Professional Java Mobile Programming

Ronald Ashri
Steve Atkinson
Danny Ayers
Constantinos Hadijis
Mårten Haglind
Rob Machin
Nadia Nashi
Bill Ray
Richard Taylor
Chanoch Wiggers

Wrox Press Ltd. ®

Professional Java Mobile Programming

Published by Wrox Press Ltd,
Arden House, 1102 Warwick Road, Acocks Green,
Birmingham, B27 6BH, UK
Printed in the United States
ISBN 1-861003-89-7

Trademark Acknowledgements

Credits

Authors
Ronald Ashri
Steve Atkinson
Danny Ayers
Marten Haglind
Rob Machin
Nadia Nashi
Bill Ray
Richard Taylor
Chanoch Wiggers

Additional Material
Costas Hadjisotiriou

Category Manager
Viv Emery

Technical Architects
Gregory Beekman
Chanoch Wiggers

Technical Editors
Tim Briggs
Mankee Cheng
Tabasam Haseen
Matthew Moodie
Mohammed Rfaquat

Author Agent
Velimir Ilic

Project Administrators
Simon Brand
Chandima Nethisinghe
Claire Robinson

Technical Reviewers
Terry Allen
Ronald Ashri
Steven Atkinson
Danny Ayers
Yogesh Bhandarkar
Jeremy Crosbie
Michael J Fichtelman
Pascal Van Geest
Costas Hadjisotiriou
Ethan Henry
Kamran Kordi
Rob Machin
James Maidment
Piroz Mohseni
Faisal Nazir
James Scheinblum
Keyur Shah
Gavin Smyth

Production Manager
Simon Hardware

Production Coordinator
Mark Burdett

Production Assistants
Abbie Forletta
Paul Grove
Natalie O'Donnell

Cover
Dawn Chellingworth

Index
Indexing Specialists

Proof Readers
Fiver Locker
Miriam Robinson
Dianna Skeldon
Chris Smith

About the Authors

Ronald Ashri

Ronald Ashri is a postgraduate researcher at the University of Southampton, working in the "Intelligence, Agents and Multimedia" group that is part of the Department of Electronics and Computer Science. His research is focused on infrastructures for agents in heterogeneous network environments, which range from the typical PC to Palm-sized devices and mobile phones. This work is jointly funded by British Telecom and the University of Southampton. Other interests include the use of Jini network technology for the support of agent-based systems.

Earlier, Ronald worked for a short period at Adastral Park, BT's research center, focusing on security issues for mobile agents. He graduated from Warwick University with a First Class Honours degree in Computer Systems Engineering.

Many thanks to the great Wrox team for all the hard work they've put into making this book a reality. Special thanks to Katia, Andreas, Photini, Daniel, Carla and Titta.

Steve Atkinson

Steven is a software architect and technical consultant with over 10 years experience. He has worked with clients from a range of industry sectors including the financial, public, telecommunications and broadcasting sectors both in the UK and, more recently, Silicon Valley. His particular area of expertise includes the management of complex data and the integration of database technology with internet/intranet technology.

Danny Ayers

Danny Ayers has a 20-year history of trying to make the best of inhuman operating systems, which in recent years have been less of an issue thanks to the Java language. Professionally he has been primarily involved with networking technologies, and as an independent information engineer he is currently working on tools that may help in the development of a semantic web. His other interests include music, woodcarving and travel. His personal web site may be found at http://www.isacat.net.

Thanks to my wife Caroline for being incredibly tolerant during my bad code days.

Constantinos Hadjisotiriou

After a diverse career start, that included Mechanical and Environmental engineering, FORTRAN programs and database management, Costas moved to writing WAP application software and content. He is currently working on image viewers for wireless devices, and industrial software for PDAs.

I have been teleworking since 1999 from Spain and I recommend everybody do the same. I can be contacted via http://www.anadysis.com.

Mårten Haglind

Mårten Haglind has been a Java Enterprise developer and architect since 1997 and is one of the founders of Haglind & Thörnblad Soft Solutions - a Swedish consulting and instruction company. His experience includes J2EE development and consulting for European and US companies. He currently is focused towards server-side web development, but has been around the block with Java. In addition to programming and mentoring, he teaches Java training courses for Learning Tree International, and is involved as technical editor and course author within both their Java and WAP curriculum.

Mårten would like to thank his family for their continued support. Special thanks goes to the honorable Chad Darby for introducing me to the publishing world.

Rob Machin

Rob graduated from Durham University in 1994 with a First in Mathematics and Philosophy. He has since discarded his studies of Leibniz, Voltaire and Sartre to work for both management consultancy firms and software houses.

Rob is a specialist in n-tier systems architecture and new technologies, developing component-based software using both Smalltalk and Java. He is currently working as a Technical Architect with the Concise Group Ltd., advising organizations, particularly within the global financial markets and investment-banking sector, on the use of Java 2 Enterprise Edition technology in their e and m-commerce strategies.

Thanks to Mo, Nell and Dora, without whom his chapters would not have been completed.

Nadia Nashi

Nadia Received an exemption from the Part One Exam in Architecture by the Royal Institute of British architects in 1990. A First class Post Graduate Diploma in architecture form Greenwich university in 1993 and an MSC in Software Engineering from Westminster University 1994. She taught Computer Aided Design at Greenwich University before becoming a full-time Object Oriented design analyst, java consultant and software developer.

Bill Ray

Bill worked at Swisscom (formally Swiss Telecom), world leaders in mobile interactive applications, designing their Interactive Television set-top box. Responsible for security and video application development, he was able to participate in the early stages of both JavaTV and the JMF. Having established a reputation for high security application development (and paranoia), Bill then set up Network 23 to provide interactive video application consultancy.

Richard Taylor

Richard Taylor is an independent mobile internet developer specializing in multi-device portals, and content conversion specialist. He has coded, team-led and project-managed his way through every computing fad since 1982, concentrating on object-orientation for the past 8 years. In his spare time, Richard dreams of Laverda motorcycles with perfect electrics. During this book it was the battery, the regulator, the brake micro-switch and fuse connection that failed. When he wins the Lottery, he is going to buy the marque. He can be reached at rct@poqit.com.

Many thanks to the following gurus. They not only offer excellent solutions, but they give advice and fix bugs overnight for complete strangers. Who could ask for anything more: Hiram Chirino – jBoSSMQ, Stefan Haustein – kAWT, Silvano Maffeis – Softwired.

Last, but never least, to Julia, Alys and Jamie ("Why don't you get a proper job?" "Like what?" "Well, a pilot. Or a fireman...").

Chanoch Wiggers

Chanoch Wiggers is a technical architect with Wrox Press Ltd. Over the last year, Chanoch has concentrated on mobile applications, especially with regards to J2ME, and on the emerging web services technologies. He specializes in java technology, specifically with regards to extending enterprise applications to the device. Chanoch enjoys riding his bike to work and back and the occasional weekend off.

Thanks go to Barbara, my wife, for your patience. Apologies to everyone who made me promise to have my picture on this book. Maybe next time....

PROFESSIONAL JAVA MOBILE

TOOL SELECTION

WEB TABLET
MOBILE PHONE
MP3 PLAYER
DIGITAL CAMERA
PALMTOP
LAPTOP

NOW SELECTED

PALMTOP

Table of Contents

Introduction 1

J2ME and Java for Mobile Devices 1

What's Covered in this Book 2

Who Should Use this Book 2

What You Need to Use this Book 2

Conventions 3

Customer Support 4

Source Code and Updates 4
Errata 4
P2P Online Forums 5

Chapter 1: Java in the Mobile Arena 7

The Growth of Device-Based Technology 8

Why J2ME and Why Now? 9
Benefits of Local Computing Power 10

The Evolution of J2ME 11

Sun, Stealth, and Brainstorming the Future 11
The Green Project and the *7 11
Oak, aka Java 12
FirstPerson Inc. and Television Set-Top Boxes 13
The Web and a Breakthrough Announcement 13
Duke and Computing Mascots 13
Historical Perspective on J2ME and Device-Based Systems 13
The Evolution of Java from Applets to Java 2 Enterprise Edition 13
Java Technology Editions 14
Java 2 Enterprise Edition (J2EE) 14
Java 2 Standard Edition (J2SE) 14
Java 2 Micro Edition (J2ME) 14

The Anatomy of J2ME 15

The Virtual Machine Layer 16
The Configuration Layer 16
The Profile Layer 16

Current J2ME Implementations and SDKs **17**
Java Card 17
Java Card Networking 18
EmbeddedJava 19
The Connected Limited Device Configuration 20
The K Virtual Machine (KVM) 21
The Mobile Information Device Profile (MIDP) 22
Scope of the MIDP 23
How the CLDC, KVM, JAM and MIDP Coexist on a Device 23
MIDlets 24
The Sun MIDP Toolkit and Emulation Environment 26
The Personal Digital Assistant (PDA) Profile 27
Palm and the Spotless Implementation 27
The Connected Device Configuration (CDC) 30
From PersonalJava to the Personal Profile (JSR62) 30
Personal Profile Extensions 31
Other Planned CDC-Based Profiles 32
The Symbian EPOC Java SDK 32

Other Technologies Impacting Mobile Java Development **34**
Carrier Technology: GSM to 3G Networks 34
2G: GSM (Global System for Mobile Communication) 34
2.5G: GPRS (General Packet Radio Service) 35
3G (3rd Generation) 35
Wireless Networking Technology 36
Bluetooth 36
Wireless LAN – 802.11b and HIPERLAN 38
Infrared 40
Considerations for Developers 40
The Move to 3G 40
The Move to Local Wireless Networks 41
J2ME and Browser Technologies 41
Wireless Application Protocol (WAP) 41
Jini and the Jini Surrogate Architecture 43
Location Hiding and Network Transparency 43
Jini 43
Jini Surrogate Architecture 43
Java Embedded Server Technology 44
The Open Services Gateway Initiative (OSGi) 45

How J2ME will Develop – The Java Community Process **46**
Initiation 46
Community Draft 46
Public Draft 47
Maintenance 47

Summary **47**

Chapter 2: Architecting and Developing J2ME Software **51**
Architectural Patterns 51
Model, View, Controller (MVC) Architectures 52

Component-based Systems for Devices 53
Improvements in Mapping the Business Problem-domain 53
Resilience to Changes in Requirements 53
Improved Scalability 54
Debugging, Safety and Resilient Systems 54

Linking to the Server **54**
Client-Server Deployment Models 55
Stand-alone Client Application Model 55
Moving Towards Client-Server 55
2-tier Client-Server Application Model 55
Moving to a More Scaleable System Architecture 56
The Java 2 Platform, Enterprise Edition 58
A Typical J2ME/J2EE Architecture 60
Service-oriented Architectures 61
Integration of Component-based Systems 61
Example of the Benefits to J2ME 62
Network Transparency and Distributed Systems 62
The Way Forward 63

J2ME Architecture Case Study **63**
Initial Requirements-gathering and Analysis Workshop 64
Clarification of Requirements 66
Use-case Diagrams 67
Use-case Sign-off Meeting 68
Initial Work and System Development 69
Business Model Class Diagrams 69
Screen Design for Timesheet Entry 70
The Initial Proposed Architecture 72
Component Diagrams 73
Sequence Diagrams 73
Planned System Structure 74
Analysis Workshop for the Initial Architectural Designs 75
Target Devices 77
Security 77
Java Application Server Load and Responsibilities 77
Creating a Mobile Application's Infrastructure 78

Summary **78**

Chapter 3: CDC and the Foundation Profile **81**

What We Will Need **82**

Connected Device Configuration **83**

Foundation Profile **84**
The CVM 84

Supported Classes **85**
Removed Classes and Methods 85

Additional Classes Defined in the Foundation Profile **86**
java.text.resources 86

javax.microedition.io 86
An Example of UDP Programming 89
The Remaining Classes 90

Implementing the GCF **90**
Socket Programming Revision 90
Implementing the socket and serversocket Connections 91
com.wrox.cdc.j2se.socket.Protocol 93
com.wrox.cdc.io.j2se.serversocket.Protocol 94

Developing Java Applications for the Foundation Profile **96**

A Basic Socket Application **96**
The SimpleServer Class 97
The SimpleClient Class 98
POP3 100
An E-mail Client 102
The Message Class 103
The MailStore Class 104
The PopReader Class 105
The MailClient Class 109
Testing The Application 112

A Simple POP3 Server **113**
The PopServer Class 114
Testing the PopServer Class 119
The PopProxy Class 120

Other CDC Profiles **121**
The RMI Profile 121
The Personal Profile and PersonalJava 122

Summary **122**

Chapter 4: Introducing the KVM, CLDC, MIDP, and other Profiles **125**
What is a Connected Limited Device? 126
The K Virtual Machine (KVM) 126
The Classfile Pre-Verifier 128
Sandbox Security 128
KVM Implementations 129
Porting the KVM 129

KVM Utilities **129**
The Java Application Manager (JAM) 130
Java Code Compact 130
The Connected Limited Device Configuration 130
CLDC Classes 131
Upward Compatibility – The CLDC and CDC Compared 132
Downward Compatibility – The CLDC, JavaCard and EmbeddedJava Compared 132
Minimum Hardware Requirements 133
Minimum Software Requirements 134

MIDlets 134
A Simple MIDlet 135
Compiling and Running the Example MIDlet 137
Instructions for J2MEWTK Stand-alone Build 138
Instructions for Forte – J2MEWTK Build 140
Instructions for Palm MIDP Installation 140
Installing a Sample Application 140
The MIDlet Suite in Detail 142
The MIDlet Classes and Resource Files 142
The MIDlet Application Descriptor File 142
The MIDlet Suite Manifest File 143
MIDlet and MIDlet Suite Attributes 144
Launching and Controlling the MIDlet 146

CLDC Development Environment **147**
CLDC and KJava Download 148
Testing the Environment 148
Build a Sun Example KJava Application 149
Pre-verify the Classes 149
Test the Application 149
Running on the Palm 149
Create a Palm Executable 150
Test the KJava Palm Application 150
Note on Palm Connectivity: 151
kAWT Download 151
Sample kAWT Application 152
Building Our Sample kAWT Application 153
Testing the kAWT Palm Application 154

Summary **154**

Chapter 5: MIDP & CLDC API **157**

The MIDP User Interface **157**

The Base UI Class: Displayable **159**

High-Level UI Events **160**

High-Level API **161**
TextBox 162
Lists 164
Alerts 167
Forms 169
Interactive Forms 173
Tickers 175

Low-Level UI API **176**
Low-Level UI Events 176
Low-Level Graphics 178
Canvas Size 179
The Graphics Class 179
Low-Level UI Methods Not Guaranteed to be Portable 192

The MIDP Record Management System 193
Record and Record Store Creation 193
Record Retrieval 196
Modifying and Deleting Records 197
Deleting a Record Store 198
Filtering and Sorting Records 198
RecordListener Events 201
Synchronization Support 201

CLDC and MIDP System Properties 201
CLDC System Properties 201
MIDP System Properties 202

Networking 203

Timers 206

Starting a MIDlet from Another MIDlet 208

Downloading Images 208

Decreasing the Size of an Application 209

The Missing Pieces – What MIDP Doesn't Cover 209
System-level APIs 209
Application Delivery and Management 209
Security 209
User Interface 210

Debugging 211

Development Platforms, Emulators and Tools 211

Alternative KVMs 212

3rd Party Class Libraries 212

General KVM Resources 213

The Future 213

Summary 214

Chapter 6: A Complete Mobile Application – The Contact Database **217**
Contact Application Requirements 218
User Requirements Overview 218
User Requirements in Detail 219
Separating Business Methods From Low-Level Implementation 219
The Database Format and Contact Class 220
The 'Glue' Layer and Supporting Classes 222
Database – Design and Classes 222
User Interface – Design and Classes 228
The Business and Control Class – ContactApp 233
The MIDP Port 242
Starting the MIDP Application 251

The Palm/KJava Port 252
KJava-Specific Classes 253
Starting the KJava Application 265
Improvements to the Application 266
Summary 266

Chapter 7: Converting Applications To CDC and CLDC 269

Summary Of Differences 270
The Connected Device Configuration 270
Connected Limited Device Configuration 272

JavaCheck 272

PersonalJava Emulation Environment 274

Example Conversion 274
Using JavaCheck 287
Interpreting the Results 290
Designing a New GUI 291
Running the Application on the Device 293

Porting to CLDC 294
Networking 294
The CLDC Application 295
Other Options 303

Summary 303

Chapter 8: Synchronization 307

The Palm Pilot 308
Different Types Of Synchronization 308
Mirror Image 308
One Directional Transfer 309
Transaction Based 309
Back-Up 311
Problems With Synchronization 311
All About Protocol 312
Creating the Protocol 312
Writing The Code 321
The Server 321
The Client 326
Using Synchronization Code 330

Synchronization With Palm Pilot Devices 331
HotSync Technology 331
SyncBuilder 332

SyncML: A New Standard 332

Summary 333

Chapter 9: The JavaPhone API 335

History of the JavaPhone API 335

What We Cover in this Chapter 336

The JavaPhone Architecture 336

The JavaPhone Packages in Details 337

JTAPI Core and JTAPI Mobile 337

Address Book 338

User Profile 338

Calendar 338

Datagram 338

Power Management 338

Power Monitor 338

Install 338

Communication API and SSL 338

Overview 339

Developing Applications with the JavaPhone API 339

The Emulator File System 339

Running Java Applications 340

The AIF Builder Tool 340

Event Handling in JavaPhone applications 341

Programming Concepts of the JavaPhone API 341

JTAPI Core 341

Address Book 344

User Profile 351

Calendar 352

JTAPI APIs 354

Datagram 355

Power Monitor 360

Capabilities 361

The Symbian SDK 362

Symbian's JTAPI Implementation 362

JTAPI Under the Emulator 363

Summary 364

Chapter 10: Java Card Applications 367

What is a smart card? 368

Why Smart? 369

Language or Operating System? 369

Multos 369

Windows for smart cards 370

API Overview 370

ISO7816 and Friends 370

Talking to a smart card 371

Listening to the Card 372

smart card Communication Example 372

smart card File Structure 372

The smart card File System 373

Example Commands 374

Creating Our Own Commands 376

Using a smart card with Java **377**

Programming the Card **378**

Using the Java Card Development Kit 381

Testing the Installation 382

Developing the Application 383

Applet Lifecycle 383

Example Java Card Applet 384

Compiling the Example 386

Converting the Example 386

Running the Example 388

Working with the Example 390

Working with a Real smart card **390**

Deploying Applications **391**

Summary **391**

Chapter 11: Interfacing with Servlets and Other Elements of J2EE **393**

What is Java 2 Enterprise Edition? **394**

Distributed Applications **394**

Communication 395

RMI and RMI-IIOP 396

Location 403

JNDI 404

Front-end Logic and Presentation 405

Server Connectivity 406

Basic J2ME Communication Example – Reading a File from a Web Server **407**

Deployment 411

Servlet Example I – Filtered Reading From a File 414

The Client Side 414

The Server Side 416

Servlet Example II - Writing to a File 418

The Client Side 418

The Server Side 420

Database Access via a Servlet 422

The Client Side 422

The Server Side 425

EJB Access via a Servlet 429

Advanced Connectivity Topics 430

User Feedback 430

Predictive Loading of Data 431

Resuming Uploads and Downloads 431

Data Formats 431

JSP 431

Alternative Communications 432

JMS 432

JavaMail 432

The Future 433

Summary 433

Chapter 12: Messaging and JMS 435

What is an Asynchronous Message? 436
The Short Message Service 436

Types of Asynchronous Messages 437
Store and Forward Networks 437
Example 438
Tracking IP Addresses 439
Contacting Mobile Devices 440
Datagram-Based Messages 441
When to use Datagram Communications 441
Example Datagram Code 441

The Auction House 444

MOM Basics 446

JMS 446
JMS Messaging Models 448
Point-to-Point Messaging 448
Publish-and-Subscribe Messaging 449
Senders and Receivers 449
The JMS Message 450
Sending a Message 452
Administered Objects 453
Connecting to a Messaging Server 454
Sessions 454
Sending (Point-to-Point Model) 455
Receiving (Point-to-Point Model) 455
Publishing (Pub/Sub Model) 456
Subscribing (Pub/Sub model) 456
The Auction House Revisited 457
AuctionHouse JMS Client 461
Bidder JMS Client 466
The Bidder User Interface 469
Running the Auction House Application 475
Example Palm iBus//Mobile Application 476
Code Changes 477
Obtaining and Installing the iBus Server 480
Obtaining and Installing the iBus//Mobile Gateway and Libraries 481
Testing the Basic iBus Installation 481
AuctionHouse-specific iBus Configuration 481
Palm Simulators 482
kAWT Installation 482
Building the JMS Clients 482
Running the Palm JMS Client 484
The Compromises 485
Is It a Success? 486
The MIDP Client 486
Enhancements 491
Other JMS Features 492
The Future 492

Summary 493

Chapter 13: Developing a J2ME Application for the MIDP and the Palm 495

Designing for Portability 495
Designing for Small Devices 496

The Towers of Hanoi 496
The J2ME Wireless Toolkit 496
The Palm OS Emulator 497
The User Interface Design 497
The Screen Navigation 498
The Screen Design 498
The Welcome View 499
The Main Game View 499
The Game Over View 500
The Help View 501
The About View 501
Commands and Action Keys 502
The Application Design 502
High-Level Design 502
Detailed Design 503
The Implementation 505
Shared Implementation 506
MIDP Implementation 513
Palm OS Implementation 523

Summary 534
References 534

Chapter 14: Case Study: A Portable Expert 537

Initial Analysis 537
Production Rules 538
Our Requirements for the Model-View-Controller System 538
System Overview 539
From Abstract to Concrete 539
Sample Knowledge 540
Knowledge Base Classes 541
The KHashTable Class 542
The FactBase Class 543
The Fact Class 543
The Condition Class 543
The Rule Class 543
The KnowledgeBase Class 545
The Engine Class 546
Using XML for Knowledge Persistence 546
Reading a Palm Memo 548
I/O Classes 548
The PdbReader Class 548
The KMemoReader Class 549
Extracting the Data 550
SAX 551
XML Classes 552
User Interface 560
Trouble-Shooting the Knowledge Bases 561
Summary 561
Further Reading 561

Chapter 15: Case Study: Mobile Positioning 565

Mobile Location Solutions 565
Location 566
Mobile Positioning 566
Network-Oriented Positioning 566
Future Developments of Network-Oriented Positioning 567

Developing Mobile Positioning Solutions 568
Handset-Oriented Positioning 569
Ericsson MPS 569
Mobile Positioning Protocol, version 3.0 570
Using the SDK 573

Hotel-Search 574
Use Cases 575
Architecture 575
Orion Application Server 576
JRun 576
MPS SDK 3.0 577
Application Tier 579
Enterprise JavaBeans 580
Presentation Tier 590
HTML 590
WML 598
Installation & Deployment 607
Summary 611

Chapter 16: Security 613

Setting the Scene 614
What is a Secure System? 614

Secure Computer Systems 615
Detection and Prosecution 615
Prevention 615
Access Control 616
Confidentiality 616
Integrity 616
Availability 616

Security in the Context of Mobile Devices 616
Access Control Problems and Solutions 617
Digital Attacks on Wireless Networks 617
Security in the Context of Mobile Code 618

Securing Java for Mobile Devices 620
J2SE Security Architecture 620
Bytecode Verifier 621
Class Loader 621
Security Manager and Access Control 622

J2ME Security Architecture 623
Changes from J2SE to the CDC 623
Changes from the J2SE to the CLDC 624
Security in the MID Profile 624
Security for Network Communication 625
Public Key Infrastructure 625

Securing the Communication Channels 626
WAP Security Problems 627
Possible Solutions 627
Security Considerations 628
Mobile Electronic Transactions Initiative 628
Summary 629

Chapter 17: Designing J2ME Software 631

What is Design? 631
Sponsors 632
Project Management 632
Developers 632
Sales Staff 633
The Purchasing Agent 633
End-users 633
Support and Maintenance Staff 633
System Design versus System Architecture 633
Building a House, Building a City 634
Managing Complexity 634
User-Centered Design 635
Short and Long-term Memory and Conceptual Models 636
Click Depth and Small Screens 636
Making a Mobile System Explorable 637
Design to Cope with Error 638
Reverse Usability 638
Infrastructure and System-level Design 639
A Programmer's Perspective on J2ME 639
Construction of J2ME-based Applications 639
Techniques and Tricks 639
Elaboration – Patterns and Metaphors for Software Design 640
Introduction to Component-based Software Design 640
Components versus Objects 641
Perceived Problems with Component-based Systems 641
Performance 641
Security 642
Reliability 642
Device-based Programming 642
Considerations for Device-based Systems 642
Timeliness 643
Feedback to the User and System Responsiveness 643
Predictability, Correctness, and Robustness 644
Fault Tolerance and Safety 644
Optimization and Correctness in Design 644
Get the Infrastructure Right Early On 645
Use the Techniques of Code Refactoring Throughout the Coding Process 645
Optimize at a Code-level Late in the Development Cycle 647

Analyze to Determine Where You Really Need to Optimize 647
Improving Startup Time versus Execution Speed 647
Improve Network Performance versus Result Processing 647
Improving Performance versus Reducing Memory Usage 647
Don't Over-optimize or Optimize Unnecessarily 648
Optimize the Object Creation Process 648
Reuse Existing Objects 648
Make Use of Object Pools 648
Caching Information on the Device 648
Assert Yourself 650
Refactor with a Plan 651
Code Defensively 651

J2ME Coding Standards **651**

Availability of Developers **652**

Factors for Small Connected Systems **652**
The Continuing Need for Client-side Computing 652
Disconnected Access 653

What Sort of Applications Benefit from J2ME? **653**

Summary **654**

Appendix A: API Comparison **655**

Appendix B: The CLDC Classes **689**

java.io Package **689**

java.util Package **692**

java.lang Package **694**

Appendix C: The MIDP classes **701**

javax.microedition.lcdui Package **701**

javax.microedition.midlet Package **744**

javax.microedition.rms Package **747**

Appendix D: javax.microedition.io Package **759**

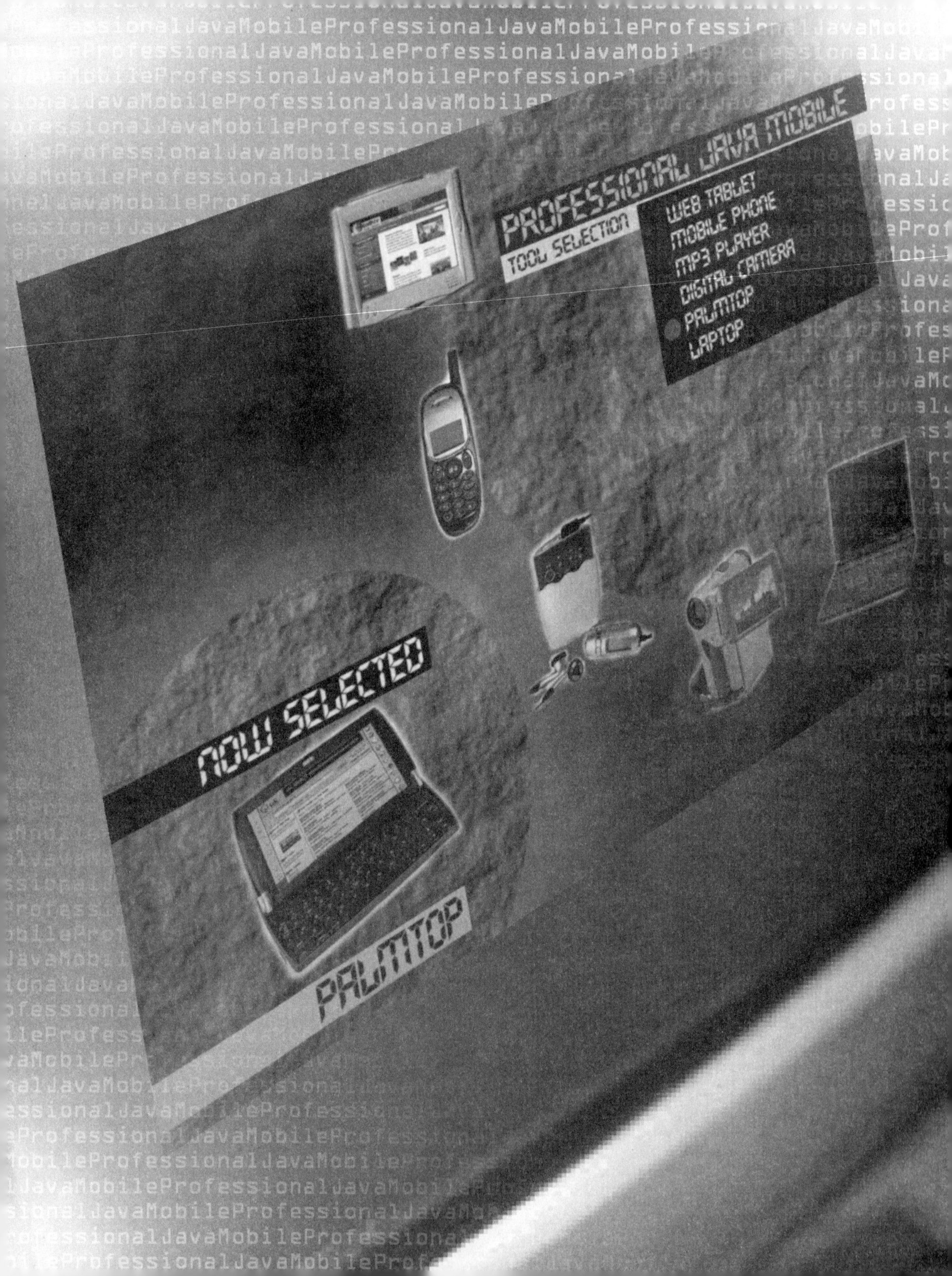
PROFESSIONAL JAVA MOBILE
TOOL SELECTION
WEB TABLET
MOBILE PHONE
MP3 PLAYER
DIGITAL CAMERA
PALMTOP
LAPTOP
NOW SELECTED
PALMTOP

Introduction

J2ME and Java for Mobile Devices

The introduction of J2ME into the Java programming world marks a return to the design roots of Java – providing small connected devices with a stable, secure, and flexible environment for application development. Obviously, the capability of devices is a factor in their ability to utilize the functionality available through the Java platform and therefore merely implementing Java 2 is impractical and will not gain much support.

The efforts on the part of Sun Microsystems and related partners in providing a Java platform for mobile devices hold true to the requirements for compatibility with standard and enterprise applications while limiting excess functionality, and have brought about an exciting new world of application development, where context and language awareness become much more important, and applications have a more direct impact on users. The J2ME platform represents Sun's entry into the real-time, consumer-device, and embedded markets and allows the developer access to these previously closed markets.

This book is an introduction to these new technologies, showing both the new APIs and attempting to document the changes in thinking required for developing applications in severely restricted devices. We also show the potential for powerful, timely solutions to the problems raised by the devices' limitations.

What's Covered in this Book

In this book we will be covering the fundamentals of programming mobile devices using Java. Specifically we will be looking at:

- J2ME and the idea of profiles and configurations
- The Connected Device Configuration (CDC) and the Foundation Profile
- The Connected Limited Device Configuration (CLDC) and the Mobile Information Device Profile (MIDP)
- Developing applications for a variety of devices, using a number of different profiles and configurations
- Converting existing J2SE applications to the CDC and CLDC
- Synchronization issues, that is, how to keep mobile applications consistent with their desktop counterparts
- The JavaPhone and JavaCard APIs that further specialize J2ME
- Integrating J2ME with J2EE APIs, specifically JMS and servlets
- Security
- Design and architecture considerations
- Case studies that show advanced uses for the APIs described throughout the rest of the book

Who Should Use this Book

This book is for professional Java developers who need a comprehensive explanation of J2ME and how it can be used to solve computing problems. It will also be of interest to developers who have a good working knowledge of Java, and some familiarity with J2EE. It provides developers with core J2ME theory alongside practical case studies exemplifying real-world uses of J2ME.

What You Need to Use this Book

All of the APIs, and most of the toolkits, used in this book are available from Sun. Below is a list of those toolkits and applications that you will need to run the code in certain chapters.

The J2ME Wireless Toolkit

The J2ME Wireless Toolkit is a set of tools that provides Java developers with the emulation environment, documentation and examples needed to develop CLDC/MIDP, compliant applications:

- http://java.sun.com/products/j2mewtoolkit/

Palm OS Emulator

Many of the examples in this book require a Palm OS Emulator (POSE), or a physical Palm, to run. The emulator can be downloaded from the Palm developers' web site, but requires a ROM image to run. This can be obtained from the same web site, or downloaded from a Palm unit:

- http://www.palmos.com/dev/
- MIDP for Palm OS – http://java.sun.com/products/midp/palmOS.html

PPP Server

Many of the examples in this book require you to set up a client-server relationship with a Palm device. This is not a problem if you are using the POSE, but a real Palm device requires some way of accessing the network. The easiest way to do this is via the Point-to-Point Protocol (PPP):

- For Windows download a PPP server such as Mochasoft PPP: http://www.mochasoft.net
- For Linux configure a PPP server: http://www.linuxports.com/howto/ppp/

JBoss Application Server

JBoss is an open source J2EE server that will be used as the server side in our integration chapters:

- http://www.jboss.org

iBus//Mobile 2.0.1 from Softwired Inc.

iBus//Mobile is a mobile JMS client suitable for installing on Palms and other wireless devices. We will use it to link to a desktop server using JMS:

- http://www.softwired-inc.com

Conventions

We have used a number of different styles of text and layout in this book to help differentiate between the different kinds of information. Here are examples of the styles we use and an explanation of what they mean:

Code has several styles. If it's a word that we're talking about in the text, for example when discussing a Java `StringBuffer` object, it's in this style. If it's a block of code that you can type as a program and run, then it's in a gray box:

```
public void close() throws EJBException, RemoteException
```

Sometimes you'll see code in a mixture of styles, like this:

```
<?xml version 1.0?>
<Invoice>
   <part>
      <name>Widget</name>
      <price>$10.00</price>
   </part>
</invoice>
```

In cases like this, the code with a white background is code we are already familiar with; the line highlighted in gray is a new addition to the code since we last looked at it.

Advice, hints, and background information comes in this type of font.

Important pieces of information come in boxes like this.

Bullets appear indented, with each new bullet marked as follows:

- **Important Words** are in a bold type font
- Words that appear on the screen, in menus like File or Window, are in a similar font to that you would see on a Windows desktop
- Keys that you press on the keyboard like *Ctrl* and *Enter*, are in italics

Customer Support

We've tried to make this book as accurate and enjoyable as possible, but what really matters is what the book actually does for you. Please let us know your views, either by returning the reply card in the back of the book, or by contacting us via e-mail at feedback@wrox.com.

Source Code and Updates

As you work through the examples in this book, you may decide that you prefer to type in all the code by hand. Many readers prefer this because it's a good way to get familiar with the coding techniques that are being used.

Whether you want to type the code in or not, we have made all the source code for this book available at our web site at the following address:

http://www.wrox.com

If you're one of those readers who likes to type in the code, you can use our files to check the results you should be getting – they should be your first stop if you think you might have typed in an error. If you're one of those readers who doesn't like typing, then downloading the source code from our web site is a must!

Either way, it'll help you with updates and debugging.

Errata

We've made every effort to make sure that there are no errors in the text or the code. However, to err is human, and as such we recognize the need to keep you informed of any mistakes as they're spotted and corrected. Errata sheets are available for all our books at http://www.wrox.com. If you find an error that hasn't already been reported, please let us know.

Our web site acts as a focus for other information and support, including the code from all our books, sample chapters, previews of forthcoming titles, and articles and opinion on related topics.

P2P Online Forums

Join the mailings lists at http://www.p2p.wrox.com for author and peer support. Be confident that your query is not just being examined by a support professional, but by the many Wrox authors and other industry experts present on our mailing lists.

PROFESSIONAL JAVA MOBILE
TOOL SELECTION
WEB TABLET
MOBILE PHONE
MP3 PLAYER
DIGITAL CAMERA
PALMTOP
LAPTOP
NOW SELECTED
PALMTOP

1

Java in the Mobile Arena

Computers have evolved and their usage has increased at a dramatic rate, since the 1940s. From the era of huge house-sized mainframes, to the world of the personal computer and the now ubiquitous computing power facilitated by handheld devices, we have seen the computer age bring sweeping changes to the way that people conduct their business and, increasingly, go about their everyday lives.

Like computing in general, Java has also experienced an extremely rapid and dynamic growth, albeit only over the last 6 years. As a programming environment, it has been adopted by the industry at an astonishing rate. Originally developed for programming consumer devices, and later hailed as a way to make the World Wide Web more dynamic, Java did not succeed in its promise of transforming the Web with graphical applets, but hit the jackpot on the server-side with the triumphant Java 2 Enterprise Edition specification.

Java is now, arguably, the most flexible and comprehensive development tool available to programmers in this newly connected world. In which case, Java is ready to return to its roots – the device. It has the capacity to enable the era of pervasive computing by connecting and developing the capabilities of the device. This is the type of Java that is the subject of this book – **Java 2 Micro Edition** (**J2ME**) – Java in one's pocket, in one's car, and on one's television, to name but a few.

In this introductory chapter we'll look at:

- The explosion of devices
- Why J2ME is a good thing
- Java's history and why J2ME gives some people a sense of déjà vu
- Splitting J2ME into virtual machines, configurations, and profiles
- Other device technologies that will influence J2ME
- How the parts of J2ME are developed

The Growth of Device-Based Technology

As improvements in wireless networks, device technology, and device-based services have materialized over the last few years, we have witnessed an explosion in the usage of mobile technology. This growth trend looks set to continue.

J2ME is uniquely placed to take advantage of this phenomenal growth in device-based computing:

- Currently 72% of all Fortune 1000 businesses use Java. (Gartner Group)
- In 2001, 25% of all workers will be mobile (in other words, will spend a significant portion of their working day out of a fixed line PC/office environment), and 30% of IT budgets will be for mobile workers. (Gartner Group)
- There are currently over 300 million mobile phone subscribers, growing at a rate of 50% per annum (PC growth globally is now only about 20% per annum and falling). (Gartner Group)
- Smart phone penetration into the mobile phone market is predicted to be 8% in 2000, 22% in 2001, 50% in 2002 and 85% in 2003. (Gartner Group)
- The number of people using mobile data services in 1999 was 31.7 million. This is expected to grow to 1.187 billion by 2005. (ARC Group)
- It is predicted that the m-commerce market in Europe will be worth 5 billion Euros during 2000, rising to over 20 billion Euros by 2003. (Gartner Group)
- By 2003, over 50% of Internet access will be via non-PC-based access mechanisms. (Meta Group)

No single channel will dominate over the next five years, and consumers will use a wide range of devices from televisions, fixed line PC connections, and mobile connections to the Internet to obtain information, make decisions, and purchase goods and services, to name just a few online activities.

Companies need to develop a multi-channel approach to this marketplace, aiming to reach out to their customers and offer both time- and location-related services via a number of digital and electronic conduits. For example, we can imagine the scenario where an office dweller receives an alert that their favorite pizza company has a special happy hour offer on now, and information that the 'The Matrix III' is playing at the cinema as they drive past on their way home.

The construction of these location and time-based services will prove a significant technical and engineering challenge for hardware manufacturers and telecommunications companies. Once this location and temporal information is available to customize and personalize the nature of data and services however, the rewards to consumers and businesses will be profound.

In parallel with this explosion of mobile technology, improvements in device technology may prompt a surge in the uptake of the use of smart devices. These devices and commodity items range from automobiles to household appliances, potentially offering many improvements and changes to the way people conduct their business and personal lives.

Why J2ME and Why Now?

In order to cope with this growth in mobile technology, and to facilitate its continuance, appropriate and useful services must be developed to run on these new and more powerful devices and the networks that connect them. The complexity of software development for consumer electronics and wireless devices has increased dramatically over the last decade. Conventional software development approaches are now becoming unworkable in the device environment due to the following factors:

- The increasing capabilities of the hardware
- The growth in the need to network-enable these devices
- The resulting increase in the complexity of the code
- The need for portability as components within the devices (such as the processor, graphics card or communication mechanisms) are swapped around
- The increased pressure of short times to market for products and device-based software

J2ME exists, and evolves, in a technology space created by the convergence of three key enablers:

- Improved mobile carrier networks that can support packet-switched protocols and Internet protocols, such as HTTP
- The development of device technology to allow sophisticated services to be implemented on the device
- The ubiquity of the Java programming language

This convergence has created an ideal environment to introduce the benefits of Java to the device software market:

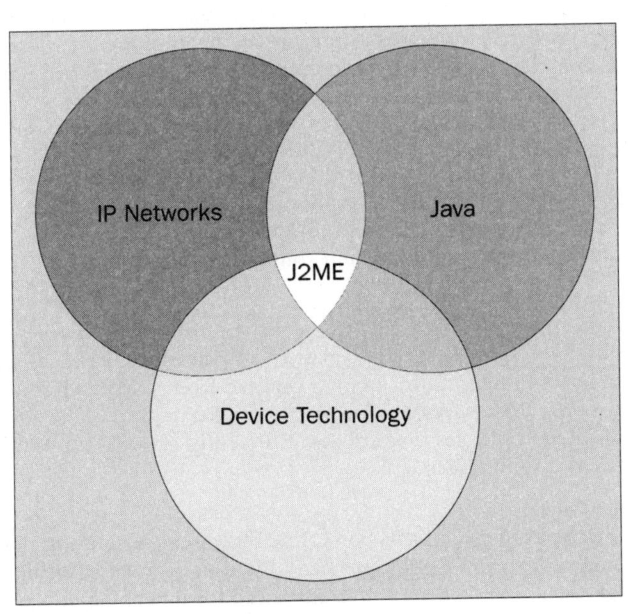

J2ME provides some unique benefits to the device market, and these include:

- Dynamic delivery of new software, and software updates for existing components, to the device after it is shipped to the user (often referred to as 'post issuance application updates')
- Cross-platform compatibility
- Enhanced user-experience – the ability to write more sophisticated user interfaces than is possible through a browser-based application
- Access to applications locally whilst disconnected from, or with limited connectivity to, the network
- Security, safety and reliability
- Extensive class libraries to support development

Another important and key benefit of Java technology for these devices is the increasing availability of developer skills and talent. Traditional embedded and device-based applications have demanded a level of experience and programming specific to systems-level and device-level coding that is both hard to find, and expensive to develop.

Java as a skill set, while in high demand in the marketplace, is relatively easy to find compared to embedded and device-oriented development skills in C/C++ and Assembler, especially when these developers must also be capable of network-enabling the device-based applications they construct. With a large body of programmers that are eager to acquire new Java skills, it should be relatively easy to attract, develop, and retain programming talent in this area.

Benefits of Local Computing Power

Browser-based applications can often be an extremely useful and adequate solution for constructing software for devices, allowing web access with no, or minimal, device configuration. It is sometimes valuable to be able to run an application on the device however, by executing code on the device's processor and using the device's resources to run the application locally. This local computing power can complement the thin-client approach, and often win over a purely browser-based solution, for the following reasons:

- **Flexibility**
  Local software utilizes the computing power available on the device and enables a greater degree of flexibility. Using J2ME provides programmers with the tools necessary to take advantage of local computing power and operating system features. J2ME can then provide a raft of benefits over a purely thin-client solution.
- **User Experience**
  Browser-based applications can often lack the sophistication in user interface capabilities that can be offered by applications taking greater advantage of the local operating systems capabilities. This sophistication can be leveraged from J2ME, whether it is through a graphical user interface on a higher-end device, or from better manipulation of device-side services. J2ME can also reach devices that are not sufficiently capable to run a micro-browser, but need a non-graphical user interface.
- **Disconnected Operation**
  As we will discuss throughout this book, J2ME provides a solution that can run on or off the network, something browser solutions can only do in a very unsophisticated way with local browser page caching. J2ME can offer advanced local caching of state to provide truly mobile applications, being able to take advantage of today's carrier networks, and the capabilities of **3rd Generation** (**3G**) networks, as they appear.

- **Security**
  J2ME is able to offer an enhanced security model over a purely browser-based solution. Despite the challenges of securing device-based operations, with J2ME a device can be made secure for mobile use, as it is capable of advanced tamper-proofing and both local and network-based security.

- **Connectivity**
  Developing for the constrained systems targeted by J2ME this offers many similar challenges to those met by client-server developers. Thought must be given on how to balance the use of local computing power, the much greater resources and power on the server side, and making the most efficient use of the network. J2ME also offers a powerful mechanism for interacting with distributed systems, accessing information from the network and even creating powerful local networks of interacting device-based services, with the planned widespread availability of **Wireless LAN** and **Bluetooth** technologies. We will look further into 3G networks, Wireless LAN, and Bluetooth later on in this chapter.

The Evolution of J2ME

J2ME offers a development tool for consumer devices that addresses a design and architecture problem domain that is quite different from that of the traditional fixed network of computers, Internet-enabled by Java 2 Enterprise and Standard Editions.

That Java fits well into its new role on the device should come as no surprise, since Java, despite its current success on the server-side, was developed with the device market in mind.

Sun, Stealth, and Brainstorming the Future

In January 1991, Sun convened a brainstorming meeting in Aspen, Colorado including such luminaries as Bill Joy, James Gosling, Mike Sheridan, and Patrick Naughton. The project, initially named **Stealth** by Scott McNealy, was tasked with looking into the future of computing, and creating products that would ride the next wave of device technology. By April 1991, the project team had moved out of Sun premises to Menlo Park, and the project was renamed the **Green** project.

*The Green Project and the *7*

The Green project had a wide brief: the aim was to create a hardware platform, operating system, programming language, and graphical user interface. To complicate matters even further, the team used new technology at every turn.

The ***7** (pronounced star-seven) device – named after the call pick-up dial function on the project office's phone system – was conceived as a revolutionary personal assistant that could be used, and customized, for a wide range of consumer-oriented and household tasks such as:

- TV/VCR remote control
- Agenda and contact management
- Checkbook balancing and personal finance management
- Distributed whiteboard for meetings
- TV program guide

The *7 had an impressive hardware specification, even when compared to most of today's PDA devices. It included a SPARC RISC CPU, a 16-bit 240x240 pixel color display, touch screen input, and both wireless networking and infrared communication capabilities.

A version of UNIX that could run in less than 1MB of memory was developed for the device, and a file system was developed to take advantage of the *7's flash memory.

The most important aspect of the *7 was its ability to load new software dynamically, and therefore become a general purpose, configurable device. This is unlike the static configuration of most consumer-oriented devices that have entirely pre-installed software and fixed-functions.

The *7 was an incredibly ambitious project, but the stellar skill set and dedication of the entire Green team ensured that it was finished and demonstrated to Sun's management in early September 1992.

> *For a little more on the history of the Green project have a look at* *http://java.sun.com/people/jag/green/index.html.*

Oak, aka Java

James Gosling initially began to develop software for the *7 with C++. However, it quickly became apparent that the traditional memory allocation and buffer overflow problems associated with C++ would become especially frustrating and annoying when creating programs for the new device.

Gosling wanted to create a language that would be less error-prone, and therefore more productive for developers cutting code on the *7. He also aimed to maintain execution efficiency, and actually increase effectiveness for the network-centric and distributed computing requirements for this new device.

The key requirements that Gosling had in mind, entailed creating a language that was:

- Network aware
- Distributable
- Secure and reliable
- Platform independent
- Multi-threaded
- Capable of dynamic loading
- Small in size
- Simple and familiar to programmers

Gosling took a practical approach to the design of this new language, which he named **Oak** after a tree he could see from his office window. Part of this approach to the language's design involved analyzing and cannibalizing several pre-existing programming languages, all of which had strengths that the new language, Oak, could draw on and exploit. These languages included Simula and Smalltalk for their object-orientation and productiveness, Fortran for its handling of numeric data, C for its ability to program at systems level, and C++ to provide a familiar syntax.

However, Gosling wanted to craft a programming language that, unlike these traditional languages, would be network-aware and capable of containing distributed computing concepts from the ground up.

The Oak language was renamed Java due to a trademark problem with the name Oak; Java was chosen after a visit to a local café, and is not, as often suggested, an acronym standing for Just Another Vague Acronym.

FirstPerson Inc. and Television Set-Top Boxes

Shortly after that first demonstration to Sun's management, the Project Green team was incorporated in November 1992 as **FirstPerson Inc.** and began to look for commercial opportunities for the *7 development.

Between 1993 and 1994, FirstPerson Inc. targeted the interactive television and set-top box market, but after failing to win a large project from Time Warner – losing out to alternative technology from Jim Clark's SGI – concluded that the market was not yet sufficiently mature to support their endeavors.

The Web and a Breakthrough Announcement

FirstPerson Inc., and Patrick Naughton in particular, were becoming more focused on the World Wide Web and browser technology. To Naughton, this was an ideal opportunity for Java and FirstPerson Inc., with the web representing a diverse client-base crying out for the multi-platform, architecture-neutral, reliable, and secure applications development environment that had been created for the *7.

Naughton and his colleague Jonathan Payne developed a web browser called WebRunner written entirely in Java. This eventually became the HotJava browser, which had the peculiar feature of being able to dynamically download small programs, known as **applets**, from a server as needed, which could then be executed within the browser.

The major breakthrough for Java, came at the Sun World Conference in May 1995, when John Gage and Marc Andreessen announced that the Netscape browser, at the time the most popular browser available, would support Java and the applet technology.

Duke and Computing Mascots

As well as Java, Green's legacy to the IT world also includes the concept of computing mascots. Joe Palrang created Duke as the mascot of the Green team, and he has continued to be used for all Java products, becoming an enduring symbol for the Java language. Mascots have since become a regular feature of trendy technology: Duke's descendents include the Linux penguin, Tux and FreeBSD's daemon. More on the Java mascot can be found at http://java.sun.com/features/1999/05/duke_gallery.html.

Historical Perspective on J2ME and Device-Based Systems

As we have seen, J2ME is not so much as a departure from, or a new avenue for, Java technology, but rather a return to one of the central driving issues in the original development of the Java language. Java and related technologies, such as **Jini**, focus strongly on the empowerment of service-based computing and J2ME enables the development of device-based services.

The Evolution of Java from Applets to Java 2 Enterprise Edition

Following the announcement from Netscape, Java was initially heralded as a way to make the web dynamic by providing applet-based services and solutions. However, applets were slow to launch, suffered from browser incompatibilities, and subsequently failed to catch on. Competing technologies from Microsoft and various graphics packages that enabled dynamic pages, such as Shockwave and Flash, have proved to be more popular.

Despite this setback, Java's unique attributes enabled it to make a transition from the client-side to the server-side, where APIs from **JDBC** to **JavaMail** allowed it to revolutionize the way server-side and distributed computing software were constructed. The same benefits that Java brought to the device, particularly its suitability for writing small and portable components, were essential to the new breed of n-tier applications that were being constructed to replace the client-server model. Java was simpler to use than CORBA, and ideal for creating server-based, thin-client applications.

Java 2 Enterprise Edition (J2EE) took the computing world by storm during 1998/99, and has now established itself in the marketplace. At the expense of other distributed component communication technologies, such as CORBA and DCOM, J2EE has become the leading technology for constructing server-side applications.

Java Technology Editions

Even though Java has come to dominate server-side computing, Sun has recognized that one (rather large) size does not fit all, and has split the Java technology into three editions. Each edition is aimed at a specific area of computing technology, and each will be used to develop different types of application.

Java 2 Enterprise Edition (J2EE)

Java 2 Enterprise Edition is used to create large and powerful server-side and distributed applications, based on application server products such as BEA WebLogic and IBM Websphere. J2EE has been successful in repositioning Java for high-end server development, providing a Java solution in areas of computing previously dominated by C++ and CORBA. J2EE forms the basis of many new thin-client applications for the Web.

Java 2 Standard Edition (J2SE)

The Standard Edition of the Java 2 platform is aimed at the desktops of PCs and workstations. It contains highly portable graphics libraries and other features necessary for writing Internet-based and network-based client applications as stand-alone clients and as applets.

Java 2 Micro Edition (J2ME)

The Micro Edition of Java 2 is the newest addition to the family, and provides a mechanism for delivering Java solutions to resource-constrained consumer and mobile devices.

The Java family of editions, **virtual machines, configurations**, and **profiles,** can be best understood by the following visual model, which provides the context needed for a good understanding of the discussions that follow in the rest of this chapter:

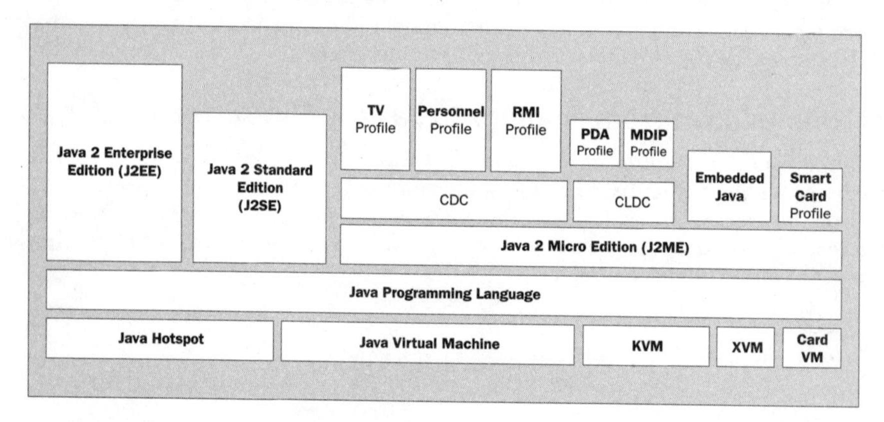

This diagram shows the relative positions of the current Java 2 elements. It also shows how the underlying virtual machine and Hotspot technologies are positioned within this family.

The Anatomy of J2ME

J2ME is not, and never will be, one solid specification dictating functionality and demanding capabilities from a target device. In fact, J2ME has been designed specifically to avoid this particular trap, and has adopted a modular extensible mechanism for embracing new device technology, while maintaining service levels and supporting code on pre-existing device types. J2ME itself, as a family of specifications, is designed to support ever larger families of devices flexibly.

This modular architecture splits the service provided by the Java environment into three separate layers:

- Virtual machines
- Configurations
- Profiles

The configuration layer is probably the most pivotal. Configurations are specified against a **class** of device; it is the configuration that defines which profiles are available, and that mandates the specific level of functionality and service from the virtual machine. The three layers can be thought of as a stack sitting on top of the device's operating system, which matches a machine's capabilities:

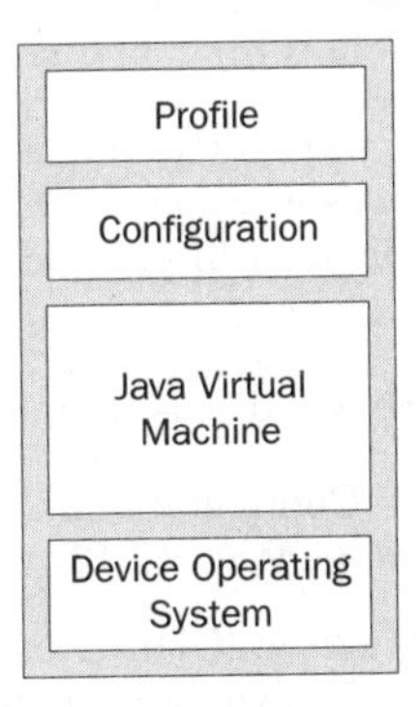

This modularity is vital to the success of the J2ME model, and to any device-centered development project, for the following reasons:

- There are a large number of possible delivery target devices for any particular application
- Device technology is constantly shifting and improving
- There is a diverse range of devices classified under the banner of consumer electronics, PDAs, and smart phones with a wide range of capabilities
- As the network application environment develops and matures, the software targeting these devices will also need to be capable of development and growth

The Virtual Machine Layer

The virtual machine used in J2SE and J2EE has grown in size since the days of Oak, and in an attempt to improve performance, its physical memory size and associated resource requirements have increased beyond the reach of all but the most powerful devices. This is especially true of the Hotspot technology used to deploy many enterprise Java applications.

In J2ME the virtual machine implementation is closely tied to the configuration that it runs. This enables J2ME to make use of far more minimal virtual machines. The virtual machine can control the way in which it implements certain features, for example, how it implements **garbage collection**, to ensure an optimum environment. These virtual machines need only provide for the underlying calls required by the particular configuration for which it is providing services, and can be customized for the underlying host operating system and device hardware to ensure that performance does not suffer unnecessarily.

The Configuration Layer

A configuration defines the minimum set of Java Virtual Machine (JVM), Java language and class library features for a particular 'horizontal' class of device – for example, groups of devices with a certain functional and capability profile.

To avoid excess fragmentation within the J2ME family, the number of configurations will be kept intentionally small (at present just two configurations are defined), and specific markets and device types will be addressed at the profile level. Having a smaller number of basic configurations, reduces the need for numerous and complex virtual machine implementations, enabling a relatively small number of configurations to serve a large number of profiles, and hence, a large number of device markets.

The configuration layer remains largely invisible to the users of a device, but is one of the most significant features to the system software developer. The configuration layer provides a contract to the suppliers of profiles, ensuring that the device will provide a certain basic set of Java libraries and services.

The two J2ME configurations, which we will discuss in the remainder of the chapter, are:

- The Connected Limited Device Configuration (CLDC) – used for quite limited devices, for example, smart-phones and simple, low-end PDAs.
- The Connected Device Configuration (CDC) – used for more capable devices with better network connectivity, for example, Symbian EPOC communicators

The focus of this book is on the former configuration, CLDC, although there are a couple of chapters in which we take a look the CDC.

The Profile Layer

A profile defines a set of APIs, and is available for a particular configuration. It provides functionality for a specific range of devices within a vertical functional market or device-type family – for example, smart phones.

The profile layer is perhaps the most visible element of the J2ME hierarchy and defines the device in the eyes of application developers. It provides a contract between the supplier of software and services for the device and the device itself, and tends to contain Java APIs that are far more domain-specific than those specified in the configuration.

A profile represents a complete specification against which application software may be developed without the need for any additional Java libraries or technology. All J2ME software is targeted against a particular profile and guarantees that certain mandatory device capabilities and behaviors will be present, and that the Java libraries and APIs will also run on the device. Portability within J2ME, refers to portability within a particular profile.

It should be noted that while a single device model will typically only support a single configuration, it could, and often will, support numerous profiles.

Current J2ME Implementations and SDKs

J2ME has attracted, a wide base of cross-industry support with mobile communications device manufacturers among the keenest to deliver early access implementations for their flagship products. Companies whose implementation or SDK we will be considering in this chapter, and indeed in greater detail throughout the whole book, are:

- Java Card
- EmbeddedJava
- The Connected Limited Device Configuration (CLDC)
- Sun's MIDP Toolkit
- Palm KVM (Spotless) implementation
- The Connected Device Configuration (CDC)
- Symbian's EPOC PersonalJava implementation

Note that we will not cover any setup or installation for these SDKs, as this will be covered in later chapters.

Java Card

Java Card technology represents the smallest version of the Java language and environment for use on extremely constrained smart card-based devices, such as:

- GSM SIM cards
- Smart cards (for example modern credit cards, such as VISA and MasterCard)
- Java Ring devices

These smart cards all contain an integrated circuit board, which houses a small microprocessor and some memory for the card to store information (typically 1Kb RAM, 16Kb EEPROM and 24Kb ROM). Using Java Card on these devices has several major advantages for smart card manufacturers, namely:

- Multiple applications can be run safely and securely on a single card
- These applications can be updated after the card is issued
- Java network-based applications can be easily accessed and utilized

- Application development in Java is easier than traditional development
- Applications can be easily ported to new types of card that support the Java Card platform
- The Java Card platform has attracted support and interest from all smart card manufacturers and, as such, is likely to add increased longevity to applications developed for this platform, compared to native applications

The Java Card environment consists of two parts; one that runs on the card itself and the other that executes off the card, usually on the card reader. Many memory and resource-intensive processes, such as class loading and bytecode verification, are run on the off-card virtual machine as it is less constrained than the on-card virtual machine.

The Java Card specification defines a runtime environment that, besides the language subset, specifies the communications protocol, security protocols, card memory usage, and an application execution model.

The Java Card environment supports only a minimized subset of the Java language, and these supported features are as follows:

- `boolean`, `byte`, and `short` primitives (32-bit `Integer` and `int` support is optional)
- One-dimensional arrays
- Java packages, classes, interfaces and exceptions, and object-oriented features (for example, inheritance, overloading, polymorphism, and scope)

Unsupported features include the following:

- All large primitive data types: `long`, `double`, `float`
- `char` primitive, `Character` and `String` classes
- Multi-dimensional arrays
- Dynamic class loading
- Security manager
- Garbage collection and finalization
- Threads
- Object serialization
- Object cloning

Java Card Networking

The communication process between the card and host (reader) is slightly unusual in that unlike TCP/IP the protocol is **half duplex**, which means that information can be transmitted from the host to the card or from the card to the host, but not both at the same time. This mode of operation is also known as **two-way alternate**.

TCP/IP, on the other hand, operates as a ***full duplex protocol****, also known as* ***two-way simultaneous****.*

The packet communication model between the card and the host is based on the exchange of Application Protocol Data Units (APDU) data packets. An APDU message contains either a command or a response to a command. The message system follows a master-slave model with the Java Card environment on the card playing the role of slave, executing the command APDUs sent from the host, as shown below:

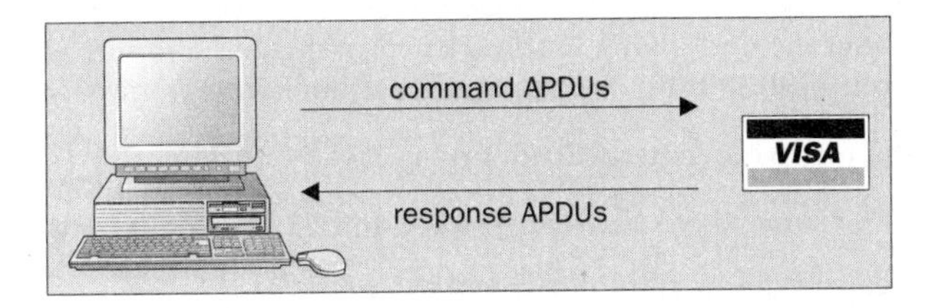

The Java Card programming and architecture model is covered in greater depth in Chapter 10, and is a highly interesting exercise in minimalism, relevant to all J2ME programmers. Full coverage of programming with Java Card is given in Java Card Technology for Smart Cards, by Zhiqun Chen (Addison Wesley, 2000, ISBN 0201703297).

EmbeddedJava

EmbeddedJava supports developers of specialized device applications who need a dedicated Java environment to create applications that do not necessarily require end-user interaction. The EmbeddedJava environment is highly configurable, and can be optimized by the device vendor to contain a specific subset of the Java language with a specialized virtual machine. This allows vendors to leave out classes and virtual machine features that are not needed, therefore optimizing the environment, while still benefiting from developing application software for their platforms in the Java language.

EmbeddedJava is optimized for environments with no need of GUIs, and that may have, at best, an intermittent networking facility. The EmbeddedJava environment contains the following layers, and can be customized, as can be seen from the configurable elements marked on the diagram:

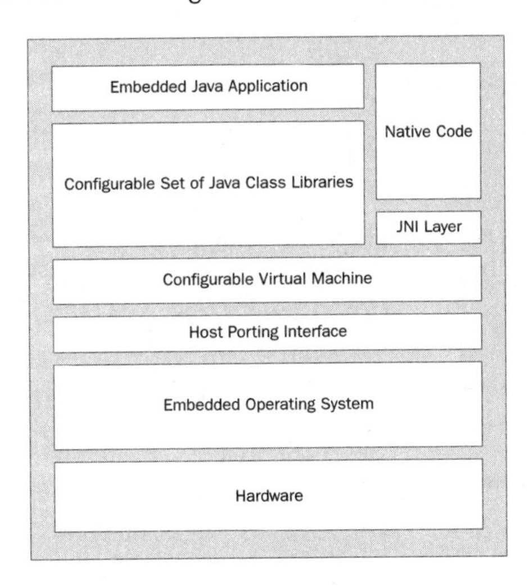

The CLDC

The CLDC has been designed to service small devices with limited resources and limited network connectivity, from mobile communication devices, point-of-sale terminals and even consumer appliances. The devices targeted by the CLDC, tend to be manufactured in high numbers, and so their unit cost must be kept to a minimum. The CLDC specification favors generality as a conscious effort to avoid assumptions about the underlying hardware within this basic remit. The CLDC is the result of the work of the expert group JSR30 (http://java.sun.com/aboutJava/communityprocess/search.html).

The CLDC makes the following general assumptions about the client device:

- Memory size of somewhere between 160 Kb and 512 Kb, split between ROM and RAM
- Low power consumption and limited battery life
- Limited, perhaps intermittent, network connection at 9600 bps or less
- Constrained, or perhaps even complete lack of, user interface

A further note on memory issues, is that the non-volatile memory size of the CLDC implementation, including its associated virtual machine, should come to not more than 128 Kb. It is also assumed that there is an additional amount of volatile heap space of at least 32 Kb for the application's execution.

The CLDC identifies, and defines, a certain basic version of Java's functionality:

- Core Java libraries (`java.lang.*` and `java.util.*`)
- I/O and networking
- Security (classfile verification and sandbox)
- Internationalization

However, the CLDC itself refrains from defining the following, leaving the implementation to its various profiles:

- User interface support
- Event handling
- High-level application model
- Persistence support

The CLDC provides almost full Java language support, with just the following exceptions:

- No floating point numbers (`float` and `double`)
- No Java Native Interface (JNI)
- No reflection
- No thread groups or daemon threads
- No weak references
- No user-defined class loaders
- No finalization (that is, no `Object.finalize()` method)
- No RMI
- No serialization

The CLDC provides only partial support for the subclasses of `java.lang.Error`, but generally supports exception handling. This is, in part, due to the fact that error handling is expensive and entails a significant overhead, but it is mainly because error recovery is highly device-specific. This error handling would be difficult to abstract sufficiently, in order to effectively define the necessary support at the configuration level.

The CLDC also introduces a new package to the Java language: `javax.microedition.*`. Special mention should be made of this package, which includes the `javax.microedition.io.Connector` class. In order to replace the rich I/O and networking provided to J2SE and J2EE by the `java.io.*` and `java.net.*` packages, this class provides a lighter weight model for obtaining connections. Sub-classes of the `Connector` class can be implemented (in profiles) to return a number of connections types.

An example of obtaining various connections using the `Connector` class is shown below:

```
Connector.open("http://www.concisemobile.com");    // HTTP

Connector.open("socket://127.0.0.1:9000");         // socket

Connector.open("file:concisewrox.doc");            // files
```

Note that the CLDC itself does not define any protocol implementations, but relies on its profiles to supply the necessary infrastructure.

The K Virtual Machine (KVM)

There is currently a wide gap in the JVM technology market between the virtual machine implementations for J2SE and PersonalJava (now a CDC profile), and the minimal virtual machine used in the Java Card platform.

The standard JVM of J2SE and PersonalJava is too large to run on many consumer devices and even most sophisticated smart phones. The Java Card virtual machine also lacks the basic functionality to take advantage of these types of consumer device hardware.

The virtual machine and the configuration need to be uniquely designed, therefore, to capture and maximize the capabilities of the underlying device. This was the motivation for the creation of new, minimal virtual machines with full Java bytecode support, like that used in the **Spotless** implementation, which we will discuss later, and the KVM itself.

> *The 'K' originally stood for Kuaui – a codename for the project, and not Kilobytes as you might expect. However, this soon changed to Kilo as it refers to the fact that the virtual machine was designed to only take up a matter of Kilobytes of memory, rather than the much larger amount required for the other available JVMs.*

The KVM is a highly portable JVM designed from the ground up for use on devices that have limited memory and network connectivity, such as smart phones and personal digital assistants (PDAs). It is designed specifically to be:

- Used on small devices – the virtual machine memory footprint is around 40 Kb to 80 Kb
- Easily portable to new devices
- As complete and high performance as possible

The KVM is currently closely linked to the CLDC and is often used to deploy the CLDC on devices. However, the CLDC is not tied to a particular virtual machine, and in future it is likely that the CLDC will utilize other JVMs.

Running KVM Applications

The KVM can be run in a number of different ways:

- From the command line on desktop implementations (emulators and SDKs).
- On host operating systems such as the PalmOS and EPOC. The KVM can be configured to launch in the same way as native applications.
- For other devices that do not have this sort of operating system, the KVM provides an implementation of the Java Application Manager (JAM). The JAM serves as an interface between the application launching requirements of the host operating system and the JVM.

The Mobile Information Device Profile (MIDP)

The MIDP specification was developed as JSR37 by the Mobile Information Device Profile Expert Group (MIDPEG), which includes representatives from companies such as America Online (AOL), Ericsson, Nokia, NTT DoCoMo, Palm, Research in Motion, and Symbian.

It was the first profile designed to run on top of the CLDC, and the main goal was to achieve simplicity rather than completeness. The openness of the environment was crucial in attracting the third-party device vendors and developers, essential to the platform's success in the marketplace.

The MIDP defines a profile for hardware that should have the following minimum specification:

Hardware Aspect	Minimum Specification
Display	Screen-size: 96x54 pixels
	Display depth: 1-bit
	Aspect ratio: approx. 1:1
Input by one of	One-handed ITU-T phone keypad
	QWERTY keyboard
	Touch screen
Memory	128 Kb of non-volatile memory for the MIDP components
	8 Kb of non-volatile memory for application-created persistent data
	32 Kb of volatile memory for the Java runtime heap
Networking	Two-way, but possibly intermittent network connection with limited bandwidth

The MIDP also defines some additional packages to those found in the CLDC:

Package	Purpose
`javax.microedition.rms`	Defines an interface to the MIDP's simple record-oriented storage mechanism, called the Record Management System (RMS)
`javax.microedition.midlet`	Defines the MIDP application model, the MIDlet
`javax.microedition.io`	Extends the generic `Connection` framework of the CLDC to include networking capabilities for HTTP
`javax.microedition.lcdui`	Provides a basic set of user interface components for use in MIDlets

Scope of the MIDP

The devices targeted by the MIDP potentially have a wide range of capabilities. Certain items of functionality however, are considered to be within the scope of the MIDP specification – these include:

- Persistent storage mechanism
- Timers
- Networking with HTTP (extends the CLDC `Connection` framework)
- User interface components
- Application life cycle

By the same token, the expert group defining the MIDP considered certain aspects to be outside the scope of the specification, as too device-specific:

- System-level APIs
- Application delivery and management
- Low-level security (the MIDP does not extend the scope of the CLDC in this area)
- Application-level security
- End-to-end security

J2ME, and therefore MIDP devices, provides the ability of not only loading new applications to the device, but also to upgrade the J2ME environment by downloading new class libraries and components of the Java environment. This is the responsibility of the JAM.

How the CLDC, KVM, JAM and MIDP Coexist on a Device

The mobile device and its operating system support both native applications and the basic networking interfaces, which enable the CLDC and JAM to be ported to the device.

Once the basic virtual machine and CLDC configuration are present, the MIDP and any additional CLDC-compliant profiles may be installed to give a fully functioning J2ME-compliant environment, necessary for the execution of non-device-specific J2ME-based applications.

This diagram shows the logical layout of operating system and infrastructure software on a device aimed at a MIDP device:

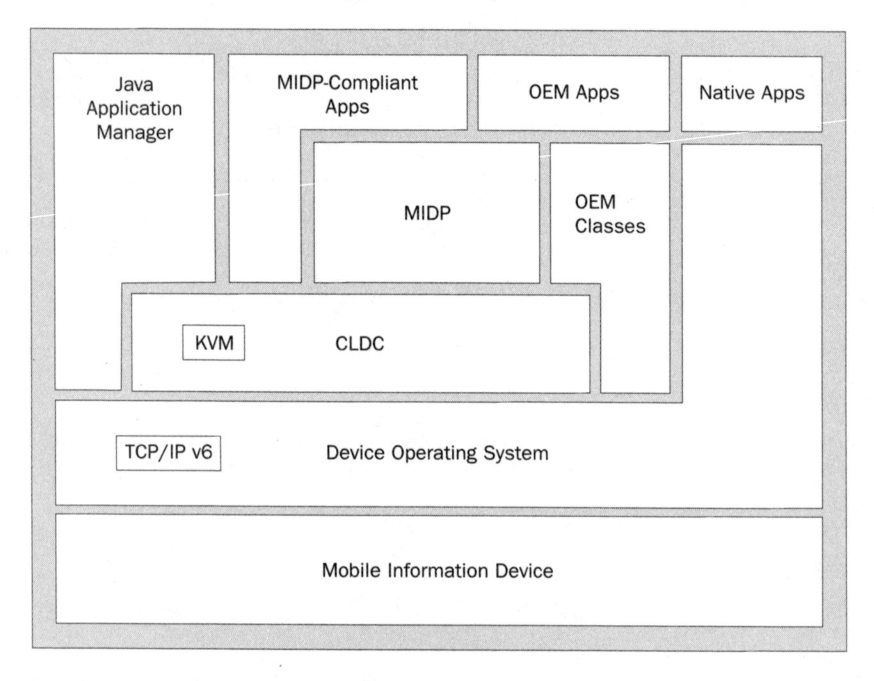

MIDP-based applications take the form of MIDlets, small applications written to include a specific life-cycle interface, specialized for small resource-constrained devices. We shall now take a look at MIDlets to give a feel for the examples that follow.

MIDlets

The MIDP defines an application model optimized to share the limited resources of the underlying device between multiple MIDP applications (MIDlets).

The MIDlet Life Cycle

The interaction of MIDlets with the MIDP application manger requires certain application life cycle criteria to be met by the MIDlet. The MIDlet must define the following methods:

Method	Function
`startApp()`	Acquires resources and starts the MIDlet execution
`pauseApp()`	Releases certain resources and ensures the MIDlet becomes quiescent, or inactive
`destroyApp()`	Releases all the resources held by the MIDlet, destroys any threads and ends all activity

The MIDlet is initially in a **paused** state when the device is turned on:

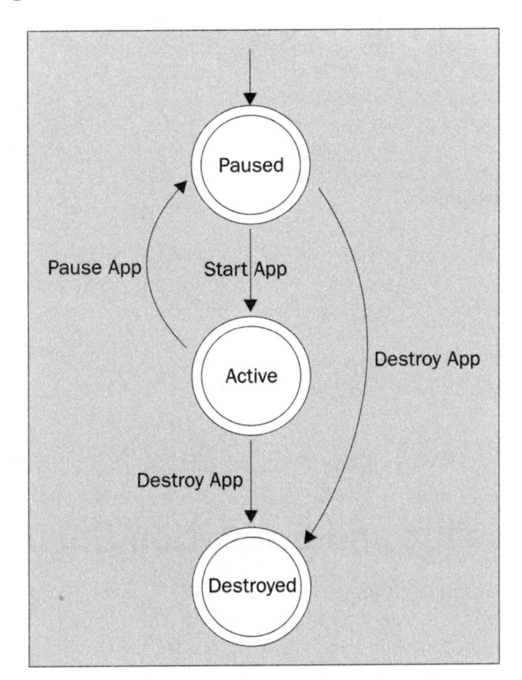

An Example

Now that we have reviewed the evolution of J2ME and considered the CLDC and MIDP combination in some detail, we are ready to take a look at some code aimed at the MIDP.

The example we will look at is a version of the ubiquitous 'Hello World' application. This simple MIDlet shows a label carrying the traditional message, and an exit option to cease execution. We will then take a look at how the MIDlet might look running on a variety of devices from the Sun Microsystems development environments. These will be explained fully in Chapter 4.

The following source code is named, unsurprisingly, `HelloMidlet.java`:

```
import javax.microedition.midlet.MIDlet;
import javax.microedition.lcdui.Command;
import javax.microedition.lcdui.CommandListener;
import javax.microedition.lcdui.TextBox;
import javax.microedition.lcdui.Displayable;
import javax.microedition.lcdui.Display;

public class HelloMIDlet extends MIDlet implements CommandListener {
  private Command exitCommand;
  private Display display;

  public HelloMIDlet() {
    display = Display.getDisplay(this);
    exitCommand = new Command("Exit", Command.SCREEN, 2);
  }
```

```
    public void startApp() {
      TextBox txtHello = new TextBox("Hello World", "", 20, 0);

      txtHello.addCommand(exitCommand);
      txtHello.setCommandListener(this);
      display.setCurrent(txtHello);
    }

    public void pauseApp() {}

    public void destroyApp(boolean unconditional) {}

    public void commandAction(Command c, Displayable s) {
      if (c == exitCommand) {
        destroyApp(false);
        notifyDestroyed();
      }
    }
  }
```

The Sun MIDP Toolkit and Emulation Environment

Sun have provided a development environment and toolkit, aimed at producing MIDlet code. The **Wireless Toolkit** has the ability to run a number of simulators, for example with our MIDlet composed above, the environment might provide the following simulations:

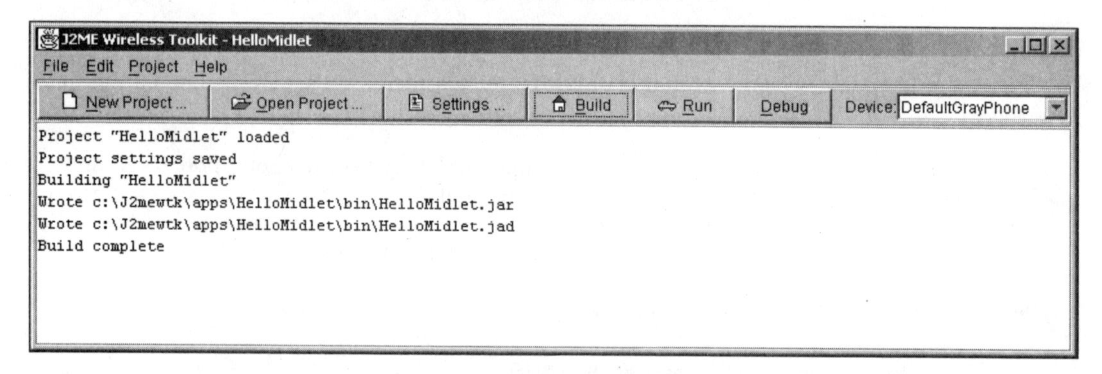

The phone and pager shown, are the ones provided 'out-of-the-box' by the Wireless Toolkit environment and give an indication of how the MIDlet might look on a number of different types of device and smart phones, ranging from a low-end Internet phone (similar to the first generation of WAP devices), to a more advanced smart phone, and even how the MIDlet might look on pager technology:

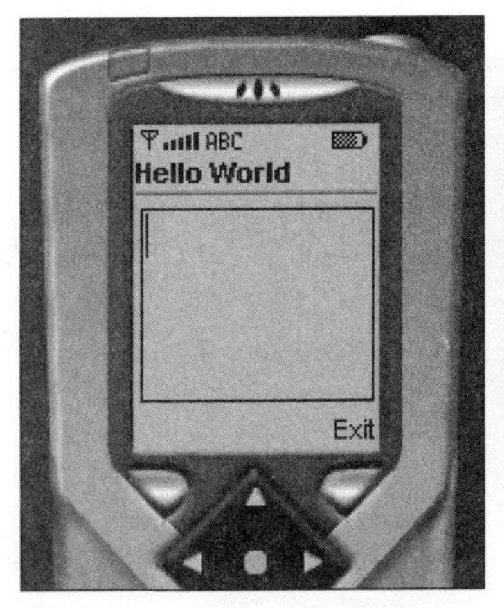

The Personal Digital Assistant (PDA) Profile

Several manufacturers, notably Palm Computing and Symbian, have spearheaded the development of J2ME on PDA devices. The Personal Digital Assistant (PDA) Profile specification (JSR75), being developed by an expert group led by Palm, will be designed to sit on top of the CLDC. It is aimed specifically at PDA-type devices such as those manufactured by Palm, whose capabilities exceed those of MIDP devices.

A typical device within this profile will:

- Have not less than 512 Kb of memory, and no more than 16 Mb.
- Usually be battery operated.
- Have reasonable GUI capabilities (20000 pixels), and allow for both character-based input and a pointing device. Will support a version of the AWT.

Palm and the Spotless Implementation

PalmSource'99 and JavaOne'99 were watershed events for J2ME, as they featured the high profile launch of the **Spotless Palm** environment. Most of the hype surrounding this event was generated by the highly successful Spotless application model, which enabled application developers to create simple Java applications for the Palm platform.

The Spotless environment grew out of the Sun Laboratory's **Kanban** project (http://www.sun.com/research/kanban/), aimed at creating a highly optimized, pure Java environment for constrained devices – initially the PalmOS computing platform. The KVM was developed as a direct descendant of the Spotless JVM implementation, and the Palm device has continued to play a central role in the development of J2ME.

Despite the fact that the Palm KVM implementation has continued to support the Spotless application model, it is unlikely that the **Spotlet** model will continue to be developed as the J2ME platform matures. It is probable that, on all future CLDC/MIDP implementations, MIDlets will replace the Spotlet model, which is currently still supported on the Palm KVM.

The KVM is still supported on the Palm, even though it was not really designed for such powerful devices. However, both Palm and Sun are committed to the KVM, and they will release an implementation of the MIDP on Palm while the PDA Profile is developed. This is due to the huge, and continuing, response from Palm users generated since PalmSource'99.

Here is another simple 'Hello World' application, this time aimed at the KVM implementation running on the Palm platform. These applications can be tested on Palm devices and on the Palm OS Emulator (POSE) emulation environment. This piece of code is named `HelloPalm.java`:

```
import com.sun.kjava.Spotlet;
import com.sun.kjava.Graphics;
import com.sun.kjava.Button;

public class HelloPalm extends Spotlet {

  /*
   * Maintain the graphics handle as a singleton pattern (see "Design
   * Patterns", Addison-Wesley for more details).
   */
  static Graphics g = Graphics.getGraphics();

  // The close button
  Button butClose;

  /*
   * This main method creates the HelloPalm Spotlet, then registers its
   * event handlers.
   */
  public static void main(String[] args) {
    HelloPalm helloPalm = new HelloPalm();
    helloPalm.register(NO_EVENT_OPTIONS);
  }

  // Default constructor creates and renders the GUI components.
  public HelloPalm() {

    // Create the close button
    butClose = new Button("Close", 130, 140);
    paint();
  }

  private void paint() {
    g.clearScreen();
    g.drawString("Hello World", 60, 80);
    butClose.paint();
  }

  // Handle the pen down event on close button.
  public void penDown(int x, int y) {
    if (butClose.pressed(x, y)) {
      System.exit(0);
    }
  }
}
```

This code can be tested in the KVM test environment from the command line, as shown below (note the OS specific parameter in the second and third commands), but the exact details of how this is achieved is explained later on in Chapter 4:

```
javac -classpath <CLDC_HOME>\bin\kjava\api\classes HelloPalm.java

<CLDC_HOME>\bin\<OS>\preverify -classpath<CLDC_HOME>\bin\kjava\api\classes
;.;<JAVA_HOME>\jre\lib\rt.jar HelloPalm

<CLDC_HOME>\bin\kjava\<OS>\kvmkjava -classpath <CLDC_HOME>\bin\kjava\api\
classes;output HelloPalm
```

Or alternatively, it can be tested directly on the POSE that has the KVM Palm implementation installed. Again the details of how to do this are beyond the scope of this introduction chapter, so the procedures for doing this are explained later on:

The Connected Device Configuration (CDC)

The CDC of J2ME accommodates a more powerful device than the CLDC, and also one that has a better connection to the network. The CDC is specified in JSR36, and is aimed at the following device type:

- 512 Kb ROM minimum
- 256 Kb RAM minimum
- Connectivity to a network
- Support for a full JVM implementation

The user interface for a CDC device is not specified exactly and can range in sophistication from a fully-fledged GUI, through devices with a relatively constrained user interface, to no user interface at all.

A typical example of a CDC-type device could be an Internet screen phone, a television set top box, or a high-end connected personal organizer (such as an EPOC Communicator device).

From PersonalJava to the Personal Profile (JSR62)

The **Personal Profile** is an extension of the technology formally known as PersonalJava into the world of J2ME. A Personal Profile-based device fits broadly into the following range:

- 2.5 Mb of ROM
- 1 Mb RAM
- Robust connectivity to a network
- A graphical user interface capable of running applets and AWT components

Comparison of PersonalJava to the Personal Profile

The Personal Profile and CDC replace the PersonalJava environment as shown in the diagram below:

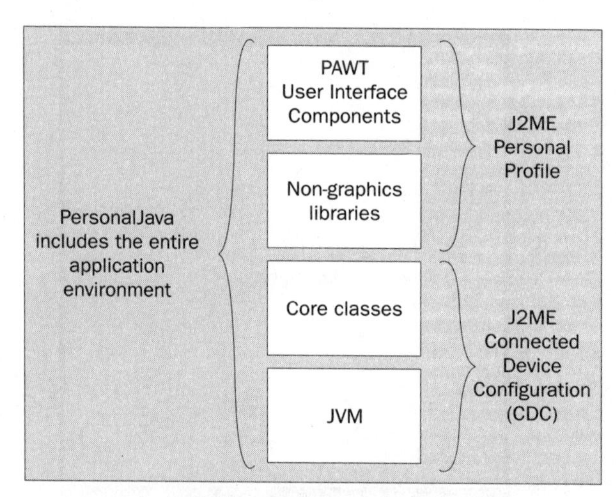

Personal Profile Extensions

The Personal Profile environment excludes certain niche functionality from the main specification. This missing functionality will be accommodated by a raft of new APIs, which define functionality focused on specific industry and functional needs. By way of example, we shall consider the JavaPhone API and the JavaTV API.

JavaPhone API

The JavaPhone API is a vertical extension to the main Personal Profile and consists of two separate profiles, one targeting Internet screen phones and the other targeting smart phones/communicators. The API is designed specifically to provide access to the functionality unique to wireless smart phones and Internet screen phones.

Particular focus is on the telephony functions needed by applications on devices of this nature, which include:

- Direct telephony control
- Datagram messaging
- Access to address book and calendar information
- Access to user profile information
- Power monitoring and management
- Application installation mechanisms

The JavaPhone API provides an optimal environment for the secure delivery of content and services to telephony-enabled devices.

JavaTV API

The JavaTV API is another extension to the Personal Profile, which provides development and deployment platform for emerging interactive television services – remember this is, in part, what Java was designed for in the first place.

The API provides developers with the tools needed to develop applications for the demanding real-time display of context-sensitive information to add value to television programming and advertising. Examples of this might include the ability to initiate e-commerce transactions from television advertisements, and interactive communication with television programs such as chat shows, game shows, and sports presentations.

The JavaTV API is being designed to provide access to the functionality unique to a digital television receiver. This functionality includes:

- Audio/video streaming
- Control over certain television mechanisms
- Access to service information data
- Tuner control for channel changing
- On-screen graphics control to allow for user interaction

Other Planned CDC-Based Profiles

The CDC provides a powerful and function-rich environment for higher-end devices and appliances. As more industry expert groups are formed, a number of device profiles will appear. At present, the most important profiles under discussion are the **Foundation Profile** (JSR46), and the **RMI Profile** (JSR66), which are discussed briefly below. More detailed coverage can be found in Chapter 3.

Foundation Profile (JSR46)

The Foundation Profile extends the CDC for devices that need a rich set of networking functionality but have little or no need for a graphical user interface. It provides a useful basis for devices needing network connectivity, where there might be a requirement to implement additional profiles to provide graphics capabilities or other specialized features.

A typical Foundation Profile device would have the following hardware characteristics:

- A minimum of 1 Mb ROM
- A minimum of 0.5 Mb RAM
- Network connectivity
- No GUI support (unless supplied by an additional profile)

Java RMI Profile (JSR66)

The Remote Method Invocation (RMI) Profile is intended to work in tandem with the Foundation Profile to supply additional support for inter-application messaging using RMI over TCP/IP.

The Symbian EPOC Java SDK

Symbian has been an early adopter of Java, and has a full implementation of PersonalJava for the EPOC operating system. The EPOC v5 Java environment is based on the on the JDK 1.1.4 platform. However, with EPOC v6, there are substantial developments in EPOC support for recognized J2ME configurations and profiles.

The 'Hello World' application for EPOC, like most PersonalJava code, is actually identical to a simple J2SE application and, unlike the other examples seen so far, is quite capable of execution under J2SE:

```
import java.awt.Frame;
import java.awt.Label;
import java.awt.Button;
import java.awt.GridLayout;
import java.awt.event.ActionEvent;
import java.awt.event.WindowEvent;
import java.awt.event.WindowAdapter;
import java.awt.event.ActionListener;

public class HelloEpoc extends Frame implements ActionListener {
  public static void main(String[] args) {
    HelloEpoc hello = new HelloEpoc();
  }

  private HelloEpoc() {
    Label lblHello = new Label("Hello World");
    Button butClose = new Button("Close");
```

```
        butClose.addActionListener(this);
        this.addWindowListener(new WindowAdapter() {
          public void windowClosing(WindowEvent e) {
            System.out.println("Closing");
            System.exit(0);
          }
        });
        this.setLayout(new GridLayout(2, 1, 10, 10));
        this.add(lblHello);
        this.add(butClose);
        this.pack();
        this.show();
      }

      public void actionPerformed(ActionEvent e) {
        dispatchEvent(new WindowEvent(this, WindowEvent.WINDOW_CLOSING));
      }
    }
```

This `HelloEpoc.java` example can run quite happily under J2SE. For example, here is how it will look running under the Microsoft Windows J2SE implementation:

The same code can also run, without recompilation, under the EPOC Java simulator:

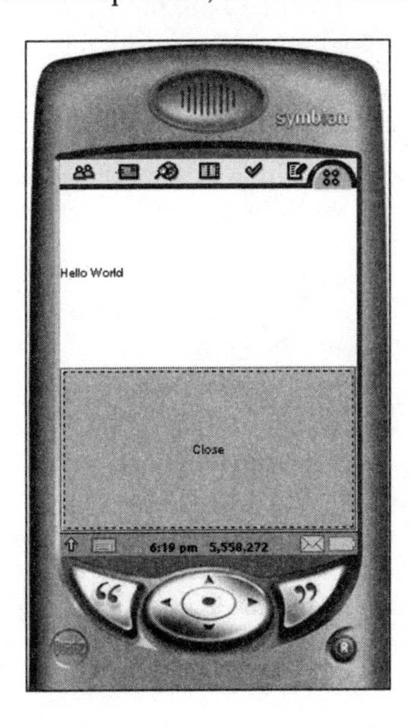

Symbian plan to continue support for Java in the EPOC v6 release, and also, their Quartz and Crystal specifications include support for the J2ME CDC and Personal Profile. Symbian's Quartz specification is intended to run on smart phones while the Crystal specification is intended for higher-end personal communicators. Symbian also plan to include support for applications designed for the MIDP in EPOC v6.

Other Technologies Impacting Mobile Java Development

There is a huge amount of innovation occurring in the mobile, wireless, and device-based computing arena at the present time. This innovation is occurring as a triad of technologies develops and matures, namely:

- Carrier technology and wireless networks
- Device technology
- Mobile Java and thin-client technology

Good J2ME programmers will need to maintain their knowledge of other technologies and developments in the field of mobile computing, in order to prepare for, and take advantage of, these innovations as they occur. It is essential, for the better understanding of J2ME, to be able to place these technologies within their wider historical context.

Carrier Technology: GSM to 3G Networks

Wireless voice and data transmission technologies are naturally of paramount importance to the widespread availability and uptake of all wireless services.

For mobile technology to succeed, wireless networks will need to provide organizations with reliable, high bandwidth wireless networks across their operational territories, and individuals with the freedom to roam over national and international boundaries, without complex billing, device configuration, or service level fluctuations.

The evolution of network technologies may be the single most important factor in determining the success of J2ME and wireless data services, for without the appropriate network coverage – both in terms of bandwidth and reliability of service coverage – wireless data solutions are not viable.

The success of J2ME is intrinsically bound up with the successful evolution of mobile and wireless networks. The intensity with which the mobile and cell phone network operators are striving to achieve dominance is evidence of the critical advantage which success in this area could bring.

2G: GSM (Global System for Mobile Communication)

GSM operates in the 900MHz and the 1800MHz (1900MHz in the US) frequency band, and is the current standard in Europe and most of the Asia-Pacific region. More than 50% of the world's mobile phone subscribers are connected via GSM networks. In the US there are only about 5 million GSM users (this compares to well over 200 million outside the US), with a high proportion of pre-GSM devices still in use.

GSM is a common standard throughout Europe, with seamless international roaming now taken for granted between European countries. This high level of market penetration – reaching 50% or more in most major European countries – has enabled a range of innovative services to become commercially viable in the region.

2½G: GPRS (General Packet Radio Service)

GPRS brings the benefits associated with packet switched protocols to wireless networks. Principally this is the ability to maintain a connectionless communication with the server, enabling near instant access to data networks without having to re-establish a connection each time it is lost.

GPRS works by bundling GSM channels together, allocating them to those users of the network that require higher bandwidth. Since users who pay for more bandwidth are given a larger slice of what is available, and the underlying supply from the network cell is limited therefore, the greater the number of users connecting via GPRS, the lower the average data rate each user will enjoy.

This means that in practice, while transmission speeds of up to 171Kbit/s are theoretically possible, actual transmission rates will be reduced to lower than 30Kbit/s as the user base grows, and the ratio of network cells to users falls.

A basic 'always on' GPRS service has been available in Japan for over a year through the NTT DoCoMo iMode service. GPRS has been the critical success factor for a wide range of services in Japan, and communication technology experts predict that this will also be the case in Europe and the US, once GPRS is established in those regions.

3G (3rd Generation)

3rd Generation networks will bring a higher data transmission rate to wireless networks, and, unlike GPRS, which builds on the existing GSM technological legacy, requires a new network infrastructure to be created.

It is predicted that NTT DoCoMo will launch the world's first commercial 3G network in Japan before the end of 2001. Some time after this date, the European network will roll out across each country, led by the Nordic countries; estimates for the timescale for the delivery of this network range from 2002 to 2005. The pan-European network can be reasonably expected, then, to have reached completion by the end of 2005, while coast-to-coast US availability may lag behind Europe by around 2 years.

The quality of 3G services outside major urban centers in the US is likely to be very varied. UMTS (Universal Mobile Telecommunications System) requires a considerably higher number of base stations to provide coverage over a given area than GSM or other 2G systems. Within a country such as the US, with enormous open spaces, this may lead to significant gaps between areas of high population density.

3G is the next major generation of wireless technology on the carrier side. The official headline figure for data transmission rates is often quoted as 2Mbit/s. This is unlikely to be achieved initially, however, since this figure assumes a stationary user with exclusive usage of a base-station; a more realistic transmission rate for 3G networks is expected to be a maximum of 384Kbit/s in metropolitan areas.

It should be noted that while the term UMTS is often used as an umbrella description for 3G standards, globally the standard for 3G is the IMT-2000 (International Mobile Telecommunications 2000).

It is forecast that there will be three optional harmonized modes within the overall IMT-2000 standard:

- W-CDMA (Wideband Code Division Multiple Access) for Europe and the Asian GSM countries
- Multi-carrier CDMA for North America
- TDD (Time Division Duplex)/CDMA for China

It should eventually be possible to use a single mobile handset/device across these modes within the overall IMT-2000 standard. However, this is unlikely to be achieved for several years, however.

Wireless Networking Technology

Local Area Wireless Network Technology will change the way we use device-based technology, both in the office, in the home, and on our journeys between the two. These new, short-lived, fluid wireless network structures will have a huge impact on the way J2ME developers design, and use, external services, both from other devices and servers. An understanding of the workings of and developments in wireless network technologies is a critical skill for Java developers to acquire.

We will take a brief look at three wireless networking technologies that are currently under development, and will be used in order to create local short range and powerful networks for collaborative device-centric networks:

- Bluetooth
- Wireless LAN (802.11b and HIPERLAN)
- Infrared

Bluetooth

Bluetooth (http://www.bluetooth.com) represents a new standard for cable-free networking. Bluetooth was created with the specific intention of delivering dynamic, collaborative, and short-lived networks between communities of devices.

Bluetooth was originally developed by Ericsson, and to ensure cross-manufacturer co-operation and agreement, the Bluetooth Special Interest Group was founded in February 1997 with IBM, Intel, Nokia and Toshiba as founder members. The group now includes over 2000 members developing the Bluetooth specification, and producing inter-operable Bluetooth devices.

The Bluetooth Network Model

Bluetooth-enabled devices are designed to detect and communicate dynamically with compatible devices within their communication range of around 10 meters (32 feet), using the radio frequency range around 2.4GHz. The 2.4GHz–2.4835GHz range is freely available internationally, known as the 2.4GHz Industrial-Scientific-Medical (ISM) band, and we shall encounter it again with 802.11b Wireless LAN and Home RF.

When Bluetooth-enabled devices encounter each other, they are able to establish a temporary network known as a Personal Area Network (PAN) or a Piconet. The devices can be securely identified, and devices known to each other can seamlessly connect to share services. Furthermore, unknown devices can be configured with security levels to offer services, or wait for manual confirmation before authorizing a new device's connection to the Piconet.

The Bluetooth Communication Model

The Bluetooth network uses a master-slave model, with one device playing the role of **master**, and coordinating the communication of the other **slave** devices in the Piconet.

Bluetooth devices will be required to operate in a noisy radio environment. Dealing with data-collision and preventing security violations and snooping on Piconet communication are key issues.

The military have had to deal with these same issues in many of their radio frequency communication models for much of the last century; the techniques used in both Bluetooth and Wireless LAN are therefore able to take advantage of the maturity of military research in this area.

Spread Spectrum Communication Techniques

Bluetooth and Wireless LAN both use techniques in their communication models to spread a single message exchange over a range of frequencies. There are two main types of Spread Spectrum techniques used in commercial wireless communication:

- Frequency-Hopping Spread Spectrum (FHSS) – used in Bluetooth
- Direct Sequence Spread Spectrum (DSSS) – used in 802.11b Wireless LAN

Spread Spectrum techniques use a range of frequencies to transmit a single message. Since the mechanism for performing this frequency spreading is known only to the sender and intended recipient of the message, it makes the transmission more difficult to block and intercept. Furthermore, unless one knows how to follow the message across the range of frequencies used for its transmission, it will appear to be more like noise than a coherent message.

The differences between these techniques center on how the spread of frequencies is utilized in the sending of a message. The Frequency-Hopping Spread Spectrum model used by Bluetooth splits up the message for transmission into packets, and uses an algorithm to 'hop' between frequencies when transmitting these packets over the Piconet. Only other members of the Piconet will be able to follow these hops; to other devices who are in range, but who are not members of the Piconet, the transmission will have the appearance of white noise.

Benefits of Bluetooth

Bluetooth promises a potential revolution in mobile technology, by allowing high-bandwidth temporary and dynamic networking of devices:

- High data transfer rates of approximately 1 Megabyte per second (Mbps)
- Low power consumption and efficient power-saving modes
- Secure, robust communication even in **noisy** environments, using military standard FHSS techniques
- Inter-operability using the globally available 2.4GHz ISM transmission range, and manufacturers cooperating with the Bluetooth Special Interest Group to ensure device level compatibility
- Ease of use, through low maintenance and seamless connectivity via dynamically formed Piconets

Development of Java APIs for CLDC devices to interface with Bluetooth is the responsibility of JSR 82. It will look at support for the RFCOMM protocol for stream connections, and the higher-level OBEX protocol for synchronization and file transfer.

Wireless LAN - 802.11b and HIPERLAN

Wireless LAN is intended to assist and augment the existing wired infrastructural models connecting Ethernet LANs. We will concentrate our discussion on the 802.11b standard, as this is by far the most widely understood in the market place; the HIPERLAN (High Performance Local Area Network) specification supported by the European Telecommunications Standards Institute (ETSI) has yet to make a commercial impact.

The technology is based upon the 802.11 standard developed by the Institute of Electrical and Electronic Engineers (IEEE) and products developed to this standard are checked by the Wireless Ethernet Compatibility Alliance (WECA) for cross manufacturer inter-operability compliance.

802.11b Wireless LAN transmission rates can reach up to 11 Mbps, and options for providing 40-bit and 128-bit encryption are available.

The 802.11b Wireless LAN Communication Model

Like Bluetooth, 802.11b Wireless LAN operates in the 2.4GHz frequency range, and uses a Spread Spectrum technique to reduce the effect of radio noise and increase the security of transmissions. However, 802.11b uses the Direct Sequence Spread Spectrum (DSSS) method. Despite this, the effect is the same and the pseudo-random transmissions are hard to detect, decrypt, demodulate, and block.

Wireless LAN Network Topologies

Wireless LANs can be operated in two main configurations:

- Fixed infrastructure topologies
- Ad hoc topologies

Fixed Infrastructure Topology

In a fixed infrastructure topology, specialized wireless nodes, called microcells or access points, broadcast and receive messages within a fixed area (for example, within areas of an office building). The network infrastructure can take care of forwarding functions, and can control the interaction with the wired LAN.

These access points, shown in the diagram, provide a bridge to the organization's wired network infrastructure:

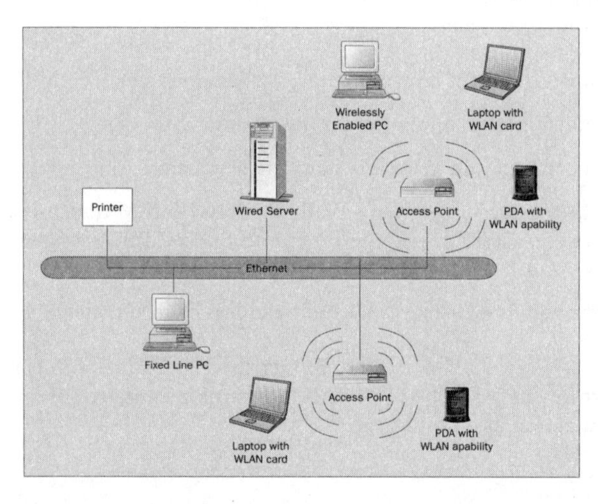

Ad hoc topology

An ad hoc, wireless network relies on each node to communicate in a peer-to-peer fashion to establish a wireless network:

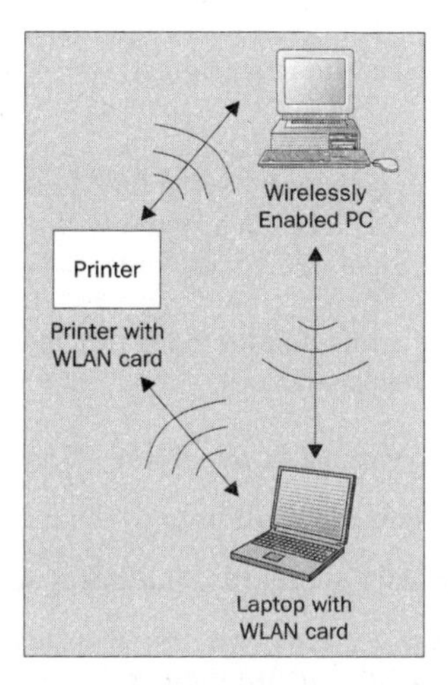

The complexity of each node must be slightly higher, since each node must be capable of independent control and operation.

This is the style of network topology favored by Bluetooth, however, it is unlikely that future wireless LAN infrastructures will be exclusively designed with only one of these topologies allowed. A mixed infrastructure made up of core fixed infrastructural areas and functions linked to a wired network, will be complemented by ad hoc and collaborative local structures, to give an optimal working environment.

Bluetooth versus 802.11b

Much has been made of the competition between 802.11b Wireless LAN and Bluetooth, but in reality the technologies, while overlapping, are intended for different markets.

Wireless LAN is already enabling organizations to replace their wired LAN infrastructure with cable-free wireless environments that will reduce maintenance overheads and allow their employees to benefit from modern open plan and hot-desk workplaces without the need for re-cabling, or docking stations for laptops.

Bluetooth, once enabled devices become more widely available, will open up the world of personal and consumer devices, allowing us to create networked clusters of collaborative devices dynamically and robustly. This will potentially enable new and exciting cross-device consumer services and information exchange, in the following areas:

- Seamless data synchronization
- Network and Internet access
- Peripheral connectivity
- Robust and efficient data exchange

The scope of these technologies does, of course, extend beyond this, and as software technologies like Jini and XML-based web service protocols mature and are adopted in this space, there is sure to be innovation in this area.

Infrared

Infrared communication (via Infrared Data Association (IrDA) compatible mobile phones) is utilized by many PalmOS, Pocket PC, and EPOC devices to connect to the Internet. IrDA is also used by these devices in order to exchange information (such as agenda and contact details) install software, and to communicate with network services, such as printers.

Infrared-enabled devices are shipping to consumers in huge numbers: it is estimated that over 247 million IrDA-enabled devices will be purchased in 2001, including 12 million Palms, and 9.3 million printers (IrDA.org figures). However, IrDA is a severely limited networking technology compared to the other wireless networking technologies we've just seen.

Infrared networking is likely to decline as Bluetooth replaces it as the communication model of choice in consumer devices for the following reasons:

- It has a short transmission range (a maximum of 5 meters)
- It relies on line of sight communication between a maximum of two devices
- Suffers from lack of industry standards, and has many proprietary implementations
- Does not deal will effectively with noisy environments – just try beaming information with a Palm in a crowded airport departure lounge

Considerations for Developers

The migration from circuit-switched protocols such as GSM to packet-switched mechanisms, such as GPRS and 3G will have a profound effect on the convenience, usability, and reliability of wireless network data services. Likewise, the move to cable-free local area networks and Piconets, will have a profound influence on the way we use computing power.

The 'always on, always connected' and push capabilities that packet data networks allow, will make it possible to reliably deliver key alerts and messages via J2ME-enabled mobile devices which are far more sophisticated than the simple pagers currently used for this purpose.

Developers using J2ME for the construction of device-centered application services should endeavor to pursue a strategy that, as far as possible, isolates the backend architecture and business-logic application layers of mobile solutions from the carrier protocol used to transmit the information through the airwaves.

The Move to 3G

Given the uncertainties surrounding the availability of increased bandwidth and the cost of using these services, developers should adopt a habit of considering bandwidth a scarce resource. Any development work tied exclusively to a particular carrier technology immediately limits the scope of its deployment, as well as imposing an external longevity constraint that may be costly to overcome in the future.

Building bloated applications that require large amounts of data to be transmitted on the assumption that increased bandwidth will soon come to the rescue (a sort of bandwidth equivalent of Moore's Law) is a high-risk strategy that could lead to user disappointment. While it is clearly desirable to start planning for increases in bandwidth, J2ME-based developments should be designed and built around bandwidth needs and connection models that can be supported by existing carrier technologies.

The Move to Local Wireless Networks

The benefits of Personal Area Networks and configuration-free, cable-free, environments will lead to whole new areas of computing and types of devices. As mobile devices start to communicate dynamically with each other and their surrounding networks, we will be able to create new types of services and communication that may, in turn, be as revolutionary as email was to the Web.

The move to local and personal wireless networks, connecting to the global Internet, with the additional possibility of location-based services made possible by the carrier networks and Global Positioning Service (GPS) systems, will be a turning point for mobile technology.

Applications that use wireless networks to communicate with the server can do so on three levels:

- The ability to transmit (and receive) information (such as contact address details and position)
- The ability to synchronize device and server datastores
- The ability to network interactively with a server-side framework to create a client-server application

As wireless network technology improves, we shall see a movement from the first two levels to the third, which will require a further level of programmatic complexity from application designers, but which will ultimately lead to more usable and useful applications.

J2ME and Browser Technologies

There are two major competing Wireless Markup Language standards, for developing thin-client browser-based applications, namely:

- The Wireless Application Protocol (WAP)
- NTT DoCoMo's iMode Technology

While iMode is prevalent in the Japanese market, WAP is still struggling to make inroads in Europe and North America. Although it currently lacks the innovation of the NTT DoCoMo's services, and suffers from the predominantly GSM-based European network constraints, with the advent of 2.5G and 3G Networks and WAP 2.0's new features, WAP will become more usable and better services should be possible.

J2ME and WAP overlap in the smart phone and PDA market, with WAP targeting devices that have a permanent connection to the network and applications that can be entirely server-based, and with J2ME targeting the needs of applications and devices that rely on an element of local computing power.

Wireless Application Protocol (WAP)

WAP encompasses a range of protocols from the network transport layer through to a presentational language and a scripting language designed to deliver web-based information to mobile computing devices in a wireless environment as efficiently as possible.

The WAP Forum (http://www.wapforum.org) defines and evolves the WAP standards. These protocol standards include a range of technologies necessary for the complete support on interactive web access via a wireless device. The standards include:

- A programming model that is based on the existing World-Wide-Web programming model.
- A presentational language called Wireless Markup Language (WML) that draws inspiration from HyperText Markup Language (HTML) and the HTML Basic specification from the World-Wide-Web Consortium (W3C), and Handheld Device Markup Language (HDML) originally developed by Phone.com. WML is designed as an eXtensible Markup Language (XML) compliant grammar.
- A scripting language called WMLScript based on ECMAScript (the same scripting language JavaScript is based on).
- A lightweight protocol stack that reduces the amount of network traffic. This protocol stack replaces TCP as used in the standard WWW model, with either a Wireless Datagram Protocol (WDP) or User Datagram Protocol (UDP) transmission protocol.
- A framework for interacting with the functionality of the wireless device, the Wireless Telephony Applications (WTA) interface.

What Needs Does WAP Address?

WAP was designed to address the needs of wireless, thin-client web-access in permanently connected environments that have the following properties:

- Less bandwidth
- Higher latency
- Less connection stability
- Less predictable availability

The device WAP targets are mass-market, handheld, wireless devices of the sort targeted by J2ME. Such devices present multiple challenges to the designer of applications and a more constrained environment when compared to a desktop PC.

These devices tend to have:

- More limited computational capabilities
- Less memory and stack space
- More restricted power consumption
- Smaller displays
- A variety of input devices, from keyboards and touch screens to voice input

Both J2ME and WAP can offer complementary services to the device. WAP opens up these devices to flexible browser-based Internet access, whereas J2ME offers these devices the ability to run more specific and powerful application software that can both interact with the network and maintain a certain level of usage when the network is unavailable.

Jini and the Jini Surrogate Architecture

The aims of Jini gel seamlessly with the collaborative networking structures that we hope will arise with the advent of 3G and the improved personal wireless networking technologies we have been discussing during this chapter.

Traditional client-server design metaphors have relied on a network environment that is effectively hard-wired and static. This assumption is barely true in a normal LAN environment, and is certainly a false axiom in a mobile, wireless, environment where the components, devices and services on the network continuously come and go.

Jini, then, is an attempt by Sun to embrace this fluid network structure and to design pervasive computing application frameworks that can actually take advantage of this situation.

Location Hiding and Network Transparency

Jini differs from many programming tools for distributed components in that it fully acknowledges and embraces the concept of remoteness. Programmers using remote components in many programming metaphors are shielded from the remote nature of the system elements they manipulate.

Jini forces the programmer to confront remote services, accessible only via the network, and to treat them as transitory things that must be discovered, used via a temporary lease, and then returned.

Component and service location cannot be ignored in mobile systems, as they have been in many traditional client-server implementations; services that are not locally available to the device must be treated as remote and temporary in order to avoid disappointment at run-time.

Jini

Jini can be used to create software that can enable federations of devices, connected via a local or personal area network such as Bluetooth or Wireless LAN (802.11), to form impromptu, collaborative, and unified software systems. Jini builds on Java technology to deliver and extend the benefits of the Java language into the construction of these services, and contains five key concepts:

- Discovery
- Lookup
- Leasing
- Remote events
- Transactions

Jini enables the creation of spontaneous communities of services providers and consumers using the processes of service **Discovery** and **Lookup** (akin to the directory services used in RMI). **Leasing** ensures that the temporary nature of dynamic network environments is embraced, and the Jini community can recover from the loss of a service. Finally, **Remote Events** and **Transactions** allow services to notify consumers of those changes, and allow the safe exchange of information.

Jini Surrogate Architecture

Jini requires the use of RMI, and so at present cannot run on the majority of devices. This is because RMI is only contained within J2ME as a profile of the CDC, intended for high-end and powerful devices.

To ensure that Jini can encompass as many device types as possible (able to support elements as simple as ROM-based devices to full-blown J2ME-based implementation) the 'surrogate' architecture is being developed to enable small devices connected via resource-limited networks to participate in Jini device federations:

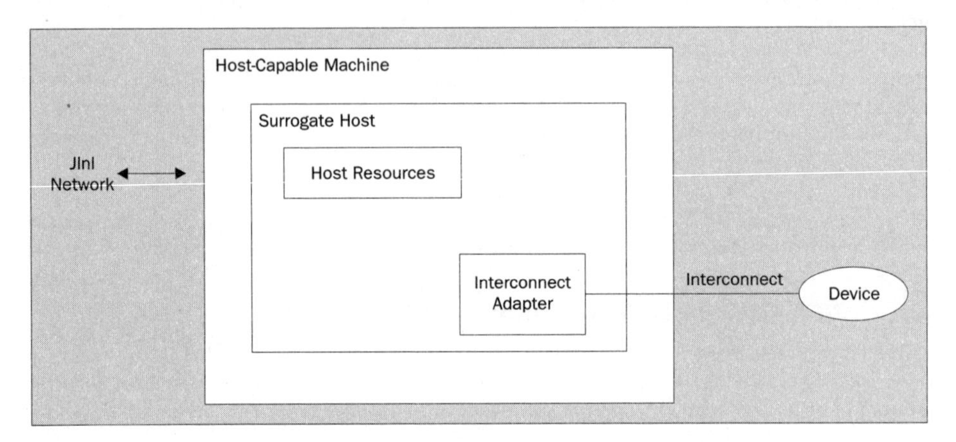

The surrogate architecture insulates a device incapable of supporting Jini itself from the Jini community, while allowing it to share and expose services via the surrogate.

The main requirements for the surrogate architecture are:

- Device-type independence
- Network-type independence
- Preserve Jini's plug-and-work philosophy

For more information on using Jini, see Professional Jini by Sing Li, (Wrox Press, 2000, ISBN 1861003552) and Core Jini 2nd Ed. by W. Keith Edwards (Prentice Hall, 2000, ISBN 0130894087).

Java Embedded Server Technology

Java Embedded Server (JES) is a technology used to create Home Gateways. A Home Gateway controls the household appliance network, and its connection to the Internet.

In the home of the future, networked appliances will combine alarm clocks, coffee machines, televisions, cars, refrigerators, air conditioning systems, and telephones together into dynamic and collaborative service-based networks. Combining these networks with an Internet-based control paradigm will offer consumers powerful new ways to use technology to enhance their daily lives – and present some interesting security concerns.

Java Embedded Server technology can be used to create Home Gateways that could provide:

- Device integration and e-commerce services – for example, to allow your alarm clock to inform your coffee machine when to make your coffee, or to provide your refrigerator with the ability to order more milk from the online supermarket
- Device monitoring services to allow for alerts and remote administration of your home appliances – for example automatically contacting support engineers regarding the status of your washing machine in the event of a device experiencing a failure

Some of these uses may sound fairly clichéd, but ubiquitous device integration with the Web using standard protocols and Java programming models allows all sorts of remote sensing and automated control applications to be constructed. Also of interest in the same application area is the Brazil project (http://www.sun.com/research/brazil/), which "extends the endpoints [of the network] to new applications and smaller devices".

The Home Gateway uses Java Embedded Server technology to link devices together into local networks and to link these networks to the Internet and external service providers, creating a focal point for enterprises and service providers to deliver services to client devices. The Home Gateway serves two primary roles:

- As a hub to manage the intelligent appliances connected to the home network
- As a gateway for communications between the home appliances and the Internet

JES is a technology that is closely related to the Jini technology framework, and can enable spontaneous Jini-enabled device federations to locate and utilize external services via the gateway.

Developers can construct new services by building on the core functions of the Java Embedded Server, allowing companies to quickly respond to changing client requirements, and provide new software and services on the devices over the network.

The JES architecture allows for services to be incrementally developed and added:

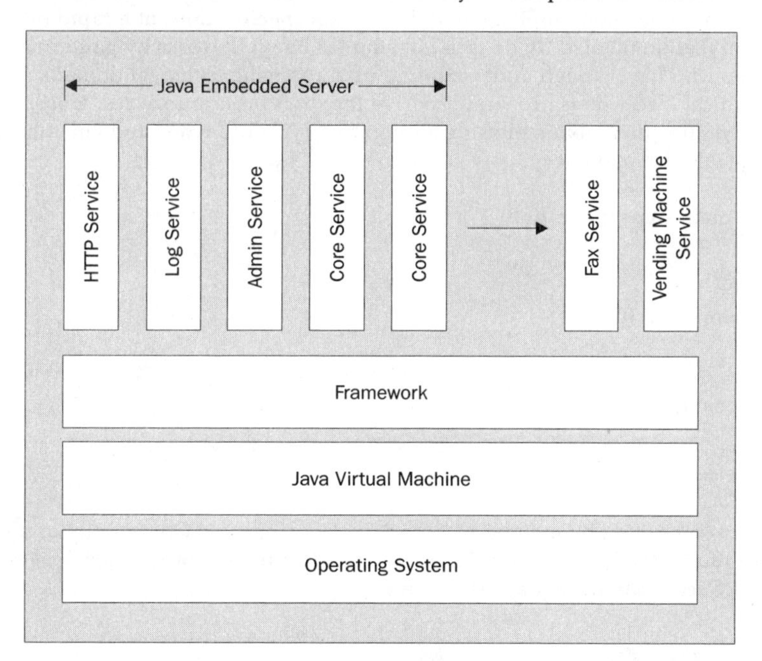

The Open Services Gateway Initiative (OSGi)

"The Open Services Gateway Initiative (OSGi) is an industry group working to define and promote an open standard for connecting the coming generation of smart consumer and small business appliances with commercial Internet services." (source: http://www.osgi.org).

The OSGi specification is being designed as an open standard to utilize Java technology in creating gateway services, enabling appliance and device-based services to be deployed and managed via a central residential, or small-office, gateway hub.

OSGi delivers an open, common architecture for interested service providers, developers, and appliance manufacturers, and forms the basis of the Java Embedded Server and Home Gateway technologies. The JES framework is compliant with the Open Services Gateway Initiative (OSGi) 1.0 framework.

How J2ME will Develop – The Java Community Process

Sun Microsystems, with extensive industry collaboration, has determined both the scope for, and pace of development of, J2ME. Sun has concentrated on the process of how J2ME will be developed, and has allowed industry leaders and experts to create flexible groups of like-minded individuals and companies to develop and define general configurations, and specific profiles, of J2ME.

This innovative and dynamic approach to creating specifications is rooted in Sun's Java Community Process (JCP), in contrast to the traditionally staid and bureaucratic, committee-led standards development cycle.

While having a strong technical committee can lead to innovation, the lessons learnt with CORBA show that it is important to develop, implement and maintain specifications at a rapid rate, if they are to compete in a market dominated by de-facto standards. The JCP works by gathering together a group of industry experts who have a deep understanding of the specific technical domain, and have a strong individual technical lead who is prepared to drive the draft process forward. Consensus around the contents of the draft is then obtained using an iterative cycle of review and amendment, among an ever-widening technical audience.

There are four main steps in the JCP:

- Initiation
- Community draft
- Public draft
- Maintenance

Initiation

A specification is submitted by members of the Java Development Community to the Process Management Office (PMO) within Sun Microsystems. It is then approved for development and is assigned a Java Specification Request (JSR) code.

Community Draft

An expert group is formed to develop a first draft of the specification. The expert group uses the feedback it obtains from the community draft phase to refine the specification, and decide with the Executive Committee (EC) who guide the development of Java Technology, whether this specification is ready to proceed to public draft.

Public Draft

The draft specification can then be reviewed by anyone who has an Internet connection, and wishes to comment or participate on the draft. The expert group uses this feedback to further refine the specification. Finally, the leader of the expert group oversees the development of the reference implementation and the associated compatibility-testing suite. Once the Executive Committee is satisfied, the specification receives final approval.

Maintenance

Once the specification is complete, the specification, reference implementation, and compatibility tests are maintained and developed in response to the need for clarification, enhancements and revisions to the specification and technical environment.

J2ME is undergoing a huge amount of innovation as a result of numerous expert groups working to develop new profiles and configurations, and refine those currently available, as we've seen in the number of references to JSRs in this chapter.

Summary

As we have discussed, J2ME offers a development tool for consumer devices that addresses a design and architecture problem-domain that is quite different from that of the traditional fixed network of computers, Internet-enabled by Java 2 Enterprise and Standard Editions.

J2ME is a very exciting set of technologies, providing the ability to use Java to extend the capabilities of the device, thus bringing Java back to its design roots. J2ME extends Java to the embedded and mobile device market, which includes consumer electronic devices as diverse as refrigerators and PDAs. Wireless device types, operating systems, and networks will proliferate, but Java provides a unifying programming platform.

Java device configurations are expected to run to hundreds of millions of devices (Symbian predicts that high end Wireless Information Devices (WIDs) will top 100 million by 2003). Java's acceptance in the wireless market will also hinge on enterprises extending their J2EE applications to cater for these new devices, and on content providers using Java-based services, and on-device applications, for which portability is key.

J2ME provides a host of benefits to the programmer approaching the development of device-based systems. As we have investigated in this chapter, and will explore throughout the rest of this book, Java enables programmers to take advantage of key factors to expand the capabilities of devices, namely:

- Code and component level portability
- Software reuse
- Language and environment simplicity
- Safety, reliability and security
- Availability of developers
- Longevity of software as the hardware develops

Unlike most traditional device programming paradigms, using Java creates an environment in which complexity can be managed, problem-domains can be modeled with objects, and functionality can be captured, using Java's high-level features.

J2ME furthers Sun's stated goal of making data services available to anyone, anywhere, at any time, on any device. At a code level, Java provides programmers with an environment in which device-based, and real-time, issues can be dealt with effectively. J2ME is therefore, set to revolutionize the way in which application software can be written for consumer devices.

J2ME enables device vendors, and developers, to target particular device configurations and profiles, ensuring that the highest level of portability can be achieved for application software in this difficult and varied market space. At the same time, J2ME allows programmers to fully exploit good OO programming language patterns and practices.

Mobile computing creates a unique set of challenges that are centered around the balancing of resources and code execution between device and server, network and device security, and creating useful services, tailored for the device and network environment. J2ME is ideally placed to take full advantage of this new environment, and enables programmers to build sophisticated services for these new mobile network and device infrastructures, in a familiar and productive Java-based environment.

PROFESSIONAL JAVA MOBILE
TOOL SELECTION
WEB TABLET
MOBILE PHONE
MP3 PLAYER
DIGITAL CAMERA
PALMTOP
LAPTOP
NOW SELECTED
PALMTOP

2

Architecting and Developing J2ME Software

Implementing a J2ME-based architecture must be carried out within the context of good project management control, and with the support of proper development processes and controls. Once however, the environmental factors have been resolved, the architect must actually create a device-based system. There are several sources of inspiration that can be drawn on as the initial plans for the system are laid.

Architectural Patterns

The architect must be aware of the use of design patterns and higher-level architectural models and patterns that have been developed to promote clear architectural modeling. This enables the architect to rapidly construct a framework for the type of systems that are being developed.

The architect of J2ME systems can benefit from the client-server development tradition. Although most of the devices being considered in this book have resources falling far short of those employed in most PC client-server deployments, and have a less reliable connection to the network, many of the same techniques will apply. For example:

- Caching certain information locally
- Minimizing network traffic
- Separating the business, presentation and persistence logic

The devices are however, also targeted by a new generation of more interactive and collaborative system models. The development of distributed architectures of loosely coupled collaborative server-side systems to provide services will also be a major source of ideas and models for the J2ME software architect.

As well as the growth in n-tier server-side technology, the growth of Jini technology, to promote short-lived collaborative networks, will provide possible models and influencing factors for the architecture of J2ME-based systems. Considerations from these architecture camps will include:

- Loosely-coupled services
- Robust message exchange
- Collaborative and dynamic look-up and use of available services
- Thin-client interaction
- Reliance on the server-side for intensive processing
- Designing for systems with limited resources

The final, and perhaps most directly relevant area of inspiration for J2ME-architecture, will come from the existing device-based application development community. The majority of existing embedded and device-based developments are performed in C/C++ and perhaps Assembler. However, the following success factors for these systems apply equally well to the world of J2ME:

- Performance
- Resilience and robustness
- Efficiency

Model, View, Controller (MVC) Architectures

The single most widely-used architectural concept for client server development is the model-view-controller (MVC) architecture. This model of system design has migrated to server-side systems, and provides a generic model for the separation of responsibility within an application for the business model, presentation and user interaction, and persistence and control of the objects of the system:

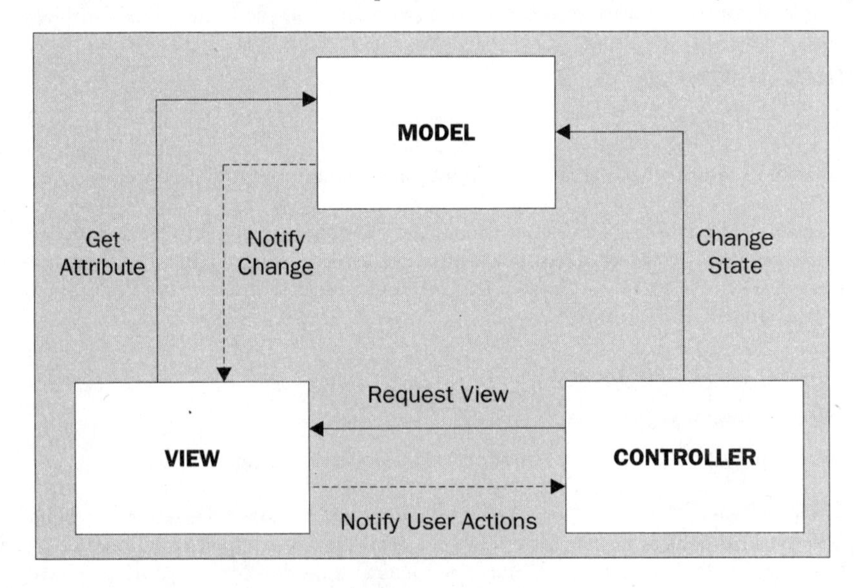

The Model

The model encapsulates the business model of the system and responds to requests from the view regarding its state. It also provides an interface to the controller to allow changes to that state.

The View

The view is responsible for rendering the information from the underlying application model, and interacts with the user, passing on user actions and input to the controller for processing.

In a J2ME application, the view will be located primarily on the device, but may also have elements with a servlet layer to assist in the rendering of data and results.

The Controller

This layer encapsulates the application's behavior, and allows the view to interact with the model to change the state of the application. The controller is specialized to a particular type of interaction with the system and there may be many controllers with a particular application.

Component-based Systems for Devices

Component-based systems assist the developers in rapidly modeling and mapping the business problem domain into an abstract framework, against which code can be implemented to fulfill the requirements.

Using components facilitates the construction of new applications from an existing toolbox of well-designed and mature units. Components also allow the discrete testing and implementation of system functionality, as we will discuss in Chapter 17. The architect of device-oriented systems can leverage the following advantages of component-based design:

- Improvements in mapping the business problem-domain and creating technical abstractions of business processes.
- Resilience to, and accommodation of, changes in requirements, in the underlying object model.
- Improvements in the area of component re-use.
- Improved scalability.
- Easier to debug, safer systems that possess increased resilience.

Improvements in Mapping the Business Problem-domain

J2ME enables a component-based, and object-oriented, approach to the modeling of system objects based on user requirements and business entities.

This alignment, between the problem domain at the business level and the components of the system at a technical level, brings dramatic improvements in the mapping of the software to the business solution, and enables rapid validation of the software and its architecture.

Resilience to Changes in Requirements

The cohesion of the software model to the business domain, increases the ease with which changes to the business model and scenarios can be reflected in the design.

For mobile systems that may undergo changes in both business functionality and deployment during the project life cycle, as the opportunity for the use of mobile technology develops and the underlying device technology develops and matures, the flexibility enabled by component-based systems will become invaluable.

J2ME components allow the investment in development to be discretely encapsulated, and selectively re-used and redeployed in new situations as necessary without recompilation and redesign, in a way that cannot be achieved through simple re-use at a code level.

Improved Scalability

The scalability of many device-to-server systems is limited by problems associated with a lack of clear separation between business and technical functionality, or between the model, view, and controller. Many systems solve these problems by introducing ad hoc **fix** code, which will limit the future deployment of the software.

A clear separation in layer responsibility coupled with the discrete componentization of the problem domain functionality should help to ensure that the system can be incrementally scaled without the nasty discovery of assumptions and 'gotchas' in the deployment model.

Debugging, Safety and Resilient Systems

J2ME allows for the construction of device-based software that interacts via clear component interfaces. This separation enables assertion and debugging techniques to be applied to validate the correctness of the elements of the system rapidly.

J2ME's garbage collection ensures that accidental memory leaks should be rare, and the error and exception checking in Java allows for the careful capture and handling of erroneous behavior.

Java provides a safe, resilient and debuggable environment for the construction of device-based software, in a way that has never been achieved with the traditional C, C++, and Assembler programming environments on these platforms.

Linking to the Server

One of the most important technologies for the vast majority of J2ME projects will be server-side connectivity, and the technology that can add most value on the server-side is, without a doubt, Java 2 Enterprise Edition (J2EE).

In developing applications to run on device-based clients, the balance between stand-alone and server-side processing, and the quality and quantity of server interaction needs to be carefully considered.

Client-Server Deployment Models

In deploying applications to the J2ME platform, the architect has the following deployment models to utilize:

- Standalone application
- 2-tier client-server model
- 3-tier client-server model
- n-tier client and server model
- Collaborative, dynamic, distributed, model

Stand-alone Client Application Model

This model is the simplest for developers and also the least functional and least useful. The application runs, interacts with, and stores information solely on the device, with no requirement for server-side connectivity or interaction.

Moving Towards Client-Server

The choice of client-server architecture is determined by the needs of the project, and the architect must balance the need for simplicity of design with the scalability that can be achieved by the expense of additional complexity in the system deployment model.

Limited forms of client-server interaction could even include the exchange of information with a desktop (server in this context) machine, for the purposes of data synchronization.

2-tier Client-Server Application Model

The first type of client-server architecture involved separating the permanent data storage from the application code, typically by the use of server-side relational databases.

In this model, the PDA would contain all the user interface and application logic, relying on the server solely as a datastorage and retrieval mechanism:

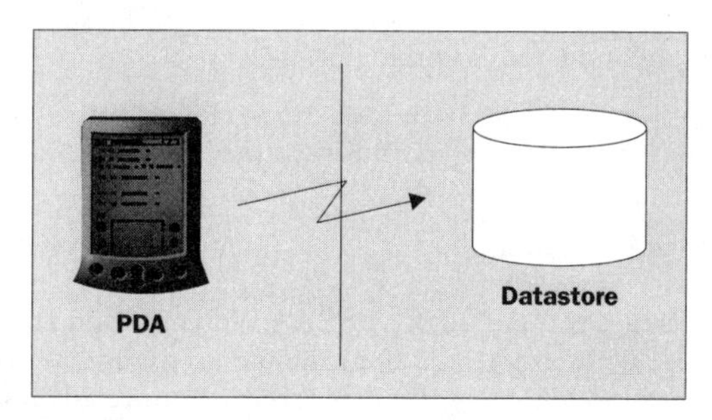

This architecture was responsible for the client-server revolution. At a departmental level, with datastorage restricted to the local area network and a limited number of users, it can meet the necessary performance criteria.

However, as the number of users increases, this model has difficulty scaling. The metaphor for database access via an independent connection, directly from the client device, is not particularly efficient – either in terms of network traffic, or the cost in database access licenses – and this architecture has now fallen out of favour, being replaced by the 3 and n-tier models we are about to encounter.

Moving to a More Scaleable System Architecture

Moving to three tiers, moves the model and application logic from the client device to a server-side execution model. This sharing of processing power and server resources amongst the clients can be a critical step in increasing scalability.

Code complexity should not be significantly effected by this move, since once the crucial step of separating out the user-interface, application logic and data storage elements of the system has taken place in the code – using good object-oriented design practices, and particularly with the use of the Model-View-Controller architectural pattern – the physical separation of these elements onto individual machines should then be a more trivial step:

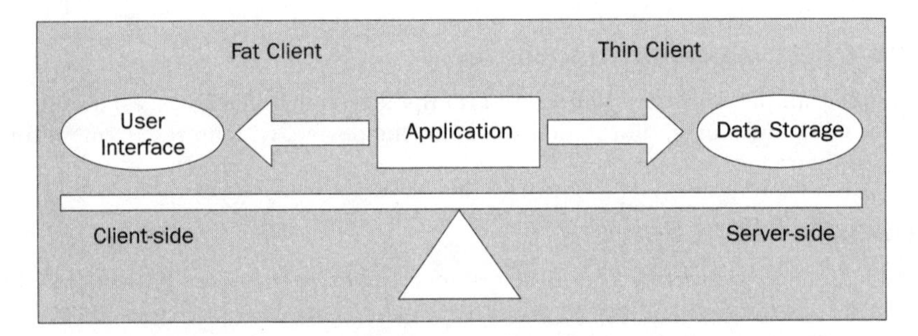

The balancing of this workload is a key decision in the application design. The need for the speed that can be obtained from local validation of user input against business rules, must be balanced with the robustness, reduced bandwidth usage, and improved data security, of performing such validation on the server.

3-tier Client-Server Application Model

The simplest type of architecture making use of server-side processing of business rules is the 3-tier model. This model moves the application business logic to a middle tier, and leaves the responsibilities as:

- **Client Tier**
  Interface and simple input validation logic
- **Middle Tier**
  Application business model logic and business process logic
- **Database Tier**
  Persistent storage of information

The development of application servers, such as BEA's Weblogic and IBM's Websphere has greatly increased the number of projects exploring 3-tier architectures, and has created secure, robust and highly scaleable Java application logic engines that can drive more advanced architectural approaches.

The diagram shows a typical 3-tier architecture with the user-interface logic on the device, the session and business logic on to the middle-tier application server, and the third, or back-end, tier for the storage of the applications data:

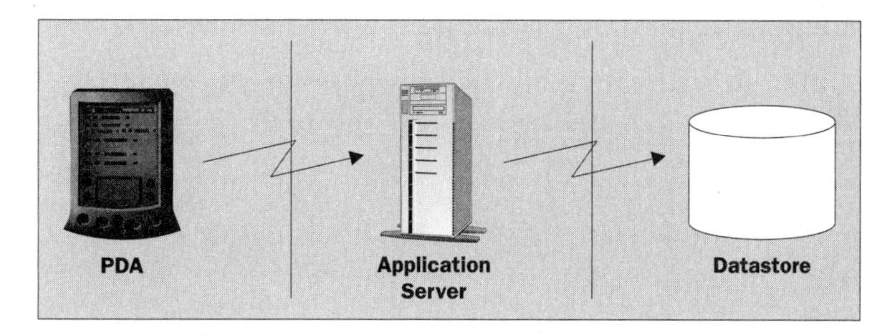

n-tier Client-Server Application Model

Once the move to a 3-tier model has occurred, the next logical step is the use of additional server-side tiers. Essentially, the user-interface logic remains on the client, but the middle-tier and backend data storage may now be split over several different physical servers, locating the business processing near the particular datastorage area, and publishing the function as an independent service.

In this architectural model, the client talks to a primary application server that may utilize other back-end application servers to provide specific functions and may use data from several back-end databases:

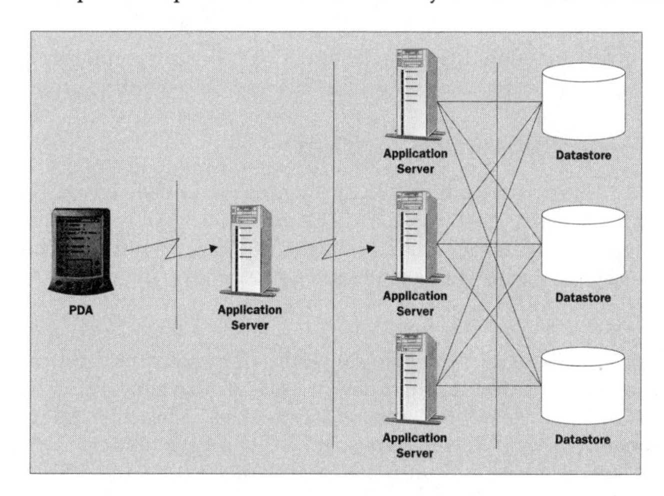

This model has the advantage of being extremely scaleable and, by componentizing the core business functionality into separate logic layers whose services can be more easily re-used, the model remains flexible as the system develops.

Collaborative, Dynamic, Distributed Application Model

The n-tier model discussed above is essentially a 'hard-wired' model, with the services statically linked together and expected to provide a certain level of services to the network. With the promise of dynamic local area networks, and technologies for devices that will better enable the exploration, discovery and utilization of new network services, such as Jini, a more collaborative application model is emerging.

In the collaborative network shown below, the distinction between client and server is blurring, and all that remains are federations of suppliers and consumers of services, who publish and subscribe to these services over a network. This network may well be temporary, short-lived, and fluid, with new services becoming available, and others leaving the group:

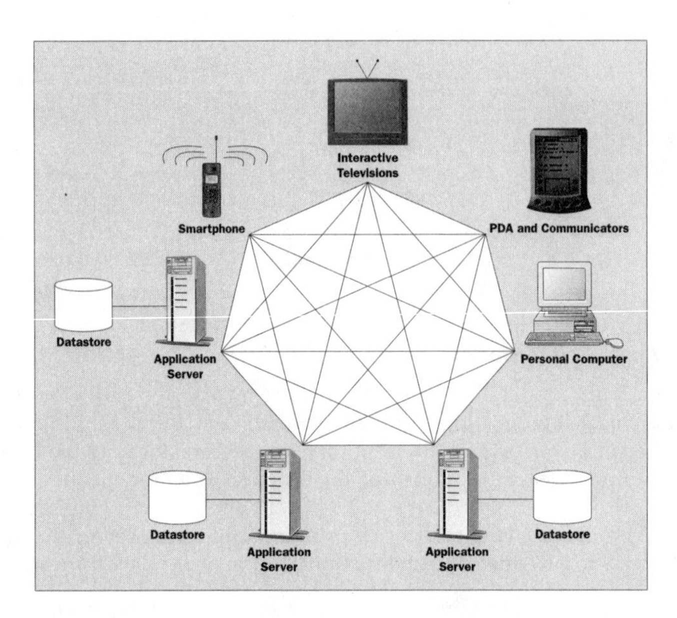

By promoting server-side functionality as services that can be independently accessed and utilized, by local software agents, the architect can design flexible and collaborative application structures.

The Java 2 Platform, Enterprise Edition

Using Java and Java 2 Enterprise Edition (J2EE) requires the knowledge of a number of best practices to develop and deliver component-based systems. Mobile developers should accumulate a considerable amount of knowledge and experience in delivering such systems, and work hard to communicate this knowledge, and transfer the value of this experience through the development of such enterprise scale software systems.

J2EE defines an industry standard for developing multi-tier enterprise applications. J2EE simplifies enterprise applications by establishing their design on a standardized, modular component environment, providing 'out-of-the-box' support for databases, directory services, and many other enterprise environments, as we shall see. The scope of Java development is extended firmly into the enterprise arena, making real Sun's promise of "Write Once, Run Anywhere" in a world where "the network is the computer".

From the JDBC API for database access, through CORBA technology for access to existing enterprise systems and resources, to a robust security model, J2EE builds on the ideas behind the Java language and provides added value to distributed computing projects. It also adds full support for Enterprise JavaBeans, Java servlets, and XML technology.

Enterprise JavaBeans - Creating a Middleware Standard

Portability, scalability and legacy integration are essential attributes for almost all enterprise development projects. J2EE enables the construction of such systems.

Portability and scalability are also important for the long-term value of n-tier applications, providing the basis for flexible and well-designed server-side application frameworks. Applications must be capable of scaling from relatively simple working prototypes to fully functional 24x7x365 component-based software structures, capable of supporting a robust and flexible user access model over the Internet.

Application servers provide a robust environment for hosting the business logic of enterprise systems, leveraging the transactional capabilities of RDBMS's and providing out-of-the-box persistence capabilities for EJB objects. Device-based applications may well need to utilize the functionality of server-side systems; this can be done most easily by implementing systems based on a J2EE-compliant Application Server, such as BEA's Weblogic or IBM's Websphere.

Basic Elements of J2EE

The diagram below provides a visual representation of the API areas covered by J2EE:

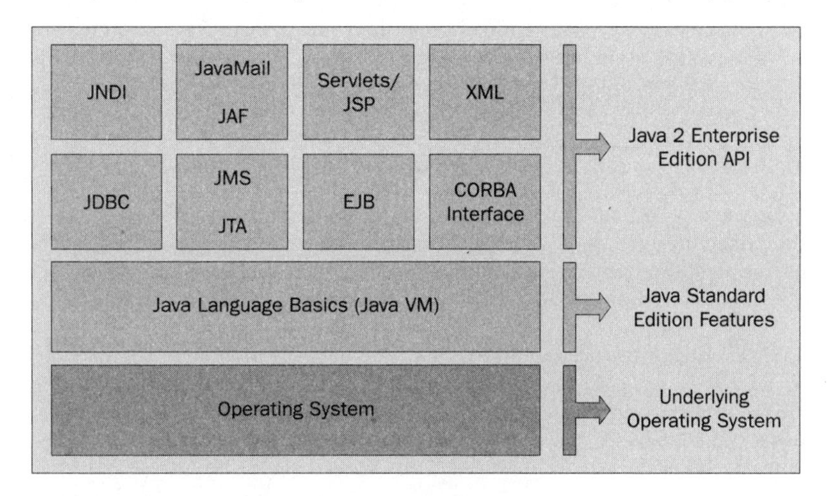

In addition to providing support for Enterprise JavaBeans, Java servlets that can model the server-side business entities and process, and the interaction with the client devices, the J2EE specification also defines a number of standard APIs to assist in the development of distributed software systems:

- **JDBC API**
  JDBC enables applications to access and manipulate data held in relational databases, and other data repositories. J2EE includes version 2.0 of the JDBC specification – the latest implementation of the JDBC API.

- **Java Naming and Directory Interface API**
  The Java Naming and Directory Interface (JNDI) API provides a simple mechanism for Java-based software to access and manipulate information held in directory services, such as LDAP servers and file systems.

- **JavaMail API**
  The JavaMail API provides the ability to access the functionality of Internet messaging and email systems, including SMTP, POP3, and IMAP4 servers. This can be useful in order to send to provide administrator alerts, order confirmations for e-commerce systems and other information via email.

- **CORBA Compliance**
  J2EE supports JavaIDL and RMI-IIOP, which are both CORBA-compliant technologies. JavaIDL has been designed to allow Java-based software to interact directly with CORBA-compliant systems. RMI-IIOP combines Java Remote Method Invocation API (RMI) with CORBA's Internet Inter-ORB Protocol (IIOP), creating a specification which allows vendors to leverage existing CORBA legacy systems and middleware products into the Java world.

- **Java Transaction API**
  Java Transaction API (JTA) provides a way for multiple J2EE-based components to participate in and manage a single, or multiple transactions.

- **Access to XML Documents**
  XML is a meta-markup language, capable of providing the definition of new data description grammars. J2EE provides access to a set of APIs to manipulate XML documents in different ways.

- **Java Message Service**
  Java Message Service (JMS) defines a standard mechanism for components to send and receive messages asynchronously, using message queuing middleware products such as IBM's MQSeries, for example.

A Typical J2ME/J2EE Architecture

The balance between server-side processing and client responsibility is a delicate one, and has a huge impact on the system design, and potentially, on both the performance and usability. J2ME allows for the creation of sophisticated client-side application logic and user interfaces that are limited only by the capabilities of the device, and the imaginations of the architect and programmers.

A typical J2ME-J2EE architecture would include the following elements:

- **Backend systems and services**
  This layer manages the provision of access to backend storage, mail servers and other pre-existing systems.

- **Java application server**
  The application server layer will include:

  - **Persistent/Entity objects**
    Entity Enterprise JavaBeans can be used to model the business objects and entities of the business system, in order to handle their persistence.

  - **Session objects**
    The session beans (session Enterprise JavaBeans) model the business processes, and hide the complexity of the underlying entity business model from the high level layers. This layer can also manage interaction with certain non-persistence related backend systems such as the interaction with the mail server to send and receive email, and securely maintain user-session information.

  - **The servlet layer**
    The servlet layer should be used to model the presentational and data format issues that are needed to respond to the device's request for information or services.

- **Web server**
  The web server is a thin layer in this system that marshals the requests from, and the responses to, the devices.

- **Client device resident layer**
  The client device must handle user-interface issues, handle input validation, and potentially cache information for off-line usage. Potential client-side solutions must be evaluated within the context of a distributed architecture. In simple systems – such as many of those currently dealt with in WAP systems – this client layer can be handled by a micro-browser.

The responsibility for managing and initiating interaction with the server, and the handling of information pushed to the device, must all be carefully considered in the formulation of the final design plans for the client-side layer. The final system structure for a client device interacting with an enterprise Java backend might look something like this:

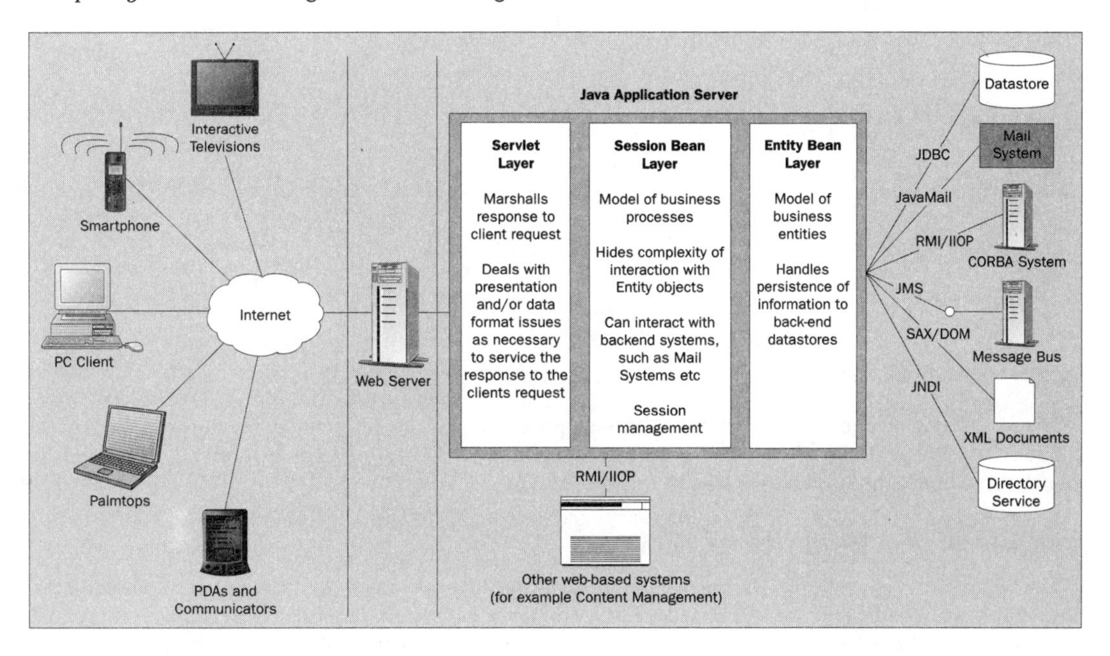

Service-oriented Architectures

Making effective use of J2ME on the server-side can be assisted by separating and layering the client and server layers in a manner that reduces the knowledge that client-side systems need to have of the server-side services. Hiding the physical complexity of the server-side and presenting the client layer with a simple and logical view of the services available, can simplify the job of the J2ME programmer.

The role of the architect is to design these service-oriented architectures, which will provide the necessary logical or service-oriented view of the distributed server-side system's functionality.

Integration of Component-based Systems

Once an organization has begun to componentize its main business applications, it is ready to move to the next level, and produce a service-oriented architecture as the foundation of future development and systems integration.

A service-oriented architecture uses an integration layer to hide the complexity of the backend systems, behind a simple logical view. This view provides the programmer using the system with a simple view of the services available; enabling the rapid and safe utilization of these services, while hiding the intrinsic complexity of the separate physical backend systems ultimately providing them:

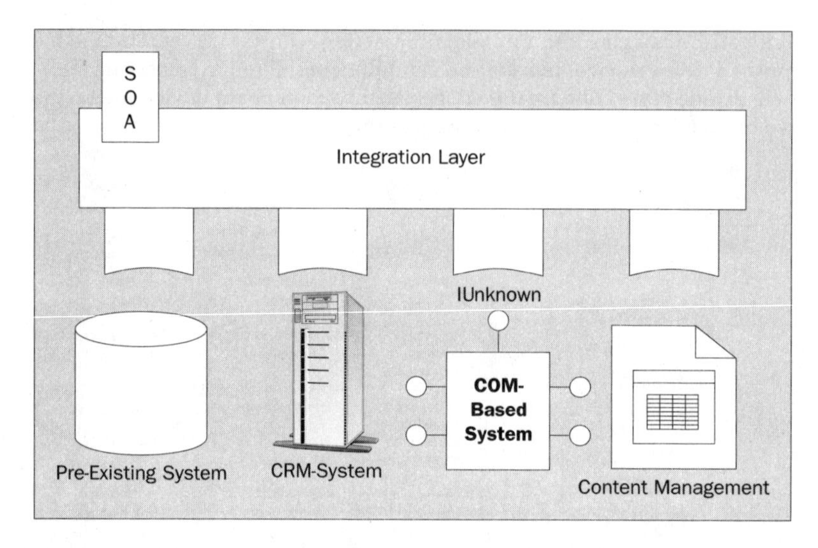

These sorts of designs enable straight-through processing to be performed on the underlying business functions, and will allow the rapid construction of new applications that can quickly be scaled to cope with new deployment scenarios and user requirements.

Example of the Benefits to J2ME

The job of a device-side programmer can be made significantly easier by being able to use service-oriented architecture on the server-side. For example, a J2ME application constructed on top of a server-side integration layer designed to trade stock could affect several traditional systems. The invoice settlement and stock management systems, as well as the salesman's commission information, and the CRM system, could all require data updates as the result of a single purchase; the integration layer should mask the potential physical isolation of these systems.

An integration layer enables the developer of device-based services to ignore the complexity of designing applications when communicating with a distributed server-side environment. By using the abstraction enabled by the integration layer, the burden of knowing details of the implementation of the physical service, is removed from the designer of applications. This can promote a better component design-philosophy, and potentially increase component boundary observance, thereby increasing the value of component in terms of reusability.

Network Transparency and Distributed Systems

While the location of a component may be an element of complexity that can be hidden by a service-oriented architecture, the location of components in a distributed system need not be hidden, and the remote component will often need to be treated very differently to local ones.

Creating a service-oriented architecture does not necessarily mean that the physically separate nature of the components and services are hidden, but rather that by moving the burden of service complexity away from the developers, and into a single component-based interface, planning, security, and scalability would move to being a function of the application architecture and infrastructure teams, rather than an element managed by the programmers in their code.

Dealing with remote components in a collaborative network structure is a new challenge for software developers, and one that technologies such as Jini are being designed to combat. For more information on distributed software architecture, see the "The eight fallacies of distributed computing" identified by Peter Deutsch (http://www.java.sun.com/people/jag/Fallacies.html). The following eight false assumptions can all potentially cause serious problems in distributed systems:

- The network is reliable
- Latency is zero
- Bandwidth is infinite
- The network is secure
- Topology doesn't change
- There is one administrator
- Transport cost is zero
- The network is homogeneous

Programmers using J2ME can benefit from the ideas behind straight-through processing and service-oriented architectures, and can learn from distributed programming technologies such as Jini.

Using these ideas can increase the potential for advancements and re-use in server-side components and services, enabling organizations to take advantage of improvements in server-side technology and security, without re-developing entire applications.

The Way Forward

While a certain amount of diversity in development approach should be expected and even embraced within the overall framework, organizations should define rules and standards for component boundaries, and develop standards for the construction of future component systems to ensure easy inter-operability. This type of integrated architecture will allow the rapid construction of device-oriented applications that can quickly be scaled and refactored at the server-side to cope with new deployment scenarios and user requirements.

J2ME Architecture Case Study

To give a practical example of applying the techniques of software architecture, we are going to look at the requirements and initial design ideas for a time-sheet entry system where there was a specific requirement for mobile data entry.

Our imaginary company had a problem with time recording; since the highly mobile workforce is typically out of the office, or in meetings, they had difficulty completing and submitting paper timesheets, or using the company's ageing client-server system, which was only available on PCs in the office, to meet a 11am deadline on Monday morning. This led to the late submission of many time sheets, which caused friction with and problems for the internal accounts department.

The majority of the company's consultants already used a Personal Digital Assistant (PDA), therefore the company was keen to provide time-sheet entry via this type of device. As examined in this chapter, it was the job of the architect to harvest and analyze the requirements for the system, and to ensure that the application designed and implemented would meet these needs.

In this case study we will look at:

- Following a requirements review and a design review for a hand-held-based system
- Considering choices and trade-offs in designing for the chosen platform/device
- Refactoring designs and balancing client and server load

Initial Requirements-gathering and Analysis Workshop

Key system sponsors for the project include:

- The head of the consultancy division, whose team would be the primary users
- The head of development, who would be responsible for developing the system
- The Finance Director, whose budgetary control the project would be developed within, and whose accounts staff would administer and support the new system

An initial requirements gathering workshop was held for the first timebox of the project and the participants were as follows:

- Architect (meeting chair)
- Business Analyst (scribe)
- Project Manager
- Senior Developer
- Finance Director
- Business user (a consultant who will remotely use system)
- Accounts Clerk

This workshop identified and documented the current processes, and began to harvest and document the requirements for the new system, using UML to document the use-cases and business scenarios described during the day.

The basic requirement was for all users to enter timesheets weekly between 11-12am on Monday, with Accounts issuing a general reminder e-mail at 4pm on Fridays, followed by a targeted one at 1pm on Monday. The timesheets needed to be approved by line managers, checked by an individual within the Accounts department, and processed before close of play on Monday evening.

Historically, users usually aimed to enter timesheets between 4.30-6pm on a Friday or 9-11am on a Monday. The reminder e-mails from the Accounts department were generated automatically via Lotus Notes. The timesheets were either submitted to line managers on paper-based forms, or e-mailed as Excel spreadsheets. Some, but not all, employees working in the head office used a client-server application written in Visual Basic to submit their timesheets.

The timesheets were approved by line managers in the format that they were received in (i.e. via all 3 methods), and passed onto Accounts where they were furiously checked by a clerk and entered into the Visual Basic application if they had been submitted via email or on paper. The clerk strove to complete this before their departure on Monday evening (which should have been at 6pm, but was frequently later due to last minute submission of timesheets), so that the timesheets could be processed as a batch job at 10pm.

The entire scope of the new system was conceived as providing both a web-based and a device-based interface to the timesheet entry. However, the pressing need and initial scope was to provide a mobile interface to the system, which would allow consultants to enter their timesheets 'on the road'. Automating the entry and approval processes would also free up the accounts clerk's time to process expense claims instead, which funnily enough always made it in on time.

Existing Timesheet System

It was agreed at the workshop that the existing system should run in tandem with the new system, at least initially, since it had some advanced approval, user administration, and reporting functionality that would take time to re-develop.

The old system was designed for Windows 3.11, in Visual Basic, against an Oracle database. The Oracle database would be shared with the new system, producing an element of data integrity testing and risk that the architect and project manager needed to quantify.

The system had been updated to a Win32-based version of Visual Basic, but the code had not been written using the object-based features of Visual Basic, nor were the system components exposed as COM components in any meaningful way. The version of Oracle had recently been updated from Oracle 7 to Oracle 8.1.6, and the database schemas were provided at the workshop.

Fortunately, the old system was seen by all in the company as having being a good investment, that had now run its course, particularly because its usage was erratic, not even being used by all those employees based at head office. With no egos to bruise in changing technology, and phasing out the existing system, the workshop could progress relatively smoothly and avoid the politics, and arguments, often associated with changing and replacing technology.

SPIN® Questioning

During the workshop, the architect and business analyst spent the majority of the time using the techniques of SPIN® questioning. Originally, SPIN® techniques were developed, primarily by Neil Rackman at Huthwaite Research, as a methodology for salesmen to uncover the needs of clients in order to make a successful sales-pitch.

The use of SPIN®-based questioning techniques to obtain information, however, need not be limited to a purely sales-led scenario, and here the approach was used to identify, clarify and separate the business requirements of the system, in order that they could be documented in use-case diagrams and templates. The most important tools used at the workshop were the whiteboard and marker. Much of the time was spent discussing and delimiting the boundaries of the system.

As an example of scenario development and problem definition, the following exchange was recorded:

Situation Question:
(Architect) How are timesheets currently submitted?

(Accounts Clerk) Well, they should be submitted using the internal timesheet application. Some of the people who are based here do use this, but others use spreadsheets which they send via email, or the paper-based form which was used before the internal system was ever developed. The consultants on the road tend to e-mail spreadsheets. The line managers approve the timesheets in whatever format they receive them, and then forward them to me for processing.

Problem Question:
(Architect) What are the difficulties you experience with consultants submitting time sheets to their line managers working in the head office?

Implied Need:
(Finance Director) Well, our consultants must be accurate in reporting their time to ensure accurate billing, and so obviously cannot fill in time sheets in advance. This means that they have quite a tight time period between Friday afternoon and Monday morning to enter and submit the timesheets. However, we want to keep our management reporting and billing cycles as short as possible, so don't want to give them more time to submit them.

The difficulty is physically transmitting the timesheet: if it's posted on a paper form, it can take days to arrive and is cumbersome to input for the accounts clerk, and the Excel spreadsheets means that the consultant must be able to access e-mail, which is not always convenient when they are on the road. We certainly don't want to purchase laptops purely for people to enter timesheets. Coming into the office specifically to use the internal timesheet application is clearly inefficient, and is often out of the question, since some of our consultants can be based anywhere in the country.

Implication Question:
(Architect) Presumably where the timesheets are submitted on paper and via e-mail they need to be keyed into the timesheet application once they have been authorized and reached the Accounts department?

(Finance Director) Yes. In theory, if the timesheet has been received as an e-mailed spreadsheet, then it can be copied and pasted rather than re-keyed, but in practice this can often take as long. With the new system we want to have just the one way of submitting timesheets, which will of course make submission easier to track, and also to remove the burden of keying in data from the Accounts team.

Needs-payoff Question:
(Architect) So am I right in thinking that one of the things you want to achieve with the new system, is to increase the ease and efficiency by which by all remote workers enter timesheets, by allowing them to directly enter timesheets into the system using a mobile device?

(Finance Director) Yes: it's important that the new system is easy for the mobile consultants to use, and if the line managers receive the timesheets in a consistent fashion, it will also make it easier and quicker for them to authorise them. However, we must ensure that the system is secure as well as easy to use.

For more information on SPIN®, see SPIN Selling by Neil Rackman (Gower Publishing Limited, 1995) ISBN: 0070522359.

Clarification of Requirements

These questions established a context for the project manager, senior developer, and architect, to begin to think about how the usage scenarios could be met technically and where further clarification was needed in the requirements in order to be able to implement the system.

It was also noted that there were various non-functional requirements for the system – such as the appropriate level of security, raised as a concern by the Finance Director that would need to be investigated and addressed.

The timesheet system contained an element of workflow in the validation and processing of the employee timesheets. In the first timebox it was decided to concentrate on the development of the client-device frontend to the system, and on developing the necessary J2ME and J2EE infrastructure. However, automating the workflow elements of processing the timesheets was noted and was planned, by the team, to be addressed in a future timebox.

Use-case Diagrams

Use cases were used in the workshop to capture the functionality and scenarios required by the business users. The business scenarios initially identified as use-cases for the system were:

- Login
- Create a timesheet
- Review/Edit a timesheet
- Approve a timesheet
- Process a timesheet

It was determined that the following basic functionality was required for the system to be a success:

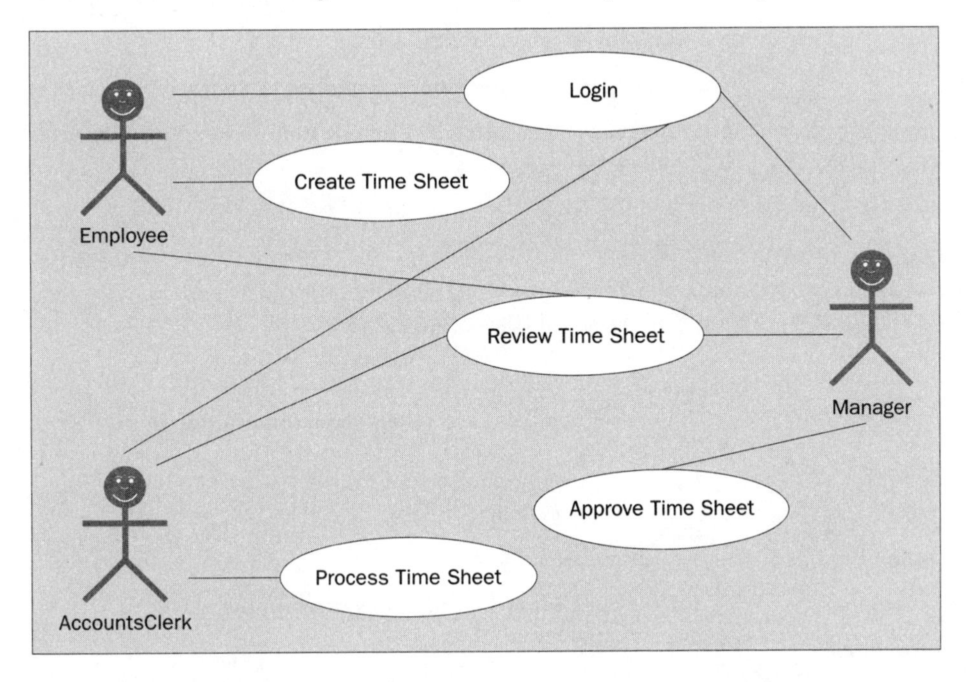

This diagram (produced by the architect after the workshop from the whiteboard notes) shows all of the use-cases identified for the basic time-sheet entry system, and which actors (users) of the system needed to be able to perform which use-cases.

Use-case Templates

The use-case template contained the elaboration of the use-case diagram, and was used to clarify and elaborate on the diagram, enabling the developers to ensure that the behavior of the system was correct and would meet the user acceptance criteria.

A template was developed for each use-case. The use-case template for the login process was as follows:

USE CASE NAME:	LOGIN
Iteration	First Draft
Summary	User logs on to the system
Actor	All
Basic Course of Events	User starts system and is presented with a challenge for user name and password User enters user name/password combination User session created (role in system defines next screen) one of: Employee can enter timesheet Manager can either enter a timesheet or approve timesheet Accounts clerk, can either enter a timesheet, approve or process timesheets
Alternative Paths	3a. User not registered, session not created, user must register to use system 3b. Password expired/incorrect
Exception Paths	1a. User unable to use system, user contacts administrator
Extension Points	
Triggers	User starts system
Assumptions	User registered to use system
Preconditions	User not already logged onto system on the device
Postconditions	
Related Business Rules	Register user with system
Author	Rob Machin, The Concise Group Ltd
Date	24th February 2001

Use-case Sign-off Meeting

A further short workshop with the business sponsors, and an expert user, was sufficient to sign-off the diagrams and templates, defining the scope of the first timebox.

The priority associated with each use-case was also discussed, and agreement using the MoSCoW (Must have, Should have, Could have, Won't have) Rules was obtained as follows:

USE CASE NAME:	MOSCOW PRIORITY
Login	Must Have
Create Timesheet	Must Have
Review Timesheet	Must Have
Approve Timesheet	Could Have: can be performed by existing system
Process Timesheet	Won't Have (Yet): performed by existing system

The project manager and architect were then able to begin to consider the approximate costing for each use-case. With this information, budget, timing, and development resource could all be planned. From these figures, an initiation proposal could be constructed for the project and agreed with the business sponsor.

Initial Work and System Development

In order to begin developing the system, the architect considered three areas of design:

- The business model needed to be developed from the signed-off use-cases
- The screen design and work-flow needed to be modeled and mapped (possibly with the assistance of a graphic designer)
- The system structure and model needed to be planned

Business Model Class Diagrams

At this point it was necessary to define the business entities that would support the business processes and logic of the system. These business objects would eventually be implemented using Enterprise JavaBeans, or as ordinary Java objects. The issues facing mobile systems will be raised later in the case study; however, it is worth noting at this point that the architect needed to consider how, if Enterprise JavaBeans are used on the server, a series of 'lighter' objects will be used, and potentially cached, on the client.

The existing database design had to be accommodated in this new system, since the same database backend would be used by both systems. This meant that the `TimeSheet` and `TimeDetail` storage structure was essentially dictated to the architect; it was also clear that it made sense to re-use as much of the control structure (`Calendar`) and `SystemUser` information as possible.

Further investigation of the pre-existing Visual Basic application at an informal workshop with the development team revealed that the `SystemUser` tables contained sufficient information to fit, without change, into the new business model outlined in the diagram below. However, the calendar information for the timesheets was hard-coded into the system as being based on a Monday to Sunday working week.

After consideration and a review of the requirements with the expert business user and business analyst, it was decided that:

- Timesheets were only to be 'open' for editing in the current week, and a certain number of weeks in the future
- For weeks that had been processed completely, or that were far in the future, timesheet entry would be 'closed' (this would prevent any historical revisionism or optimism on the part of consultants)

The existing database did not store calendar or permission information (but hard-coded this logic) therefore, the architect noted that some new tables for calendar information and permission information would have to be added to the existing database, and arranged for this to be discussed with the Oracle DBA and Visual Basic Project Lead. This would need to be done carefully to avoid conflicts with, and re-development of, the existing system.

This diagram shows the main business object classes of the system as designed by the architect:

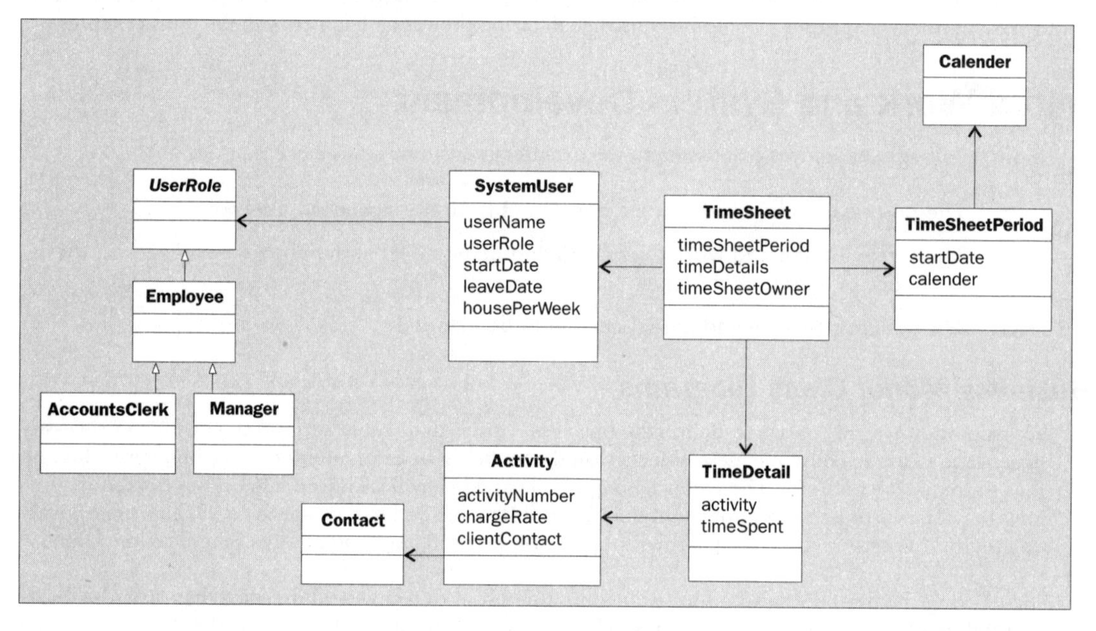

A `TimeSheet` class is for a particular `TimeSheetPeriod` (usually a calendar week) and a specific `SystemUser`, it contains a collection of `TimeDetail` objects. Each `TimeDetail` is for a particular `Activity`. The `SystemUser` has a `UserRole`, that defines the user as an `Employee` and perhaps more specifically as an `AccountsClerk` or a `Manager`.

Screen Design for Timesheet Entry

Screen design and usability were important issues for the company, and an initial user interface design exercise – using a simple graphics package – was undertaken to give the business users a feel for how the eventual system might look and feel. The design of the user-interface kept in mind the need for simple and 'touchable' screens. Since the planned deployment platform for the time-sheet application was a Palm/Pocket PC/Symbian device, having controls that could respond to touch-screen 'clicks' and minimizing the level of user-input was imperative.

The screen design concentrated on capturing the logical flow of user input for the time-sheet:

- Selecting the individual whom the time-sheet is for
- Selecting the week the time-sheet is for

Then inputting the time details:

- The activity code to book the time against
- A brief description of the work done
- The number of hours spent during the week on the activity

Recording hours spent day by day was agreed to be a *Won't Have Yet* for timebox one, although the users could enter multiple entries for any particular activity, and enter a text description specifying the precise date if they wished.

Touchable Screen Design

The issues with designing the look and feel for PDA-type devices, needed to be given careful consideration. For example, Palm recommends that the controls within a page be operable by a user with his or her fingers, in case the Palm pen has been lost.

The graphical designer for the system worked with the architect to produce some mock screens to help the team visualize potential interface options and issues. The time-sheet entry screens, shown below, were felt to provide the minimum amount of input to enable the viewing and entry of timesheets:

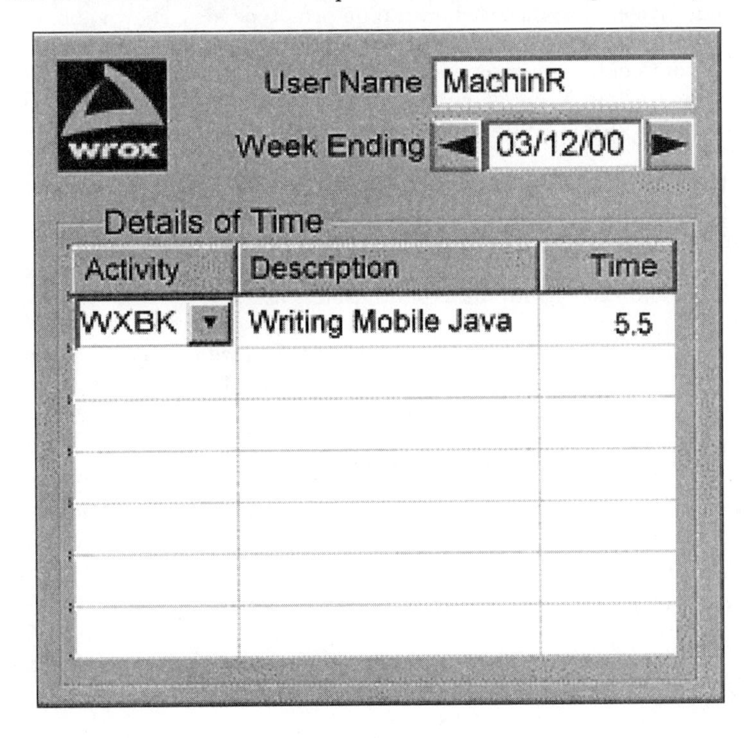

There was some debate over the entry of the user name on this form. The current user was identifiable by the current session, created for the user when they logged in, and this user name was only needed if the viewing or entry of timesheets other than your own was a requirement of the system. This problem was referred to the user representative on the project team, and it was decided that the user name should be kept here as it was common practice for people to enter and view other people's timesheets. The architect, project manager, business analyst, and expert business user, did not consider it necessary to refer this matter to the project steering committee consisting of the three departmental heads.

The team identified some basic screen flows to demonstrate possible user-interface issues:

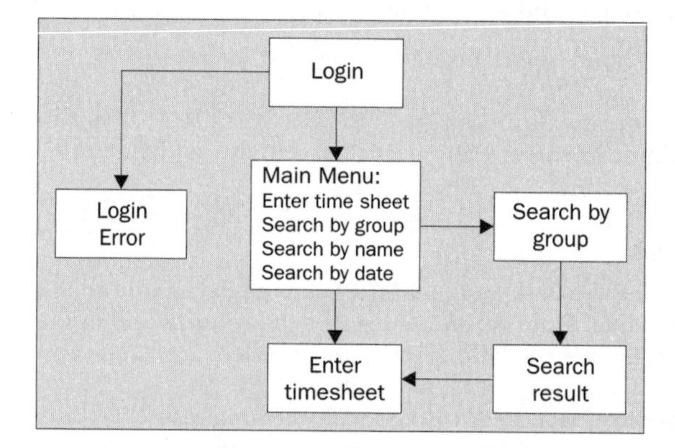

A design workshop was held with the original requirements workshop participants to select designs and agree upon interface issues. The final demonstration prototype idea, shown above, was constructed in advance, and presented to the group to give an idea of how the system might look and feel. This prototype, the first draft of the use-cases and the initial architectural designs were circulated to the participants after the workshop, and taken to a final meeting with the project sponsors, in order to obtain consensus on the project scope and boundaries to initiate the project.

Various user interface concepts were reviewed at the workshop and certain ideas, such as the use of drop down lists for permitted `Activity` objects, in order to reduce user input, and reduce the chance of mis-typing activity codes (it was recognised that there was a far greater risk of this if data was input while the user was on the move between appointments rather than seated at a desk), were advocated by the senior developer and agreed with the expert business user.

The expert business user and business analyst were then tasked with completing a user acceptance-testing plan, to ensure that the system produced could be validated as having fulfilled the identified requirements.

The Initial Proposed Architecture

The architect worked from the identified use-cases and business model to produce:

- System component diagrams
- Sequence diagrams
- System structure diagrams (not UML)

Component Diagrams

This diagram shows the major system and sub-system components within the application. Certain elements, such as User, Calendar and Activity, were independent of the TimeSheet model and could therefore be re-used in future application development projects within the organization:

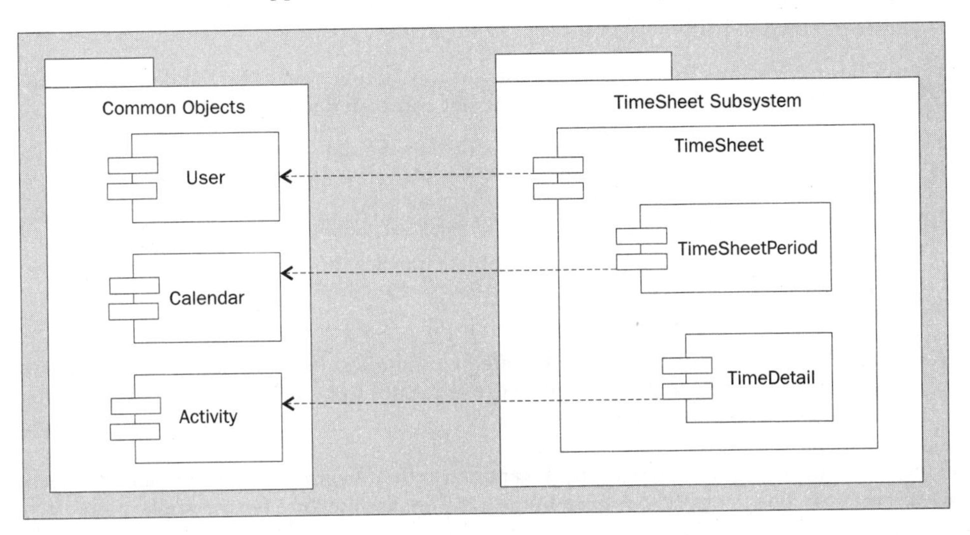

Sequence Diagrams

Sequence diagrams were developed by the architect to allow the developers to better their understanding the order of events and actions in the system, and to develop their ideas of the required controller objects and infrastructural code.

For example the `UserSettingsManager` controls requests to retrieve the settings within the system for a particular user, such as the activities they are permitted to enter time against – information retrieved via the `ActivityManager`:

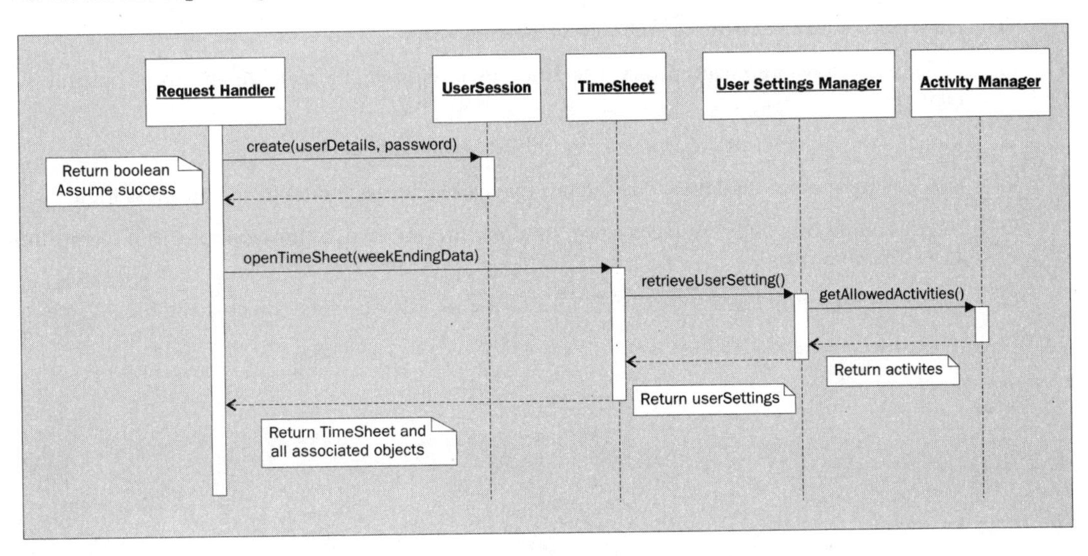

This diagram shows the flow of events and actions as a user creates a time sheet:

- First the user must attempt to log into the system by entering a vaild username and password combination
- Assuming this login is successful, a `UserSession` object is assigned to the user to manage their interaction with the server
- An `openTimeSheet()` request is made for a particular week (by the default the last outstanding week that the individual has not completed a timesheet for)
- This attempts to find a time sheet for the use for the date specified, if none is found a new `TimeSheet` is seamlessly created
- To create a `TimeSheet` certain settings must be queried
- A controller `UserSettingsManager` object queries the `ActivityManager` for the employee to determine which activities the `SystemUser` is permitted to include in their `TimeDetails`
- The `TimeSheet` object and the associated configuration objects are populated with the required information, and returned to the requestor to be passed to the device-side cache where they can be processed

From the use case and sequence diagrams, the architect could begin to perform functional decomposition on the scenarios of the system, to create an initial software architecture for the application.

Planned System Structure

UML is not designed to notate software architecture specifically, so the diagrams used to communicate this element of the software's design did not conform to a particular notation standard, but nonetheless were constructed to be as clear as possible.

The diagram below shows the architect's initial attempt to construct a system that would:

- Utilize the backend datastore from the initial system
- Utilize the client's existing web-based security systems
- Make use of Java application server technology to construct a scalable and robust business logic tier
- Manage the `UserSession` objects in a robust server side cache
- Use a request handling layer to create an extensible access model to the business logic
- Isolate the mobile delivery mechanism to allow for alternative (for example, thin client) future access mechanisms
- Utilize a client-side layer for the user interface and a device-side object cache for off-line use

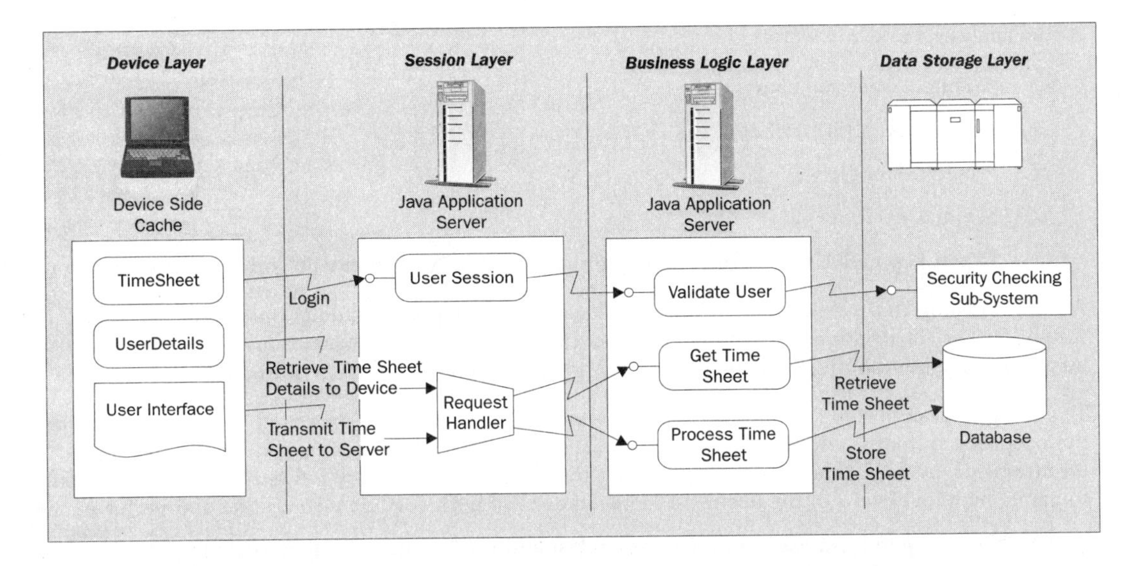

Analysis Workshop for the Initial Architectural Designs

The design analysis workshop used the SAAM (Software Architecture Analysis Method)-cycle to analyze the initial proposed architectures and design plans. SAAM is a process that can be applied to software architectural designs in order to validate their system model. It can be applied to both proposed software development plans and potential software acquisitions.

SAAM uses a scenario-based approach, relying heavily on end-user involvement to provide information on the business requirements, and a strong technical knowledge and prototyping validation model to verify the correctness and fitness of architectural design plans. It is an iterative process that cyclically refines designs throughout the design phase of a timebox, and can be used during the entire development cycle by the architect to adapt and validate the continuing implementation and Quality Assurance processes.

Each SAAM cycle is made up of six distinct phases:

- Describing the candidate architecture that is to be reviewed and validated
- Developing the use-case scenarios with the business users
- Classifying the scenarios in relation to the proposed design
- Performing an evaluation of the architecture in relation to the scenarios
- Checking for and revealing scenario interaction within the design
- Performing an overall evaluation of the architecture in relation to the findings of the business review, taking into account fitness for business purpose in terms of both functional and non-functional requirements (such as robustness)

At the analysis workshop were:

- Architect (meeting chair)
- Business Analyst (scribe)
- Project Manager
- Senior Developer
- Expert Business User assigned to help project team (a consultant who will remotely use system)

At the workshop the architect talked through the architecture plan and diagrams, as shown above, and sought further clarification and detail on the use-case scenarios, working with the expert business user, business analyst and project manager to record and understand any new requirements.

The design was analyzed and the scenarios classified in relation to the proposed design, to ensure that the proposed system would meet the functional requirements. At this point, non-functional requirements, performance levels, security, and the cost of the necessary infrastructure were discussed to gain a high-level view of the businesses expectations of both cost, time-to-market and performance.

The architect, expert business user, business analyst and project manager then had the task of:

- Performing an evaluation of the architecture in relation to the scenarios
- Checking for and revealing scenario interaction within the design

Once this had been completed, and any initial patches to the design plans had been made on the whiteboard, the workshop had the task of validating the architecture and focusing on the technical implementation issues. This involved the architect and senior developer working to perform an overall evaluation of the architecture in relation to the findings of the business review, taking into account fitness for business purpose in terms of both functional and non-functional requirements (such as robustness and mobile issues).

For example, an issue picked up and raised by the senior developer from the design plans was as follows. With to regard the `Activity objects` that each `SystemUser` can access, are these 'permissions' associated with:

- Each `TimeSheet`? – which seems unlikely
- Each `SystemUser`?
- A `TimeSheetPeriod` combined with a specific `SystemUser` object's permission?

This question was answered by the expert user, who felt that the system should restrict or permit access to the activities on a per `SystemUser` basis as in the initial plans. However, this permission is time-sensitive, since people join and leave project teams and therefore `TimeSheetPeriod` will need to be closed once processed. So, the last option was the correct behavior.

The senior developer and architect also wanted to spend some time on the following issues in the workshop:

- Agreement on initial target devices
- Considerations for mobile devices
- Server versus client load, responsibilities and caching
- Choice of server-side platform

For more information on the use of SAAM within the design and development process, see Software Architecture in Practice, *by Len Bass, Paul Clements and Rick Kazman (Addison Wesley Longman Publishing Co, 1998) ISBN: 0201199300.*

Target Devices

The expert business user and business analyst had done some initial research before the workshop. Of those consultants who already owned a PDA (over 60%), the usage of different models was divided up as follows:

Device Type	Percentage
Palm V and Vx	50%
Palm III models	20%
Palm Pilots & Professionals	10%
Psion Series 3	5%
Psion Revo	5%
Psion Series 5x	5%
Pocket PC-based devices	5%

Since the overwhelming number of consultants already possessed, and were comfortable with using, Palm-based devices, and 50% possessed Palms capable of supporting a Java virtual machine therefore, it was decided for the purposes of the pilot, to support only the Palm platform in timebox 1.

The choice of application development model for the Palm using J2ME was either the Spotless or the MIDP/CLDC model. The architect felt that it would be better to structure the application for the MID profile on the Palm as well as the other platforms (on which the Spotless implementation was not available) since this would promote portability to other MIDP devices in future timeboxes.

Security

It was agreed at the initial requirements workshop to allow the user to enter a plain text username and password to be checked against the values stored in the current Oracle database. However, this was identified as an area of concern, given that the system was going to be available outside of the company's LAN security infrastructure.

The architect raised security as an issue at the subsequent sign-off meeting and ensured that at a later stage, proper attention would be paid to methods for securing this system, and utilizing good Internet security practices. Initially the project manager scheduled this for a workshop at the end of timebox 1, with work expected to commence on the security issues as part of timebox 2.

Java Application Server Load and Responsibilities

The architectural plans relied on a robust and scalable middle-tier layer utilizing Java application server technology. The choice of application server provider was open at the stage of the architectural review: the company is an important Oracle customer, and one option was therefore to use Oracle Java database and application server features for the new system. However, the company management was concerned with using a single vendor solution, and it was therefore decided to allow three industry-leading vendors, Oracle, BEA and IBM, to bid.

The main processing and object creation load in the software architecture was placed firmly on the server-side, with the client relying on the server to create objects, caching and manipulated these objects locally on the device, before returning the objects to the server-side to be validated and persisted in to the Oracle database.

Creating a Mobile Application's Infrastructure

There is no silver bullet solution to the problems of moving from fixed network, powerful PC clients to mobile devices and wireless networks; there are, however, certain concrete steps that can be taken by development teams to maximise their chances of success. The management of tactical solutions in this area is critical, and allowing the organic growth of wireless systems, which develop on an ad hoc basis according to differing requirements, could lead to poorly architected systems that may not always be secure or scaleable.

Organizations must think strategically in order to be able to cope with the exponential growth in demand for the core functions that employees and clients will demand as this technology matures. By ensuring that the backend systems are developed to a high standard, best of breed, scaleable, and secure architectures can be created to deliver the content to any client platform irrespective of the client technology, including thin-client, J2ME and J2SE applications.

An Effective Pervasive Computing Environment

The strains placed on development projects from supporting wireless technology projects, will prove to be an acid test of many web development and design methodologies. However, viewed and executed correctly, the development and expansion of the mobile platform will allow many organizations to gain a competitive edge in an already technically-aware market.

The role of an architect in this environment should also include monitoring and reviewing:

- How applications and components are developed and deployed
- How change control is implemented in the component deployment and development process
- How the component framework services will be monitored in real-time, including:
  - Component heartbeats
  - Logging of component problems, exceptions and failures
  - Automatic restart of component systems on failure
- How security policies are decided and implemented, and how intrusion is detected
- How policies are monitored and reviewed, and how processes are audited

Summary

Architecture holds a key position in the decision-making process for software development. If important choices on design, and development, can be caught and analyzed early, prior to implementation, costly mistakes, and fruitless prototyping implementations, including technological blind alleys, can be avoided. Architecture is, therefore, an element of the development cycle of device-based applications that can yield a high cost/benefit return.

As we have discussed, close attention must be paid to the balance between client and server processing. J2ME-based software must be designed to take advantage of new devices and delivery channels as they emerge, whilst ensuring that a stable, scalable, robust and secure server-side infrastructure is created.

In this chapter, we have explored how new techniques in the gathering and recording of requirements can be combined with an iterative, timeboxed approach to the development process to enable a more responsive and flexible approach to project delivery. J2ME is such a new and rapidly evolving technology therefore, architects and project managers will need to be particularly careful in designing and planning potential solutions, while of course leveraging the lessons learnt from the client-server and distributed computing traditions where appropriate.

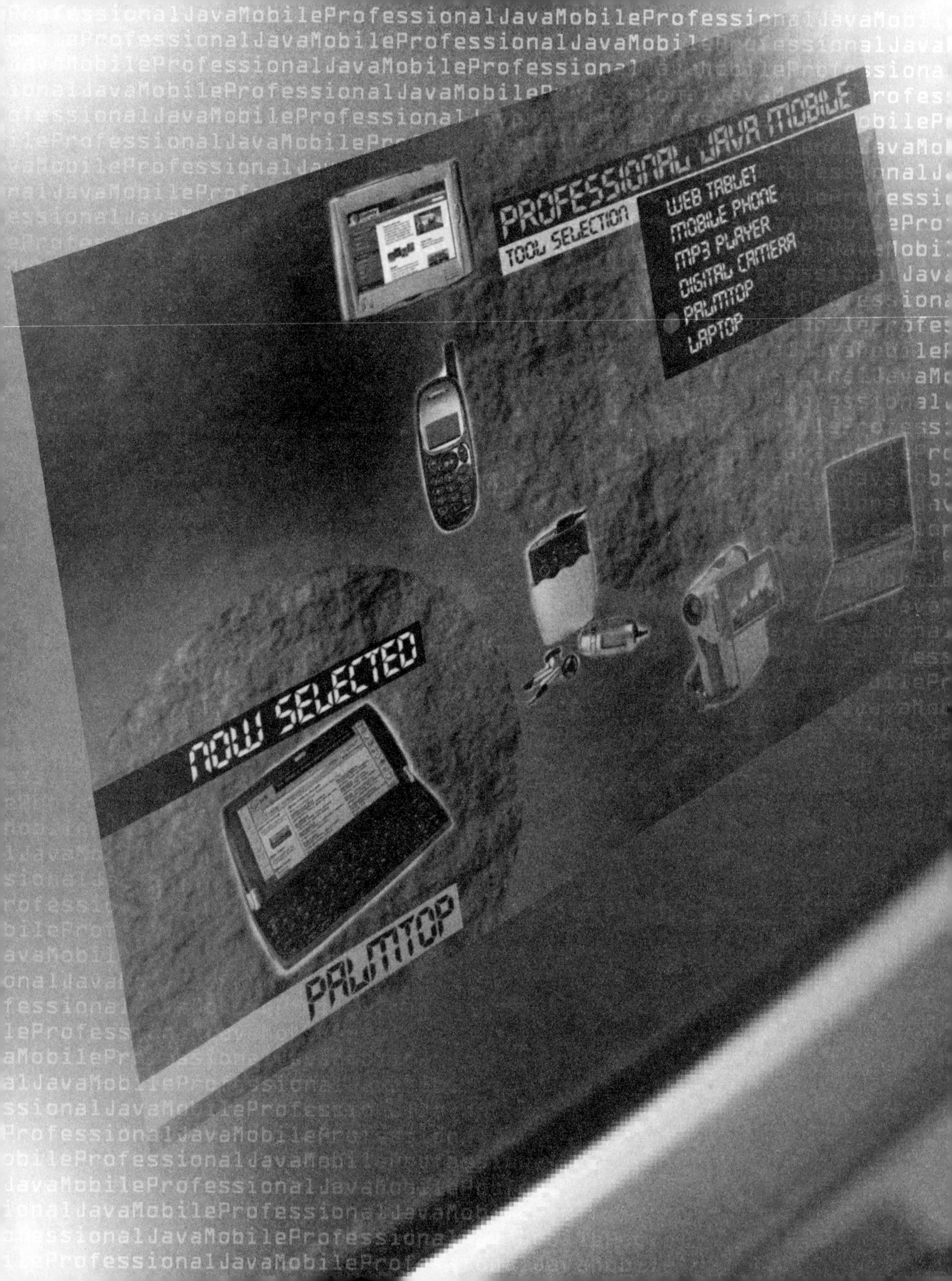

PROFESSIONAL JAVA MOBILE

TOOL SELECTION

WEB TABLET
MOBILE PHONE
MP3 PLAYER
DIGITAL CAMERA
PALMTOP
LAPTOP

NOW SELECTED

PALMTOP

3

CDC and the Foundation Profile

In addition to devices usually associated with mobile technology, namely mobile phones, PDAs, and Palmtops, Sun (together with several licensees), saw a niche for devices with greater capabilities than small devices but that do not have the full processing and memory capabilities of PCs.

Examples include, consumer devices, such as web phones (a web based, packet switched telephone service), web pads (dedicated HTML browser devices), set top boxes that allow access to the Internet on TVs, and digital television set top boxes. In addition, all these devices have some type of network access. In fact, we should begin to see a thread running through developments in Java and its associated technologies, which is echoed in the examples above: connectivity.

With the advances made in radio modems, and other networking technologies, it is almost inconceivable that any given device will not have access to TCP/IP or similar network services within the next few years. This extends to devices such as in-car navigation systems and position based services, set top boxes including audio and video systems, and home application servers for small devices, to name but a few.

In all these cases, resources and capabilities are less constrained than small mobile devices, such as mobile phones. Memory is more plentiful, we can expect a file system for storing data, and processing power may be equivalent to that of a PC, especially when freed from the full-featured, but ponderous software usually provided with PCs.

The specification for Java programming on these types of devices, defined in the Connected Device Configuration (CDC), assumes that they vary greatly in form factor, and in addition are somewhat limited in memory, processing power, and input capabilities, and therefore cannot afford to carry excess capabilities. For this reason, this configuration exploits the full potential of the J2SE platform without making assumptions, such as the availability of a graphic user interface, or requiring RMI support.

The Foundation Profile implements the minimum specification set out by the CDC. It does not specify a GUI, but merely incorporates the file handling, networking and basic Java language classes, together with Java's security model and exception handling.

When the Foundation Profile specification was written, it was understood that it would also act as a base that other profiles would build on to add a GUI, RMI support, or another desired functionality depending on requirements. In this chapter, we will discuss the functionality provided by the Foundation Profile, as specified by the Connected Device Configuration that it implements, and both the current and potential devices for this configuration.

We will look at the types of applications that are possible with it, and implement a networking application that illustrates its usage. We will also look at the Personal Profile and its relation PersonalJava as the natural extension to the Foundation Profile. To summarize, we discuss:

- The Connected Device Configuration (CDC).
- The Foundation Profile – the base profile for all other CDC Profiles.
- The CVM – a C language Virtual Machine, which is a Linux and VxWorks implementation of the VM for the CDC.
- We discuss the classes from the J2SE that are supported, and those classes that have been removed in this configuration.
- We next define and explain the new connectivity classes provided in the CLDC originally and made available in the CDC. These classes provide a new way of accessing resources both locally and over a network. We will also minimally implement these classes for Windows and the J2SE for those readers that do not have Linux systems.
- We will end by developing a POP3 Proxy Server that will show both the similarities and differences in developing applications for the CDC and the Foundation Profile and by discussing future profiles for the CDC.

What We Will Need

There are two platforms for which implementations of the CDC and the Foundation Profile are available: Linux (Red Hat 6.2 or later on Intel) and VxWorks (also on Intel). We will need to build the distributions for these in order to test the code in the latter part of the chapter. The distributions include a VM for CDC devices, called the CVM, which is available as C source with appropriate make files for compilation, an optimized run time environment and supporting classes for the Foundation Profile and the Micro Edition IO API for generic connections. We will learn about this API in this chapter.

Download both the CDC and Foundation Profile, which can be found at http://www.sun.com/software/communitysource/j2me/cdc/download.html. The installation is fairly straightforward; instructions are given in the guide to build first the CDC and then the Foundation Profile. There is also a suite of classes provided to test the installation. One quirk, which we encountered in the installation process for Linux, is that we needed to create a /linux sub directory in the JDK 1.3 installation and move the contents of the /jdk1.3 folder into it for the installation to be able to find `javac`.

Once the CDC and Foundation Profile are installed, compilation of class files must be done with the bootclasspath pointing at the `cdc.jar`. This JAR file defines both the optimized runtime classes and the new networking classes for the J2ME platform. (We will talk about these later in this chapter.). This can be done by using the `-bootclasspath` switch on the command line. So to compile a class named `myClass.java` we would therefore enter the following command:

```
javac -bootclasspath <path_to>/cdc.jar myClass.java
```

Where <path_to> is the path to the relevant JAR file. For those readers that do not have access to Linux or the VxWorks platform, we will provide a minimal implementation that will allow us to run the examples nevertheless.

The classes can then be run using:

```
<path_to>/cvm -Djava.class.path=<path_to_code_folder> myClass.java
```

We will provide further information on compiling and running the examples in this chapter as needed.

Connected Device Configuration

The Foundation Profile is a Connected Device Configuration. Recall that this configuration sits between the full J2SE platform and the Connected Limited Device Configuration (CLDC) with the more specialized configurations such as EmbeddedJava (eJava) and JavaCard at the far end.

The Foundation Profile defines a basic functionality for high-end J2ME consumer and embedded devices. These devices require a platform that is scalable and robust, and that has a good security model, which is available through the J2ME platform. The reasons for these are two fold; first, in order to facilitate automatic software updates and repair, and second a good security model makes accepting downloadable third party software viable.

CDC devices must have:

- A minimum of 512Kb ROM
- A minimum of 256Kb RAM
- Connectivity, whether IR, serial, bluetooth, or an as-yet-unrealized networking technology
- A complete implementation of the Java Virtual Machine according to the JVM specifications, 2nd edition
- Any degree of user interface including none

The CDC addresses issues such as application lifecycle, the security model and code installation, as well as the supported APIs.

Specifically, this configuration seeks to differentiate itself from the CLDC in several ways. Devices that match the CLDC are even more restricted in terms of memory, processing power, and input capabilities, although they may have equivalent network access in terms of reliability and bandwidth. Second, CDC devices have a file system. In addition, the CDC supports a full implementation of the JVM, which in comparison to lesser devices includes full garbage collection, all of the data types, and class file verification and so on. (The CLDC is discussed in full in Chapters 4 and 5).

Another important point, is that this configuration specifies the full security model as defined in the Java 2 Standard Edition platform. Unlike the CLDC, which uses a sandbox model reminiscent of the applet security model, this configuration includes the full `java.security` package and its subpackages.

As with the CLDC, the intention is to keep applications upwardly compatible through the platforms, making applications designed for these devices possible to run on J2SE and J2EE systems, although there is a proviso in the form of a new set of classes that provide an abstraction of connections. These file are defined in the `javax.microedition.io` package and will be discussed in more detail later in the chapter.

Foundation Profile

As mentioned above, the Foundation Profile is an implementation of the CDC that does not add functionality to it, but is only a base profile that provides all the basic classes and functionality required by the intended devices. The Foundation Profile is a subset of the J2SE APIs for running on small devices with a network connection, and the target devices for this profile match those that do not require GUI functionality.

One potential device for this profile is the Consumer Application Server concept. The Consumer Application Server acts as an application proxy host for CLDC and other CDC devices in the home gathering, deploying, and uninstalling applications, and acting as a proxy server for devices. The application server would also schedule software updates.

In fact, looking around at currently planned devices, the proxy server is the most visible of these devices. Proxy services may extend to acting on behalf of electric, gas, and other meters to enable reporting of readings over a network, automatic downloading or recording of audio and video content to disk, tape or CD and facilitating switched language tracks for movies. Another possibility may be monitoring and reporting pay-per-view service usage, such as pay-per-view sport and movie channels, which may also extend to audio services in the future. For numerous examples of the possibilities, go to http://www.windriver.com.

The Foundation Profile requires the following minimum characteristics for a device that uses it:

- 1024Kb ROM
- 512Kb RAM
- Network connectivity

In both cases, memory requirements do not take into account the requirements of installed applications. Keeping the minimum requirements minimal, allows manufacturers to specify devices more closely to their intended function. In addition, there is no GUI functionality and, if required, this would need to be added, by the definition of a new profile that incorporates the Foundation Profile.

Command line interaction is still available although whether this would be available to the user is doubtful. Note that the assumption is that the user will often be unfamiliar with PCs and may not even be aware of the proxy that the device represents.

The CVM

The CVM is a Virtual Machine written in C. It forms the basis for J2ME devices that implement the CDC and its profiles. As standard J2SE VMs do not scale well to these types of devices, a completely new implementation of the Virtual Machine was written, providing a small footprint (40% when compared to the classic JDK VM), reliable Java 2 1.3 VM with an optimized interpreter. This includes method calls and return value optimisations and synchronization. The CVM also includes native thread support, RMI, reflection, serialization, and JNI support for those profiles that have those functionalities.

For the more technically minded, the CVM includes optimized (generational) garbage collection, although this can be replaced with an alternative garbage collector. The VM also runs in its own separate memory space to the system memory. Part of the optimisation has been the inclusion of "pre-loaded" classes – classes that are provided as byte code in ROM memory and do not need to be interpreted.

In addition, Sun have also thought about the stability of the VM and have added protection against memory starvation, running out of file descriptors, stack usage limiting, and the CVM also checks for and tries to reduce C recursion.

Continuing in the same vein, there has been some work towards enabling porting of the CVM, including providing clear and well documented interfaces and code that is port aware. On the whole, this information is likely to be of interest to only a few. It does appear that Sun have tried to provide a good quality VM for this platform.

Supported Classes

The classes included from the J2SE platform come from the following packages. They define, in order, input and output functionality, the basic Java language classes, interaction with the garbage collection, reflection, arbitrary point math, networking classes, and finally, security:

- `java.io`
- `java.lang`
- `java.ref`
- `java.reflect`
- `java.math`
- `java.net`
- `java.security`

Additional classes defined in the Foundation Profile are in the following packages:

- `java.text.resources`
- `javax.microedition.io`

The AWT and Swing classes are not included as previously mentioned (no GUI), nor are the applet and beans packages. Once again, we have already mentioned that `java.rmi` and all subpackages have been removed. Besides these, `java.sql`, `javax.accessibility`, `javax.naming`, `javax.transaction`, the sound classes introduced in Java 2, and finally, the CORBA classes are also not present.

Removed Classes and Methods

As far as classes that have been removed are concerned, these will be summarized below. Deprecated classes and methods have on the whole been removed. An exception to this seems to be the `java.security.acl` package, which has not, despite the fact that it has been superceded by the current security system as introduced in Java 2.

Among the classes that will not be missed are two classes that were deprecated in JDK1.1 for assuming that bytes can represent characters adequately, these being `LineNumberInputStream` and `StringBufferInputStream`. In any case, operations such as these should be done with an appropriate reader.

Another class that has been removed is the `BigDecimal`, which has disappeared without trace. This class provided basic arithmetic functions and scale manipulation with control over rounding, but is not present.

Additional Classes Defined in the Foundation Profile

The following classes fall into two areas. The first of these, in the `java.text.resources` package, was specifically introduced in the Foundation Profile, while the second, specifically concerned with access to resources, was originally introduced in the CLDC and are available in the CDC. We discuss them in order.

java.text.resources

`java.text.resources` has a single class `LocaleData` that contains a single method `getAvailableLocales()` that returns a list of the available locales. This method is actually only included to provide a way to list available resources, although this will, in all likelihood, change. Locale specific resources (language resources for the internationalization of applications) should be accessed via resource bundles in any case.

Unfortunately, whole localization issues are likely to be of high concern in this Profile, discussion of this topic is out of scope for this chapter. In any case, the techniques for accessing locale specific resources are the same as those for the J2SE platform.

javax.microedition.io

The classes defined in the `javax.microedition.io` package are collectively known as the **Generic Connection Framework** (GCF). These classes provide simplified IO and networking capabilities with an easy programming interface, using a `Connection` as the basic concept. (Do not confuse this with the `java.sql.Connection` interface.)

This package was originally designed with a view to allowing CLDC devices access to networking services that are more simplified, abstracted, and that also have a reduced footprint. In addition, as there should be upward compatibility through the platforms, this will need to be added to the J2SE and J2EE platforms.

The `microedition.io` package abstracts IO to connections. A `Connection` represents a connection to a resource, whether this represents a file, datagram connection, or a content source, such as a web site. The interface hierarchy is as follows:

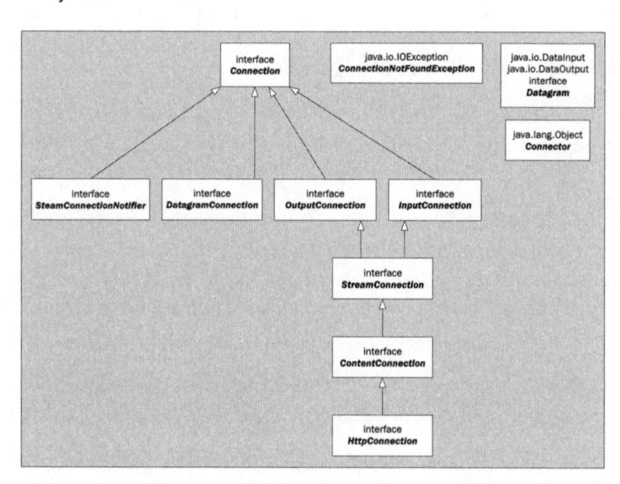

To summarize, the behavior of each interface is as follows. Connection itself only defines one method, close(), that releases the resource bound to this connection. InputConnection and OutputConnection represent stream connections for input and output respectively, and StreamConnection implements them both for two-way communication.

ContentConnection extends StreamConnection to represent HTTP connections specifically and includes methods for accessing the character encoding type, length and content-type of the content at the URI specified for the connection.

In order to get a connection to a resource, we use the Factory class Connector, which will return an implementation of Connection or one of its sub interfaces.

Something to be aware of, is that the CDC does not provide these implementations. It only provides the ability to create diverse types of connections in a profile. Furthermore, profiles are not forced to implement all these different and useful connections above. The MIDP, which sits on the CLDC, for example, only has an HTTP connection – it has no datagram, comms ports or file connections. The Foundation Profile defines a Socket, ServerSocket, and http:// connection in addition to the file:// and datagram:// connections that the CDC defines.

We open a connection using the open() method of the Connector class, which has the following format:

```
Connection conn = Connector.open("<scheme>:<address>;<parameters>")
```

The <scheme>:<address>;<parameter> string conforms to the Uniform Resource Indicator (URI) syntax defined in the IETF standard RFC2396 (http://www.ietf.org/rfc/rfc2396.txt) of which URL is a subset.

For example, here is a code snippet to create an HTTP connection that creates an InputConnection:

```
InputConnection in =
                (InputConnection) Connector.open("http://www.poqit.com");
```

We can open a client socket in one of the following ways:

```
StreamConnection sc =
        (StreamConnection)Connector.open("socket://212.67.193.221:9000");
StreamConnection sc =
     (StreamConnection)Connector.open("socket://www.someserver.com:9000");
```

The Connector class attempts to instantiate a class whose name is Protocol, in a package that is the sum of the value of the microedition.protocolpath system property, the platform, and the scheme given in the connection string. If microedition.protocolpath is not set Connector uses com.sun.cdc.io by default. For example, the handler for file connections on CDC devices is called com.sun.cdc.io.j2me.file.Protocol.

This class then attempts to connect using the connection string. Implementers who wish to add connection types may do so, in which case microedition.protocolpath can be changed to reflect the path to the handler class. It should be noted that an implementation that adds a scheme not specified by the CDC and the Foundation Profile could reduce its portability.

There are three forms of the `Connector.open()` method. The simplest takes a connection string as shown above, in which case it will create a connection with read and write capabilities by default which does not throw exceptions in case of a (connection specific) timeout. The other two forms also accept a mode attribute that takes the value READ, WRITE, or READ_WRITE, and that presumably helps minimize wasting resources on the unused parts of the connection, as well as checking for incorrect usage, when the connection only supports one-way communication. It should be noted that this mode is a hint to the handler and it may ignore it.

In addition to the mode attribute, the third form also takes a parameter, `timeouts`, which indicates that the calling code can handle timeout exceptions and its syntax is as shown:

```
static Connection open(String name, int mode, boolean timeouts)
```

If `timeouts` is set to `true`, then the scheme may throw an `InterruptedIOException` when it detects a timeout condition. Again, this flag is only a hint to the handler and there is no guarantee that such exceptions will be thrown. The default is `false`, indicating that exceptions will be thrown. Note that the timeout period is scheme specific, and there is no way of setting it – only the original implementers
can specify its value.

Therefore connections are usually opened to gain access to an input or output stream four convenience functions are provided when we merely need to access an input or output stream:

```
static DataInputStream openDataInputStream(String name);
static DataOutputStream openDataOutputStream(String name);
static InputStream openInputStream(String name);
static OutputStream openOutputStream(String name);
```

For example:

```
InputStream in = (InputStream) Connector.open("http://www.poqit.com");
```

Could also be written as:

```
InputStream in = Connector.openInputStream("http://www.poqit.com");
```

`InputConnection` and `OutputConnection` simply describe the functionality provided by the `openXXXStream()` helper methods above for input and output streams respectively. `StreamConnection` implements the combination of the two to represent two-way communication.

We would cast the return value from `Connector's open()` method to `StreamConnectionNotifier` when the connection represents a `ServerSocket`. `ServerSocket` has a single method `acceptAndOpen()` that waits for a client to request a connection on the relevant port, and returns a `StreamConnection` that represents the server side of that connection.

`DatagramConnection` and `Datagram` allow us to send UDP data packets across the network. Data packets are currently most often used in network management applications, but could be used for automated utility meter readings, or to facilitate automated traffic reports through data gathering devices, to name just a few possibilities. The key here is that these usage scenarios do not require guaranteed timely receipt of packets, if a meter reading is missed we can get the next one, we can interpolate any missing data from a traffic camera, and so on.

In fact, the potential for data packet communications with J2ME devices is so great that we will look at some examples of programming with datagrams in more detail.

An Example of UDP Programming

For J2ME, a new interface, `Datagram`, has been defined which holds the destination address and a buffer containing the data to be sent/received from that address. `Datagram` extends both `java.io.DataInput` and `java.io.DataOutput` for reading and writing primitive Java data types. These classes provide convenience methods for reading and writing of data to the datagram buffer, for example, `readBoolean()` to read a boolean type, and `writeChars()` to convert an array of type `char` to binary data, and so on.

We'll start by creating and sending a few datagrams, and then see how a server would capture datagrams. To send or receive datagrams, we must first create a datagram connection:

```
DatagramConnection datagramConn =
        (DatagramConnection)Connector.open(datagram://127.0.0.1:3456);
```

The values provided by the connection string will be used as the default values for the `DatagramConnection` methods that return a datagram. In this example, unless otherwise specified, data will be sent to machine 127.0.0.1 on port 3456. Omitting an IP address will cause the connection returned to be a server-side one, meaning that the server will wait for datagrams and can then process them, responding if necessary.

Once we have a connection, we can create datagrams with it:

```
Datagram datagram = datagramConn.newDatagram(256)
```

This will allocate a buffer of 256 bytes and create a datagram with it. The `newDatagram()` method is overloaded to allow an alternative address and port to be specified, pass a buffer to store the data in, or both. We can also check the nominal and maximum buffer lengths (in the Sun implementation these values are the same).

The datagram is initialized with an address and port number to reply to; the address will be the one associated with the localhost (this is how the address is assigned in multi-homed systems), and binds the socket to any available port.

All that is left to do now is to set and send our data. This could be as simple as the code below although it is left up to you to decide how useful this particular packet is:

```
byte[] bytes = "keep alive".toBytes();
datagram.setData(bytes, 0, bytes.length);
datagramConn.send(datagram);
```

We assign a buffer with a "keep alive" message, set the data for the datagram, and send it. `sendData()` takes a byte array, the offset into the array and the length of the data.

Effectively, that is all that it takes to program with datagrams. The datagram also has accessor methods for retrieving the sender's address and port number to reply to, and the offset into the buffer of the data and the length of the data.

To create a server that will receive this packet, we would create a datagram connection set to server mode and call the `receive()` method on it. This will bind a socket to the port provided and the server will then wait for a packet, blocking execution until one is received. For this example, we would bind the server to port 3456:

```
DatagramConnection datagramConn =
        (DatagramConnection)Connector.open(datagram://:3456);
```

Providing no IP address sets this connection to server mode, notice that the colon preceding the port number is still required, we call the receive method giving a datagram as the argument. We can then access the data in the datagram:

```
Datagram datagram = datagramConn.newDatagram(256)
datagramConn.receive(datagram);
// do something with datagram here
```

Datagram programming also allows us to broadcast data packets to multiple clients by binding the connection to a multicast port, a port that is reserved for client groups. Clients can join this group and receive any packets sent to the group and port combination.

The Remaining Classes

Besides the previously mentioned interfaces there are two other, `RandomAccessConnection`, for accessing random access memory, and `DirectoryConnection`, which represents directory services type metadata stores.

Implementing the GCF

At the time of writing, the only available implementations of the CDC, Foundation Profile, and the GCF, are for VxWorks and Linux. There are no plans to implement this profile for any other system. In order to allow readers with other systems to try out the examples in the remainder of the chapter, we will implement classes for the socket and server socket connections.

Socket Programming Revision

Before we begin, we should briefly review our knowledge of socket programming. We can create a (client) socket connection by constructing a socket and passing it the server name and port number to connect to, as follows:

```
Socket sock = new Socket("www.poqit.com", 80);
```

`Socket` also has constructors to connect to a specific IP address by passing an `InetAddress` object that allow us to bind the client socket to a specified local port and address.

When we have connected to the server, we can begin to communicate with it. For example, to retrieve the home page at http://www.poqit.com, we would now establish a writer to the output stream accessible via the socket and print `GET /` to it:

```
PrintWriter out = new PrintWriter(socket.getOutputStream(), true);
out.println("GET /");
```

The code above would then return output such as the following:

```
<!doctype html public "-//w3c//dtd html 4.0 transitional//en">
<html>
<head>
   <meta http-equiv="Content-Type" content="text/html; charset=iso-8859-1">
   <meta name="Author" content="Richard Taylor">
   <meta name="GENERATOR" content="Mozilla/4.6 [en-gb]C-CCK-MCD NetscapeOnline.c
o.uk  (Win98; I) [Netscape]">
   <title>Poqit Home</title>
</head>
<body text="#999999" bgcolor="#000000" link="#999999" vlink="#999999" alink="#99
9999" nosave>
etc etc
```

Of course, in order to be able to read this, we will also need to obtain an input stream with the `getInputStream()` method from which we can construct a `Reader` of some sort. In addition to communicating using character readers/writers, we can also communicate with other streams, allowing us to send or receive serialized classes, and send binary data (including images and audio).

All these can throw a variety of exceptions, including `UnknownHostException`, `IOException`, and `SecurityException`. `UnknownHostException` occurs if the hostname cannot be resolved by the network, `IOException` covers a variety of circumstances where connection is not possible. `SecurityException` occurs if a `SecurityManager` is installed and is blocking access to the port being connected to.

To construct a server side socket, we instantiate a `java.net.ServerSocket` passing the port number to listen on:

```
ServerSocket server = new ServerSocket(110);
```

The code snippet above would bind a server socket to port 110 on the default IP address for the system. We accept requests by clients to connect by calling the server socket's `accept()` method. This will block execution until a client connects and will return a `Socket` that represents the client:

```
Socket client = server.accept()
```

Once we have a client, we can call the `getInputStream()` and `getOutputStream()` methods as above to communicate with the client.

Implementing the socket and serversocket Connections

If you remember, we can substitute our classes for the default classes provided by Sun by changing the `microedition.protocolpath` system property, in this case to `com.wrox.cdc.io`. In addition, since the platform we are developing on is J2SE, we should also change the `microedition.platform` system property to "j2se". Therefore, the complete package names for our socket and serversocket implementations are `com.wrox.cdc.io.j2se.serversocket` and `com.wrox.cdc.io.j2se.socket` respectively. When it comes to testing and deploying the application on a J2ME system, we can omit the system property command line to run the code without problems.

In order to implement the socket and serversocket connections, we will need to extract the interfaces for the `javax.microeditions.io` package, the connector class and `ConnectionNotFoundException` from the Foundation Profile download. The classes we need are all in the `/cdfoundation/src/share/classes/cldc` directory. Copy the contents of the `/javax` sub-folder into the download code folder for this book, and make them available in the classpath.

In addition, add the folders necessary to create the implementing classes. The directory structure should look like the following:

```
/javax
      /microedition
                   /io
/com
    /wrox
         /cdc
             /io
                /j2se
                     /socket
                     /serversocket
```

In order to provide our own implementation of the socket connections, we will also need to modify the `Connector` class to remove a dependency on the rest of the Linux implementation. The change is as follows:

- The `Connector` class imports an interface called `com.sun.cdc.io.ConnectionBaseInterface`. This import will need to be changed to `import com.wrox.cdc.io.ConnectionBaseInterface`, a class that we will go on to implement next.

> **Modifying the files from the Sun Microsystems download is subject to the license for the download.**

The code for the `ConnectionBaseInterface` class is very simple:

```
package com.wrox.cdc.io;

import java.io.IOException;
import javax.microedition.io.Connection;

public interface ConnectionBaseInterface {
  public Connection openPrim(String name, int mode,
                             boolean timeouts) throws IOException;
}
```

It merely defines the `openPrim()` method used to open the connection. The files can now be compiled using the following command from the download code folder for this chapter:

```
javac -classpath . javax/microedition/io/*.java
```

We will now implement the socket and serversocket `Protocol` classes.

com.wrox.cdc.j2se.socket.Protocol

The Protocol for socket connection is a simple wrapper for a socket. We must implement ConnectionBaseInterface and StreamConnection interfaces:

```
import javax.microedition.io.Connector;
import javax.microedition.io.Connection;
import javax.microedition.io.StreamConnection;

import com.wrox.cdc.io.ConnectionBaseInterface;

import java.net.Socket;
import java.io.DataOutputStream;
import java.io.OutputStream;
import java.io.DataInputStream;
import java.io.InputStream;
import java.io.IOException;

public class Protocol implements ConnectionBaseInterface, StreamConnection {

  static final int READ = 1;
  static final int WRITE = 2;
  static final int READ_WRITE = (READ | WRITE);
  Socket socket;
```

To implement ConnectionBaseInterface, we have to implement its openPrim() method. We begin by checking that the connection string is of the form "//host:port" by checking for the double forwards slash and colon characters:

```
  public Connection openPrim(String name, int mode,
                             boolean timeouts) throws IOException {

    int colon = name.indexOf(":");

    if (!name.startsWith("//") || colon == -1) {
      throw new IllegalArgumentException("Invalid connection string" + name
                                        + "usage \"//hostname:port\"");
    }
```

If they do not, we throw an IllegalArgumentException. We next parse the connection string for the host and port number. If there is no host specified, this indicates that a server socket is required; we construct a serversocket connection and return it:

```
    String host = name.substring(2, colon);
    int port = Integer.parseInt(name.substring(colon + 1));

    if (host.equals("")) {
      return new com.wrox.cdc.io.j2se.serversocket.Protocol().openPrim(name,
              READ_WRITE, false);
    }
```

Otherwise, we create a socket with the host and port number given and return it to the calling method:

```
      socket = new Socket(host, port);
      return this;
    }
```

The next method is one that does not immediately spring to mind. Therefore the implementation of a server socket is going to be a wrapper of the com.net.ServerSocket class, its acceptAndOpen() method will simply call the ServerSocket's accept() method. However, this will return a java.net.Socket so we must have a way to wrap the java.net.Socket that is returned so that it exposes the StreamConnection and ConnectionBaseInterface interfaces. We can do this by specifying an open() method that simply assigns the socket passed to it to the socket property of this class:

```
    public StreamConnection open(Socket socket) throws IOException {
      this.socket = socket;
      return this;
    }
```

The remaining methods are implementations of the Connection.close() method and StreamConnection's methods, getInputStream(), getOutputStream(), getDataInputStream(), and getDataOutputStream():

```
    public InputStream openInputStream() throws IOException {
      return socket.getInputStream();
    }

    public OutputStream openOutputStream() throws IOException {
      return socket.getOutputStream();
    }

    public DataInputStream openDataInputStream() throws IOException {
      return new DataInputStream(socket.getInputStream());
    }

    public DataOutputStream openDataOutputStream() throws IOException {
      return new DataOutputStream(socket.getOutputStream());
    }
    public void close() throws IOException {
      socket.close();
    }
  }
```

com.wrox.cdc.io.j2se.serversocket.Protocol

This class is simpler to implement. Let's see first how we would create a server socket connection with the GCF. We can instantiate a server socket using the "socket://" or "serversocket://" protocol strings. Notice that we do not specify an IP address or host name to bind to; this is how the socket connection implementation can tell it's a server socket. A typical connection string is as follows:

```
  StreamConnectionNotifier conn = Connector.open("socket://:110");
```

This will bind a server socket to port 110 on the device and returns a StreamConnectionNotifier that represent this connection. Once we have a StreamConnectionNotifier, we can listen for client requests by calling the acceptAndOpen() method, which will return an StreamConnection representing the client.

We will, therefore, implement the javax.microedition.io.StreamConnectionNotifier interface, which represents a server side connection, as well as our ConnectionBaseInterface. The server property represents the java.net.ServerSocket that will be bound to the local port:

```
package com.wrox.cdc.io.j2se.serversocket;

import java.io.IOException;

import java.net.Socket;
import java.net.ServerSocket;

import javax.microedition.io.Connection;
import javax.microedition.io.StreamConnection;
import javax.microedition.io.StreamConnectionNotifier;

import com.wrox.cdc.io.ConnectionBaseInterface;

public class Protocol implements ConnectionBaseInterface,
                                 StreamConnectionNotifier {

  ServerSocket server;
```

The first method to implement is the openPrim() method. We check that the connection string passed in has been formed properly, and that no IP address is specified, and then read in and parse the port number to bind to. An improperly specified port number will throw a NumberFormatException, we cannot recover from this, we throw an IllegalArgumentException.

Otherwise, assuming all is well, we construct a ServerSocket with the given port number, assign it to the server property for this object, and return the object to the calling method:

```
  public Connection openPrim(String name, int mode,
                             boolean timeouts) throws IOException {
    if (!name.startsWith("//:")) {
      throw new IllegalArgumentException("Connection string must start"
                                         + " with
                                             [server]socket\"//:\"port");
    }

    int port = 0;

    try {
      port = Integer.parseInt(name.substring(3));

    } catch (NumberFormatException nfe) {
      throw new IllegalArgumentException("Invalid port number" + port);
    }

    server = new ServerSocket(port);
    return this;
  }
```

That completes the implementation of ConnectionBaseInterface's methods. We must now implement StreamConnectionNotifier's acceptAndOpen() method. We call ServerSocket's accept() method to bind to the specified port. This will block execution until a client connects. When a client does connect the accept() method will return a socket that represents that client to us. We construct a socket Connection around the socket (to present the socket through the GCF API) and return it to the calling method. The client can then call the getInputStream() and getOutputStream() methods to communicate with the client:

```
    public StreamConnection acceptAndOpen() throws IOException {
      Socket client = server.accept();
      return new com.wrox.cdc.io.j2se.socket.Protocol().open(client);
    }

    public void close() throws IOException {
      server.close();
    }
  }
```

This is a somewhat simplistic implementation of these classes and, in addition to missing basic error checking and is architecturally rather naïve, however it should achieve its aim of allowing us to simply test applications on platforms that do not have an implementation of the CDC and Foundation Profile for them. From now on, the code will assume that these classes are available on the classpath for systems that require them. Now that we have an "implementation" available, we can go on to develop a networking application for it.

Developing Java Applications for the Foundation Profile

One of the most powerful application types that we can develop with the classes available must be the networking application. The networking classes give us abstracted access to networking with the Socket and related classes that we can utilize to communicate with other applications over TCP/IP or similar network. The Generic Connection Framework adds a level of abstraction to this.

A socket represents a communication point between two machines, and may be used for application-level or machine-level communications. For our purposes, we can illustrate this best with application level protocols such as HTTP or POP.

Although these protocols are easiest to visualize, they only illustrate the possibilities. We will see how we can write a server that will respond to clients, processing their requests, and responding accordingly. We can use this to write a WAP browser for our client, or e-mail enable a CLDC or CDC device, but we can also use them for our own means. Applications that utilize streaming capabilities for audio and video will use sockets to communicate with the content server.

A Basic Socket Application

In order to get a familiarity with socket programming in the CDC, we will write a simple client and server pair that will talk to each other across socket connections. We will implement the server first, get it running, and then write a client that it can talk to.

The SimpleServer Class

Here is a simple server that waits for a connection to be made to it and returns a greeting:

```
import javax.microedition.io.StreamConnectionNotifier;
import javax.microedition.io.StreamConnection;
import javax.microedition.io.Connector;

import java.io.PrintWriter;
import java.io.IOException;

public class SimpleServer {
```

We have hard coded the port number to connect to, 4000 in this case. We open a `StreamConnectionNotifier` with the connection string "`socket://:4000`". The socket connection handler sees that there is no host name specified, and passes this to the serversocket handler. We call `acceptAndOpen()` on the server and execution will then block until a connection is made:

```
  public static void main(String[] args) {
    int port = 4000;

    try {
      StreamConnectionNotifier serverConnection =
        (StreamConnectionNotifier) Connector.open("socket://:4000");
      StreamConnection clientConnection = serverConnection.acceptAndOpen();
```

When one is made, we will be handed a `StreamConnection` that represents the client, we construct a `PrintWriter` to the output stream to the client and print "Hello there" to it. The stream will automatically flush when it is closed, but it is good manners to do so ourselves:

```
      PrintWriter out =
        new PrintWriter(clientConnection.openOutputStream());

      out.println("Hello there!");
      out.flush();
      out.close();
    } catch (IOException ioe) {
      System.out.println("IOException: " + ioe);
    }
  }
}
```

That is all there is to it. All of the above methods can throw an `IOException`, so we must wrap the calls in a `try...catch` block. Assuming that both the `javax.microedition.io` and our `com.wrox.cdc.io` package and subpackages are compiled and on the classpath, we can compile the code above with the following command:

```
javac -bootclasspath <path_to>/cdc.jar SimpleServer.java
```

This will compile the code with the J2ME implementation classes rather than the J2SE classes that are the default. Once the code is compiled, run the server as follows:

```
<path_to>/cvm -Djava.class.path=<path_to_code>  SimpleServer
```

Where <path_to_code> is the path to your code directory.

On systems that do not have the Foundation Profile implementation, this will be a little different. Compiling is more like you would expect:

```
javac SimpleServer.java
```

Then running the server would be as follows:

```
java -Dmicroedition.platform=j2se -Dmicroedition.protocolpath=com.wrox.cdc.io
SimpleServer
```

These additional command line arguments will make sure that `Connector` searches for the handlers for `com.wrox.cdc.io.j2se.<protocol>.Protocol` instead of the default. The server will bind to the port and wait for a client to make a connection.

The SimpleClient Class

Writing the client is very simple. Its code does not differ greatly to the code above. By default, we create a socket connection to the localhost on port 4000. The client also takes one or two optional command line arguments; the first specifies the host to connect to. The second argument can optionally specify an alternative port number to connect to. This will be useful in a moment when we begin to implement the main application in this chapter.

Once we have a connection, we open an input stream to it and construct a reader around it. We have chosen a `BufferedReader` because it is more efficient than other readers, and because it provides the `readLine()` method for reading input. Notice that we only read one line, the client assumes it will only ever connect to our own server, which only ever output the single line "Hello there" before severing the connection and exiting:

```
import javax.microedition.io.StreamConnectionNotifier;
import javax.microedition.io.StreamConnection;
import javax.microedition.io.Connector;

import java.io.InputStreamReader;
import java.io.BufferedReader;
import java.io.IOException;

public class SimpleClient {

  public static void main(String[] args) {
    int port = 4000;
    String host = "localhost";

    if (args.length > 0) {
      host = args[0];
    }
    if (args.length > 1) {
      port = Integer.parseInt(args[1]);
    }

    try {
      StreamConnection serverConnection =
```

```
          (StreamConnection) Connector.open("socket://" + host + ":" + port);

        BufferedReader in =
          new BufferedReader(new InputStreamReader(serverConnection
            .openInputStream()));

        String reply = in.readLine();
        System.out.println(reply);
        in.close();

      } catch (IOException ioe) {
        System.out.println("IOException: " + ioe);
      }
    }
  }
```

We can compile this code as above. For J2ME systems this will be:

```
javac -bootclasspath <path_to>/cdc.jar SimpleClient.java
```

J2SE systems will require a normal compilation:

```
javac SimpleClient.java
```

Once the code is compiled, we can run the client:

```
<path_to>/cvm -Djava.class.path=<path_to_code>  SimpleClient
```

Or:

```
java -Dmicroedition.platform=j2se -Dmicroedition.protocolpath=com.wrox.cdc.io
SimpleClient
```

On J2SE, this should give the following output:

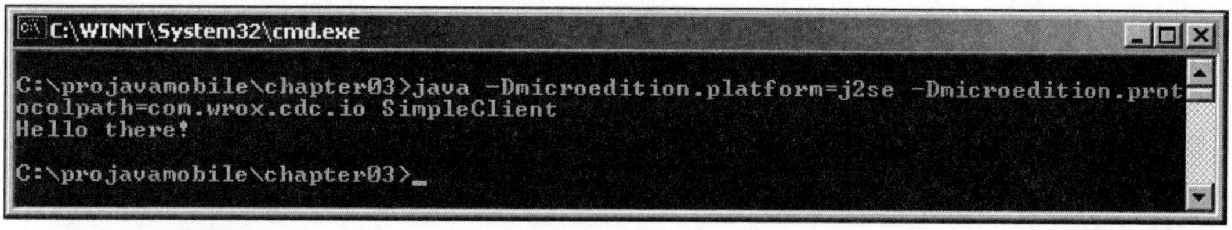

```
C:\WINNT\System32\cmd.exe

C:\projavamobile\chapter03>java -Dmicroedition.platform=j2se -Dmicroedition.prot
ocolpath=com.wrox.cdc.io SimpleClient
Hello there!

C:\projavamobile\chapter03>_
```

Let's move on to develop an application that shows the power of the GCF.

POP3

POP3, the Post Office Protocol, is an electronic mail protocol currently enjoying its third incarnation. It defines a set of commands for querying a POP3 server for information on e-mail it is holding for a specified user. All of these commands, and the responses to them from the server, are in plain text, and this is what makes POP so simple to use.

A typical POP session is as follows:

```
> USER Fruitcake
+OK
> PASS cherry
+OK
> STAT
+OK 1 1960
RETR 1
+OK 1960
Reply-To: "Orange Popsicle" <popsicle@mypopserver.com>
From: "Orange Popsicle" <popsicle@mypopserver.com>
To: "Fruit Cake" <fruitcake@someotherserver.com>
Subject: this message
Date: Fri, 1 Dec 2000 15:02:24 -0000
MIME-Version: 1.0
Content-Type: text/plain;
      charset="iso-8859-1"
Content-Transfer-Encoding: 7bit
X-Priority: 3
X-MSMail-Priority: Normal
X-Mailer: Microsoft Outlook Express 5.50.4133.2400
X-MimeOLE: Produced By Microsoft MimeOLE V5.50.4133.2400

Hello there!

popsicle

DEL 1
+OK
QUIT
```

What is happening here? after connecting to the POP server, we must authenticate ourselves. This is what we are doing in the first and second lines:

```
> USER Fruitcake
+OK
> PASS cherry
+OK
```

`USER` is a command specifying the user whose account is being queried. `PASS`, is the password for that user. The command and any additional content passed to the server are separate by white space. Each time the server should respond with the message `+OK`. An invalid account user name or password will return an error message such as the following, and in some cases the server will also sever the connection:

```
-ERR authorization failed
```

We can now query the status of the mailbox with the `STAT` command:

```
> STAT
+OK 1 1960
```

The server again responds with `+OK`, followed by the number of messages and the size in bytes of the messages. We can retrieve any messages using the `RETR` command followed by the index number of the message (starting with 1), which we did with the next command:

```
> RETR 1
```

The server will return a `+OK` message that specifies the size of the individual mail followed by the e-mail itself in text format. If we so choose, we can then delete the message from the server before quitting the session, with the `DEL` and `QUIT` commands.

In fact, the server always responds affirmatively with a `+OK` if there were no problems. Possible error messages include attempting to query the server without first authenticating, and attempting to use a non-existent command, which will give error messages such as the following:

```
> STAT
-ERR authorization first

> BOOGIE
-ERR unimplemented
```

The full list of commands and their meanings is provided below:

Command	Meaning
`USER`	User name.
`PASS`	Password.
`APOP`	User and password encrypted.
`QUIT`	Quit the session.
`STAT`	Query the mailbox status – this will return `+OK` followed by the number of messages and then the size in bytes of those messages.
`LIST`	Mailbox summary.
`RETR`	Retrieve a message, the message is specified giving an index number beginning with message 1.
`DELE`	Delete the message specified by the index number following the command.
`NOOP`	Do nothing.
`RSET`	Resets the session.
`TOP`	The number following this command specifies the number of lines from the top of the message to retrieve.
`UIDL`	Retrieves a unique ID for the message.

By default, POP services are available through port 110 as we saw earlier. Essentially, that is all there is to it. Using this information we can now go on and implement our POP aggregator.

We have chosen to implement a POP3 application here, as it illustrates the power of networking applications simply. For this example, we will need access to a POP service, such as most ISPs provide. In addition port 110 must be open to contact the mail server on. Corporate networks often have a firewall that will make this port unavailable unless it is opened by a network administrator. Internet access using dial up access should present no problems.

*nix systems (and their users) are usually more aware of such things as ports and networking and should have no problem with this. Either way, we can user `SimpleClient` to check this very simply by running the following command:

```
<path_to>/cvm -Djava.class.path=<path_to_code>  SimpleClient mail.mailserver.com
110
```

Or:

```
java -Dmicroedition.platform=j2se -Dmicroedition.protocolpath=com.wrox.cdc.io
SimpleClient mail.mailserver.com 110
```

Where `mail.mailserver.com` is the name of a POP3 mail server. This will attempt to connect to the specified mail server on port 110 and read the response. A successful connection will return a `+OK` message. Any other message indicates that there was a problem connecting. This may be simply because the computer is not connected to the Internet, and will also occur if a firewall is preventing access to the appropriate port:

An E-mail Client

The first application we will construct is an e-mail client. The client will read any mails on the mail server for an e-mail account and store them locally.

> **Note that in order to make the application more easily understood much of the standard error checking has been taken out. We do not recommend that you use this on a production system.**

The application also contains several `System.out` commands that report on the progress of the application, whether they are useful or not obviously depends on what user interface is available.

The design revolves around the class `MailClient`. Three other classes will also be needed; `MailStore` that will be used to store any e-mails transferred, and `PopReader` that connects to the server and reads any e-mails on it. Here is a UML diagram that shows the package relationships:

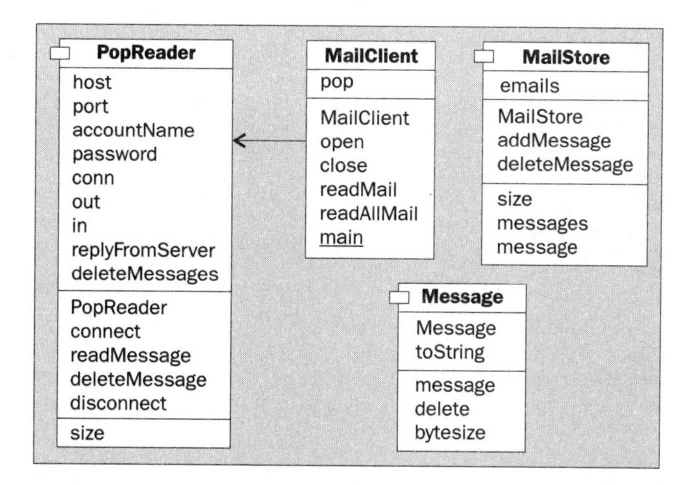

We also have a `Message` class, with methods to return the sender, intended recipient, date, and any other headers we wish to make available to users and other applications. For this example, it is sufficient for the class to store its e-mail as a string. A flag is also included to indicate when the e-mail has been deleted and a value `bytesize` that indicates the size of the string in bytes.

The details of each e-mail account are specified on the command line when the application is started. Here is an example:

```
<path_to>/cvm -Djava.class.path <path_to_code> TestMailClient pop.mypopserver.com
110 fruitcake cherry
```

Here `pop.mypopserver.com` is the mail server address, `110` is the port number, `fruitcake` is the account name and `cherry` is the password.

The Message Class

We will implement each class in order of complexity. `Message` is shown below. The constructor accepts the string, which represents the e-mail in its entirety including headers and body and does not attempt to separate out the various headers and so on.

We assign the message to the `message` property and calculate the size of the message in bytes:

```
package com.wrox.pop3;

public class Message {
  String message;
  boolean deleted;
  long bytesize;

  public Message(String message) {
    this.message = message;
    this.deleted = false;

    // the number of characters times 2 (unicode=2 bytes per character)
    this.bytesize = message.length() * 2;
```

```
  }

  public String toString() {
    return message;
  }

  // accessors, mutators
  public String getMessage() {
    return message;
  }

  public void setMessage(String message) {
    this.message = message;
    this.bytesize = message.length() * 2;
  }

  public boolean isDeleted() {
    return deleted;
  }
  public void setDeleted(boolean deleted) {
    this.deleted = deleted;
  }
  public long getBytesize() {
    return bytesize;
  }
}
```

To compile this class enter the following commands. In J2ME:

```
javac -bootclasspath <path_to>/cdc.jar Message.java
```

J2SE systems will require a normal compilation:

```
javac com/wrox/pop3/Message.java
```

The MailStore Class

`MailStore` is a simple wrapper for a `Vector` that stores each message in this application, all the e-mails are kept in memory; if the device is switched off or crashes, all e-mails would be lost. Implementing a standalone store may be preferable as it would be responsible for persisting e-mail, and may even run continuously in the background, more appropriate for 'always on' devices.

Here is the code for `MailStore`, which again is very simple:

```
package com.wrox.pop3;

import java.util.Iterator;
import java.util.Vector;

public class MailStore {
  Vector emails;
```

```
    public MailStore() {
      emails = new Vector();
    }

    public int getSize() {
      return emails.size();
    }

    public Iterator getMessages() {
      return emails.iterator();
    }

    public void addMessage(Message message) {
      emails.add(message);
    }

    public Message getMessage(int index) {
      return (Message) emails.elementAt(index - 1);
    }

    public void deleteMessage(int index) {
      Message message = (Message) emails.elementAt(index - 1);
      message.setDeleted(true);
    }
  }
```

Compile this class as above. Note that there is a conversion between the zero based index that vectors use and POP3, which uses an index that begins with 1 (in the last two methods). `deleteMessage()` sets the deleted flag to true for the appropriate message.

It would also be better to write an interface that defines the functionality given above and to return an object that implements `MailStore` via a Factory class. A Factory class is a class that instantiates another class and is a way of loosely coupling applications from classes or services that may need to change in the future for business or operation requirement. `javax.microedition.io.Connector` is a Factory class. Calling its `open()` method can instantiate different classes depending on a combination of command line options, system variables and other conditions.

Creating a Factory class that can be configured to return different instances would allow us to change the storage method, perhaps to use a third party application, or to make a persistent store, and would make interfacing to other application simpler. For the purposes of illustrating the relevant concepts, this class is quite sufficient, and in any case making these changes is straightforward.

The PopReader Class

The next class we must develop is `PopReader`, which establishes a connection to the mail server and allows reading of e-mails. Notice that this is the first class to use the GCF classes:

```
package com.wrox.pop3;

import java.io.InputStreamReader;
import java.io.BufferedReader;
import java.io.PrintWriter;
import java.io.IOException;
import javax.microedition.io.Connector;
```

```
import javax.microedition.io.StreamConnection;
import java.net.UnknownHostException;
import java.util.StringTokenizer;

public class PopReader {

  private String host;
  private int port;
  private String accountName;
  private String password;

  StreamConnection conn;

  // size of inbox
  int size;

  PrintWriter out;
  BufferedReader in;

  String replyFromServer;
  public boolean deleteMessages;
```

The class properties are as follows: `host`, `port`, `accountName`, and `password` are the host name and port of the mail server, and the mail account name and password. No surprises there. `conn` is the socket connection to the server.

`size` is the number of messages on the server for this account, there is a `PrintWriter` for writing to the server and a `BufferedReader` for reading its replies, and `replyFromServer` holds the reply as a `String`. `deleteMessages` is a flag that indicates if the e-mails should be deleted from the server when they have been read.

The constructor takes the account details and a the value of `deleteMessages`, and these are used to initialize the corresponding properties:

```
  public PopReader(String host, int port, String accountName,
                   String password, boolean deleteMessages) {
    this.host = host;
    this.port = port;
    this.accountName = accountName;
    this.password = password;
    this.deleteMessages = deleteMessages;

    String prevLine;
    String replyFromServer;
    boolean endOfMail = false;
  }
```

There are four methods that do all the actual work of reading e-mails, namely `connect()`, `disconnect()`, `readMessage(,)` and `deleteMessage()`. We will discuss them next.

The connect() method

Connect establishes a connection to the server by opening a socket to it:

```
public void connect() throws UnknownHostException, IOException {

  // open socket connection to mail server
  conn = (StreamConnection) Connector.open("socket://" + host + ":"
                                           + port);
  System.out.println("Connection to host open.");
```

We pass the server and port to the constructor for `Socket`. Notice that any `IOException` raised is passed to the calling method, which is better equipped to handle any interaction with the user.

Once the socket is open, we can open the input and output streams:

```
  // open input and output readers
  out = new PrintWriter(conn.openOutputStream(), true);
  in = new BufferedReader(new InputStreamReader(conn.openInputStream()));
```

The second argument to `PrintWriter`, is a `boolean` that sets automatic flushing of the output buffer when `println()` is used for output. We can now read the acknowledgement from the server that it is ready for authentication:

```
  // reading server connect message
  System.out.println(in.readLine());
```

The `readLine()` method returns when a carriage return, line feed or a combination of the two is received. One way that we could improve this class, would be to extend `Thread` and write a `wait()` method that causes the current thread to sleep until a response is received. This would free up the system while we are waiting for a message across potentially slow connections and servers, but is left as an exercise for the reader.

Assuming that we have successfully established a connection, we must now authenticate ourselves. We send the `USER` and `PASS` commands to the server and read the responses:

```
  // send username and password and request status.
  // assumes account details are correct
  out.println("USER " + accountName);
  System.out.println("Sent user name");

  replyFromServer = in.readLine();

  out.println("PASS " + password);
  System.out.println("Sent password");
  replyFromServer = in.readLine();
```

We now send a `STAT` message requesting the status of the account:

```
  out.println("STAT");
  System.out.println("Sent STAT message");
  StringTokenizer st = new StringTokenizer(in.readLine());
  System.out.println("read message from server");
```

Remember that the response will be something along the line of:

+OK 1 1960

Where the second part is the number of messages and the third, the size of the messages in bytes. We use the `StringTokenizer` class to retrieve the number of messages with the following code:

```
        // fast forward past "+OK" and read the number of emails in inbox
        st.nextToken();
        size = Integer.parseInt(st.nextToken());
        System.out.println(size);
    }
```

The first token is of course the `+OK` message; we fast forward past it and parse the next token for an integer value of the number of messages on the server. We now have open input and output streams, the number of messages on this server.

The readMessage() Method

We now need to read each message, the `readMessage()` method accepts an integer representing the message number of the e-mail to read. We check this number to check we are not attempting to read an invalid message and define a `StringBuffer` to hold the message. To retrieve a message from the server, we send it the `RETR` command together with the message number to be read, for example the following command returns the first message:

RETR 1

The server responds with another `+OK` message specifying the size of the e-mails about to be sent, which we throw away. We now read the first line of the e-mail:

```
    public Message readMessage(int message) throws IOException {
      if (message > size) {
        return null;

      }
      StringBuffer sb = new StringBuffer();

      out.println("RETR " + message);
      in.readLine();    // read the acknowledgement of this command
```

We continue to read from the server until it sends a line-feed, ".", line-feed sequence, which signifies the end of the mail. `readLine()` does not, however, append the new line to the string that it reads and so any line feed on its own on a line in the e-mail will be an empty string when read. We therefore, test for the reply from the server being an empty string, following by ".", and ending with an empty string. Notice that the condition in the while loop, depends on both the last two lines being true and the current read line being a line-feed:

```
      String lineBeforeThat = replyFromServer;
      replyFromServer = in.readLine();    // read +OK message

      String lastLine = replyFromServer;    // assign it to last line
      replyFromServer = in.readLine();    // read first line of message
```

```
    while (!(lineBeforeThat.equals("") && lastLine.equals(".")
            && replyFromServer.equals(""))) {
      sb.append(replyFromServer + "\r\n");

      lineBeforeThat = lastLine;
      lastLine = replyFromServer;
      replyFromServer = in.readLine();
    }
    sb.append(replyFromServer + "\r\n");
    if (deleteMessages == true) {
      deleteMessage(message);
    }

    return new Message(sb.toString());
  }
```

If the boolean `deleteMessages` value set in the constructor is set to true, we call the `deleteMessage()` method, which we will see next. Deleting a message from the server is very simple:

```
  void deleteMessage(int message) {
    out.println("DELE " + message);
  }
```

Once we are done with the server, we must disconnect from it. We do this by closing the input and output streams represented by the `Reader` and `Writer`, again responsibility for dealing with any exceptions is left to the calling method here:

```
  public void disconnect() throws IOException {
    out.println("QUIT");
    out.close();
    in.close();
    conn.close();
  }
```

`getSize()` is simply a helper method/accessor and returns the number of e-mails on the server:

```
  public int getSize() {
    return size;
  }
}
```

The final class is the mail aggregator itself. We now have all the components required to read from a mail server. We have the `PopReader`, which has methods for reading from a server, and we have the `Message` class and a `MailStore` to store messages in.

The MailClient Class

`MailClient` can be run as a standalone, in which case it will read all the e-mails for the specified account and output them to the screen or can be used as part of another application. We will begin by outlining its various methods and properties:

```
import java.io.IOException;

import java.util.Vector;
import java.util.Iterator;

import com.wrox.pop3.Message;
import com.wrox.pop3.MailStore;
import com.wrox.pop3.PopReader;

public class MailClient {

  PopReader pop;
```

We define a single class property, `PopReader` that we will use to connect to a mail server and read the mail from it. The constructor takes the required account details and constructs a `MailClient` with them. We also provide two (aliased) method for opening and closing the connection to the server, `open()` and `close()`. Both methods simply pass on any `IOException` up the stack:

```
  // this class can be used standalone or as an object
  // if used as standalone, it will read all the emails and output them to
  // the screen.

  public MailClient(String host, int port, String accountName,
                    String password, boolean deleteMessages) {
    pop = new PopReader(host, port, accountName, password, deleteMessages);
  }

  public void open() throws IOException {
    pop.connect();
  }

  public void close() throws IOException {
    pop.disconnect();
  }
```

`MailClient` provides two methods for reading mail. The `readMail()` which reads the message whose index number is given as an argument, and `readAllMail()` which iterates through every message reading it, and returns an `Iterator` of `Message` objects representing the e-mail received:

```
  public Message readMail(int i) throws IOException {
    return pop.readMessage(i);
  }

  public MailStore readAllMail() throws IOException {
    MailStore messages = new MailStore();

    // retrieve all emails
    System.out.println("Retrieving all emails:");
    for (int i = 1; i <= pop.getSize(); i++) {
      messages.addMessage(readMail(i));
    }

    return messages;
  }
```

Remember that we iterate through messages 1 to `getSize()` not from zero as is usual in java, as the POP3 protocol specifies index number that begin at 1. Finally, we will write the code for the `main()` method that will test the code out. `main()` accepts the account details as command line arguments, we parse the arguments for correct usage, and then call the constructor for this class:

```
  public static void main(String[] args) {

    int numArgs = args.length;

    if (!(numArgs >= 4)) {
      System.out
        .println("Usage: java MailClient host port username password [delete]");
      System.exit(-1);
    }

    String host;
    int port;
    String accountName;
    String password;
    boolean deleteMessages;

    host = args[0];

    try {
      port = Integer.parseInt(args[1]);

    } catch (NumberFormatException nfe) {
      throw new IllegalArgumentException("Invalid port number, type"
                                         + " \njava MailClient for usage
instructions");
    }

    accountName = args[2];
    password = args[3];

    deleteMessages = false;

    if (numArgs == 5) {
      deleteMessages = args[4].equalsIgnoreCase("delete");
    }

    MailClient mailclient = new MailClient(host, port, accountName,
                                           password, deleteMessages);
```

When we have an instance of `MailClient`, we can `open()` the connection, read the all mails on the server and then close the connection, outputting any read mail to the screen before execution ends:

```
    Iterator iter = null;
    try {
      mailclient.open();
      iter = mailclient.readAllMail().getMessages();

    } catch (IOException ioe) {
      System.out.println("IOException: " + ioe);
      System.exit(-1);
```

```
        }
        finally {
          try {
            mailclient.close();
          } catch (IOException ioe) {
            System.out.println("IOException: " + ioe);
            System.exit(-1);
          }   // nothing we can do here
        }

        while (iter.hasNext()) {
          System.out.println((Message) iter.next());
        }
      }
    }
```

Compile the application as before. You may or may not have noticed, but this class is not included in the com.wrox.pop3 package, as it does not implement any POP functionality. The best place for this class is probably in the parent directory for the package root (/com), which will be the folder for this chapter in the code download.

Testing The Application

We can test this application using the following command as follows:

```
<path_to>/cvm -Djava.class.path=<path_to_code> MailClient pop.apopserver.com 110
fruitcake cherry
```

Or for J2SE systems:

```
java -Dmicroedition.platform=j2se -Dmicroedition.protocolpath=com.wrox.cdc.io
MailClient pop.apopserver.com 110 fruitcake cherry
```

Here is an example of a session. MailClient first retrieves the account details, and then connects to the mail server. The server in question is my own, hence the rather dubious credentials:

```
java -Dmicroedition.platform=j2se -Dmicroedition.protocolpath=com.wrox.cdc.io
MailClient localhost 110 fruitcake cherry delete
```

MailClient indicates when the socket connection is open, and at each stage of authentication:

```
C:\WINNT\System32\cmd.exe - java -Dmicroedition.platform=j2se -Dmicroedition.protocolpath=com....
Connection to host open.
+OK hello from PopServer
Sent user name
Sent password
Sent STAT message
read message from server
3
```

As we can see, there are three e-mails on this account. Here is an example of the retrieved messages:

```
From:him
To:fruitcake
Subject:this and that

Hello you, it's me, love him.

.

+OK
From:me
To:fruitcake
Subject:me

Did you get the email that him sent you, you?

.


+OK
From:her
To:fruitcake
Subject:him

you, me asked me to ask you whether you have heard from him recently and whether you got
the email from me?

.
```

The output leaves a little to be desired but you should get the picture. We have now got a working e-mail client using the GCF. As can be seen, it is quite straightforward to implement these types of networking solutions and the result is only a little different than the J2SE version would be. In fact, we can further extend this application without too much trouble to allow another client network access to read these e-mails.

We will implement a POP3 server that will make available the e-mails; any external, authorized device could then use this server. In fact, we would be defining a POP proxy. Let's go ahead and see how we might implement this solution.

A Simple POP3 Server

We can implement a server using a serversocket connection. In order to write a POP3 server, we will implement the following subset of the POP3 commands, namely: `USER`, `PASS`, `STAT`, `RETR`, `RSET`, `NOOP`, `QUIT`, and `DELE`. The server will have a `MailStore` passed to it in the constructor that holds the e-mails, together with a user name and password to authenticate against. Here is the updated class diagram:

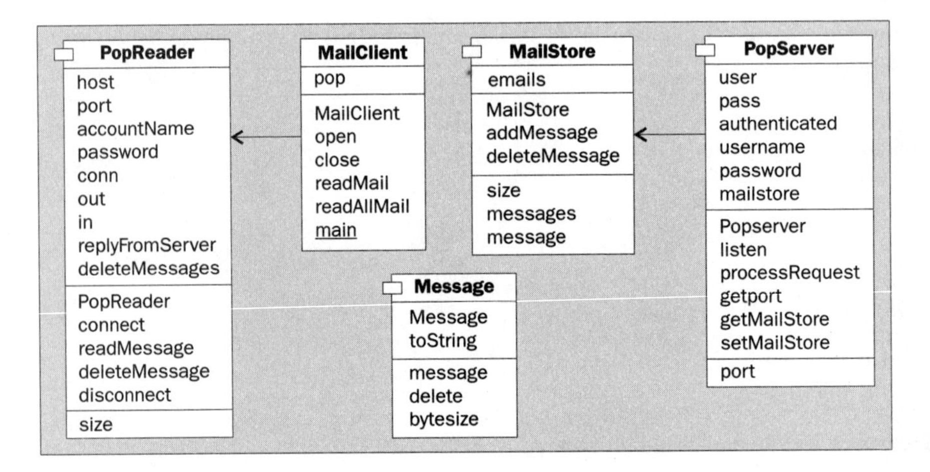

The PopServer Class

Let's see what we have got so far:

```
package com.wrox.pop3;

import javax.microedition.io.StreamConnectionNotifier;
import javax.microedition.io.StreamConnection;
import javax.microedition.io.Connector;

import java.io.BufferedReader;
import java.io.InputStreamReader;
import java.io.PrintWriter;
import java.io.IOException;

import java.util.StringTokenizer;

public class PopServer {

  // user entered values
  String user;
  String pass;

  boolean authenticated;

  // check values
  String username;
  String password;

  int port = 110;    // default port number

  MailStore mailStore;
```

`user` and `pass` represent the values input by the client using the `USER` and `PASS` POP3 commands, we check these values against the `username` and `password` properties set in the constructor. In addition, `authenticated` is used to check that the user has signed in before releasing any information on the e-mails in the server. The port to bind to on the local host is set by default to 110, we can make an alterations to this through an accessor/mutator pair.

In addition to the user name and password, the constructor also takes a `MailStore`, preloaded with e-mails:

```
// constructor
public PopServer(MailStore mailStore, String username, String password) {
  this.mailStore = mailStore;
  this.username = username;
  this.password = password;
}
```

The controlling method in this class is `listen()`, it instantiates the serversocket connection, binds it to the specified port, and then waits for a client to connect. When we have a client connection, we can obtain a `PrintWriter` and a `BufferedReader`, acknowledge the client with a message, and read any requests from the client.

We use the `readLine()` method to read the client request. Please note, that this method also blocks execution until a new-line is encountered so this would be better implemented as a threaded application, which would also allow us to specify a timeout period for client requests:

```
// main work method
public void listen() {

  // request from client
  String request;
  StreamConnection conn = null;

  try {
    StreamConnectionNotifier server =
      (StreamConnectionNotifier) Connector.open("socket://:" + port);

    // bind to port and wait for client request
    StreamConnection client = server.acceptAndOpen();

    // get a Reader and a Writer
    PrintWriter out = new PrintWriter(client.openOutputStream(), true);
    BufferedReader in =
      new BufferedReader(new InputStreamReader(client.openInputStream()));

    // acknowledge connection to client and read client request
    out.println("+OK hello from PopServer");
    System.out.println("+OK ");

    request = in.readLine();
```

The server checks each request for the `QUIT` command, and breaks the loop if so. Otherwise, it passes the request to its helper method `processRequest()`:

```
    while (true) {
      // case sensitive for QUIT command
      if (request.equals("QUIT")) {
        out.println("+OK");
        break;
      }
```

```
        // process request and print reply to client
        out.println(processRequest(request));

        // read next request
        request = in.readLine();
      }

      // close all resources on error.
      out.close();
      in.close();
      client.close();
      server.close();
    } catch (IOException ioe) {
      System.err.println("IOException occurred: " + ioe);
      System.exit(-1);
    }
  }
```

Remember that the `listen()` method has already dealt with the `QUIT` command. We will need to deal with the remaining commands. We begin by parsing the request from the client for the command and any further details made available by the client.

Next, we will instantiate a string tokenizer. Each POP3 command should have no more than two tokens, the command and one parameter maximum, so we can safely ignore further values. We check how many have been passed, save the first token in a local variable named `command`, and if there is more than one token, save the second in the `parameter` variable. If the server is passed a blank line, it will return an error message:

```
  // helper method to process client requests
  String processRequest(String request) {
    String command = "";
    String parameter = "";

    StringTokenizer tokenRequest = new StringTokenizer(request);
    int numberTokens = tokenRequest.countTokens();

    if (numberTokens > 0) {
      command = tokenRequest.nextToken();
    } else {
      return "--invalid request--";
    }
    if (numberTokens > 1) {
      parameter = tokenRequest.nextToken();
    }
```

Let's deal with each command in the order they are likely to appear, and finish off any that have not been implemented at the end. First `USER`. This may be the simplest to implement, we assign the value in parameter (empty string if none is passed) to the `user` property for this class, set the return message to `+OK` and return:

```
    // USER name in request
    if (command.equals("USER")) {
      user = parameter;
      return "+OK";
```

Next, the client should authenticate the user with the PASS command. We assign the value of parameter to the pass property and check that the user and pass values passed match those of our check values. If so authenticated is set to true, we return the value +OK. Any other result returns an error message to the client:

```
    // PASSword in request
  } else if (command.equals("PASS")) {
    pass = parameter;

    if (user.equals(username) && pass.equals(password)) {
      authenticated = true;
      return "+OK";
    } else {
      return "--authentication failed--";
    }
```

From now on, all commands should check that the authentication stage has been carried out before processing. Except for this check, the STAT command needs to return the number of messages in the MailStore and their size:

```
    // client requesting mail box STATus
  } else if (command.equals("STAT")) {
    if (!authenticated) {
      return "--authentication first--";
    }
    int size = 0;
    for (int i = 1; i <= mailStore.getSize(); i++) {
      size += mailStore.getMessage(i).getBytesize();
    }
    return "+OK " + mailStore.getSize() + " " + size;
```

The RETR command requires the index number of the message as a parameter. We need to parse the passed parameter for the message number, which we check against the limits of the mail store before retrieving the message. Notice the check to see if the user has deleted the message:

```
    // RETRieve a message
  } else if (command.equals("RETR")) {
    if (!authenticated) {
      return "--authentication first--";
    }
    int index = Integer.parseInt(parameter);

    if (index < 0 || index > mailStore.getSize()) {
      return "--usage RETR messagenumber': invalid message number entered";
    }
    Message message = mailStore.getMessage(index);
    if (message.isDeleted()) {
      return "--message has been deleted--";
    }
    return "+OK\r\n" + message + "\r\n.\r\n";
```

DELE is more or less identical, barring a couple of essential differences:

```
      // DELete message request
    } else if (command.equals("DELE")) {
      if (!authenticated) {
        return "--authentication first--";
      }

      int index = Integer.parseInt(parameter);

      if (index < 1 || index > mailStore.getSize()) {
        return "--usage 'DELE messagenumber': invalid message number entered";
      }

      mailStore.deleteMessage(index);
      return "+OK message deleted";
```

NOOP only really makes sense in a threaded environment but here goes:

```
      // NOOP
    } else if (command.equals("NOOP")) {
      return "+OK";
```

RSET resets the session for the user, in this context this means undoing any deletions. We iterate through each message in the mail store and make sure they are not deleted. Any other command will return an error message as shown at the end:

```
      // ReSET the session - undeleted any messages.
    } else if (command.equals("RSET")) {
      for (int i = 1; i <= mailStore.getSize(); i++) {
        mailStore.getMessage(i).setDeleted(false);
      }
      return "+OK";
    }

    return "-- unimplemented command--";
  }
```

All we have to do now, is write a couple of accessors and mutators, and this class is finished. Beware that it may not make sense for some properties to be modifiable via mutators:

```
  // accesssors, mutators
  int getPort() {
    return port;
  }

  MailStore getMailStore() {
    return mailStore;
  }

  public void setPort(int port) {
    this.port = port;
```

```
  }

  void setMailStore(MailStore mailStore) {
    this.mailStore = mailStore;
  }
}
```

Testing the PopServer Class

We can test this class with the following code for the `PopServerTest` class, it is by no means a rigorous test class nor is it meant to be; unit testing as well as thorough system testing is recommended. It is merely a means to test the class generally works before we combine it with the previous client:

```
import com.wrox.pop3.MailStore;
import com.wrox.pop3.Message;
import com.wrox.pop3.PopServer;

class PopServerTest {

  public static void main(String[] args) {
    if (args.length < 2) {
      System.out.println("usage: java PopServerTest username password");
      System.exit(-1);
    }

    String username = args[0];
    String password = args[1];

    MailStore ms = new MailStore();
    ms.addMessage(new Message("From:him\r\nTo:you\r\nSubject:this and "
                              + "that\r\n\r\nHello you, it's me, love him."));
    ms.addMessage(new Message("From:me\r\nTo:you\r\nSubject:me\r\n\r\nDid"
                              + " you get the email that him sent you, you?"));
    ms.addMessage(new Message("From:her\r\nTo:you\r\nSubject:him\r\n\r\n"
                              + "you, me asked me to ask you whether you"
                              + " have heard from him recently and whether"
                              + " you got the email from me?"));

    PopServer server = new PopServer(ms, username, password);
    System.out.println("Preparing to start server...");
    server.listen();
  }
}
```

Here we are constructing the previous messages by hand and loading them into a mail store, that is then passed to an instance of `PopServer`. Compile and run this in the command line before running the `MailClient` application on it. First, we will need to open a shell to run the server on, and run the following command:

```
<path_to>/cvm -Djava.class.path=<path_to_code> PopServerTest fruitcake cherry
```

Or:

```
java -Dmicroedition.platform=j2se -Dmicroedition.protocolpath=com.wrox.cdc.io
PopServerTest fruitcake cherry
```

The server will return the following message:

```
Preparing to start server...
```

Open an additional shell (or command line window) and enter the following:

```
<path_to>/cvm -Djava.class.path=<path_to_code> MailClient fruitcake cherry delete
```

Or:

```
java -Dmicroedition.platform=j2se -Dmicroedition.protocolpath=com.wrox.cdc.io
MailClient localhost 110 fruitcake cherry delete
```

You should get output that is identical to that above.

The PopProxy Class

We should now combine the two applications to create a POP3 proxy:

```
import com.wrox.pop3.MailStore;
import com.wrox.pop3.PopServer;

import java.util.Iterator;

import java.io.IOException;

class PopProxy {

  public static void main(String[] args) {

    int numArgs = args.length;

    if (!(numArgs >= 4)) {
      System.out
        .println("Usage: java PopProxy host port username password [delete]");
      System.exit(-1);
    }

    String host = args[0];
    int port = Integer.parseInt(args[1]);

    String username = args[2];
    String password = args[3];

    boolean mailDeleted = false;

    if (numArgs == 5) {
```

```
        mailDeleted = args[4].equals("delete");
      }

      MailClient mailClient = new MailClient(host, port, username, password,
                                             mailDeleted);

      // read every mail
      System.out.println("Retrieving mail...");
      MailStore mailStore;
      try {
        mailStore = mailClient.readAllMail();

        // setup server
        PopServer server = new PopServer(mailStore, username, password);

        // uncomment the following line to change the port the server binds to
        // server.setPort(someportnumber);
        server.listen();
      } catch (IOException ioe) {
        System.out.println("IOException " + ioe);
        System.exit(-1);
      }
    }
  }
```

Assuming that we have some e-mails in the account specified at the command line, running this file should give us a server on the localhost containing the e-mails from them. This example has the server set to not to delete the e-mails on the various accounts, so you can run this example as many times as you wish.

From the examples that we have seen above, and if you have any familiarity with socket programming, you will realize that the differences are minimal. For those of you that have not, of the seven classes that we have created, only two needed to change, `PopServer` and `PopReader`. Even then, the changes were quite minimal. Therefore, you should find that networking applications are as easy to create for the CDC profile, as they are for anything else. Otherwise, the platform is much the same, and so programming for it should not present you with any major surprises, only fun.

Other CDC Profiles

As explained previously, as well as standing as a profile in its own right, the Foundation Profile will also act as a base for other profiles that add functionality to it. Two such profiles already exist; the RMI Profile and the Personal Profile, and we will deal with these next. Unfortunately, they are both in their early stages, so information is a little scarce.

The RMI Profile

The RMI profile JSR 066, currently in the last stages of public draft, adds RMI functionality to the Foundation Profile. It has been put forward mainly by companies wishing to implement Jini in small mobile devices. The RMI profile includes JRMP (wire protocol), marshaled object support, the activator interface, and client side activation protocol and remote object lookup, etc.

This profile quotes the following minimum requirement for devices that implement it:

- 2.5Mb ROM
- 1Mb RAM
- TCP/IP network connectivity
- A full implementation of the Foundation Profile

As usual this does not include application memory requirements.

The RMI profile does not include tunneling remote method calls, so no RMI through firewalls and via proxies. There is no multiplexing protocol to allow multiple logical data streams to be multiplexed into a single data stream. In addition, this specification removes deprecated method, classes and interfaces from the J2SE specification and the stub skeleton protocol and compiler deprecated in the JDK1.1.

The Personal Profile and PersonalJava

The **Personal Profile** has its roots in **pJava**, also known as **PersonalJava**, a runtime environment targeted at devices whose users have minimal or no computer experience. This includes such devices as set top boxes for digital and satellite TV. It also has hardware assumptions, namely 2Mb ROM and 1-2Mb RAM not including application memory requirements.

pJava includes the applet, `java.awt` and bean packages, in addition to the Foundation Profile classes (in retrospect). It does not include the full `util` package, nor does it include RMI. From the AWT classes, several are optional including `Frame`, non modal `Dialog` objects and classes related to `Menu`. In addition some classes have been modified and methods removed.

While lightening the footprint requirement for these devices, this selective inclusion of some classes has caused portability problems. The J2ME specification supports incremental deployment of the JVM and upwardly compatible applications, but the inability to determine at design time the presence of classes severely limits rapid development and deployment.

The Personal Profile does not include any optional packages, or classes; it does represent the next generation of pJava and therefore applications designed for pJava will operate on Personal Profile devices.

Summary

In this chapter, we have covered the basis for a Connected Device Configuration and its base profile, the Foundation Profile. Devices covered by the Connected Device Configuration are usually fixed point devices with a fair amount of resources, both in memory and in processor power terms, and are therefore more able than small mobile devices normally associated with the J2ME platform.

We discussed briefly the classes included in the Foundation Profile, including the new `Connector` and `Connection` abstraction for data streams of all kinds. A `javax.microedition.io.Connection` covers a multitude of data sources including stored in a file, via a socket, or as datagrams. We will see more of the `microedition.io` package in later chapters on the Connected Limited Device Configuration.

We then went on to develop one of the possible applications of a CDC, a proxy server. In this case we developed a POP3 proxy, utilizing some of the new classes available in the `microedition.io` package. The theme of the J2ME and the Java language generally is connectivity, therefore applications that are network focused, or at least network enabled, will become increasingly commonplace as they are in the J2SE platform.

PROFESSIONAL JAVA MOBILE
TOOL SELECTION
WEB TABLET
MOBILE PHONE
MP3 PLAYER
DIGITAL CAMERA
PALMTOP
LAPTOP
NOW SELECTED
PALMTOP

4

Introducing the KVM, CLDC, MIDP, and other Profiles

The three most common categories of mobile device are the mobile phone, the pager, and the relatively low cost Personal Digital Assistant (PDA). Currently, only a handful of these devices are extensible, in other words we can add our own applications to them, and the applications are usually proprietary to that device, or family of devices.

The sheer volume of this market immediately makes it an attractive area for third party mobile Java solutions, but the two features that are an advantage to the user – size and simplicity – can also create a difficult platform for development and run-time, whether using Java or other languages.

In this chapter, we introduce the Java 2 Micro Edition (J2ME) technology for 'limited' devices. First we discuss what 'limited' device means. We then move on to explain the following J2ME building blocks required for CLDC and MIDP development:

- K Virtual Machine (KVM), and how it compares with the standard JVM
- Connected Limited Device Configuration (CLDC), its features and limitations, and how it fits between the CDC and embedded Java solutions
- Mobile Information Device Profile (MIDP), its capabilities and limitations

We describe how these three major building blocks fit together to create a development and run-time platform for Java on the smallest of devices.

A simple Hello World example MIDP application is introduced and instruction is given on how to run the application on Sun's J2ME Wireless Toolkit. We also describe how to run Sun's CLDC examples on a Palm Emulator (POSE).

By the end of the chapter, you will have created and run your first MIDP application, and will have a feel for the differences between J2ME development and the more familiar J2SE platform.

To make best use of this chapter, the only prerequisite is some exposure to standard Java programming. You need no knowledge of PDA development, or any mobile phone development kits.

What is a Connected Limited Device?

In Sun's definition, a 'connected limited' device is a personal item such as a PDA that has these characteristics:

- Simple user interface (compared to a desktop user interface)
- Minimal memory budget (160Kb – 512Kb, exclusive to the Java platform)
- Low bandwidth connectivity (often 9600 bps or less)
- Intermittent network connectivity

Sun also considers the limited device to be a personal item, such as a mobile phone, which is probably battery-powered. The restrictions do not affect the architecture of a J2ME application, but will affect the possible functionality included in an application.

Let's quickly take these limited device features in turn so as to get an idea of their effect on our potential application. The simple user interface means we cannot guarantee that the device will have such sophistication as multiple windows, mouse-like pointers, drag and drop capabilities, a QWERTY keyboard, and so on.

If we are to make our application run on many devices seamlessly, then our user interface has to be simple, very simple. As we see later though, Sun has allowed for several levels of user interface complexity to avoid unacceptable compromises such as running a 90x50 phone application on a PDA with a 320x240 screen.

The minimal memory budget will limit our ability to save large objects and data, but this goes hand-in-hand with the minimal user interface. The intermittent low bandwidth connectivity means we cannot guarantee real-time flow of data to and from the server side, so we need to offer temporary storage for any user-entered data in the application until the user can synchronize it with a server. Of course, we must still be careful however, not to break the memory limitations mentioned above.

Subsequent chapters give examples of how to develop applications with the tight restrictions of the CLDC and it is surprising, given the restrictions, what can be developed. First of all, however let's discover more about the building blocks we need to develop for our mobile devices.

The K Virtual Machine (KVM)

The KVM is a subset of the standard Java Virtual Machine (JVM), with its main goal being to fit into a small memory footprint (it is only 40-80Kb in size), while still being as complete and fast as possible. 'K' originally stood for Kuaui – a codename for the project, and not Kilobytes as you might expect. The KVM is suitable for both 16-bit and 32-bit devices.

Currently, the limited device configuration and the KVM are intimately related, and CLDC will **only** run on top of the KVM, although this is likely to change in the future. Note that the KVM is bundled with the CLDC for this reason, which can confuse developers looking for a separate KVM download.

Since the KVM is a subset of Sun's standard JVM, its features are best described by what has been left out and why, compared to official Virtual Machine Specification (Lindholm and Yellin, 1996):

Feature	Reason
No Java Native Interface	Memory size (security model-related)
No user-defined class loaders	Memory size (security model-related)
No reflection	Memory size (security model-related)
No threads groups or daemons	Memory size (security model-related)
No finalization	Memory size (security model-related)
No float or double	No hardware support can be assumed
Minimal Error classes	Memory size, and minimal error-handling capability on real devices

The missing functionality is related to either memory size or hardware support for floating point arithmetic, with a lot of classes being absent because of the limited security model of the KVM. As you can see, the standard security model is a major omission, and again, this is down to the code size of all the supporting classes. However, there is still a perfectly safe security model in J2ME, the 'sandbox' model, as we see later. What has been lost is functionality such as the ability to create your own class loaders, and so on. Here is a little more detail on each of the major omissions:

- **Java Native Interface**
  The JNI was eliminated for security reasons – the sandbox security model assumes the set of native methods must be closed.

- **User-Defined Class Loaders**
  The CLDC has a built-in class loader that cannot be overridden, replaced, or reconfigured by the user, for security purposes.

- **Reflection**
  The lack of reflection in the CLDC eliminates serialization, RMI, JVMDI, and JVMPI.

- **Thread Groups and Daemon Threads**
  Although the CLDC and JVM support multi-threading, there is no built-in support for multi-thread control. It is possible to perform such control, but we must implement that functionality ourselves within our own application.

- **Finalization**
  There is no `Object.finalize()` method in the CLDC specification. If the cleaning up of a specific resource is required then one alternative is to use the 'finally' part of a try/catch/finally statement.

- **Weak References**
  The J2ME `java.lang.ref` package allows applications to interact with the garbage collector, and allows more flexible caching of data. However, on a mobile device of already limited resources, this sort of flexibility is not of great value, and so has been removed from the CLDC. It also decreases memory requirements.

- **Float and Double**
  It is assumed that no floating-point hardware will be available on the device for the support of float or double data types because of the limitations of handheld devices. However, software libraries that implement this functionality are available from www.jscience.net. Other vendors will no doubt offer competitive libraries in the future.

- **Error Classes**
  The majority of error classes have been removed not only to save memory, but also because handheld devices often have fairly primitive native error handling abilities. They usually rely on a soft reset, reboot, or battery removal (hello, Nokia 7110 owners) when a failure occurs.

- **Security**
  Security in J2ME can be considered at two different levels:

  - Virtual machine level security (classfile pre-verification)
  - Application level security (sandbox model)

At the virtual machine level, the security aspects are focused on denying applications access to 'forbidden' memory areas, and so forth on the target device. In J2ME, as in J2SE and J2EE, the low-level mechanism to make sure downloaded Java bytecode doesn't access memory outside of the Java heap is the classfile verifier.

The Classfile Pre-Verifier

In standard J2SE and J2EE implementations class verification is performed on the target device, but takes a significant amount of memory (Sun estimates that it needs a minimum of 50Kb code space and 30-100Kb of dynamic memory when running).

Therefore, in J2ME, a new verifier has been written, taking only 10Kb of code, and typically less than 100 bytes of dynamic memory. However, each class now needs to contain special attributes to identify it to the new verifier on the target device. The pre-verifier tool supplied with the J2ME Wireless Toolkit adds these attributes.

There is a small increase in the size of a class (~5%), but the classes are backwards compatible, and so can still be loaded by J2SE and J2EE verifiers. The use of the pre-verifier also decreases application startup time on the target device. For more detail on exactly what steps the pre-verifier and verifier take, see the CLDC specification 1.0, Section 5.3.

Sandbox Security

At the higher application level, even though the class files have been verified, it would still be possible for rogue applications to access resources such as the network, file systems, and the target device's hardware features. The comprehensive security aspects of the standard JVM cope with these potential digressions by implementing a security manager, which is called whenever access to these external resources is required. However, this implementation is a severe problem in terms of code and dynamic memory requirements, so the KVM reverts to a simple closed 'sandbox' security model.

The sandbox model trades memory resources against flexibility, in particular, the ability to define class loaders, or libraries with native functionality. In particular, the loss of the reflection classes means Java RMI is not available on a KVM platform – you must 'roll your own' remote invocation.

In addition to the simplified security, it is not possible to override the CLDC system classes with alternate implementations, for obvious reasons. For example, downloaded MIDP classes could pose as the real CLDC core classes, and do malicious damage.

The CLDC specifications state that a more comprehensive security model can be introduced in later versions, so Sun has left the door open for the future when mobile devices are powerful enough to run the standard J2SE security model.

KVM Implementations

The reference KVM 1.0 from Sun is written in C, allowing it to be easily ported to the majority of mobile and embedded hardware. In many cases, this simply means compiling it on the target platform. Sun has created implementations for Windows, Solaris, and Palm.

Other implementations are available from vendors such as Motorola, and RIM. As expected, the non-reference implementations perform rather better than Sun's KVM, which was written to demonstrate J2ME capabilities rather than as a commercial product.

Since all these alternatives must adhere to the KVM/CLDC specification our application can be run on all of them without modification:

- The Motorola development kit can be obtained from: http://www.motorola.com/idendev/
- The RIM development kit can be requested from: http://www.rim.net

Porting the KVM

To give a greater understanding of the reasons why CLDC and MIDP have certain features and limitations, you may wish to investigate how to port the KVM. Sun supplies the KVM source with the KVM/CLDC reference implementation. It comes with comprehensive documentation so you are at liberty to examine, and port, the code yourself.

We will not go into detail here about porting the KVM, other than to say that the source is ANSI C (except the 64-bit integer arithmetic), so any standard compiler will suffice for compilation. Sun documentation states that the KVM has been built and tested successfully on machines using 32-bit pointers using the following compilers:

- MS Visual C++ 6.0, on Windows 98 and Windows NT 4.0
- GNU C compiler on Solaris
- Metrowerks CodeWarrior Release 6 for Palm
- Sun DevPro compiler 4.2 on Solaris

Refer to the KVM Porting Guide supplied with the reference CLDC 1.0 for more details.

KVM Utilities

There are a few reference utilities supplied with the KVM, in particular the Java Application Manager, and the Java Code Compact. Although we may not need these utilities as a third-party developer, knowing what they do and why they have been supplied will help us understand the J2ME platform.

The Java Application Manager (JAM)

Not all limited devices have a native way of launching and managing applications. This is true of the majority of mobile phones and pagers, whereas PDAs, such as Palms and Psions, are already designed to be able to find and start third-party applications.

For those devices with no application launcher, the reference KVM implementation includes the JAM. The JAM expects to be able to fetch the applications in JAR format from a network or storage protocol. The most likely route is naturally HTTP, alternatives include cable, and Over The Air (OTA) downloads.

The JAM reads the JAR, and discovers the main class by reading its manifest file (see the **MIDlet Suite Manifest File** section below). It also checks the application is compatible with its versions of the CLDC and MIDP, and then launches the KVM passing it the main class to run.

The JAM can also be used to update applications by version number comparison, and obviously it can halt and remove applications too. The JAM uses the application descriptor file (see the **The MIDlet Application Descriptor File** section below) to load/update applications, and also check there is enough free memory for the new application to run.

Java Code Compact

The Java Code Compact (JCC) utility, supplied with the KVM reference implementation, allows Java classes to be linked to the KVM directly, to give quicker application startup times. JCC takes Java classes and creates a C file from them, suitable for compiling and linking into the KVM. Unless you are providing the KVM yourself, of course, this option will not be available to you.

The Connected Limited Device Configuration

Now we understand the underlying layer of the KVM and its capabilities, we can move up an invisible layer, to the CLDC, remembering that the CLDC and KVM are intimately connected.

The figure below shows how the various components of J2ME for limited devices fit together. As you can see, the CLDC/KVM layer covers the operating system from the point of view of any Java application:

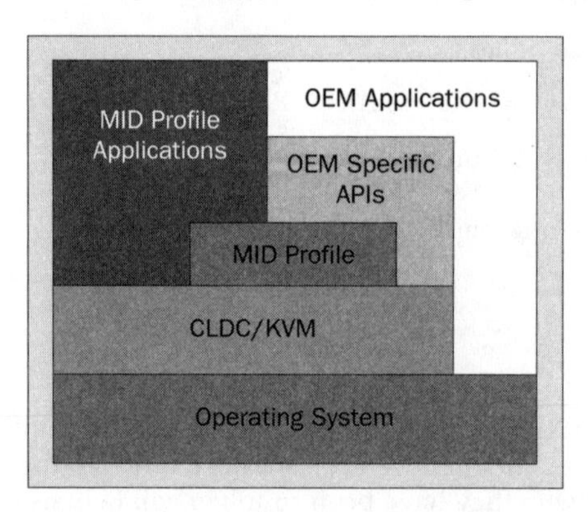

The CLDC must make minimal assumptions about the underlying device capabilities because configurations are horizontal building blocks – in fact it actually leaves the MIDP to tackle these problems, as we shall see later. In effect, a configuration mandates the minimum classes that we can develop with, which gives us the maximum portability across the wide range of devices.

Currently there is only one major version of the CLDC specification available, version 1.0 from Sun. However, there are three implementations – for the Windows, Solaris, and Palm platforms. The minimum requirement of a target device to run the CLDC 1.0 is 128Kb of non-volatile memory, plus a minimum of 32Kb of volatile memory for the Java runtime and object memory. The required volatile memory needed depends on the target platform and its role, for example dynamic loading of applications will require more volatile memory to hold those new classes than pre-loading an application into non-volatile memory.

In keeping with its minimalist nature, the CLDC specification makes no promises about real-time performance, the scheduling of applications, or even whether downloaded applications are stored persistently. Note that this may make the CDLC inappropriate for applications such as games on some platforms, as sufficient performance cannot be guaranteed (despite the device having the potential to deliver the required performance).

One vital point about any CLDC implementation, is that there are no optional features – a vendor must satisfy the complete CLDC specification, however they decide to implement it for their platform. That's good news for developers.

CLDC Classes

Although, in general, the CLDC is subset of J2SE it does include a handful of new classes, covered in the previous chapter, which generalize the standard Java networking and I/O classes. The new classes merely serve to unify external connection classes, and thus decrease the memory footprint. We will review which standard J2ME classes are available in the CLDC specification, and see the effect that any missing classes have. There are five categories of CLDC classes, the first three being inherited from J2SE:

- Core classes (`java.lang` and `java.util` subsets)
- I/O
- Internationalization
- Networking
- Properties

There is no device-specific functionality in the CLDC, such as user interface or event handling. This is strictly the responsibility of a profile such as the MIDP, simply because there is such a range of different mobile devices that it is not possible to squeeze all UI functionality into a unified, yet compact, set of classes. Hence, the need for the classes to be split into two sets of APIs, with the second set, profiles, being stacked upon the CLDC.

We give no code examples for the J2SE-derived classes, since these are covered in Chapter 3. However, for new features such as the CLDC-specific classes there are code snippets in Chapter 3, and a comprehensive CLDC-compliant application is created in the Chapter 6.

Upward Compatibility – The CLDC and CDC Compared

The CDC (Connected Device Configuration) has already been covered in Chapter 3 and we are concentrating on the CLDC here. However, let's compare the two configurations in case we wish to write a cross-configuration application and need to understand what limits CLDC imposes upon us compared with other J2ME configurations.

Sun defines the CDC and CLDC as being suited to devices with these resources and capabilities, with the ability to support these libraries:

Resources/capabilities	CLDC	CDC
Memory	128K minimum	256K RAM 512K ROM
Network Connection	Intermittent, unreliable	Intermittent, unreliable
Security	Sandbox	Full security manager

Java Libraries		
AWT	No	Yes
RMI	No	Yes, optional
JDBC	No	Yes, optional
JNI	No	Only with optional RMI profile
Applets	No	Yes

Although there are many 'standard' functions missing from the CLDC, a vital point to note is that the CLDC is a complete subset of the CDC. This means CLDC code is portable upwards to more powerful J2ME configurations without any effort. As we see later, though, this is not true of the MIDP component, which is why Sun chose to split the J2ME platform into configurations and profiles.

Downward Compatibility – The CLDC, JavaCard and EmbeddedJava Compared

There is very little in common between the CLDC and the next Java levels 'down', in other words Java on smart cards and EmbeddedJava. These lower layers are written for use by firmware developers in closed environments, at the lowest level of hardware, and with no user interface. Here they can use Java as a device-independent language, but they will always need to use native libraries specific to their hardware. The CLDC on the other hand, is devoted to allowing third party developers to abstract out any hardware dependence.

More information is available from their respective home pages:

- http://www.java.sun.com/products/embeddedjava/
- http://www.java.sun.com/products/javacard/
- The Mobile Information Device Profile

The Mobile Information Device Profile (MIDP) version 1.0 is currently the only official Java Community Process CLDC-based profile available (see http://java.sun.com/products/midp/ for details). As described earlier, a profile sits above both the CLDC and KVM, and defines a set of functionality and APIs expected from a 'vertical' set of devices, i.e. closely related devices, but from different vendors.

In MIDP's case, the target devices are mobile phones, pagers, simple wireless PDAs, and of course hybrid combinations of these devices. One thing to beware of, as we see later in MIDP's definition of the minimum hardware capabilities, is that using MIDP on advanced PDAs will not be always be appropriate. Developers (and users) will find some functionality either unavailable, or just too simple to do justice to the device. In these cases, consider another profile that better matches your application.

The MIDP API is limited to those APIs that were considered absolute requirements to achieve broad portability. These APIs are:

- Application (i.e. defining the semantics of an MIDP application and how it is controlled)
- User interface
- Persistent storage
- Networking
- Timers

The MIDP assumes a narrower set of hardware capabilities than the CLDC, especially in the areas of user interface and storage features, or rather, lack of features.

Minimum Hardware Requirements

The **minimum** hardware characteristics required of a Mobile Information Device (MID) that supports the MIDP are:

- 96 by 54 pixels of 1 bit depth, with approximately square pixels
- User input by one of the following methods:
  - 'one-handed' keyboard (i.e. a phone keypad)
  - 'two-handed' keyboard (i.e. a QWERTY keyboard)
  - touchscreen
- 128 Kb of non-volatile memory for the MIDP components
- 8 Kb of non-volatile memory for persistent application data
- 32Kb of volatile memory for the Java runtime
- Wireless network connectivity, possibly intermittent

Note that the MIDP specification doesn't actually insist on these characteristics, only that a MID should normally have these. Realistically speaking however, these are the characteristics of many of today's devices, and in future will no doubt come to include hardware such as wireless watches. From the minimum hardware specification, we can see there is allowance in the MIDP for persistent storage of data the application creates, and for various methods of input. We will discuss these later.

There is no support for voice input or output. This will presumably appear in a future profile, when such devices are more common.

Minimum Software Requirements

Much of the **minimum** expected software characteristics of a Mobile Information Device (MID), reflect the above hardware, and are:

- A kernel (i.e. host OS) to manage the hardware, and run the KVM
- The ability to read and write to non-volatile memory
- Time and date support, for time-stamping data, and also for running timers
- The ability to write to a bit-mapped display
- The ability to accept user input
- The ability to control the application lifecycle (i.e. find, launch, stop, and so on.)

Again, the MIDP specification doesn't actually mandate these software characteristics, but we can see from the list that there is nothing out of the ordinary about the requirements, so you can assume all these features will be on real MIDP devices.

The time and date support in the MIDP is worth noting. It is there to make sure that synchronization of any persisted data can be performed, by using time and date comparison.

MIDlets

With the basic classes and interfaces being supplied by the CLDC, the MIDP concentrates on a higher level of classes, such as support for the user interface, persistent storage, timers, and also the implementation of the `Connection` network interface specified in the CLDC.

An application that only uses the MIDP and CLDC APIs is known as a **MIDlet**, and must extend the `javax.microedition.midlet.MIDlet` class. One or more MIDlets can be grouped together, along with other shared resources, into a **MIDlet suite**, which is held in single standard Java JAR file, suitable for downloading onto a target device, whether by wire or over the air.

The MIDP specification does not define the way in which the MIDlet suite JAR file is loaded onto a target device, or how the loaded MIDlets are found, started, stopped, and removed from the device. It also makes no attempt to say how the device should react on encountering an application error. This has all been carefully avoided, because of the wide variation of host OS methods to perform such operations making it impossible to abstract out the application management. However, Sun does supply the JAM as an example application manager for device vendors.

The MIDlet suite's JAR file contains:

- One or more MIDlet classes, plus any classes shared between them
- Any additional resource files, containing strings, images, and other application data
- A manifest file

In addition to the JAR file, an **application descriptor** file must also exist, which holds attributes describing the MIDlet, and MIDlet suite. We cover the content of the application descriptor and manifest files later.

A Simple MIDlet

We will now step through a very basic MIDlet, `IntroMIDlet1`, which is naturally the industry standard Hello World application. This is a slight adaptation of the `HelloMIDlet` supplied with Sun's MIDP 1.0 reference implementation. We will cover the minimal code required to build `IntroMIDlet1` from the UI manager, through displaying data, to event handling for the 'Exit' button. Note that all the source code for this, and the other examples in the book can be downloaded from the Wrox website: http://www.wrox.com.

The application displays:

- A text entry box, with a title of 'IntroMIDlet1' and default text of 'Hello World', which we can enter, and edit, up to 256 characters
- An 'Exit' button

Here is the complete MIDlet class. As you can see, there are only three methods required to create our basic MIDlet:

- The constructor
- `startApp()`
- `commandAction()`

In the constructor, the MIDlet's `Display` object is initialized by calling `Display.getDisplay()`. The `Display` class is the MIDlet's UI manager. There is only one `Display` instance per MIDlet, and the MIDlet uses it to decide which `Screen` to display (see the next chapter for a definition of `Screen`):

```
import javax.microedition.midlet.MIDlet;

import javax.microedition.lcdui.Command;
import javax.microedition.lcdui.CommandListener;
import javax.microedition.lcdui.Display;
import javax.microedition.lcdui.Displayable;
import javax.microedition.lcdui.TextBox;
import javax.microedition.lcdui.TextField;

public class IntroMIDlet1 extends MIDlet implements CommandListener {
  private Command exitCommand;
  private Display display;

  public IntroMIDlet1() {
    display = Display.getDisplay(this);
```

The second step taken in the constructor is to create an `EXIT` command, ready for attaching to our subclass of `Screen`. A `Command` represents an action that the user may select. We set its priority to 1, in other words most important, to hint that we would like this visible on a soft button, or other navigation device, rather than somewhere in a menu. Note that we don't actually say **what** the `Command` is going to do at this stage, although we do indicate that it is of type Command.EXIT and has a label 'Exit'. The behaviour associated with the command is defined later as we shall see:

```
    exitCommand = new Command("Exit", Command.EXIT, 1);
  }
```

The startApp() method is called whenever our MIDlet becomes the device's current process. In the case of the reference implementation device, this is only called once, straight after the constructor method. Note that a device's application manager calls startApp() and pauseApp() whenever it schedules the MIDlet to be the current process, or a background process, respectively. Therefore, once startApp() is called, we know it is worth displaying something:

```
    public void startApp() {
```

In this case, that 'something' is a TextBox, a subclass of Screen. A TextBox takes up the whole of the MIDlet's drawing area. Its constructor takes four arguments, the title of the TextBox, the default string to display, the maximum number of characters it will hold, and finally, the type of input it will allow. We set the text to 'Hello World' and allow the user to type up to 256 characters of any kind:

```
        TextBox t = new TextBox("IntroMIDlet1", "Hello World", 256,
                                TextField.ANY);
```

Although we have created a TextBox, our 'Hello World' text is still not visible to the user. We need to take three more steps. The first of these is to associate the EXIT command with the TextBox, so it is displayed along with 'Hello World':

```
        t.addCommand(exitCommand);
```

Next, we have to make sure there is a method to deal with the Exit button when the user selects it. As you can see from the class definition, we have implemented the CommandListener interface, in other words we have defined a commandAction() method. t.setCommandListener(this) specifies that the command listener is our own MIDlet instance:

```
        t.setCommandListener(this);
```

Finally, we make the TextBox, and the associated EXIT command, visible. All we need to do is call display.setCurrent(t) to display the TextBox:

```
        display.setCurrent(t);
    }
```

At this point, the application draws our TextBox and sits patiently waiting for user input, or the Exit key to be pressed. We don't have to explicitly catch key inputs or re-draw the screen whenever the user types/deletes/moves – the TextBox class does all this for us. However, we do need to catch the EXIT command action, and make sure the application exits.

The commandAction() method is called whenever a command event occurs. There's only one possible command – EXIT, so there's no point checking the values of the c and s parameters:

```
    public void commandAction(Command c, Displayable s) {
```

The notifyDestroyed() method tells the application management software that the MIDlet has finished executing, and that any associated resources can be released:

```
        notifyDestroyed();
    }
```

The `pauseApp()` method is called whenever the MIDlet becomes a background process. This is also left empty, because we have not used any resources that need freeing up. Again, it must be defined because the MIDlet interface defines it:

```
    public void pauseApp() {}
```

The `destroyApp()` method is called just before the MIDlet is stopped. In this case it is left empty, because there is no cleaning up required, but we need to define it to satisfy the MIDlet interface:

```
    public void destroyApp(boolean unconditional) {}
  }
```

We have only used high-level UI APIs, so there is no need to initiate any paint, repaint, scroll, or other low-level call. It is the responsibility of the MIDlet high level UI classes to do any re-drawing and scrolling on our behalf. We define the features and differences between the high-level and low-level UI APIs in the next chapter.

One thing to consider is whether we should place code in the constructor or in the `startApp()` method. Placing it in the constructor means our `TextBox` is created once only, but it does use up resources regardless of whether the `IntroMIDlet1` is the current application or not.

If we place it in `startApp()`, every time the `IntroMIDLet1` application is woken up/brought to the front then the `TextBox` is created, which saves on resources when the application is not the foreground application, but uses processing power every time the user brings the application to the foreground. There is no right way, since both memory and CPU are constrained on MIDs. We must experiment and find out what works best for the user and the device.

Of course, for devices that make no attempt to schedule multiple applications, the choice doesn't matter because both the constructor and `startApp()` will only ever get called once. Sun's J2ME Wireless Toolkit runs MIDlets in this fashion, but it is a dangerous assumption to make of the myriad of devices that may be available in next few years. It's a limitation of the J2ME Wireless Toolkit that's worth remembering when testing MIDlets on it.

Compiling and Running the Example MIDlet

The tools supplied with the basic MIDP and CLDC reference implementations are quite limited in their functionality, so we recommend the J2ME Wireless Toolkit (J2MEWTK) as your MIDP build tool, and we have based the following instructions on it. It's a nice lightweight way of building and testing a MIDP project, and does a great job of auto-creating MIDlet suites and associated files (see later).

It comes complete with the CLDC and MIDP libraries, utilities, Javadoc documentation and example MIDlets, runs on any Java 1.3 enabled platform, and it is free. It also integrates with Sun's free Forte for Java Community Edition Integrated Development Environment. Download it from http://java.sun.com/products/j2mewtoolkit/download.html.

There are also some optional resources you will definitely find useful:

- A real MIDP device. At the time of writing (May 2001) there are precious few MIDP devices. However, Sun has just released its beta Palm MIDP implementation:
    - Sun Palm MIDP – http://java.sun.com/products/midp/. This will also work with a Palm emulator available for download from http://www.palmos.com/dev/.
    - Motorola have released two handsets through Nextel, the Motorola/iDEN i85s and i50sx. They are currently only available in the US, although the Accompli 008 is due to be released in Europe and Asia by Summer 2001: http://www.motorola.com/idendev/.
    - There are also J2ME phones available in Japan, but they appear to be just CLDC-compliant, and not MIDP-compliant. They are based on a Sun specification named KittyHawk, which suggests they probably aren't MIDP. A development toolkit is available from LG, but currently there is minimal documentation in English: http://java.ez-i.co.kr.
    - A list of current J2ME devices is maintained at http://www.JavaMobiles.com.
- A J2ME-compatible IDE, usually complete with an emulator:
    - Sun Forte: http://www.sun.com/forte/ffj/. The J2MEWTK integrates seamlessly with this IDE.
    - Metrowerks CodeWarrior: http://www.metrowerks.com (used by Motorola as their IDE for iDen development).
    - Zucotto: http://www.zucotto.com. Version 2 has supporting for debugging, and also a useful image creation tool.
    - RIM: http://www.rim.net.

We now describe the three most likely methods of testing our example MIDP application:

- Building and running with the J2MEWTK, stand-alone
- Building and running with the J2MEWTK, integrated with Forte for Java Community Edition
- Building by J2MEWTK, and installation and testing on the Palm MIDP implementation

Note the CLDC/KVM and MIDP come precompiled as executable emulators for Windows, but if we are developing on another platform, then we'll need a make utility, such as GNU make, to compile the emulator before we can test our own applications.

Instructions for J2MEWTK Stand-alone Build

The J2ME Wireless Toolkit is available from the Sun Java web site. It comes with both CLDC and MIDP API documentation to help develop applications, but it doesn't contain the MIDP or CLDC specifications. They are available separately from Sun's Java web site.

Build a Sample Application

We will now create a sample application using the `IntroMIDlet1` from above. Start the KToolbar toolkit up by executing `ktoolbar.exe` from the `bin` directory of the toolkit's root directory (or from the Start menu on Windows). Create a new project and name it IntroMIDlet1. Specify `IntroMIDlet1` as the MIDlet class name and click on Create Project:

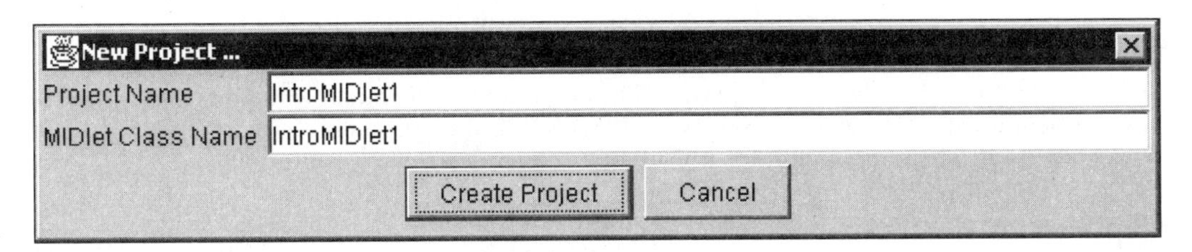

The properties for this project are then displayed. Accept the defaults (press OK) and return to the main toolkit window. If everything has gone according to plan we should be presented with the locations where our application's source and resource files should be placed:

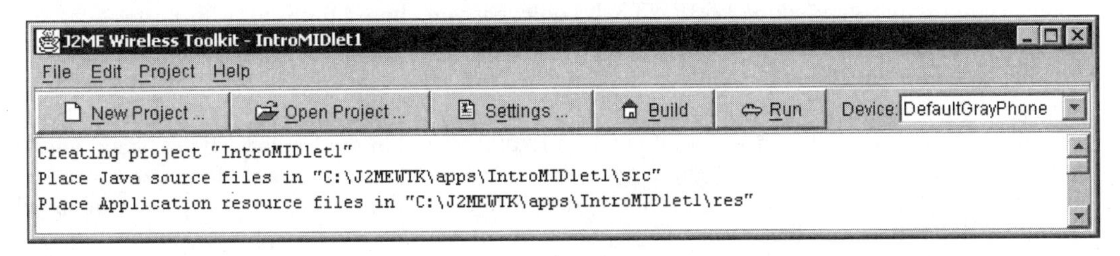

The next step is to place the `IntroMIDlet1` source file in the directory specified (and optionally, an icon file. This needs to be placed in the `src` directory). Now click Build:

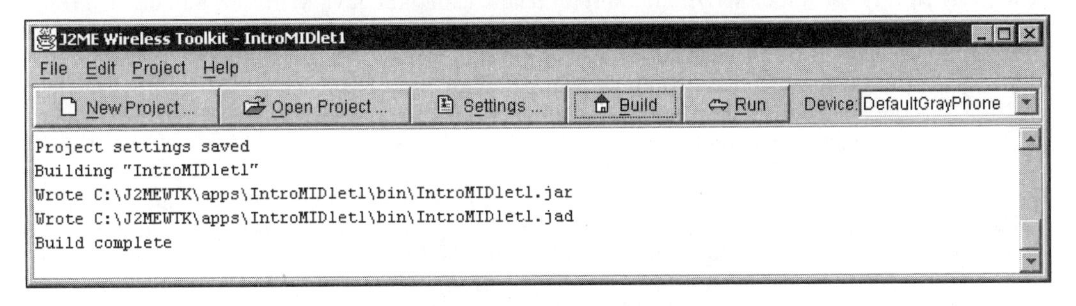

The final step is to run the project, which should produce the following emulator window:

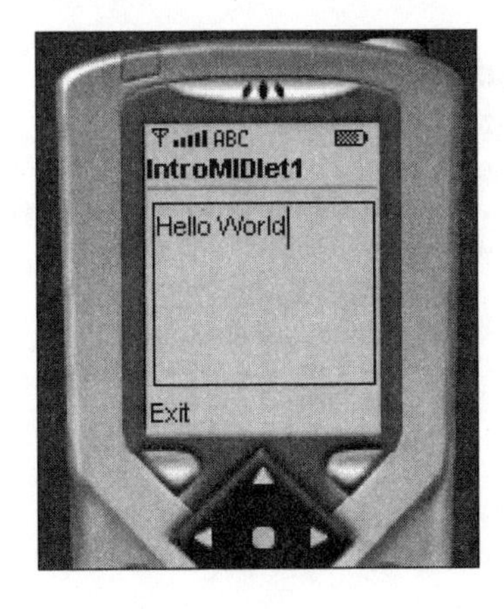

Instructions for Forte – J2MEWTK Build

Follow these if you downloaded the J2MEWTK, and want to integrate it with Forte. We will not attempt to describe the Forte IDE in any depth as the J2MEWTK has a good user guide explaining the basics, as does Forte itself. The Forte user guide is available at http://www.sun.com/forte/ffj/.

Download

Note that you must install Forte first, in order for J2MEWTK to integrate with it. Follow the standard Sun installation instructions, and select the 'Integrated with Forte' choice when installing J2MEWTK. Also, avoid spaces in the directory names, because the pre-verifier cannot cope with them.

You can still use the stand-alone J2MEWTK KToolbar if you choose to install with Forte.

Build a Sample Application

When starting Forte, you will be asked whether you want to set the compiler type to J2ME Wireless compiler. Answer yes and then select File | Open Explorer, and in the Explorer [Filesystem] window open the IntroMIDlet1 directory. Right mouse click over RunAll at the bottom of the example listing, and select the Execute option. After some seconds or minutes, depending on your processor speed, the example MIDlet suite will be compiled and presented on a pseudo-mobile phone, as above.

Instructions for Palm MIDP Installation

The Early Access version of MIDP for Palm (officially called the Java Manager, but also referred to as JavaMgr and sometimes midp4palm) is available from the Sun Java developer web site at the time of writing. You must register to access downloads from this site: http://developer.java.sun.com/developer/. The MIDP for Palm is expected to be publicly available by the time this book is published, so also check the public MIDP site for a released version: http://java.sun.com/products/midp/.

The beta includes the Palm MIDP implementation plus a very useful GUI-based utility for converting a MIDlet suite into a Palm-compatible `.prc` file. This PRC Converter Tool is much easier to use than the command line `MakePalmApp` equivalent supplied with the CLDC reference implementation (see later).

The PRC Converter Tool requires JDK 1.3, but also note that the minimum Palm OS version is 3.5, so we'll need a reasonably up-to-date Palm device or ROM image for our emulator.

Installing a Sample Application

First, we must install Java Manager on the Palm. Having unpacked the zip file, install the `MIDP.prc` file on the Palm device or emulator, and then start the PRC Converter Tool by running the `converter.bat` batch file. Note that it expects us to have an environment variable called `JAVA_PATH` set up, pointing to our JDK installation. The PRC converter tool requires two files to convert a MIDlet suite – the `MIDlet.jad` file and the matching JAR file:

Assuming that we have installed the J2MEWTK on Windows, click on the folder button, and navigate to the directory <J2MEWTK_install_dir>\apps\IntroMIDlet1\bin\.

Here we will find the IntroMIDlet1.jad file that the converter needs to build our MIDlet suite. The JAR is also there, but the converter tool doesn't bother showing it – it expects the JAR to be in the same directory.

Highlight the IntroMIDlet1.jad file, and press the 'Convert' button. The generated PRC file will be written to the same directory. Finally, load the IntroMIDlet1.prc file onto your Palm/emulator, and start the MIDlet by selecting its icon. Here is the result:

If we click on the JavaMgr icon instead, we can access the preferences, both basic and advanced. Probably the most useful preferences are:

- In the basic preferences – enabling the networking
- In the advanced preferences – saving the output is a useful, if simple, debugging feature, since it captures `System.out` and `System.err`
- In the advanced preferences – enabling debugging, which causes the application to pause until a debugging program makes contact with the Palm/emulator

The MIDlet Suite in Detail

So far, we have assumed that we just need to create a J2MEWTK project, place the source file in the right directory, and crank the handle. However, for your own projects, you need to fully understand the content and attributes of a MIDlet suite, because there are several other files supporting the Java code. We now describe those extra files and attributes.

The MIDlet Classes and Resource Files

There is no limit to how many MIDlets there can be in a MIDlet suite, other than the target device's memory capacity, of course. MIDlets in a MIDlet suite can interact with each other, and MIDlets have access to any class in their MIDlet suite, allowing code sharing. Interestingly, MIDlets can actually start other MIDlets in their suite, but it's not clear what effect this might have on a device's application manager.

MIDlets can only exit by calling the `MIDlet.notifyDestroyed()` method. They should not use `java.lang.System.exit()` or `java.lang.runtime.exit()`, as these will both cause a `SecurityException` as defined in the MIDP specification.

The resource files are accessed using `java.lang.Class.getResourceAsStream(String name)`. There is no connection support for the 'file' protocol in the MIDP therefore, this is the only method of accessing files, and these files must be local to the device, i.e. in the MIDlet suite's JAR file. This method also allows access to the MIDlet suite's manifest file, as well as other resource files.

The MIDlet suite descriptor file can also be read using `javax.microedition.midlet.MIDlet.getAppProperty()`. You cannot read the MIDlet's class files however, as they are executable only.

The MIDP does not override any CLDC classes, and similarly, your classes cannot override any of the MIDP or CLDC classes, because of the security implication mentioned earlier in this chapter.

The MIDlet Application Descriptor File

The application descriptor allows the target device's application management software to decide if the MIDlet suite is suitable for the device, and also allows custom configuration parameters to be supplied without needing to modify the JAR file.

The Java application descriptor (JAD) allows the device to check these before loading the suite, all of which may cause the load to be aborted:

- The configuration version (does the application need a more recent version than that on the device?)
- The profile version (is a more recent version required?)
- The size of the MIDlet suite (is there enough memory on the device?)
- The version of the suite (is there a newer version already resident?)

Note that there is no attempt to suggest the minimum dynamically allocated memory this application needs, so even if the application manager does check the MIDlet suite size, it's a pretty meaningless test if the application tries to grab 500Kb when it starts up. All the check does is to make sure that the application can be stored persistently, but not necessarily run.

The JAD file is read before the much larger MIDlet suite JAR, containing the manifest file, is read in, which is why it needs to be a separate file. The application descriptor points to the MIDlet suite JAR, containing the manifest file, with its `MIDlet-Jar-URL` attribute.

The advantage of having a separate JAD file is that, if the device was a wireless one, the application manager only needs to download the few tens of bytes of the JAD to discover whether a full load of the JAR file is worthwhile.

Note that the application descriptor file MIME type is `text/vnd.sun.j2me.app-descriptor` *and the filename suffix must be:* `.jad`*.*

Here is an example application descriptor file:

```
MIDlet-1: IntroMIDlet1, IntroMIDlet1.png, IntroMIDlet1
MIDlet-Jar-Size: 1140
MIDlet-Jar-URL: IntroMIDlet1.jar
MIDlet-Name: IntroMIDlet1
MIDlet-Vendor: Sun Microsystems
MIDlet-Version: 1.0
```

The application descriptor file can be found in the bin directory of your J2MEWTK project. It was automatically generated by the toolkit. You can change its contents by selecting the 'Settings...' button in the toolbar.

The MIDlet Suite Manifest File

The manifest file describes the contents of the MIDlet suite JAR file and being part of that JAR file, is loaded onto the target device after the application descriptor file has been read. The application management software and the MIDlets can access it at a later date.

The manifest file can be found in the `bin` directory of your J2MEWTK project. It is also auto-generated by the toolkit.

Here is an example of a minimal manifest file:

```
MIDlet-1: IntroMIDlet1, IntroMIDlet1.png, IntroMIDlet1
MIDlet-Name: IntroMIDlet1
MIDlet-Vendor: Sun Microsystems
MIDlet-Version: 1.0
MicroEdition-Configuration: CLDC-1.0
MicroEdition-Profile: MIDP-1.0
```

MIDlet and MIDlet Suite Attributes

The attributes found in the application descriptor and the manifest file can be split into two types – those describing the MIDlet suite, and those describing an individual MIDlet.

All attributes are simple name-value pairs of unlimited length, with both the name and value being case-sensitive. Each attribute must appear on its own line, with the name and value being separated by a colon. Any white space is ignored. Here is an example attribute:

```
MIDlet-Name: Mancala
```

The table below lists which attributes can appear in the manifest file, and which methods can appear in the deployment descriptor, and whether they are mandatory, optional, or just not valid.

Be aware that the attribute naming is unintuitive, with MIDlet suite attributes names confusingly starting with "`MIDlet-`" rather than something more descriptive like "`MIDletSuite-`".

The MIDlet suite attributes (**not** the specific Midlet attributes) are:

MIDlet Suite Attribute	Description	In manifest file?	In descriptor file?
`MIDlet-Name`	The name of the MIDlet suite that identifies the MIDlets to the user.	Mandatory	Mandatory
`MIDlet-Version`	The version number of the MIDlet suite. The format is `major.minor.micro` as described in the JDK Product Versioning Specification. The application management software can use this attribute for installation and upgrading purposes, as well as for user information.	Mandatory	Mandatory
`MIDlet-Vendor`	The vendor of the MIDlet suite.	Mandatory	Mandatory

MIDlet Suite Attribute	Description	In manifest file?	In descriptor file?
`MIDlet-Icon`	The name of a Portable Network Graphics (PNG) 1.0 image file within the JAR file used to represent the MIDlet suite. It should be used when the application management software displays an icon to identify the suite.	Optional	Optional
`MIDlet-Description`	The description of the MIDlet suite.	Optional	Optional
`MIDlet-Info-URL`	A URL for information further describing the MIDlet suite.	Optional	Optional
`MIDlet-Jar-URL`	The URL from which the MIDlet suite JAR file can be loaded.	No	Mandatory
`MIDlet-Jar-Size`	The number of bytes in the MIDlet suite JAR file.	No	Mandatory
`MIDlet-Data-Size`	The minimum number of bytes of persistent data required by the MIDlet suite. This is merely a hint, rather than a guarantee of resource. The default is zero.	Optional	Optional
`MicroEdition-Profile`	The J2ME profile required in order to run this MIDlet suite, (e.g. "`MIDP-1.0`"). This is compared with the system property `microedition.profiles`.	Mandatory	No
`MicroEdition-Configuration`	The J2ME configuration required to run this MIDlet suite, (e.g. "`CLDC-1.0`"). This is used to compare with the system property `microedition.configuration`.	Mandatory	No
`xxxx`	Custom attributes specific to the MIDlet suite. They must not begin with the string "`MIDlet-`".	Optional	Optional

Note that the three attributes which are mandatory in both the manifest and application descriptor file – `MIDlet-Name`, `MIDlet-Version` and `MIDlet-Vendor` – must be identical in both files, otherwise the MIDlet suite will not be loaded by the target device's application management software. If any other attributes are defined in both the application descriptor and the manifest file, then the value in the application descriptor file always takes precedence.

There is only one attribute specific to an individual MIDlet, but it holds three pieces of information about the MIDlet:

- MIDlet name
- MIDlet icon name
- MIDlet class name

An example MIDlet attribute is:

```
MIDlet-1: IntroMIDlet1, IntroMIDlet1.png, IntroMIDlet1
```

MIDlet Attribute	Description	In manifest file?	In descriptor file?
`MIDlet-<n>`	The name, icon, and class of the nth MIDlet in the JAR file separated by commas.	Mandatory	No

The MIDlet identifier, <n>, must start at 1, and increment without gaps in the numbering, in other words 1,2,3.

The name is used to identify this MIDlet to the user. The icon is the name of the PNG image file within the JAR for the icon of the nth MIDlet. The class is the name of the class extending the MIDlet class for the nth MIDlet, and it must have a public no-argument constructor, so the application manager on the target device can launch the MIDlet.

Note that the first two parts of the `MIDlet-<n>` attribute are optional – we can just leave them blank, with the requisite number of comas, but with only the class defined as a minimum. Omitting either may leave your user in the dark about what application is about to launch, though, since there is no guarantee which of the title or the icon a device will use to display to the user. The J2MEWTK, for example, displays both.

Launching and Controlling the MIDlet

When a MIDlet suite is installed onto a target device, the contents of its JAR file are kept in persistent storage, ready for the user to launch the MIDlet.

MIDlets are launched via the application manager on the MID, which creates an instance of the MIDlet using its public no-argument constructor. From then on, control of the MIDlet is via the `java.microedition.midlet.Midlet` class lifecycle methods; `startApp()`, `pauseApp()`, and `destroyApp()`. Note that these functions mean that a MIDlet may be running, but in the 'background', i.e. it is not the current application. `pauseApp()` and `startApp()` move the MIDlet from foreground to background mode, and vice versa.

It is worth re-iterating that `startApp()` and `pauseApp()` may be called many times during the lifetime of an application, as the application manager makes the MIDlet the current process, and then changes it to a background task. You may prefer to think of `startApp()` as `resumeApp()` to remind you of its true function.

CLDC Development Environment

There is one other avenue of development worth exploring, and that is the plain CLDC development environment. The CLDC reference implementation does **not** include MIDP, but it does support two pseudo-profiles worth investigation:

- KJava, available with the CLDC Palm overlay from Sun's CLDC download pages
- kAWT, available from http://www.kawt.de

KJava runs on both Windows, Solaris and Linux machines, and also on Palm OS, with the 'overlay' classes. It was Sun's first UI for CLDC testing, and was released so that developers had some way of running CLDC code in the absence of any working MIDP devices back in mid-2000. Sun supplied additional native methods in the UNIX and Windows KVM, outside of the CLDC specification, to demonstrate the CLDC working on a Palm-like interface.

Sun are expected to withdraw KJava as official profiles, such as the MIDP, become available for each of the above platforms, but we still include it here for interest. It is currently still useful for writing and testing CLDC on non-Windows platforms.

kAWT (KVM Abstract Windowing Toolkit) is a lightweight implementation of Sun's standard Abstract Windowing Toolkit (AWT). It is not a Sun product, but can be downloaded free for evaluation purposes. kAWT is of great interest, since it is very likely to be similar to the yet-to-be-released PDA Profile (PDAP).

It supports a higher level of user interface than MIDP because it has windowing features, and is far better suited to the Palm platform. Not only that, but because it is a sub-set of AWT you can write J2ME code that will run on a desktop with minimal or no porting. (However, because it's only a sub-set, do not necessarily expect to be able to just copy over any of your existing AWT applications.)

Just to re-iterate, these two libraries are not as a result of official profiles like MIDP, and therefore not a supported part of J2ME. In the absence of the official PDA Profile (PDAP), KJava, kAWT, and the imminent Palm MIDP are the only Palm development choices currently available for CLDC. It's also worth noting that the PDAP page states that it will support a sub-set of AWT, so if you are tempted to start developing for the Palm, then kAWT is likely to be the closest match of all the graphics libraries to the future PDAP.

For exploring these Palm-based examples, we will need:

- Palm CLDC overlay (Sun)
- A Palm OS device, or Palm emulator (Palm)
- If you use the Palm emulator, you also need a Palm ROM 'image'

We will need to join the Palm Developer Program to obtain the Palm emulator, but this can be done online at http://www.palmos.com/dev/. The Palm emulator is known as the Palm Operating System Emulator (POSE), and emulates a Palm by reading a copy of a Palm OS ROM. It does **not** come with one as default, but you can obtain a Palm ROM image by either requesting one (from Palm at the same web site as above), or by downloading one from a real Palm device.

CLDC and KJava Download

The CLDC (currently at v1.0.2) can be downloaded from the Sun Java web site. The download comes complete with the CLDC specification and API documentation and KJava API, plus porting notes for the KVM. Note that to run CLDC on a Palm, or Palm emulator (POSE), you must also download the Palm 'overlay', which includes KJava, since you need some way of viewing your application output. Unzip the standard CLDC zip, move into the `j2me_cldc` directory, and then unzip the Palm zip right on top of it.

Testing the Environment

Once we have downloaded and unzipped the CLDC, we can test our environment by running one of the pre-compiled examples.

```
cd {cldc_install_dir}
.\bin\kjava\win32\kvmkjava -classpath bin\kjava\samples\classes pong.Pong
```

Here we have started the pre-compiled KVM supplied with CLDC 1.0.2, and told it to load and run the example Pong. We should see a simple application like this:

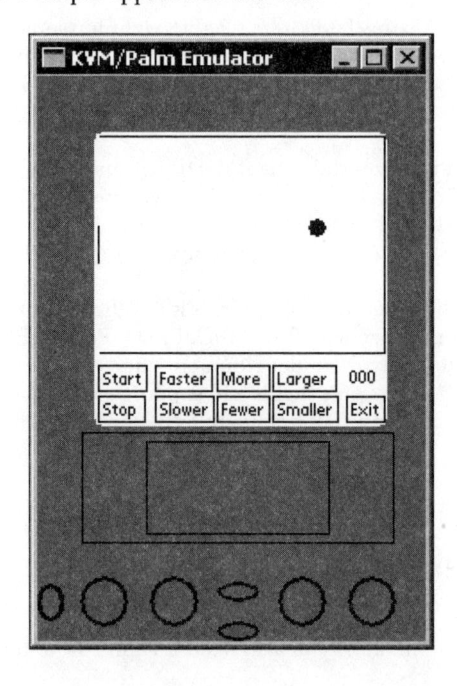

Now let's move on to build an application from scratch.

Build a Sun Example KJava Application

Having proved that we have a suitable environment for developing CLDC-based applications, we'll now build some sample source by hand. The main point to note here is that rather than using standard Java libraries, we must direct the Java compiler to use only the CLDC-supplied ones, which are a subset of the standard J2SE equivalents. This is done with the `-bootclasspath` directive. Our generated classes are placed in the `tmpclasses` directory (which must be created by hand), ready for pre-verifiying:

```
cd {cldc_install_dir}\samples
mkdir tmpclasses
javac -g:none -d tmpclasses -classpath tmpclasses;..\bin\kjava\api\classes  -
bootclasspath ..\bin\common\api\classes src\pong\Pong.java src\pong\PongBall.java
```

Pre-verify the Classes

To pre-verify the classes we run the preverify utility supplied with CLDC. The pre-verified classes are placed in the `classes` directory. Again the directory name is arbitrary, and must be created first:

```
mkdir classes
..\bin\win32\preverify -d classes -classpath
tmpclasses;..\bin\kjava\api\classes;..\bin\common\api\classes pong\Pong
pong\PongBall
```

The application is now ready for testing. Note that for CLDC-based applications, we don't need to create a manifest file or an application descriptor, because there is no application manager (like the JAM) running to load the application.

Test the Application

Type in the following on the command line to invoke the application:

```
cd {cldc_install_dir}\samples\classes
..\..\bin\kjava\win32\kvmkjava pong.Pong
```

If the application didn't work, then you should double-check that your `bootclasspath` points to the CLDC classes, and also have a look in the `tmpclasses` and `classes` directories to make sure the Java compiler and pre-verifier did their jobs correctly.

Running on the Palm

First, download the separate Palm overlay from the Sun Java web site (http://java.sun.com/products/cldc/).

Unzip the overlay files, and let them overwrite your existing CLDC reference implementation files. Don't worry, we can still run the CLDC demos on Windows as well as the Palm.

The Palm overlay contains:

- KVM Palm executable
- KVM utils executable
- Pre-built .prc executables

Create a Palm Executable

To create a Palm executable from your Java class files, you must use the MakePalmApp utility supplied by Sun with the CLDC. Again, this is not part of the official CLDC specification, but an add-on convenience utility.

MakePalmApp is supplied as Java source, so we need to build it from the source before we can use it. To build the MakePalmApp itself:

```
cd {cldc_install_dir}\tools\palm\src
javac palm\database\*.java
jar cvf classes.jar palm\database\*.class palm\database\*.prc palm\database\*.bmp
```

Now we have to build a JAR of our application classes (this is optional, but simplifies builds where there are multiple Java files to be made into a single application):

```
cd {cldc_install_dir}\samples\classes
jar -cvf pong.jar pong\*.class
```

Finally, we will create the Palm executable. Note that MakePalmApp, not the Java interpreter, uses the classpath, classes\pong.jar, in the last line, below. When setting the classpath for the Java interpreter, we must place it **between** the java command and palm.database.MakePalmApp, or define it earlier like we have in the third line:

```
cd {cldc_install_dir}\samples
mkdir bin
set CLASSPATH=..\tools\palm\src\classes.jar; %CLASSPATH%
java palm.database.MakePalmApp -v -version "1.0" -icon icons\pong.bmp -
bootclasspath ..\bin\common\api\classes -classpath
..\bin\kjava\api\classes;classes\pong.jar -o bin\Pong.prc pong.Pong
```

The Palm executable (Pong.prc) will be found in the samples\bin directory.

Test the KJava Palm Application

Note: if you are going to test on the POSE rather than a real Palm device, when first creating a new emulator configuration, a .psf file, you must allocate at least 2Mb (8Mb is better) of RAM for the emulator; otherwise you will get the following error:

```
"KVM" 1.0 reports "MemoryMgr.c", Line:3527, MemMove to NULL"
```

Once the emulator, or device, is running:

- Load the KVM.prc and KVMUtils.prc onto the emulator
- Load the Pong.prc application, and start it

This is the result:

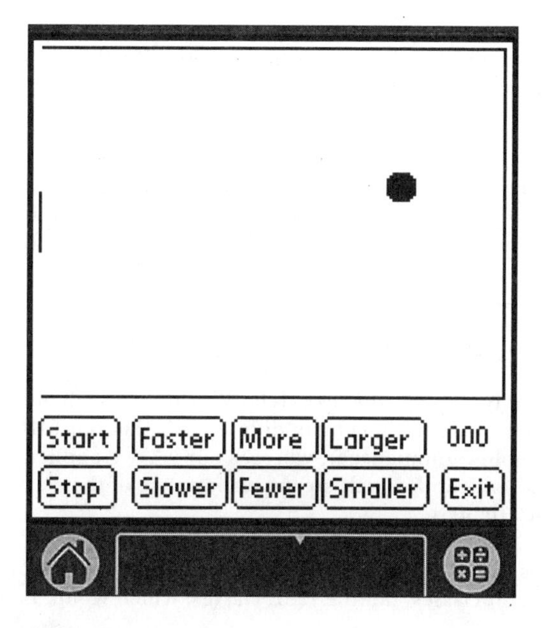

Note on Palm Connectivity:

If you write code examples that use the CLDC `Connection` classes, you must:

- Add the `-networking` option to `MakePalmApp`. Not doing so will result in the following error: `Could't init InetLib -1, 1290.`
- Secondly, within the Palm emulator, go to Settings | Properties and in the Communications section, select 'Redirect the NetLib calls to the host TCP/IP'.
- Last of all, in your URLs, use the pseudo protocol 'testhttp' rather than 'http', since most Palms don't have 'http' support. 'testhttp' has been included by Sun in the `KVMUtils.prc`.

All the above networking configuration advice applies equally to the Palm emulator. The one exception is if you are using (or emulating) a Palm VII/VIIx, which of course does have native HTTP support.

kAWT Download

In this section, we show how to build a kAWT 'Hello World' example using the CLDC Windows implementation. Again, kAWT requires a Palm or the POSE to run, and we must load Sun's `KVM.prc` and `KVMUtils.prc` onto the emulator, or device, of course, in addition to our own executable Palm application.

The kAWT classes comes in several flavors:

- Basic UI Java class files, `kawt.zip`, for compiling into your own application
- UI files complete with networking and IO classes, `kawt_io_net.zip`, again for compiling into your own application

- A basic Palm pdb file, kawt_pdb.zip, containg only UI classes, ready for loading straight onto a Palm or Palm emulator
- A Palm pdb file complete with UI, networking and IO classes, kawt_io_net_pdb.zip, also for loading onto a Palmor emulator
- Color Palm-compatible equivalents of the above

We need to download one set of plain Java class files for compilation, and also the equivalent .pdb to load onto the Palm, along with our own application. We recommend getting the kawt_io_net.zip and kawt_io_net_pdb.zip because these will allow us to use the CLDC Connector classes in later chapters.

Sample kAWT Application

This kAWT example simply displays the text Hello World, in an editable form, on a Palm, with an Exit button, just like our MIDP example. The main difference is that our MIDP application is under the control of the device's application manager, whereas our kAWT application is a standard Java executable with a main() method, which creates an instance of the class. Both classes have some form of catching action events, though.

The first thing to note, is the importing of the java.awt and java.awt.event UI libraries. These are actually kAWT's equivalent classes, not the J2SE classes, which would be much larger:

```
import java.awt.Button;
import java.awt.Frame;
import java.awt.Panel;
import java.awt.BorderLayout;
import java.awt.FlowLayout;
import java.awt.TextField;

import java.awt.event.ActionListener;
```

Next, our class, HelloWorldkAWT, extends Frame and implements ActionListener. These can be considered to be the equivalents of our MIDP class definition, earlier:

```
public class HelloWorldkAWT extends Frame implements ActionListener {
```

In the HelloWorldkAWT constructor, we create two Panel objects. The 'main' Panel, which occupies the top of the screen, contains our TextField with the default text of Hello World. The second Panel, buttonBar holds a single 'Exit' button. These Panel objects are used just for a layout control – there are several alternative ways of laying out graphics components (using a LayoutManger – we chose BorderLayout which works on four points of the compass(North, South, East, West) and also has a center):

```
  public HelloWorldkAWT() {
    Panel mainPanel = new Panel(new BorderLayout());
    add("Center", mainPanel);
    Panel buttonBar = new Panel(new FlowLayout());
    add("South", buttonBar);

    TextField tf = new TextField("Hello World");
    mainPanel.add(tf);
```

As well as adding the button to the `buttonBar`, we also call `addActionListener(this)` so that when the user presses **Exit** our `actionPerformed()` method is called, and this exits the application:

```
      Button b = new Button("Exit");
      b.addActionListener(this);
      buttonBar.add(b);
    }
```

The only other thing to do is actually make the UI components visible, by calling `show()`, which is methods inherited from `Frame`:

```
    public static void main(String[] args) {
      HelloWorldkAWT hw = new HelloWorldkAWT();
      hw.show();
    }

    public void actionPerformed(ActionEvent e) {
      System.exit(0);
    }
  }
```

Building Our Sample kAWT Application

Build the kAWT example in the same manner as the Pong KJava example, but include the kAWT libraries in place of the KJava ones when building and creating an executable.

Here is a batch file to compile, preverify, and make a Palm application. It assumes that the kAWT classes (held in the `kawt_io_net.zip` download) have been unzipped into the directory `{cldc_install_dir}\bin\kawt\`, and also that the `HelloWorldkAWT.java` source code has been placed into the directory `{cldc_install_dir}\samples\src\HelloWorldkAWT\`:

```
cd samples

REM create the temporary directories (do only once)
REM mkdir tmpclasses
REM mkdir bin
REM mkdir classes

REM compile the class
javac -g:none -d tmpclasses -classpath tmpclasses;..\bin\kawt -bootclasspath
..\bin\common\api\classes src\HelloWorldkAWT\HelloWorldkAWT.java

REM preverify the class
..\bin\win32\preverify -d classes -classpath
tmpclasses;..\bin\kawt;..\bin\common\api\classes HelloWorldkAWT

REM make the palm app
set CLASSPATH=..\tools\palm\src\classes.jar; %CLASSPATH%
java palm.database.MakePalmApp -v -version "1.0" -icon icons\HelloWorldkAWT.bmp -
bootclasspath ..\bin\common\api\classes -classpath ..\bin\kawt;classes -o
bin\HelloWorldkAWT.prc HelloWorldkAWT
```

Testing the kAWT Palm Application

Test the kAWT application exactly the same way as the KJava sample, but first unzip `kawt_io_net_pdb.zip`, and load the `.pdb` onto the Palm. Load your application, run it, and this is the result:

Summary

We have learnt about the core building blocks of the limited device J2ME, and how they fit together:

- The virtual machine, KVM
- The configuration, CLDC
- The profile, MIDP

We have seen that the CLDC support for 'standard' J2SE classes is limited, like the devices it will run on, but that it is still quite feasible to develop for families of devices at a high level of abstraction.

An example MIDP application was introduced, the Sun Wireless Toolkit was used to build and run it, and in conjunction with the build, we have also understood the need for application descriptors, and how MIDP applications are likely to be dynamically loaded by device's application managers.

Finally, we covered the two pseudo-profiles available for CLDC, in particular, kAWT, which is likely to be very similar to the PDA Profile, and how we can run examples on a Palm

Next, we cover the MIDP's user interface capabilities in greater depth.

PROFESSIONAL JAVA MOBILE

TOOL SELECTION

WEB TABLET
MOBILE PHONE
MP3 PLAYER
DIGITAL CAMERA
PALMTOP
LAPTOP

NOW SELECTED

PALMTOP

5

MIDP & CLDC API

In this chapter, we'll look in more detail at the MIDP API, and also cover the areas of the CLDC API that differ from standard J2SE. Specifically we will cover:

- The MIDP user interface.
- The MIDP support for data storage, the Record Management System (RMS).
- MIDP and CLDC system properties.
- Networking and connectivity. How to connect to a remote server.
- Timers.
- Debugging MIDlets.
- Toolkits, emulators and other J2ME resources.
- Limitations of MIDP. What can't you do?
- The future of J2ME. It's a young standard. Will it survive?

The MIDP User Interface

Probably the most limited aspect of a 'limited' device is that of the user interface (UI). With a minimum of 96x54 pixel display and a minimum of 15-20 keys, the MIDP specification had to ignore the existing standard Java user interfaces and define its own.

In particular, the MIDP user interface shares no code with either the Abstract Windowing Toolkit (AWT) or Swing, found on most Java desktop systems. The main reason for not using these existing UI libraries, is because they expect Windows, Icons, Mouse, and Pointer (WIMP) resources to be available.

Many MIDs have one or two of these WIMP features at most, but even then these features are likely to be limited when compared to their desktop equivalent. A secondary reason was the large memory requirements that the dynamic objects created by a WIMP user interface require.

As a result, there is no facility for multiple windows in the MIDP user interface, and none of the accompanying features, such as overlaps, minimizing, maximizing, and resizing. Instead, there is a `Displayable` class, which occupies the whole available screen area available to the MIDP application.

Interestingly, this does **not** necessarily mean the device's whole screen. The device may reserve space for an entry area or perhaps menus. It is possible to envisage an application manager of, say, a Palm-sized device, that allocates a quarter-screen sized window per MIDlet, that would give a pseudo-windows feel to the UI, but nevertheless, it would still **not** allow:

- Cutting and pasting between MIDlets
- Windowing within a MIDlet's `Screen` area
- Other expected windowing features

Although these windowing features are not an integral part of the MIDP, third party UIs, most notably the kAWT, are available, and provide most sophisticated user interaction. If we know that our target devices are always the top-end MIDs, then we should consider using these alternative libraries, but keep an eye on the issue of portability.

The MIDP user interface classes, found in the library `javax.microedition.lcdui`, can be considered on two levels, the high-level UI API, and the low-level UI API.

The high-level part of the UI API gives the maximum portability for our application. It specifies **what** functionality exists (say a selectable list), rather than specifying **how** that function is displayed (a pull down list, a new screen for the list, or maybe a single sideways-scrollable line of choices). For those familiar with WAP devices, this high-level API is the equivalent of the diverse look and feel of select/option lists found on `Phone.com` (for example Motorola, Alcatel, Siemens,and so on) and Nokia phones – all the devices perform the same function when supplied with the same WML code, but their implementations can look completely different.

In the high-level APIs, the implementation of the UI functionality is left to host OS methods, as are implementations of implicit features such as scrolling and navigation. The high-level APIs cannot access the physical attributes of the target devices, for example accessing key '2'.

The low-level part of the UI API provides very little protection from the physical device's capabilities. This gives us the chance to manipulate individual pixels, keys, listen for specific events, and so on. This has both advantages and disadvantages, as we will see later, and purposefully sails between the Java mantra of 'Write Once, Run Anywhere' and writing pure device-specific applications.

> **Some parts of the low-level UI API are not guaranteed to be portable. Even though our application may be MIDP-compatible, we still have no guarantee that it will run on another MIDP-compliant device if we use certain low-level methods. However, the API does provide methods for feature-checking prior to using such methods.**

We detail these low level messages below, but remember the extreme device limitations forced the MIDP designers to break the 'Write Once, Run Anywhere' rule, so either avoid using those methods if we want our application to be completely portable, or always use the feature-check methods first.

Before we dive into more detail, here are the user interface classes in Unified Modelling Language (UML) format, as a reference for the first half of this chapter:

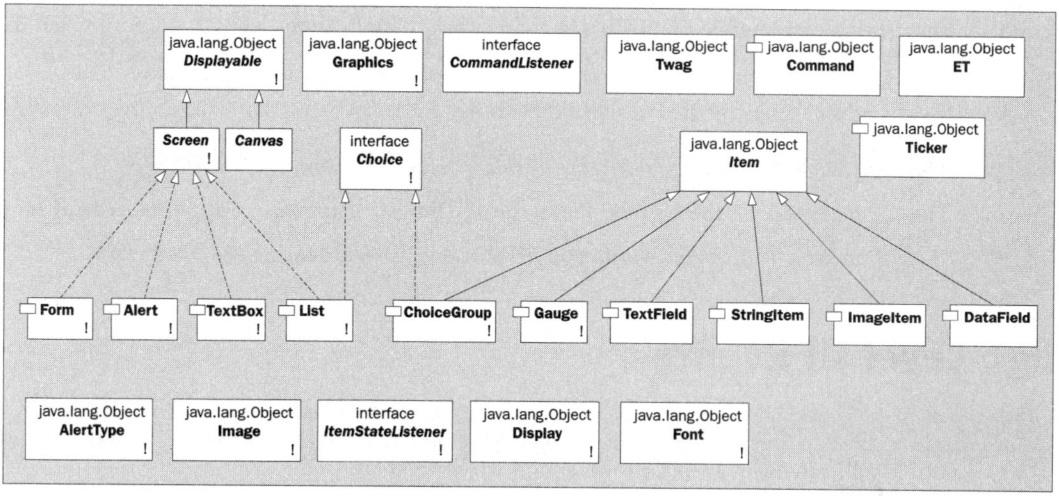

The Base UI Class: Displayable

The MIDP UI has no concept of a moveable, resizable Window, as we outlined previously. Instead, the MIDP has a simple fixed-size screen, represented by the `Displayable` object. Our application can have many `Displayable` objects, but only one `Displayable` is visible, or current, at any one time, and the user can only access items on that `Displayable`. If the MIDlet is running in the foreground, then the MIDlet's current `Displayable` handles all events that occur as the current application is being used.

Note that there is no guarantee of the dimensions of a `Displayable` in pixels – it may not be the whole screen of the target device. That decision is left to the target device's application manager, and there is no way we can offer 'hints' as to the size we might like., we can interrogate the size, However as we see later, and once the `Displayable` has been created, its size is guaranteed not to change.

If you are familiar with WAP devices, then a `Displayable` is a little like a WML card, and your application can have many cards, but just like WML, only one will be visible at any time.

The `Displayable` class has two abstract sub-classes:

- `Screen`
- `Canvas`

`Screen` is the base class for all the high-level UI `Displayable` objects, which are sub-divided into generic screens (`Form`) and pre-defined screens (`Alert`, `List` and `TextBox`). A `Form` allows us to build application-specific `Screen` objects, whereas the pre-defined `Screen` objects are complete in themselves, and cannot have other UI components added to them.

`Canvas` is the base class for all low-level UI `Displayable` objects. It allows direct manipulation of the target device's user interface, with such features as key mapping, double buffering, and clipping. In combination with the `Graphics` class, `Canvas` offers a basic set of graphics commands including lines, rectangles, arcs, filled shapes, and simple font rendering.

The applications that will use `Canvas` are typically games and other graphically intensive applications such as location/direction maps. `Canvas` also allows us to tailor applications to fit the device's screen, since it can be interrogated for its dimensions. When we consider that MIDP is now available for Palm, and yet can also run on a mobile phone, it's likely that many applications will use `Canvas` to get the best from each device.

A `Displayable` can check whether it is visible with the `isShown()` method. This only returns true if:

- The `Displayable` object's MIDlet is running in the foreground
- The `Displayable` is the current one in the MIDlet (set using `Display.setCurrent()`)
- No system screens (in other words non-MIDlet ones) are obscuring the `Displayable`

High-Level UI Events

Like the display classes, UI events can be split into high-level and low-level events, but in this case both types of `Displayable`, `Canvas` and `Screen`, can have high-level events associated with them., only a `Canvas` can handle low-level events However and we cover these later.

The high-level events (or actions and commands, as they are also called in the MIDP documentation) are represented by the`Command` class. There are four steps to create and handle a `Command`. First, the `Command` is assigned a label, a type, and a priority:

```
exitCommand = new Command(label, type, priority);
```

The `label` is displayed to the user when the command is rendered on the display.

The `type` must be one of: `BACK`, `CANCEL`, `EXIT`, `HELP`, `ITEM`, `OK`, `SCREEN`, and `STOP`. This set is defined in the `Command` class, and is not extensible. Also note that only `SCREEN` guarantees to display the label we defined – all the other types can be overridden if the implementation wishes to do so.An example might be a mobile phone mapping 'OK' to a hard button, or a jog dial press.

The assumption is that many devices will have icons or pre-defined strings and buttons more familiar to the user of that type of device. We should use the type `SCREEN` when we want to guarantee that the label will appear on the screen, in some form. In effect, it's the generic type for any other label outside of the above set.

Finally, the `priority` gives a hint to the application as to which commands to give priority to when displaying them, and possibly how to display them. For example, the highest priority commands should be immediately visible to the user, whereas lower priority ones could be held in a `Menu` or `Option` choice which the user must open to get to them.

All three pieces of command information are normally used as hints by the specific implementation of `Command` objects on the target device, rather than by our code, although we can read the label, type, and priority information if need be.

The second step in the process is to associate the `Command` with a `Displayable` by using the `addCommand()` method. Here is a snippet of the `IntroMIDlet1` example from the previous chapter, where the variable '`t`' is a `TextBox`, with which we associate our `Command`:

```
exitCommand = new Command("Exit", Command.EXIT, 1);
t.addCommand(exitCommand);
```

We have created a command labeled 'Exit' which is of type EXIT, which will be enabled when t is the current Displayable. Note that common Command actions can be shared between Displayable objects, which decreases application size.

The third step is to define the code that actually handles the Exit button event when it occurs. This is performed by the CommandListener class. In our case, we will make our MIDlet class handle the events, by defining that it implements the CommandListener interface:

```
public class IntroMIDlet1 extends MIDlet implements CommandListener {
...
}
```

Our MIDlet just needs to define a commandAction() method to implement the CommandListener interface. All our commandAction() needs do is stop the application by calling notifyDestroyed(). This tells the application management software that the MIDlet has finished being used, and that any associated resources can be released:

```
public void commandAction(Command c, Displayable s) {
    notifyDestroyed();
}
```

The parameters passed in to commandAction() are the Command that was fired off, and the Displayable that was associated with it. In our case there has only been one Command defined, exitCommand, so we can just assume any event received by the MIDlet closes the application.

Also left out, but good practice in more complex applications, is the call to the MIDlet's destroyApp() method to free up any resources used before informing the application management software that the MIDlet has finished running., in this application However there is no specific cleanup (like saving data) required.

Finally, we need to connect together the Displayable and the CommandListener (in our case, the MIDlet itself, in other words 'this') that will handle its events. This normally happens wherever the Displayable is first setup (in the startApp() method in our IntroMIDlet1 example):

```
t.setCommandListener(this);
```

Now, whenever an exit event occurs, the commandAction() method of our IntroMIDlet1 gets called. There is no way of removing a CommandListener from a Displayable, but we can replace any existing one with a subsequent call to setCommandListener().

High-Level API

Now we understand the basic framework of the high-level APIs, we can go into more detail on both the high- and low-level APIs of the MIDP. Here we just cover selected aspects of the APIs. The full list of classes, methods and parameters can be found in Appendix C.

First, we'll look at the four sub-classes of the high-level `Screen` class:

- `TextBox`
- `List`
- `Alert`
- `Form`

As we have already used a `TextBox` in our `IntroMIDlet1` example from the previous chapter, we will examine this first. We then cover the `List` class, and use its features to build the skeleton of a demo application for the rest of the high-level API, by creating a `List` to hold the choice of demos a user can choose from.

Keep in mind that we do not have control over the look and feel of these components, only what function they perform. If we need a specific look to a `List`, or `TextBox` for example, then we must 'roll our own' versions using the low-level UI, and check that they work satisfactorily on all possible target devices.

TextBox

Our `IntroMIDlet1` application had a `TextBox` that allowed the user to enter and edit text. On creation, the application can set the title, the initial contents, and ask for a certain (maximum) size of text box:

```
TextBox t = new TextBox("IntroMIDlet1", "Hello World", 256, TextField.ANY);
```

The final parameter, is for 'input constraint', in other words. restricting the text input to a certain type, potentially with some formatting. Examples are telephone numbers, numeric-only input, or maybe a URL. The different types of input constraint are defined in the `TextField` class and are referenced by `TextBox`. Our application will accept any sort of user input, but these are the full set of constraints:

- `ANY`
  Any user input is accepted.
- `EMAILADDR`
  A valid email address is accepted.
- `NUMERIC`
  Integer values accepted, including negative ones. A decimal point is not allowed – very annoying.
- `PASSWORD`
  The real input characters will be masked by an alternative characters, such as asterisks. Can be combined with other constraints, for example `PASSWORD | NUMERIC` for a 4 digit PIN.
- `PHONENUMBER`
  Region-specific phone characters only (for example numeric input plus a '+' prefix). May be linked to a native dialing application, but this is implementation-dependent.
- `URL`
  A valid URL is accepted
- `CONSTRAINT_MASK`
  A mask for checking which constraint is currently in use. Use in conjunction with `getConstraints()`. It cannot be used for setting the constraint, and will throw an `IllegalArgumentException` if used as such.

Strangely, there are no constraints defined to limit entry to ALPHABETIC, ALPHANUMERIC, or CURRENCY, which would have all been very useful.

How these different constrained fields are presented is entirely down to the implementation, so we can assume nothing about look and feel. For example, the J2MEWTK v1.0.1 separates out the first two sets of 3 digits of a phone number:

```
    TextBox t = new TextBox("Phone Number:", "01295690746", 20,
  TextField.PHONENUMBER);
```

This is rendered on-screen as:

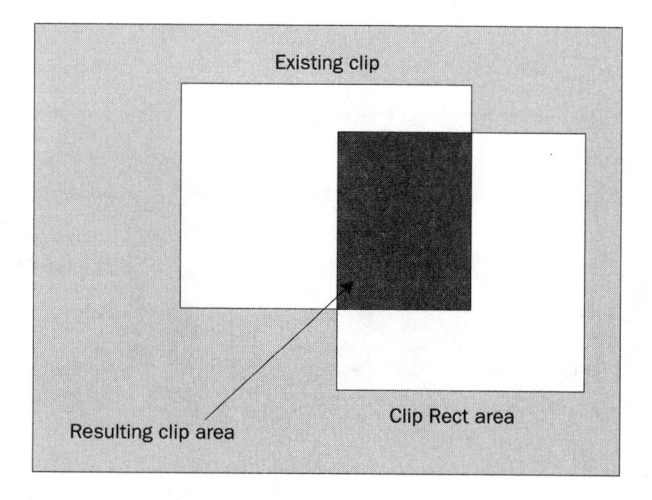

It is Also worth noting is that only the individual characters are validated, not the input string as a whole entity. For example, in the J2MEWTK v1.0.1 again, EMAILADDR constraint allows user entry such as 'fred@home@silly@com', which is not a valid email address. It is still up to our application to check the complete user input is valid, so both EMAILADDR and URL constraints, in particular, are of very limited use.

Once our TextBox, t, has been created, the current Displayable (of which TextBox is a subclass) can be set using the Display.setCurrent() method:

```
  display.setCurrent(t);
```

As we have used high-level APIs only, at this point (assuming our MIDlet is the foreground process, and not obscured by a system screen) the TextBox becomes visible without any more coding, because all high-level MIDP Screen objects paint the display for us.

There are a couple of points to be amde about the text. We can specify the initial String as null if we want the TextBox to be empty, and also the maxSize, 256 in our example, is not guaranteed – it's merely a request. We should check the actual size allocated with getMaxSize() after creation, if our application is dependant on a specific size being available.

Lists

A `List` is, not surprisingly, a list of choices, from which the user can pick and choose from.

`List` implements the `Choice` interface, which allows the creation of textual or image lists. `Choice` allows us to add, remove, and generally manipulate the choices in the `List`.

There are three types of `List`:

- `MULTIPLE`
- `EXCLUSIVE`
- `IMPLICIT`

Here is what they look like in the J2MEWTK v1.0.1:

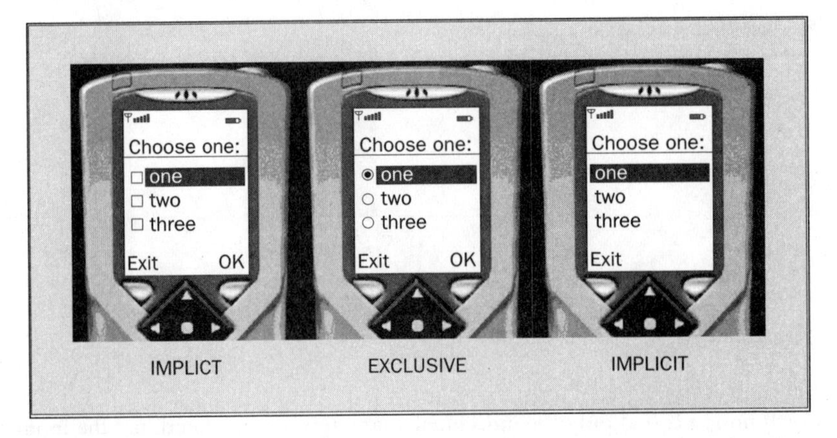

`MULTIPLE List` objects allow a user to select one or more choices by toggling the value of the choice between selected and not selected. Being a high-level feature the application never gets to know what the user is doing until they explicitly press an 'OK', 'Finish' or similar button, therefore generating a `Command` event, to signify they have finished their choosing. Until that point, the MIDP classes are responsible for scrolling, refreshing, and all other rendering of the screen. This is similar to check boxes in HTML.

`EXCLUSIVE List` objects are similar, but instead of allowing multiple selections, only one choice can be selected at any one time, with the previous choice being toggled off when a new choice is made. This is similar to radio buttons in HTML.

`IMPLICIT List` objects act slightly differently. As soon as a choice is selected, a `Command` event is fired. One example of an implicit `List` is a simple menu where we make a choice, and go directly to that chosen item, via the event-handling code in the matching `CommandListener.commandAction()` method. As can be seen in the screen shot above, there is no need for an 'OK' command.

This is a particularly attractive UI feature for mobile devices, because it minimizes the user input to just scrolling and selection (assuming the `List` isn't a huge length, of course). There is no HTML-equivalent to implicit lists in HTML, a list of hyperlinks is not a bad analogy, but in J2ME the application is responsible for rendering the new screen after the user has selected his choice.

Lists may be a choice of text labels, or images to display to the user, with the image being optional. The implementation of MIDP on the target device must decide how to display these (both, text only, or maybe image only if it's present), but a wise assumption is that the images, if any, should be compact both for memory and screen area considerations.

To demonstrate lists and commands, we will expand on `IntroMIDlet1` from the previous chapter, and give the user a choice between two applications. We will use an `IMPLICIT` List to present the user with their initial choices. There are only two demos initially, but we will be adding to this framework through the rest of this chapter. The first demo is just our standard Hello World `TextBox`, and the second is Bye World that just exits us from the application. Note that all the source code from this chapter can be downloaded from the Wrox web site: http://www.wrox.com.

Here is the revised source, complete with additional event-handling for our `IMPLICT List`:

```
import javax.microedition.midlet.MIDlet;
import javax.microedition.lcdui.*;

public class HighLevelMIDlet extends MIDlet implements CommandListener {
  private final static String hw = "HelloWorld";
  private final static String gw = "ByeWorld (=exit)";

  private Command exitCommand;
  private Command backCommand;
  private Display display;
  private Ticker ticker;
  private TextBox t;
```

We now create each `Command` in the constructor because one (`exitCommand`) gets shared between two screens (the initial `List` and our Hello World `TextBox`). We have also added a 'Back' button to helloWorld() to allow navigation back to the main list of demos:

```
  public HighLevelMIDlet() {
    display = Display.getDisplay(this);
    // create our commands
    exitCommand = new Command("Exit", Command.EXIT, 1);
    backCommand = new Command("Back", Command.BACK, 1);
  }
```

Our initial list has choices added to it using the `append()` method. The `null` parameter is where we would specify the optional image or icon for this choice.

Our event handler now needs to handle 3 events:

- A list selection
- An `EXIT` event
- A `BACK` event

```
  // display the menu of demos
  public void startApp() {

    // Create a List Screen, and make it IMPLICIT, i.e. like a menu
```

```
    List list = new List("Choose a Demo", Choice.IMPLICIT);
    // add choices to the menu
    list.append(hw, null);
    list.append(gw, null);

    // only add the EXIT command, because BACK make no sense
    list.addCommand(exitCommand);
    list.setCommandListener(this);

    display.setCurrent(list);
  }

  // display some text, and allow user any sort of input
  public void helloWorld() {
    t = new TextBox(hw, "Hello World", 256, TextField.ANY);
    t.addCommand(exitCommand);
    t.addCommand(backCommand);
    t.setCommandListener(this);
    display.setCurrent(t);
  }
```

we have a single event handler and several events (a choice in the `List`, and the `BACK` and `EXIT` commands)therefore we need to distinguish between them. They don't all need to share this code – we could create multiple `CommandListener` objects, one for each type of event, but for this demonstration, it's not necessary. It also potentially saves memory by not creating multiple event-handling objects, but this saving needs to be balanced against decreased execution speed(with the multiple tests for each event), and the readability and maintainability of the code.

In the `commandAction()` method, we can find the index in the `List` if the choice the user made, and from that, we can examine the label of the choice made, and decide what action to take. We still have to handle the `BACK` and `EXIT` commands, as well as the `IMPLICIT List` choices:

```
  public void commandAction(Command c, Displayable s) {
    // check if the event was from a List object
    if (c == List.SELECT_COMMAND) {
      // It was, so look at the labels to find out which choice the user made
      List l = (List) s;
      String choice = l.getString(l.getSelectedIndex());

      if (choice.equals(hw)) {
        // display Hello World TextBox
        helloWorld();
      }
      if (choice.equals(gw)) {
        // stop the MIDlet
        notifyDestroyed();
      }
    }
    int commandType = c.getCommandType();
    // if BACK was pressed, then take user back to the List menu
    if (commandType == Command.BACK) {
      startApp();
```

```
      }
      // if EXIT as pressed, then stop the MIDlet
      if (commandType == Command.EXIT) {
        notifyDestroyed();
      }
    }

    // not needed
    public void pauseApp() {}

    // not needed
    public void destroyApp(boolean unconditional) {}
  }
```

This gives the following output:

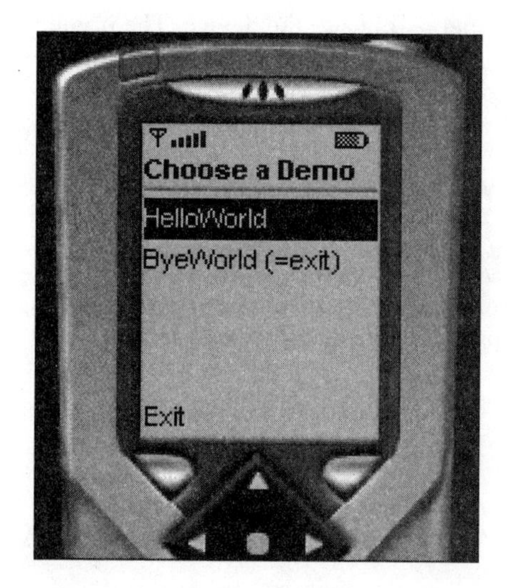

For each of the other high-level UI demos below, we add another entry into the initial `List` of demos, and of course, add matching event-handling code to re-direct control to that demo in `commandAction()`.

Alerts

An `Alert` is another pre-defined `Screen` useful for informing the user of a warning, error, or exception in the application.

The difference between `Alert` and other `Screen` objects, such as `TextBox` and `List`, is that it has a timeout associated with it. we Also, cannot add `Command` objects or `CommandListener` objects to an `Alert` – they will throw an `IllegalStateException`. (This is a rare case of inherited functionality being ignored, and breaking the object-oriented rule of not inheriting a class unless all the methods make sense in the subclass).

The basic alert types, defined in the associated AlertType class, are:

- ALARM
  Alerts the user of an event that they asked notification for ("Meet Mike at 10am.")
- CONFIRMATION
  Confirms user actions ("Record deleted.")
- ERROR
  An error occurred ("Could not download latest prices.")
- INFO
  Information only, no error, or danger ("Visit our website for the latest updates.")
- WARNING
  Warn the user of a potentially dangerous operation ("Deleting whole database – are you sure?")

All these types of Alert are merely hints to the device. The implementation will decide how to render each type of Alert.

Alert implementations can also give a spontaneous audible alert, using AlertType. playSound(), but there is no control over which AlertType the implementation decides is worthy of an audible warning as well as a textual one – it is, as usual, implementation-dependent. Neither can the type of sound be specified.

The Alert timeout can be set to the number of milliseconds that the alert is to be displayed for, or set to Alert.FOREVER if we want our user to dismiss the Alert, having presumably read it. We add the following lines in the appropriate places to our HighLevelMIDlet:

In our class definition:

```
private final static String alerts = "Alert";
```

In startApp(), we add:

```
  list.append(alerts, null);
```

A new method to display our Alert:

```
// display an alert for 2 seconds only
public void alertTest() {
  Alert a = new Alert("NASDAQ Alert:");
  a.setTimeout(2000);
  a.setType(AlertType.WARNING);
  a.setString("ECsoft stocks up 23% to $12.87");

  display.setCurrent(a);
}
```

Finally, in commandAction(), we add one more test, to see if 'Alert' is the choice in our list of demos:

```
    if (choice.equals(alerts)) {
      alertTest();
    }
```

Now run `HighLevelMIDlet`, and we will see three choices, "HelloWorld", "ByeWorld", and "Alert". Selecting "Alert" will display:

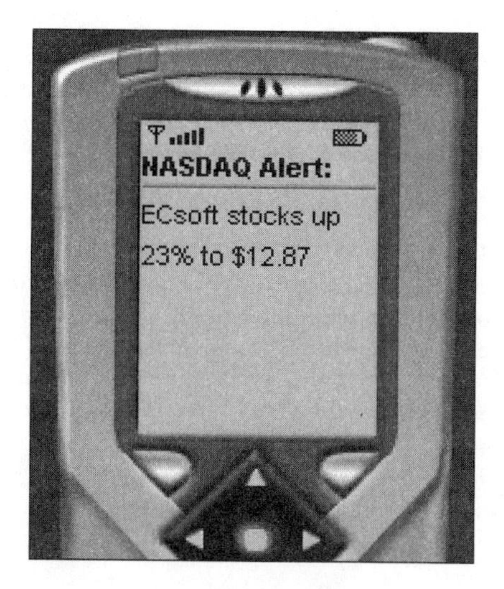

If our alert content is too long to fit into the display and requires scrolling, then the timeout is disabled, and the `Alert` becomes modal, i.e. the user must dismiss it. The target device's implementation will decide how to allow the user to dismisss it.

Forms

Form objects allow the application to define the content of a `Screen`, but because a `Form` is part of the high-level API the target device's implementation handles all layout, scrolling, and movement through the `Form`. In particular, there is no guarantee that the contents of a `Form` all appear in a single scrollable screen – the implementation may decide that certain input items require their own temporary 'popup' screen, just while the user is focused on that part of the form. This has parallels in the WAP world, where some browsers, notably Nokia, use the same mechanics to capture user input.

All objects that appear within a `Form` inherit from the abstract `Item` class. `Item` only specifies that there is a label associated with each `Item`, and it is up to each implementation how to identify each `Item` in a `Form` using this label. The following `Item` implementations are available for adding to a `Form`:

- `ChoiceGroup`
- `DateField`
- `Gauge`
- `ImageItem`
- `StringItem`
- `TextField`

The `ChoiceGroup Item` is very similar to `List`, but exclusively for embedding in a `Form`. Both inherit the majority of their functionality from the `Choice` class, giving single and multiple choices options, with both textual and image choices possible. The major difference is that we cannot create an `IMPLICIT ChoiceGroup` (yet another breaking of the OO rule of not inheriting a class unless all the methods/attributes make sense in the subclass):

```
ChoiceGroup cg = new ChoiceGroup("Hotel facilities:", Choice.MULTIPLE);
cg.append("pool", null);
cg.append("gym", null);
cg.append("golf", null);
```

`DateField` allows the user to enter and edit a time, a date, or both. Note there is no read-only version of `DateField`, so if we just want to display the date and time, we must use a text string instead.

```
DateField df = new DateField("Remind me at:", DateField.TIME);
```

`Gauge` is a horizontal bar graph, and can optionally be altered by the user. The graph's range of values spans from 0 to an application-defined maximum integer value, and the application is responsible for scaling the data to that range. The application can alter the value of the graph during display (for example for a progress indication). If the gauge is interactive, the implementation decides how to offer the user the ability to increase/decrease the value (up and down buttons, for example).

```
Gauge(String label, boolean interactive, int maxValue, int initialValue)
```

For example, here is a gauge with a range of 10, and an initial value of 5, or halfway.

```
Gauge g = new Gauge("Marks out of 10:", true, 10, 5);
```

`ImageItem` allows an immutable image (that is, one that cannot be changed) to be displayed in a `Form`. Interestingly, and unlike other `Item` implementations, `ImageItem` has a comprehensive set of layout hints, but the target implementation can still decide to ignore these. The image format must be Portable Network Graphics (PNG) version 1.0.

An alternative text description can be specified for an `ImageItem`, for cases where the device's memory capacity is exceeded, and it cannot manage to render the `ImageItem`. It is not mandatory for the device to display the string instead, though. `Form` also offers a convenience method for adding images, `append()`, when no layout hints are needed:

```
ImageItem(String label, Image img, int layout, String altText)
```

Here's an example:

```
    Image i = Image.createImage("/images/JJ.png");
    ImageItem ii = new ImageItem("JavaJeans", i, LAYOUT_DEFAULT, "JavaJeans");
```

`StringItem` is for adding read-only text to a `Form`. Again, `Form` offers a convenience method to do this too, also called `append()`, but taking a String as the argument.

```
StringItem si = new StringItem(null, "Copyright JJ Inc, 2001");
```

`TextField` lets us add editable text to a `Form`, and shares the same functionality as `TextBox`, in other words it has some minimal control over user input, with restrictions such as numeric-only input, telephone numbers, email addresses, or URLs:

```
TextField tf = new TextField("Special delivery instructions:", "Type here", 256,
TextField.ANY);
```

Here is our example Form rendered using the MIDP 1.0 reference implementation from Sun. It comprises of, (in top-to-bottom order):

- An ImageItem ("JavaJeans"), obtained from a PNG file which is in the MIDlet suite's JAR file
- A string added directly to the Form (Welcome to JavaJeans Mobile)
- An EXCLUSIVE ChoiceGroup (small, medium, large)
- A DATE_TIME DateField
- A TextField (Special delivery instructions plus input area)
- A StringItem (Finally – do you like our site?)
- A Gauge

Here is what the top of the form appears like in the J2MEWTK:

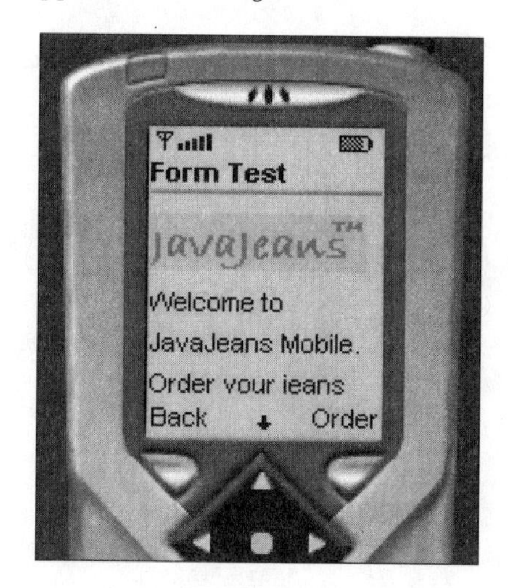

Here is the source code that creates the Form above. Again, we've added the Form code to HighLevelMIDlet.

In the class definition:

```
private final static String forms = "Form";
private Command orderCommand;
```

In the constructor:

```
orderCommand = new Command("Order", Command.EXIT, 1);
```

In startApp():

```
list.append(forms, null);
```

A new method to handle the selection of "Form" from the demos list:

```
    // Display a complex Form
    public void formTest() {

      // Create a ChoiceGroup with three entries, equivalent to a List
      ChoiceGroup cg = new ChoiceGroup("Choose a size:", Choice.EXCLUSIVE);
      cg.append("small", null);
      cg.append("medium", null);
      cg.append("large", null);

      // Create a date and time Field
      DateField df = new DateField("Deliver when:", DateField.DATE_TIME);

      // A TextField entry area, equivalent to a TextBox
      TextField tf = new TextField("Special delivery instructions:",
                                   "Type here", 256, TextField.ANY);

      // A gauge or bar chart
      Gauge g =
        new Gauge("Finally - please tell us your opinion of our site:", true,
                  10, 5);
```

We will try and create an image that we will fetch from the MIDlet JAR file in this instance. If for some reason it fails, we just display some alternative text:

```
      Image i = null;
      String il = null;
      try {
        i = Image.createImage("/images/JJ.png");
      } catch (java.io.IOException ioe) {
        il = "<logo>\n";
      }
      ImageItem ii = new ImageItem(il, i, ImageItem.LAYOUT_DEFAULT,
                                   "JavaJeans");

      // A plain text string
      StringItem si = new StringItem(null, "\nCopyright JJ Inc, 2001");
```

Here, we create the `Form`, and add every `Item` to it. They are displayed in the order that they are added to the `Form`:

```
      Form f = new Form("Form Test");
      f.append(ii);
      f.append("Welcome to JavaJeans Mobile. Order your jeans now!");
      f.append(cg);
      f.append(df);
      f.append(tf);
      f.append(g);
      f.append(si);
```

We also need some Command buttons, like the 'Order' command to allow the user to navigate around:

```
    f.addCommand(backCommand);
    f.addCommand(orderCommand);
    f.setCommandListener(this);

    display.setCurrent(f);
  }
```

Finally, the test for "Form" in commandAction():

```
      if (choice.equals(forms)) {
        formTest();
      }
```

For demonstration purposes this is fine, but for real-world applications take care not to force too much information into a single Form, because it may become unwieldy on a real device. One way to do this would be to split the form up into several smaller forms that the user navigates between.

Interactive Forms

Unlike choices in a List (that is, except an IMPLICIT List), an interactive Item can cause an event when its value is changed by the user. Form has an associated ItemStateListener in addition to the standard CommandListener of a Screen to handle these Item events. This gives us a novel dynamic form capability that we can feedback to the user what effect their choices and input is having.

In our example, we track a Gauge Item, and change its label depending on the value the user sets the gauge to. This is all done in the itemStateChanged() method that we write to satisfy the ItemStateListener interface.

Here is the additional code needed to catch events from the Guage Item on our Form. First, we change our HighLevelMIDlet to implement the ItemStateListener interface:

```
public class HighLevelMIDlet extends MIDlet implements CommandListener,
ItemStateListener {
  ...
```

```
  private final static String interactive = "Interactive Form";
```

We add another entry into the list of demos the user can pick from:

```
    list.append(interactive, null);
```

Then Add the test for the interactive form demo in commandAction():

```
      if (choice.equals(interactive)) {
        interactiveTest();
      }
```

We copy the `formTest()` method to `interactiveTest()`, and just a line to associate an event listener `ItemStateListener` to it:

```
    f.setItemStateListener(this);
```

Finally, in our MIDlet, we implement the item event-handling method. As the `ItemStateListener` only gets an `Item` passed in, we have to decide which `Item` on the `Form` caused an event, by checking it class instance:

```
  public void itemStateChanged(Item i) {
    // we're only interested in Gauge Item events
    if (i instanceof javax.microedition.lcdui.Gauge) {
      int gaugeValue = ((Gauge) i).getValue();
```

Once we have ascertained it's the `Gauge`, we can just give the user some feedback by modifying its label:

```
      // if user moves Gauge above 5, she likes us
      if (gaugeValue > 5) {
        i.setLabel("Oh good :-)");
      }

      // if user moves Gauge to 5, she's ambivalent
      if (gaugeValue == 5) {
        i.setLabel("So-so :-|");
      }

      // if user moves Gauge below 5, she's not impressed
      if (gaugeValue < 5) {
        i.setLabel("Oh dear :-(");
      }
    }
  }
```

All the code can be found in the `HighLevelMIDlet` example source.

Here's the output, showing the feedback of Oh good :-) and Oh Dear:-(:

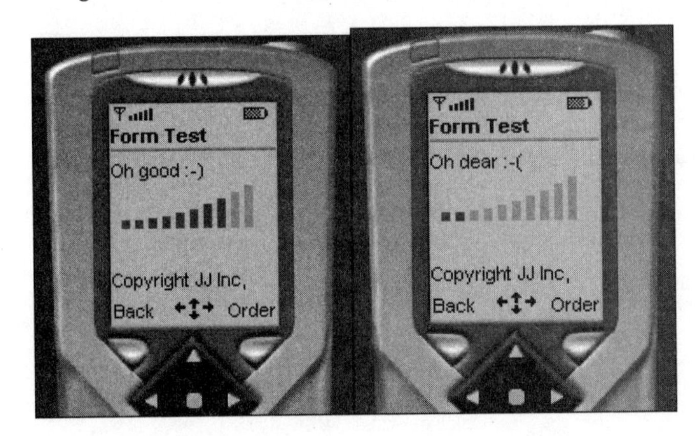

Item objects cannot be shared across forms. If we wish to share some entered data, it needs to be kept in our own Java class (or a MIDlet storage class, see later), and, at some later point, we can synchronize that data between the two Items on the different Form objects.

A useful facet of the high-level APIs, mentioned earlier, is that any changes to the display contents, like Form, are reflected immediately on the screen, so if we remove an Item, the Form is refreshed without additional code.

Tickers

All high-level Screen objects can have a Ticker attached to them. A Ticker is a ticker-tape of continually scrolling horizontal text, not a timer. A Ticker can be shared across many screens, which is useful if a certain piece of information needs to be visible at all times.We will construct the Ticker in the constructor, so it can be shared across several screens:

```
    private final static String tickerTape = "Ticker";
```

```
    private Ticker ticker;
```

```
      ticker = new Ticker("NASDAQ Latest: Sun shares up this morning, Apple's are
  cooking slowly, and no sight of Dawn raids.");
```

In startApp():

```
    list.append(tickerTape, null);
```

We display Hello World, but this time, complete with Ticker, having changed the helloWorld() method to accept a Boolean input parameter, tickerOn. If tickerOn is true, then we display a ticker:

```
    public void helloWorld(boolean tickerOn) {
      t = new TextBox(hw, "Hello World", 256, TextField.ANY);
      t.addCommand(exitCommand);
      t.addCommand(backCommand);
      t.setCommandListener(this);
      if (tickerOn) {

        // add ticker to screen
        t.setTicker(ticker);
      }
      display.setCurrent(t);
    }
```

If the `Ticker` String is changed, then that new String is immediately shown in place of the old one, if the `Ticker` is visible.

Low-Level UI API

Having met all the high-level components of the MIDP UI, we can now move on to the low-level components and see what features they offer.

The low-level UI API consists of both events and graphical commands. Just to recap, some of these features **will not necessarily be portable across different devices. Well written code will check first, before assuming** any specific functionality is available. This will be discussed in more detail later in this chapter.

All low-level user interface calls must run on a `Canvas` – the low-level `Displayable` equivalent of `Screen`. `Canvas` is an abstract class, though, and we must implement several of the methods ourselves, in particular the event handling methods. All the low-level MIDP graphics drawing commands are encapsulated in the `Graphics` class.

Note that we can mix a `Canvas` and a `Screen` in a single application – a `List` might select a particular part of a game, and a TextBox could hold the top scores, etc. The distinction between high-level and low-level UI is for simplified understanding only, they are still both part of the same library., it is important to note However that we **cannot** mix the high and low-level UI APIs together on the same `Screen`.

Low-Level UI Events

All the low-level events are handled by the current visible `Canvas`, and are serial, in other words they cannot interrupt each other but rather get handled in the order they were fired off by the application. In other words any method handling an event will complete its operations and return before the next queued event is dealt with. (There is actually one exception to this serial calling rule – and that is the `serviceRepaints()` method, which makes sure the screen content is up to date, always gets satisfied immediately. We cover this later.)

Here is the full list of low-level events:

- `showNotify()`
- `hideNotify()`
- `keyPressed()`
- `keyRepeated()`
- `keyReleased()`
- `pointerPressed()`
- `pointerDragged()`
- `pointerReleased()`
- `paint()`
- The `CommandListener.commandAction()` method

The `showNotify()` event occurs just before a `Canvas` becomes the current visible `Displayable`. In the reference implementation this method is empty, but can be overridden by a subclass, and allows initialization of resources such as timers and animations.

The `hideNotify()` event occurs shortly after the `Canvas` stops being the current `Displayable`, and allows the application to pause those same timers and animations that would unnecessarily waste processing power when the `Canvas` isn't visible.

The `keyPressed()`, `keyRepeated()`, and `keyReleased()` events are all keypad/keyboard related. All keys in MIDP devices have an associated key code, but they assume only a keypad exists, in other words the key codes are limited to the 0-9, # and * keys (`KEY_NUM0`, `KEY_NUM1`, and so on). Other MIDP devices with QWERTY keyboards will also define key codes for the extra keys, but we cannot assume these alphabetic keys will be on all MIDP devices. A surprising omission is that multiple alphabet mappings on keypads, such key 2 mapping to the characters A, B, and C, are not handled by the low-level APIs – it is up to the application to associate three rapid presses of key 2 to the letter 'C'.

For games applications, use the `getKeyCode()` and `getGameAction()` methods to find out if that key has been mapped to a certain functionality, like UP, DOWN, LEFT, RIGHT, and FIRE. The target implementation decides which keys map to these gaming functions, making sure our game will still be usable on any MIDP keypad or keyboard. The mapping does not change whilst the application is running, therefore we just need to call `getKeyCode()` once per gaming key during initialization to build a simple key map:

```
void initGame() {
  up = getKeyCode(Canvas.UP);
  down = getKeyCode(Canvas.DOWN);
  fire = getKeyCode(Canvas.FIRE);
}

public void keyPressed(int keyCode) {
  switch(keyCode) {
    case up:
      ...
      break;
    case down:
      ...
      break;
    case fire:
      ...
      break;
    default:
      break;
  }
```

`keyRepeated()` is not guaranteed to be implemented on all devices. We must first check using `hasRepeatEvents()` before using it.

`Canvas` has no default code for these keyboard/keypad methods, so to capture these events we need to override them.

`pointerPressed()`,`pointerDragged()`,`pointerReleased()` are all pointer-related events, and as such not mandatory on MIDP devices. Use the `hasPointerMotionEvents()` to decide whether the target device supports these. All three methods return the (x,y) coordinates of the pointer, relative to the upper left corner of the `Canvas`. Again we need to override these methods in our code because they are, by default, empty.

Canvas is a Displayable,therefore it can have high-level Command events associated with it, just like all the high-level API Screen objects.

Low-Level Graphics

Although the low-level graphics allow an application the flexibility to access individual pixels on the Canvas, all repainting of that Canvas must be performed by our application.

To repaint a Canvas, the application must call Canvas.repaint(), even though the application's actual drawing code is held within the method paint(). Applications must not call paint() directly – the target device's implementation will do this for the application, but not necessarily immediately, and sometimes several repaint() calls may be made before a paint() occurs. This potentially allows the implementation to give better graphics performance by combining multiple repaint() calls together.

Here is a simple Canvas class that we have created inline in the code as an internal class, rather than create a separate class file for it. It draws a black square 20*20 pixels in size, in the center of the screen area available to this MIDlet:

```
public void drawRectangle() {

  c = new Canvas() {

    protected void paint(Graphics g) {
      int width = c.getWidth();
      int height = c.getHeight();
      g.setGrayScale(0);
      g.fillRect(((width / 2) - 10), ((height / 2) - 10), 20, 20);
    }

  };
  c.addCommand(exitCommand);
  c.addCommand(backCommand);
  c.setCommandListener(this);
  display.setCurrent(c);
}
```

As can be seen, we have only implemented one of the methods described above, paint(), and what is more, we do not call it directly ourselves. We leave that to the underlying implementation to do by calling display.setCurrent().

The application can However flush any repaint() calls by calling the serviceRepaints() method. serviceRepaints() will block until the paint() method has completed. On return from a serviceRepaints() call, the Canvas is guaranteed to be up to date. ServiceRepaints() is the one exception to the serialized events rule mentioned earlier. This means we can call it from within an event handling routine such a commandAction, and it will complete first, and not be queued up.

One point to be aware of is that it is possible to become deadlocked if the code calling serviceRepaints() holds a lock which paint() also requires. If this is the case then paint() will sit waiting forever to get that lock, and will never return. Use callSerially() instead to avoid this potential deadlock.

Our `paint()` code should assume nothing about the existing contents of the screen, because there is no guarantee that the last `Graphics` commands have not been overridden since by another application that wishes to alert the user or interrupt in some way. On a MIDP of course, interrupts are quite likely, for example phone calls, SMS messages, and paging messages. Therefore, always redraw the complete screen in the `paint()` method.

Canvas Size

The origin of the `Canvas` coordinate system is at the top left of the display, being 0,0, with positive values being right for x, and down for y, like so:

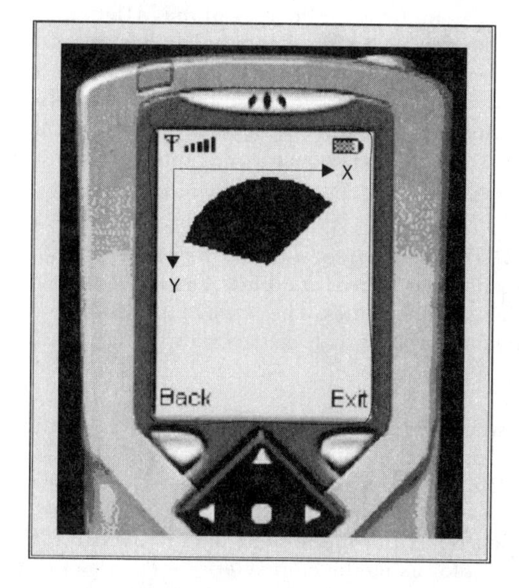

The maximum size of the `Canvas` can be obtained from the`Canvas.getWidth()` and `Canvas.getHeight()` methods. Although the width and height are not guaranteed to be the same every time an application starts up, once the application has started, width and height will not change, even for newly created `Canvas` objects.

Pixels are not guaranteed to be exactly square, but MIDP implementations are required to cope with any transformation should the device's pixels be significantly non-square. Therefore, our applications do not need to compensate for this, and we can assume approximately square pixels. This is particularly important for maintaining the correct aspect ratio of images, for example, which is a problem in the WAP device world.

Let's now review each of the drawing features found in the `Graphics` class.

The Graphics Class

All the low-level drawing commands are called from within the `Graphics` class, an instance of which is always passed into the `paint()` method of `Canvas`. This is the only way to obtain a `Graphics` object whose destination is the display, and that `Graphics` object is not available outside the `paint()` method. Reference to this `Graphics` instance should not be kept, because calls to it once `paint()` returns are undefined.

The only other way to obtain a Graphics object is from a mutable (in other words writeable) image using Image.getGraphics(). Mutable images can be created off-screen as drawing areas, as we will see later.

There is also support for double-buffering of the display memory, but since not all devices can support this functionality, check first, by using the Canvas method isDoubleBuffered(). Double-buffering lets us draw off-screen, by using additional memory, and then swap out the existing screen contents for the new ones we have drawn.

This technique is vital for animations, because drawing direct to the display memory can cause annoying flashes as the display updates. In the case of the MIDP, the implementation takes care of any extra memory needed for the double buffer. Any drawing commands we call are rendered in off-screen memory, and when repaint() is called, the off-screen contents are displayed in place of the current screen contents, rather than have to be drawn, pixel by pixel, potentially causing screen interference. Of course, we have no guarantee that the target device has the processing power to make this happen sufficiently quickly to avoid screen interference.

Grayscale and Color

Although most current MIDP devices are not color, the MIDP pixels are defined as 24-bits, with 8 bits for each primary component – red, green, and blue. The implementation then maps those pixel values onto the target display's actual capabilities. The Display class allows us to determine whether color, grayscale or monochrome is supported with the Display.isColor() andDisplay.numColors() methods.

The current color is set using setColor(int RGB) where RGB is in the hexadecimal format 0x00RRGGBB, or setColor(int red, int green, int blue), where each primary color's value must fall between 0 and 255, or 0xFF in hexadecimal.

getColor() returns the current color in the same 0x00RRGGBB format, or we can find the individual components of the current color using the convenience methods of getBlueComponent(), getGreenComponent(), and getRedComponent().

The grayscale can be set between 0 and 255 using setGrayScale(), and the current value can be retrieved using getGrayScale(). It is possible to mix color and grayscale commands, with the implementation being responsible for interpreting the best match as it converts between the two. Here's an example where our variable, shade, would be expected to be set to a gray value of approximately 170:

```
SetColor(0xff8080);
int shade = getGrayscale();
```

Note that on a monochrome LCD the grayscale value of 255 is white, in other words, an absence of a pixel, and 0 is a filled in black pixel.

There is no concept of foreground and background colors in the MIDP, only of the current color, which is what all the graphics commands use to render with. Also, there is no way of color mixing or blending because MIDP offers is no way of reading the color value of an individual pixel – a disappointing limitation which will hopefully be offered in future versions of the MIDP profile.

Fills

Three basic shapes can be solid filled – rectangles, arcs, and round cornered rectangles. Using `fillRect(int x, int y, int width, int height)` is the most efficient way of clearing or initializing the display. Below is some code that will draw a square at the center of the screen:

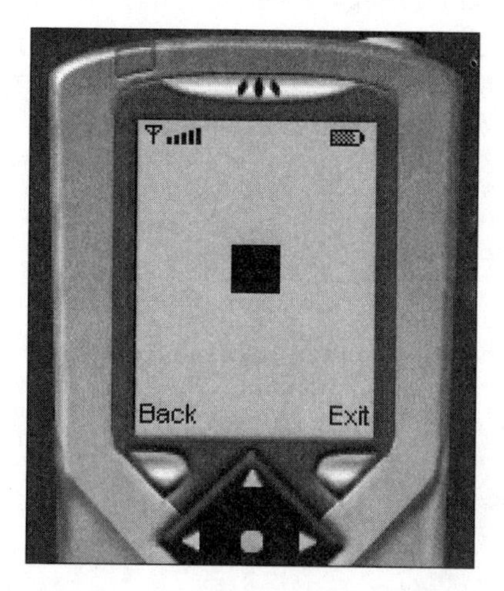

We have built up a low-level equivalent to our previous example, `HighLevelMIDlet`, so all of the demo `List` and event-handling code will be familiar to us. What have changed are the demo methods themselves, in this first case `drawRectangle ()`. Here is the code for our low-level demo, `LowLevelMIDlet`:

```
import javax.microedition.midlet.MIDlet;
import javax.microedition.lcdui.Canvas;
import javax.microedition.lcdui.Command;
import javax.microedition.lcdui.CommandListener;
import javax.microedition.lcdui.Form;
import javax.microedition.lcdui.Graphics;
import javax.microedition.lcdui.Display;
import javax.microedition.lcdui.Displayable;
import javax.microedition.lcdui.List;
import javax.microedition.lcdui.Choice;
import javax.microedition.lcdui.Image;

public class LowLevelMIDlet extends MIDlet implements CommandListener {
  private final static String rect = "Rectangle";

  private Command exitCommand;
  private Command backCommand;
  private Display display;
  private Canvas c;

  public LowLevelMIDlet() {
    display = Display.getDisplay(this);
    // create our commands
    exitCommand = new Command("Exit", Command.EXIT, 2);
    backCommand = new Command("Back", Command.BACK, 2);
  }
```

The `startApp()` method initializes the list:

```
public void startApp() {
  // Create a List Screen, and make it IMPLICIT, i.e. like a menu
  List list = new List("Choose a Shape", Choice.IMPLICIT);

  list.append(rect, null);

  // only add the EXIT command, because BACK make no sense
  list.addCommand(exitCommand);
  list.setCommandListener(this);

  display.setCurrent(list);
}
```

Here is the important part. The `drawRectangle()` method creates an anonymous instance of a `Canvas` object and specifies its `paint()` method. If we knew we were going to draw repeatedly into the same `Canvas`, then we might create a more long-lasting one in the constructor, or `startApp()`:

```
// display a rectangle, plus two commands, BACK and EXIT
public void drawRectangle() {

  c = new Canvas() {

    protected void paint(Graphics g) {
      int width = c.getWidth();
      int height = c.getHeight();
      g.setGrayScale(0);
      g.fillRect(((width / 2) - 10), ((height / 2) - 10), 20, 20);
    }

  };
  c.addCommand(exitCommand);
  c.addCommand(backCommand);
  c.setCommandListener(this);
  display.setCurrent(c);
}
```

The remaining methods in `LowLevelMIDlet` are not new. They simply implement the required `pauseApp()`,`destroyApp()`, and `commandAction()` methods:

```
// not needed
public void pauseApp(){}

// not needed
public void destroyApp(boolean unconditional) {}

// The event handler now needs to cope with 3 events:
// - a list selection
// - an EXIT event
// - a BACK event
public void commandAction(Command c, Displayable s) {

  // check if the event was from a List object
```

```
      if (c == List.SELECT_COMMAND) {
        // It was, so look at the labels to find out which choice the user made
        List l = (List) s;
        String choice = l.getString(l.getSelectedIndex());
        if (choice.equals(rect)) {
          // draw a rectangle
          drawRectangle();
        }
      }

      // if BACK was pressed, then take user back to the List menu
      if (c.getCommandType() == Command.BACK) {
        startApp();
      }

      // if EXIT as pressed, then stop the MIDlet
      if (c.getCommandType() == Command.EXIT) {
        notifyDestroyed();
      }
    }
  }
```

Arcs in MIDP may be circular or elliptical, and are filled using `fillArc(int x, int y, int width, int height, int startAngle, int arcAngle)`. The arc is filled like a pie chart, from the arc-ends to the center of the arc, and not direct from arc-end to arc-end. The `startAngle` is relative to 3 o'clock, in other words a `startAngle` of 0 means the arc has a horizontal edge to the right of the circle's center, and even stranger, the `arcAngle` rotates 'backwards' or anti-clockwise from the `startAngle`, clockwise rotation can However be specified with a negative `arcAngle` value:

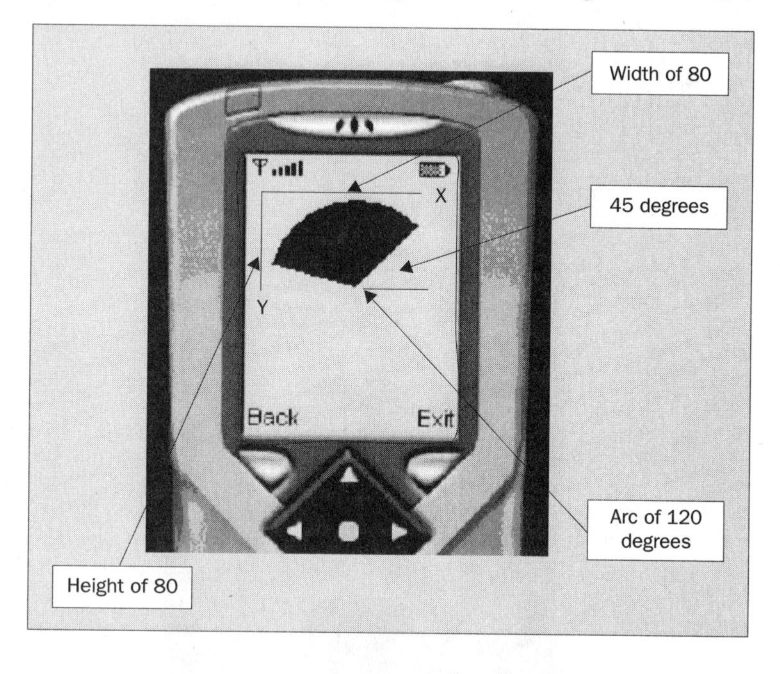

Here's an example where we've specified a `startAngle` of 45 degrees from 3 o'clock, and an `arcAngle` of 120 degrees. The filled arc is part of a solid circle that is 80 pixels in width and height, offset from the top left corner by (10,10) or rather, the invisible box bounding the circle has an origin of (10,10):

```
public void drawArc() {

  c = new Canvas() {

    protected void paint(Graphics g) {
      g.setGrayScale(0);
      g.fillArc(10, 10, 80, 80, 45, 120);
    }
  };
  c.addCommand(exitCommand);
  c.addCommand(backCommand);
  c.setCommandListener(this);
  display.setCurrent(c);
}
```

To create a filled circle (or ellipse), just set `startArc` to 0, and an `arcAngle` of 360 degrees, like so:

```
public void drawCircle() {

  c = new Canvas() {

    protected void paint(Graphics g) {
      g.setGrayScale(0);
      g.fillArc(10, 10, 80, 80, 0, 360);
    }
  };
  c.addCommand(exitCommand);
  c.addCommand(backCommand);
  c.setCommandListener(this);
  display.setCurrent(c);
}
```

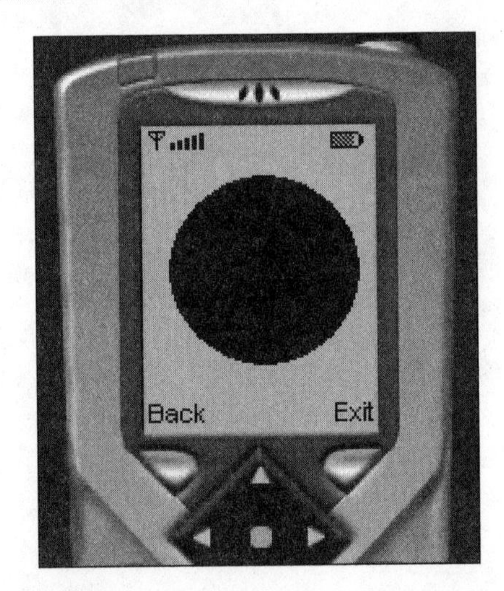

Round cornered rectangles can be filled using `fillRoundRect(int x, int y, int width, int height, int arcWidth, int arcHeight)`, where `arcWidth` and `arcHeight` define the width and height of the arc that defines the rounded corners. Again, here is an example:

```
public void drawOblong() {

  c = new Canvas() {

    protected void paint(Graphics g) {
      g.setGrayScale(0);
      g.fillRoundRect(10, 10, 80, 80, 40, 40);
    }
  };
  c.addCommand(exitCommand);
  c.addCommand(backCommand);
  c.setCommandListener(this);
  display.setCurrent(c);
}
```

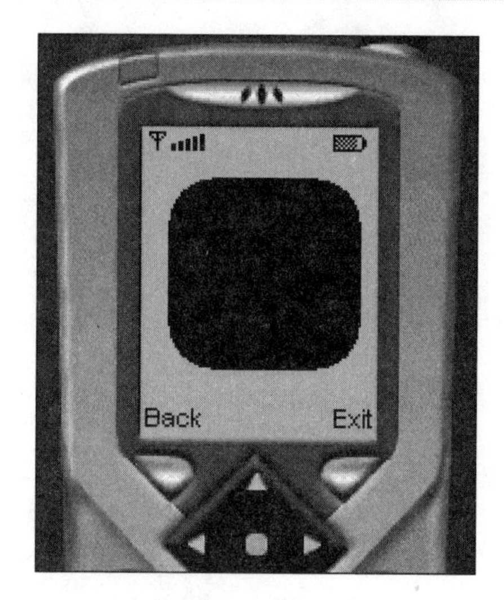

Clipping

Clipping describes the process of limiting the area in which graphics commands have an effect. It's useful when we only want to update part of the screen, but we want to use standard commands that might waste processing power overwriting areas that are already correctly rendered.

The clipping area is always rectangular, and is set using `setClip(int x, int y, int width, int height)`. The values of the current clip can be obtained using `getClipX()`,`getClipY()`,`getClipHeight()` and `getClipWidth()`. Here is an example, where our original black filled rectangle drawn, with `drawRectangle()`, is now clipped to only the part of the rectangle within the top left quadrant of the `Canvas`:

```
protected void paint(Graphics g) {
  int width = c.getWidth();
  int height = c.getHeight();
  g.setGrayScale(0);
  g.setClip(0, 0, (width / 2), (height / 2));
  g.fillRect(((width / 2) - 10), ((height / 2) - 10), 20, 20);
}
```

The clip area, (0,0) to the center of the `Canvas`, is shown as a solid outline. The original black-filled square is shown as a dotted outline, and the actual rendered area is the black top left corner of the original square:

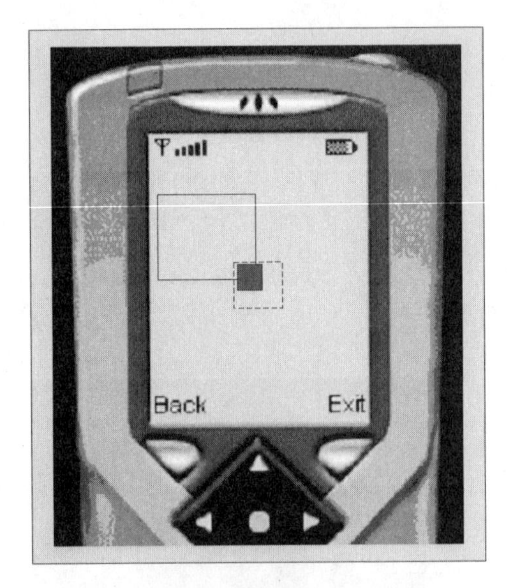

We can also specify a new clipping area by using `clipRect (int x, int y, int width, int height)`. The new clipping area is the intersection of the existing clip rectangle with the one specified in `clipRect()`:

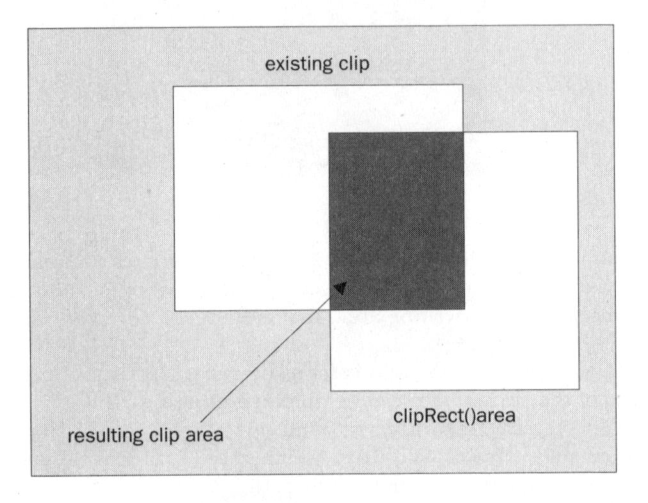

Lines and Outlines

Simple 1-pixel width lines are drawn using the `drawLine(int x1, int y1, int x2, int y2)` method in `Graphics`. There are two styles of lines, `SOLID` or `DOTTED`, which are set by calling `setStrokeStyle()`, and can be read using `getStrokeStyle()`. The length of the filled parts of the line, and the gaps between them, is implementation-dependant. Strokes styles can only be applied to lines and outlines, not to solid fills or text rendering. For `SOLID` lines, both end-points are guaranteed to be drawn, but there is no similar guarantee for `DOTTED` lines.

In addition to lines, the outlines of rectangles, arcs, and rounded rectangles, can also be drawn using `drawRect()`,`drawArc()`, and `drawRoundRect()`. They have exactly the same parameters as their solid fill equivalents. When combining outlines and solid fills, their respective outlines will not match if we use the same height and width variables, because the `drawRect()` always has a greater depth, by 1 pixel. An extreme example that demonstrates this is, the following two very thin rectangles that share the same parameters, one is a fill, one a draw:

```
public void draw2Rect() {

  c = new Canvas() {

    protected void paint(Graphics g) {
      g.setGrayScale(0);
      g.setStrokeStyle(Graphics.SOLID);
      g.fillRect(25, 20, 50, 1);    // fills 1 horizontal row of pixels
      g.drawRect(25, 30, 50, 1);    // draws 2 horizontal rows of pixels
    }
  };
  c.addCommand(exitCommand);
  c.addCommand(backCommand);
  c.setCommandListener(this);
  display.setCurrent(c);
}
```

This is standard graphics functionality, because (assuming pixel blending is allowed – it isn't in MIDP) it makes sure that fills always butt together if they share co-ordinates, but it can be confusing if you haven't programmed graphical components before. Here is the result:

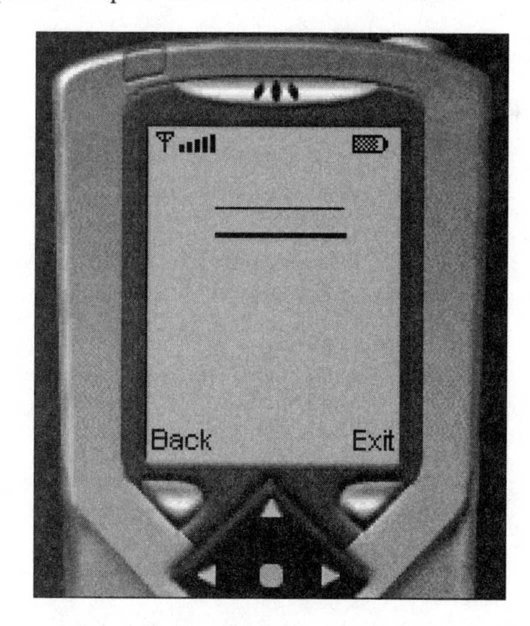

This mismatch is also true of arc and rounded rectangle outlines too, when compared with their solid fill equivalents.

Text

Fonts are implementation-dependant and cannot be defined by our application. However, the target device tries its best to match our request with its closest font. Our application can specify one or more of these font attributes:

- face(FACE_SYSTEM, FACE_MONOSPACE, FACE_PROPORTIONAL)
- size(SIZE_SMALL, SIZE_MEDIUM, SIZE_LARGE)
- style(STYLE_PLAIN, STYLE_BOLD, STYLE_ITALIC, STYLE_UNDERLINED)

We can mix styles together (bold + italic), but only one each of face and size can be specified when requesting a font from the implementation.

To obtain a font, use the static method `Font.getFont(int face, int style, int size)`, as there is no `Font()` creation method. Then use `Graphics.setFont(Font font)` to set the current font. If called with a null parameter, then the default font is used, in other words the equivalent longhand statement would be `setFont(Font.getDefaultFont()).Graphics.getFont()` returns the current font being used, not a new `Font`, as might have been expected.

If the implementation cannot match our request for a specific font exactly, it delivers its closest match. We can check the actual attributes of the returned font using `Font.getSize()`, `Font.getStyle()` and `Font.getFace()`.

All text is rendered relative to an 'anchor' point. The anchor point is defined as a combination of a horizontal constant (`TOP`, `BASELINE`, `BOTTOM`) plus a vertical constant (`LEFT`, `HCENTER`, `RIGHT`), and generally describes a point on the bounding box of a character or array of characters. Here are the possible anchor points for an example text string (in a font and size unlikely to be found on a MIDP, currently):

To render one or more characters, we specify the coordinates of the anchor point, and which anchor point around the characters we want to use. For example, the code below will draw the string with the top left corner of the 'b' positioned at (20,20) on the display:

```
public void renderText() {

  c = new Canvas() {

    protected void paint(Graphics g) {
      g.drawString("big pebbles", 20, 20, g.TOP | g.LEFT);
    }
  };
```

```
        c.addCommand(exitCommand);
        c.addCommand(backCommand);
        c.setCommandListener(this);
        display.setCurrent(c);
    }
```

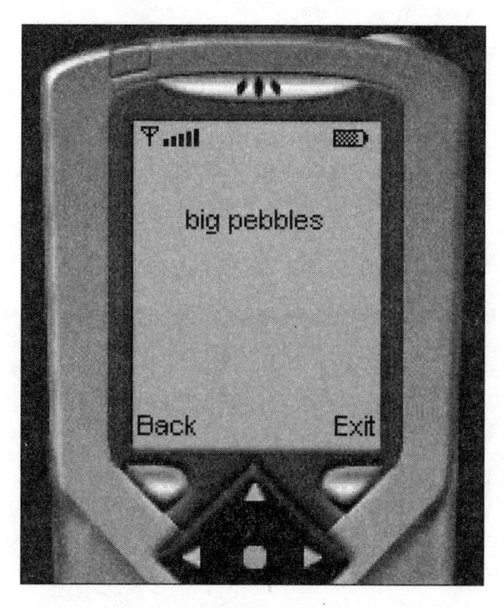

Similarly, for an anchor point of `TOP|RIGHT`, which position the top right corner of the 's' at (50,50):

```
    g.drawString("big pebbles", 50, 50, g.TOP|g.RIGHT);
```

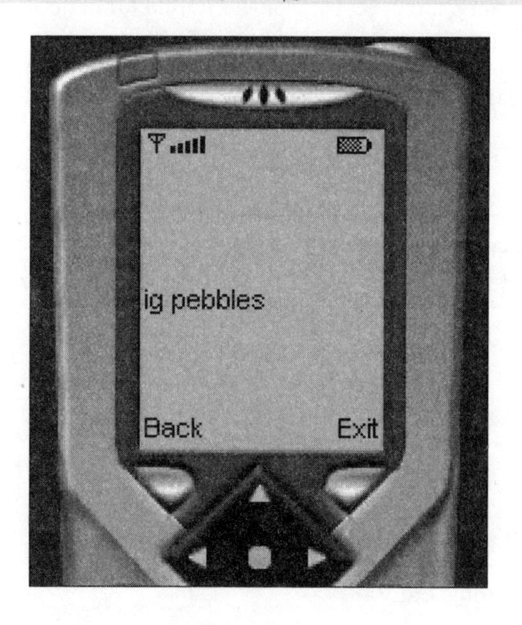

The only relative anchor point constant – the position of the baseline – can be determined using `getBaselinePosition()`, which returns the distance in pixels that the baseline is from the top edge of the bounding box.

There are four methods to render text:

- `drawChar()` renders a single character
- `drawChars()` renders an array of chars
- `drawString()` renders a String
- `drawSubstring(java.lang.String str, int offset, int len, int x, int y, int anchor)` renders a substring by specifying the starting point in the string, plus the number of characters to render

All methods take a text parameter, (x,y) coordinates and an anchor point.

The `Font` class allows us to find the width, in pixels, of a string or sub-string in the current font, using the methods `stringWidth(String)` and `substringWidth(String, int, int)`. We can use this to position our text accurately, and also to check that the complete string will be visible on the display. The width-checking equivalents for characters and character arrays are `charWidth(char)` and `charsWidth(char[], int, int)`.

There is minimal support for internationalization (or I18N as it is called, because there are 18 characters between the 'I' and the 'n') in the MIDP. The underlying CLDC libraries have no support for `java.text`, only for translating Unicode characters to bytes and vice versa, using the `InputStreamReader` and `OutputStreamWriter` classes.

If we truly need I18N, but we don't know which device the user will access our application with, then we are probably not going to be able to create a unified solution on such limited devices. Partial I18N solutions may be possible by downloading new font glyphs onto the device, but this is liable to suck up excessive amounts of memory and processing power if used for multiple fonts for multiple languages. Away from the rendering problems of fonts, devices should present locale-specific calendars, times, phone numbers, and so on without us needing to intervene, which may help in some applications, but we need to use the high-level UI classes for these.

Images

All images are rendered on a `Canvas` using `drawImage(Image img, int x, int y, int anchor)` using the anchor point in a similar manner to the text rendering methods. The only notable differences are that an image can specify `VCENTER` as the vertical constant in the anchor point, but it cannot specify `BASELINE`, since an image has no baseline. The following code centers an image at (50,50):

```
protected void paint(Graphics g) {
  Image i;
  try{
    i = Image.createImage("/myImage.png");
    g.drawImage (i, 50, 50, g.VCENTER | g.HCENTER);
  } catch(java.io.IOException ioe){System.out.println(ioe);}
}
```

We can also create an off-screen image, and using the `Image.getGraphics()` method to obtain a `Graphics` object, which lets us draw into that image, and at some later date, place the image we created/modified onto a visible `Canvas`.

Here's an example of drawing a white circle into an off-screen `Image` of size 50x50, and then placing the off-screen image into a `Canvas`. We then set the `Canvas` as the current `Screen`, therefore making the `Image` visible:

```
Image i;
public void drawOffScreenImage() {
  // create an empty Canvas
  MyCanvas c = new MyCanvas();
  // create a mutable Image
  i = Image.createImage(50,50);
  // get its Graphics
  Graphics g = i.getGraphics();
  // fill with black
  g.setColor(0,0,0);
  g.fillRect(0,0,c.getWidth(),c.getHeight());
  // draw a white circle
  g.setColor(255,255,255);
  g.fillArc(10, 10, 30, 30, 0, 360);
  // attach some commands
  c.addCommand(exitCommand);
  c.addCommand(backCommand);
  c.setCommandListener(this);
  // by calling setCurrent() the Canvas' paint() method is run,
  // causing the image to be copied into the Canvas, thus making it visible
  display.setCurrent(c);
}

class MyCanvas extends Canvas {
  protected void paint(Graphics g) {
    g.drawImage(i, 0, 0, g.TOP|g.LEFT);
  }
}
```

The result looks like this:

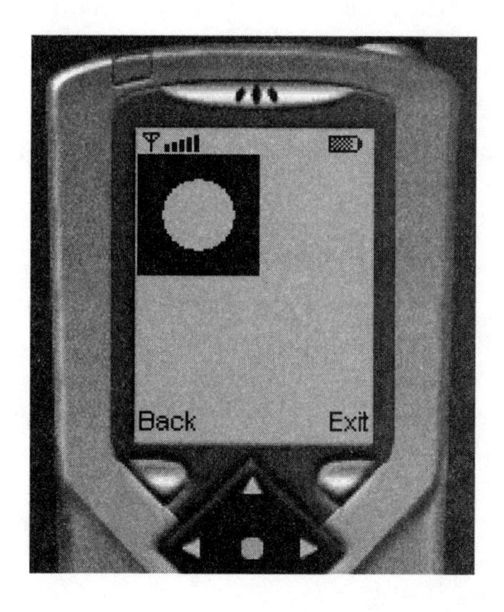

Translations

The origin of the coordinate system can be moved from its default of top left, around the display using `translate(int x, int y)`. Applying a second `translate()` is cumulative, so the following examples both have the same effect of moving the origin to (70,80):

```
translate(70,80);
```

And:

```
translate(50,50);
translate(20,30);
```

Thereafter, any drawing is relative to the point 70,80. We can find the current origin by using `getTranslateX()` and `getTranslateY()`, and, if need be, reset the origin back to (0,0) with:

```
translate(-getTranslateX(),-getTranslateY());
```

Low-Level UI Methods Not Guaranteed to be Portable

Just to sum up this low-level section, the following methods are not guaranteed to be available on all MIDP devices. We have listed the check method we should call before using them:

Potentially Unsupported Methods	Check Method
Any that assume color or grayscale	`Display.isColor()`
Any that assume color or grayscale	`Display.NumColors()`
Any assuming animation capabilities	`Canvas.isDoubleBuffered()`
`Canvas.keyRepeated()`	`Canvas.hasRepeatEvents()`
`Canvas.pointerPressed(int, int)`	`Canvas.hasPointerEvents()`
`Canvas.pointerReleased(int, int)`	`Canvas.hasPointerEvents()`
`Canvas.pointerDragged(int, int)`	`Canvas.hasPointerMotionEvents()`
`Font.getFont(int face, int style, int size)`	`Font.getSize()`, `Font.getStyle()`, `Font.getFace()`, and optionally for checking styles, `Font.isBold()`, `Font.isItalic()`, `Font.isPlain()`, `Font.isUnderlined()`

Now we have explored the differences between the high-level and low-level UI, and seen how different types of applications can use the different features, let's move on to cover the rest of the MIDP functionality, in other words data storage, networking, timers, and finally, general system properties.

The MIDP Record Management System

The MIDP defines an API for simple persistent storage, called the Record Management System (RMS). The RMS is not a database as such, and has no relational or SQL functionality – it simply allows the controlled saving and retrieving of arrays of bytes, called records.

The RMS is composed of zero or more records held in a record store, with each MIDlet suite having zero or more record stores. The record stores are not available to other MIDlets outside its suite, but they can be shared by all the MIDlets within the suite. The mechanism of persistent storage is under the control of the implementation, so there is no guarantee that a record store will survive through battery changes, failures, or reboots, or even if there is actually enough non-volatile memory available.

The RMS has no built-in record-locking, so it is the application's responsibility to implement any locking mechanism, and make sure that multiple applications or threads pointing to a single record do not accidentally overwrite that record's value without the application controlling the writes.

However, the RMS does guarantee to serially write the records, and not be interrupted halfway through by a subsequent write, so at least the store will not be corrupt with a mix of data. The solution is have a single application (or method) in the MIDlet suite that controls the RMS, and any other MIDlets that want to store and retrieve data should do so via this one route, therefore ensuring atomicity of RMS access.

All RMS classes are held in the `javax.microedition.rms` package.

Record and Record Store Creation

A record store is opened using `javax.microedition.rms. openRecordStore(String recordStoreName, boolean createIfNecessary)`. The record store's name must be 32 characters or less, and must be unique within its MIDlet suite. If the `createIfNecessary` flag is set to `true`, the record store should be created as well, if is does not already exist. To close a record store, use `closeRecordStore()`. If a store is opened twice, it must be closed twice too. To permanently delete a whole record store, use `deleteRecordStore(String)`. Only closed stores can be deleted.

The records themselves are a simple array of bytes, so there is no `Record` class defined in MIDP. Records can be added to an open record store with `addRecord(byte[] data, int offset, int numBytes)`.

In our first RMS example, `RMSMIDLet1`, we save three names to a newly created record store, and then list the names out in a `Form`. We have listed the full application below, for completeness:

```
package com.poqit.j2me;

import javax.microedition.midlet.MIDlet;
import javax.microedition.lcdui.CommandListener;
import javax.microedition.lcdui.Command;
import javax.microedition.lcdui.Display;
import javax.microedition.lcdui.Displayable;
import javax.microedition.lcdui.Form;

import javax.microedition.rms.RecordStore;
import javax.microedition.rms.RecordStoreFullException;
import javax.microedition.rms.RecordStoreNotOpenException;
import javax.microedition.rms.RecordStoreException;
import javax.microedition.rms.RecordEnumeration;
```

```
public class RMSMIDlet1 extends MIDlet implements CommandListener {

  // the name of our record store
  private static final String recordStoreName = "names";

  private Command exitCommand;
  private Display display;
  private RecordStore recordStore;

  public RMSMIDlet1() {
    display = Display.getDisplay(this);
    exitCommand = new Command("Exit", Command.EXIT, 1);
  }
```

The `RecordStore`, `rs`, is created on the very first call to `startApp()`, as well as being opened. Note that we must catch a `RecordStoreException`, in case there is some problem, such as the store being full, or (if we did not set `createIfNecessary` flag to true) the store isn't found. For any subsequent times that we start the application (or whenever it becomes the foreground process), the record store is just opened, not created.

Note that we always create new 3 records whenever `startApp()` is run. This is obviously not desirable in a real application, because, as we will see, we get multiple records being created whenever it is run or comes to the foreground. However, it makes our application as simple as possible:

```
  public void startApp() {
    try {
      recordStore = RecordStore.openRecordStore(recordStoreName, true);
    } catch (RecordStoreException rse) {
      System.out
        .println("startApp: couldn't create or open the record store.");
      System.out.println(rse);
      rse.printStackTrace();
      notifyDestroyed();
    }
```

Having opened the record store, we then call `createRecords()` to create 5 simple records – each containing a person's name:

```
    // create some records
    createRecords();
```

Finally, we display the records by calling `viewRecords()`, which we discuss below:

```
    // now retrieve them
    viewRecords();
  }
```

To create the records, we call `RecordStore.add()`, being sure to catch the potential exception again:

```
public void createRecords() {
  String name1 = "Julia";
  String name2 = "Jamie";
  String name3 = "Alys";
  String name4 = "Richard";
  String name5 = "09234234";
  try {
    byte[] b = name1.getBytes();
    recordStore.addRecord(b, 0, b.length);
    b = name2.getBytes();
    recordStore.addRecord(b, 0, b.length);
    b = name3.getBytes();
    recordStore.addRecord(b, 0, b.length);
    b = name4.getBytes();
    recordStore.addRecord(b, 0, b.length);
    b = name5.getBytes();
    recordStore.addRecord(b, 0, b.length);
  } catch (RecordStoreException rse) {
    System.out.println("save: couldn't save records to store.");
  }
}

public void viewRecords() {
  Form form = new Form("All Records:");

  RecordEnumeration enum = null;
  try {
    enum = recordStore.enumerateRecords(null, null, false);
    byte[] b;

    // iterate through the records one by one, until none are left
    // the Enumerator throws an exception when it can't find another record
    while (enum.hasNextElement()) {

      // get the record's bytes
      b = enum.nextRecord();

      // add it to our Form
      form.append(new String(b) + "\n");
    }
  } catch (RecordStoreException rse) {
    System.out
      .println("viewAllNames: record store doesn't exist or is not open");
    notifyDestroyed();
  }

  // release the Enumerator
  enum.destroy();

  form.addCommand(exitCommand);
  form.setCommandListener(this);

  display.setCurrent(form);
}
```

```
    // free up the record store whilst we are not the foreground app.
    // startApp will open it again when we come back to the foreground.
    public void pauseApp() {
      try {
        recordStore.closeRecordStore();
      } catch (RecordStoreException rse) {
        System.out.println("pauseApp: couldn't close record store.");
        notifyDestroyed();
      }
    }

    // close the record store for the last time, before exiting
    public void destroyApp(boolean unconditional) {
      try {
        recordStore.closeRecordStore();
      } catch (RecordStoreException rse) {
        System.out.println("destroyApp: couldn't close record store.");
        notifyDestroyed();
      }
    }

    public void commandAction(Command c, Displayable s) {
      notifyDestroyed();
    }
  }
```

The only other point to note is that because opening a record store is a potentially resource-hungry process, we should consider closing the store whenever we are not the foreground application, and also close it when exiting the application. Therefore, our `pauseApp()` and `destroyApp()` methods close the store, again catching any exceptions.

Record Retrieval

There are two ways to retrieve records:

- By their unique ID, which acts as the primary key
- By an enumerator, `RecordEnumerator`

The record IDs start at 1 and increment by one for every new record added to the record store. When records are later deleted, the 'empty' ID is not made available to the next new record added to the record store. Therefore, we cannot rely on IDs being contiguous. However, we can find out how many records have been created in the total lifetime of the record store, by using the method `getNextRecordID()`, and we may find the ID useful as a guide to chronological order of the records.

To retrieve a specific record:

```
byte[] myBuffer = recordStore.getRecord(id);
```

Alternatively, we can copy the record contents directly into a buffer at a certain offset. Here we copy the first record into the 100th byte of myBuffer, onwards:

```
byte[] myBuffer = new byte[200];
int offset = 100;
int id = 1;
int bytesCopied = rs.getRecord(id, myBuffer, offset);
```

If we intend to retrieve many records, then the most efficient way is using RecordEnumerator. RecordEnumerator shares similar features to the standard java.util.Enumeration, but curiously doesn't actually sub-class it. Here is a comparison of methods:

- hasNextElement() is equivalent to Enumeration.hasMoreElements()
- nextRecord() is equivalent to Enumeration.nextElement()

To retrieve our records, we use these two basic methods in viewRecords(), and work through the record store, adding each name retrieved into a Form, which is then displayed:

```
while(re.hasNextElement()) {
  b = re.nextRecord();
  form.append(new String(b) + "\n");
}
```

In addition to these two most frequently used methods, we can also work backwards through a record store with hasPreviousElement() and previousRecord(). We can also ask for the total number of records by calling numRecords().

RecordEnumerator can optionally keep its index into the records up to date even while records are being added or deleted. This comes with a caveat However that there may be a performance hit. The alternative is to call rebuild() whenever we know we have added records, and want the RecordEnumerator updated. The RecordEnumerator can be reset at any time by calling reset(), and iterate through the complete record set again.

Modifying and Deleting Records

Records are modified by calling setRecord(), passing in the new contents of the record as a byte array, like so:

```
name = "Fred";
byte[] myBuffer = name.getBytes();
int offset = 0;
recordStore.setRecord (recordID, myBuffer, offset, myBuffer.length);
```

Note that the offset is the point to start reading data from myBuffer, **not** the position to store into the existing record.

To delete a record, simply call:

```
recordStore.deleteRecord(recordID);
```

In the Wrox downloads (http://www.wrox.com), you will find an expanded example of RMSMIDlet1 (imaginatively named RMSMIDlet2), which allows us to not only to enter new names into a database, but also to change them and delete them from the record store. The code is too long to reproduce here, but the majority of it is straightforward, once we understand the basics described above.

Deleting a Record Store

A complete record store can be deleted using deleteRecordStore(), as long as it is closed first:

```
recordStore.closeRecordStore();
RecordStore.deleteRecordStore(recordStoreName);
```

Also note that with the J2MEWTK, all record stores can be deleted from the KToolbar, by using the File | Clean Database command.

Filtering and Sorting Records

In combination with RecordEnumerator, we can implement the RecordFilter interface to define a filter to find selected records from the record store, and also the RecordComparator interface which allows us to define a record-sorting algorithm. To use RecordFilter we must implement the matches() method, which returns true or false depending on whether we consider the record passed in matches our criteria or not. To use RecordComparator, we must compare the two records passed in, and sort them according to our criteria into one of the following results:

- EQUIVALENT
  Considered to be equal in order, although they do not necessarily have to be the same record
- FOLLOWS
  The first record follows the second record in order
- PRECEDES
  The first record precedes the second record in order

To get the RecordEnumerator, we pass in our previously created RecordFilter and RecordComparator, like so:

```
Boolean keepUpdated = false;
RecordEnumeration enum = recordStore.enumerateRecords(myFilter, myComparator,
keepUpdated);
```

The RecordEnumerator, instead of enumerating through every record, now only enumerates through those that match the criteria set by our implementations of RecordFilter and RecordComparator. It is the RecordStore that calls our RecordFilter and RecordComparator code, we do not call it ourselves.

Here is an extract from RMSMIDlet3, where the content of our record store can be filtered and sorted by various means. Note that we must first create the record store, by using either RMSMIDlet1 or RMSMIDlet2, and that to share a record store, these MIDlets must all be in the same MIDlet suite.

First, the RecordFilter implementation, which in our case, filters records by comparing the first character of each record to a low and high character. For example, if the low character is 'a' and the high' character is 'l' then passing a record in containing "Alys" will return true, whereas "Richard" will return false. Here is the call to create our implementation of RecordFilter:

```
RecordFilter rf = new Filter('a', 'l');
```

The `Filter` class itself:

```
class Filter implements RecordFilter {
  private char lowChar;
  private char highChar;

  // Constructor
  Filter(char low, char high) {
    lowChar = Character.toLowerCase(low);
    highChar = Character.toLowerCase(high);
  }

  // returns true if the record is within in the filter range low->high
  public boolean matches(byte[] record) {

    // convert to String first, because that's how we saved it.
    String str = new String(record).toLowerCase();
    char ch = str.charAt(0);
    if (ch < lowChar) {
      return false;
    }
    if (ch > highChar) {
      return false;
    }
    return true;
  }
}
```

The `RecordComparator` has a similar algorithm, but it compares the two records together rather than with 'high' and 'low' characters:

```
RecordComparator rc = new Comparator();
```

```
class Comparator implements RecordComparator {
  public int compare(byte[] record1, byte[] record2) {
    String str1 = new String(record1).toLowerCase();
    String str2 = new String(record2).toLowerCase();
    if (str1.compareTo(str2) < 0) {
      return RecordComparator.PRECEDES;
    }
    if (str1.compareTo(str2) > 0) {
      return RecordComparator.FOLLOWS;
    }
    return RecordComparator.EQUIVALENT;
  }
}
```

Here are the methods to create and use the filter and comparator. The first method, `viewOptions()` gives the user a choice of actions to perform on the records, such as filtering only records between 'a' and 'l'. Each option maps to `viewAll()`, `viewA_L()` and `viewM_Z()`, and `viewOther()` respectively. Each of the specific `viewXXX()` methods creates a `RecordFilter` and a `RecordComparator`, and then passes them into the generic `view()` method.

`view()` simply iterates through all records and displays them on a `Form`, but this time passes a filter and a comparator into the `recordStore.enumerateRecords()` call:

```
    public void viewOptions() {
      List list = new List(MAIN_LIST_TITLE, List.IMPLICIT);
      list.append(VIEW_ALL, null);
      list.append(VIEW_A_L, null);
      list.append(VIEW_M_Z, null);
      list.append(VIEW_OTHER, null);
      list.addCommand(exitCommand);
      list.setCommandListener(this);
      display.setCurrent(list);
    }

    // view all entries, unsorted
    public void viewAll() {
      RecordFilter rf = null;
      RecordComparator rc = null;
      view(rf, rc);
    }

    // view all entries between "a" and "l"
    public void viewA_L() {
      RecordFilter rf = new Filter('a', 'l');
      RecordComparator rc = new Comparator();
      view(rf, rc);
    }

    // View all entries between "m" and "zzzzzzzzz"
    public void viewM_Z() {
      RecordFilter rf = new Filter('m', 'z');
      RecordComparator rc = new Comparator();
      view(rf, rc);
    }

    public void viewOther() {
      RecordFilter rf = new NegativeFilter('a', 'z');
      RecordComparator rc = new Comparator();
      view(rf, rc);
    }

    public void view(RecordFilter rf, RecordComparator rc) {
      Form form = new Form("Result:");

      RecordEnumeration enum = null;
      try {

        // need to keep tabs the record ID for each name in the ChoiceGroup
        recordIDs = new int[recordStore.getNumRecords()];
        int i = 0;

        // pass the filter and comparator in to the Enumerator.
        // The Enumerator does all the filter and comparator calls for us, so
        // we need never call them directly.
        enum = recordStore.enumerateRecords(rf, rc, false);
```

```
        byte[] b;
        while (enum.hasNextElement()) {
          recordIDs[i] = enum.nextRecordId();
          b = recordStore.getRecord(recordIDs[i]);
          form.append(new String(b) + "\n");
          i++;
        }
      } catch (RecordStoreException rse) {
        System.out.println("View: record store doesn't exist or is not open");
        notifyDestroyed();
      }
      enum.destroy();

      form.addCommand(exitCommand);
      form.addCommand(backCommand);
      form.setCommandListener(this);
      display.setCurrent(form);
    }
```

RecordListener Events

Records being added, changed, or deleted can fire off an event, just like commands and `IMPLICIT` lists. These `RecordStore` events can be caught by the application by using the `RecordListener` class. The `RecordListener` is enabled by the `addRecordListener(RecordListener)` method in `RecordStore`. This could be an interesting way of communicating between MIDlets in the same suite. One MIDlet could 'listen' for record store changes made by the other MIDlets in the suite, and act upon that information.

As with other MIDP events, the `RecordListener` events are guaranteed to occur in the order they are fired off.

Synchronization Support

Record stores (but not individual records) are time-stamped when saved or modified, and also the record store's version number is incremented on each change. Both these features are vital if the data stored is to be synchronized at some later date, with say, a central server, or maybe a user's desktop PC. The relevant methods are `getVersion()`, where the version increments by one on every change, and `getLastModified()`, which returns the date in milliseconds from midnight GMT 1st Jan 1970, as per the standard Java `System.currentTimeMillis()` method.

CLDC and MIDP System Properties

We cover these at this point, because one of the main uses of system properties is as HTTP header information, which we discuss in the next section.

CLDC System Properties

Although the CLDC does not support the standard J2SE `java.util.Properties` class, it does have a subset of the functionality for accessing system properties. All the system properties are accessed by a call to the `java.lang.System.getProperty()` method, as per normal J2SE operations.

The CLDC standard system properties are:

- `microedition.platform`
  The name of the host platform or device, and its default is `null`. The vendor of the platform is expected to set this property. This is vital for MIDlets that want to use vendor-specific extensions, and need to be able to positively identify their platform, rather than guessing from information such as screen size and other low-level UI hints.
- `microedition.encoding`
  The default character encoding, and is set to "`ISO8859_1`".
- `microedition.configuration`
  The name and version of the supported configuration, with the default being "`CLDC-1.0`".
- `microedition.profiles`
  A space-separated list of profiles supported by this device, and it has a default value of `null`. Any profiles sitting upon the CLDC are expected to set this property (see below).

Note that profiles based upon the CLDC can define additional system properties. MIDP does just this, as we now see.

MIDP System Properties

The MIDP system properties are also accessed using the `java.lang.System.getProperty()` method, and are:

- `microedition.locale`
- `microedition.profiles` – overrides CDLC value of `null`

`microedition.locale` defines the current locale of the device, and must contain:

- `language`
- `country code`
- `variant`

The language code must be a two lower case letters as defined by ISO-639, and the country code as two upper case letters, as defined by the standard ISO-3166.

Examples of valid values for locale are:

- `en-US`
- `en-GB`
- `fr-FR`
- `fr-CA`

`microedition.profiles` is set to "`MIDP-1.0`" or a similar version String, rather than being null as the CLDC defaults it to.

Networking

We introduce the MIDP `HttpConnection` class here for completeness, but do not attempt to give full coverage at this point. Chapter 11 contains several working examples of how to connect MIDP devices to servlet-enabled web servers.

Although the CLDC defines several interfaces devoted to connectivity and networking in its `javax.microedtion.io` package, it implements none of them, leaving this job to the profiles that sit on top of it. In the MIDP's case, only a subset of the HTTP 1.1 protocol is actually mandatory, and that is using `javax.microedition.io.HttpConnection`.

It is a subset, and not a complete implementation because only the following requests are allowed:

- `GET`
- `POST`
- `HEAD`

These three requests probably cover 99% of all our requirements, However so the limitation is not great. As with everything else in the MIDP, the actual implementation details are of no consequence to us as third party developers. Whether the device uses an underlying WAP protocol, or full-blown end-to-end TCP/IP does not affect how we code our applications. All we know, and trust, is the ability to connect to a remote service using the HTTP 1.1 protocol. Therefore, we will only concentrate on the `HTTPConnection` class, which specifically supports HTTP.

First, we need to open the `HttpConnection`, and specify which request type we are using – `GET` in this case:

```
HttpConnection hc = (HttpConnection)Connector.open(url);
hc.setRequestMethod(HttpConnection.GET);
```

Also, before moving to the 'Connected' state, as the MIDP specification calls it (in other words causing our request to be sent to a server), we can optionally set some request parameters, such as '`User-Agent`', which the receiving server may use to tailor its reply to us (such as delivering a shorter reply because it recognizes it is a limited J2ME device that is requesting):

```
String profiles = System.getProperty("microedition.profiles");
String configuration = System.getProperty("microedition.configuration");
hc.setRequestProperty("User-Agent", "Profile/"
                                    + profiles
                                    + " Configuration/"
                                    + configuration);
```

There is no limitation to the headers we can define, other than the MIDP specification recommends that the '`User-Agent`' and '`Content-Language`' header should contain the values of the system properties `microedition.profiles`, `microedition.configuration`, and `microedition.locale`.

Last of all, we activate the `Connection` (the 'connected' state) by opening the `InputStream`:

```
InputStream is = hc.openInputStream();
```

Calling `openInputStream()` causes our request to be sent, and we are now ready to read the result sent back by the server. If we can read the length of the reply (by calling `getLength()`, which reads the 'content-length' HTTP header), then it's possible to read all the reply characters in one go. If the length is not known, then we read a character at a time, until –1 is returned, indicating the end of the reply stream, building up the reply in a `StringBuffer`, character by character:

```
InputStreamReader isReader = new InputStreamReader(is);
int len = (int)hc.getLength();
if(len > 0) {
  char[] chars = new char[len];
  int actual = isReader.read(chars);
  // convert chars to a String
  result = new String(chars);
} else {
  StringBuffer sb = new StringBuffer();
  int ch;
  while((ch = is.read()) != -1) {
    sb.append((char)ch);
  }
  result = sb.toString();
}
```

We can also read the content type and headers of the reply too, rather than just reading only the data part of the HTTP message.

In our example below, an extract from the `ReadFile` example in **Interfacing With Servlets** (Chapter 11), we request a file from an HTTP server. The only parameter passed in to this `read()` method is the URL of the text file (it can be an HTML file, for example). We are passed back a String containing the contents of the requested file, or possibly an error String. The only additions to the code snippets above, are copious amounts of exception handling, to try and diagnose any problems, and also to make sure that all resources are guaranteed to be closed on exiting the method:

```
public final static String read(String url) {
  String result = null;
  HttpConnection hc = null;
  InputStream is = null;
  try {
    hc = (HttpConnection)Connector.open(url);
    hc.setRequestMethod(HttpConnection.GET);
    // Send a little information about ourselves.
    String profiles = System.getProperty("microedition.profiles");
    String configuration = System.getProperty("microedition.configuration");
    // Optional, so may be null.
    String locale = System.getProperty("microedition.locale");
    // Set by the device manufacturer. May be null.
    String platform = System.getProperty("microedition.platform");
    hc.setRequestProperty("User-Agent", "Profile/"
                                  + profiles
                                  + " Configuration/"
                                  + configuration);
    if(locale != null) {
      hc.setRequestProperty("Content-Language", locale);
    }
    if(platform != null) {
```

```
        hc.setRequestProperty("J2ME-Platform", platform);
      }
      // find the length of the text file
      try {
        is = hc.openInputStream();
        InputStreamReader isReader = new InputStreamReader(is);
        int len = (int)hc.getLength();
        if(len > 0) {
          char[] chars = new char[len];
          int actual = isReader.read(chars);
          // convert chars to a String
          result = new String(chars);
        } else {
          StringBuffer sb = new StringBuffer();
          int ch;
          while((ch = is.read()) != -1) {
            sb.append((char)ch);
          }
          result = sb.toString();
        }
      } catch(IOException e) {
        result = "Could not open the InputStream. Check your web server is
running.";
      } finally {
        if(is != null) {
          is.close();
        }
        if (hc != null) {
          hc.close();
        }
      }
    } catch(IOException ioe) {
      result = "The URL is probably invalid";
    }
    return result;
  }
```

When we wish to POST data to the server rather than receive it, we need to send that data in addition to setting the request type and any request parameters. We do this by obtaining an OutputStream by calling HttpConnecton. openOutputStream().

Again, from an example in Chapter 11, this time ServetWrite, we show the write() method, which takes a URL – the target servlet in this case, and also a String to send to that servlet as the data part of the POST request:

```
  public final static String write(String url, String content) {
    String result = null;
    try {
      boolean status = false;
      HttpConnection hc = null;
      InputStream is = null;
      OutputStream os = null;

      hc = (HttpConnection)Connector.open(url);
      hc.setRequestMethod(HttpConnection.POST);
```

```
String profiles = System.getProperty("microedition.profiles");
String configuration = System.getProperty("microedition.configuration");
String locale = System.getProperty("microedition.locale");
String platform = System.getProperty("microedition.platform");
hc.setRequestProperty("User-Agent", "Profile/"
                                    + profiles
                                    + " Configuration/"
                                    + configuration);
if(locale != null) {
  hc.setRequestProperty("Content-Language", locale);
} if(platform != null) {
  hc.setRequestProperty("J2ME-Platform", platform);
}
try {
  os = hc.openOutputStream();
  OutputStreamWriter osWriter = new OutputStreamWriter(os);

  osWriter.write(content);
```

Having built and sent the request, we can optionally open the `InputStream` and read the servlet's reply – maybe a status message confirming success. This code is identical to the `GET` request reading code above.

Timers

MIDP has the same `Timer` support as J2SE/JDK 1.3 (but not earlier versions) – in other words the `java.util.Timer` and `java.util. TimerTask` classes. `Timer` and `TimerTask` allow an application to schedule a background task to occur, either once only, or at regular set intervals. `Timer` is the scheduling class, and the application implements the `TimerTask` interface to define what code runs whenever the scheduled event occurs.

Here is a simple example of a news flash, which pops up a `Form` for a few seconds, and disappears again. When initially started, the first `Form`, `form`, created in `startApp()` stays visible for twice the `Timer` delay, then the correct timing kicks in. Here is the code:

```
package com.poqit.j2me;

import java.util.Timer;
import java.util.TimerTask;
import javax.microedition.midlet.MIDlet;
import javax.microedition.lcdui.Form;
import javax.microedition.lcdui.Canvas;
import javax.microedition.lcdui.Command;
import javax.microedition.lcdui.Display;
import javax.microedition.lcdui.Displayable;
import javax.microedition.lcdui.CommandListener;

public class TimerMIDlet extends MIDlet implements CommandListener {
  private Command exitCommand;
  private Display display;
  private Timer tim;
```

```
    public TimerMIDlet() {
      display = Display.getDisplay(this);
      exitCommand = new Command("Exit", Command.EXIT, 1);
    }

    public void startApp() {
      Form form = new Form("Some Little App");
      form.append("An application minding its own business...");
      form.addCommand(exitCommand);
      form.setCommandListener(this);
      display.setCurrent(form);
```

The call to `createNewsFlash(form, 30, 4000)` creates a `Form` which pops up every 4000 milliseconds, and stays visible for 30% of those 4000 milliseconds, or approximately 1 second:

```
      createNewsFlash(form, 30, 4000);
    }
```

`createNewsFlash()` creates a `Timer` and two associated `TimerTask` helpers, one to make the newsflash `Form` visible, and the other to revert to the original `Displayable`. The `TimerTask` implementation, `Flash`, simply sets the current displayable and returns immediately:

```
    public void createNewsFlash(Displayable displayable, int percentVisible,
                                long period) {

      Form newsFlash = new Form("News Flash!");
      newsFlash.append("This newsflash lasts for approximately "
                   + percentVisible * period / 100000 + " seconds.");
      Flash newsFlashOn = new Flash(newsFlash);
      Flash newsFlashOff = new Flash(displayable);
      tim = new Timer();
      tim.schedule(newsFlashOn, period, period);
      tim.schedule(newsFlashOff, period + (percentVisible * period / 100),
                  period);
    }

    class Flash extends TimerTask {
      Displayable currentDisplayable;

      Flash(Displayable displayable) {
        currentDisplayable = displayable;
      }

      public void run() {
        display.setCurrent(currentDisplayable);
      }
    }

    public void commandAction(Command c, Displayable s) {
      notifyDestroyed();
    }

    public void pauseApp() {
      tim.cancel();
```

```
    }

    public void destroyApp(boolean unconditional) {
      tim.cancel();
    }
  }
```

Starting a MIDlet from Another MIDlet

For some reason, many developers seem interested in starting another MIDlet off from the current one. This can be done, as we show below. However, because the subsequent MIDlets we fire off are in our control and not the device's application manager, it is probably of limited use, unless we intend writing our own application-manager-in-a-MIDlet.

Here is a `commandAction()` method that creates a new MIDlet instance (which must be one in the same MIDlet suite), and then exits, leaving the new MIDlet, `midlet2`, running in its place:

```
  public void commandAction(Command c, Displayable s) {
    try {
      Class midletClass = Class.forName("LowLevelMIDlet");
      LowLevelMIDlet midlet2 = (LowLevelMIDlet)midletClass.newInstance();
      midlet2.startApp();
    } catch(Exception e) {
        e.printStackTrace();
    } finally {
        notifyDestroyed();
    }
  }
```

Downloading Images

Downloading images is no different to loading any other data. Only PNG images are currently supported by MIDP, though. Here is basis for a method to load image data from a web server, and create an image from that loaded data, using `createImage()`. This code can be placed into the networking examples earlier in the chapter. It plugs in just after we have opened the `HttpConnection`, `hc`. Note that we use a `DataInputStream` rather than an `InputStreamReader`:

```
    int len = (int)hc.getLength();
    if(len > 0) {
      byte[] imageData = new byte[len];
      is = hc.openInputStream();
      dis = new DataInputStream(is);
      // load the image data in one go
      int actual = dis.read(imageData);
      loadedImage = Image.createImage(imageData, 0, (int)len);
    }
```

Decreasing the Size of an Application

Code obfuscators, which are normally used to make Java applications difficult to copy, are useful tools for decreasing an application's size. They change our (usually) readable variable names into shorter, obscure names, which take up less space. Examples include, HashJava (http://www.meurrens.org/ip-Links/java/codeEngineering/blackDown/hashjava.html) The JBuilder IDE (http://www.borland.com/jbuilder/) and Jshrink (http://www.e-t.com/jshrink.html).

The Missing Pieces – What MIDP Doesn't Cover

Several areas of limited mobile device functionality are not covered by the MIDP, usually because it is impossible to enforce adherence to these functions across such a diverse set of devices. We cover these, and any effect they may have.

System-level APIs

At some level, the MIDP designers needed to draw a line across what functionality could be assumed, and what could not. For example, voice input and output, available on some PDAs and mobiles, has purposefully been left out because although there are system-level APIs available, they are not universal enough to abstract that layer out into the MIDP. The same goes for power management functions, telephony functionality, mobile Subscriber Identifier Module (SIM) access and interacting with common OEM software such as WAP browser functions, such as setting and getting bookmarks.

As devices improve in functionality, and new features appear, we can expect the profiles to incorporate these changes over the years, or have completely new profiles defined for those new devices.

Application Delivery and Management

The method of application delivery to the target device, and management after being loaded, are both outside the scope of the MIDP specification, even though at the CLDC level, the JAM has been offered as an example solution by Sun. In particular, there is no guarantee that across devices they all have any form of application filing system (as opposed to RMS) to help store and manage the application.

Another unknown, is whether the device is capable of receiving a complete application across an intermittent network connection. If the application size is relatively large in comparison to the intermittent nature of the connection, it may be only be possible to load some applications via a more reliable connection such as a serial communication port or infra-red link.

Security

The MIDP specifies no additional security above and beyond that of the CLDC, and certainly doesn't attempt to define any end-to-end security at the application level. If a secure transport or connection is needed, then we must either use a third party set of libraries, or create our own (see the resources at the end of this chapter for SSL libraries).

There is nothing stopping OEMs, third parties or ourselves from adding libraries above and beyond those that are specified in the MIDP, to cover the areas above. Of course, our application may no longer run on all devices unless the library supplier can port them over to all the other devices we intend to target.

It is generally unlikely that more limited devices (phones, for example) will allow third-party developers to add libraries onto their devices. Although some open-platform devices are starting to be announced, most will be limited to the providers choice of libraries, even if MIDlets can be freely added to the device.

User Interface

Having specified a low-level user interface, the MIDP offers reasonable control over the user interface. The MIDP specification does not However define the exact look of the high-level user interface. This could well lead to a repeat of the WAP 'browser wars', for example one type of phone implements a `TextBox` as a direct user input area, whereas a second phone type defines it as a completely new `Screen`. This is inevitable, and the MIDP team was probably right to avoid it, but nevertheless, it could be a problem in future, because our application doesn't quite appear as we might expect on all platforms. We could write our application to support several look-and-feels, but the downside will be application bloat, which may use more storage than the device (or user) thinks is acceptable.

If we need a user interface widget that appears the same on all devices, then we must create our own using the low-level `Canvas` and `Graphic` classes, or use a third party UI library, but expect higher memory usage and decreased performance.

There is no support for reading pixel values, which can be quite a limitation when doing advanced graphics. It means that we need to remember what values we wrote into the display, which is very CPU and memory-intensive.

Another missing feature is the ability to mix the high and low-level UI APIs together on the same `Screen` – maybe something like an application-defined `Item` would have allowed low-level graphics to be added to a `Form`.

One last omission to note, is that there is no equivalent of a pseudo-hyperlink in a Form, in other words a simple piece of highlighted text that allows the user to navigate elsewhere, rather than having to use a soft key. A disappointing omission that could have easily been incorporated by the definition of a new `Item` type. We could attach a `Listener` so that whenever a user pressed the device's default 'OK/Accept' button the application could decide what action to take.

An example of this missing feature, is when scrolling through a long list, and we don't really want to display all 100 entries, just the first 10. We can add a hyperlink `Item` at the bottom of the `Form`, and let the user decide whether they want to view the next 10 by selecting the hyperlink.

With the current `Form` and `Item` implementations, we must define a 'More...' command. This solution often means additional key presses for the user, because the 'More...' doesn't necessarily get mapped direct to a soft key, but to a `Menu`, which the user has to go in to, and then select it before they can get the next 10 items. This is very tedious for the user, and a waste of processing power. One of the greatest challenges of mobile applications is that the user is often naturally in a rush – they want the information immediately, and pressing just one more time may be enough to put them off our application.

In effect, adding a hyperlink `Item` would improve the user interface of any MIDP device because instead of just having two navigation choices available from single key presses (in other words the commands mapped to the expected standard of two soft keys), we have three. The 'OK/Accept' button becomes the third navigation choice. In terms of user acceptance, many mobile phones already work this way – the 'OK/Accept' button is the default navigation choice.

We have now covered MIDP in enough detail to understand it well, so we will move on to debugging, development environments, the limitations of MIDP, and finally the future of MIDP.

Debugging

Debugging is not easy in J2ME. Currently, most toolkits have no support for it, since the code runs on a remote unit, in early 2001, However Sun did release the KVM Debug Wire Protocol (KDWP) specification, which defines how a CLDC device can support source code-level debugging. The specification is included with the CLDC 1.0.2 download, and is directed towards IDE vendors and manufacturers, rather than developers.

In the words of the specification, "The KVM Debug Wire Protocol (KDWP) is the protocol that is used for communication between a Debug Agent (DA) and a CLDC-compliant J2ME Java Virtual Machine (usually KVM)."

However, because of memory constraints on J2ME devices, support for the standard JVMDI (Java Virtual Machine Debug Interface) and the full JDWP (Java Debug Wire Protocol) is not possible. Instead, KDWP was created. It is a subset of JDWP, and is derived directly from the JDWP Specification (see http://java.sun.com/products/jpda/doc/jdwp-spec.html).

At the time of writing, only Zucotto's Whiteboard 2.0 IDE appeared to support full MIDP source code debugging (see below), although RIM will follow soon, as no doubt, will other vendors, until IDEs start supporting KDWP However, we currently have to revert to using good old-fashioned methods such as `System.out.println()` statements, which we can optionally wrap up to make sure they do not reach the final production version of our application.

If we have lots of generic (CLDC) code, then we could consider writing a J2SE equivalent front-end (simulator) that exercises our code on a standard Java 2 platform before moving it over to running on J2ME. Hallvard Trætteberg has already written one at http://www.idi.ntnu.no/~hal/development/palm/PalmApp.html.

Development Platforms, Emulators and Tools

At the time of writing, there are still very few alternative MIDP implementations to Sun's J2MEWTK. There are only two at the present moment, although RIM are part-way MIDP-compatible at this time, and intend to be fully compliant before releasing their commercial version (see below):

- Motorola & Metrowerks CodeWarrior – http://www.metrowerks.com/products/palm. Motorola have specified Metrowerks as the IDE for their MIDP-compliant iDEN platform.
- Zucotto – http://www.zucotto.com
- RIM – http://developer.rim.net.

  The early access RIM SDK is currently based upon a mixture of CLDC, MIDP database classes, and also RIM-specific UI classes. There is no HTTP support yet, but the intention is to become fully MIDP compatible, which will add this functionality.

  The one major advantage over the Sun reference MIDP (and J2ME Wireless Toolkit) implementation is the ability to debug applications. The RIM IDE and simulator, are tightly integrated, letting us set breakpoints, single step through our code, skip over method calls, check variable values, set watches on those values, and so on. It's relatively lightweight in processor requirements, and kicks off rather quicker than, say, Forte or Borland, should we be running on something less than an 800-1000Mhz PC.

Once this becomes fully MIDP-compatible, this could be an invaluable toolkit.

Other platforms that worth considering for CLDC development, rather than MIDP, are:

- Borland JBuilder Handheld Express (Palm only) – Can be found at http://www.borland.com/jbuilder/hhe.
- LG Telecom – A Korean mobile operator, which has LG and British Telecom (BT) as major shareholders. They also manufacture (or rather LG manufactures) mobile devices, such as the I-Book. LG's devices use Sun's KittyHawk library. Their ez-java emulator is available from http://java.ez-i.co.kr. Currently, there is minimal English documentation.
- Color KVM – The Color KVM is a useful port of the standard CLDC KVM for the Palm, to include greyscale and color support. Can be found at http://www.kawt.de.
- KVM for Linux – The KVM has been ported to the Linux platform, since Sun only offered Windows and Solaris reference implementations, http://www.extreme-java.de.

Alternative KVMs

There are several alternative 'micro' JVMs to Sun's reference implementations, but because of the additional `Connection` classes defined specifically for CLDC, none currently support J2ME. Features to look for, are increased performance over Sun's reference implementation, and also better support for specific platforms, such as Palm and PocketPC. Of course, the down-side is that if we do use the extra functionality, we need to be sure to isolate it in our application for removal/ replacement in future versions.

3rd Party Class Libraries

There is already a raft of third party libraries available for the CLDC and MIDP, which in itself is very encouraging. Here is a selection of the most useful and adventurous libraries currently available:

- AWT User Interface – http://www.kawt.de.
- SSL – http://playground.sun.com/~vgupta/KSSL.
- JDBC driver – http://www.alphaworks.ibm.com/tech/jdbcme.
- Floating point math – http://www.jscience.net.
- XML parsers:
  - KXML – http://www.kxml.org.
  - TinyXML for KVM – http://www.gibaradunn.srac.org/tiny/index.shtml.
  - NanoXML – http://nanoxml.sourceforge.net.
- Java Message Service – http://www.softwired-inc.com. See Chapter 12.
- 3D wireframe – http://www.sourceforge.net/projects/j3dme.
- SOAP – http://www.ksoap.org. Currently experimental.

General KVM Resources

Here are two very good general sources of KVM information and links to the latest and greatest tools, applications, and tutorials, plus a new site listing all mobile devices that support Java, along with their capabilities:

- http://www.microjava.com
- http://www.billday.com/j2me
- http://www.JavaMobiles.com

The Future

The future for running applications on mobile devices is rosy, for several reasons:

- Mobile connectivity is always intermittent, promoting the need for local data stored on the device, but with the ability to synchronize, when possible, with a back-end server.
- Having said that the connectivity is intermittent, when one does open a connection, its speed will improve in the next few years. Downloading a 50Kb Java application will therefore be seconds not minutes.
- The increase in mobile phone usage alone, plus the need to personalize them with applications, games, and other goodies, will provide enough impetus to the J2ME bandwagon.
- WAP phones, although currently getting some bad press, will whet the appetite of some users, who will want rather more sophisticated devices.

In addition, there are the Java-specific advantages too:

- In general, developers do not want to learn another language. Java now has great enough market penetration that it's easier to find Java engineers than, say, Palm developers. This makes development companies' lives much, much easier. Better still, those skills are transferable within the development company.
- The back-end server mentioned above, for synchronization and so forth, is more and more likely to be written in Java, so the temptation is to use the same skills on the front-end too, with easy choices of connection libraries straight out of the box. Expect to see SyncML, with both Java and non-Java support libraries on every connected limited device within 5 years.
- Nobody likes rewriting code once for Palm, once for Psion, once for Pocket PCs, and so on. In combination with this will be the proliferation of third party libraries, already appearing, which create a very strong argument for using Java.
- Mobile processing power is always increasing, therefore eating away at the 'Java is slow' complaint.

This doesn't mean, that J2ME will be the hot item of 2001 for end users, however. The combination of affordable but powerful mobile devices plus good bandwidth, is still probably another 1 or 2 years away for the average pocket. However, it is time to start laying the foundations, to make sure that there is content available at the same time as the devices. In Europe, the most useful WAP applications were between 6 to 9 months behind the handset delivery dates, which, along with some poorly thought out marketing, led to a temporary blip in its popularity. Let's hope J2ME avoids that situation.

Why might J2ME fail? there are a few possibilities, but most seem unlikely at the moment:

- If hardware performance grows really quickly, then J2SE may become the de-facto standard for mobile devices, completely missing out J2ME. This is a distinct long-term possibility, but not for several years – at least 3, maybe even 5. The beauty is that a well-written J2ME application should be portable quickly to J2SE, so even if this does happen, the investment is not lost.
- One of the hardware manufacturers could dominate the field and force a specific toolkit in place of J2ME. Only Palm seems likely, and even then sales of mobile phone numbers dwarf Palm sales.
- The current Configurations and Profiles are not sufficient for gaming (no alpha blending on the MIDP, for example), and this could be the first big market – teenage mobile users. If a console company offers an open mobile gaming library/API, J2ME's first-to-market advantage could be lost.

Summary

We have seen how the three layers, the KVM, the CLDC, and the MIDP together form a cohesive, but imperfect, subset of Java 2. It is an excellent set of compromises that match the same compromises made by mobile device manufacturers, in terms of limited screen, keyboard, memory, and processor power. There are still some improvements we would like to see, though, especially for games developers, and better networking support for sockets and datagrams in MIDP specifically.

Having understood a lot of the details behind the three component layers, we need to consider the following points before we dive into writing an application aimed at a limited device, be it Java or another language:

- Is our application going to make sense, to the user, on such a limited device?
- Can we actually implement the user interface we want, with all the given limitations?
- Will it fit into the device's memory?
- Will it run fast enough on all devices?
- Are we expecting too much of the wireless networking interface (will it be as reliable as we wish)?
- Do we want to run our application on many different profiles, and devices?

In the next chapter, we develop a larger, and more realistic application, applying some design effort to separate out the CLDC code from the MIDP (and other profiles), and apply our new found knowledge of the APIs.

PROFESSIONAL JAVA MOBILE

TOOL SELECTION

WEB TABLET
MOBILE PHONE
MP3 PLAYER
DIGITAL CAMERA
PALMTOP
LAPTOP

NOW SELECTED

PALMTOP

6

A Complete Mobile Application – The Contact Database

To demonstrate CLDC and MIDP development, we will create a simple contact database application that can be extended in many ways. It also demonstrates porting to mobile Java platforms other than the MIDP, and last of all it allows us to take a further step and integrate with server-side applications which we will consider further in Chapter 11.

In developing this application our aims are the:

- Abstraction of the user interface layer
- Abstraction of the database layer
- Practical use of the CLDC and MIDP
- Practical use of KJava by porting the MIDP code to this alternative pseudo-profile
- To create a design that allows future interfacing with servlets

The first two points allow us to move MIDP and KJava -specific peculiarities into maintainable classes specific to each platform, leaving us with generic re-usable classes. These re-usable classes are CLDC code only, and thus, can be used with other future J2ME profiles as they appear.

To compile and run all three of the application variants described here, you need the following downloads and tools. All have been introduced in the previous two chapters, so there should be no surprises:

- MIDP – www.java.sun.com/products/midp
- CLDC and Palm overlay - www.java.sun.com/products/cldc

- A Palm device or Palm emulator. If you choose to use the emulator, you must also obtain a ROM image, either from an actual Palm or by requesting one from Palm – http:/www/palm.com
- An IDE (or two). We used the Java Wireless Toolkit (J2MEWTK), a standard Java editor and some batch files.

Contact Application Requirements

Although a contact database or address book written in Java replicates functionality that many mobile phone, pagers, and PDAs already have, very few, if any can share their data with devices different to themselves. Most operate on the principle of a back-end synchronizing application doing all the data conversion for them, with the hope that the sync software supports all devices a user owns.

Our simplistic approach, is that the data is independent of the device; a more complex approach would be to tailor the data to each device, but still deliver it from a common source. This is outside the scope of a single chapter, though. If we can get contact data uploaded from our mobile device to a central server in a device-independent format, then we can share the resulting data with other devices that can then download it. Better still, because we can read, add, and edit entries whilst disconnected it's rather more useful than, say, a pure 'thin-client' WAP browser-based solution, where the accessing, editing, and updating can only be performed while online, which is inconvenient, because of the connection delay and cost of the phone call.

We start with the MIDP port first because, at the time of writing, it's the only official profile for the CLDC. It is also the lowest common denominator, so whatever we can run on a MIDP device is portable to a more powerful device such as a Palm or Psion. We then follow up the MIDP port by illustrating the same functionality using Sun's example Palm KJava libraries.

User Requirements Overview

These are the functions that we want our application to perform:

- View a list of contacts (an implicit requirement)
- Create a new contact
- Read the details of an existing contact
- Modify an existing contact
- Delete a contact
- Delete all contacts

We would also like to share our contacts with other devices and people, so we also need to be able to:

- Upload contacts to a server
- Download contacts from a server

Let's now develop each of these requirements a little further and explore the database design we need to create to hold our contacts.

User Requirements in Detail

Here are the descriptions of each of the possible functions that the user can perform with our contact application:

- **View contacts**
  The user will want an overall view of their contacts and to be able to pick one out for reading, modifying, or deleting. This will be the most used function of the application, since it will be the first screen. Users need to be able to find a contact with the minimum of key presses.
- **Read contact**
  This will be the second most used function of the application, and so, must always be a visible option. This an important consideration for MIDP devices with only two or three soft menu buttons. The user also needs to be able to browse through the contacts, by various means. In our simplified application, the only way a user can select a contact is from a simple unsorted list.
- **Create contact**
  The user must be able to create new contacts. The information entered is simply a list of text fields. We have chosen to allow the entry of name, phone, e-mail, web address, and postal address.
- **Modify contact**
  The user can modify any or all fields in the contact.
- **Delete contact**
  Any contact can be deleted.
- **Delete all contacts**
  All contacts are deleted, leaving a clean database. Useful for starting afresh, if the user wants to download a completely different database from the server.
- **Upload contacts**
  The user can upload all contacts from their target device to a central server.
- **Download contacts**
  The user can download new contacts from a remote server to their device.

Separating Business Methods From Low-Level Implementation

As mentioned earlier, one of our main development aims is that our business logic is separated from the nitty-gritty of the database and user interface specifics of the particular platforms we choose to port to. In other words, when we port to another platform, we want to do it with minimum effort.

In addition, if we design a generic database schema, we can also avoid re-writing a reasonable amount of the middle tier of our application.

To illustrate our layers and for reference throughout the chapter, here are our main classes and how they build up upon MIDP, CLDC and the underlying host operating system. We have also considered kAWT implementation in this diagram:

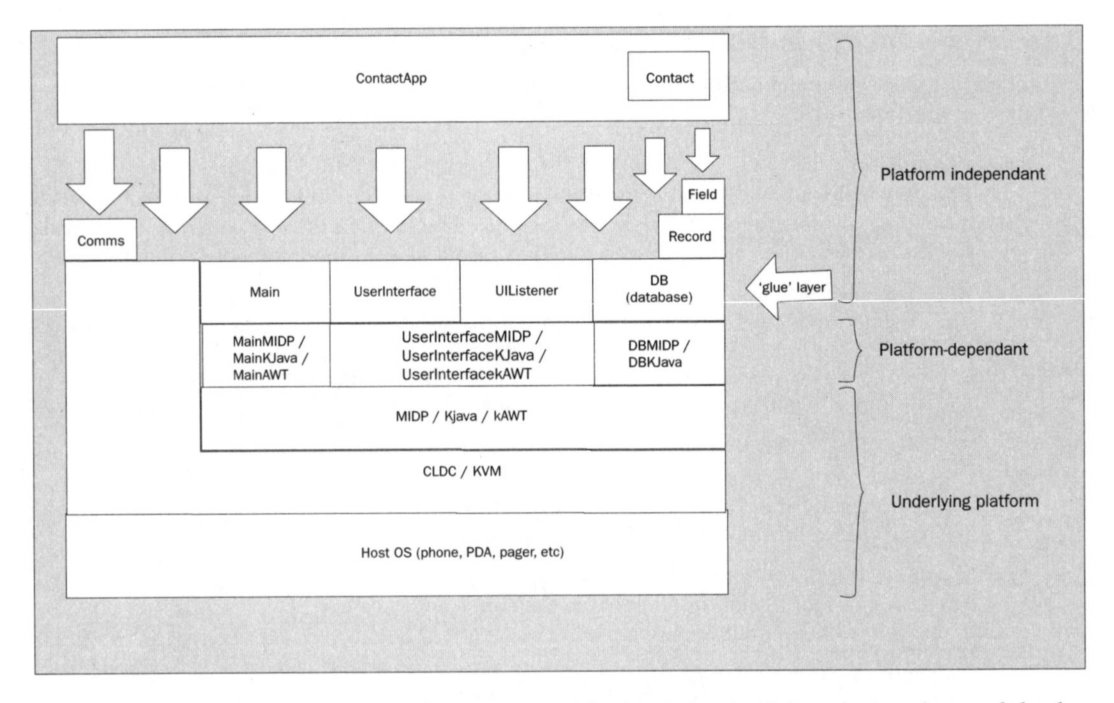

From this, we can see that there is no direct access of the lower levels of the user interface and database by our 'business' logic, contained within the `ContactApp` class. All access to profile-specific code is via the 'glue' layer, but the business classes might still need the CLDC for the basic Java libraries.

We will look at each of the areas in the above figure, in the order in which they were designed in the application:

- The database format and `Contact` data class
- The abstract 'glue' layer and supporting classes such as `Record`
- The business/application classes, ie. `ContactApp`
- The platform-dependent instances of the glue layer, for example, `MainMIDP`, and so on

Note that when creating a layered system such as this, you will most likely find yourself designing and developing from both ends of the diagram, until you meet in the middle. By developing both top-down and bottom-up, you can be sure that the underlying resources are actually capable of doing what your application wants to achieve. An example of this occured during our application development and this marked the inherent differences between the KJava and MIDP databases, which led to some modifications to our high-level functionality.

The Database Format and Contact Class

Each contact database contains zero or more contacts and nothing else. On the target device, we do not care what form the contact database takes, as long as we can access and save our contacts. However, on a central server, we do have to specify a format, so we have chosen a simple comma-separated (CSV) text file, with each contact taking a single line and each piece of contact data being separated by commas. Here is an example:

```
Richard Taylor,+447971000000,rct@poqit.com,www.poqit.com,1 Mow Meadows
Grace Charlesworth,01613390000,,,2 Mow Meadows
Joe Public,02071230000,joep@nowhere.com,www.nowhere.com,12 Acacia Avenue
```

Every contact database entry, known as a record, is made up of the following fields of information:

- Name
- Phone
- E-mail
- Web address
- Address

All pieces of data, or fields, within a contact entry, are free-form text in our application. In a commercial application, we would validate the users' data, such as checking for valid email addresses, valid URLs and so on. We know at some point these contact records will be uploaded and downloaded from a central server and we also know the majority of the data is strings rather than numeric, therefore we have chosen a human-readable format for the contact record. With CSV format we can also move the data into a SQL database on the server with ease.At this point you might be thinking why not use XML to describe the data, rather than an anonymous CSV file. This is an excellent idea, especially since several XML parsers are available for CLDC, but we have limited space to describe our application and have decided to go for a simpler, but less robust, solution.

The Contact Class

The `Contact` class is a very simple class that just defines the names of the data fields (used for labelling them in the user interface, later) and the order that the fields appear in the CSV file. The order is used both by the `Record` class, described later, and in the CSV above:

```
package com.poqit.j2me.common.app;

public class Contact {
  public static final String NAME = "n";
  public static final String PHONE = "p";
  public static final String EMAIL = "e";
  public static final String WEB = "w";
  public static final String ADDRESS = "a";

  public static final String[] fieldNames = {
    NAME, PHONE, EMAIL, WEB, ADDRESS
  };

  public static final int NAME_INDEX = 0;
  public static final int PHONE_INDEX = 1;
  public static final int EMAIL_INDEX = 2;
  public static final int WEB_INDEX = 3;
  public static final int ADDRESS_INDEX = 4;
}
```

The names of the fields (`NAME`, `PHONE`, and so on) are short to minimise wasted space on MIDP devices. We could sub-class `Contact` for each profile or pseudo-profile and put more descriptive strings in, depending on screen space.

The 'Glue' Layer and Supporting Classes

The 'glue' classes separate the business classes from the platform-specific classes, yet they are still platform-independent themselves. Interestingly enough, the connectivity to a back-end server is shared code, but is only dependent on the CLDC, not MIDP or KJava, so we don't need to port that to each platform.

It is worth noting that none of the 'glue' layer classes can be instantiated – they're all interfaces or abstract classes and every port must have its equivalent concrete class such as `MainMIDP`, `DBMIDP`, and `UserInterfaceMIDP` for MIDP, and so on. This is not true of the supporting classes like `Record`, though. There is no `RecordMIDP` equivalent that we need to create, because a `Record` is completely independent of the profile.

We will cover the database-related classes first, then the user interface classes and finally, the `Main interface`.

Database – Design and Classes

Now we know the sort of data we want to store and retrieve we can define some database-related classes, `Field`, `Record`, and `DB`.

The Field Class

The `Field` class represents a single piece of data in the contact record, such as an e-mail address. It holds a String name and a String value and has methods to access these:

```
package com.poqit.j2me.common.database;

public class Field {
  private String name;
  private String value;

  public Field(String initialName, String initialValue) {
    name = initialName;
    value = initialValue;
  }

  public String getName() {
    return name;
  }

  public String getValue() {
    return value;
  }

  public void setName(String newName) {
    name = newName;
  }

  public void setValue(String newValue) {
    value = newValue;
  }
}
```

Note that the name of the `Field` is not part of the information in the server's CSV file, only the values are saved. The `Field` names and the order they are saved in the CSV file are defined by the `Contact` class.

The Record Class

To describe a single record in the database, we will define a generic `Record` class, which is independent of the underlying platform. First, here are the class variables:

```
package com.poqit.j2me.common.database;

import java.util.Hashtable;
import java.util.Enumeration;

public class Record {
```

The `PK` and `NO_PK` are modes that we describe below, and are used when converting the `Record` to a byte array:

```
  public static final int PK = 0;
  public static final int NO_PK = 1;
```

`SEPARATOR` is the separator character used in our CSV file. If we thought we might use a different separator at some future date, we could pass this as a variable into the relevant methods below, rather than define it rigidly:

```
  public static final char SEPARATOR = ',';

  private int primaryKey;
  private String[] fields;
```

The `primaryKey` is a unique integer that we can identify this `Record` by. Note that it is not saved to the server's CSV file, because we cannot guarantee it to be unique across multiple devices – it is only unique in the database this `Record` is held in, that is the target device's database. All new `Records` have a primary key of -1 when constructed, whereas a `Record` saved to the database will leave a positive integer value.

We have three constructors. The first constructor creates an empty `Record`. The field count is passed in to make sure the `fields` String array is the right size before any data is placed in it. The `fields` String array holds just the field values, that is the contact name, phone number, e-mail, etc and not the field names themselves. The primary key is initialized to -1 using `setPK()`:

```
  public Record(int fieldCount) {
    fields = new String[fieldCount];
    setPK(-1);
  }
```

The second constructor creates a `Record` from a `Field` array that is passed. Again, note the field names are ignored:

```
public Record(Field[] initialFields) {
  fields = new String[initialFields.length];
  for (int i = 0; i < initialFields.length; i++) {
    fields[i] = initialFields[i].getValue();
  }
  setPK(-1);
}
```

The third constructor creates a `Record` from a byte array. The byte array is in the form of comma-separated variables. Normally, in J2SE, we would use the `StringTokenizer` here to parse and extract each piece of information between the commas, but it is not in the CLDC API:

```
public Record(byte[] b, int mode) {
  int fieldsStart = 0;
```

The mode parameter passed in is one of `PK` or `NO_PK` as mentioned earlier. This allows us to parse byte arrays containing a primary key as well as the `Field`. This constructor can then be used for two purposes:

- `Record` creation using a byte array read from the underlying database (primary key is in array)
- `Record` creation using a byte array received from a remote web server (no primary key)

The primary key is always the first item in the byte array if it is present:

```
  if (mode == NO_PK) {
    setPK(-1);
  } else {

    // primary key is first piece of data
    while (b[fieldsStart++] != SEPARATOR) {

      // loop, counting up to first separator
    }
    setPK(new String(b, 0, fieldsStart - 1));
  }

  // find fields - there's always one more than there are separators
  int numFields = 1;
  for (int i = fieldsStart; i < b.length; i++) {
    if (b[i] == SEPARATOR) {
      numFields++;
    }
  }
  fields = new String[numFields];

  StringBuffer fieldValue = new StringBuffer();
  int fieldIndex = 0;
  for (int i = fieldsStart; i < b.length; i++) {
    if (b[i] == SEPARATOR) {
      fields[fieldIndex] = fieldValue.toString();
      fieldValue = new StringBuffer();
      fieldIndex++;
```

```
      } else {
        fieldValue.append((char) b[i]);
      }
    }

    // the last field value
    fields[fieldIndex] = fieldValue.toString();
  }
```

Once a Record has been created, we cannot add fields or set field values, only read them. This is an easy functionality to add, but as we didn't need it here, we've left it out.

We then have some primary key accessor methods. As we see later, the MIDP and Palm database both reference records by integer IDs, as opposed to our Strings identifier – name, so we have to save the primary key within the record itself. We access its value with the getPK() and setPK() methods.

A primary key value of -1 shows the record has never been saved to a database. The primary key has no real use outside the target device, since different database formats use different methods to assign their primary keys, and therefore, we cannot rely on the value assigned to it as being unique, or even valid. For example, the MIDP database classes always increment to give the next record identifier, regardless of whether there is a 'spare' record ID where a record has been deleted, whereas the Palm database 'packs' the records in whenever a record is deleted. Due to these differences, we choose not to save the primary key when saving databases to a remote server:

```
  public int getPK() {
    return primaryKey;
  }

  public void setPK(int value) {
    primaryKey = value;
  }

  public void setPK(String value) {
    primaryKey = -1;
    try {
      primaryKey = Integer.parseInt(value);
    } catch (NumberFormatException nfe) {

      // fall through leaving key as -1.
    }
  }
```

The index of the fields is defined in the Contact class, above. To read the value of single fields from the record, such as the contact name, getFieldValue() is used, passing in the index of the field we want:

```
  public String getFieldValue(int fieldIndex) {
    return fields[fieldIndex];
  }
```

The reverse of creating a record from a byte array is the getBytes() method, which converts a record to a byte array. Again, the mode decides whether the record's primary key should also be placed in the byte array or not, depending on whether we use the resulting byte array for sending to a database, or a server:

```
public byte[] getBytes(int mode) {
  String fieldName;
  String fieldValue;
  StringBuffer buffer = new StringBuffer();

  // primary key
  if (mode == PK) {
    buffer.append(Integer.toString(getPK()) + SEPARATOR);
  }

  // fields
  for (int i = 0; i < fields.length; i++) {

    // convert field value to bytes
    fieldValue = fields[i];
    if (fieldValue == null) {
      fieldValue = "";
    }
    buffer.append(fieldValue);
    if (i != (fields.length - 1)) {
      buffer.append(SEPARATOR);
    }
  }
  return buffer.toString().getBytes();
}
```

Finally, the `Record` class has a convenience function, `toFields()`, for use by the user interface layer. It creates an array of `Field` objects for the user interface to display, with each `Field` holding a name String and a value String. `Record` doesn't know the field names therefore we must pass them in using the `Contact` class, for example:

```
record.getFields(Contact.fieldNames)
```

Here is the `getFields()` method:

```
  public Field[] getFields(String[] fieldNames) {
    String fieldName;
    String fieldValue;

    int numFields = fieldNames.length;
    Field[] fieldArray = new Field[numFields];

    for (int i = 0; i < numFields; i++) {
      fieldName = fieldNames[i];
      fieldValue = fields[i];
      if (fieldValue == null) {
        fieldValue = "";
      }
      fieldArray[i] = new Field(fieldName, fieldValue);
    }
    return fieldArray;
  }
}
```

The DB Class

Now we have a Record definition, we can define the generic database class, which we have called DB (to avoid clashing with the previously defined Database class in KJava). All our generic database methods can be rolled into this one class – both database-related methods and record-related calls. Each port that we do needs to subclass this abstract DB class.

The database-related methods required are:

- open() – open the contacts database, creating it if necessary
- close() – close the database
- clear() – clear all records from the database

The record-related methods are:

- readRecords() – read all records in the database
- readRecord() – read a single record, identified by either a value in a particular field or by primary key
- writeRecord() – write a record to the database, creating it if necessary, or overwriting the existing record
- deleteRecord() – delete a record identified by either a value in a particular field, or by primary key
- numRecords() – return the number of records in the database
- merge() – merge multiple records into the database from a byte array passed in

DB has only one class variable, the name of the database, which is set at the very start of the application, in MainMIDP or MainKJava. The majority of methods are abstract, with only one method, merge(), having a significant body of code:

```
package com.poqit.j2me.common.database;

public abstract class DB {
  protected String dbName;

  public abstract void open();
  public abstract void close();
  public abstract void clear();
  public abstract int numRecords();
  public abstract Record[] readRecords();
  public abstract Record readRecord(String value, int fieldIndex);
  public abstract Record readRecord(int primaryKey);
  public abstract boolean writeRecord(Record record);

  public void deleteRecord(String name, int fieldIndex) {
    deleteRecord(readRecord(name, fieldIndex));
  }

  public void deleteRecord(Record record) {
    deleteRecord(record.getPK());
  }

  public abstract void deleteRecord(int primaryKey);
```

merge() is exclusively for creating and adding records to the database from the byte array passed in from the server. It identifies records by looking for '\n' (each record in the CSV file is terminated by the carriage return+newline characters \r and \n). Once it has found that character, it creates a sub-array containing a single record's data (leaving out the terminating characters), and then creates a record from that sub-array:

```
    public boolean merge(byte[] b) {
      int startIndex = 0;
      int len = b.length;
      byte[] dst;
      for (int i = 0; i < len; i++) {
        if (b[i] == '\n') {
          dst = new byte[i - startIndex];
          System.arraycopy(b, startIndex, dst, 0, i - startIndex - 1);
          writeRecord(new Record(dst, Record.NO_PK));
          startIndex = i + 1;
        }
      }
      return true;
    }
  }
```

User Interface – Design and Classes

The platform-independent portion of the user interface is defined within the UserInterface and UIListener interface definitions.

The functionality closely maps that of the MIDP, because it is the simplest of all our user interfaces. If we were to start with one of the more sophisticated user interfaces, then we would be in danger of assuming functionality that just wouldn't be available in the MIDP when we came to port to it.

UserInterface deals only with the output side of the user interface, whilst UIListener specifies which methods will receive user input. Here is the name of the package we have created and placed our relevant classes in and therefore must import:

```
package com.poqit.j2me.common.ui;

import com.poqit.j2me.common.database.Field;
```

The first three static variables are common button labels, which are likely to be used in any application, not just a contact database, so we share them here:

```
public interface UserInterface {

  // Generic button/soft key labels
  public static final String EXIT_AL = "Exit";
  public static final String BACK_AL = "Back";
  public static final String OK_AL = "OK";
```

The next two, NAMEEND_ST and VALUEEND_ST are just separation characters needed to make the displaying of fields look tidy, one placed at the end of the field names and one placed at the end of the field's value:

```
public static final String NAMEEND_ST = ": ";
public static final String VALUEEND_ST = "\n";
```

The four possible application states help the user interface identify which type of screen is active, and therefore, what actions are valid at the current time. There are four types, which could be expanded for more complex applications:

- `SIMPLEACTION` – the default, in other words none of the other three choices below (viewing the details of a record is an example)
- `SINGLEEDIT` – the user is editing a single text field (entering a database name for example)
- `MULTIEDIT` – the user is editing multiple text fields (creating a new record, or modifying one)
- `LIST` – the user is viewing a list (of contacts)

```
// possible application states
public static final int SIMPLEACTION = 0;
public static final int SINGLEEDIT = 1;
public static final int MULTIEDIT = 2;
public static final int LIST = 3;
```

There are four basic display methods in `UserInterface`:

- `displayFields()` – display an array of `Fields`, such as the contents of a record, as delivered by `Record.toFields()`
- `editField()` – display a single editable `Field`, such as the name of a database to download
- `editFields()` – displays multiple editable `Field` objects, such as the contents of a record
- `displayList()` – display a list of selectable Strings, e.g. a list of contacts

```
public void displayFields(String title, Field[] nameAndValue,
                          String[] actionLabels);
public void editField(String title, Field field, String[] actionLabels);
public void editFields(String title, Field[] nameAndValue,
                       String[] actionLabels);
public void displayList(String title, String[] choice,
                        String[] actionLabels);
```

All four display methods require a title for the screen and an array of action label Strings to be passed in. These are used to create buttons, or menu entries for each screen. An example call might be one that allows the user to edit a contact:

```
   ui.editFields("Contacts:", record.getFields(Contact.fieldNames), new
String[]{UserInterface.BACK_AL, UserInterface.OK_AL});
```

Here the title is "Contacts:" and the only two buttons available to the user are "Back" and "OK". The field names and values are retrieved by calling `record.getField()` as previously mentioned.

In addition, there are three miscellaneous supporting methods:

- `setUIListener()` – the vital link between the user interface and the class that handles any user actions. Whenever a button or soft key is pressed, or an edit field is "OK"ed, something has to perform some work on that information. For our application it is whichever `UIListener` class has been registered with the `UserInterface`.
- `close()` – gives the application a chance to clean up any user interface resources, if need be.

```
  public void setUIListener(UIListener uil);
  public void close();
}
```

The UIListener Interface

`UIListener` is also a straightforward interface. Classes interested in catching and handling user input must implement it, that is business classes like `ContactApp`. These classes must also register themselves with `UserInterface` to receive user input events, by calling `setUIListener()`:

- `actionListener()` – called whenever a button, soft key, or menu item is pressed. The text label of the button is passed in as an argument.
- `listListener()` – called whenever an item in a `List` is selected, such as a list of contacts. The name of the item is passed in as an argument.
- `editListener()` – receives a single string of text entered by the user. This is used when the user enters a single text field, in other words a database name.
- `multiEditListener()` – receives an array of strings entered by the user. This is called whenever the user edits multiple fields, in other words a contact.

```
package com.poqit.j2me.common.ui;

import com.poqit.j2me.common.database.Field;

public interface UIListener {
  public void listListener(String name);
  public void actionListener(String action);
  public void editListener(String inputText);
  public void multiEditListener(Field[] fields);
}
```

The Main Interface

Our `Main` interface which every application port needs to implement, is very simple:

```
package com.poqit.j2me.common;

import com.poqit.j2me.common.database.DB;

public interface Main {
  public void exit();
  public DB getDB();
}
```

It only has two reasons for existence. The first is to allow applications to exit in a platform-independent manner, by calling `exit()`. This helps us hide the MIDP's missing `System.exit()` method. The second is to serve up a reference to the contacts database for business classes to gain access to.

The Comms Class

The `Comms` class is not really a 'glue' class in our application, since all ports use the same code. However, other platforms are likely to need slightly different communications, so we have separated it out into its own class.

The `Comms` class has two methods:

- `downloadAndMerge()` – downloads the contents of a specified CSV file to the target mobile device and merges the records into the existing database
- `upload()` – uploads the whole of the target device's database up to a named file on the server

The Palm CLDC has no `HttpConnection` implementation we avoid using the `HttpConnection` class and use the `ContentConnection` class instead. The main losses of functionality when using `ContentConnection` instead of `HttpConnection` are:

- No way of setting the request type. At the server end it defaults to `GET`.
- No way of reading the HTTP status code returned.
- No headers can be set, such as content type and so on.

We will not discuss the `Comms` class in detail here, as Chapter 11 details several code examples very similar to this class. There are only two points of interest where `Comms` differs to those examples. Firstly, in the `downloadAndMerge()` method, `database.merge()` is called once the complete byte array has been received from the server:

```
        status = database.merge(data);
```

Secondly, in the `upload()` method, after each record has been written to the output stream, carriage return and newline characters must be added to the output stream to make each record appear on its own line in the resulting file:

```
            os.write(ra[i].getBytes(Record.NO_PK));
            os.write('\r');
            os.write('\n');
```

Here is the `Comms` class code:

```
package com.poqit.j2me.common.comms;

import com.poqit.j2me.common.database.Field;
import com.poqit.j2me.common.database.Record;
import com.poqit.j2me.common.database.DB;
import java.io.InputStream;
import java.io.OutputStream;
import java.io.IOException;
import javax.microedition.io.Connector;
```

```
import javax.microedition.io.ContentConnection;

public class Comms {
  public static boolean downloadAndMerge(String baseURL, String filename,
                                         DB database) {
    boolean status = false;
    ContentConnection c = null;
    InputStream is = null;
    try {
      byte[] data;
      c = (ContentConnection) Connector.open(baseURL + filename);
      is = c.openInputStream();
      int len = (int) c.getLength();
      if (len > 0) {
        is = c.openInputStream();
        data = new byte[len];
        int actual = is.read(data);
      } else {
        StringBuffer sb = new StringBuffer();
        int ch;
        while ((ch = is.read()) != -1) {
          sb.append((char) ch);
        }
        data = sb.toString().getBytes();
      }
```

The `database.merge()` method is called when the complete byte array has been received from the server:

```
      status = database.merge(data);
      if (is != null) {
        is.close();
      }
      if (c != null) {
        c.close();
      }
    } catch (IOException ioe) {
      System.out.println("download: IOException");
      ioe.printStackTrace();
    }
    finally {}
    return status;
  }

  public static boolean upload(String baseURL, String filename,
                               DB database) {
    boolean status = false;
    try {
      ContentConnection c = null;
      InputStream is = null;
      OutputStream os = null;

      c = (ContentConnection) Connector.open(baseURL + filename);
      os = c.openOutputStream();
```

```
      // write records out, one by one, having converted them to byte
      //arrays,
      // AND added a carriage return and newline to separate records.
      Record[] ra = database.readRecords();
      for (int i = 0; i < ra.length; i++) {
```

The following code writes records out, one by one, having converted them to byte arrays:

```
        os.write(ra[i].getBytes(Record.NO_PK));
        os.write('\r');
        os.write('\n');
      }

      is = c.openInputStream();
      int len = (int) c.getLength();
      if (len > 0) {
        byte[] data = new byte[len];
        int actual = is.read(data);
        status = true;
      } else {
        StringBuffer sb = new StringBuffer();
        int ch;
        while ((ch = is.read()) != -1) {
          sb.append((char) ch);
        }
        status = true;
      }
      if (is != null) {
        is.close();
      }
      if (os != null) {
        os.close();
      }
      if (c != null) {
        c.close();
      }
    } catch (IOException ioe) {
      System.out.println("IOException");
      ioe.printStackTrace();
    }
    return status;
  }
}
```

The Business and Control Class – ContactApp

Moving up a layer to the pseudo-business classes, `ContactApp` class is our business and control class. `ContactApp` has no knowledge of the underlying profile, and relies only on our 'glue' classes and the CLDC for its basic Java classes. It only specifies **what to do** (create a record, display its contents, and so on) and **not how** it is done. It has parallels with the MIDP high-level UI classes in this respect, but extends that behaviour to the database layer too.

Rather than being pure business logic, it combines both logic and UI calls together, since this application has minimal business logic. In a more complex system, it would be wise to partition the business logic and the user interface methods. So, in addition to creating a contact record, it causes the edit user interface to be drawn and also accepts user interface events or 'callbacks'. The user interface layer lets the application know that certain events have happened, like button presses and selections made from a list, which the application must react to.

The only public methods ContactApp are list(), called just once on startup by the main application (MainMIDP or MainKJava)and the four UI listeners that are defined by the UIListener interface that ContactApp implements. All the other methods are private to ContactApp, and are fired off by the user as they use the application.

We now step through the ContactApp source, section by section, with the relevant commentary above each section:

```
package com.poqit.j2me.common.app;

import com.poqit.j2me.common.Main;
import com.poqit.j2me.common.ui.UIListener;
import com.poqit.j2me.common.ui.UserInterface;
import com.poqit.j2me.common.database.DB;
import com.poqit.j2me.common.database.Field;
import com.poqit.j2me.common.database.Record;
import com.poqit.j2me.common.comms.Comms;
import java.io.IOException;
```

ContactApp implements the UIListener, with its four listeners, so it can process user input received from the UserInterface object:

```
public class ContactApp implements UIListener {
```

The screen IDs help ContactApp work out what state it is in. For example, when a generic "Back" button is pressed, we need to know where to go back to, since it's dependent on where we are. The ID, set every time a screen is drawn, helps us do that:

```
    // screen IDs
    public static final int CONTACT_LIST = 0;
    public static final int CONTACT_NEW = 1;
    public static final int CONTACT_DETAILS = 2;
    public static final int CONTACT_EDIT = 3;
    public static final int DB_DOWNLOAD = 4;
    public static final int DB_UPLOAD = 5;
```

The action labels are additional buttons to those defined in the generic UserInterface class, but the ones here are more specific to the contact database application, which is why we separated them out:

```
    // action labels
    private static final String NEW_AL = "New";
    private static final String DELETE_AL = "Delete";
    private static final String EDIT_AL = "Edit";
    private static final String DOWNLOAD_AL = "D'load";
    private static final String UPLOAD_AL = "U'load";
    private static final String DB_CLEAR_AL = "Clear";
```

The screen titles are just text strings displayed by each screen, to tell the user where they are in the application:

```
//screen titles
private static final String CONTACTLIST_ST = "Contacts:";
private static final String CONTACTNEW_ST = "Enter new contact name:";
private static final String CONTACTDETAILS_ST = "Contact details:";
private static final String CONTACTEDIT_ST = "Edit contact:";
```

The DOWNLOAD_PATH and UPLOAD_PATH are used when communicating with the remote server, to upload and download contacts. There are two parameters in the URL, for helping the remote server (a servlet in our case, as you will see later on in this chapter):

- service – tells the servlet whether the request is a GET or a POST, in effect. We cannot use HTTP requests, because our CLDC ports do not have an HttpConnection implementation – the nearest we have is a ContentConnection.
- filename – the file name to load or save to on the server.

```
  private static final String DOWNLOAD_PATH =
"http://localhost:8080/projavamobile/servlet/ContactServlet?service=download&filen
ame=";
  private static final String UPLOAD_PATH =
"http://localhost:8080/projavamobile/servlet/ContactServlet?service=upload&filenam
e=";

  // For CLDC on a Palm, i.e. KJava
  // private static final String DOWNLOAD_PATH =
"testhttp://localhost:8080/contacts/ContactServlet?service=download&filename=";
  // private static final String UPLOAD_PATH =
"testhttp://localhost:8080/contacts/ContactServlet?service=upload&filename=";
```

Also note that there are two versions, because the CLDC Palm overlay, which we port to later, requires us to use 'testhttp' rather than 'http' as the protocol. This is because, in general, most Palms do not have HTTP support. Instead, Sun included pseudo-HTTP classes in their CLDC Palm overlay for us to use, but named it differently to distinguish it from any Palm equivalents.

Finally, we have some class variables:

- main – holds the reference to the Main that called us, so we can get a reference to the (already) open database, by calling Main.getDB() and also to exit the application correctly, by calling Main.exit().
- ui – holds the reference to the one and only UserInterface object, so we can register our four UI listeners with it, and also call its methods to draw our screens.
- currentScreen – holds the ID of the screen we're on, for navigation purposes.
- selectedRecordPK – holds the currently selected record's primary key. It saves time searching through the database.

```
private Main main;
private UserInterface ui;
private int currentScreen;
private int selectedRecordPK;
```

The following constructor merely initializes the `main` and `ui` variables:

```
public ContactApp(Main main, UserInterface ui) {

  // remember the main reference, so we can exit
  this.main = main;

  // remember the ui reference, so we can draw
  this.ui = ui;
}
```

The `list()` method displays a list of all contacts in the database by using `readRecords()` to fetch them all and showing each contact's name field. "Exit", "New","Download", "Upload", and "Clear" buttons/menus/soft keys are also displayed for user navigation. Once the list is displayed, the application waits for user input. The user can either select one of the contacts, or press a button/menu/soft key:

```
public void list() {
  String[] choice;

  Record[] contacts = main.getDB().readRecords();
  if (contacts != null) {
    int len = contacts.length;
    choice = new String[len];
    for (int i = 0; i < len; i++) {
      choice[i] = contacts[i].getFieldValue(Contact.NAME_INDEX);
    }
  } else {
    choice = new String[0];
  }
  ui.displayList(CONTACTLIST_ST, choice, new String[] {
    UserInterface.EXIT_AL, NEW_AL, DOWNLOAD_AL, UPLOAD_AL, DB_CLEAR_AL
  });
  currentScreen = CONTACT_LIST;
  ui.setUIListener(this);
}
```

Also note, that we set the UI listener to be ourselves in `list()`, rather than in the constructor, because for MIDP, we cannot guarantee we are in the foreground until `list()` is called, since it's called from `startApp()`.

The `details()` method displays the entire details of contact selected from the contact list, and adds Action buttons/soft key to allow the user to edit or delete that contact:

```
private void details() {
  details(main.getDB().readRecord(selectedRecordPK));
}
```

```
private void details(String name) {
  details(main.getDB().readRecord(name, Contact.NAME_INDEX));
}

private void details(Record record) {
  ui.displayFields(CONTACTDETAILS_ST,
                   record.getFields(Contact.fieldNames), new String[] {
    UserInterface.BACK_AL, EDIT_AL, DELETE_AL
  });
  selectedRecordPK = record.getPK();
  currentScreen = CONTACT_DETAILS;
}
```

`create()` creates a new contact and tells the interface to draw the empty contact as a set of editable fields. The resulting newly edited contact is handled by the `multiEditListener()` method:

```
private void create() {
  Record record = new Record(Contact.fieldNames.length);
  selectedRecordPK = record.getPK();
  ui.editFields(CONTACTEDIT_ST, record.getFields(Contact.fieldNames),
                new String[] {
    UserInterface.BACK_AL, UserInterface.OK_AL
  });
  currentScreen = CONTACT_NEW;
}
```

The `edit()` method displays the entire contact as editable fields. When the user finishes editing and assuming s/he presses "OK", rather than "Back", the edited contact is handled by the snappily-titled `multiEditListener()` method, below:

```
private void edit() {
  Record record = main.getDB().readRecord(selectedRecordPK);
  ui.editFields(CONTACTEDIT_ST, record.getFields(Contact.fieldNames),
                new String[] {
    UserInterface.BACK_AL, UserInterface.OK_AL
  });
  currentScreen = CONTACT_EDIT;
}
```

The following method converts the edited `fields` to a `Record` and saves the edited, or new contact `Record` to the database and returns to the updated contact list:

```
private void editOK(Field[] fields) {
  Record record = new Record(fields);
  record.setPK(selectedRecordPK);
  main.getDB().writeRecord(record);
  list();
}
```

`delete()` deletes the currently selected contact from the database and returns the updated contact list:

```
private void delete() {
  main.getDB().deleteRecord(selectedRecordPK);
  list();
}
```

The `download()` method allows the user to enter the name of the file to download from the remote server:

```
private void download() {
  ui.editField("Download database", new Field("DB to load:", ""),
               new String[] {
    UserInterface.BACK_AL, UserInterface.OK_AL
  });
  currentScreen = DB_DOWNLOAD;
}
```

`downloadOK()` requests records to be downloaded and merged from a remote server. The name of the file on the server to fetch from, was entered in `download()`:

```
private void downloadOK(String name) {
  if (Comms.downloadAndMerge(DOWNLOAD_PATH, name, main.getDB())) {
    System.out.println("Downloaded OK");
  } else {
    System.out.println("dbDownloadOK: couldn't download the database.");
  }
  list();
}
```

This next method allows the user to enter the name of the file to save the current database contents to on the remote server. The actual uploading occurs in `uploadOK()`, below, which is called by `editListener()`:

```
private void upload() {
  ui.editField("Upload database", new Field("DB to save to:", ""),
               new String[] {
    UserInterface.BACK_AL, UserInterface.OK_AL
  });
  currentScreen = DB_UPLOAD;
}
```

The `uploadOK()` method sends the entire database contents to a remote server. The name of the file the server saves to was entered in `upload()`:

```
private void uploadOK(String name) {
  if (Comms.upload(UPLOAD_PATH, name, main.getDB())) {
    System.out.println("Uploaded OK");
  } else {
    System.out.println("uploadOK: couldn't upload the database.");
  }
  list();
}
```

This next method simply clears all records from the database. We don't check with the user, but obviously a commercial application would need to pop up some double-check mechanism at this point:

```
    private void dbClear() {
      main.getDB().clear();
      list();
    }
```

`listListener()` is the event handler for when a contact is selected from the contact list. It calls details, which displays the contact's details, read-only:

```
    public void listListener(String name) {
      details(name);
    }
```

The `actionListener()` method is the event handler for all button/soft key presses:

```
    public void actionListener(String action) {
      if (action.equals(UserInterface.BACK_AL)) {
        if (currentScreen == CONTACT_EDIT) {
          details();
        } else {
          list();
        }
      } else if (action.equals(EDIT_AL)) {
        edit();
      } else if (action.equals(DELETE_AL)) {
        delete();
      } else if (action.equals(DOWNLOAD_AL)) {
        download();
      } else if (action.equals(UPLOAD_AL)) {
        upload();
      } else if (action.equals(DB_CLEAR_AL)) {
        dbClear();
      } else if (action.equals(UserInterface.EXIT_AL)) {
        main.exit();
      } else if (action.equals(NEW_AL)) {
        create();
      } else {
        list();
      }
    }
```

The `editListener()` method is the event handler for when a single edit field is successfully completed:

```
    public void editListener(String inputText) {
      if (currentScreen == DB_DOWNLOAD) {
        downloadOK(inputText);
      } else if (currentScreen == DB_UPLOAD) {
        uploadOK(inputText);
      } else {
        list();
```

```
        }
    }

    public void multiEditListener(Field[] fields) {
        editOK(fields);
    }
}
```

Now we have covered all the profile-independent classes. Just before we look at the three ports of the 'glue' classes, starting with the MIDP variants, we'll just cover tha last piece of shared code – our servlet, which processes the download and upload requests from our J2ME client.

The ContactServlet Class

We have included the `ContactServlet` class so the complete client-server source is in one chapter. Chapter 11 covers servlets in greater detail therefore However, we only skim the code here. We will highlight the differences between the examples here and in Chapter 11 and invite you to return later, once you have read it.

The source code of `ContactServlet` is below:

```
import javax.servlet.http.HttpServlet;
import javax.servlet.http.HttpServletRequest;
import javax.servlet.http.HttpServletResponse;
import java.io.IOException;
import java.io.PrintWriter;
import java.io.BufferedReader;
import java.io.FileReader;
import java.io.FileOutputStream;

public class ContactServlet extends HttpServlet {
  private final static String DATABASE_DIRECTORY = "C:\\contacts\\";

  protected void doGet(HttpServletRequest request,
                       HttpServletResponse response) throws IOException {
    PrintWriter out = response.getWriter();
    String filename = request.getParameter("filename");
    if (filename == null) {
      System.out
        .println("ContactServlet.doGet() error: no filename parameter in URL");
      return;
    }
    String service = request.getParameter("service");
    if (service == null) {
      System.out
        .println("ContactServlet.doGet() error: no service parameter in URL");
      return;
    }
    if (service.equals("upload")) {
      doPost(request, response);
    }

    BufferedReader file = null;
    try {
      file = new BufferedReader(new FileReader(DATABASE_DIRECTORY
```

```
                                                + filename + ".csv"));
        String line;
        while ((line = file.readLine()) != null) {
          out.println(line);
        }
      } catch (IOException ioe) {
        System.out.println("ContactServlet.doGet() IOException");
        ioe.printStackTrace();
      }
      if (file != null) {
        file.close();
      }
      out.close();
    }
    public void doPost(HttpServletRequest request,
                       HttpServletResponse response) throws IOException {
      PrintWriter out = response.getWriter();
      String filename = request.getParameter("filename");
      if (filename == null) {
        return;
      }
```

We do not bother trying to tell the J2ME client whether there is a failure, other than the length being 0. We do print out an error to `System.out` for our debugging, though. Obviously, a commercial application deserves some error-handling to inform the user what is going on, but realistically, real devices **will** have `HttpConnection` implemented, so we haven't put much effort here in implementing error-passing back to the client. Refer to the servlets chapter to see the status being passed back using `HttpConnection`:

```
      // find the length of the data to be saved.
      int len = request.getContentLength();
      if (len < 1) {
        System.out.println("ContactServlet.doPost() error: file length="
                           + len);
        out.close();
        return;
      }
      try {
        FileOutputStream fos = new FileOutputStream(DATABASE_DIRECTORY
                                                + filename + ".csv");
        byte[] data = new byte[len];
        request.getInputStream().read(data, 0, len);
        fos.write(data);
        fos.close();
      } catch (IOException ioe) {
        System.out.println("ContactServlet.doPost() IOException");
        ioe.printStackTrace();
      }
      out.close();
    }
  }
```

The major difference between the examples in Chapter 11 and this servlet is that the J2ME client is forced to use `ContentConnection` rather than `HttpConnnection`, since the CLDC Palm overlay does not support `HttpConnection`. Therefore, there is no value in returning HTTP status codes, such as 200 (OK) or 404 (not found) and so on, since the J2ME client cannot access them.

The MIDP Port

All the above classes are shared across all three of our ports. The next classes are for the MIDP port only and deal with the specifics of three MIDP areas:

- Database
- User interface
- Starting the application

We start with the database classes.

The DBMIDP Class

Our MIDP specific database class is snappily named `DBMIDP`, and it extends our abstract `DB` class, fleshing out the empty the method calls defined in there. Here is the code for `DBMIDP`, split into methods, with commentary following each method:

```
package com.poqit.j2me.midp.database;

import com.poqit.j2me.common.database.DB;
import com.poqit.j2me.common.database.Field;
import com.poqit.j2me.common.database.Record;
import javax.microedition.rms.RecordStore;
import javax.microedition.rms.RecordEnumeration;
import javax.microedition.rms.RecordStoreException;

public class DBMIDP extends DB {
```

`DBMIDP` has a single private reference to a MIDP `RecordStore`, which is initialized by calling `DBMIDP.open()`, as we shall see later:

```
    public RecordStore recordStore;
```

The `RecordStore` is identified using a String name, as described in the previous chapter on the MIDP API. The constructor stores the database's name for use by `open()` later. It does not open the database, because we would prefer to call `open()` within the MIDP `startApp()`, rather than create a `DBMIDP` object every time the application came to the foreground:

```
    public DBMIDP(String name) {
      recordStore = null;
      dbName = name;
    }
```

We do not have a `create()` method for the database, only an `open()` method, which opens the database, creating it if necessary, because there are no situations when we ever want to create a database without subsequently opening it. If `recordStore` is non-null, it must have been previously opened and we don't bother opening it again. This avoids us getting mixed up and opening the database more times than we close it, which would cause the database to stay open:

```
public void open() {
  if (recordStore != null) {
    return;
  }
  try {
    recordStore = RecordStore.openRecordStore(dbName, true);
  } catch (RecordStoreException rse) {
    System.out.println("open: couldn't open database.");
  }
}
```

Closing the database requires a call to the MIDP `closeRecordStore()` method. If `recordStore` is already null it must have been previously closed:

```
public void close() {
  if (recordStore == null) {
    return;
  }
  try {
    recordStore.closeRecordStore();
  } catch (RecordStoreException rse) {
    System.out.println("close: couldn't close database.");
  }
  recordStore = null;
}
```

The `clear()` method removes all records from the database. It is the responsibility of the calling class to check with the user that this is all right to do. We do not bother checking, but a commercial application obviously would:

```
public void clear() {
  try {
    close();
    RecordStore.deleteRecordStore(dbName);
    open();
  } catch (RecordStoreException rse) {
    System.out.println("clear: database couldn't be cleared");
  }
}
```

The next method simply returns the number of records in the database:

```
public int numRecords() {
  try {
    return recordStore.getNumRecords();
  } catch (RecordStoreException rse) {
    System.out
      .println("numRecords: database doesn't exist or is not open");
    return 0;
  }
}
```

`readRecords()` enumerates through the complete database, and builds an array of `Record` objects to pass back to the calling method. This is used to display a list of contacts to the user:

```
public Record[] readRecords() {
  Record[] entries = null;
  if (numRecords() == 0) {
    entries = new Record[0];
    return entries;
  }

  RecordEnumeration enum = null;
  try {
    entries = new Record[recordStore.getNumRecords()];
    int i = 0;
    enum = recordStore.enumerateRecords(null, null, false);

    byte[] b;
    while (enum.hasNextElement()) {
      int recordID = enum.nextRecordId();
      b = recordStore.getRecord(recordID);
      Record r = new Record(b, Record.PK);
      entries[i] = r;
      i++;
    }
  } catch (RecordStoreException rse) {
    System.out
      .println("readRecords: database doesn't exist or is not open");
    enum.destroy();
    return null;
  }
  enum.destroy();
  return entries;
}
```

We do not bother to limit the number of records read in for our example application, but this feature would be needed for a real MIDP device, since scrolling through more than 10-15 contacts would be tedious. A search facility would also be a useful addition, as would a sorting algorithm. MIDP does offer such features, using `RecordFilter` and `RecordComparator` classes, but the CLDC Palm overlay does not, so for simplicity we chose not to use the MIDP features:

```
public Record readRecord(String value, int fieldIndex) {
  boolean found = false;
  byte[] b;
  Record record = null;
  RecordEnumeration enum = null;
  try {
    enum = recordStore.enumerateRecords(null, null, false);
    while (enum.hasNextElement()) {
      int recordID = enum.nextRecordId();
      b = recordStore.getRecord(recordID);
      record = new Record(b, Record.PK);
      if (record.getFieldValue(fieldIndex).equals(value)) {
        found = true;
        break;
```

```
            }
          }
        } catch (RecordStoreException rse) {
          System.out
            .println("readRecord: database doesn't exist or is not open");
        }
        enum.destroy();
        if (!found) {
          return null;
        }
        return record;
      }
```

readRecord() has two variations – by name and by primary key. The find by name is rather slow, since it must work through the database until it finds a match with the name String passed in, whereas the read by primary key can directly read the desired record. The MIDP RMS offers no other way to find a specific record other than by primary key. We pass a field index in (in other words Contact.NAME_INDEX) and the value to find a match for:

```
      public Record readRecord(int pk) {
        try {
          return new Record(recordStore.getRecord(pk), Record.PK);
        } catch (RecordStoreException rse) {
          System.out.println("readRecord: record with pk of " + pk
                             + " doesn't exist");
          return null;
        }
      }
```

The writeRecord() method allows the storing of both new records and the overwriting of existing records. A primary key value of -1 means that the record is a new one that has not previously been saved to the database. Any other value is a valid primary key. Notice that MIDP database records start from 0, not 1, unlike the Palm database records:

```
      public boolean writeRecord(Record record) {
        boolean status = false;
        try {
          if (record.getPK() == -1) {
            record.setPK(recordStore.getNextRecordID());
            byte[] b = record.getBytes(Record.PK);
            recordStore.addRecord(b, 0, b.length);
          } else {
            byte[] b = record.getBytes(Record.PK);
            recordStore.setRecord(record.getPK(), b, 0, b.length);
          }
          status = true;
        } catch (RecordStoreException rse) {
          System.out.println("writeRecord: couldn't save record to database.");
          System.out.println(rse);
          rse.printStackTrace();
        }
        return status;
      }
```

Deletion of records is by primary key value, but our abstract DB class has convenience methods that allow deletion by name and record:

```
    public void deleteRecord(int primaryKey) {
      try {
        recordStore.deleteRecord(primaryKey);
      } catch (RecordStoreException rse) {
        System.out.println("deleteRecord: record couldn't be deleted");
      }
    }
  }
```

The UserInterfaceMIDP Class

As we saw from the previous chapter where we covered the MIDP UI API, the MIDP user interface is considerably simpler than a PDA user interface such as a Palm. This is the main reason we stuck with simplistic lists, edits, and buttons for our abstract UserInterface class, with no windows, tabs, pull-downs, and so on.

Over and above the UserInterface methods defined previously, all our user interface ports (the MIDPand KJava) must also catch user events and pass them on to the handler specified using the setUIListener() method, which is the generic ContactApp class in our application.

We'll now describe each method in UserInterfaceMIDP, the MIDP implementation of the UserInterface interface.

In UserInterfaceMIDP, there are two class variables, one to hold a reference to the Display object (passed in by MainMIDP, as we see later) and also one to reference the UIListener class, which is ContactApp in our application:

```
package com.poqit.j2me.midp.ui;

import com.poqit.j2me.common.ui.UserInterface;
import com.poqit.j2me.common.ui.UIListener;
import com.poqit.j2me.common.database.Field;
import com.poqit.j2me.midp.MainMIDP;

import javax.microedition.lcdui.Command;
import javax.microedition.lcdui.TextField;
import javax.microedition.lcdui.TextBox;
import javax.microedition.lcdui.Form;
import javax.microedition.lcdui.List;
import javax.microedition.lcdui.Display;
import javax.microedition.lcdui.Displayable;
import javax.microedition.lcdui.CommandListener;

public class UserInterfaceMIDP implements UserInterface, CommandListener {

  private static Display display;
  private UIListener eventListener;
```

In the constructor we must get a reference to the MIDP Display object. Only MIDlets can obtain this, which is why we must get it through the MainMIDP reference passed in:

```
public UserInterfaceMIDP(MainMIDP main) {
  display = Display.getDisplay(main);
}

public void close() {}
```

Displaying a list of contacts is a matter of creating a `List.IMPLICIT` and adding each contact to that `List`. The `IMPLICIT` mode allows the user to select an item from the list without any further input, in other words it treats the `List` as a menu. The choice String array passed in (by `ContactApp`) is composed of the name fields for all contacts in the database. `addActions()` creates soft buttons or a menu, depending on the amount of screen space (and the device, of course). Finally, the `List` component is dispayed on the screen by calling `setCurrent()`:

```
public void displayList(String title, String[] choice,
                        String[] actionLabels) {
  List l = new List(title, List.IMPLICIT);
  addActions(l, actionLabels);
  for (int i = 0; i < choice.length; i++) {
    l.append(choice[i], null);
  }
  display.setCurrent(l);
}
```

Here is what `displayList()` looks like using various emulators from the J2ME Wireless Toolkit:

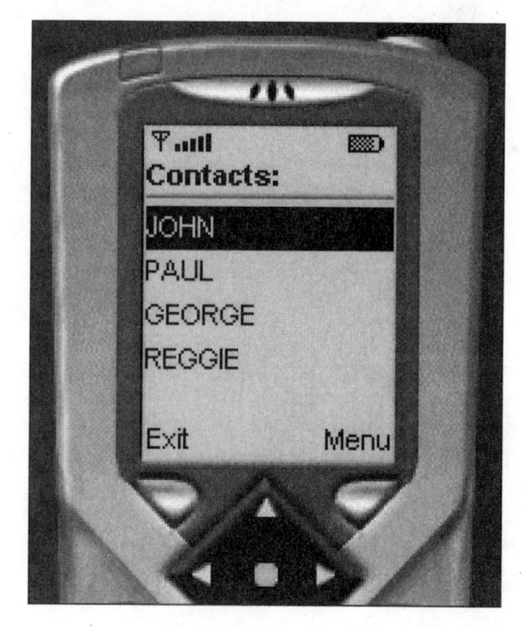

The fields of a contact are displayed using the `displayFields()` method. A `Form` is created to hold a list of the name and value of each `Field` separated by some static Strings, `NAMEEND_ST`, and `VALUEEND_ST`, for readability:

```
public void displayFields(String title, Field[] fields,
                          String[] actionLabels) {
  Form f = new Form(title);
  addActions(f, actionLabels);

  for (int i = 0; i < fields.length; i++) {
    f.append(fields[i].getName());
    f.append(UserInterface.NAMEEND_ST);
    f.append(fields[i].getValue());
    f.append(UserInterface.VALUEEND_ST);
  }
  display.setCurrent(f);
}
```

Here is the resulting screen:

A single editable `Field` is displayed using the `editField()` method. Here, we just need a single MIDP `TextBox`, plus any actions – "Back" and "Save" in this case. We have left the `TextBox` length as an arbitrary value, but might want to limit this depending on which other platforms are being ported to. Setting the value using a properties file, (included in the MIDP suite) is a flexible solution:

```
public void editField(String title, Field field, String[] actionLabels) {
  TextBox t = new TextBox(field.getName(), field.getValue(), 128,
                          TextField.ANY);
  addActions(t, actionLabels);
  display.setCurrent(t);
}
```

The corresponding screen will look like this:

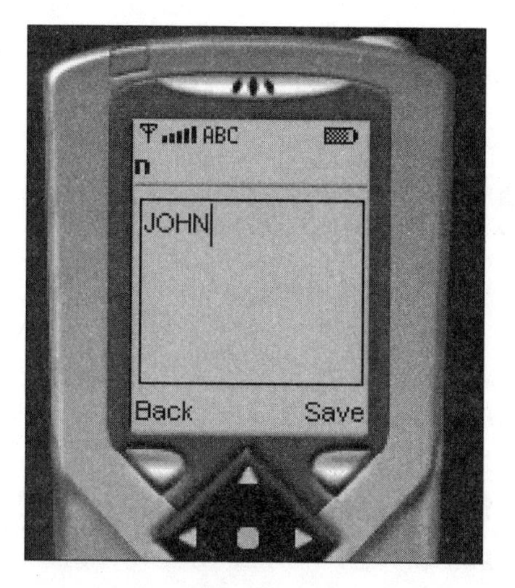

Multiple edit `Field` objects are displayed similarly, but using multiple `TextField` objects within a `Form` rather than a `TextBox`, since `TextBox` is a `Displayable` and not an `Item`:

```
public void editFields(String title, Field[] fields,
                       String[] actionLabels) {
  Form f = new Form(title);
  addActions(f, actionLabels);

  for (int i = 0; i < fields.length; i++) {
    TextField tf = new TextField(fields[i].getName(),
                                 fields[i].getValue(), 128,
                                 TextField.ANY);
    f.append(tf);
  }
  display.setCurrent(f);
}
```

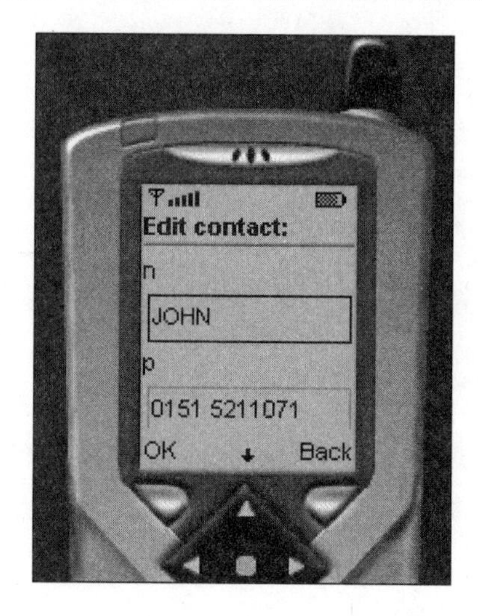

The private method addActions() adds multiple button/soft keys to the Displayable and sets the CommandListener to be our UserInterfaceMIDP instance, that is this:

```
private void addActions(Displayable currentDisplayable,
                        String[] actionLabels) {
  int num = actionLabels.length;
  for (int i = 0; i < num; i++) {
    Command newCmd = new Command(actionLabels[i], Command.SCREEN, 1);
    currentDisplayable.addCommand(newCmd);
  }
  currentDisplayable.setCommandListener(this);
}
```

UserInterfaceMIDP implements CommandListener, therefore we must define a commandAction() method to catch all MIDP user input events. This is where all the initial event-handling is performed, that involves:

- Working out which Displayable caused the event
- Extracting the data from the user interface Displayable in a form suitable for passing to a higher-level class
- Calling the correct method of the registered UIListener instance

One of the vital things here, is that the data being passed to the UIListener class (ContactApp) is generic – we cannot pass any MIDP-specific data like the Displayable itself up into the higher-level class because it is platform-independent. The three steps above apply to all the ports we implement:

```
public void commandAction(Command c, Displayable s) {
  if (c == List.SELECT_COMMAND) {
    List l = (List) s;
    String name = l.getString(l.getSelectedIndex());
    eventListener.listListener(name);
  } else {
    String action = c.getLabel();
    boolean eventFound = false;
    if ((s instanceof Form) && (action.equals(OK_AL))) {
      Form f = (Form) s;
      Field[] fields = new Field[f.size()];
      for (int i = 0; i < f.size(); i++) {
        Field field = new Field(((TextField) f.get(i)).getLabel(),
                                ((TextField) f.get(i)).getString());
        fields[i] = field;
      }
      eventListener.multiEditListener(fields);
      eventFound = true;
    }
    if ((s instanceof TextBox) && (action.equals(OK_AL))) {
      eventListener.editListener(((TextBox) s).getString());
      eventFound = true;
    }
    if (!eventFound) {
      eventListener.actionListener(action);
    }
  }
}
```

Last of all, is the very short method that sets the `UIListener`:

```
    public void setUIListener(UIListener uil) {
      eventListener = uil;
    }
  }
```

This is the same code for all platforms, but because `UserInterface` is forced to be an interface rather than an abstract class, we must add it to every implmentation of `UserInterface`. As we see later, we could not allow `UserInterface` to be an abstract class because the other ports of the `UserInterface` interface have to extend other platform-specific UI classes and Java's single inheritence model does not allow us to extend two concrete classes

Starting the MIDP Application

Last but not least, we somehow have to start the application. This step is also platform-specific of course, but much of the code is shared and could potentially be abstracted out. The class is called `MainMIDP`.

The initialization of the application takes the following steps, regardless of the platform:

- Create the user interface
- Open the current contact database, creating it if it doesn't already exist
- Create our business object, `ContactApp`, passing in a reference to ourselves, so that it can call `Main.exit()` at some point later in time
- Finally, display the list of contacts from the database

The MainMIDP Class

Here is the source for the `MainMIDP` class:

```
package com.poqit.j2me.midp;

import com.poqit.j2me.common.Main;
import com.poqit.j2me.common.app.ContactApp;
import com.poqit.j2me.common.database.DB;
import com.poqit.j2me.midp.database.DBMIDP;
import com.poqit.j2me.midp.ui.UserInterfaceMIDP;

import javax.microedition.midlet.MIDlet;
```

We'll see that in the constructor below, all but the last step is performed. This is because MIDP has a `startApp()` method, which is called whenever our MIDlet becomes the device's current process, so it is there that we call the method to display the contact list:

```
public class MainMIDP extends MIDlet implements Main {
  private DB database;
  private ContactApp app;

  public MainMIDP() {
    UserInterfaceMIDP ui = new UserInterfaceMIDP(this);
    database = new DBMIDP("names");
```

```
    app = new ContactApp(this, ui);
  }

  public void startApp() {
    database.open();
    app.list();
  }
```

The pauseApp() and destroyApp() methods close the database to free up resources, while the application is in the background:

```
  public void pauseApp() {
    database.close();
  }

  public void destroyApp(boolean unconditional) {
    database.close();
  }
```

The exit() method allows business objects to force the application to stop. This is only necessary because MIDP applications cannot call System.exit(), they must call notifyDestroyed() instead, and adding this method to our business classes would have made them platform-specific. Therefore, we wrapped notifyDestroyed() up in the exit() method instead, this does However cause all the other ports to define an exit() method too. The exit() method is defined in the generic Main interface, which all our 'main' application starting classes must implement:

```
  public void exit() {
    database.close();
    destroyApp(true);
    notifyDestroyed();
  }
```

Finally, getDB() returns a reference to the database, for the business class to use:

```
  public DB getDB() {
    return database;
  }
}
```

The Palm/KJava Port

The KJava user interface and database classes are bundled with the CLDC Palm overlay, but are not actually part of the specification. KJava exists to allow developers to exercise the CLDC code on a real hardware platform, and also to a lesser extent, to demonstrate how to write database and user interface classes for the Palm.

It is extremely likely that KJava will disappear from the CLDC bundle once the PDA Profile Reference Implementation is available, especially with the recent arrival of MIDP for Palm, for our purposes, However it is a useful platform for demonstrating porting of our application.

KJava-Specific Classes

The KJava-specific classes all adhere to our common interfaces and abstract classes, which are:

- DB
- UserInterface
- Main

The DBKJava Class

The main effort of porting the application to KJava went into disguising the differences between the Palm database and the MIDP database:

```
package com.poqit.j2me.kjava.database;
import com.poqit.j2me.common.database.DB;
import com.poqit.j2me.common.database.Field;
import com.poqit.j2me.common.database.Record;

import com.sun.kjava.Database;

public class DBKJava extends DB {

  private static int creatorID = 0x54616374;
  private Database database;
```

Palm databases are identified by a unique ID known as a `creatorID`, (in comparison to the String name used by MIDP databases). These various `creatorID` are maintained by Palm, and you can choose a unique one of your own online at the Palm developers web site, http://www.palmos.com/dev/. We have used the value 0x54616374, which translates to "Tact" in ASCII, as in Con-Tact. Also, in place of MIDP's `RecordStore`, we use KJava's `Database` class as our database instance.

The constructor is identical to the MIDP version and just saves the database name:

```
  public DBKJava(String name) {
    database = null;
    dbName = name;
  }
```

The `open()` method creates a new instance of the KJava `Database` class, using our unique `creatorID`, plus a `typeID` (we chose 0). This does not create a `Database`, it only opens it if it already exists. Therefore, having found it doesn't exist, we must open the `Database` object again, which requires having to call `Database.create()`:

```
  public void open() {
    if (database != null) {
      return;
    }
    int typeID = 0;
    database = new Database(typeID, creatorID, Database.READWRITE);
    if (!database.isOpen()) {

      // not open, so it needs creating
```

```
            Database.create(0, dbName, creatorID, typeID, false);

            // now try and open it again
            database = new Database(typeID, creatorID, Database.READWRITE);
        }
        return;
    }
```

Closing the database is as simple as the MIDP implementation. Again we set `db` to null to signal to any subsequent `open()` calls that there is no database currently open:

```
    public void close() {
        if (database == null) {
            return;
        }
        database.close();
        database = null;
    }
```

Unlike a MIDP `RecordStore`, we cannot delete a KJava `Database`, so we revert to deleting records, one by one. Note that the Palm database re-indexes its records after every deletion, so all we need do is delete the first record, until there are no records left:

```
    public void clear() {
        int num = numRecords();
        for (int i = 0; i < num; i++) {

            // NOTE: Palm DB 're-indexes' records after every delete, so just keep
            // deleting the 'first' record.
            database.deleteRecord(0);
        }
    }
```

Reading all records in the database is a straightforward loop, which is similar to that of the MIDP implementation, because both databases deliver database records as byte arrays. Therefore, we can just re-use our `Record` constructor that takes a byte array as the constructor parameter to convert between the database format and our desired `Record` class:

```
    public Record[] readRecords() {
        int num = numRecords();
        Record[] entries = new Record[num];
        byte[] b;
        for (int i = 0; i < num; i++) {
            b = database.getRecord(i);
            Record r = new Record(b, Record.PK);
            entries[i] = r;
        }
        return entries;
    }
```

KJava's `Database.getNumberOfRecords()` returns -1 rather than 0 when there are no records. We hide this peculiar result in `numRecords()` and return the more conventional 0 instead:

```
public int numRecords() {
  int num = database.getNumberOfRecords();
  if (num == -1) {
    num = 0;
  }
  return num;
}
```

Reading a single record is slightly different to the MIDP implementation because of the way the two database libraries signal that a record has not been found. MIDP throws an exception, `RecordStoreException`, which we must catch, whereas the KJava equivalent simply returns a null byte array. In both cases, we choose to return a null `Record` as the simplest way of indicating failure:

```
public Record readRecord(int pk) {
  byte[] b = null;
  b = database.getRecord(pk);

  if (b == null) {
    System.out.println("record not found");
    return null;
  }
  return new Record(b, Record.PK);
}
```

Reading a single record in KJava using a String name is similar. As with the MIDP implementation, we have no way of searching cleanly by name; we have to trawl through every record until we find the matching one. We pass a field index in (being `Contact.NAME_INDEX`) and the value to find a match for:

```
public Record readRecord(String value, int fieldIndex) {
  int num = numRecords();
  byte[] b;
  for (int i = 0; i < num; i++) {
    b = database.getRecord(i);
    Record r = new Record(b, Record.PK);
    if (r.getFieldValue(fieldIndex).equals(value)) {
      return r;
    }
  }
  return null;
}
```

Writing a record in KJava is nearly identical to the MIDP version, with the KJava write methods also being named `addRecord()` and `setRecord()`. Both KJava and MIDP versions take a byte array as the data input. The only difference is that the MIDP versions have additional parameters: one to specify the length of the data being saved, and an offset into the array (if not all of the byte array is to be saved).

The other noteable divergence is that in MIDP we use `getNextRecordID()` to discover the primary key value for a new record, whereas in KJava we can simply use the value returned from `numRecords()`. This is because Palm records are always contiguous, starting from 0, regardless even of record deletions:

```
public boolean writeRecord(Record r) {
  if (database == null) {
    return false;
  }

  boolean status = false;
  if (r.getPK() == -1) {
    int num = numRecords();
    r.setPK(num);
    byte[] b = r.getBytes(Record.PK);
    return database.addRecord(b);
  } else {
    byte[] b = r.getBytes(Record.PK);
    return database.setRecord(r.getPK(), b);
  }
}
```

Deleting a record in KJava has an important difference compared to the MIDP deletion. In the Palm database, the record IDs are always contiguous; in other words, if there are 20 records, their indices are always 0 to 19. If the user deletes the 5th record, then the indices are all changed so that the range of record IDs becomes 0 to 18. MIDP, on the other hand, always increments the index for every new record, with no attempt to fill in the 'gaps', or pack the records down into the lowest possible set of IDs.

The MIDP implementation suits our application perfectly, because even if the user does delete records, we don't need adjust the value of the primary key in our records, because it always matches the record's index, for the Palm implementation However, we have a problem. Whenever a record is deleted, we have to adjust every record's primary key that is 'above' the deleted record. Here's an example.

Before any deletions happen in this fresh database, there are 4 records:

```
0,Richard Taylor,+447971000000,rct@poqit.com,www.poqit.com,1 Mow Meadows
1,Grace Charlesworth,01613390000,,,2 Mow Meadows
2,Joe Public,02071230000,joep@nowhere.com,www.nowhere.com,12 Acacia Avenue
3,Fred Bloggs,05557754554,fred.bloggs@somewhere.com,www.fredb.com,
```

After deleting the second record, this is what the records would look like:

```
0,Richard Taylor,+447971000000,rct@poqit.com,www.poqit.com,1 Mow Meadows
2,Joe Public,02071230000,joep@nowhere.com,www.nowhere.com,12 Acacia Avenue
3,Fred Bloggs,05557754554,fred.bloggs@somewhere.com,www.fredb.com,
```

Now, we need to make sure the primary keys are contiguous by decrementing all those above the deleted record, so both Joe's and Fred's primary keys now look like this:

```
0,Richard Taylor,+447971000000,rct@poqit.com,www.poqit.com,1 Mow Meadows
1,Joe Public,02071230000,joep@nowhere.com,www.nowhere.com,12 Acacia Avenue
2,Fred Bloggs,05557754554,fred.bloggs@somewhere.com,www.fredb.com,
```

Our KJava `deleteRecord()` method implements this functionality, below:

```
    public void deleteRecord(int primaryKey) {
      database.deleteRecord(primaryKey);
      int num = numRecords();
      if (num == 0) {
        return;
      }
      byte[] b;
      for (int i = primaryKey; i < num; i++) {
        b = database.getRecord(i);
        Record r = new Record(b, Record.PK);
        r.setPK(r.getPK() - 1);
        writeRecord(r);
      }
    }
  }
```

The UserInterfaceKJava Class

The KJava user interface is no more sophisticated than MIDP, except in its support for pen events and key events. We use pen events for detecting whether a button was pressed, an item was selected from a list and for scrolling. More graphically-oriented applications could make rather more use of the pen events (`penUp`, `penDown`, and `penMove`), but since we need to be compatible with the MIDP functionality, we have chosen not to. Key events are for keyboard (or a keyboard equivalent) input and again we don't try and perform anything spectacular, we just gather the input as it's entered.

Where MIDP gave no choice of layout, KJava gives us the chance to position our GUI components, but does not have any sort of layout manager, so we must do all the work ourselves. We have decided to define a very simple screen layout with three areas:

- Title
- 'Main' area
- Button bar

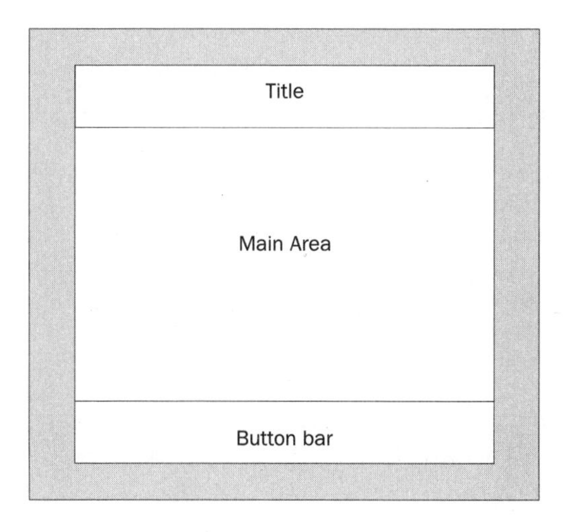

The 'Main' area holds the specific component for each of the screens, such as the list of contacts, the editable contact and so on.

Here is the `UserInterfaceKJava` source:

```
package com.poqit.j2me.kjava.ui;

import com.poqit.j2me.common.ui.UIListener;
import com.poqit.j2me.common.ui.UserInterface;
import com.poqit.j2me.common.database.Field;
import com.sun.kjava.Graphics;
import com.sun.kjava.Button;
import com.sun.kjava.List;
import com.sun.kjava.SelectScrollTextBox;
import com.sun.kjava.ScrollTextBox;
import com.sun.kjava.Spotlet;
import com.sun.kjava.TextField;
import com.sun.kjava.TextBox;
```

First note, that we extend the `Spotlet` class, the KJava user interface class.
Then we define the screen dimensions. We've assumed a standard Palm screen of 160 x 160 pixels and then extrapolated from that to give each of the three screen areas a certain amount of space (`BUTTONHEIGHT`, `TITLEHEIGHT` and `MAINHEIGHT`):

```
public class UserInterfaceKJava extends Spotlet implements UserInterface {

  // Useful screen dimensions. Assumes a standard Palm screen.
  public static int MINX = 0;
  public static int MINY = 0;
  public static int MAXX = 159;
  public static int MAXY = 159;
  public static int CHARHEIGHT = 10;

  // All these are derived from above dimensions.
  public static int SCREENWIDTH = MAXX - MINX;
  public static int SCREENHEIGHT = MAXY - MINY;
  public static int BUTTONHEIGHT = CHARHEIGHT * 2;
  public static int TITLEHEIGHT = ((CHARHEIGHT * 3) / 2);
  public static int EDITHEIGHT = ((CHARHEIGHT * 3) / 2);
  public static int MAINHEIGHT = MAXY - MINY - BUTTONHEIGHT - TITLEHEIGHT;

  private int UImode;
```

`eventListener` is our usual registered listener that deals with all our events, i.e. `ContactApp`:

```
  private UIListener eventListener;
```

`g` is our graphics context,and we only need to set it once, so it's a static variable:

```
  private static Graphics g = Graphics.getGraphics();
  static SelectScrollTextBox sstb;
  static TextField tf;
  static TextField[] tfa;
```

The labels of `TextField` are not retrievable, so we save them in `tfaLabels`, since we need to identify each `TextField` in some way:

```
static String[] tfaLabels;
```

Finally, we have a `Button` array and again, a matching button label array:

```
Button[] buttons;
String[] buttonLabels;
```

The constructor for a `Spotlet` is minimal. All we need to do is specify that we want the input focus, so we get pen down events, key presses, and so on:

```
public UserInterfaceKJava() {
  register(NO_EVENT_OPTIONS);
}

public void close() {}

public void displayList(String title, String[] choice,
                        String[] actionLabels) {
```

The class variable `UImode` is set to `LIST` at the start of the method and helps us identify which user interface component caused events later on in our event-handling method.
When displaying the list, we also need some buttons for navigation and a title for the screen. As with the MIDP implementation, the `actionLabels` are the button labels for this screen:

```
  UImode = LIST;
  addTitle(title);
  addActions(actionLabels);

  StringBuffer sb = new StringBuffer();
  for (int i = 0; i < choice.length; i++) {
    sb.append(choice[i] + "\n");
  }
```

For displaying a list of contacts (or databases), KJava has the `SelectScrollTextBox` class, which, as its name suggests, is a scrollable text box, with which the user can click in to select one of the items in the text box. The selection is always a complete line in the text box – we cannot select words, or arbitrary parts of the text box contents. Therefore, we need to build our list of contacts with one per line, and then place them all into the `SelectScrollTextBox`. Calling the `paint()` method causes it to be drawn

We also choose to use a class-wide reference, `sstb`, for the `SelectScrollTextBox`, because we have the luxury of knowing that only one `SelectScrollTextBox` is ever used at one time. Therefore, we can directly access `sstb` from the event handling method (`penDown()`) and see which contact the user selected:

```
  sstb = new SelectScrollTextBox(sb.toString(), MINX, MINY + TITLEHEIGHT,
                                 MAXX, MAINHEIGHT);
  sstb.paint();
}
```

We use such class-wide references for all of our low-level KJava-specific user interface components. In a more complex application, you might want to consider a more robust way of preserving the application's state.

The size of the list is set to be the majority of the Palm screen with the constant values MINX, MINY, MAXX, and MAXHEIGHT. MAXHEIGHT is less than MAXY to make space for buttons at the bottom of the screen (the drawing origin is the top left corner of the screen, like MIDP), and CHARHEIGHT allows space for the title. Here is the resulting screen:

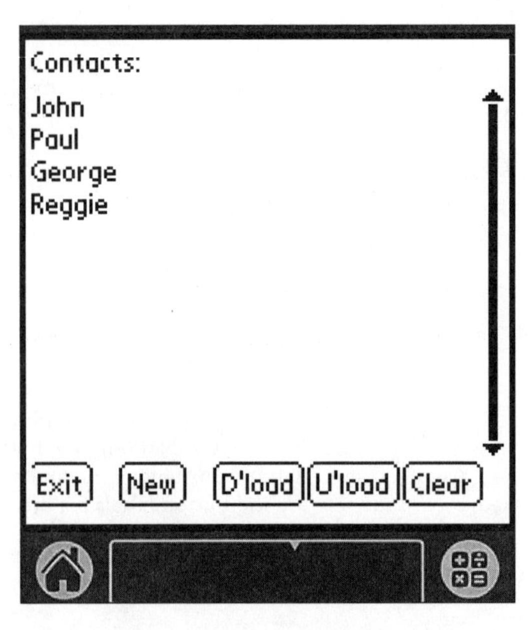

Displaying the fields of a contact is done by displaying a single TextBox, with all the details of the contact or database inside it. We do this by appending the field names and values together, separated as usual by the pre-defined NAMEEND-ST and VALUEEND_ST Strings for readability:

```
    public void displayFields(String title, Field[] fields,
                              String[] actionLabels) {
      UImode = SIMPLEACTION;
      addTitle(title);
      addActions(actionLabels);

      int num = fields.length;
      StringBuffer sb = new StringBuffer();
      for (int i = 0; i < num; i++) {
        sb.append(fields[i].getName());
        sb.append(UserInterface.NAMEEND_ST);
        sb.append(fields[i].getValue());
        sb.append(UserInterface.VALUEEND_ST);
      }
      ScrollTextBox stb = new ScrollTextBox(sb.toString(), MINX,
                                            MINY + TITLEHEIGHT, MAXX,
                                            MAINHEIGHT);
      stb.paint();
    }
```

Here is the result:

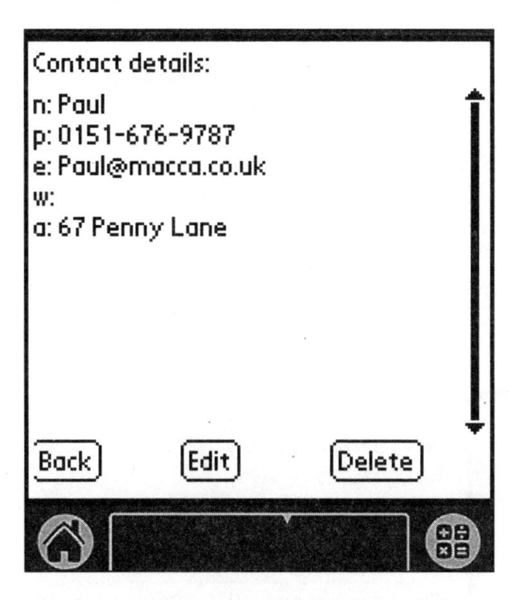

To edit a field, KJava offers a `TextField` that allows text entry:

```
public void editField(String title, Field field, String[] actionLabels) {
  UImode = SINGLEEDIT;
  addTitle(title);
  addActions(actionLabels);

  tf = new TextField(field.getName(), MINX, MINY + TITLEHEIGHT,
                     SCREENWIDTH, EDITHEIGHT);
  tf.setText(field.getValue());
  tf.paint();
}
```

Editing multiple fields is the same, except the fields must be placed down the main area of the screen. It isn't possible to identify the `TextField` after their creation (except by a reference to the object of course), because although we can set the title of a KJava `TextField`, we cannot read that title at a later date. Therefore, we remember the titles by holding them in a String array named `tfaLabels`, which is used later in the event-handling method `penDown()`:

```
public void editFields(String title, Field[] fields,
                       String[] actionLabels) {
  UImode = MULTIEDIT;
  addTitle(title);
  addActions(actionLabels);

  int num = fields.length;
  tfa = new TextField[num];
  tfaLabels = new String[num];
  for (int i = 0; i < num; i++) {
```

We can then set the input focus to the first `TextField` on the screen, (with `setFocus()`) and remember which `TextField` is the 'current' one, in other words, the one with the focus:

```
      tf = new TextField(fields[i].getName(), MINX,
                             MINY + TITLEHEIGHT + (EDITHEIGHT * i),
                             SCREENWIDTH, EDITHEIGHT);
      tf.setText(fields[i].getValue());
      tf.paint();
      tfa[i] = tf;
      tfaLabels[i] = fields[i].getName();
    }
    tf = tfa[0];
    tf.setFocus();
  }
```

Here is the multi-edit screen:

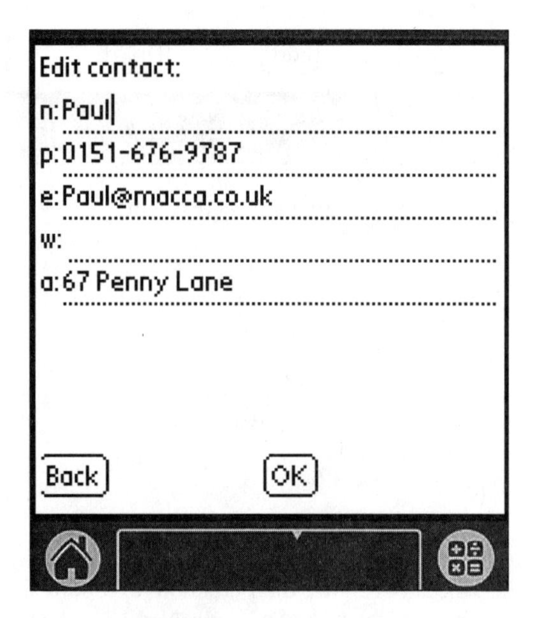

Adding a title to each screen is very straightforward:

```
  private void addTitle(String title) {
    TextBox tb = new TextBox(title, MINX, MINY, MAXX, TITLEHEIGHT);
    tb.paint();
  }
```

Adding the buttons to the screen is also simple, with some crude computation to space the buttons out roughly equal across the bottom of the screen. This works only for small numbers of buttons and with sensible length labels:

```
  private void addActions(String[] actionLabels) {
    buttonLabels = actionLabels;
    int num = actionLabels.length;
```

```
      buttons = new Button[num];
      for (int i = 0; i < num; i++) {
        buttons[i] = new Button(actionLabels[i], (MAXX * i) / num,
                                MAXY - BUTTONHEIGHT + 3);
        buttons[i].paint();
      }
    }

    public void setUIListener(UIListener uil) {
      eventListener = uil;
    }
```

Where our MIDP implementation of `UserInterface`, `UserInterfaceMIDP`, implemented `CommandListener`, we have to extend the `Spotlet` class, which is one of the reasons we couldn't have `UserInterface` as an abstract class. A `Spotlet` can draw to the screen and can receive user input events. There are several event-handling methods defined for a `Spotlet`, but only two are of interest to us; `penDown()` and `keyDown()`.

`penDown()` is by far the most complex handler because, as with the MIDP implementation, we must be able to:

- Work out which GUI component caused the event
- Extract the data from the user interface component in a form suitable for passing to a higher-level class
- Call the correct method of the registered UIListener instance, passing it the platform-neutral user data

`penDown()` is only passed the co-ordinate data where the pen was pressed down, whereas the MIDP helped us a little by delivering the object that caused the event. So in `penDown()`, to link a raw pen press to an object, we have to do some testing to find out exactly what caused the event. Each component of interest to us has a method to test whether the pen press was within its boundaries, so we must test every component on the screen, until we find which one it was. Here are some examples of how the components tell us about events:

- For buttons, `Button.pressed(x,y)` returns true if the pen was within its boundaries
- For lists/menus, `SelectScrollTextBox.getSelection(x,y)` returns the item selected as a String if the pen was within its boundaries
- For input fields, `TextField.pressed(x,y)` returns true if the pen was within its boundaries

We combine this with `UImode` to decide which `UIListener` method to call and with what data from the user interface:

```
    public void penDown(int x, int y) {
      boolean found = false;
      int num = buttons.length;
      for (int i = 0; (i < num) &&!found; i++) {
        if (buttons[i].pressed(x, y)) {
          g.clearScreen();
          if (UImode == MULTIEDIT) {
```

The swapping of input focus is indicated to the user by a blinking caret. Whenever an edit is completed with "OK "or "Back", we must stop the caret from flashing, with `tf.loseFocus()`:

```
            tf.loseFocus();
            if (buttonLabels[i].equals(OK_AL)) {
              Field[] fields = new Field[tfa.length];
              for (int j = 0; j < tfa.length; j++) {
```

Here we can note the retrieving of the button labels from the `tfaLabels` array when the user presses "OK" after having edited a contact:

```
                Field field = new Field(tfaLabels[j], tfa[j].getText());
                fields[j] = field;
              }
              eventListener.multiEditListener(fields);
              found = true;
            }
          }
          if (UImode == SINGLEEDIT) {
            if (buttonLabels[i].equals(OK_AL)) {
              tf.loseFocus();
              eventListener.editListener(tf.getText());
              found = true;
            }
          }
          if (!found) {
            eventListener.actionListener(buttonLabels[i]);
            found = true;
          }
        }
      }

      if (!found) {
        if (UImode == LIST) {
          String name = sstb.getSelection(x, y);
          if (name != null) {
            g.clearScreen();
            eventListener.listListener(name);
            found = true;
          } else {
            sstb.handlePenDown(x, y);
            found = true;
          }
        }
        if (UImode == MULTIEDIT) {
          int count = tfa.length;
          for (int i = 0; i < count; i++) {
            TextField temptf = tfa[i];
            if (temptf.pressed(x, y)) {
              tf.loseFocus();
              temptf.setFocus();
              tf = temptf;
            }
          }
        }
      }
    }
```

The `keyDown()` method is very simple, because KJava does all the key input handling for us and also because we only have one possible source of key input. The single source of key input is the `TextField` that currently has focus. We just need to call its `handleKeyDown()` method and it will refresh the screen for us, drawing/deleting characters as necessary:

```
    public void keyDown(int keyCode) {
      tf.handleKeyDown(keyCode);
    }
  }
```

Starting the KJava Application

Starting the KJava application differs from the MIDP application in only very small areas, namely:

- `DBKJava` database object are instantiated instead of `DBMIDP` ones
- A `UserInterfaceKJava` object is created instead of a `UserInterfaceMIDP` one
- The initial displaying of the initial contact list occurs within the constructor, since there is no `startApp()` equivalent in KJava
- We have a static `main()` method to start the application

The MainKJava Class

Here is the code for the `MainKJava` class:

```
package com.poqit.j2me.kjava;

import com.poqit.j2me.common.Main;
import com.poqit.j2me.common.app.ContactApp;
import com.poqit.j2me.common.database.DB;
import com.poqit.j2me.kjava.database.DBKJava;
import com.poqit.j2me.kjava.ui.UserInterfaceKJava;

public class MainKJava implements Main {
  private DB database;
  private ContactApp app;

  public MainKJava() {
    UserInterfaceKJava ui = new UserInterfaceKJava();
    database = new DBKJava("names");
    database.open();
    app = new ContactApp(this, ui);
    app.list();
  }

  public void exit() {
    System.exit(0);
  }

  public DB getDB() {
    return database;
  }

  public static void main(String[] args) {
    MainKJava app = new MainKJava();
  }
}
```

Improvements to the Application

As usual, there are a host of improvements and extensions we could make to this application, but which space doesn't allow us to cover here. Here are a few ideas for more advanced features, if we wish to build on our application:

- **Memory space and performance**
  There is no attempt to tune the application for minimal memory space or maximum performance. Both will affect each other too, with high performance code often taking more space up (loop unrolling and so on). For a discussion of performance and size tuning and design, refer to Chapter XX.
- **Database Sharing**
  Multiple MIDlets or Java mobile apps could share this database (we need to make sure our target devices will support multiple database access, of course).
- **Filtering and Searching**
  Contacts should be sorted alphabetically rather than in the order they were created. Filtering and searching of contacts is necessary for commercial applications. Using a 'folder' analogy for storing contacts is another form of filtering. The MIDP defines some useful filtering and sorting interfaces such as `RecordComparator` and `RecordFilter`, which are used by `RecordStore.enumerateRecords()`.
- **Merging of databases**
  True merging of databases would be very useful. This is probably best done on the target device because any changes will not be permanent until uploaded by the user. This may require significant memory requirements, of course. SyncML is a useful standard to investigate.
- **Data validity**
  The validity of entered data, such as URLs and e-mail addresses, is not checked.
- **Controlling databases**
  A server-side administration application to control all the uploaded databases would be an option. Also some controlled personal folder mechanism to keep them in would be necessary for privacy, if this is to be used on the open Internet.
- **Comma checking**
  No checking for commas during data entry to make sure user does not enter those characters.
- **XML use**
  Use XML for the data format, for greater simplicity of back-end and third party interfacing.

Summary

We have seen how to separate our CLDC business logic from the underlying profile, such as MIDP. Following on from there, we saw how to build a generic application that could be run on three mobile platforms, by separating the CLDC-based business logic from the underlying profile.

We also learnt the differences between the various profiles and pseudo-profiles, particularly in the areas of event handling, screen layout, user input, and database (persistence) capabilities. We also saw how large a simple application can become when it needs to be layered to promote portability.

With the additional functionality the application deserves, plus rigourous error handling, it is obvious that J2ME applications may run on relatively simple devices, but this does not mean the code is also simple. In fact the reverse is often true, since gaps in the API need to be plugged by your own code, thus increasing the application size.

Having worked through at least some of the code, you now have enough knowledge to write your own applications and recognize some of the potential pitfalls and complications of writing mobile applications for MIDP and CLDC.

PROFESSIONAL JAVA MOBILE
TOOL SELECTION
WEB TABLET
MOBILE PHONE
MP3 PLAYER
DIGITAL CAMERA
PALMTOP
LAPTOP
NOW SELECTED
PALMTOP

7

Converting Applications To CDC and CLDC

The subject of porting applications to different platforms should be something the Java developer doesn't have to worry about. After all, the mantra of Java is 'Write Once, Run Anywhere'. However, back in the real world, not all computers are made equal, and it was inevitable that there would have to be different versions of Java to work in all these different environments.

In this chapter, we'll be looking at how applications developed for full-scale Java (J2SE) can be moved onto platforms supporting the various J2ME profiles. Working through an example of such a conversion, we'll look at the tools and techniques available to smooth the transition from desktop to device, and how a Java application can truly be said to be platform independent.

There are two aspects to getting application to work on any platform, the first is purely practical, while the second is more application specific. The Java 2 Standard Edition has various APIs available to programmers, and some of these are not available to those working on smaller implementations. While some of these APIs can be replaced, some require a rethink of what task the API is performing, and how that task can be performed without it.

The second issue is the user interface, reduced resolution and color depth may make our user interface impractical, and certain user interactions, such as `MouseEntered` events, just don't happen in a pen driven interface, or key input only environment. We will be looking at both these issues, and while we may make clear indications of how to deal with missing APIs, alterations to the user interface will be largely dependent on the application in question.

Summary Of Differences

While there have been several alterations deep within the JVM, to make it operate more effectively with limited resources, they are of limited interest to the application developer. Improvements in garbage collection were also essential to get applications working, they don't actually affect the way in which code is written, Java provides insulation for the developer from these issues. Finally, while code optimization techniques may differ between profiles, and the desired results may also be different, initial application development can ignore these issues.

Initially, we are only interested in how we have to change our programming style, and our programs, to make them work effectively, later we will have to spend time considering how our application may change due to its environment. Mobile users may have different requirements, and we may decide that our application architecture is inappropriate to a device of limited processing power, but these changes will depend on the application being ported.

The Connected Device Configuration

Programming to the CDC currently makes use of the PersonalJava Specification, which is very close to what we think of as desktop Java. Here, programmers with more experience of Java, have a real advantage as the APIs available to the PersonalJava developer are basically those available under Java 1.1. Anyone with experience developing with Java before version 2, will have no problem working with PersonalJava. Indeed, applications originally developed for Java 1.1 should generally work under PersonalJava without any alteration at all.

However, programmers used to the later versions of Java should still have little problem getting used to the limitations of PersonalJava, as long as they are prepared to do without some of the tools they are more familiar with.

The big difference is the lack of the Swing API. Unlike the AWT, the basic GUI framework of Java 1.1, Swing uses so called "lightweight" components created in Java. To understand this difference, it's important to understand how the AWT operates; when an AWT component, such as a `Button`, is created, the JVM makes a request of the resident operating system for a `Button` component, this is known as a "peer component".

Practically, this means that when an AWT component is used in a Java application, executing on a Microsoft Windows platform, it shows a button that is and looks like all other Microsoft Windows buttons. The same application running on an Apple Mac computer will produce a button that is, and looks like, a Mac button.

This was considered a positive advantage, as it meant that Java applications always looked like native ones, and also that such peer components executed very fast, being developed as part of the operating system. Unfortunately, as operating system developers added new GUI features, such as the Tabbed Panel and Tree View, it was impossible for the AWT to access these new components, without being constantly updated.

Developers also found that they wanted more control over the appearance of GUI widgets, and looked for a consistent "Java" look-and-feel. The solution was found with the development of the Swing API. This API is developed completely in Java (and therefore considered lightweight as it makes no use of underlying OS components) and provides much more flexibility to the developer. Application developers working on desktop applications now have the choice of the native-looking, fast, AWT, and the flexible, pretty, Swing. Unfortunately, no such choice exists for the J2ME developer.

It should be noted, that as Swing was developed completely in Java, it should be possible to move the Swing APIs to run in a J2ME environment. While Swing cannot be used with the CLDC, it will run (just about) on CDC devices. However, attempting this does show the real speed advantages of using native peer components, as Swing operation on currently available CDC devices is far too slow for reasonable usage.

we have no Swing therefore, though the AWT is available in full, and can be used to provide just about any interface imaginable. Developers desiring features absent from the AWT, but popular in Swing (such as tabbed panels) may find free implementations available that were designed prior to the release of Java 2 platform. Developing these GUI widgets, was a very popular hobby until Sun bundled them into Java, and while they were designed to run on Java 1.1.x, they will work well on a PersonalJava platform. Remember to consider speed issues when working with any lightweight components.

However, one aspect the developer should be aware of is that local implementations of familiar AWT components may look very different on some devices. The replacement of AWT components with appropriate local peers was always intended for desktop systems, though the uniformity of most desktop systems has reduced the impact considerably.

Considering the placement of AWT `Menu` objects, may provide some guidance in this. On most platforms the `Menu` object, which is associated with a `Frame`, will appear at the top of that `Frame`, however, on the Apple Macintosh platform, such `Menu` objects always appear at the top of the screen. This was very much the original idea of the AWT, but it is particularly pertinent when porting applications to a wider style of devices.

In PersonalJava, this process is known as Truffle, and Sun provide an implementation called Touchable to give us some idea what to expect. Touchable is designed for touch screen devices, and as such has very large buttons and scroll bars.

Of course, none of this should be any problem to the Java programmer. Older Java applications, developed without Swing or other lightweight APIs, have always appeared different on different platforms, making use of native peer components. This process makes the application interface consistent with the operating system that the user is familiar with, rather than being imposed by the programmer.

Most of us However would like slightly more control over how our applications appear. This is still possible through the use of lightweight components, though care should be taken if this route is to be followed. Lightweight components have a greater processing and memory overhead than their AWT equivalents as already discussed.

While it is in the area of Graphical User Interfaces that Java 2 had the most impact to the platform, some other useful classes were added, and which are absent from PersonalJava. While a full listing is given in the appendices, it is worth further enlarging on a few important ones.

First, the `Collections` added in Java 2 are missing. While these can always be implemented, this represents considerable work if applications make extensive use of them. The list of missing classes includes the `ArrayList` and `Stack` objects so useful in Java 2.

Also missing are Java2D and the drag & drop functionality from J2SE. While useful (these two APIs can be implemented using the AWT) this task is not easy, and in the case of drag & drop, of dubious usefulness in devices. Finally, the Java Sound API is also absent, though the static `Applet.getAudioClip` may be available (depending on the implementation).

Connected Limited Device Configuration

The Connected Limited Device Configuration is far further removed from J2SE than the CDC. Rather than try to implement parts of the J2SE JDK in limited form, it takes a different approach in that it throws out the bits that just aren't going to work with severely limited resources. The big differences for the application programmer are the complete lack of AWT and the different networking classes. The network classes are discussed in Chapter 3 on the CDC and the Foundation Profile, so here we'll concentrate on the lack of AWT.

The AWT is a wonderful thing, and losing it, along with its event model, does change programs a great deal, as we will see. Implementations of the CLDC make various AWT components available to us, though they operate in a different way to what most developers are familiar with. We can find a lot more detail in Chapter 5, but it's worth remembering that the pixel positioning will limit our applications to a particular resolution, if not a particular device. The following example shows how a `Button` is instantiated with two additional parameters, giving the size of the button in pixels:

```
quit = new Button("Quit", 135, 145);
```

Programmers with long memories might recognize the event mechanism of CLDC as being similar to the one used on Java 1.0. The main difference is that in Java 1.0 all events are handled in the same method of a single class, meaning that enormous `if{}else{}`chains are needed to work out what action has occurred. As actions included `MouseMove` events, this had a massive impact on program execution speed and responsiveness.

The big problem with the event model from Java 1.0 was that it got overly complex with large applications and became quite unmanageable. On a CLDC device we shouldn't be developing large applications, and the simplicity of having all events handled in one place can be useful. Here is an example:

```
public void penDown(int x, int y) {
   if(insert.pressed(x,y)) {
   ...// do stuff
   } else if(quit.pressed(x,y)) {
   ...// do stuff
   } else if(last.pressed(x,y) && current > 0) {
   ...// do stuff
   } else if(next.pressed(x,y) && current < (slides.size()-1)) {
   ...// do stuff
   } else if(delete.pressed(x,y)) {
   ...// do stuff
   }
}
```

At times a `case` statement may be more appropriate.

JavaCheck

In order to ease the porting of J2SE applications to PersonalJava, Sun provide software known as JavaCheck (http://java.sun.com/products/personaljava/javacheck.html), which can analyze Java class files and report on their use of unsupported APIs. It's available free as a download and is written completely in Java. For Microsoft Windows users there is a self extracting archive available, though note that you will also have to download at least one specification file to be able to use the software (at the moment there is only one specification available, as we shall see in the following examples).

Most of the supported classes are pretty obvious, as applications using Swing aren't going to work. As we shall see, JavaCheck is not a perfect tool, but it can highlight less obvious APIs, like the use of the `Collections` APIs, which might not be obvious to the developer.

While JavaCheck is a useful tool for checking compatibility, it is important to understand the status of PersonalJava platforms, and the level of support they can provide. There are four possible states for an API to be within PersonalJava:

- `Supported`
  The API or class is available in all PersonalJava compatible platforms.
- `Unsupported`
  The API or class is not supported so find some other way to do it.
- `Modified`
  The API or class is available, but isn't quite the same as in J2SE. An example would be `Frame`, which is the same except that we can only have one of them.
- `Optional`
  Support for this is optional.

While the first three states are pretty obvious, it is the fourth state that can give problems. Several APIs are considered optional, and from the developer's point of view it makes things very complicated. There is no guarantee that an application that makes use of optional components will work on every PersonalJava-compatible platform, and in some ways it negates the usefulness of having the standard. Experience has shown, so far, that all implementations are supporting all the optional components, but there is no way of telling if this will continue.

There are obvious exceptions, in that a device without a screen is not going to have any use for the AWT. At the time of writing the only commercially available platforms for PersonalJava deployment are EPOC, WinCE/PocketPC, and VxWorks, and all these implementations have all of the optional components available, so it seems likely that this trend will continue.

JavaCheck will test an application, using the `.class` files, and report on which of the above areas the APIs and classes fail. It can also provide some hints as to how to approach dealing with the modified or unsupported calls.

JavaCheck checks against different versions of Java, and it does this through specification files. When you download JavaCheck, you also download the appropriate specification file for the version of Java your target device has implemented.

Unfortunately, at the time of writing (and for the last few years), the only specification file available is for PersonalJava, so it's ideal if we are developing for CDC devices, but less than useful for anything else. It is envisioned that licensees of Java will produce specification files for their devices, though no one has yet done so.

There are different versions of JavaCheck for the different versions of PersonalJava, though in theory it should just be a matter of using a different specification file. There are two ways of running JavaCheck once it's installed, either using a GUI or from the command line. The GUI version is, obviously, easier to understand and is easily run by entering:

```
java JavaCheckV
```

The V at the end of the name denotes that we are running the visual version (with its GUI). It is also possible to specify the CLASSPATH from which to operate:

```
java -cp c:\JavaCheck\bin\JavaCheck.jar JavaCheckV
```

Running the command line version is slightly more complicated, as we have to specify everything using one line. We will be looking at examples of both modes of operation in the next section.

PersonalJava Emulation Environment

Sun provide an emulation environment for testing PersonalJava applications. This is very useful and can be downloaded from http://www.javasoft.com/products/personaljava/pj-emulation.html.

Once you have downloaded it, you will be offered several options. Although several of these regard the version you wish to develop for, which will depend on our intended target platform, there is also the question of interface. As already noted, PersonalJava implementations are encouraged to modify the AWT components to suit the device they are porting to, and this can have quite an impact on the appearance of your application.

Sun provides one alternative interface, the Touchable interface, which is designed for touch screen devices, and this is the default interface provided to licensees. The idea is that the licensees will either use Touchable, or develop their own interface elements (or provide access to peer objects, as J2SE does) and provide these as plug-ins for the emulator. Unfortunately, no one is yet doing this, though the lack of deployed PersonalJava implementations may have an impact on this.

If your intended target device has a novel interfacing system, then it might be worth trying to get the emulator configured to use widgets appropriate to that device, if the device manufacturer is prepared to make them available. However, there is no substitute for testing on the device itself, which should always be the final stage.

Once downloaded, the emulator is easy to use, though you might have to reset your CLASSPATH and JAVA_HOME variables to get it working. See the accompanying User Guide for detailed instructions.

Example Conversion

The application we will be converting is a basic slide editor, which enables the user to enter new slides, delete slides and move around their list. It also synchronizes with a client device, so we'll be porting the code to that client device. We'll start by looking at the source code for the example, developed using Java Platform 2 Standard Edition (J2SE), then consider what changes we are going to need to get the same application working on the various J2ME configurations.

The application is based around three windows, which are shown below; the one on the top shows the current slide, and offers navigation controls for moving around the presentation. The second window shows the titles of all the slides in the current show. From this window, we can delete the highlighted slide, insert a new slide, and synchronize this slide show with another device. The synchronization process is used in the following chapter, so we won't dwell on it here.

The third window is only visible when inserting a slide, and allows the user to add lines to the new slide. It was developed to allow drafting of presentations, and while it lacks many features, it will serve as a useful example for our purposes:

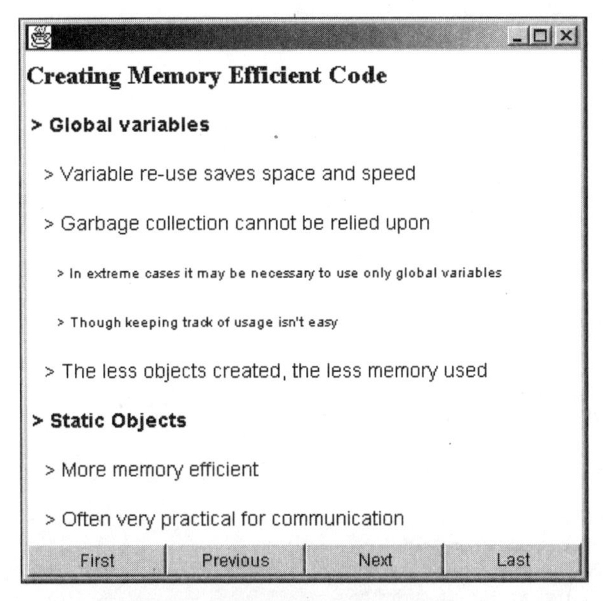

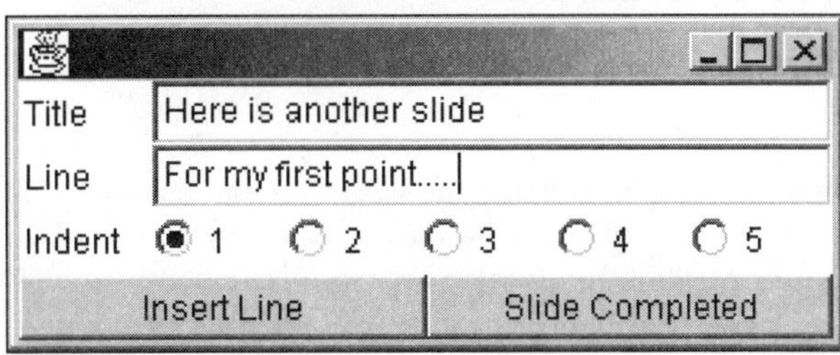

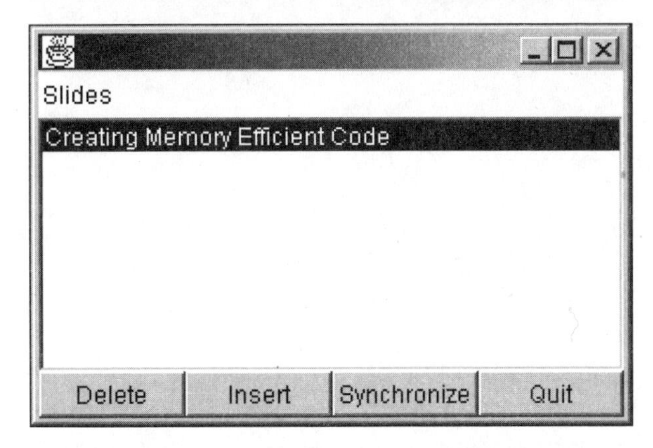

It's not necessary to go through this applications source code in detail now, but we will be referencing parts of it in the following sections, so it's worth understanding how it operates.

The SlideEditor Class

Here we declare the `SlideEditor` class. A vector is used to store the `Slide` objects we'll be working with. Note that if we had used a different `Collection`, such as an `ArrayList` or `Stack`, we would have had more problems moving to J2ME as it doesn't support these collections:

```
import java.io.OutputStream;
import java.io.InputStream;
import java.io.IOException;

import java.net.Socket;
import java.net.ServerSocket;

import java.util.Vector;
import java.util.Date;

import java.awt.BorderLayout;
import java.awt.Button;
import java.awt.Frame;
import java.awt.GridLayout;
import java.awt.Label;
import java.awt.Panel;

import java.awt.event.ItemListener;
import java.awt.event.ItemEvent;
import java.awt.event.MouseListener;
import java.awt.event.MouseEvent;

public class SlideEditor implements MouseListener, ItemListener {
  Vector slides;
```

In the next code segment, we will be creating the various variables we are going to be using, including the confirmation dialog and the frame used for editing slides being inserted. Note that we are using the `List` class from the AWT, which has to be specified in full to avoid conflict with the Swing object of the same name:

```
  Button first, previous, next, last;
  Button delete, insert, quit, insertLine, complete, sync;
  Date d = new Date();
  Slide sl;
  int current = 0;
  SlideViewer disp;
  java.awt.List titles = new java.awt.List();
  Confirmation confirm;
  Frame f;
  Insertion myInsertion;
  String CRLF = new Character((char) 13).toString()
              + new Character((char) 10).toString();
  Panel slideDisplay;
  Status update;
```

The `main()` method simply instantiates a non-static instance of itself:

```
  public static void main(String args[]) {
    SlideEditor mySlideEditor = new SlideEditor();
    mySlideEditor.go();
  }
```

The `go()` method builds the GUI and adds the various listeners, as well as creating a sample slide to display. If this was a real application we would certainly want some persistence (the ability to load and save slides), and probably some way of exporting them into a standard format, however, this example will serve our purposes. It also calls the `buildList()` method, which builds the list of slide titles used in the second window. Notice that the `myInsertion` variable is instantiated using a reference to the `SlideEditor`, as well as the display window:

```
   public void go() {
      first = new Button("First");
      previous = new Button("Previous");
      next = new Button("Next");
      last = new Button("Last");
      delete = new Button("Delete");
      insert = new Button("Insert");
      quit = new Button("Quit");
      insertLine = new Button("Insert Line");
      complete = new Button("Insert Slide");
      sync = new Button("Synchronize");

      first.addMouseListener(this);
      previous.addMouseListener(this);
      next.addMouseListener(this);
      last.addMouseListener(this);
      delete.addMouseListener(this);
      insert.addMouseListener(this);
      sync.addMouseListener(this);
      quit.addMouseListener(this);
      titles.addItemListener(this);

      f = new Frame();
      f.setLayout(new BorderLayout());
      Panel masterControl = new Panel();
      masterControl.setLayout(new GridLayout(1, 4));
      masterControl.add(delete);
      masterControl.add(insert);
      masterControl.add(sync);
      masterControl.add(quit);

      f.add(masterControl, BorderLayout.SOUTH);
      f.add(new Label("Slides"), BorderLayout.NORTH);
      f.add(titles, BorderLayout.CENTER);
      slides = new Vector(0, 1);
```

Here we will be stipulating the text that will be displayed in the slide:

```
      sl = new Slide("Creating Memory Efficient Code");
      sl.addLine(1, "Global variables");
      sl.addLine(2, "Variable re-use saves space and speed");
      sl.addLine(2, "Garbage collection cannot be relied upon");
      sl.addLine(3,
                "In extreme cases it might be necessary to use only global
variables");
      sl.addLine(3, "Though keeping track of usage isn't easy");
      sl.addLine(2, "The less object created, the less memory used");
      sl.addLine(1, "Static Objects");
```

```
    sl.addLine(2, "More memory efficient");
    sl.addLine(2, "Often very practical for communication");
    sl.setTime(d.getTime());
    slides.addElement(sl);
    disp = new SlideViewer(first, previous, next, last);
    disp.setVisible(true);
    disp.viewSlide((Slide) slides.elementAt(current));
    myInsertion = new Insertion(this, disp);
    confirm = new Confirmation(f, this);
    update = new Status(f);
    buildList();
```

```
    f.pack();
    f.setLocation(450, 20);
    f.setSize(300, 200);
    f.setVisible(true);
  }
```

The `changeOfSlide()` method is called whenever a slide is changed, for whatever reason. It uses the variable `current` to retrieve the slide from the `Vector` and passes it to the display window. It also makes sure that the right title is highlighted in the titles window:

```
  public void changeOfSlide() {
    disp.viewSlide((Slide) slides.elementAt(current));
    titles.select(current);
  }
```

The following method is called whenever a slide is added or removed from the show. It passes each slide title to the titles window for display:

```
  public void buildList() {
    titles.removeAll();
    for (int i = 0; i < slides.size(); i++) {
      titles.add(((Slide) slides.elementAt(i)).getTitle());
    }
  }
```

Events are handled here, with `changeOfSlide()` being called at the end to ensure that the display is consistent:

```
  public void mouseClicked(MouseEvent e) {
    if ((e.getComponent() == next) && (current < (slides.size() - 1))) {
      current++;
    } else if ((e.getComponent() == previous) && (current > 0)) {
      current--;
    } else if (e.getComponent() == last) {
      current = slides.size() - 1;
    } else if (e.getComponent() == first) {
      current = 0;
    } else if (e.getComponent() == quit) {
      System.exit(0);
    } else if (e.getComponent() == delete) {
```

```
      confirm.setVisible(true);
    } else if (e.getComponent() == insert) {
      myInsertion.reset();
      myInsertion.setVisible(true);
    } else if (e.getComponent() == sync) {
      syncUpWithMirror(slides);
      buildList();
    }
    changeOfSlide();
  }
```

The rest of the methods necessary for any class implementing the `MouseListener` interface:

```
  public void mousePressed(MouseEvent e) {}
  public void mouseReleased(MouseEvent e) {}
  public void mouseEntered(MouseEvent e) {}
  public void mouseExited(MouseEvent e) {}
```

The following method is called when the user clicks on a slide title in the titles window, and causes that slide to be displayed in the display window:

```
  public void itemStateChanged(ItemEvent e) {
    current = titles.getSelectedIndex();
    changeOfSlide();
  }
```

The `deleteSlide()` method also checks that the currently selected slide isn't beyond the number of slides in the show (as would be the case if the last slide has been deleted):

```
  public void deleteSlide() {
    slides.removeElementAt(current);
    if (current >= slides.size()) {
      current = slides.size() - 1;
    }
    buildList();
    changeOfSlide();
  }
```

The next method inserts a slide at the current position (directly after the currently displayed slide):

```
  public void insertSlide(Slide s) {
    current++;
    slides.insertElementAt(s, current);
    buildList();
    changeOfSlide();
  }
```

These subsequent methods are used when the Synchronize button is pressed, and are used to synchronize the slide show with one located on another device. They are examined in detail in the next chapter, so we won't bother with the details here:

```
    public void syncUpWithMirror(Vector v) {
    }
    private String getResponse(InputStream is) throws IOException {
    }
    private Slide receiveSlide(InputStream is) throws IOException {
    }
    private void sendSlide(Slide s, OutputStream o) throws IOException {
    }
}
```

This marks the end of the `SlideEditor` class.

The Insertion Class

This class makes up the third window, and is only made visible when the user is editing a slide to be inserted into the show:

```
import java.util.Date;

import java.awt.BorderLayout;
import java.awt.Button;
import java.awt.Checkbox;
import java.awt.CheckboxGroup;
import java.awt.Frame;
import java.awt.GridLayout;
import java.awt.Label;
import java.awt.Panel;
import java.awt.TextField;

import java.awt.event.MouseListener;
import java.awt.event.MouseEvent;

class Insertion extends Frame implements MouseListener {
```

The slide being edited is held in this variable:

```
    Slide working = new Slide();
```

These slides make up the GUI components of the third window:

```
    Button insert, complete;
    TextField textInput = new TextField();
    TextField titleInput = new TextField();
    CheckboxGroup indent = new CheckboxGroup();
```

Next, we store a reference to the object that created this instance (`parent`), in order to be able to call methods of that object when we need to. We also store a reference to the display window (`picture`), so we can show the slide as it is edited:

```
    SlideEditor parent;
    SlideViewer picture;
```

Our constructor is used to create the layout of the window. Note that this window will not be visible when instantiated, only when `setVisible()` is called will this window become available to the user:

```
Insertion(SlideEditor r, SlideViewer s) {
  parent = r;
  picture = s;
  Panel indents = new Panel();
  indents.setLayout(new GridLayout(1, 5));
  working.setTitle("Slide under construction");
  indents.add(new Checkbox("1", indent, true));
  indents.add(new Checkbox("2", indent, false));
  indents.add(new Checkbox("3", indent, false));
  indents.add(new Checkbox("4", indent, false));
  indents.add(new Checkbox("5", indent, false));
  insert = new Button("Insert Line");
  complete = new Button("Slide Completed");
  setLayout(new BorderLayout());
  Panel headings = new Panel();
  headings.setLayout(new GridLayout(3, 1));
  headings.add(new Label("Title"));
  headings.add(new Label("Line"));
  headings.add(new Label("Indent"));
  Panel dataEntry = new Panel();
  dataEntry.setLayout(new GridLayout(3, 1));
  dataEntry.add(titleInput);
  dataEntry.add(textInput);
  dataEntry.add(indents);
  Panel buttons = new Panel();
  buttons.setLayout(new GridLayout(1, 2));
  buttons.add(insert);
  buttons.add(complete);
  insert.addMouseListener(this);
  complete.addMouseListener(this);
  Panel input = new Panel();
  input.setLayout(new BorderLayout());
  input.add(headings, BorderLayout.WEST);
  input.add(dataEntry, BorderLayout.CENTER);
  add(buttons, BorderLayout.SOUTH);
  add(input, BorderLayout.CENTER);
  setSize(300, 120);
  setLocation(450, 250);
}
```

Once a line of a slide has been entered we clear the text entry boxes:

```
public void reset() {
  textInput.setText("");
  titleInput.setText("");
}
```

Here we are catching the events:

```
public void mouseClicked(MouseEvent e) {
```

First, we check to see which button has been pressed, and if it's the **Insert** button, we add a line to the slide. The `addLine()` method takes two parameters, the indent and the contents. The indent is calculated by the check box selected, which is converted into an `int` through a complex conversion:

```
      if (e.getComponent() == insert) {
        if (!textInput.getText().equals("")) {
          working
            .addLine(new Integer((indent.getSelectedCheckbox()).getLabel())
              .intValue(), textInput.getText());
          textInput.setText("");
        }
        working.setTitle(titleInput.getText());
        picture.viewSlide(working);
      }
```

If the slide is complete, we add an identification string, and the current time (used for synchronization) and call the `insertSlide()` method of the `SlideEditor` object. This last step is done this way to ensure the overall application can update the various parts of the GUI to reflect the additional slide:

```
      if (e.getComponent() == complete) {
        setVisible(false);
        working.setID();
        Date d = new Date();
        working.setTime(d.getTime());
        parent.insertSlide(working);
      }
    }
```

The following methods are required by classes that implement the `MouseListener` interface:

```
    public void mousePressed(MouseEvent e) {}
    public void mouseReleased(MouseEvent e) {}
    public void mouseEntered(MouseEvent e) {}
    public void mouseExited(MouseEvent e) {}
  }
```

The SlideViewer Class

The constructor for this class again creates the GUI objects for this window, and also creates `Font` objects to be used for the display of differently indented elements of the slides. Note that the buttons displayed in this window, are actually created in the `SlideEditor` class, and the events associated with those buttons, are dealt with in that class:

```
  import java.awt.Frame;
  import java.awt.Font;
  import java.awt.Panel;
  import java.awt.Button;
  import java.awt.GridLayout;
  import java.awt.BorderLayout;
  import java.awt.Label;

  class SlideViewer extends Frame {
```

```
    Font indent1, indent2, indent3, indent4, indent5;
    Panel controls, SlideDisplay;
    SlideViewer(Button b1, Button b2, Button b3, Button b4) {
      controls = new Panel();
      controls.setLayout(new GridLayout(1, 4));
      controls.add(b1);
      controls.add(b2);
      controls.add(b3);
      controls.add(b4);
      indent1 = (new Font("SansSerif", Font.BOLD, 14));
      indent2 = (new Font("SansSerif", Font.PLAIN, 14));
      indent3 = (new Font("SansSerif", Font.PLAIN, 12));
      indent4 = (new Font("SansSerif", Font.PLAIN, 10));
      indent5 = (new Font("SansSerif", Font.PLAIN, 8));
      setLayout(new BorderLayout());
      SlideDisplay = new Panel();
      setLocation(20, 20);
      setSize(400, 400);
    }
```

This following is the one and only method of this class, which displays the slide handed to it as a parameter:

```
    public void viewSlide(Slide s) {
      removeAll();
      SlideDisplay.removeAll();
      Label l = new Label(s.getTitle());
      String pad = "";
      SlideDisplay.setLayout(new GridLayout(s.countLines() + 1, 1));
      l.setFont(new Font("Serif", Font.BOLD, 18));
      SlideDisplay.add(l);
      for (int i = 0; i < s.countLines(); i++) {
        l = new Label();
        switch (s.getLine(i).value) {
        case 1:
          l.setFont(indent1);
          pad = "";
          break;
        case 2:
          l.setFont(indent2);
          pad = "  ";
          break;
        case 3:
          l.setFont(indent3);
          pad = "    ";
          break;
        case 4:
          l.setFont(indent4);
          pad = "      ";
          break;
        case 5:
          l.setFont(indent5);
          pad = "            ";
          break;
        }
```

```
        l.setText(pad + " > " + s.getLine(i).text);
        SlideDisplay.add(l);
      }
      SlideDisplay.doLayout();
      add(SlideDisplay, BorderLayout.CENTER);
      add(controls, BorderLayout.SOUTH);
      show();
    }
  }
```

The Slide Class

This class is used to represent a `Slide`, and consists almost solely of mutator methods and instance variables:

```
import java.util.Vector;

class Slide {
```

The next lines of code making up the slide, are stored in a `Vector`. Again, use of an `ArrayList`, or `Stack`, would have caused us problems later, as these aren't supported in J2ME. The `seen` variable is used in synchronization to record if this slide has been synchronized:

```
  String ID;
  String title;
  long creationTime;
  Vector lines;
  boolean seen;
```

The `Slide` class also creates, and stores, a unique reference created from the first three characters of the title and its own hashcode. This is used in synchronization to gain a unique reference to the slide:

```
  Slide() {
    lines = new Vector(0, 1);
    seen = false;
  }
  Slide(String topLine) {
    title = topLine;
    lines = new Vector(0, 1);
    ID = title.substring(0, 3) + new Integer(this.hashCode()).toString();
    seen = false;
  }
  public void setTitle(String t) {
    title = t;
  }
  public void setID() {
    ID = title.substring(0, 3) + new Integer(this.hashCode()).toString();
  }
  public void setID(String s) {
    ID = s;
  }
  public void addLine(int indent, String text) {
    Line l1 = new Line(text);
    l1.value = indent;
```

```
      lines.addElement(ll);
    }
    public String getTitle() {
      return title;
    }
    public int countLines() {
      return lines.size();
    }
    public Line getLine(int i) {
      return (Line) lines.elementAt(i);
    }
    public long getTime() {
      return creationTime;
    }
    public void setTime(long l) {
      creationTime = l;
    }
    public String getID() {
      return ID;
    }
  }
```

The Line Class

This class stores the individual lines of the slide, consisting of a string and the amount the line is to be indented:

```
  class Line {
    public int value;
    public String text;
    Line(String s) {
      text = s;
      value = 0;
    }
    Line(int v, String s) {
      value = v;
      text = s;
    }
  }
```

The Confirmation Class

One instance of this class, is used to provide a confirmation dialog when the user wishes to delete a slide:

```
  import java.awt.Button;
  import java.awt.Panel;
  import java.awt.Dialog;
  import java.awt.Label;
  import java.awt.Frame;
  import java.awt.GridLayout;
  import java.awt.BorderLayout;

  import java.awt.event.WindowListener;
  import java.awt.event.WindowEvent;
  import java.awt.event.MouseListener;
  import java.awt.event.MouseEvent;
```

```
class Confirmation extends Dialog implements WindowListener, MouseListener {
  Button OK, cancel;
  SlideEditor parent;
```

The constructor creates the GUI framework, but nothing is displayed until the `setVisible()` method is called. The `WindowListener` interface is used so that the dialog can be closed in a manner appropriate to the OS being used (clicking on the top right corner of the window, in the case of a Microsoft Windows platform):

```
  Confirmation(Frame f, SlideEditor p) {
    super(f, "Delete?", true);
    parent = p;
    add(new Label("Please Confirm Deletion"), BorderLayout.NORTH);
    Panel options = new Panel();
    options.setLayout(new GridLayout(1, 2));
    OK = new Button("Delete");
    cancel = new Button("Cancel");
    options.add(OK);
    options.add(cancel);
    OK.addMouseListener(this);
    cancel.addMouseListener(this);
    add(options, BorderLayout.CENTER);
    setSize(200, 70);
    setLocation(120, 185);
    addWindowListener(this);
  }

  public void windowOpened(WindowEvent e) {}

  public void windowClosing(WindowEvent e) {
    setVisible(false);
  }

  public void windowClosed(WindowEvent e) {}
  public void windowActivated(WindowEvent e) {}
  public void windowDeactivated(WindowEvent e) {}
  public void windowIconified(WindowEvent e) {}
  public void windowDeiconified(WindowEvent e) {}
  public void mousePressed(MouseEvent e) {}
  public void mouseReleased(MouseEvent e) {}
  public void mouseEntered(MouseEvent e) {}
  public void mouseExited(MouseEvent e) {}
  public void mouseClicked(MouseEvent e) {
    if (e.getComponent() == cancel) {
      setVisible(false);
    }
    if (e.getComponent() == OK) {
      parent.deleteSlide();
      setVisible(false);
    }
  }
}
```

If the window is closed, or Cancel is pressed, the dialog simply vanishes. If OK is pressed, then it calls the `deleteSlide()` method of the `SlideEditor` object passed into the constructor.

The Status Class

This last class is only used during synchronization, where it shows the current progress of the synchronization process:

```
import java.awt.Label;
import java.awt.Frame;
import java.awt.Dialog;
import java.awt.BorderLayout;

class Status extends Dialog {
  Label message;
  Status(Frame f) {
    super(f, "Status", false);
    setLayout(new BorderLayout());
    message = new Label();
    add(message, BorderLayout.CENTER);
    setSize(220, 50);
    setLocation(480, 95);
  }
  public void setMessage(String s) {
    message.setText(s);
    repaint();
  }
}
```

Clearly, this application isn't going to work on a hand-held, since the number of frames is going to cause problems. We will start by getting it working on a Pocket PC device, which supports CDC (PersonalJava), and then have a go at trying to get it working on a CLDC device, using the KVM. So, our first stage is therefore to use JavaCheck to see what classes might give us problems.

Using JavaCheck

We run JavaCheck with the following command:

```
java -cp c:\JavaCheck\bin\JavaCheck.jar JavaCheckV
```

This will give us the following screen:

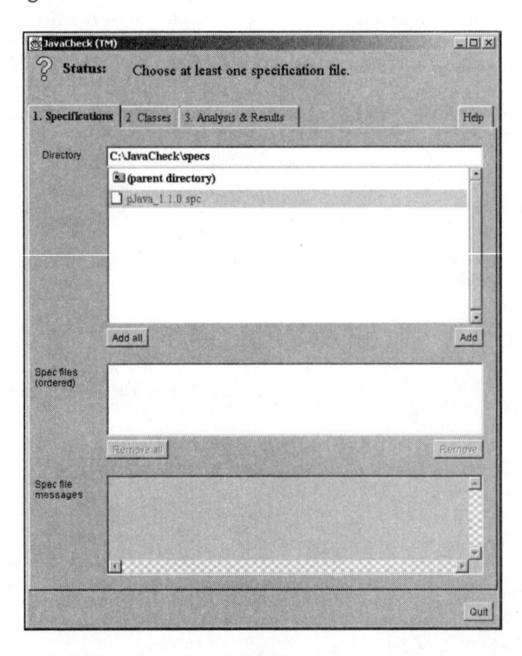

Note the top line, which clearly states that we must select a specification file before continuing. We have downloaded one specification so we select it and add it to the list at the bottom. Then we can move to the next tab, and select our classes to analyze:

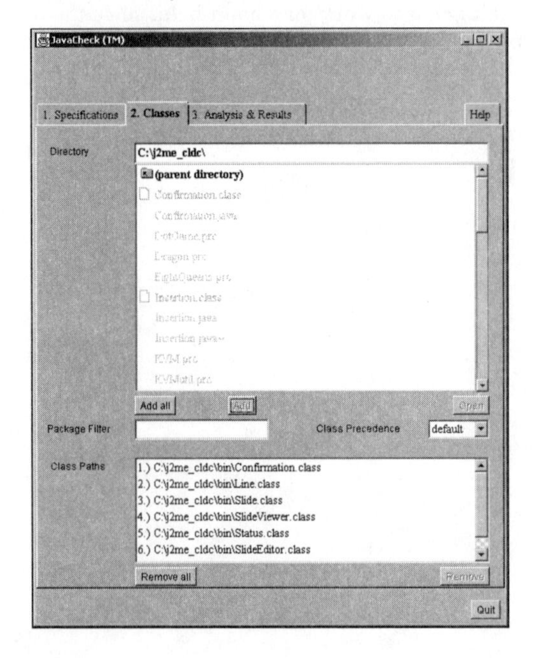

We have to select all the `.class` files that make up our application, and add them to the list at the bottom. Once we've added all the classes we want to analyze, we can move on to that stage. The process is actually very fast, and presents a list of classes that may have problems:

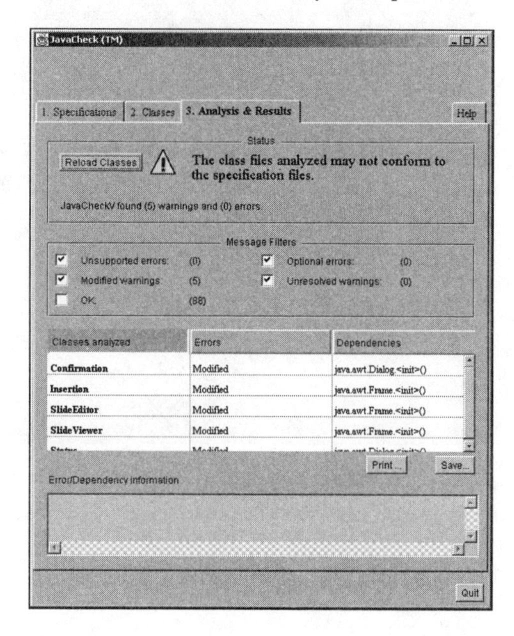

As we can see, there is a list of classes that have references to classes not in the core PersonalJava specification, though they are only warnings as the classes we've used, are `Modified` rather than `Unsupported`. By clicking on the **Dependencies** column we can actually get more details of the modifications that have been made to this class and why they might cause problems with our application.

We could have equally used JavaCheck from the command line, by entering the following command:

```
C:\WINNT\system32\cmd.exe

C:\JavaMobile\chapter07>java -cp c:/JavaCheck/bin/JavaCheck.jar JavaCheck -i c:/
JavaCheck/specs/pJava_1.1.0.spc -classpath .
The class files MAY NOT CONFORM to the specified platform.

C:\JavaMobile\chapter07>_
```

The parameter `-i` gives the location of the specification file, and `-classpath` is used to tell the application where to look for files to analyze. In this case, it has been set to "`.`", the current directory. As this produces somewhat unhelpful output, we'll try it again, this time with the `-v` parameter, for a slightly more verbose response:

```
C:\WINNT\system32\cmd.exe

C:\JavaMobile\chapter07>java -cp c:/JavaCheck/bin/JavaCheck.jar JavaCheck -i c:/
JavaCheck/specs/pJava_1.1.0.spc -classpath . -v
  JavaCheck: Version 3.0m, (10/16/98), Copyright 1994 - 1998, Sun MicroSystems,
 Inc.

Package Precedence = default.
Classpath = ..
JavaCheck: Loaded platform specification: PersonalJava (version 1.1).

   Warning: class "Confirmation" depends on Modified method "java.awt.Dialog.<in
it>()".
   Warning: class "Insertion" depends on Modified method "java.awt.Frame.<init>(
)".
   Warning: class "SlideEditor" depends on Modified method "java.awt.Frame.<init
>()".
   Warning: class "SlideViewer" depends on Modified method "java.awt.Frame.<init
>()".
   Warning: class "Status" depends on Modified method "java.awt.Dialog.<init>()"
.

    7 classes were loaded.

     JavaCheck:  93 dependencies were analyzed.
     JavaCheck:  0 dependencies were Unsupported.
     JavaCheck:  0 dependencies were Optionally implemented.
     JavaCheck:  5 dependencies were Modified.
     JavaCheck:  0 dependencies were Unresolved.
     JavaCheck:  88 dependencies were OK.

    0 error(s), 5 warning(s) found.
The class files MAY NOT CONFORM to the specified platform.

C:\JavaMobile\chapter07>_
```

The results are obviously the same regardless of which method is used to obtain them, though the visual interface gives us more information. Using the command line is particularly difficult on Windows machines, particularly with the limits imposed on the length of possible commands with Windows '95 and '98. It's relatively simple to reduce the command length, either through the use of a batch file (or UNIX shell script), or by specifying the `CLASSPATH` before running the command.

Interpreting the Results

So, JavaCheck has analyzed our application and the only thing we might have a problem with is the use of several frames. We should have already been aware of this as PersonalJava usually only allows one frame per application (and one dialog per frame). Even if we were unaware of this, it is obvious that our application is going to have to have a different GUI purely from a screen size point of view.

The fact that no other problems have emerged is not remarkable. The APIs for PersonalJava cover the vast majority of J2SE, and as long as the programmer avoids Swing and certificate management stuff there isn't any real reason why our application shouldn't work without modification, other than to the GUI. Once we get used to working in PersonalJava we shouldn't hit any problems at all.

Given this therefore, is JavaCheck actually useful? Well, not really just yet. If we were porting a large application it might be useful, but we really shouldn't be porting large applications onto small devices anyway.

As mentioned, it can be useful for finding obscure APIs introduced in Java 2, and will prove more useful to the inexperienced developer. With practice, the J2ME developer should know what APIs are available, rendering JavaCheck redundant unless porting to a different implementation is required. Such a case will only arise if more specification files are created, and while Sun officially support JavaCheck and talk about future releases, nothing has been seen for several years. It is, however, very possible that as more J2ME devices become available, JavaCheck may prove an invaluable tool for ensuring portability.

As we shall see, porting the application to CLDC devices is much more difficult, and there isn't a JavaCheck to tell us where we might have problems. Hopefully, there will be more specification files in the future, making JavaCheck the essential resource it wants to be, but for the moment it's a useful shortcut, though not actually essential.

Designing a New GUI

Currently our application has three frames, and we can only use one, so the first thing is to start thinking about how we can combine the frames into a single frame. This isn't as big an issue as it might be, as we can use panels within the frame.

The target device has a resolution of 320x240, this is pretty standard for hand-held devices, and while we hope our application will work at other resolutions (and it should), it is impossible to design a GUI which will look perfect at any resolution. The first thing to do is to create a new master frame, then change all the other frames into panels, then add them to the master frame. This actually only requires very small changes to our code:

```
    public void go() {

  ...

      f = new Frame();
      f.setLayout(new BorderLayout());
      Panel masterControl = new Panel();
      masterControl.setLayout(new GridLayout(1, 4));
      masterControl.add(delete);
      masterControl.add(insert);
      masterControl.add(sync);
      masterControl.add(quit);
```

The first additional segment of code simply creates a new panel (called `slideList`), and puts everything that is to be used in the master frame (`f`) into the new one. The second additional block then adds the new frame to `f` along with the slide display panel:

```
      slideList = new Panel();
      slideList.setLayout(new BorderLayout());
      slideList.add(masterControl, BorderLayout.SOUTH);
      slideList.add(new Label("Slides"), BorderLayout.NORTH);
      slideList.add(titles, BorderLayout.CENTER);

  ...

      disp = new SlideViewer(first,previous,next,last);
      disp.setVisible(true);
      disp.viewSlide((Slide)slides.elementAt(current));
      myInsertion = new Insertion(this, disp);
      confirm = new Confirmation(f,this);
      update = new Status(f);
      buildList();

      f.add(slideList, BorderLayout.NORTH);
      f.add(disp, BorderLayout.CENTER);
      f.pack();
      f.setLocation(450, 20);
      f.setSize(240, 320);
      f.setVisible(true);

    }
```

Additional code is also needed when the Insert button is pressed. The following code inserts the `Insertion` panel into the south of the master frame (`f`):

```
    } else if (e.getComponent() == insert) {
      f.add(myInsertion, BorderLayout.SOUTH);
      f.show();
      myInsertion.reset();
      myInsertion.setVisible(true);
    }
```

This additional method is called when the new slide is completed and removes it:

```
  public void insertSlide(Slide s) {
    f.remove(myInsertion);
    f.show();
    current++;
    slides.insertElementAt(s, current);
    buildList();
    changeOfSlide();
  }
```

Finally, we have the changes necessary to the class declarations themselves. Here is the amendment required for the `Insertion` class:

```
class Insertion extends Panel implements MouseListener {
```

The change required for the `SlideViewer` class is as follows:

```
class SlideViewer extends Panel {
```

Given our restricted screen space, it also makes sense not to add the `Insertion` panel until it is needed, for example, to give more space for viewing.

Using the PersonalJava emulation environment, we can be confident that the application will work properly:

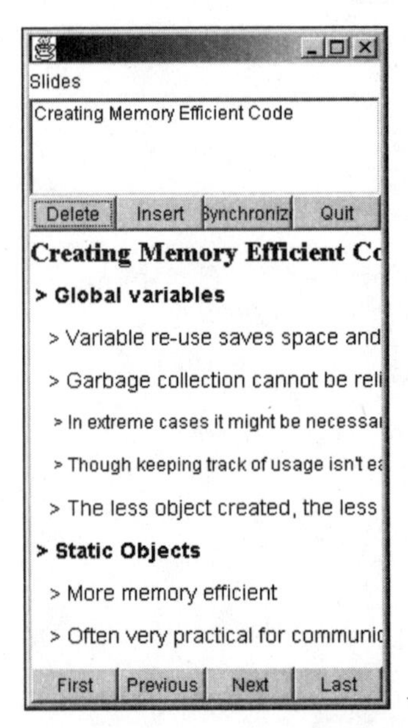

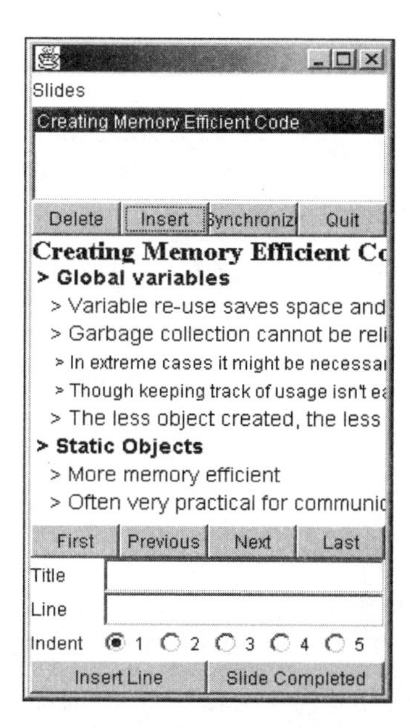

However, if we try to run the application using the emulated environment with the Touchable AWT implementation, it will fail because `Dialog` is not supported by Touchable. This is an important aspect to PersonalJava development. While our application may be PersonalJava-compliant, it is not guaranteed to run on any PersonalJava implementation. Having said that, this interface really wouldn't work on a touch screen device, so we would need to do quite a lot of porting work anyway.

Running the Application on the Device

Installing the PersonalJava environment (virtual machine) onto your device, will differ widely depending on the device. Quite possibly by the time this book has gone to print it, will already be installed on the device when it is purchased, but for the moment we'll probably need to download a PersonalJava Runtime Environment from the Javasoft web site, http://www.javasoft.com/products/personaljava. Installing is, again, very device-specific, but once installed things should get a little more standard.

Taking a Pocket PC as an example, there are a few things the emulator couldn't show us. We will lose some space due to the Start bar at the top of the screen, and the pen input configuration on the bottom right will cover one of the buttons. The worst bit is when we actually try to use it – the handwriting recognition pop-up will obscure everything being written. This is fairly typical of the problems of moving applications onto non-desktop devices, and applies no matter what language we are using.

Obviously, this application is going to need quite a lot more work to make it usable on this device, though it may already be very functional on other PersonalJava-compatible devices. The question of the feasibility of true platform independence may be asked, and it is certainly arguable that it isn't possible to produce any application that will work on every device. Hopefully, this example has shown that porting an application onto a CDC platform isn't actually very hard, with a minimum of modification to our normal Java code required.

All we have done is ported the application so it works on a CDC device, we may well wish to do further work on usability, particularly given the further restrictions on the PocketPC platform (such as the pop-up writing area). This will depend on the application, and who will be using it. We are also going to have to change the synchronization methods (see the next chapter for more details of this).

Porting to CLDC

Getting our application to run on a CLDC device is going to be much more of a challenge. With no AWT at all, and different networking classes, we will need to basically re-think how our application is going to be used.

There is no JavaCheck specification file for CLDC, at the time of writing, so there is no way to automatically check classes for compatibility. Having said that though, it again tends to be easier than expected. Most APIs and classes are supported, assuming we discount the AWT and networking stuff. Multi-dimensional arrays are out, but as Java arrays are not 'proper', we really should be using `ArrayList` and `Vector` (in the case of J2ME applications) instead. Most applications won't actually be technically too difficult to port, but designing the GUI and making everything fit together can be a nightmare.

Networking

Though the networking section of the `SlideEditor` example above was left out, the areas we are interested in, which are different on CLDC devices, are very simple and worth showing here.

The following code is standard J2SE socket communications. The first line creates a `CRLF` String for later use. The second opens a server socket, or daemon, listening for connections on socket 8001 (chosen at random for our communications protocol). `server.accept()` is a blocking action which will cause the application to wait for an external connection and return a `Socket` object from which we can then create streams to write to and read from.

Writing is very simple indeed, as can be seen, while reading is slightly more complicated and taken care of in another method called `getResponse()` which takes an `InputStream` as a parameter:

```
String CRLF = new Character((char)13).toString() + new
                                              Character((char)10).toString();
ServerSocket server = new ServerSocket(8001);
Socket sock = server.accept();
InputStream in = sock.getInputStream();
OutputStream out = sock.getOutputStream();
out.write(("+OK Waiting For First Slide" + CRLF).getBytes());
String received = getResponse(in);
```

We can see that the application opens a server socket and sends some data, before getting a response. The variable `CRLF` is used to hold the characters ASCII 13, 10 respectively, and we have decided that in our protocol these should be sent at the end of every line. The `getResponse()` method looks like this:

```
private String getResponse(InputStream is) throws IOException {
   int i = 0;
   String s = "";
   while(i != 10) {
      i = is.read();
      if(i != 10) {
         s = s + new String(new Character((char)i).toString());
      }
   }
   return s.substring(0, s.length()-1);
}
```

This should be pretty clear; it collects characters from the incoming stream (which is passed as a parameter) until it gets an ASCII 10, then it returns what it received.

If we look at the corresponding code to do the same thing on a CLDC device that supports network connections, it would appear as follows:

```
myDaemon = (StreamConnectionNotifier)Connector.open("socket://:1710");
mySocket = (StreamConnection)myDaemon.acceptAndOpen();
i = mySocket.openInputStream();
o = mySocket.openOutputStream();
in = new InputStreamReader(i);
out = new OutputStreamWriter(o);
out.write("+OK Waiting For First Slide" + CRLF);
String received = getResponse();
```

It should be noted, that the CLDC specification makes no demands of devices to support TCP/IP networking, and additionally the `Connector` object may not be happy to provide such a network object. Making truly device independent code is difficult in such circumstances, though it would probably be appropriate to present an error message in such circumstances.

While all CLDC implementations will have a `Connector` object, it may be restricted to other forms of communications, such as serial or infrared connections. As we have stated, this application is intended for the Palm Pilot device using the KVM, so we can assume that network connectivity is available. As more profiles emerge, we will have to be careful how we advertise our applications and their compatibility.

All network objects within CLDC come from a single static `Connector` object, and we can create our streams from that object, or at least, from the object, which the `acceptAndOpen()` method returns. Notice that none of the objects are created within the same code. This is because they are all global objects, rather than being created whenever this method is run. It saves on memory and lightens the load on the garbage collector. We've also used an `OutputStreamWriter` in this example for simplicity.

The `getResponse()` method is actually identical, except that it doesn't take a parameter, it uses a global variable instead, for efficiency.

So, while the actual methods of creating the stream may be different, once they are created they are therefore identical and can be used in exactly the same way. For more information about networking in CLDC please see Chapters 3 and 5.

The CLDC Application

Now we are going to look in detail at our application and see what amendments we are going to have to do to get it working on a CLDC device. We will be working with a Palm Pilot device, as these are easily the most common CLDC devices available at the time of writing:

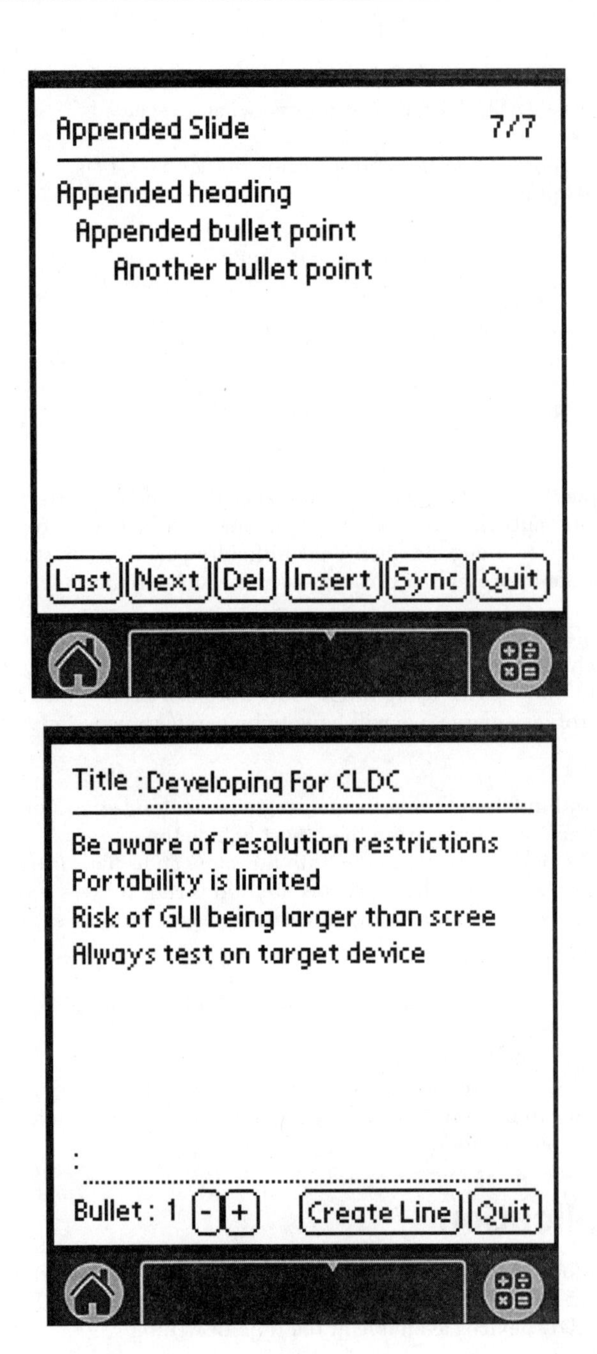
Appended Slide
7/7
Appended heading
Appended bullet point
Another bullet point
Last
Next
Del
Insert
Sync
Quit
Title :
Developing For CLDC
Be aware of resolution restrictions
Portability is limited
Risk of GUI being larger than scree
Always test on target device
:
Bullet : 1
-
+
Create Line
Quit

The following classes are identical to the ones used in the other versions of the application, and demonstrate that many classes won't need any modification to work on CLDC devices:

```
class Slide {}
class Line {}
```

All KVM applications must extend `Spotlet`, you can also glimpse the classes we will need to implement this port pf the application:

```
import java.io.OutputStreamWriter;
import java.io.OutputStream;
import java.io.InputStreamReader;
import java.io.InputStream;
import java.io.IOException;

import javax.microedition.io.Connector;
import javax.microedition.io.StreamConnection;

import java.util.Date;
import java.util.Vector;

import com.sun.kjava.Button;
import com.sun.kjava.TextField;
import com.sun.kjava.TextBox;
import com.sun.kjava.Graphics;

import com.sun.kjava.Spotlet;

public class Mirror extends Spotlet {
```

The user interface to our application is going to have to be rethought, as there isn't space for the three windows of our original application, and even the replaceable panels we used for the CDC version aren't going to work given that the AWT is missing from the CLDC.

We are therefore reduced to using what is basically a single canvas with pixel positioning of components. That's not to say that we can't use AWT components, (as these are provided by the implementation) but without using panels, they work in a very different way.

The first thing we have to do is get a reference to our global `Graphics` object, which we will be using for all graphical screen drawing in this application:

```
  static Graphics g = Graphics.getGraphics();
  Vector slides;
  Slide sl;
  int current;
```

We also define a global variable that contains a number of spaces, and use these to pad our text and ensure that it is displayed cleanly:

```
  String space = "                        ";
```

All our Button objects are global and are owned by this class. The CLDC profile only has a single Namespace in each class, making scope redundant. While this can be confusing to developers used to using scope to define identical variables, given the size of application expected to be developed on the CLDC platform, it is actually less of an issue than it might appear.

As events are only delivered to this class (as an extension of Spotlet), it makes sense to do the entire GUI work here. CLDC components won't draw themselves, so when we want our button to appear on the screen we have to ask it to draw itself:

```
Button last, next, delete, quit, confirm, insert, sync, createLine, up,
        down;
```

As all our events are going to be handled from here, we need to keep track of what mode the user is in. These Boolean values are used for that:

```
boolean deleting = false;
boolean inserting = false;
Date d = new Date();
int indent = 1;
```

The following TextBox is going to be used when the user is entering a new slide. It will display the level of indent of the next line to be added. The additional parameters are included to specify its location on the screen. This is specified in pixels from the top left corner of the screen.

It should be noted that using pixel positioning limits the application to working within a particular resolution of device. Although it should operate on devices with a greater resolution, it might look a little odd. It would, obviously, be possible to create our own lightweight components that resized themselves to suit the resolution being used (or purchase such a set of components, such as the kAWT), though there would be a considerable processing overhead.

Notice that in addition to stating where it should appear on the screen (35, 145), we also state its size in pixels (10, 10):

```
TextBox bullet = new TextBox("", 35, 145, 10, 10);
```

The titleEntry variable will hold the title of a new slide being created:

```
TextField titleEntry = new TextField("", 27, 5, 125, 10);
```

The following variable holds the text for a line being entered. It should be noted that none of these text elements are displayed until the user enters the Insert mode.

In limited memory environments there is an advantage in using global variables, especially where the performance of the garbage collection is suspect (as it always is in new implementations). Making the variables static can be even more efficient in terms of memory usage, but can also lead to more confusion for the developer as each variable name can only be used once:

```
TextField lineEntry = new TextField("", 5, 128, 145, 10);
StreamConnection c;
InputStream i;
OutputStream o;
InputStreamReader is;
OutputStreamWriter os;
String s = "";
String store = "";
int h1 = 0;
int h2 = 0;
String CRLF = new Character((char) 13).toString()
            + new Character((char) 10).toString();
```

Getting our application looking right, is only the start of the process, as we also need to make sure it responds to our users. This process is very different from that used in other versions of Java, having more in common with JDK 1.0 than anything else.

As our application extends `Spotlet`, we can use the `register()` method of `Spotlet` to make our new instance the focus of events. This action will also have the effect of removing any other `Spotlet` objects, which had previously registered to receive event notifications. The parameter is one of two possibilities: `NO_EVENT_OPTIONS` and `WANT_SYSTEM_KEYS`. The latter option should be used if we are interested in intercepting system key presses.

We use the following `main()` method, which instantiates an instance of the current application and registers it to receive events:

```
public static void main(String args[]) {
  (new Mirror()).register(NO_EVENT_OPTIONS);
}

Mirror() {
  last = new Button("Last", 2, 145);
  next = new Button("Next", 27, 145);
  delete = new Button("Del", 55, 145);
  insert = new Button("Insert", 77, 145);
  sync = new Button("Sync", 108, 145);
  quit = new Button("Quit", 135, 145);
  up = new Button("-", 45, 145);
  down = new Button("+", 55, 145);
  createLine = new Button("Create Line", 80, 145);
  confirm = new Button("Confirm Deletion", 40, 80);
  bullet.setText("1");
  slides = new Vector(0, 1);
  sl = new Slide("Slide 1");
  sl.addLine(1, "The First Line");
  sl.addLine(2, "The Second line");
  sl.addLine(2, "The Third line");
  sl.addLine(2, "The Fourth Line");
  sl.setTime(d.getTime());
  slides.addElement(sl);

  sl = new Slide("Slide 2");
  sl.addLine(1, "The 1st Line");
```

```
    sl.addLine(2, "The 2nd line");
    sl.addLine(2, "The 3rd line");
    sl.addLine(2, "The 4th Line");
    sl.setTime(d.getTime());
    slides.addElement(sl);

    sl = new Slide("Slide 3");
    sl.addLine(1, "Line 1, Slide 3");
    sl.addLine(2, "Line 2, Slide 3");
    sl.addLine(2, "Line 3, Slide 3");
    sl.addLine(2, "Line 4, Slide 3");
    sl.setTime(d.getTime());
    slides.addElement(sl);

    sl = new Slide("Slide 4");
    sl.addLine(1, "Top Line");
    sl.addLine(2, "Not Top Line");
    sl.addLine(2, "Not Bottom Line");
    sl.addLine(2, "Bottom Line");
    sl.setTime(d.getTime());
    slides.addElement(sl);

    sl = new Slide("Slide 5");
    sl.addLine(1, "The First Line of Slide 5");
    sl.addLine(2, "The Second Line of Slide 5");
    sl.addLine(2, "The Third Line of Slide 5");
    sl.addLine(2, "The Fourth Line of Slide 5");
    sl.setTime(d.getTime());
    slides.addElement(sl);
    paintSlides();
  }
```

We aren't interested in the keys, and will only be using pen events, but we are still required to have a `keyDown()` method. This is because the KVM is not able to pass events on to GUI components, and if we want a user to be able to enter text into a `TextArea`, then we need to pass any key presses on to those components:

```
  public void keyDown(int k) {
    if (titleEntry.hasFocus()) {
      titleEntry.handleKeyDown(k);
    }
    if (lineEntry.hasFocus()) {
      lineEntry.handleKeyDown(k);
    }
  }
```

The Boolean variable `inserting` is used to keep track of when the user is adding a slide to the show. When a user enters `insert` mode some of the buttons change their labels, and have different effects. Therefore, we check for the current mode before looking at the individual buttons:

```
  public void penDown(int x, int y) {
    if (inserting) {
```

We use the `pressed()` method of the `Button` object, which will return a boolean value indicating if the x and y coordinates handed to it fall within its screen area. Obviously, it is very possible to end up with a very large `penDown()` method, consisting of a long line of `if` statements, which may not be efficient. However, it should be noted that the KVM is only intended for limited application development, and should our interface start to get so complex we have difficulty dealing with the events generated, then we ought to perhaps consider a more fuller Java platform, such as the CDC:

```
      if (quit.pressed(x, y)) {
        sl.setTitle(titleEntry.getText());
        lineEntry.setText("");
        titleEntry.setText("");
        inserting = false;
        d = new Date();
        sl.setTime(d.getTime());
        sl.setID();
        current++;
        slides.insertElementAt(sl, current);
        paintSlides();
      }
      if (titleEntry.pressed(x, y)) {
        titleEntry.setFocus();
      } else {
        titleEntry.loseFocus();
      }
      if (lineEntry.pressed(x, y)) {
        lineEntry.setFocus();
      } else {
        lineEntry.loseFocus();
      }
      if (up.pressed(x, y) && indent > 1) {
        indent--;
        bullet.setText(new Integer(indent).toString());
        bullet.paint();
      }
      if (down.pressed(x, y) && indent < 5) {
        indent++;
        bullet.setText(new Integer(indent).toString());
        bullet.paint();
      }
      if (createLine.pressed(x, y) && indent < 5) {
        sl.addLine(indent, lineEntry.getText());
        indent = 1;
        bullet.setText("1");
        lineEntry.setText("");
        paintInsertion();
      }
    } else {
      if (insert.pressed(x, y)) {
        sl = new Slide();
        inserting = true;
        paintInsertion();
      }
      if (quit.pressed(x, y)) {
        System.exit(0);
      }
```

```
      if (last.pressed(x, y) && current > 0) {
        current = current - 1;
        paintSlides();
      }
      if (next.pressed(x, y) && current < (slides.size() - 1)) {
        current = current + 1;
        paintSlides();
      }
      if (delete.pressed(x, y)) {
        confirm.paint();
        deleting = true;
      }
      if (sync.pressed(x, y)) {
        syncWithServer();
      }
      if (confirm.pressed(x, y) && deleting == true) {
        slides.removeElementAt(current);
        if (current == slides.size()) {
          current = current - 1;
        }
        deleting = false;
        paintSlides();
      }
    }
  }
```

These methods are only used for synchronization, and are examined in detail in the next chapter:

```
  private void syncWithServer() {
  }

  private String getResponse() throws IOException {
  }

  private Slide receiveSlide() throws IOException {
  }

  private void sendSlide(Slide s) throws IOException {
  }
```

The `paintInsertion()` method is used to render the screen when a user is inserting a new slide. It is called whenever a significant change has been made in the material to be displayed, and it clears the screen before laying everything out. Note that we have to clear the screen first, as objects cannot easily be removed from the display:

```
  private void paintInsertion() {
    g.clearScreen();
    g.drawLine(5, 20, 155, 20, 1);
    g.drawString("Title", 5, 4);
    g.drawString("Bullet :", 5, 145);
    bullet.paint();
    createLine.paint();
    quit.paint();
    titleEntry.paint();
```

```
        lineEntry.paint();
        up.paint();
        down.paint();
        for (int i = 0; i < sl.countLines(); i++) {
          g.drawString(sl.getLine(i).text, sl.getLine(i).value * 5,
                       (i * 12) + 25);
        }
      }
```

This method is very similar to the one above, but this time it renders the normal display, with a slide shown filling most of the screen and some control buttons below:

```
      private void paintSlides() {
        g.clearScreen();
        sl = (Slide) slides.elementAt(current);
        g.drawString(sl.getTitle(), 5, 5);
        g.drawLine(5, 20, 155, 20, 1);
        g.drawString(new Integer(current + 1).toString() + "/"
                  + new Integer(slides.size()).toString(), 140, 5);
        for (int i = 0; i < sl.countLines(); i++) {
          g.drawString(sl.getLine(i).text, sl.getLine(i).value * 5,
                       (i * 12) + 25);
        }
        insert.paint();
        sync.paint();
        last.paint();
        next.paint();
        delete.paint();
        quit.paint();
      }
    }
```

Other Options

While the KVM, and indeed the CLDC specification, does not include the AWT in any recognizable form, there are third party tools to provide much of the missing functionality. As discussed in previous chapters, the kAWT is a set of objects, which can be used in much the same way as the AWT, ensuring upwards compatibility. Even if it can't guarantee to run programs developed for the AWT, it can guarantee that applications developed using it will run on larger versions of Java.

Summary

Porting applications to other versions of Java should not be necessary. Sun have been telling us we can 'Write Once, Run Anywhere' for so long we almost believe it, but it's neither true nor desirable. The true value comes from the back of an application, beyond its network or user interface elements. As we have seen, the objects that actually do the work of the application, require almost no modification to work with every version of Java.

Creating an application for a particular platform is not particularly different or difficult. Porting an application might be considered more difficult, but with careful planning and some common sense it isn't actually as bad as it might be.

Getting J2SE applications to work under CDC shouldn't be difficult, unless a great number of Swing components have been used. It is likely that the GUI will need to be redesigned anyway, and conversion can form part of that process. JavaCheck can be used to confirm compatibility, but the PersonalJava Emulation Environment is more useful for development.

Porting to CLDC is much more difficult. These limited devices offer a different operational model and user interaction, meaning that many applications won't actually make sense on such a device. Careful consideration of how and when the application is going to be used, will be necessary before work commences.

Java isn't the solution to any porting problems, but it can make life a great deal easier.

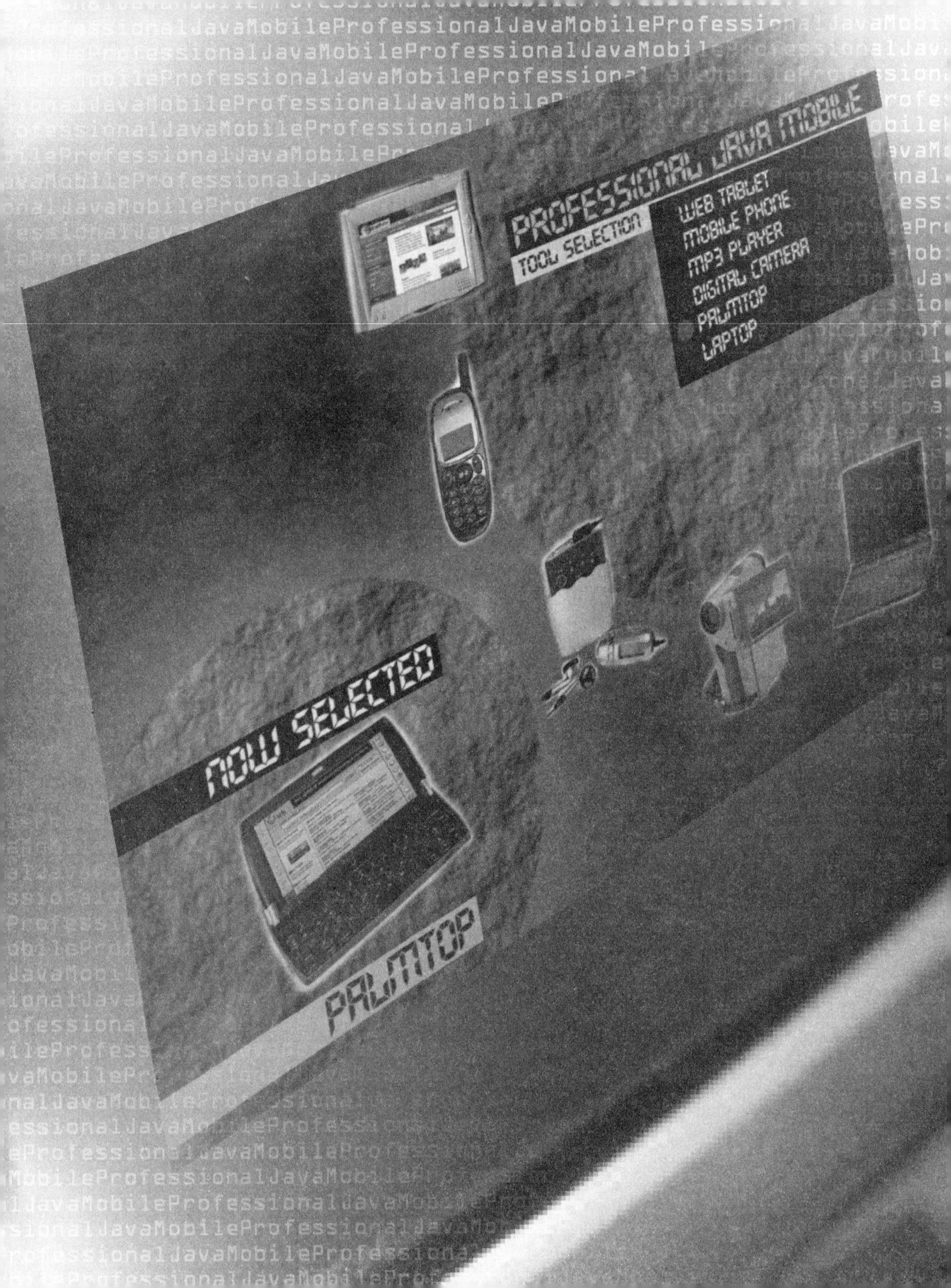
PROFESSIONAL JAVA MOBILE
TOOL SELECTION
WEB TABLET
MOBILE PHONE
MP3 PLAYER
DIGITAL CAMERA
PALMTOP
LAPTOP
NOW SELECTED
PALMTOP

8

Synchronization

We frequently talk about synchronization as though it's something basic and obvious to modern computing environments, when in fact the very need to synchronize information is very recent. It is only when there are multiple computers providing information, that we need to start to think about how we can ensure that that information is the same amongst them.

The obvious way to ensure that all computers being used for an application can see the same data, is to put that data onto a server and have the computers operate as clients to that server. This method has been used for years, and the increasing adoption of HTML interfaces for business applications is a broadening of this concept. However, while that may form an ideal situation, it is not always applicable. Some computers aren't always able to connect to their server, and even if available, connectivity costs or processing load may prove prohibitive. In the world of mobile devices, synchronization is everything.

This chapter will cover the basics of synchronization, why we do it, and when it's appropriate. We will also be discussing popular synchronization processes and the move towards standards. Working through an example of synchronization, we'll be showing some of the inherent problems, and some approaches to dealing with them. The example we will be working with demonstrates the most complex form of synchronization, and shows by its limitations how difficult it is to manage data on different devices.

It has now become obvious that mobile devices need to be able to synchronize information with desktop servers, though it wasn't always so. Early mobile computers had their own file systems and the ability to import and export data, but nothing in common with devices sold today. It was the Palm Pilot device that demonstrated that synchronization should be central to mobile computing, rather than something added onto the end.

The Palm Pilot

The Palm Pilot handheld computer changed the way we think about mobile computing. There were other devices around well before it, many of them were more powerful and flexible, some even cheaper, and it's interesting to consider what made (and continues to make) the Palm so popular, particularly as it is synchronization that lies at its core.

The Palm was never conceived as a mobile computer; in fact, it has never been advertised as such, but as an extension to a desktop computer. Its most useful ability is to allow the user to continue their work, to almost take a portion of their desktop computer with them, working on the same things they were doing, in miniature. This has been the real selling point of the Palm, and it's interesting to see other mobile device manufacturers attempt to emulate its success.

This ability to cut a chunk out of your desktop and take it with you is based on the Palm's ability to synchronize with desktop applications. When you buy one, you're not just buying a mobile Personal Information Manager (PIM), but a desktop one too. Obviously, synchronization followed to various other software packages, and now there aren't any popular PIM applications that won't synchronize their content with a Palm Pilot (at least, those running on the Windows Platform).

When we think about synchronization, we tend to think of handheld devices connecting to desktop machines, but that is far from the whole story. There are many circumstances where data in multiple places needs to be synchronized, often to allow easy integration with legacy systems, particularly databases. However, this type of operation tends to be a continuous process, ensuring that an alteration on one system is reflected as quickly as possible on the other.

While these systems are no doubt interesting, they are outside our discussion. We are talking about the necessity to consolidate information from different platforms, where many alterations may have taken place during the disconnection. A mobile device falls particularly into this area, with the necessity for frequent backups, and the desire to be able to deal with the information in a desktop environment.

Different Types Of Synchronization

While the generic term suffices for devices like the Palm, we need to think a little more deeply about what the synchronization process actually involves, and how it affects data stored by applications.

Mirror Image

A mirrored synchronization is probably the most well known, though it's also the most complicated for the programmer. It is only applicable if the user has the same application residing on both the mobile device and the desktop one. Probably the best known of these is the ubiquitous address book application. This is an example of data stored in a database that needs to be synchronized. Information is frequently entered while on the move, but is also imported from other sources (such as company phone listings) into desktop applications, and synchronization is necessary to ensure the information is always available wherever the user needs it.

It is not only databases that can be synchronized though mirror image, many applications now allow an entire file tree to be synchronized across devices, ensuring that data files, applications, and databases are all mirrored. This type of situation requires replication of the information on every device the user wishes, alterations to individual records should be reflected in all copies of the data, and new entries must be added across the instances.

This can be technically, and logistically, challenging. Imagine the situation where such an address book is synchronized, and the user makes an alteration to an entry, perhaps to add a mobile phone number. The same user, before any synchronization has occurred, then makes another alteration to the same entry on their desktop machine, perhaps updating the fax number. When synchronization occurs, the software will have to choose which version of the record is most appropriate.

If sufficiently advanced, it might recognize that one field has been added while another has been altered, but if one field has been altered on both instances of the data then there is no obvious choice. This kind of situation can be very rare, depending on the application, but it serves to illustrate how logical problems are often more important than technical ones when data is replicated in this fashion.

It is vital, when designing a synchronization process, to work out the logical flow, and understand how conflicts will be dealt with. You can be certain that conflicts will occur, and need to decide how to approach them. There is no answer that will fit every instance, and conflict resolution is central to synchronization code development.

The programmer developing a synchronization system for an application, will have to consider every eventuality, and how it will be dealt with. There are no generally applicable rules to the process. We will be looking at an example of a mirror/image synchronization application later in this chapter, which may provide some pointers to major design considerations.

One Directional Transfer

This is a much less complex situation, where data needs to be copied from one device to the other. A good example of this is where a mobile device is being used to collect information (such as stock levels), and this information is then transmitted back to a desktop machine through a synchronization protocol.

As there is no data coming from the server, the process is actually very simple, and the data then held on the server can easily be integrated into other applications. It is possible for conflicts to occur in this model, but only if the process has been badly designed. In the above example, of stock levels, there should only be one person collecting stock levels on any one stock, so no conflict should be possible (unless double-stocktaking is being done, to check the levels, in which case the conflicts are what's being searched for).

Transaction Based

This term refers to a situation where a mobile device is being synchronized with information, which is not held on the server to which it is connected. One example of this is the AvantGo application (http:\\www.avantgo.com), which downloads particular web sites to a mobile device during synchronization.

Once the synchronization process is started the AvantGo application is required to make network connections to web sites, collect the content, and translate it into a more appropriate format before sending it on to the client device. The important point here is that the AventGo application does not collect the web sites required until the synchronization process has been commenced, and the process continues until the data has been collected and formatted for sending to the device.

The same process is equally likely to apply to a situation where data is being collected from a networked database or corporate intranet. The important factor being that the server is required to perform some sort of processing on the information, such as collecting it from elsewhere, before it can be transferred to the client. These systems, from our point of view, are basically one directional once the data has been collected and organized.

Back-Up

The most basic sort of synchronization of all, this is simply the transfer of information from the client device to the safe storage of the server. Back-up is generally used when data is held in a mobile application for which there is no desktop equivalent. In these circumstances, the only desire is to safeguard the information from loss or damage.

An example may include the high-score table for a mobile game. There is little point in making such information available to desktop applications, but it may be of great importance to the owner.

Problems With Synchronization

Creating software to perform synchronization is not easy, and should not be taken lightly. There are many potential pitfalls, and the current lack of any standards can lead to the appearance of an insurmountable challenge.

Programmers should always be prepared to deal with sudden disconnection, as at any point during the process a mobile device might lose contact with its network, or the user might just get bored of waiting and pull the plug. It is, therefore, vital to ensure that any updates are fully received before being committed, and error handling should be as gracious as possible. The decision about how far to roll back after a communication failure is left to the application designer, in some circumstances it will be better to collate what data has been received, while in others it may be more appropriate to cancel the entire operation.

Mirror synchronization is only useful where there are similar, or identical, applications on two separate devices, which can be used to access the same data. While the storage formats of the two applications may differ, due to differences in their underlying architecture, the important thing is that from the user perspective they are basically the same.

In such circumstances, it will be necessary not only to transfer the data from one device to the other, but also to perform some conversion on the data to make it readable by the receiving application. This is well illustrated by the Microsoft ActiveSync application, which will synchronize files stored on a PocketPC device with a desktop running Microsoft Windows.

Files on the desktop in Microsoft Word format have to be converted to Microsoft Pocket Word so they can be edited on the handheld device, but as Pocket Word lacks many of the features of its desktop equivalent, use of these features has to be removed from the files. This can result in files being unusable, or at least, badly formatted, and is of dubious value on anything but the most basic files. This is not the fault of ActiveSync, but illustrates the problems of converting data from one format to another.

Another issue that may have to be taken into account is the capabilities of the portable device being used. It may well be that the data is of an inappropriate size for use on a handheld device. Eudora, an e-mail application from Qualcomm, nicely illustrates this. On the desktop application, e-mails may be of several megabytes in size, which obviously won't fit onto most handheld devices, therefore as part of the synchronization process, all e-mails are cut to their first 100 lines. While this limits the functionality of the handheld equivalent, it makes the application usable on devices with limited memory and processing power.

For our example, we will be using the same application we used in Chapter 7, which was ported from J2SE to CDC and CLDC. This example provides an ideal example as it clearly is the same application on either platform (or all three) and it illustrates some of the problems and complexity of working with differing data sources.

All About Protocol

The first thing to do, when designing a synchronization process, is to decide on a protocol. While there are groups working towards some standard protocols (such as SyncML, http://www.syncml.org) these are not widely used, and may in fact prove more flexible than needed (and thus more heavyweight). We will be therefore defining our own protocol.

At some point we will need to decide how our devices are going to communicate. While the protocol should be designed to be flexible about the communication channel, we will still need to decide if the devices are going to have the option to use serial communications or a network connection. While in many applications it's ideal to allow the user to choose how to connect (as the Palm Pilot synchronization does), some applications will have particular needs that require specific connectivity, while others will have no use for more options than are necessary. We will be using a TCP/IP connection for our synchronization, though our software will be designed in such a way as to make conversion to a different connection easy to accomplish.

Once these decisions have been made the actual coding can start, but be aware that you will certainly notice deficiencies in your protocol as you write the software, and ensure you have a mechanism for making alterations (particularly if you are working with a team). The protocol is the most important part of the process and should be clearly documented.

Creating the Protocol

Our example program deals with slides, and uses a `Vector` to hold all the slides in the current show. It's worth taking a quick look at the `Slide` object before we start to remind ourselves what data we are going to be synchronizing; the important sections are listed below:

Constructors	
`Slide()`	Creates an empty slide, note that a `title`, `ID` and `creationTime` must be set before the slide is used
`Slide(String topLine)`	Creates a slide with the specified title, also creates `ID`, note that `creationTime` must also be set Parameters: `topLine` – containing the title to be used

Fields	
`boolean seen`	Tag used during synchronization to mark that this slide has been checked against mirrored store

The important parts of this object, from a synchronization point of view, are the `ID` and the `TIME` variables. These record the unique identifier of the object and the time when it was last altered. They are essential if we are going to be able to synchronize two slide shows.

The `ID` is made up of the start of the slide title and a random element, which should be unique enough for our purposes, and the time is measured in milliseconds since the beginning of time (as far as Java understands it). It could be argued that this isn't unique enough, but we have to work within the limits of the environment. It might be advisable to place a marker denoting which device this slide was first created on, allowing the uniqueness to be confirmed by comparison with existing slides, or perhaps with the use of a hash table of the slide contents, but such a process is beyond the scope of our example.

Any objects we intend to mirror synchronize will have to have both these things, so that the latest version and any additions can be identified. Notice that the individual lines do not have these references, and this is because we will not be synchronizing to that level of detail. If a slide is altered on one device, then it will be reflected on the other, but if the slide is altered on both devices, with (say) a line added to the slide in both instances, then only the most recent version of the slide will be synchronized.

This is typical of the kind of compromises we have to make when designing synchronized systems. It would be possible to uniquely identify each line, with time stamps, enabling merging of two versions of the same slide, though such a process would still fail if the same line had been altered on both devices. In our example, we have decided that we will only check to the `Slide` level, and to assume that the most recent modification is the one we wish to keep, and in some rare circumstances data will be lost:

Methods	
`public void setTitle(String t)`	Sets the title of this slide. Parameters: `t` – String to set the title to.
`public void setID()`	Sets the ID of this slide to the first few letters of the title, and a random element.
`public void setID(String s)`	Sets the ID of this slide to the passed value, generally used when adding a slide received through synchronization. Parameters: `s` – Value to set the ID to.

`public void addLine`
 	
`(int indent, String text)`
	Adds a line to the slide.
	
	Parameters:
	`indent` – Value of the line, from 1 to 5. The higher the value the greater indent of the line. `text` – the text value for that line.

`public String getTitle()`	Returns the title of this slide. Returns: String containing the text of the title.
`public int countLines()`	Returns the number of lines. Returns: Number of lines on this slide, not including the title line.

Methods	
`Public Line getLine(int i)`	Returns the text of one line. Parameters: `i` – Number of line to be returned, first line has value 0. Returns: Text content of this line.
`public long getTime()`	Returns the time when this slide was last altered. Returns: Time last edited.
`public void setTime(long l)`	Set the time, normally done after slide has been altered. Parameters: `l` – Time altered, normally current time or collected during synchronization.
`public String getID()`	Gets the unique identity of this slide. Returns: Unique ID, it should be noted that uniqueness is not guaranteed but considered beyond reasonable doubt.

We can now start working out how our protocol is going to actually work. The most obvious way to achieve synchronization is to copy all the data onto the server and then check the contents there, but we won't be doing that for a couple of good reasons.

The first point is that of network bandwidth used. If the two data sources are actually already synchronized, then the entire body of data is being sent twice for no reason at all. The other impact is that of speed, and this is often particularly important when you are considering mobile devices, where bandwidth speeds may be more limited than static environments.

The less data we transmit between the pair of applications, the better, so we need to think about what information is needed to make a decision as to if the slide is already matched. We should also try to minimize the amount of processing power used on the mobile device. The host (in this context) will be the device doing most of the work, as this will generally have greater resources available to it.

While synchronization is actually a peer-to-peer arrangement, it is convenient to think of a client-server relationship when developing. Therefore, it will be the server that makes all the decisions about what slides need to be updated, and it will do this with the `uniqueID` and `creationTime`, which are both needed to be able to make a proper decision.

To start, we will sketch out a flow diagram of how we expect the protocol to work. This not only highlights the amount of work we are going to have to do, but can also be useful for spotting limitations in what is possible. Every synchronization process is a compromise, and it's important to understand where we are going to be limited in our functionality.

Here is the flow chart for the client:

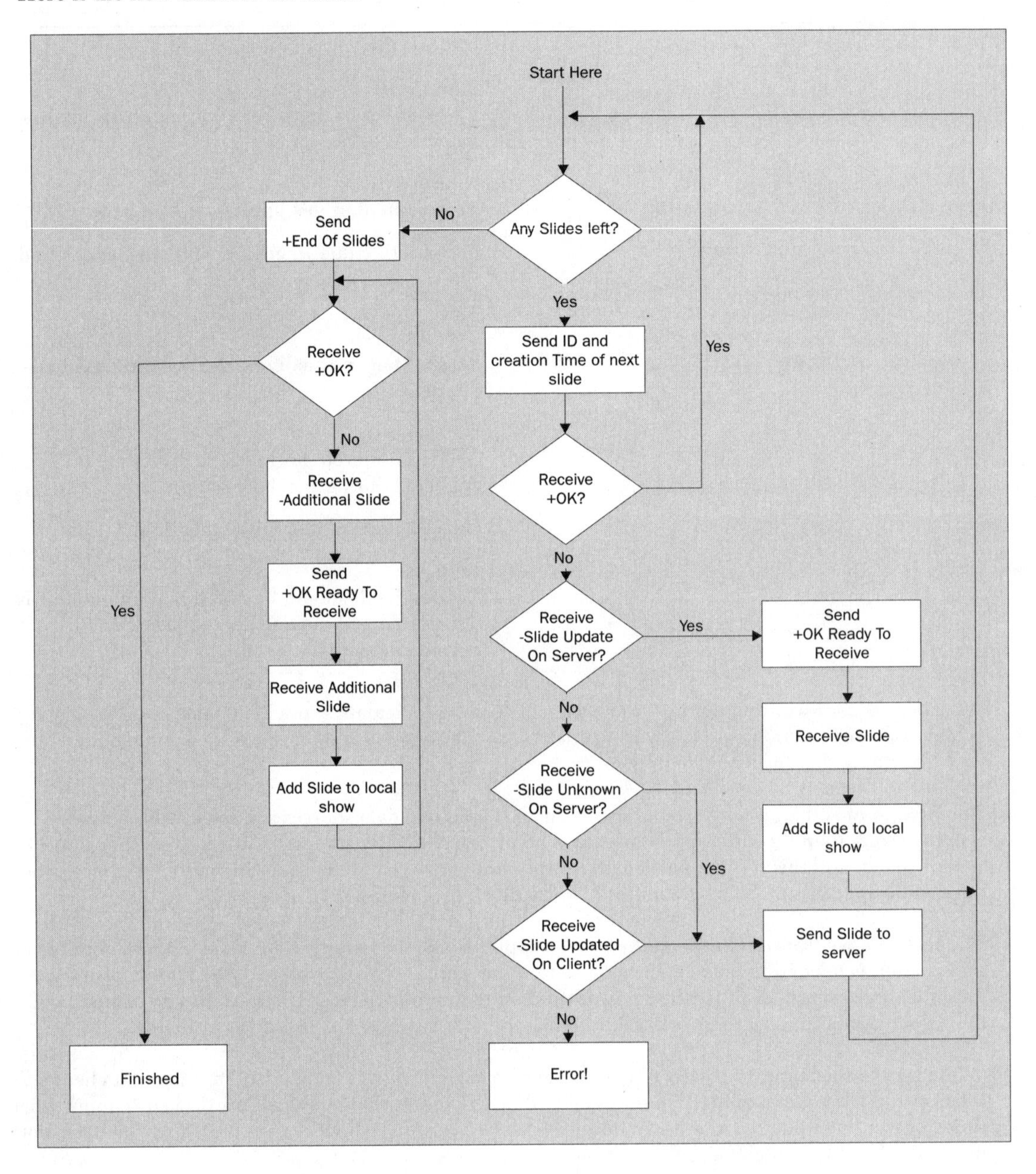

Here is the flowchart for the server:

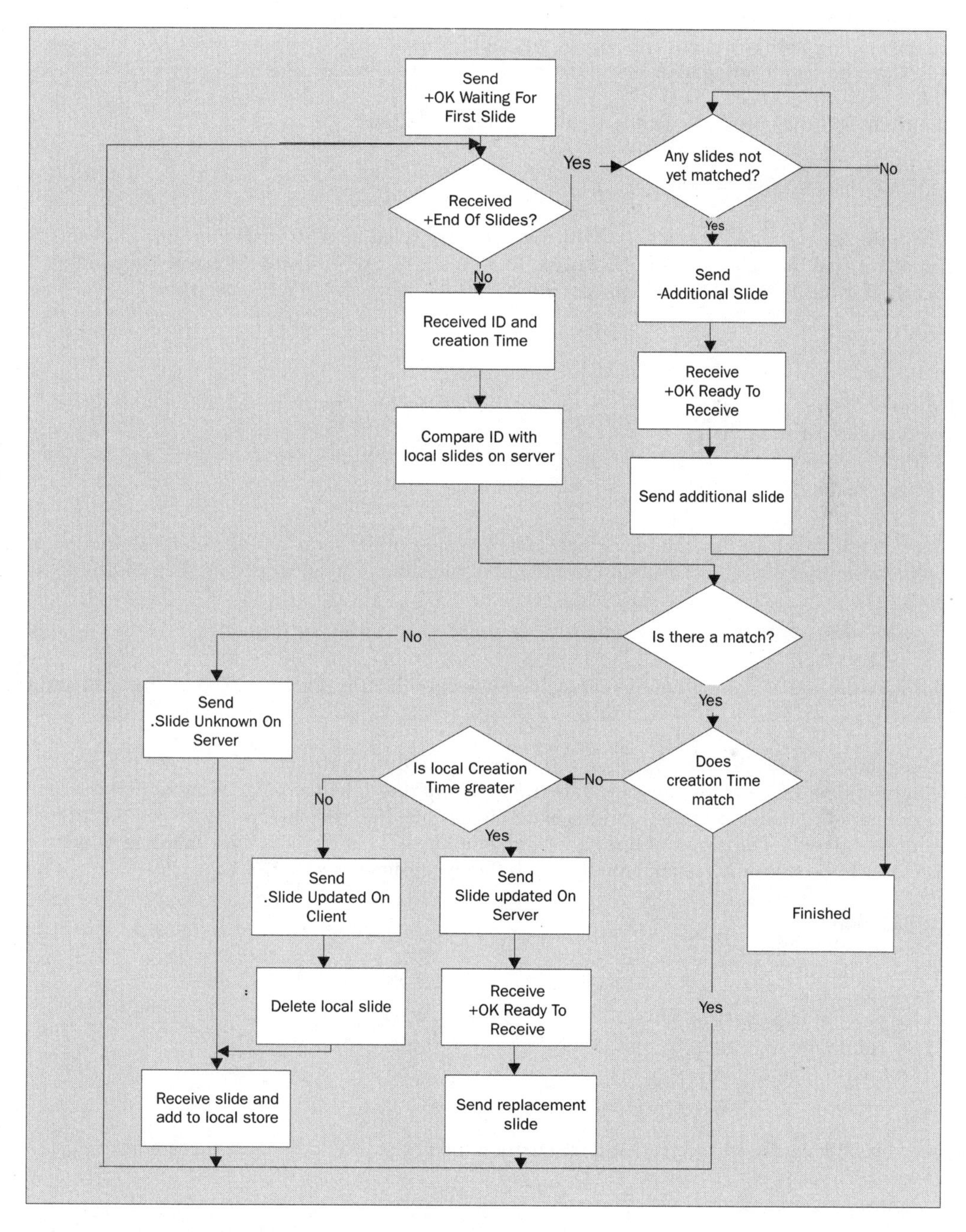

The server will start by listening on an appropriate port (1710 in this case), and wait for the client to contact it. When connected, it will respond:

```
+OK Waiting For First Slide
```

All exchanged information will be terminated with CR (ASCII 13) and LF (ASCII 10), this is in common with most Internet protocols, and ensures detection of the end of line characters. Being as we are developing both the client and server, we could actually specify whatever we liked for end-of-line indicators, but conforming to as many standards as possible makes sense.

The client will then send the details of the first slide as follows:

```
ID=xxxxxxx TIME=xxxxxxxxx
```

Where the `ID` and `creationTime` of the first slide are filled in. This is also terminated with CR LF. The server will then compare the ID against its own slides, and respond differently depending on what it finds. If it can find a match of both `ID` and `creationTime` it will response with:

```
+OK
```

However, if the server finds that its version of the slide, with a matching `ID`, has a later `creationTime` it will send:

```
-Slide Updated On Server
```

This will tell the client that the server has a later version of the slide, and the client should be ready to receive a copy of the full information contained in the slide. The client should then respond:

```
+OK Ready To Receive
```

This is so the server knows that the client acknowledges that the slide needs to be updated and is ready to receive.

The example below shows how a slide will be transmitted, with the `ID`, `creationTime` and `Title` being sent first, followed by the lines of text making up the slide itself. Each line of text is preceded with the amount of indentation required (this example has one line with indent 1 with the rest of the lines being intended 2). The group of lines are terminated by a "." alone on a line, enabling as many lines as necessary to be transmitted sequentially. Each line is terminated with CR LF:

```
ID=Sli49548
TIME=976471288600
TITLE=Slide 5
1 The First Line of Slide 5
2 The Second Line of Slide 5
2 The Third Line of Slide 5
2 The Fourth Line of Slide 5
.
```

Once the slide has been received and understood the client should confirm successful receipt:

```
+OK Slide Replaced
```

The next thing the server can respond is:

```
-Slide Updated On Client
```

This would indicate that the client has a later version of the slide, and the server would like to receive this later version to update its local slide show. The client will respond with a copy of the slide in question that the server can use to replace the one in its local store.

The last optional response from the server is:

```
-Slide Unknown To Server
```

This will be sent when the server has no reference to the slide (no matching `ID`) and therefore wishes to add the slide to its internal store. On receipt of this, the client should immediately send the slide in the same format as described above, enabling the server to create a slide object for its local store.

The actual response from the client to the last two messages is identical, as we shall see when we look at the implementation code. Once the whole slide has been sent the server will respond:

```
+OK
```

At which point the client should resume sending the `ID` and `creationTime` of the next slide.

Once the client has sent the details of every slide known it tells the server:

```
+End Of Slides
```

The server will then check if it has any slide objects of which the client is unaware, and if so it sends:

```
-Additional Slide
```

The client will notice this and respond:

```
+OK Ready To Receive
```

The server then sends the slide information in the same format used above. Once the client has received the slide it adds it to its local store and confirms receipt to the server:

```
+OK Slide Added
```

The server now can respond with another slide (using `-Additional Slide`), or can complete the exchange with:

```
+OK
```

The client will now close the connection.

Note that most of the responses start with either "+" or "-". This enables the responding application to only check the first character of the response, to know whether to continue or if more action is required. This is also typical of Internet protocols and can save processing time.

Given the above protocol, a typical exchange may look something like the following:

Server	Client
`+OK Waiting For First Slide`	
	`ID=Sli46200 TIME=976478141740`
`+OK`	
	`ID=Sli47700 TIME=976478141740`
`+OK`	
	`ID=Num48312 TIME=976478141740`
`+OK`	
	`ID=Cre48928 TIME=976478141740`
`-Slide Updated On Server`	
	`+OK Ready To Receive`
`ID=Cre48928`	
`TIME=976478141740`	
`TITLE=Creating A Protocol`	
`1 Spend time on your protocol`	
`1 Don't try to get it right first time`	
`2 Allow for later alterations`	
`1 Make sure you get peer review`	
`.`	
	`+OK Slide Replaced`
	`ID=Sli49548 TIME=976478141740`
`+OK`	
	`ID=Red2583282 TIME=976478139830`
`-Slide Updated On Client`	

Server	Client
	ID=Red2583282
	TIME=976478139830
	TITLE=Reducing Bandwidth
	1 Learn from Internet Protocols (rfc's)
	2 They were developed by large groups
	3 Never let one person develop a protocol
	1 Try to conform to as many standards as possible
	.
+OK	
	+End Of Slides
-Additional Slide	
	+OK Ready To Receive
ID=Cod82390	
TIME=976478136329	
TITLE=Coding Standards	
1 Protocols demand exact documentation	
2 With frequent examples	
2 Examples should be genuine exchanges	
3 Where possible	
.	
	+OK Slide Added
+OK	

There are a couple of things missing from this protocol, and it's important to understand what they are. Take a moment to look at the example and think about in what circumstances it would fail to work.

The protocol does not deal elegantly with two situations. The first is where a single slide has been modified on both devices, and therefore the individual elements of the slide should be checked for uniqueness to ensure the maximum data is maintained. However, the lines that make up a slide object have neither `uniqueID` nor `creationTime`, so it isn't possible to synchronize to that level and we just keep the most recently modified version.

Should there have been modifications on both devices, the result will be that the last modification made will be reflected on both devices, perhaps resulting in data loss. This is a decision made by the application designer, who has to consider the likely events and how they can be coped with.

It is very important to gain as much information as possible about how the application will be used, to work out what combination of events are allowable. Also, by being aware of the limitations of the system it will be much easier to respond to criticism (such as during a beta test) and provide a deeper (or more shallow) level of synchronization as required.

The second situation is where a slide has been deleted from one device. Being as our application doesn't actually allow the user to delete slides, it didn't get included in the protocol. However, clearly, it would be a vital component in a deployed system, and it would be necessary for the device on which the slide was deleted to maintain a reference to the `uniqueID` and an addition to the protocol to allow the transmission of that information would be necessary. This could be achieved by simply blanking he contents of slides that had been deleted, and arranging so that they were not displayed in the slide show. Such a modification could easily be done without affecting the protocol.

Writing The Code

Obviously, there are two versions of the code to perform synchronization, one for the client and one for the server. While they share elements, such as the ability to send and receive individual `Slide` objects, they are actually quite different, as they operate on different parts of the protocol. In this instance the client is also designed as a CLDC application, so it makes use of the network classes appropriate to that class of device, though that actually makes little difference to the code that carries out the protocol.

The following only shows the methods actually used for synchronization, for the rest of these applications please see Chapter 7. The code shown here for the server is taken from `SlideEditor`, while the client is `Mirror`.

The Server

The `Vector` object `v` contains the slides to be synchronized, our example application only deals with one set of slides, but this arrangement allows for expansion:

```
public void syncUpWithMirror(Vector v) {
```

These `String` objects are used to hold values relating to specific slides, so they can easily be used in comparisons:

```
    String uniqueID;
    String uniqueTime;
```

Here we are creating a `String` object to use as our end of line terminator. Note that while we use an identical construct in the client, we don't define it within the method, as defining it globally is more memory and process efficient:

```
    String CRLF = new Character((char) 13).toString()
                + new Character((char) 10).toString();
```

These Boolean values are used to keep track of if a matching slide has been found, and if a slide has been updated from the client:

```
    boolean presentOnServer, updated;
```

The update object is used to keep the user aware of what is happening, and to assure them that the application hasn't crashed:

```
    update.setMessage("");
    update.setVisible(true);
    try {
```

Here we are opening the listening socket. 1710 is just a random socket number, anything over 1024 would do fine (though, obviously, the client will have to connect to wherever the server is listening):

```
      ServerSocket server = new ServerSocket(1710);
      update.setMessage("Waiting For Connection");
```

This is our blocking action, waiting for the client to connect. In a deployable application we would probably want to put this in a separate thread, so we could provide an **Abort** button, but to keep things simple in this example we just hang the application until the client connects:

```
      Socket sock = server.accept();
```

Creating our Stream objects for communication:

```
      InputStream in = sock.getInputStream();
      OutputStream out = sock.getOutputStream();
      update.setMessage("Connection established");
```

This is the start of our protocol execution. Notice the use of the CRLF as our end of line terminator:

```
      out.write(("+OK Waiting For First Slide" + CRLF).getBytes());
```

The getResponse() method is used for all our received communication, again, in the client the received object is global:

```
      String received = getResponse(in);
      int loop = 0;
```

This loop simply loads each slide into the variable sl and sets its seen field to false, as we are starting our synchronization. The insertion of a copy of the slide, followed by the removal of the original may seem inefficient, but it is the only way to update the contents of a Vector:

```
      for (int i = 0; i < slides.size(); i++) {
        sl = (Slide) slides.elementAt(i);
        sl.seen = false;
        slides.insertElementAt(sl, i);
        slides.removeElementAt(i + 1);
      }
```

It's worth noting that in recent versions of J2SE, you can use the set() method of Vector to set the contents of a particular element, but this has not been transferred to any of the J2ME profiles, yet.

This is the start of our main loop, as our protocol states that the sequence '+End Of Slide' should only be sent when the client has sent details of every slide it is aware of. This loop should complete once for each slide on the client:

```
        while (!received.startsWith("+End Of Slides")) {
```

Here we cut the unique ID and creation time from the first line received from the client. Note that we make no attempt to confirm that the client has sent the data in the right format, relying on our `try...catch` block to manage any problems:

```
          uniqueID = received.substring(3, received.indexOf(" "));
          uniqueTime = received.substring(received.indexOf(" ") + 6);
          update.setMessage("Matching Slide : " + uniqueID);
```

This loop looks at each slide the server is locally aware of and compares it to the details sent by the client. If a matching slide is found then `presentOnServer` is set to `true`, otherwise we calculate if the sent slide is older or newer than the locally stored one and send the appropriate message. `updated` is used to remember if the slide has been updated, so even if the slide is present there is still the possibility that we might need to ask the client for a copy of it:

```
          presentOnServer = false;
          updated = false;
```

The `creationTime` is created from the string sent by the client, and converted into a long value. This is done so that it can easily be compared with the age of the slides in the local store:

```
          long creationTime = new Long(uniqueTime).longValue();
```

The following loop goes through each of the local slides, and compares it to what was sent by the client. If there is a match it compares the time stamps on what was sent and the local slide. If the local slide is newer it sends that slide to the client, if older it asks the client for a copy of the slide, and if matched it does nothing. If a slide is updated, `updated` is set to `true` to reflect this, and if any match is made then `presentOnServer` is also set to `true`. Once a slide has been found in the local store it sets the `seen` value of that slide to `true`, so we know that that slide has been seen:

```
          for (int i = 0; i < slides.size(); i++) {
            sl = (Slide) slides.elementAt(i);
            if (sl.getID().equals(uniqueID)) {
              if (!(sl.getTime() == creationTime)) {
                if (sl.getTime() > creationTime) {
                  update.setMessage("Preparing to send slide to client");
                  out.write(("-Slide Updated On Server" + CRLF).getBytes());
                  received = getResponse(in);
                  sendSlide(sl, out);
                  received = getResponse(in);
                  received = getResponse(in);
                  updated = true;
                } else {
                  update.setMessage("Preparing to receive updated slide");
                  out.write(("-Slide Updated On Client" + CRLF).getBytes());
                  sl = receiveSlide(in);
                  received = getResponse(in);
                  updated = true;
                }
              } else {
                update.setMessage("Slide Matched : " + sl.getTitle());
```

```
        }
        sl.seen = true;
        slides.removeElementAt(i);
        slides.insertElementAt(sl, i);
        presentOnServer = true;
      }
    }
```

Once we've checked against every slide, the value of presentOnServer will reflect if we found anything, and if not we have to assume that the slide is unknown to the server. The following code performs that action, and gets the slide from the client. If the slide has also not been updated, then the end of this section sends the +OK that reports this slide to have been successfully matched, and receives the details of the next slide:

```
    if (!presentOnServer) {
      out.write(("-Slide Unknown On Server" + CRLF).getBytes());
      sl = receiveSlide(in);
      sl.seen = true;
      slides.addElement(sl);
      received = getResponse(in);
    } else if (!updated) {
      out.write(("+OK" + CRLF).getBytes());
      received = getResponse(in);
    }
  }
```

The last line of the above code closes the while loop, meaning that '+End Of Slides' has been received. At this point, we go through another loop, this time looking at all the slides on the server and making sure that they have been seen.

This loop is fairly simple in that it just looks at each slide in turn, and if it hasn't been seen (if the boolean value hasn't been set to true) then it sends the details of the slide to the client. Strictly speaking, we should check that the returned value from the client is '+OK Ready To Receive', but we just assume this and rely on a communication error being thrown if there is any problem:

```
  for (int i = 0; i < slides.size(); i++) {
    sl = (Slide) slides.elementAt(i);
    if (!sl.seen) {
      out.write(("-Additional Slide" + CRLF).getBytes());
      received = getResponse(in);
      if (!received.startsWith("+")) {

        // Throw a communications error
      }
      update.setMessage("Sending Slide To Client");
      sendSlide(sl, out);
      received = getResponse(in);
      update.setMessage("Slide Updated On Client");
    }
  }
```

Once all the unseen slides have been sent to the client, we send our '+OK' message, this will let the client know that it's OK to close the connection:

```
      out.write(("+OK" + CRLF).getBytes());
      update.setMessage("Connection Closing");
```

We then shut down our side:

```
      out.close();
      in.close();
      sock.close();
      server.close();
    } catch (Exception e) {
      System.out.println("Something went wrong");
      e.printStackTrace();
    }
```

This last line removes the status window from the screen:

```
    update.setVisible(false);
  }
```

While it may only be necessary to close the `ServerSocket`, `server`, it's generally good practice to close all the streams in the reverse order they were opened. This not only provides some protection against badly implemented versions of the `Socket` peer objects, which have been known to exist, but also makes for easy to understand source code.

We then have a couple of private methods to take care of the details of communications. The first, gets the response from the client and converts it into a string. It waits for the LF character (ASCII value 10) as the end of line marker, as this character is not added to the string, the final character will be CR (ASCII 13) that will be cut off:

```
  private String getResponse(InputStream is) throws IOException {
    int i = 0;
    String s = "";
    while (i != 10) {
      i = is.read();
      if (i != 10) {
        s = s + new Character((char) i).toString();
      }
    }
    return s.substring(0, s.length() - 1);
  }
```

The next is used when a slide is going to be sent by the client to the server, and handles receiving the slide and instantiating it into an object. Note that it would be possible to use the Serialization API to send the slides, and this would be a more elegant solution, but would limit us to Java based servers and clients:

```
  private Slide receiveSlide(InputStream is) throws IOException {
    Slide newSlide = new Slide();
    String received = new String();
    received = getResponse(is);
    newSlide.setID(received.substring(3));
    received = getResponse(is);
```

```
    newSlide.setTime(Long.parseLong(received.substring(5)));
    received = getResponse(is);
    newSlide.setTitle(received.substring(6));
    received = getResponse(is);
    while (!received.startsWith(".")) {
      ;
      newSlide.addLine(Integer.parseInt(received.substring(0, 1)),
                       received.substring(2));
      received = getResponse(is);
    }
    return newSlide;
  }
```

The last private method is used to send slides to the client, when needed:

```
  private void sendSlide(Slide s, OutputStream o) throws IOException {
    o.write(("ID=" + s.getID() + CRLF).getBytes());
    o.write(("TIME=" + s.getTime() + CRLF).getBytes());
    o.write(("Title=" + s.getTitle() + CRLF).getBytes());
    for (int i2 = 0; i2 < s.countLines(); i2++) {
      o.write((s.getLine(i2).value + " " + s.getLine(i2).text
               + CRLF).getBytes());
    }
    o.write(("." + CRLF).getBytes());
  }
```

The Client

The client side of the synchronization process is very similar, though the order in which it does things is slightly different. The big difference is that the only place where the client has to make any decisions is when it sends the slide details and waits to see what the server says it should do next. This was a deliberate move to reduce the amount of processing required on the client, which may well have limited resources.

This code is written to work in the CLDC environment, and was used on a Palm Pilot device. Therefore, some of the networking classes may be unfamiliar, please see Chapter 5 for more details, though the following should be relatively easy to understand anyway:

```
  private void syncWithServer() {
    try {
```

The client only deals with one slideshow, which is contained in the global variable `slides`. This `Vector` holds all the slides being worked on, and is referenced by various methods within the application.

It is, obviously, better to adopt the approach used on the server, where the `Vector` to be synchronized is passed as a parameter, but this may use more system resources by creating an additional entry on the stack. Global variables are used a great deal more where there are limited resources, as they are often more efficient.

The CLDC does not have any AWT classes, so we make use of the global graphics object `g`, to draw directly to the screen. This is how we show the user that things are happening:

```
g.drawString("Creating Connection", 2, 132);
```

The next line opens a connection to our server. This is different to the normal networking classes, for details see Chapter 5:

```
c = (StreamConnection) Connector.open("socket://127.0.0.1:1710");
```

Next we open the communication streams, from here everything is much the same between CDC and CLDC, except where noted:

```
i = c.openInputStream();
o = c.openOutputStream();
is = new InputStreamReader(i);
os = new OutputStreamWriter(o);
g.drawString("Connection Opened" + space, 2, 132);
```

`getResponse()` is a private method used to collect responses from the server. Strictly speaking, we should check that the expected response is received, but we are depending on an exception being thrown if anything fails in the communication:

```
getResponse();
```

This loop is repeated once for each slide the client is aware of. Note that the variable `h2` is global, and is reused elsewhere in the application. This is to save resources:

```
for (h2 = 0; h2 < slides.size(); h2++) {
```

We let the user know which slide we are sending, if there are many slides, this might be a lengthy process and it's important to keep the user updated as to what is happening:

```
  g.drawString("Sending Slide " + new Integer(h2).toString() + space,
               2, 132);
```

`sl` is another global variable, used throughout the application. Here we are copying the current slide from the global vector, for ease of use:

```
  sl = (Slide) slides.elementAt(h2);
```

The next lines send the details of the current slide to the server; the client will compare the `uniqueID` and `creationTime` with the slides it has stored locally, and let us know what it would like us to do next. The response from the server is stored in another global variable store.

Note the use of the global variable `CRLF`, which is made up of the ASCII characters 13 and 10, as specified in the protocol. In the server synchronization method, we defined this as a local variable, but in the client it is a global in order to save processing time and memory:

```
  os.write("ID=" + sl.getID() + " TIME="
           + new Long(sl.getTime()).toString() + CRLF);
  store = getResponse();
```

If the server responds with '+OK', we know that it already has a copy of the slide, with the same creationTime, and therefore we can move on to the next slide in our local store. However, if it responds with anything else then we need to examine in more detail to decide what to do next.

```
          if (!store.startsWith("+")) {
```

If the server has a more recent version of the slide, it will send a '+Slide Updated On Server' command, which we should receive before getting ready to receive the details of the slide. As it happens, our application does not allow editing of slides (it's only an example), and therefore this is not a possible reaction. However, it is implemented against a more functional application being developed:

```
            if (store.startsWith("-Slide Updated On Server")) {
              g.drawString("Updating Slide From Server" + space, 2, 132);
              os.write("+OK Ready To Receive" + CRLF);
```

We receive the uniqueID, creationTime, and title at this point. The protocol states that the line sent containing the uniqueID and creationTime will have initial characters indicating what they are, and we need to cut these off. We also store the creationTime as a long variable, so we need to convert the incoming string into a long before it can be put into the slide object. We will use the private method receiveSlide() to actually do the conversion. Again, we are using the global variable sl to store the incoming slide, to save resources. Reusing variables in this way can be confusing, and care must be taken to ensure that we are only using a variable for one thing at a time, failing to do so can lead to difficult to spot bugs emerging:

```
              sl = receiveSlide();
```

As this slide is a replacement for the slide whose details we sent to the server, we need to insert it and remove the original:

```
              slides.insertElementAt(sl, h2);
              slides.removeElementAt(h2 + 1);
              os.write("+OK Slide Replaced" + CRLF);
              g.drawString("Slide Replaced", 2, 132);
```

This code will be entered if the server has never heard of the slide details we sent or has an older copy of the slide. Therefore, we must send the entire slide:

```
            } else if (store.startsWith("-Slide Unknown On Server")
                       || store.startsWith("-Slide Updated On Client")) {
              g.drawString("Sending Slide To Server", 2, 132);
```

We use the private method sendSlide() to actually do the work of sending the slide to the server:

```
              sendSlide(sl);
            } else {
              g.drawString("An Error Has Occured" + space, 2, 132);
            }
          }
        }
```

Once we have sent the details for all the slides we are aware of, we send the message '+End Of Slides', and wait to see how the server responds. If it responds with a '+OK', then the synchronization is completed, and we can close our connection, however, if it responds anything else then we need to be ready to receive additional slides. We repeat this loop as often as necessary, once for every additional slide the server wants to send us:

```
        os.write("+End Of Slides" + CRLF);
        store = getResponse();
        while (!store.startsWith("+OK")) {
          if (store.startsWith("-Additional Slide")) {
            g.drawString("Receiving New Slide" + space, 2, 132);
            os.write("+OK Ready To Receive" + CRLF);
            sl = receiveSlide();
            slides.addElement(sl);
            g.drawString("New Slide Appended", 2, 132);
            os.write("+OK Slide Added" + CRLF);
            store = getResponse();
          } else {
            g.drawString("An Error has occured", 2, 132);
            store = "+OK";
          }
        }
```

Once we have added any slides sent from the server, we close the connection and complete the synchronization:

```
        g.drawString("Closing Connection" + space, 2, 132);
        os.close();
        is.close();
        o.close();
        i.close();
        c.close();
      } catch (Exception e) {
        g.drawString("Network Error" + space, 2, 132);
      }
    }
```

We then have our private methods, used for the communications and sending and receiving slides. The first method is almost identical to getResponse() on the server, except for modifications to be less resource hungry. The use of global variables, and a global stream, can greatly reduce the processing and memory overhead:

```
    private String getResponse() throws IOException {
      h1 = 0;
      s = "";
      while (h1 != 10) {
        h1 = is.read();
        if (h1 != 10) {
          s = s + new Character((char) h1).toString();
        }
      }
      return s.substring(0, s.length() - 1);
    }
```

The next two methods are responsible for sending and receiving slides respectively, and are almost identical to their counterparts on the server:

```
private Slide receiveSlide() throws IOException {
  Slide s = new Slide();
  store = getResponse();
  s.setID(store.substring(3));
  store = getResponse();
  s.setTime(Long.parseLong(store.substring(5)));
  store = getResponse();
  s.setTitle(store.substring(6));
  store = getResponse();
  while (!store.startsWith(".")) {
    s.addLine(Integer.parseInt(store.substring(0, 1)),
              store.substring(2));
    store = getResponse();
  }
  return s;
}

private void sendSlide(Slide s) throws IOException {
  os.write("ID=" + s.getID() + CRLF);
  os.write("TIME=" + s.getTime() + CRLF);
  os.write("TITLE=" + s.getTitle() + CRLF);
  for (h1 = 0; h1 < s.countLines(); h1++) {
    os.write(s.getLine(h1).value + " " + s.getLine(h1).text + CRLF);
  }
  os.write("." + CRLF);
  g.drawString("Slide Sent" + space, 2, 132);
}
```

Overall, the client is much less complex than the server, which is appropriate. Part of the process of designing a synchronization system is to ensure that the majority of processing is taking place on the device most able to cope with it, which is, in this case, the server.

Using Synchronization Code

Hopefully you are now aware that creating code for synchronization is no easy task, but one that requires a great deal of consideration and planning, and some compromises. It is rarely possible to allow for every combination of data updates that will need to be synchronized, and some risk analysis is necessary to ensure that the best option possible is achieved.

It may be necessary to allow the user some level of control over what data source takes precedence in the case of conflicts, though this should be kept to a minimum to avoid presenting the user with too many possible options, leading them to ignore all of them. A good example of this can be seen in the Pocket PC ActiveSync software that is supplied with all PocketPC hardware. This offers the user options to make the handheld client or desktop system the final arbitrator for conflicts, or for the user to be presented with the option should such a conflict arise:

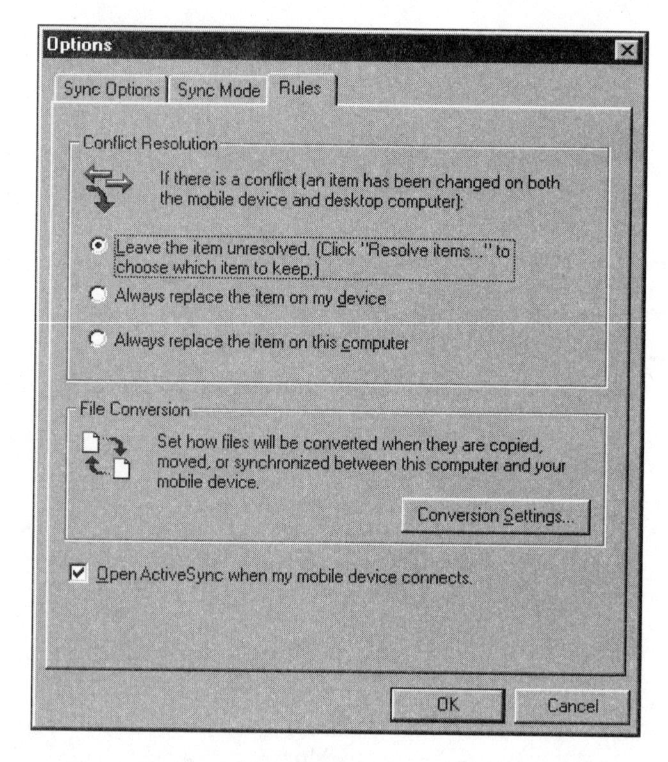

While straight backup or transaction-based synchronization may not face these issues, mirrored data will inevitably have some issues where data has been updated on both devices, and the programmer must predict what the results of such a conflict will be and decide if it is appropriate.

Synchronization With Palm Pilot Devices

A great deal has been made, in this chapter, of the synchronization capabilities of the Palm devices, and it would be remiss not to consider the particular opportunities offered by working with this platform. As has already been discussed, the Palm's success can largely be attributed to its effective synchronization process, and it is more than possible for the Java programmer to make use of the tools already developed to do this.

While Palm devices are more than able to run J2ME locally, after all, the above example was designed to run on a Palm device, it is not really in keeping with the user experience to demand that they run a custom client before running a custom server. Much better to make use of the HotSync technology available from Palm Computing (http://www.palm.com/software/) and allow the user a familiar interface and operating style.

HotSync Technology

The user of a Palm compatible device simply places their device into the provided cradle, which often provides power for recharging in addition to connectivity, and presses the single button on the cradle, to start the synchronization process. Of course, there is a requirement for software to have already been installed on the server (this is referenced to as the Desktop in the Palm documentation), and this software is always waiting for a signal from a Palm device to start synchronization.

This simplicity of use belays the complexity we have already discussed, but provides a very logical user interface which non-computer users are very happy with.

The actual code that performs the synchronization is known as a **conduit**, and there is one conduit for each application to be synchronized. Strictly speaking, the development of conduits is beyond the scope of this chapter, for several reasons that we will discuss, but as there is now a Java API available from Palm for conduit development, it seems appropriate to provide a brief overview of how such software can be developed.

The first thing to understand is that we are not talking about software that will run on the Palm device itself, only on the server. The HotSync software will call the Java conduit when the user presses the cradle button, and will then be able to interface with the data stored on the device, through the HotSync Java APIs. These APIs are not provided with the HotSync software, and must be downloaded and installed separately. It is possible to distribute copies of both the Sun Java Runtime and Palm Java Conduit APIs with your conduit, but this will increase the size of your distribution considerably.

The second factor to be aware of is that the Palm Java Conduit API is only available for Microsoft Windows operating systems. The reason for this is the necessity for native code to be developed for interfacing between the JVM and the HotSync application, which is performed using a Microsoft Windows DLL file. Palm have stated that with sufficient demand they are happy to port the native code to other operating systems, but for the moment any conduits developed in Java will only be usable on the Microsoft Windows platform.

While the above restrictions may rob Java of some of its key advantages, if your application is only ever planned to run on Palm OS based devices, synchronizing with Microsoft Windows based desktops, Java conduit development might well be appropriate. More details and documentation can be downloaded at from http://www.palmos.com/dev/tech/conduits/.

SyncBuilder

SyncBuilder is a complete replacement for the HotSync server provided by Palm Computing, written in Java. It is an open source project, working towards a completely platform independent implementation of HotSync, though for the moment it only works on Microsoft Windows and Unix platforms (including Linux).

Obviously, with the main application written in Java, the creation of Java conduits is simplified, and essential. Currently SyncBuilder provides synchronization with all the inbuilt applications and appears very stable. While still only useful for developing synchronization applications for Palm devices, it provides a more Java based environment, and the promise of platform independence in the future. More details can be found at http://stud.fh-heilbronn.de/~christ/java-link/intro.html.

SyncML: A New Standard

As should already be clear, the process of synchronization is not only about communication, but also about keeping track of alterations and updates, as well communicating those changes between clients and servers. Given the multitude of different devices, and applications, it does not make sense for every application developer to use their own proprietary standard for each application. Therefore, a number of companies have joined forces to create a standard for representing data to be synchronized, and it should come as no surprise that this format will be based on the XML standard.

While a worthy idea, there is still some question over the future of SyncML. In order to be relevant to as many applications as possible, it has become a complex standard, and may actually be overweight for some applications. With enough support, (and a nice Java API), however it should make the process of designing a synchronization protocol far less complex, as well as highlighting possible problems before the work of coding actually starts.

The specification has recently become available, and it's worth taking a moment to look at what it tries to do and how it approaches the many problems involved in synchronization. SyncML is designed to fit between the synchronization processes available on the client and server devices. That is to say, that both devices should be able to deal with SyncML, which provides a middle ground. What this means is that it is theoretically possible for devices to synchronize without any knowledge of the application or device being communicated with.

The process is quite complex, but well documented. It consists of a process of modifications made on the client being sent to the server, and the server working out the combination of modifications, and sending the results back to the client. It should be made clear that the entire data set does not need to be sent – only the modified records are communicated. This indicates that the server is responsible for all conflict resolution, which would normally be the case (as it is in our example), but with SyncML it is possible for the server to send the client notification that a conflict has taken place and let the client take appropriate action.

SyncML is still in its very early days, and while reference implementations are available for Palm, EPOC, Linux, and Microsoft Windows, it is by no means an established standard. Details may be downloaded from the SynchML web site at http://www.syncML.org.

Summary

There is a strong argument that synchronization is actually only a stopgap technology that will be redundant within years as devices are more networked and less reliant on locally stored data. We have already seen that synchronizing between two devices involves not only complex code, but also some inevitable conflicts.

As the number of devices a user wants to synchronize increases, with the addition of a work computer, home computer, PDA, and mobile phone, these problems will increase in an exponential manner. There is however, one easy way of sorting out these problems. By holding all information on a central server, and accessing it from networked devices, all the problems just go away. Of course, such a datastore will have a host of other problems, regarding multiple access and concurrent data modification, but these are more related to database issues than synchronization.

While this might appear impractical, it should be remembered that communications and connectivity are progressing very rapidly. Already, in the UK, mobile phones offering GPRS are constantly connected to the Internet by virtue of being switch on, while desktop systems using DSL can act as central repositories for all information.

Technologies such as Bluetooth may provide more connectivity for synchronization, such as the ability to have your diary updated as you walk in to work. One has to question why wait until you enter the office? When all the information is in one place, there are no conflicts or updates to worry about, the process becomes more reliable and less complex, but completely dependent on network connectivity.

However, this kind of connectivity is still restricted to limited areas, and cost may well be a feature. Even if the very idea of synchronization has a limited lifespan, such applications are going to be an essential stepping-stone to the truly connected society.

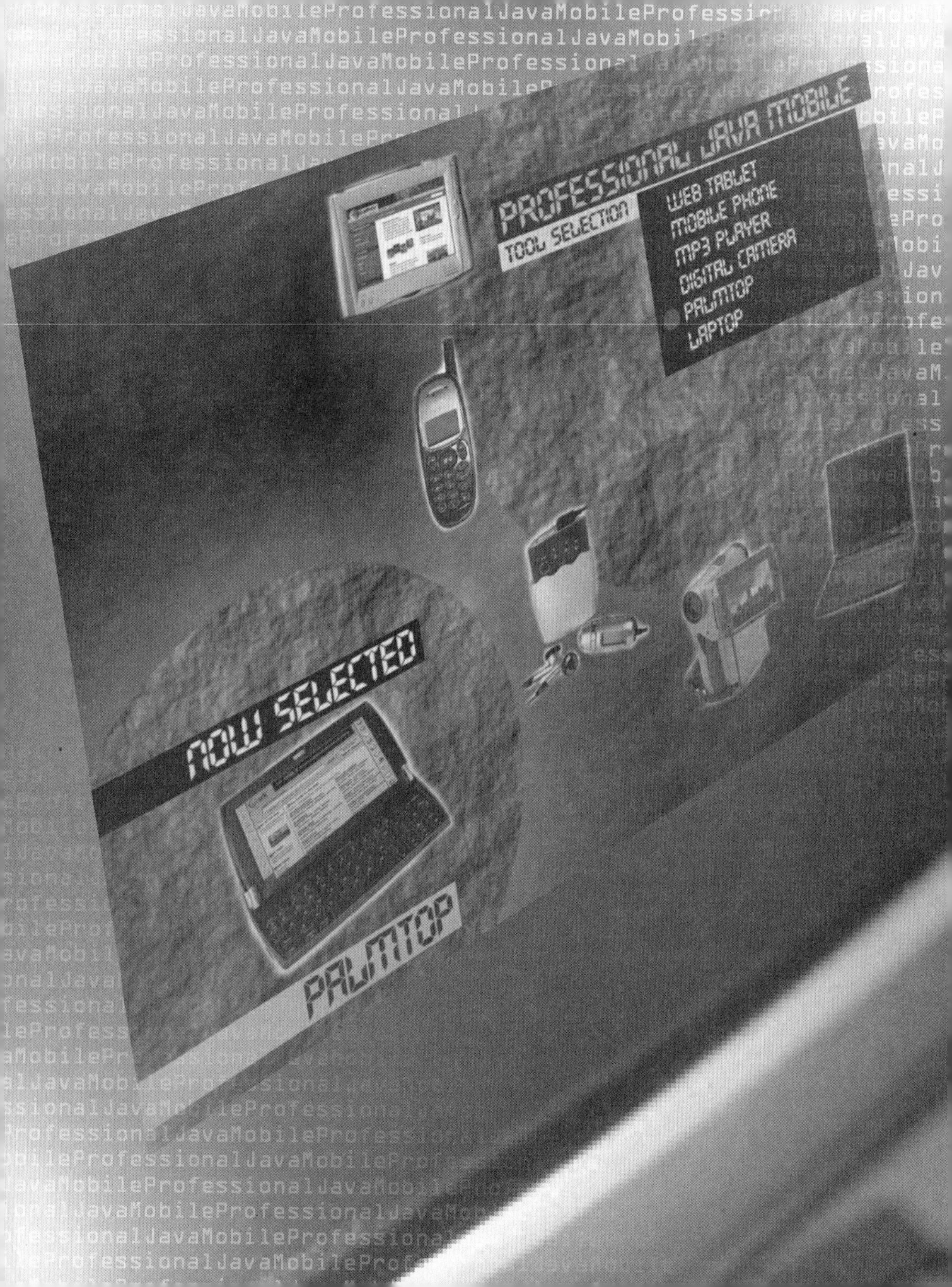
PROFESSIONAL JAVA MOBILE
TOOL SELECTION
WEB TABLET
MOBILE PHONE
MP3 PLAYER
DIGITAL CAMERA
PALMTOP
LAPTOP
NOW SELECTED
PALMTOP

9

The JavaPhone API

Increases in the sophistication of Mobile Information Devices (MIDs), and telecommunication technology are drastically changing the way we access information. It is now possible, or it will be in the very near future, to develop applications that will allow secure transactions using mobile devices (M-Commerce), the development of location based services from mobile devices which would enable us to access information about specific locations (we will see more about this in Chapter 15), and the transfer of information, for example, travel and weather alerts, news and stock information, to mobile devices.

With the introduction of third generation mobile technology (3G), **Internet screen phones** and **smart phones** are beginning to emerge. The definition is still changing but an **Internet screen phone** is a traditional phone but with a small screen and optional keyboard that provides Internet access and access to a number of Internet commerce applications and services. A **smart phone** is a mobile phone that allows Internet access and usually also address book, calendar, and notebook functionality. One of the most important areas for Internet screen phones and smart phones is the use of functionality specifically for telephony operations, and a development has been taking place in the Java community to provide precisely this functionality.

History of the JavaPhone API

The JavaPhone specification provides functionality that includes accepting and making calls, querying the handset for contact, and power information. In its full form, it allows call forwarding, access to messaging functionality, mobile radio control, and access to the components of the handset to control microphone and speaker volumes, the state of lamps (or leds), and buttons, and more.

One of the major companies that has specialized in MIDs is Symbian. Symbian is a joint venture between the major telecommunication companies including Ericsson, Nokia, Motorola, Panasonic, and Psion. During the past several years, they have developed a software platform specifically tailored towards MIDs called the Symbian platform, which is a robust 32-bit operating system with support for developing, testing, and deploying MID applications. Already, several smart phones based on the Symbian platform, are available in many countries.

The Symbian platform supports the development of applications in several languages, primarily C. However, increasingly, the language of choice for developers is Java. The Symbian platform provides a full implementation of PersonalJava, and access to telephony functionality is provided through an implementation of JavaPhone for MIDs. The JavaPhone specification was developed as part of a Java Community Process, that incorporated key leaders in the telecommunications industry, including Symbian.

This basically means that the developer is programming for Java 1.1 with an application that contains a class with a main method, etc. The JavaPhone API allows developers to develop applications that can manage telephone functionality from these PersonalJava applications (for example, make, receive or transfer calls), and enable us to access the personal information manager of a device (for example, address book, database, or calendar).

What We Cover in this Chapter

As previous chapters have dealt with programming in PersonalJava (and in the near future, its successor, the Personal Profile), we will not be looking at examples to introduce this topic. Any developer familiar with Java should be able to adjust their thinking to programming in Java1 without too much pain. Downloading and installing the SDK (which includes an emulator) is a simple and relatively painless operation. The notes quote a minimum JDK1.2.2, however the installation also downloads a separate JRE.

For an introduction to the SDK, please read the comprehensive documentation, the quick start section also includes simple instructions on writing, packaging, and deploying applications. There are several tools for generating the various deployment descriptors that are required, and for designing and packaging icons in the image format used in the Symbian phones.

In this chapter, we will discuss the JavaPhone architecture and the Symbian implementation of the JavaPhone.

The JavaPhone Architecture

The JavaPhone API represents a vertical extension to the PersonalJava platform. Coupled with PersonalJava or EmbeddedJava application environments, the API provides a mechanism for developing applications for telephony devices. The API was specifically targeted at developers and organizations developing applications for smart phones and Internet screen phones. It allows the development of functionality unique to telephony devices, and includes:

- Direct telephony control
- Datagram messaging
- Access to address book and calendar information
- Access to user profiles
- Mechanisms for application installation
- Access to system properties
- Access to power management

In the JavaPhone API, two profiles are introduced; the Smart Phone Profile and the Internet Screen Phone Profile. The diagram below shows the main components of the JavaPhone API:

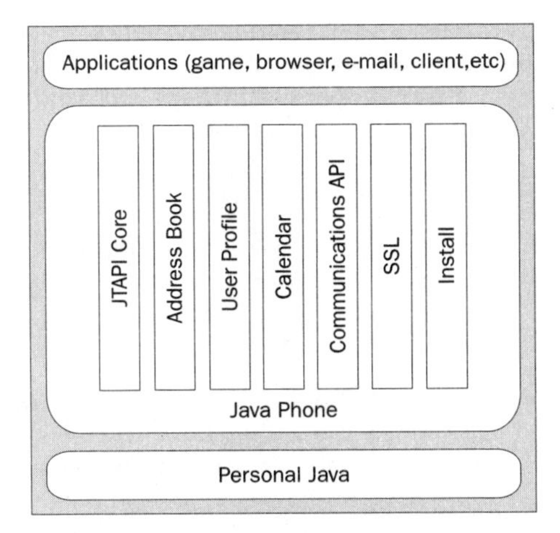

Of the packages shown above, only JTAPI Core, Address Book, and the User Profile packages are required by the Internet Screen Phone Profile. The remaining packages are optional and do not have to be fully implemented.

The Wireless Profile, as implemented by smart phones, also adds a Datagram package, JTAPI Mobile, and Power Monitoring packages, all of which are mandatory. In addition, the Calendar package is made mandatory.

The JavaPhone Packages in Detail

As can be seen in the Symbian implementation, only some of an optionally implemented package's functionality may be implemented, whether through security concerns or because the handset does not support it. The packages are further discussed below.

JTAPI Core and JTAPI Mobile

The Java Telephony API (JTAPI) is a portable API for developing telephony applications. It forms the interface between Java applications and the telephone, or telephone system implementations.

The JTAPI is functionally separated into a series of packages called capability packages. The capability package serves as an extension mechanism allowing JTAPI providers to communicate the features that are included in an implementation. JTAPI defines a `Provider` interface that provides information about what functionality is available, and includes methods for accessing references to classes that implement the desired functionality.

JTAPI Mobile groups certain functionalities together, including a `MobileRadio` provider for the Wireless Profile. Internet phones do not implement these sets of interfaces and classes. The JTAPI specifications include capability interfaces that allow applications to discover whether certain functionality has been implemented in a particular handset. In some devices though, the overhead in providing these interfaces, merely to allow capabilities to be discussed, is too high, and so certain functionality has been grouped together into `javax.telephony.mobile`. The capabilities of the providers can be tested for in these devices, using `instanceof()`.

Address Book

One of the main features of an application that implements telephony functionality is access to an address book or contacts list. Using the `AddressBook` package, an application can locate and update contact information stored in the address book, such as phone numbers and e-mail addresses. The `AddressBook` package implements the query mechanisms specified in the vCard standard. The vCard is a recognized standard for the transfer and storage of contact information.

User Profile

The User Profile package defines the objects necessary for setting and extracting the personal information for the MIDs owner, such as the user's contact information.

Calendar

Another main feature of applications for MIDs is access to calendar information. The Calendar package defines the objects necessary for accessing and adding calendar events, and deleting or modifying existing events. The Calendar package also provides access to schedule information such as to do lists and task entries.

Datagram

We have previously looked at datagrams in Chapter 3, but to summarize a datagram is a transport independent method for delivering a message. Applications send messages to and receive messages at addresses that consist of a service name and location. The device the application is executing on chooses the network device or bearer; for example, the Quartz Symbian platform specifies SMS over SMS for sending only, SMS over WDP for send/receive, and UDP over IP (send/receive).

The Datagram package defines the objects necessary to allow applications to send messages only knowing the address and data protocol. Since the API is transport independent, applications do not need to be rewritten to run on different devices.

Power Management

The Power Management package allows applications to be aware of the various states of a device, such as the device being in the sleep state, conserve state (where the device attempts to run at maximum power efficiency) and when it is in the ON mode.

Power Monitor

A `PowerMonitor` implements functionality to monitor the power level of a device. Using the `javax.power.monitor` package to which `PowerMonitor` belongs, an application can monitor the current battery level, detect whether an external power source is being used, and retrieve an estimate of the remaining power level. It can also be used to alert the application when the power is running low.

Install

The JavaPhone specification includes an interface to allow for the installation and removal of applications to/from the device. The Install package supports application packaging, installation mechanisms, and versioning.

Communication API and SSL

The Communication API and SSL packages cover serial communications. If no serial communications are available to applications, they will not be implemented.

Overview

The JavaPhone API allows MID application developers to leverage the benefits of Java and provide telephony unique functionality. In summary, the JavaPhone API provides:

- Device portability
- Lower development costs
- Application security
- Increased market availability

Writing in Java, and using the JavaPhone API, allows developers to write once and run anywhere – if a device implements the JavaPhone specification, then an application will run on that device. Using Java as opposed to the more traditional C or C++ for developing applications for MIDs will enable companies to bring applications to the market in a reduced time, and therefore at a reduced cost.

Developing Applications with the JavaPhone API

The following sections introduce the programming concepts behind developing applications for the JavaPhone specifications, and are illustrated with reference to the Symbian implementation. The examples are focused on the Quartz or Crystal device family reference designs. The reference designs and the Quartz/Crystal development kits can be downloaded from: http://www.symbiandevnet.com.

Once we have the installer application, installation is simply a matter of running it and taking a note of the directory where the SDK has been expanded. If we have chosen the default and everything has gone according to plan, we should have the following directory structure: C:\Symbian\6.0\QuartzJava\. From now on, we shall refer to this as <EPOC_ROOT> when we are describing directory structure, and %EPOC_ROOT% when referring to the CLASSPATH environment variable.

The Emulator File System

The emulator has its own file system with drives named c:, d:, and so on. While the directories and files on these drives exist as directories and files in a desktop's file system, there is no direct correspondence between them; for example, the emulator's c: drive and a PC's c: drive are not equivalent. In practice, the c: drive of the emulator corresponds to directory <EPOC_HOME>\epoc32\wins\c\ and the ROM drive corresponds to <EPOC_HOME>\epoc32\release\wins\urel\z\.

We can also add drives and make files, both inside and outside the emulators folders accessible to the emulator, using environment variables of the form _epoc_drive_**driveletter**. For example, we can set up a p: drive in the c:\projects\myapplet folder as follows:

```
set _epoc_drive_p=c:\projects\myapplet
```

Now when we run the emulator, it will have a drive p: that corresponds to the directory c:\projects\myapplet\. In addition, this can be set in the epoc.ini file found in the <EPOC_HOME>\Epoc32\Data\ directory. The following line is the default, mapping j: to <EPOC_HOME>\erj:

```
_epoc_drive_j \Symbian\6.0\QuartzJava\erj
```

The erj directory is the default directory for Java development, and is created at install time.

Running Java Applications

EPOC does not have a concept of a current working directory in the way that other operating systems do. Even though the EPOC Java VMs can be launched from a command line, they do not inherit a current directory from the shell. Instead, the current directory is set using the -cd command line option.

Unlike a desktop OS, EPOC has no environment variables, nor any meaningful notion of a registry. This causes a problem for us as Java developers, as Java Virtual Machines on other operating systems use registry or environment variables to provide a default classpath.

The emulator expects to find its classes on drives with the same directory structure as the default Java development drive, j:, mapped using the PC's _epoc_drive_j environment variable. The default classpath on the emulator consists of:

Component	Description
.	The current directory.
?:/classes	For shared classes that may be used by more than one application.
?:/ext/*	Extension packages stored in zip or JAR archives.
?:/lib/classes.zip	System packages. Contains the Java classes.

Here, the question marks refer to relative paths within the emulator's drivespace, and as a default resolve to its j: drive.

Java applications and applets may be launched on the emulator from the command prompt. For example, to run the ChangeTelCodes, class we would enter:

```
%EPOC_ROOT%\Epoc32\release\wins\udeb\pjava_g -cd
j:\projavamobile\chapter09\classes ChangeTelCodes
```

This, or any equivalent sequence of commands, launches the emulator and runs the application housed in the file j:\examples\projavamobile\chapter09\ChangeTelCodes.class in its Java run-time. The class name is case-sensitive, but paths are not. Don't worry about the specifics of this command, or compilation, as we'll look at them more carefully below, when we run examples.

The AIF Builder Tool

It may be the case that we wish to run our applications from the emulator's Launcher menu. This is accomplished by converting our Java files into an EPOC application file (.app), which, minimally contains a unique identifier (UID) to identify the program, and an application information file (.aif) to specify the application's icon and caption. Luckily, the Quartz SDK comes with the AIF builder tool that allows us to do this.

Other files may also be required, depending on the programming language: for example, for a Java application, an additional text file (`.txt`) is required to specify the command line to the JVM when the application is launched; while for OPL, the OPL source file has to modified.

Event Handling in JavaPhone applications

The JavaPhone API depends on an event mechanism to report changes to the state of various components of the telephony model. Examples of telephony events are: a call being disconnected or accepted, a connection attempt failing, the phone going off the hook, a button being pressed, etc. At each event, the application will notify the user and allow them to make choices based on the events that have occurred.

An example of state transition for a call is from the `Call.IDLE` state (the call has no connections), to `Call.ACTIVE` (the call has one or more non-disconnected connections), to `Call.INVALID` (all of the call's connections have reached the disconnected state). An event is generated for each state change. A connection will pass from being idle (not connected) to being in the process of contacting its associated address, to notifying that address that an incoming call is occurring, being connected, disconnected, etc. Different reasons for the transition of a connection between states will generate a variety of events. A full list of connection states and events is available in the JavaPhone documentations.

Now, several JTAPI events can be generated as a result of a single change in a call. For example, if a party were to hang up on a call, the connection would become dropped, the call would move from an active state to an invalid state, and so on. Before JTAPI 1.3, the event mechanism was implemented through observers. An `Observer` specifies a single method `XXXChangedEvent()`, where `XXX` specifies the type of event. This method accepts an array of events as its parameter that it then processes one at a time.

The implementing method checks what type of event each event in the array is, and conditionally executes logic according to type. The result of this is that groups of related events are processed together. In the 1.3 specifications, JTAPI past the Java 1.2 event model using event listeners. Event listeners define a method for each type of event, and each method accepts a single event as its parameter.

An application must provide code to implement the listener or observer for the relevant events, which it then registers. From then on, all events are passed to the listener or observer for that event.

JTAPI events, as we have discussed previously, are generated in groups, and so a mechanism for grouping events was devised using events called Meta events. Meta events define the beginning and end of each group of events and define the group. Examples of groups include call changes as a result of an unobtainable address, network congestion, etc. Application listeners that need finer grained event information than is available through the Meta events, are expected to store events until the Meta event has finished, and process them together in the context of the Meta event that groups them.

Programming Concepts of the JavaPhone API

We will begin investigating the APIs available through the JavaPhone specifications by covering the required packages, JTAPI Core, Address Book, and User Profile.

JTAPI Core

The JTAPI core packages (in `javax.telephony`, `javax.telephony.capabilities`, and `javax.telephony.events`) provide a model for simple telephony and methods for connecting, disconnecting, and accepting calls.

The JavaPhone specifications model a telephone call as shown below. Let's introduce some more jargon here. The telephone call is encapsulated by the `javax.telephony.Call` object. It acts as the center point for all calls. A `Call` holds a connection to every party involved in that call; so a call between two people will have two connections, a conference may have three or more parties with a connection for each party, etc.

The address of a telephone (handset, cell phone, call it what you will) is its telephone number or whatever identifier is used to connect to it. This may be an IP address or another means of identifying an end point that can be contacted using a provided service. There may be more than one address for a given telephone so a separate object models that telephone. The reverse is also true; many phones may be served by a single telephone number through a private exchange or telephone network.

The end point or target of a telephone call is called a `Terminal`. This leads to two types of relationships between `Call` and calling party. There is the logical connection, `Connection`, between the `Call` and an `Address`, and there is a physical connection, `TerminalConnection`, which is the relationship between the call and the `Terminal`.

The implementation is responsible for maintaining the relationships between addresses and terminals, and this information cannot be modified by applications. In addition, in order to facilitate efficient garbage collection, connections are mutable and cannot be reused for consequent calls, and will lose their association with both the call and the related address as soon as the connection has reached the disconnected state. A consequence of this is that once a connection is disconnected, a call cannot access the address that it was bound to through that connection. To put this another way, the application is responsible for maintaining information about what parties were involved in a call once those parties have discontinued the call:

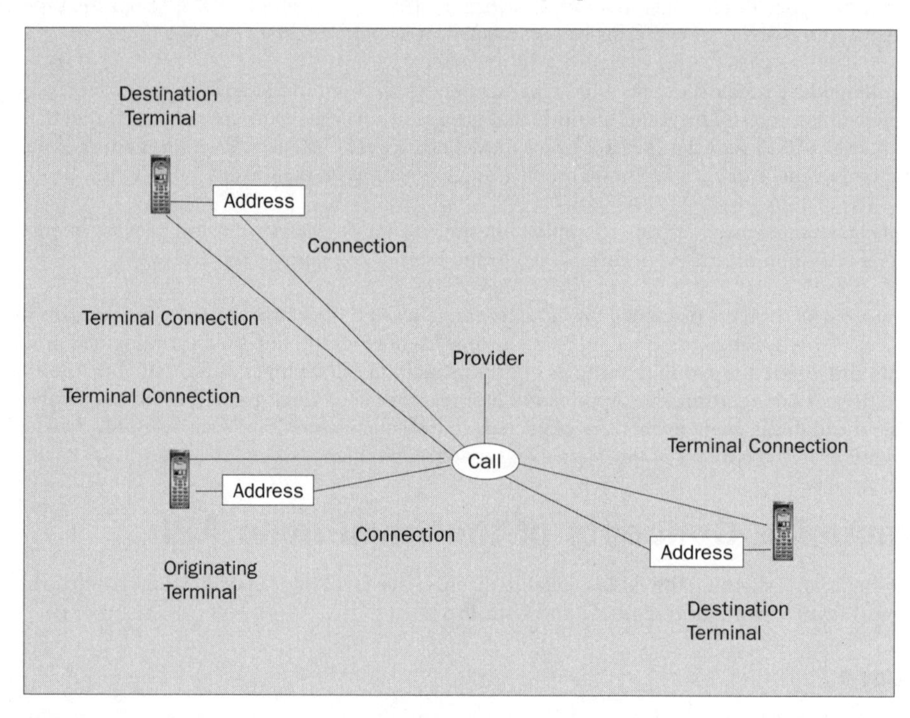

The JavaPhone specifications define a `JtapiPeer` that represents the specific (vendor) implementation of the JavaPhone specification. As with many of the other classes in these specifications, this class is managed by a factory. An application gets a reference to the `JtapiPeer` via the `JtapiPeerFactory`'s `getJtapiPeer()` method.

The `JtapiPeer` very simply gives access to the implementation's service providers. A service represents a type of supporting network. Several services may exist, in which case they are differentiated by a String name. The `JtapiPeer` has one method for retrieving the available services, `getServices()`, which returns an array of strings, and the `getProvider()` method, which takes a String parameter that is the identifier for the service. In both the `getJtapiPeer()` and the `getProvider()` methods, passing `null` to the methods will return the default.

Here is some example code to make a call using the default service:

```
public class myCaller implements CallListener {

private String telNumber;
private Provider defaultProvider;
private origAddress defaultAddress;
private origTerminal defaultTerminal;

public myCaller(String telNumber) {
  this.telNumber = telNumber;

  try {
    JtapiPeer peer = JtapiPeerFactory.getJtapiPeer(null);
    defaultProvider = peer.getProvider(null);
```

In the section of code above, we retrieve the default `JtapiPeer` for this handset and get the default provider from it. We can now get an origination address and a reference to the originating terminal from the provider. Calling the `getAddresses()` method will retrieve an array of type `Address` with a list of the addresses associated with this service. We will arbitrarily choose the first of these (blithely assuming that there is at least one) and select its first terminal. Remember that a given address may have more than one terminal associated with it:

```
    Address[] addresses = provider.getAddresses();
    origAddress = addresses[0];

    Terminal[] terminals = provider.getTerminals();
    origTerminal = terminals[0];
```

We can now go on to make the call. We create a `Call`, and attempt a connection to the telephone number specified in the constructor's parameter. An array of connections will be returned; the first represents the `Connection` from this terminal (to this `Call`), and the second is the `Connection` from the `Call` to the destination address:

```
    try {
      Call call = provider.createCall();

      Connection connections[] = call.connect(origTerminal,
                                  origAddress,
                                  telNumber);
    } catch(Exception e) {
      // deal with error here
    }
```

Typically, the class would also register a listener to listen for call events and react accordingly.

The exceptions that may be thrown by attempting to connect include attempting to connect a call when calling is not supported or is forbidden. For example, applets are not allowed access to JTAPI functionality due to the sandbox model security they are subject to. An `InvalidPartyException` is thrown if the address cannot be reached (or resolved). A `ResourceNotAvailable` is thrown, if, for example, no dial tone can be detected.

Address Book

The address book and calendar are extensions of the `javax.pim.database.Database` class in the JavaPhone specification. The `database` package defines the objects necessary to support a simple flat, non-relational, database model, and includes simple methods to add, delete, update , and traverse items stored.

The Address Book package (`javax.pim.addressbook`) models a database of contact information through the `ContactDatabase` class. The contact details are based on the vCard specification (RFC 2426). A device may contain several, named, address books. The `ContactDatabase` has static methods for creating, opening, and deleting named databases and, for retrieving a list of database names. It also has various methods for accessing the contacts stored in it.

A contact is stored in the `ContactCard` class that stores contact information in the vCard standard. A database can organize `ContactCards` into `ContactGroups`. Finally, a template for newly created cards, `ContactTemplate`, can be created, which defines default fields and default values for those fields.

The fields of a `ContactCard` are named with reference to the vCard standard. Example fields include the title field, `ContactDatabase.TITLE`, a telephone number, `ContactDatabase.TEL`, and so on. If you are familiar with the vCard specifications you will see that the fields are named as closely to their vCard equivalents as possible.

All items in a database (including contact databases) are of the super type `Item` and each item is a set of `ItemFields`. When the vCard specification specifies that a field contains a sequence of values, this is made available through the `AggregateField` class. This class takes additional indexing information when adding fields so that a sequence can be enforced. An example of a sequence of values is the name field, `N`, which requires a sequence of the family name (`FAMILY_NAME`), given name (`GIVEN_NAME`), additional names (`ADDITIONAL_NAMES`), honorific prefixes (`PREFIXES`), and honorific suffixes (`SUFFIXES`). The values in brackets represent the index values of the fields that store the respective values. The code snippet below shows how the name field would be constructed:

```
AggregateField name = new AggregateField(ContactDatabase.N);
name.addField(new ItemField(ContactDatabase.N, "McGillis"));
name.addField(new ItemField(ContactDatabase.N, "Sophia"));
name.addField(new ItemField(ContactDatabase.N, "Anne"));
name.addField(new ItemField(ContactDatabase.N, "Dr"));
name.addField(new ItemField(ContactDatabase.N, "ABC XYZ"));
```

The `addField()` method of the `AggregateField` class appends the field to the end of the aggregates sequence so index numbers are not required. In order to set a specified field, (assume we have retrieved the field previously), the following code would be required:

```
...
ItemField familyName = new ItemField(ContactDatabase.N, "McGregor");
name.setField(familyName, ContactDatabase.FAMILY_NAME);
...
```

In addition, fields that require additional parameters, can have these parameters added to them through the addParameter() method of the ItemField class, from which AggregateField inherits. An example of a field that takes additional parameters is the telephone field, which takes additional information regarding the type of number such as whether it is a work telephone number, and whether it is a fax or a voice phone.

Let's look at the possibilities by partially developing a somewhat minimal application. We will not concern ourselves with the client interaction, as it is somewhat outside the scope of the chapter and therefore not interesting to us. In any case this is the same as any PersonalJava application and therefore more or less Java1 development work which should present you with no problems.

We will begin by using the concepts we learned above to write a test class that inserts a number of contacts into a database, in this case, the default provided by the SDK:

```
import javax.pim.database.DatabaseException;
import javax.pim.database.Parameter;
import javax.pim.database.ItemField;
import javax.pim.database.AggregateField;

import javax.pim.addressbook.ContactCard;
import javax.pim.addressbook.ContactDatabase;


public class AddTelNumbers {
  private String database;

  String[] numbers = {"01222123456", "012222345678", "01203123456",
"01203345678"};
  String[] contactNames = {"Beccy", "Michael", "Kensa", "Layla"};
```

database represents the string name of the database to be opened. In this case, we will be using the default database provided by the SDK, which is empty when new. The main() method for this class merely instantiates a copy of the class and calls the addNumbers() method, which adds the numbers in the arrays above, followed by the checkUpdate() method, which extracts all contacts from the database and lists the name and telephone number for each contact returned:

```
  public static void main(String[] args) {
    AddTelNumbers addTelNumbers = new
AddTelNumbers(ContactDatabase.DEFAULT_ADDRESSBOOK);
    addTelNumbers.addNumbers();
    addTelNumbers.checkUpdate();
  }
```

The constructor accepts the name of the database to insert the numbers into:

```
  public AddTelNumbers(String database) {
    this.database = database;
  }
```

addNumbers() attempts to open the database, and assuming that it succeeds, inserts a ContactCard, per name listed, in the class variable; contactNames. The contacts entered into the database are somewhat basic. They do not contain any information except a name and one telephone number. However, you should be able to interpolate from this code:

```
public boolean addNumbers() {
try {
  ContactDatabase contactDatabase =
                          ContactDatabase.openDatabase(database);

    ContactCard contact;
    for(int i=0; i<numbers.length; i++) {
      contact = new ContactCard();
```

We insert blank strings for the fields for which we have no values:

```
      AggregateField name = new AggregateField(ContactDatabase.N);
      name.addField(new ItemField(ContactDatabase.N, ""));
      name.addField(new ItemField(ContactDatabase.N, contactNames[i]));
      name.addField(new ItemField(ContactDatabase.N, ""));
      name.addField(new ItemField(ContactDatabase.N, ""));
      name.addField(new ItemField(ContactDatabase.N, ""));
```

We add two parameters to each telephone number stating that the number is a voice line, and is a home number. You should note that unlike the vCard specifications, (which state that a number may have several assignations such as HOME, WORK, MSG, and so on, delimited by whitespace) the JavaPhone specifications mandate separate entries in the `ContactCard` with duplicate numbers for each type of telephone numbe:

```
      ItemField tel = new ItemField(ContactDatabase.TEL, numbers[i]);
      tel.addParameter(
              new Parameter(ContactDatabase.TYPE, ContactDatabase.HOME));
      tel.addParameter(
              new Parameter(ContactDatabase.TYPE, ContactDatabase.VOICE));
```

Finally, we add the name and telephone number fields to the contact card which we then add to the database:

```
      contact.addField(name);
      contact.addField(tel);

      contactDatabase.addItem(contact);
    }
  } catch(DatabaseException de) {
    System.out.println("Error adding to database:" + de);
    return false;
  }
return true;
}
```

The `checkUpdate()` method opens the database and retrieves an iterator of all the contacts in the database. For each contact, we print out the name and any home telephone numbers listed:

```
public void checkUpdate() {
try {
    ContactDatabase contactDatabase = ContactDatabase.openDatabase(database);
    javax.pim.database.Iterator contacts = contactDatabase.cardItems();
    ContactCard contact;

    while(contacts.hasNext()) {
```

```
        contact = (ContactCard)contacts.next();
      System.out.println("Contact name: " + contact.getName());
      System.out.println("Contact has the following home telephone numbers");
      String[] phone_numbers = contact.getPhoneNumbers("HOME");
      for(int i=0; i<phone_numbers.length; i++) {
        System.out.println(i + ": " + phone_numbers[i]);
        }
      } catch(DatabaseException de) {
        System.out.println("Database error:" + de);
        return;
      }
    }
  }
```

Now, let's go on to see how we can update existing contacts in the database. The addressbook API allows us to delete contacts and update information stored within it. In order to illustrate some of these concepts, we will develop a rather minimal application that will exercise some of the available methods.

The scenario is as follows: Oftel (the regulatory body for telephone industry in the UK) recently made changes to area codes in six areas. Our customers have so far failed to understand the changes and therefore correct the numbers in their address books. Unfortunately the old numbers have recently been disconnected. To target customer dissatisfaction we will distribute an application to make the changes.

For the purpose of this application we will act on the changes made to the Cardiff and Coventry areas. The Cardiff area codes changed from "01222" to "029" and local numbers now have an additional two digits at the front of the numbers; namely "20". Coventry's area code has changed from "01203" to "024" and the digits "76" have been added to the front of local numbers.

The following code shows an application, that makes the modifications to all Cardiff and Coventry telephone numbers. The class variables represent the relevant changes. The `main()` method creates an instance of this class and calls its `listContacts()` method. When all available contacts have been listed, we call the `changeCodes()` method, which will affect the changes. If the changes succeed, we will once more call the `listContacts()` method, which will list the effected classes:

```
import javax.pim.database.ItemField;
import javax.pim.database.AggregateField;
import javax.pim.database.Parameter;
import javax.pim.database.DatabaseException;
import javax.pim.addressbook.ContactDatabase;
import javax.pim.addressbook.ContactCard;

public class ChangeTelCodes {
  private String database;

  String OLD_CARDIFF_CODE = "01222";
  String NEW_CARDIFF_CODE = "029";
  String NEW_CARDIFF_DIGITS = "20";
  String OLD_COVENTRY_CODE  = "01203";
  String NEW_COVENTRY_CODE = "024";
  String NEW_COVENTRY_DIGITS = "76";
```

```
public static void main(String[] args) {
  ChangeTelCodes changeTelCodes =
                 new ChangeTelCodes(ContactDatabase.DEFAULT_ADDRESSBOOK);
System.out.println("Addressbook before amendments:");
  changeTelCodes.listContacts();

  boolean success = changeTelCodes.changeCodes();
  if(success != true) {
    System.out.println("We have a problem");
  } else {
  System.out.println("These are contacts for this database:");
    changeTelCodes.listContacts();
  }
}

public ChangeTelCodes(String database) {
  this.database = database;
}
```

Next, we open the default database. This may throw a `DatabaseException` if the `DatabaseFactory` has not been set. In the Symbian SDK, it is the `javax.pim.database.ContactDatabase` class that instantiates the factory class given by the system property `javax.pim.addressbook.factory`. In this configuration, `ContactDatabase` instantiates an instance of the factory and then delegates work for creating, opening, closing, and deleting databases to the factory.

A `DatabaseException` is also thrown if the specified database does not exist. A `java.lang.SecurityException` will be thrown if accessing the database is forbidden by the security policy for that system.

It is worth noting that the default databases are called "`DEFAULT-ADDRESSBOOK`" and "`DEFAULT-SIM`". They can also be accessed through the `ContactDatabase.DEFAULT_ADDRESSBOOK`, and `ContactDatabase.DEFAULT_SIM` static members in the Symbian implementation. Calling the no arguments `openDatabase()` method opens the default address book:

```
public boolean changeCodes() {

  ContactDatabase contactDatabase = null;

  try {
    contactDatabase = ContactDatabase.openDatabase();
  } catch(DatabaseException de) {
    System.out.println("Could not open specified database " + de);
  }
```

We retrieve all the contacts stored in the database using the `cardItems()` method as an iterator. Notice the `javax.pim.database.Iterator` class. This class works just like `java.util.Iterator` and is provided to allow Java 1.1 systems compatibility, since it was introduced in Java 2. We get all the contacts in this database:

```
  javax.pim.database.Iterator contacts = null;
  try {
    contacts = contactDatabase.cardItems();
```

thisContact represents the current card being processed. We iterate through each contact in the database, retrieving the telephone numbers for that contact. We then loop through the telephone numbers checking whether they begin with the relevant area codes, and if so, replace the number
with the correct one:

```
      ContactCard thisContact;

      while(contacts.hasNext()) {
        thisContact = (ContactCard)contacts.next();
        String thisNumber;
        String localNumber;
        ItemField[] telephones = thisContact.getFields(ContactDatabase.TEL);

  int i;
  for(i=0; i<telephones.length; i++) {
    ItemField thisTel = telephones[i];
    thisNumber = thisTel.getString();
```

We now simply check if the current telephone number begins with the Cardiff or Coventry codes, and, if so, substitute the new values:

```
  for(i=0; i<telephones.length; i++) {
    ItemField thisTel = telephones[i];
    thisNumber = thisTel.getString();

    if(thisNumber.startsWith(OLD_CARDIFF_CODE)) {
      localNumber = thisNumber.substring(OLD_CARDIFF_CODE.length());
      thisTel.setString(NEW_CARDIFF_CODE + NEW_CARDIFF_DIGITS + localNumber);

      contactDatabase.updateItem(thisContact);
    }

    if(thisNumber.startsWith(OLD_COVENTRY_CODE)) {
      localNumber = thisNumber.substring(OLD_COVENTRY_CODE.length());
      thisTel.setString(NEW_COVENTRY_CODE + NEW_COVENTRY_DIGITS +
                                   localNumber);
      contactDatabase.updateItem(thisContact);
    }
  }
```

Notice the update mechanism. The telephone number fields returned are obviously references so making changes to them updates the ContactCard. However, the contact card must be updated with the database using the updateItem().

Finally, the error checking is quite primitive and not really up to scratch for a user application:

```
        }
      } catch(DatabaseException de) {
        System.out.println("Error accessing database:" + de);
        return false;
      }
      return true;
    }
```

The `listContacts()` method quite simply opens the database, retrieves all contacts from the database, and iterates through each contact, listing the name and all telephone numbers for that contact. Note that we use the `getFields()` method that `ContactCard` inherits from `javax.pim.database.Item` rather than the `getPhoneNumbers()` method as this method, requires a telephone number type. In order to list all numbers, we would have to loop through retrieving the telephone numbers for each type:

```
public void listContacts() {
  try {
    ContactDatabase contactDatabase =
                                ContactDatabase.openDatabase(database);
    javax.pim.database.Iterator contacts = contactDatabase.cardItems();
    ContactCard contact;
    while(contacts.hasNext()) {
      contact = (ContactCard)contacts.next();
    System.out.println("Contact name: " + contact.getName());
    System.out.println("Contact has the following telephone numbers");
    ItemField[] phone_numbers = contact.getFields(ContactDatabase.TEL);
    for(int i=0; i<phone_numbers.length; i++) {
      System.out.println(i + ": " + phone_numbers[i].getString());
      }
    }
    System.out.println();
  } catch(DatabaseException de) {
    System.out.println("Database error:" + de);
    return;
  }
}}
```

Obviously, this application leaves much to be desired, including rudimentary error checking, a user interface, etc. I think, however, that it demonstrates the principles. You should note that `System.out.println()` calls will only be possible to see in the emulator when it is run from the command line. It is not meaningful to show these on the emulator in release mode, or indeed a real device and so these messages will be lost.

Running the Examples

Save the examples in the `<EPOC_ROOT>\erj\projavamobile\chapter09\src\` directory, creating the relevant folders if necessary. Also make sure you have a directory called `classes` in the `<EPOC_ROOT>\erj\projavamobile\chapter09\` directory. Remember that this structure corresponds to `j:\projavamobile\chapter09\` in the emulator's file system.

In order to compile the example we need to point the emulator's classpath to the pJava, Quartz UI, and JavaPhone classes. As it doesn't use the system classpath, we will set its classpath in the command line:

```
javac -classpath
%EPOC_ROOT%\erj\lib\classes.zip;%EPOC_ROOT%\erj\ext\qawt.jar\;%EPOC_ROOT%\erj\ext\
javaphone.jar -d classes src\*.java
```

Here `%EPOC_ROOT%` is where the emulator's directory structure begins, in other words, on a Windows installation it may be `C:\Symbian\6.0\QuartzJava\`. This command assumes that all the Java files are in a directory called `src`, and are going to be compiled into a directory called `classes`.

Once compiled, the classes need to be run in the emulator. Since Quartz uses pJava, we need to use the emulator's pJava executable. There are two versions, `pjava` (the standard executable) and `pjava_g` (the debugging version). These are located in `<EPOC_HOME>\epoc32\release\wins\urel` and `<EPOC_HOME>\epoc32\release\wins\udeb` respectively. For this example, we will use `pjava_g`.

As explained above, pathnames given to `pjava` are paths within the emulator's own drivespace; they are not system paths. Also note that because of this, the emulator has no concept of the current working directory and must be told from where to run applications, relative to its internal drive structure. With this in mind, we must run the `pjava_g` command with the `-cd` switch, which tells the emulator where the current working directory is:

```
%EPOC_ROOT%\Epoc32\Release\wins\udeb\pjava_g -cd
j:\projavamobile\chapter09\classes ChangeTelCodes
```

This will start the emulator and run the example. You should get a message, printed to the command console, that confirms the successful change in the database.

User Profile

The User Profile package allows access to the user profile of the device owner. This package allows us to extract the user profile information and, depending on security restrictions, set the profile information. The User Profile package is a very simple API and has one interface and one class `UserProfileImpl`, and `UserProfile` respectively. `UserProfileImpl` is implemented to provide the required functionality.

`UserProfile` specifies two methods, `getCurrentUser()` and `setCurrentUser()`, that can be used to get and set the current user. The methods take and/or return a `ContactCard` with the user's details. A third method, the static `setImplementation()`, can be used to set the implementation. This value may be found in the system property `javax.pim.userprofile.implementation` although `UserProfile` will check for this value and attempt to load the class in any case. `UserProfile` delegates the work to the implementation class after checking that an implementation class has been provided and loaded.

The following code shows how to use the user profile package.

```
import javax.pim.userprofile.UserProfile;
import javax.pim.addressbook.ContactCard;

public class TestUserProfile {
  public static void main(String[] args) {
    ContactCard userCard;

    userCard = UserProfile.getCurrentUser();

    System.out.println("The current user is " + userCard.getName());
    System.out.println("The user's first preferred home number is" +
                userCard.getPreferredPhoneNumbers(ContactDatabase.HOME));
    System.out.println("");

  }
}
```

The user profile is returned as a `ContactCard`. The method `getName()` is used to extract the user's name from the `ContactCard`. `getPreferredPhoneNumbers()` returns an array of type `String` of the user's preferred telephone numbers of the type given as its parameter. The types of telephone fields include `HOME`, `WORK`, `VOICE`, `FAX`, `CAR`, and so on.

Calendar

The Calendar API (`javax.pim.calendar`) allows application developers access to scheduling information on the host device. The calendar API conforms to the iCalendar specification (RFC 2445) and provides a simple and thin API to fit on a typical resource limited MID. The typical calendar functionality that can be created includes accessing, adding, deleting, and updating schedule information, and creating to-do lists.

The main classes in the `calendar` package are: `CalendarDatabase`, `CalendarEntry` and `CalendarToDo`. Supporting classes in the package include `Repeat` and `RepeatRule`. `Repeat` specifies dates on which the entry should be repeated. It accepts specific dates on which to repeat the entry it is associated with, and exceptions to the repeated dates, through the `addExceptDate()` and `addRepeatDate()` methods. It is usually easier to specify repeated occurrences of the entry using a `RepeatRule`.

The `RepeatRule` class can define from and to dates, frequency of repetition, exceptions to those rules, and more, and quite complex rules can be defined with it. An example rule might be for repetition on a weekday between 9 to 5, except for July and August.

The Symbian implementation imposes some restrictions on the types of `RepeatRule` it will support, such that there is only one repeat rule per calendar entry, explicit exception dates, and no explicit repeat dates. The following sections illustrate the use of the `Calendar` package.

Accessing Schedule Information

To access schedule information on the host device, the calendar must be opened. Schedule information can then be extracted by searching between dates and returning calendar items in the schedule. The following code illustrates how to open the calendar and extract the calendar entries:

```
CalendarDatabase localCal = null;
CalendarEntry calItem = null;

try {
  localCal = CalendarDatabase.openDatabase();
} catch (DatabaseException e) {
  System.out.println("Error opening default database!");
}

Iterator calItems = localCal.items(startDate, endDate);

while (calItems.hasNext()) {
  calItem = calItems.next();
  System.out.println("The start date for this entry is: " +
                     calItem.getStartDate());
  System.out.println("The end date for this entry is: " +
                     calItem.getEndDate());
  System.out.println("The description for this entry is: " +
                     calItem.getDescription());
}
```

In the above example, the default calendar database on the host device has opened a search made of calendar entries between `startDate` and `endDate`. An `Iterator` object is returned that contains a series of `CalendarEntryItems`. These can be iterated through and manipulated in various ways. In this example, we simply print the values to the `system.out` stream.

Adding Schedule Information

To add an entry into the calendar database, we must create a calendar item, set it with a start and an end date, and provide a description of the event. `CalendarDatabase` defines a number of fields that can be added to the entry describing the attendees to the entry (an attachment to the entry, whether a RSVP is required, etc.,) as per the iCalendar specifications.

The following code illustrates how to insert an item into the calendar database:

```
CalendarDatabase localCal = null;

 newCalItem = new CalendarEntry();

newCalItem.setStartDate(startDate);
newCalItem.setEndDate(endDate);
newCalItem.setDescription("This is an example entry");


try {
  localCal = CalendarDatabase.openDatabase();
} catch (DatabaseException e) {
  System.out.println("Error opening default database! " +
                     e.getMessage());
}

try {
  localCal.addItem(newCalItem);
} catch (DatabaseException e) {
  System.out.println("Error inserting calendar item! " +
                     e.getMessage());
}
```

In the example above an empty `CalendarEntry` is created, and populated with a start date, an end date and a description. The new `CalendarEntry` object is then inserted into the local default `CalendarDatabase`. In this case, no additional information has been provided about the nature of the entry, attendees, etc.

Creating To Do Lists

Calendar to do lists are lists of tasks to be completed by the device owner. Usually a task has an associated description, such as its due date and priority. The following code extract opens the calendar database, extracts the list of tasks, and iterates through them, printing out their due dates, descriptions, and priorities:

```
CalendarDatabase localCal = null;
Iterator toDoIterator = null;

try {
  localCal = CalendarDatabase.openDatabase();
```

```
} catch (DatabaseException e) {
  System.out.println("Error opening default database!");
}

try {
  toDoIterator = localCal.toDoItems()
} catch (DatabaseException e) {
  System.out.println("Error getting toDoItems");
}

CalendarToDo toDo = null;

while (toDoIterator.hasNext()) {
  toDo = toDoIterator.next();

  System.out.println("The to do item descriptor - " +
          toDo.getDescription());
  System.out.println("The to do item due date is  - " +
          toDo.getDueDate());
  System.out.println("The to do item priority is  - " +
          toDo.getPriority ());
}
```

JTAPI APIs

The mobile package, `javax.telephony.mobile`, covers features specific to mobile networks. Here's a summary of the principal classes and their functions:

- `MobileProvider` is the interface to the wireless implementation. This class provides methods with which to list the available `MobileNetworks`, select a `MobileNetwork`, get the current `MobileNetwork`, and get the type of network (for example, GSM, CDMA etc.).
- `NetworkSelection` provides methods to set up lists of preferred or forbidden networks and whether the network is selected automatically or manually. The `MobileNetwork` object contains all available information about the wireless network and provides methods with which to access this information.
- `MobileRadio` provides simple radio management: principally turning the radio on and off and getting the signal strength; and finally, the `MobileTerminal` interface represents a physical end point of a connection. The `MobileTerminal` extends the `Terminal` interface to provide methods to identify the terminal, such as the ID String and the manufacturer's name, together with methods to generate DTMF dial tones.

The Symbian provider implements both the `MobileProvider` and `MobileRadio` interfaces. The static `JtapiPeer.getProvider()` method returns the `MobileProvider` for this device; it must be cast to the `MobileProvider` interface. In this case, we know that the device implements these interfaces, however, the `instanceof()` checks are provided for good practice reasons:

```
Provider myProvider = JtapiPeer.getProvider(null);
if(myProvider instanceof MobileProvider) {
  MobileProvider mobileProvider = (MobileProvider)myProvider;
}
```

Or we can use:

```
Provider myProvider = JtapiPeer.getProvider(null);
if(myProvider instanceof MobileRadio) {
  MobileProvider mobileProvider = (MobileRadio)myProvider;
}
```

The JTAPI mobile package allows finely grained control of options in the implementation.

Datagram

The Datagram API provides a transport independent method of sending and receiving messages. The messages can be sent via IP or nonIP transport methods such as SMS, or GSM data. The Symbian implementation of the JavaPhone specification supports UDP over IP, SMS over GSM, and WDP over SMS.

The datagram API consists of the following classes in the `javax.net.datagram` package:

- `Address`
- `DatagramService`
- `DatagramServiceFactory`
- `Datagram`
- `DatagramNameService`.

`DatagramService` represents a datagram service for both client and server operations. A service is retrieved using the `getService()` static method. An address must be provided to retrieve the service for a specified location. Addressing is specified using URLs; the `DatagramService` class has a `parseAddress()` method that takes a string URL and returns a `java.net.datagram.Address` object. We will see more on addressing of datagrams further on in this chapter. When a service has been retrieved, datagrams may be sent and received using the `send()` and `receive()` methods respectively.

`DatagramServiceFactory` is delegated work such as creating instances of `DatagramService`, and registering and unregistering services for specified transports. When the `getService()` method of the `DatagramService` is called, each factory registered with the implementation is searched to see if it supports the specified transport, and when one is found that does, it is used to instantiate the required `DatagramService`. The factory must also support address parsing for the transport it supports, which will be delegated to it by the `DatagramService`.

Finally, `DatagramNameService` allows abstract addressing, using a server name and service. In the following section, we discuss abstract addressing in detail.

Datagram Addressing

There are two types of addressing: abstract addressing and concrete addressing. In abstract addressing, the server name and service are provided, and the implementation converts this to concrete addressing information. In the Symbian implementation, the server name is used to locate a contact card in the devices address book that contains the addressing information.

In concrete addressing, a text URL is used to represent the destination and transport for the datagram message. The `DatagramService.parseAddress(String)` is used to obtain an `Address` object that represents the relevant address.

In the following example, concrete addressing is used to obtain `Address` objects for a datagram using the SMS transport, the UDP transport, and the WDP transport:

```
import javax.net.datagram.*;
...
String contactSMS    = "sms://0123454321";
String contactUDP    = "udp://test.wrox.com:8888";
String contactWDP   = "wdpsms://0123454321"

Address dgAddressSMS     = DatagramService.parseAddress(contactSMS);
Address dgAddressUDP     = DatagramService.parseAddress(contactUDP);
Address dgAddressWDP    = DatagramService.parseAddress(contactWDP);
...
```

Note that the address representation is a URL description using the format:

```
<transport>://<address>:<port>
```

Typically, the addressing information would be provided by the application in some way or would be constructed from addressing information given by the user together with knowledge of the system in use.

In the following example, abstract addressing is used to create an `Address` object from an entry in the devices address book. A new contact is created, the name is set to "John", and a field is added to the contact that represents a datagram device. Note that the `X-DATAGRAM-` prefix is added to the name to make it available to the lookup service. The field is also modified to be the preferred method to contact the person:

```
ContactCard johnContact = new ContactCard();

johnContact.setFormattedName("John");

ItemField jMobile = new ItemField("X-DATAGRAM-Mobile", "sms://012347654321");
jMobile.addParameter(ContactDatabase.TYPE, ContactDatabase.PREF);
johnContact.addField(jMobile);
```

The local contacts database is opened using the `openDatabase()` method, the new contact is added to the database, and the lookup service is used to get the address for the datagram. Using the lookup method, the first argument is the name of the contact and the second is the preferred field – note that here the `X-DATAGRAM-` prefix is taken off:

```
ContactDatabase contacts = ContactDatabase.openDatabase();
contacts.updateItem(johnContact);

Address dgAddress = DatagramNameService.lookup("John", "Mobile");
```

Another method of `DatagramNameService`, that can be used to search the contact database for all datagram addresses for a particular service, is the `lookupAll()` method. This method returns all the addresses as an array of `Address` objects. To determine what transport methods are supported, we compare the first characters of available addresses up to the colon with the transport method names already known – `sms`, `wdpsms` and `udp` – using the `String.startsWith()` method or similar.

Sending and Receiving a Datagram

In order to illustrate the sending and receiving of datagrams, we will develop a simple client and server pair that will act as a consumer and producer of datagrams. We will begin by developing the producer.

We will need the majority of the classes and interfaces in the package, so we import the lot:

```
import javax.net.datagram.*;
import java.io.IOException;
```

`service` represents the `DatagramService` for UDP on this device. The type of service is determined by the URL string that we provide. `datagramAddress` represents the address to which we will direct the datagrams we send. `numDatagrams` is a parameter, required by the constructor for this class, that specifies the number of datagrams to send:

```
public class DatagramProducer {
  DatagramService service;
  Address datagramAddress;
  int numDatagrams;
```

We provide a `main()` method so that the producer can stand-alone:

```
  public static void main(String[] args) {
    String url = "udp://localhost:9097";
    Address address = null;
```

We attempt to parse the address, and instantiate a producer with the given address. The test code asks the producer to send 5 datagrams:

```
    try {
      address = DatagramService.parseAddress(url);
      DatagramProducer client = new DatagramProducer(address, 5);
      client.sendDatagrams();
    } catch (Exception e) {
      System.out.println("Client could not be started: " + e);
      return;
    }
  }
```

Next, we will write the constructor for this class. The constructor attempts to get a datagram sending service and will also store the various values in class variables. If it cannot retrieve a service, it will throw an `java.io.IOException`, `AddressNotSupportedException`, or `IntermittentNetworkException`. The latter two inherit from the `IOException`:

```
  public DatagramProducer(Address address,
                          int numDatagrams) throws IOException {
    this.datagramAddress = address;
    System.out.println("Client initialized");
    service = DatagramService.getService(address);
    this.numDatagrams = numDatagrams;
  }
```

The `sendDatagrams()` method is the worker method for this class. It creates `numDatagrams` (the number of datagrams) and sends them all, with the message `"Hello from Datagram client"`. In this class, each datagram is sent in intervals of a second; this is merely to illustrate a service that is spread over time. Finally, we close the service before exiting:

```
    public void sendDatagrams() {
      System.out.println("sending datagrams");
      try {
        for (int i = 0; i < numDatagrams; i++) {
          System.out.println("sent datagram: " + i);
          String message = "Hello from Datagram client";
          Datagram datagram = new Datagram(message.getBytes(),
                                           datagramAddress);
          service.send(datagram);


          try {
            Thread.sleep(1000);
          } catch (InterruptedException e) {}
        }
        service.close();
      } catch (IOException ioe) {
        System.out.println("Error sending datagrams");
      }
    }
  }
```

Let's continue by developing a consumer for these messages. The producer has the same address hard coded into it for the purposes of this example, the first few lines of code are virtually identical. The only thing to note is that rather than calling `getService()`, we call `getServerService()` to retrieve a consumer type service for datagrams:

```
  import javax.net.datagram.*;
  import java.io.IOException;

  public class DatagramConsumer {
    DatagramService service;
    Address datagramAddress;

    public static void main(String[] args) {
      String url = "udp://localhost:9097";
      Address address = null;

      try {
        address = DatagramService.parseAddress(url);
        DatagramConsumer server = new DatagramConsumer(address);
        server.receiveDatagrams();

      } catch (Exception e) {
        System.out.println("Could not start server: " + e);
        return;
      }
    }

    public DatagramConsumer(Address address) throws IOException {
```

```
      this.datagramAddress = address;
      System.out.println("Server initialized");
      service = DatagramService.getServerService(datagramAddress);
    }
```

The work method for this class is appropriately called receiveDatagrams(). We construct a buffer of the maximum size allowed by the implementation, as given by the getMaximumLength(), and use it to construct a datagram, giving the address **from** which we expect datagrams to arrive as an argument. From now on, the consumer simply loops, forever waiting for datagrams to be received and outputting the messages stored in the datagrams. To those familiar with datagram application development, this code should look virtually identical to Java 2 datagram code:

```
    public void receiveDatagrams() {
      try {

        System.out.println("Waiting for datagrams");
        byte[] buff = new byte[service.getMaximumLength()];
        Datagram datagram = new Datagram(buff, datagramAddress);

        while (true) {
          System.out.println("Receiving datagrams from " + datagramAddress);
          datagram.setLength(buff.length);
          service.receive(datagram);
          System.out.println("New datagram received: "
                             + new String(datagram.getData(),
                                          datagram.getOffset(),
                                          datagram.getLength()));
        }
      } catch (IOException ioe) {
        System.out.println("Error receiving datagram: " + ioe);
      }

    }
  }
```

Again, this code should be saved into <EPOC_ROOT>\erj\projavamobile\chapter09\src\. We cancompile the examples using the same command as the last example, and then start up the DatagramConsumer:

```
%EPOC_ROOT%\Epoc32\Release\wins\udeb\pjava_g -cd
j:\projavamobile\chapter09\classes DatagramConsumer
```

The DatagramConsumer will wait for datagrams to be sent to it:

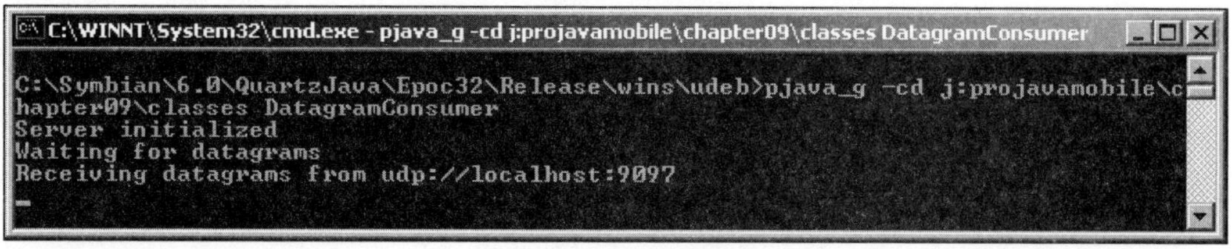

```
C:\WINNT\System32\cmd.exe - pjava_g -cd j:projavamobile\chapter09\classes DatagramConsumer

C:\Symbian\6.0\QuartzJava\Epoc32\Release\wins\udeb>pjava_g -cd j:projavamobile\c
hapter09\classes DatagramConsumer
Server initialized
Waiting for datagrams
Receiving datagrams from udp://localhost:9097
-
```

Start a `DatagramProducer` in another command window and watch as it sends datagrams, while the `DatagramConsumer` receives them:

```
C:\WINNT\System32\cmd.exe

C:\Symbian\6.0\QuartzJava\Epoc32\Release\wins\udeb>pjava_g -cd j:projavamobile\c
hapter09\classes DatagramProducer
Client initialized
sending datagrams
sent datagram: 0
sent datagram: 1
sent datagram: 2
sent datagram: 3
sent datagram: 4

C:\Symbian\6.0\QuartzJava\Epoc32\Release\wins\udeb>_
```

```
C:\WINNT\System32\cmd.exe - pjava_g -cd j:projavamobile\chapter09\classes DatagramConsumer

C:\Symbian\6.0\QuartzJava\Epoc32\Release\wins\udeb>pjava_g -cd j:projavamobile\c
hapter09\classes DatagramConsumer
Server initialized
Waiting for datagrams
Receiving datagrams from udp://localhost:9097
New datagram received: Hello from Datagram client
Receiving datagrams from udp://localhost:9097
New datagram received: Hello from Datagram client
Receiving datagrams from udp://localhost:9097
New datagram received: Hello from Datagram client
Receiving datagrams from udp://localhost:9097
New datagram received: Hello from Datagram client
Receiving datagrams from udp://localhost:9097
New datagram received: Hello from Datagram client
Receiving datagrams from udp://localhost:9097
_
```

Power Monitor

The Power Monitor API in the `javax.power.monitor` package allows applications to monitor the power level of the host device. The API specifically allows applications to:

- Retrieve current battery level
- Test whether an external power source is being used
- Estimate the remaining battery life
- Receive alerts when the battery is running low

The Power Monitor API is a very simple API and has three classes defined; `PowerMonitor`, `PowerMonitorListener`, and `PowerWarningType`. To use the power monitor functionality, first create an instance of the `PowerMonitor` class:

```
private static PowerMonitor pwMonitor;

monitor = PowerMonitor.getInstance();
monitor.addPowerMonitorListener(this);
```

The preceding code first imports the PowerMonitor package, then creates a private variable of the type, PowerMonitor. Finally a reference to the host systems PowerMonitor is created and a listener added to it. In this case, the listener will be the application instance that this code is part of. To obtain information about the battery level or power settings we simply call one of the PowerMonitor methods, as shown below:

```
int batteryRemaining  = monitor.getEstimatedSecondsRemaining();
int batteryLevel     = monitor.getBatteryLevel()
boolean extPower     = monitor.usingExternalPowerSource();
```

The variable batteryRemaining will be assigned the value of the number of seconds of remaining battery power. If an external power supply is being used, then it will be assigned Integer.MAX_VALUE. If the value is not available, -1 will be assigned.

The variable batteryLevel will be assigned an integer related to the battery level of the host device. If an external power supply is being used, then batteryLevel will again be assigned Integer.MAX_VALUE. If a value cannot be obtained, then batteryLevel will be assigned -1; otherwise a value between 0 (no life left) and 100 (fully charged) will be assigned.

The power monitor API can also be used to implement functionality to report power related events. When creating an example we must make sure that we implement the PowerMonitorListener interface; events can then be captured as follows:

```
Public void powerWarningList(int estimatedSecondsRemaining,
                              PowerWarningType warning) {
  println(  "Seconds remaining = "    + estimatedSecondsRemaining +
       " Warning = "      + warning);
}
```

The powerWarningList() method simply displays information about the power event. The power events that can occur include:

- BatteryCritical
- BatteryLow
- BatteryNormal
- CallTermination
- ExternalPowerSourceChange
- IrTermination
- NetworkTermination

Capabilities

As previously discussed, the ability of the system to carry out certain functions is determined through the capability packages. There are two types of capabilities: dynamic and static. Static capabilities define whether or not the specified capabilities are supported by the implementation. Dynamic capabilities check the state of the various objects involved, and based on this returns according to whether a method can be called safely.

Calling a method, or attempting to access functionality that is not supported, will cause one of a number of exceptions to be thrown, including `MethodNotSupportedException` and `IllegalStateException`. Example check whether an address is observable, whether calls ban be connected or connections disconnected by the application, and whether components of the device can be controlled and/or observed.

The Symbian SDK

The Symbian SDK and Quartz devices do not support the following features: The JTAPI call control and phone packages are not supported. Neither is SSL supported, this is probably because the size and price of Quartz devices does not allow this processor and power expensive functionality. Core power support is minimal (it can only tell if external power is on or not) and the install package is not supported either.

Finer grained departures from the spec are shown below; first, for the implementation itself, and then, in the SDK's ability to test certain functionality.

Symbian's JTAPI Implementation

The current JTAPI implementation does not support all of the functionality specified. For example, caller identification is not supported.

Although the Symbian implementation provides the `javax.telephony.events` package, it is not supported. The `addXXXObserver()` methods will all throw `MethodNotSupportedException`, as no observer events are ever generated.

The following table lists, in more detail, the methods not supported by Symbian's implementation:

Class	Unsupported methods
`Address`	All of the methods relating to observers are missing from the package.
`Call`	Again, methods related to observers are not supported. In addition, a deprecated method, `getCallCapabilties()`, is not supported in this release.
`Connection`	The deprecated method, `getConnectionCapabilities()`, is not supported.
`Provider`	Methods relating to observers are not supported. Deprecated capabilities methods are not supported.
`Terminal`	Methods relating to observers are not supported. Deprecated capabilities methods are not supported.
`TerminalConnection`	A deprecated method, relating to capabilities, has been removed.

Class	Unsupported methods
`MobileProvider`	The `getHomeNetwork()` and `setNetwork()` methods are not supported. This method returns the home network object; however, home networks are not supported at this time by Symbian phones.
`MobileRadio`	Functionality relating to the automatic loading of radio services at bootup is not supported. Changing the state of the radio service (from on to off, and vice versa) is not possible in this implementation. The only methods supported are setting and getting listeners for the radio provider, and getting the current radio signal strength.
`MobileTerminal`	`getSoftwareVersion()` is not supported. A `null` is returned if this method is called.

JTAPI Under the Emulator

In order to test applications, it is necessary to do some testing with a real handset. The emulator can emulate the environment, using a handset and an AT command set, to talk to a mobile phone. This allows the basic telephony functionality to be tested but does limit the scope of the emulation.

In particular, the following features are not supported: call waiting, DTMF (Dual Tone Multiplexed Frequency), tone control, signal strength monitoring, obtaining the network time zone.

Perhaps most importantly, the emulator does not always report disconnection events. Using the Vodaphone provider, we get the following results:

Disconnection reported?	Who initiates	Who disconnects	Comment
Yes	JTAPI application	JTAPI application	
No	JTAPI application	Remote party	If we look at the SH888 (cell phone) handset at this point it still has a call ongoing. It's necessary to hang up the phone to terminate the call: this action does then delivers a disconnection event to JTAPI.
No		Mobile phone	
Yes	Remote party	Remote party	

This can be summarized as follows: if the calling party is not the party that disconnects the call, a disconnection event will not be reported to the application.

Summary

The JavaPhone specifications add powerful functionality to PersonalJava applications, utilizing the telephony capabilities of implementing devices. It is expected that this functionality will eventually also be available on personal computers and such like as developers begin to use it.

We have tried to show the functionality available through the JavaPhone specifications, particularly as it relates to the Symbian implementation. We have not discussed some of the packages specified, most notably the install package. This package covers the installation and management of applications and standard extensions, but is probably deemed too unsafe, at the moment, for developers to gain access to the functionality provided by the package. Loading of applications is still available through the standard systems for installation that native applications use.

We have also not discussed the call control (advanced call features such as call forwarding, and conferencing), communications, or power management APIs, as they are not provided by the Symbian implementation at this time. As and when the JavaPhone specifications are implemented in more able devices we may begin to see these introduced into implementations. There is no doubt that as these types of devices that offer both telephony functionality and the power of the PersonalJava (and soon the Personal Profile) become more prevalent, it is likely that developments will allow them to implement more functionality as their available resources increase.

We have covered the classes that the specifications use to model a telephone call, and the programming paradigm that this introduces. We have also seen examples of codes that utilized the PIM (Personal Information Manager) functionality offered by Symbian handsets, how to access the various databases, and how to retrieve and manipulate their contents.

PROFESSIONAL JAVA MOBILE

TOOL SELECTION

WEB TABLET
MOBILE PHONE
MP3 PLAYER
DIGITAL CAMERA
PALMTOP
LAPTOP

NOW SELECTED

PALMTOP

10

Java Card Applications

Java Card was launched by Sun Microsystems in 1996, as a way of programming smart cards in Java. While theoretically Java should be usable on every conceivable platform, it isn't possible to use precisely the same language on every device as different Java Virtual Machines interpret different flavors of Java. The lack of resources on a particular platform may well make elements of Java inappropriate, and no platform capable of Java is more restrictive than the smart card.

With only a serial connection to the outside world, communication restricted to sequences of hexadecimal characters, an 8-bit processor, and a memory measured in Kilobytes, it would clearly be impossible to run anything approaching the whole Java language on such a device. Therefore, Sun Microsystems defined a restricted subset of Java for use on these cards, cutting out enormous parts of the J2SE API, to create a very simple language that could be used for creating very simple applications.

So far, Java Card has proved extremely popular, with millions of cards already deployed. The inclusion of **GSM Subscriber Identity Module** applications into Java Card compatible SIMs, has led to millions of units already being deployed in European mobile phones.

In this chapter we'll be discovering what a smart card is, and what unique functionality it can offer. We will also look at how applications can communicate with a smart card of any type, and how we can create Java applications to work with any kind of smart card. We will then move on to the card itself, seeing how Java can be used on the very smallest of embedded platforms, and this will be illustrated with an example of the kind of application that could be deployed onto a card. Finally, we'll look at how a smart card application is actually deployed.

What is a smart card?

Smart cards are defined by ISO7816; a document that describes not only the physical characteristics of the card, but also its electrical and logical properties. The full document is very dull, it is, however, available from the International Standards Organization at http://www.iso.ch.

We have to pay for a full version of the standard, although if we are planning any number of smart card application developments it's probably worth it. We will be looking at it in more detail later in this chapter.

A smart card is literally like a credit card with a chip embedded in it, some memory, some way of talking to the outside world, and perhaps a coprocessor for cryptographic functions. Not all smart cards are the size of a credit card; the most common card of all is the SIM used in GSM mobile telephone networks, but all are basically of the same construction. ISO7816 defines how flexible and robust the card must be, and how resistant to intrusion (though most manufacturers exceed these requirements by a long way):

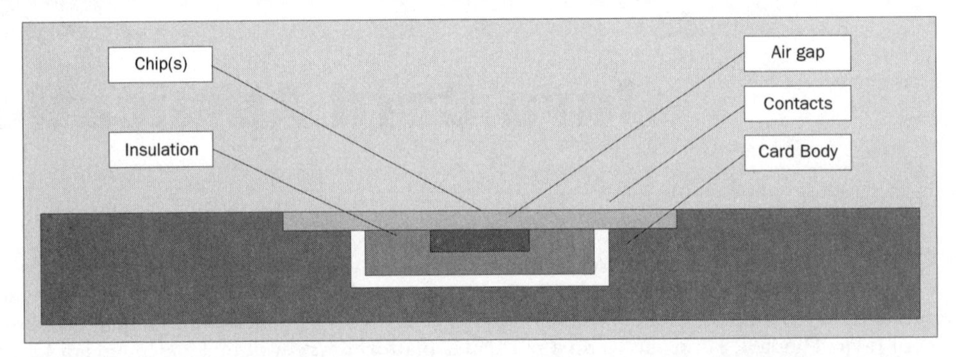

This cross section shows how a card is constructed, and how most cards communicate with the outside world. While most cards are 'wired', there are also wireless cards (the two types are also referred to as 'contact' and 'contactless' cards, respectively). Wireless cards communicate via a radio loop that also functions as an induction loop, providing the power to the chip and radio transmitter.

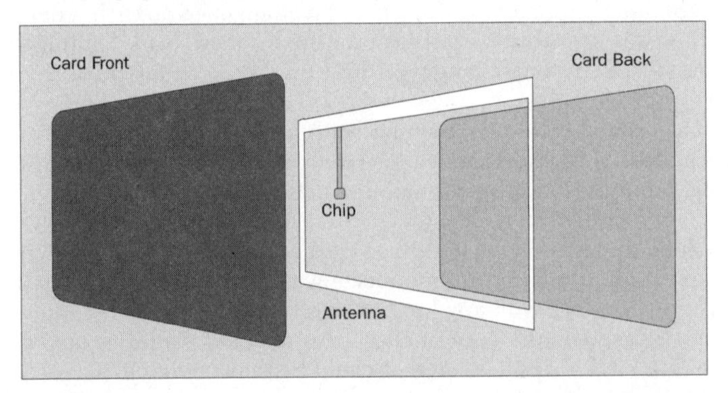

This type of arrangement is particularly well suited to transport applications, where customers only need to wave the card near the receiver to open gates into the system. Unfortunately, the range of such a card is only about 10cm, as it is limited by the ability of the base unit to induce current in the card, and the base units tend to consume a great deal of power (excessive for mobile devices). To the card application developer, the method of communication actually matters very little, as instructions are exchanged without concern for the transport mechanism.

Smart cards come in two varieties: memory cards and processor cards. While memory cards have been popular in the past for simple stored value systems, such as cards for use in public phones; it is processor cards that can provide real solutions to security issues. Obviously, only processor cards can use Java Card; memory cards aren't intended to run local applications at all.

Why Smart?

Traditional credit cards use a magnetic strip on the back to store 66 bytes of information. Holograms are used in an effort to make them harder to forge, and a nonerasable strip on the back is used to store the signature. Raised lettering on the front enables easy notation of the details using a carbon sheet.

The 66 bytes of information is frequently cited as not being enough for today's applications, though this is arguable. Kodak can store a photograph on a magnetic strip, and if the 66 bytes is simply the key to a database then the amount of storage is irrelevant. However, there are other problems with magnetic technology.

The magnetic strip on the back of a credit card is just audio tape (the same as is used on a compact audio cassette); one can even tear it off and run it past a standard tape head to 'hear' the data. More importantly, we can record on it, with someone else's data. There is no copy protection mechanism in the tape at all; a pot of glue, and a bit of tape, and we can make a card look like anyone's. This type of fraud is very common, and expensive to the card issuing companies.

Smart cards, on the other hand, have an array of physical and electrical defenses to avoid them being copied this way. The chip itself has memory areas, which cannot be written to, and may well have cryptographic keys to which even the processor does not have access. In this way, it becomes effectively impossible to copy a card.

Language or Operating System?

It is important to remember when developing applications for smart cards, that internally the card is a fully featured computer. It is roughly equivalent to a home computer 20 years ago, and it will increase in complexity with time. These devices have their own operating systems and, prior to Java Card, could only be programmed in their own proprietary assembler code. As in Java's general history, Java Card is designed to be the standard that will work on all smart cards. No matter what internal architecture the card has, it should be able to run Java Card programs, known as **applets**.

While we can consider Java Card to be a language, and a subset of Java, it is useful to consider that it is also an operating system for the card. In some cases this may be a layered arrangement, with a JVM running on top of a local OS, but often Java Card will actually fulfill the role of an operating system for the card.

Sun are not alone in realizing that the ability to develop applications for any smart card would be very valuable; and at least two standards are also being promoted, in competition to Java Card, as the solution to this problem. It's worth taking a moment to look at these other solutions in order to understand what Java Card is competing against, so we will look at Multos and Windows for smart cards.

Multos

Originally developed for the **Mondex** pre-paid card, this operating system is now available for license. It allows standard C programs to be cross-compiled for different card architectures. It also offers access to the Mondex application, which is designed for secure payment applications.

While not yet widely deployed, large-scale trials of the Mondex system have taken place, with mixed success. It has been announced that a GSM Subscriber Identity Module is available using Multos, and can theerfore provide secure payment systems for GSM mobile phones. It remains to be seen if these applications are deployed.

Mondex is 51% owned by MasterCard, and more details of which can be found at http://www.mondex.com and also at http://www.multos.com.

Visa also have their own electronic payment system, called VisaCash. However, as this is a closed, proprietary system, there is little to say about it from a development standpoint.

Windows for smart cards

While the idea of fitting the Windows operating system onto a smart card may seem unlikely, that is what Microsoft are planning. Windows for smart cards is actually both a very different system of development, and operating system, and is worth some examination.

When a user decides to develop a Windows for Smart cards application, he/she works with the development kit, using Wizards to define what they want their application to do, and what features of the operating system it will use. The Wizard then creates a customized version of the operating system which is exactly suited to the application being deployed, and nothing more. In this way it should be possible to use the most efficient operating system, without having unnecessary features taking up valuable space or processing time. Other card operating systems may be upgraded from time to time, but only Windows for smart cards can offer a customized operating system for every application.

While, at the time of writing, there are very few deployed Windows for smart cards applications, the popularity and marketing of Microsoft ensures that they will be a popular option. More details are available from http://www.microsoft.com/windowsce/smartcard/.

API Overview

Obviously, there is little point in converting most of the J2SE API; features like the AWT make little sense on a device with only serial communications. Therefore the basics of `java.lang` are available, with threading, security managers, object cloning, finalization, and large primitive data types (float, double, long, and char) all missing.

Three additional packages provide objects suitable for the kind of applications targeted at smart cards. These are `Java Card.framework`, `Java Card.security` and `Java Cardx.crypto`. Cryptography very often performs a central role in smart card applications, and it is very important that Java Card has access to these resources. The amount of cryptography available on the platform varies widely; care should be taken that applications don't depend on features that may not be available on all cards.

ISO7816 and Friends

We have already established that smart cards conform to the ISO7816 standard, and this standard is split into several sections. Not only does ISO7816 define what the card is like physically, it also defines the way in which cards are communicated with and a set of basic file manipulation commands that can be used with any compliant card.

Talking to a smart card

Communication with a smart card is normally via a serial connection to a computer, or computing device (in the case of mobile 'phone handsets). Cards are passive devices; all communication must originate with the device-based application, and consists of commands and responses from the card. It is not possible for a smart card to initiate contact.

Smart card readers conforming to ISO7816 basically, provide RS232 communications, power, and a clock signal for the card. It has been commented on, at length, that smart cards do not include their own clock, but are reliant on an external signal for this. The reason this is so controversial, is that there is potential for an attacker to slow the clock speed down (to, say, 1 tick a second) and then watch the processor gates move through an electron microscope. While this may sound far fetched, we should remember that some smart cards are proposed as replacements for cash, and must be as secure as paper money, if not more so.

Smart cards communicate at 9600 baud, which may appear slow, until we start to examine the amount of data they normally deal with. Communication between a computing device and a smart card is performed with basic commands known as **ADPU**s (Application Data Programming Units). All communication is in hexadecimal code. The following example shows the construction of a typical smart card command, which has to follow the format:

- Class of Command
- Command
- First Parameter
- Second Parameter
- Data Length
- Data

The diagram below shows the command required to perform a 'Select' action. Before anything can be done with a file, it must be selected; this includes the execution of a Java Card applet, which must be selected before any commands can be sent to it:

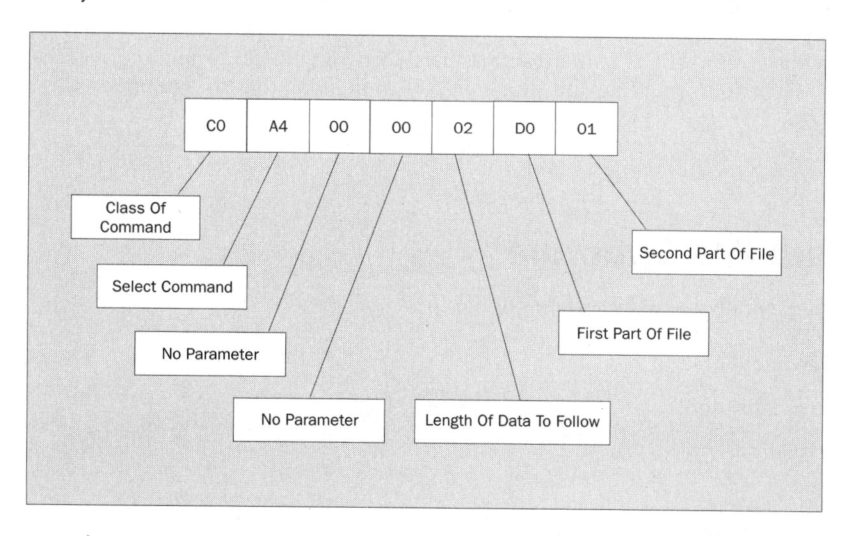

Listening to the Card

The data returned from the card is of an unknown length because there is no equivalent data length parameter defined, though each command sent to the card should have a known response. A standard 'OK' response from the card would consist of 90 and 00 (in hexadecimal), with no data. Error conditions and certain commands may give different responses, such as 9F xx when there is data to be collected from the card. The response is known as a **Status Word**, and is always 2 bytes (though these two might indicate additional data to come). The diagram below shows a length of data returned from the card, followed by a Status Word (SW1 and SW2):

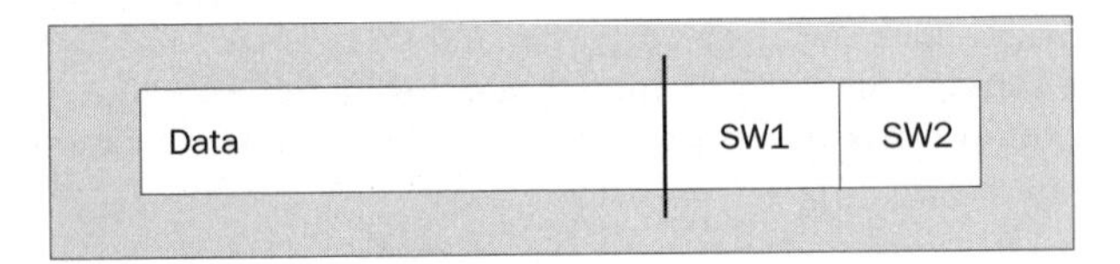

smart card Communication Example

The following example shows how communication might proceed. Note that this example could be applied to any ISO7816 smart card, and is not restricted to Java Card-compliant cards:

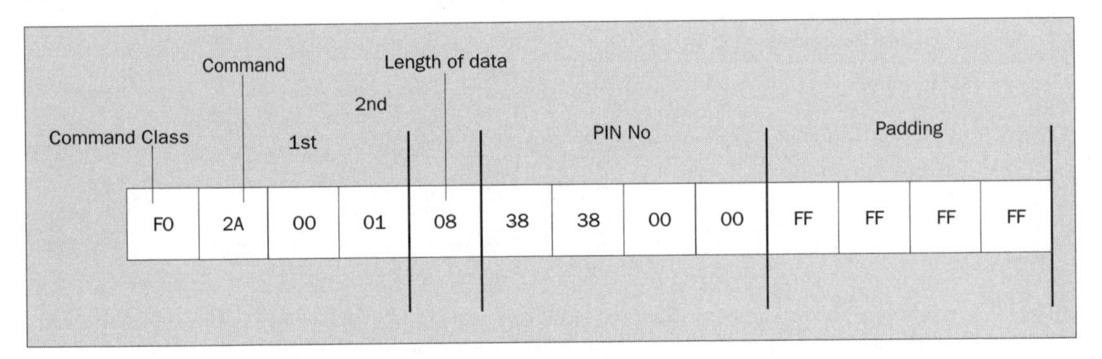

This example shows a PIN number being presented to a smart card for validation. Note that the PIN is actually 8 bytes long (commencing with the first '38'), with the final 4 being padded. The PIN used is '8800', which is sent in ASCII (American Standard for Computer Information Interchange). The response is on the following line and shows '90, 00', indicating that the command was successful:

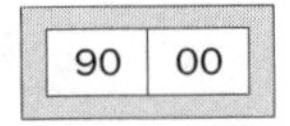

smart card File Structure

Smart cards generally have three stages of life:

- Prepersonalization
- Personalization
- Activation

Prepersonalization involves the installation of the operating system and applications. The data placed on the card during the prepersonalization phase is often written in **ROM**, rather than **FLASH** memory, and as such can never be changed. Data stored on the card that can never be altered is generally known as the **mask**, and both **soft** and **hard masks** are common.

The **hard mask** is data stored in the ROM of the card, while the **soft mask** is data stored in FLASH, which is never expected to change, and the operating system should not allow changes to be made to that data. Soft masks are particularly popular in small production runs of cards, which means anything less than a million cards, as applications can be written to the soft mask without the need to create new ROM chips with the application stored on them.

Personalization is the process whereby the card is identified to a particular user or application instance. This might involve the recording of account details or personal information, depending on what function the card is intended to fulfill.

Activation often takes place when the card is first used, or when it is issued to the customer (which might well be directly after personalization). Applications are run for the first time, and Java Card applets are no different in this respect, they should be considered analogous to the construction phase of a normal Java object.

The primary task performed in Personalization is the creation of the file system. Although Activation may well involve writing data to the various files, it is very unlikely that these will not already have been created on the card during Personalization.

The smart card File System

The file system used on a smart card is laid down by ISO7816, and is fairly basic. While the hard and soft masks are stored in a similar fashion, they are not accessible to applications running on the card. If we are familiar with common desktop computer systems should be aware of the following differences:

- All files are identified (named) as 4 bytes
- Directories are of fixed size, and known as 'Dedicated'
- Files must be 'selected' before we can do anything with them

Files on a smart card come in a variety of types:

File Type	Description
Transparent	Random access file that can contain anything.
Linear fixed files	Records of fixed length – can be navigated by moving to 'next', 'last', and so on. Examples include an address book function.
Cyclic	Fixed length records, where selecting the 'next' of the last one leads to the first. Examples include last numbers dialed.

Table continued on following page

File Type	Description
Variable length records	Not supported in GSM, a fixed length file that can contain records of uneven length. Care must be taken not to add a record that exceeds the remaining space in the file.
Incremental	Very useful files, but very small. A single byte that can be incremented or decremented by one. Useful because control can be very granular, one PIN to decrement, one to increment. Examples include pay-as-you-go applications.

Example Commands

The following command APDUs are examples of ISO7816 commands, which could be used with any compliant smart card. The sequence implemented shows a file being selected, a PIN being presented, and the file contents being read. The responses are also shown. While these commands are not directly part of the Java Card specification, they are defined by ISO7816 and are typical of the kind of communication we will be doing with a Java Card applet.

A PIN exists in an authenticated, nonauthenticated or blocked state, and once authenticated, will remain so until a different directory (or Java Card applet) is selected. When we're working with a Java Card applet, we have to take responsibility for receiving and authenticating the PIN ourselves, and we shall see how to do that in the later example:

Class of Command	Instruction	Parameters		Length	File	Name
C0	A4	00	00	02	D0	01

Select File

The "Select file" is the file creation command, as specified by ISO7816-4. C0 is the class of command, A4 being the specific command to be executed. The other parameters are not used in this command.

The length indicates how many more bytes there are in this APDU.

All file Wroxnames consist of two hexadecimal numbers up to FF FF.

The OK response:

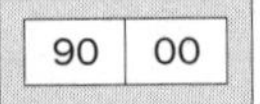

The response shows that the file has been selected.

Now we need to present our PIN to be able to read the file. It is important to identify the PIN we will be using (of the 3 that can be set in each directory), and we will be expected to know which PIN will enable us to read the file we have selected. As we have selected a linear fixed record file, the PIN we need to authenticate against will have been specified during the creation of that file:

Class of Command	Ins	Parameters		Length	Password				Padding			
F0	2A	00	01	08	38	38	00	00	FF	FF	FF	FF

Present PIN

F0 is the class of command, 2A specifically says that we are going to present a file verification PIN for authentication. The first parameter isn't used, while the second specifies that we are presenting to PIN 1, of the three possible in each directory.

The length indicates how many more bytes there are in this APDU.

The password, as ever, is presented in hexadecimal numbers representing the ASCII values of the characters making up the PIN. In this case the PIN is '8800', which translates as 38, 38, 00, 00. The PIN is 8 characters by default, so the last 4 are padded with FF. 4 digit PIN numbers are the most common, though 8 would be considerably more secure. The real reason 4 is used is due to human short term memory, which has been shown to hold between 5 and 9 items (depending on the individual), so 4 is considered simple to store in short term memory. Given that a thief is only given three opportunities to guess the number it makes little statistical difference.

The OK response:

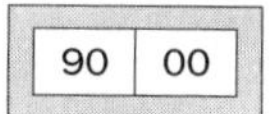

This same response tells us that the PIN has been authenticated, failure would have resulted in a 63 00, with other codes used for other problems (such as 69 81, when there is no PIN set). We should now know that we are able to read the file we have selected. This authentication will remain until power is disconnected or we select a different directory.

Now we will read some data from the file:

Class of command	Instruction	Parameters		Length
C0	B2	03	04	20

Read data

The "Read data" command is for reading data from a fixed length record file. The first two bytes are the class of command and specific instruction (C0 B2), then come the parameters. The first parameter selects a record (the third in this case), and the second defines which record to actually read (00 for the first, 01 for the last, 02 for the next from current, 03 for the previous from current, and 04 for current).

In this case the length actually refers to the amount of data wanted. We should note that this is expected to be thirty two characters in this example. The actual length of the response will depend on the data being retrieved, which must be known prior to issuing the command.

The OK response:

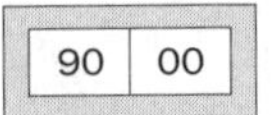

Note that we haven't actually got the data yet; all we have done is to ask the card to place the data into a buffer, from which we can then read.

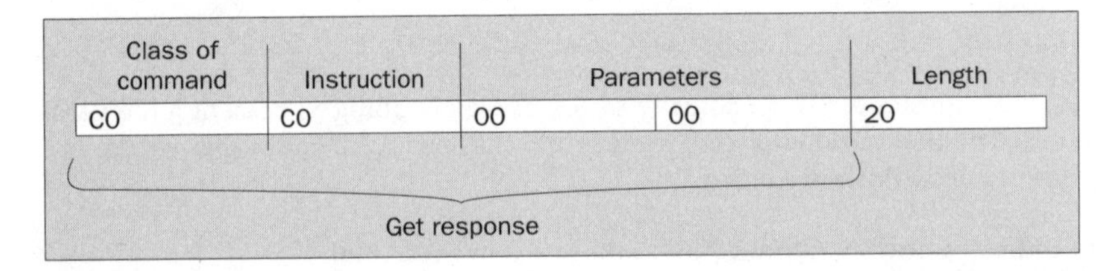

The "Get response" command is used whenever a command giving a response has been issued. We give the command asking for the response (in this case the "Read data" command above), get an OK response (90 00), and then actually ask for the data.

The length here actually refers to the amount of data required.

The response is as follows:

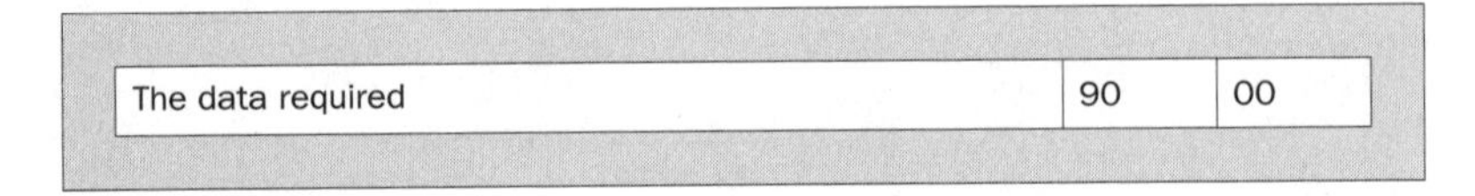

This exchange is fairly typical for a smart card application.

Creating Our Own Commands

As all communications with the smart card must be phrased in APDUs, not only will our Java Card application need to receive and interpret them, but the server-side of our application will also need to know how to work with them. Once we have decided exactly what our Java Card application is going to do, one of the first stages should be the identification of what APDUs we are going to need, and what they should be.

Using a smart card with Java

There are several ways that Java can communicate with a smart card application, though it should be remembered that all such communication will take place in the form of APDUs, as already described. These must be transmitted to the smart card through some sort of terminal device addressed through the serial port or similar.

Smart card readers are themselves complex devices, and any communication with a smart card will actually have to run through the smart card reader. Most readers will require some additional processing on the APDUs before they will pass on a command to the card. This additional processing may involve the generation, and appending, of a checksum, or even the conversion of the APDUs into an ASCII version of itself. Some readers will have multiple slots, or even another smart card built into the hardware, and, in that case commands will need to be addressed to the appropriate slot. It's not feasible to run through all the possible command structures, but they should be documented with particular smart card readers.

When a smart card reader is connected directly to a serial port, it is very possible to communicate with it through the `javax.comm` libraries. Again, the communications protocol will be documented with the specific card reader being used, though it's generally of a low speed.

The following code shows how a command can be created and sent to a smart card. In this case, the reader is a Schlumberger Reflex 60:

```
public static boolean selectFile(int fName1, int fName2) {
  return command(new int[] {
    0xC0, 0xA4, 0x00, 0x00, 0x02, fName1, fName2
  });
}

private static boolean command(int bits[]) {
  myPort.putByte(COMM, 0x67);
  myPort.putByte(COMM, 0x00);
  myPort.putByte(COMM, bits.length);
  for (int i = 0; i < bits.length; i++) {
    myPort.putByte(COMM, bits[i]);
  }
  if (myPort.getByte(COMM) != 0x62) {
    return false;
  } else {
    return true;
  }
}
```

The `selectFile()` command actually builds the APDUs, and then passes it to `command()` which formats it for the Reflex 60. In this way, it is possible to extend this object to work with different readers by over riding the `command()` method.

It should be obvious that we are able to do many things with a smart card, without using Java Card to program applications to run on the card itself. Storing information securely, electronic wallet applications, loyalty schemes, and cryptographic applications are all perfectly possible using just ISO7816 commands and a standard smart card. In fact, the lack of interesting Java Card applications has led to several competitions with prizes of up to 50,000 US dollars available for the author of the killer Java Card application.

This difficulty in making code that will work with different smart card readers has led to the establishment of OpenCard, a set of APIs designed to work with any smart card reader. OpenCard is a consortium of companies involved in smart card production or business, with a view to establishing standard Java objects that can be used to communicate with any smart card reader. However, progress on the standard has been slow, and at the time of writing, there is very little support for it. More information, and a list of supported hardware and software, can be obtained from http://www.opencard.org.

It is certain that with the growth of smart cards in a computer environment, some sort of card communications standard will be necessary. Microsoft has already launched PC/SC (Personal Computer/smart card), a library of functions available to programmers working on the Windows platform (though OpenCard could conceivably run on top of PC/SC, as it does with other APIs). OpenCard should become the standard way of communicating with smart cards, but as current support does not include Java Card-compliant smart cards, it is, for the moment, of no significance to us.

Programming the Card

Creating a Java Card application is much like creating any other application where communication is central. We start by defining what it is we want to achieve, then we work out how the various parts will communicate, and finally, we work out what objects we are going to need before writing some code.

All applications on a smart card also need an **AID** (Application Identifier). Each applet deployed on a smart card will need its own AID, as will the package they are in. AIDs are from 5 to 16 bytes long, and some care should be taken to ensure no applications on the card share the same AID, as this could have unpredictable consequences. Therefore, allocation of the first 5 bytes of the AID is done by the company, and awarded by the International Standards Organization.

The remaining bytes (up to eleven of them) are defined by the company, and should be unique for each application they develop. The parts of the AID are known as the National Registered Application Provider (RID) and the Proprietary Application Identifier Extension (PIX). Note that the word "National" is just a legacy of the system, the codes are a worldwide standard.

For the purposes of this example, we will be using fictional AIDs, but for deployed applications, a request should be made to the ISO for an AID for one's company.

AID for our package:

RID					PIX		
0xA0	0x00	0x00	0x00	0x00	0x01	0xC0	0x01

AID for our application:

RID					PIX
0xA0	0x00	0x00	0x00	0x00	0x01

There is nothing particularly significant about these numbers; they were selected pretty much at random and are easy to remember.

We are going to run through a fairly trivial example of what is possible with a Java Card applet, but it should show the process and make it obvious how the functionality could easily be expanded. We are going to create a Java Card applet that can store details of a user's e-mail account (server, username and password) and produce these on request. Therefore, having decided what the applet is going to actually do, we need to consider to what commands it will need to respond. As communication will only take place in ADPUs, we will need to define these and what responses we expect from the applet.

Select Command

This command is defined by ISO7816, notice that selecting an application is a different class of command from the file selection we saw earlier. It serves to show the format in which we record ADPUs, and these documents will prove invaluable when we come to create the other half of the software:

Class of command	Ins.	Parameters		Length	AID of application to be selected							
0x00	0xA4	0x04	0x00	0x08	0x66	0x69	0x6C	0x6C	0x73	0x00	0x00	0x02

Expected Responses

The following indicates the successful completion of the command:

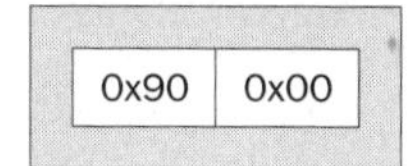

0x90	0x00

This response indicates that the command failed, because the application cannot be found:

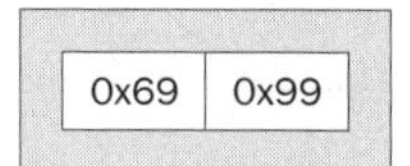

0x69	0x99

Since all incoming APDUs will be sent to the same method, it will be up to our application to work out what to do with the APDU depending on the instruction specified. These are completely up to the software developer, so here we have decided that 0x10 will mean 'get server name', 0x20 will be for the user name, and 0x30 should return the password. Note that for this simple application, we will be hard-coding the values for these things; but as we will see, adding additional commands for setting the values, and providing some sort of PIN security, is not a complex task.

Get Server Name

Class of command	Ins.	Parameters		Length	Padding							
0x00	0x10	0x00	0x02	0x02	0xFF	0xEE						

Expected Responses

The following response shows numbers representing the name of the server, equal in length to the first byte (that is, the first byte is a length indicator):

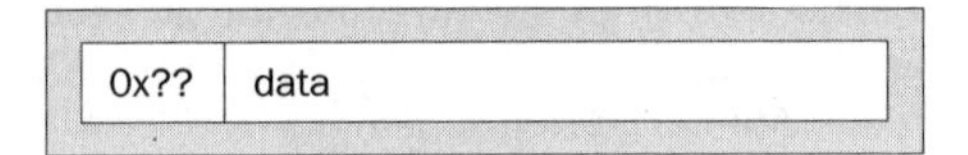

Here the status bytes indicate that the command was a success:

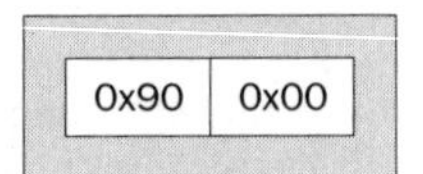

This shows the format for most of our commands, though we should still specify them in full.

Get User Name

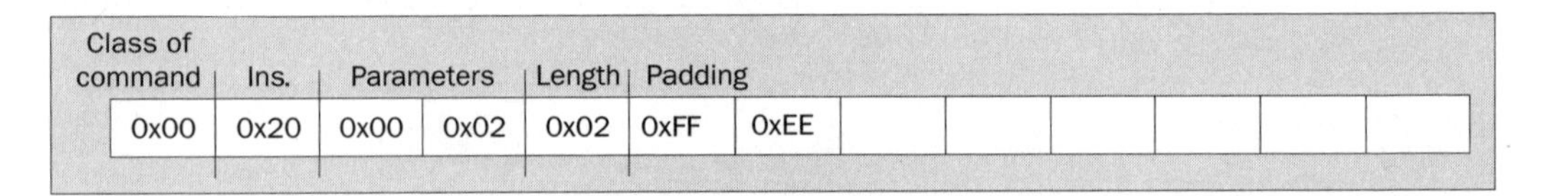

These are the numbers representing the name of the user, equal in length to the first byte:

Expected Responses

Here we have status bytes that indicate that the command was a success:

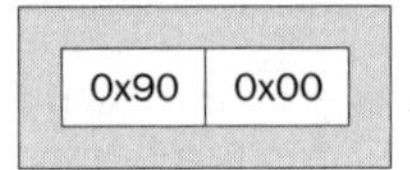

Get Password

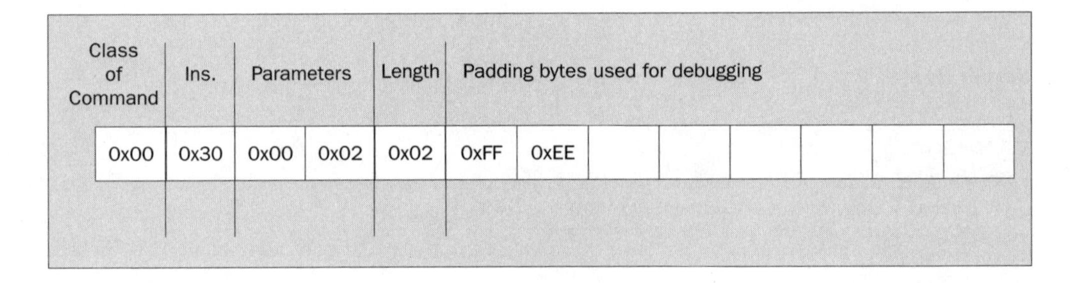

Expected Responses:

These numbers represent the password, equal in length to the first byte:

Status bytes that indicate that the command was a success:

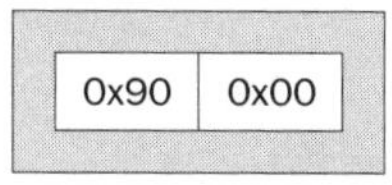

It's perfectly possible for us to define hundreds of commands and to do anything we feel inclined to, or, more realistically, anything the smart card is capable of. Beyond the basic commands already discussed, there are no standards for formatting these commands; as long as we document what we intend to do then we can create what we need.

Having decided on our commands, and the responses we expect from them, we can start to think about our application. Before we write the program, we should have an environment we can test it in, and Sun provide such an environment with the **Java Card 2.1.2 Development Kit**, which provides all the tools needed to write, and debug Java Card applications, without access to a real card. Not only is this kind of development easier in many ways (the emulator will generate errors invisible on a real card), but it is also considerably quicker; as the code is not having to be written to the soft mask every time testing is desired.

Using the Java Card Development Kit

The kit can be downloaded from the Java Card home page at http://java.sun.com/products/JavaCard/dev_kit.html. It does not come with an installer, so it should just be expanded into an appropriate directory. In our example, we are working on a Microsoft Windows computer, and we will expand it into the root of drive C. This saves a lot of typing, though it will happily sit anywhere.

Once the file has been expanded, we need to set up the working environment. We'll need two shell windows (command prompt windows), and the following settings should be applied to both. The first thing is to set up an environment variable pointing to the kit installation (which we will call <CARD_HOME>), and we must make sure these steps are followed in this order:

```
set JC21BIN=<CARD_HOME>\bin
```

UNIX users should use:

```
setenv JC21BIN=/<CARD_HOME>/bin
```

We will also need to set the `PATH` variable so that it points to the same place:

```
set PATH=%PATH%;%JC21BIN%
```

With UNIX we use the following:

```
setenv PATH ${PATH}:$JC21BIN
```

We will need to set our `CLASSPATH` variable so that it points to where we are working, in addition to the normal file containing the Java classes and the file `api21.jar`. Under Microsoft Windows, this would be done with the command:

```
set CLASSPATH=%CLASSPATH%;c:\<CARD_HOME>\lib\api21.jar
```

Testing the Installation

To ensure that the installation has worked properly, it is worth running the provided examples. These are available in the `<CARD_HOME>\samples\src\demo\` directory, and we should change to that directory before entering the following on the command line:

```
jcwde -p 9025 jcwde.app
```

This will run the emulator. The emulator uses a loopback socket connection for communications, and in this example we will listen to socket 9025 (which is the default for the other end of the connection). This should result in an output resembling the following:

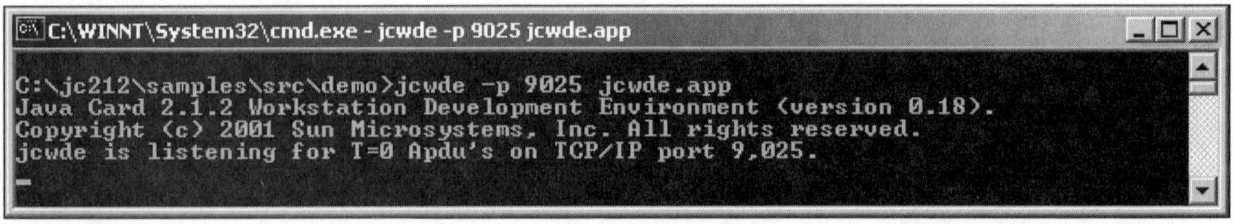

```
C:\WINNT\System32\cmd.exe - jcwde -p 9025 jcwde.app

C:\jc212\samples\src\demo>jcwde -p 9025 jcwde.app
Java Card 2.1.2 Workstation Development Environment (version 0.18).
Copyright (c) 2001 Sun Microsystems, Inc. All rights reserved.
jcwde is listening for T=0 Apdu's on TCP/IP port 9,025.
```

Now the emulator is listening, we need to provide it with some APDUs to process. These are supplied, from a new command prompt window, by an application called `apdutool`. This program takes an input file and feeds the APDU codes from that file into the emulator, and displays the output. For the demo applications, we can use the APDU script supplied, with the following command:

```
apdutool demo1.scr
```

The output should scroll up the window; and a file is provided called `demo1.scr.expected.out` which shows what output is expected from this. It's not necessary to compare every number, a glance at a few specific lines, chosen at random, should confirm that it's working. For a more complete test, we can capture the output into a file and compare the files.

A Note on Windows 9x

We will find that the output from the demonstration, and most of our applications, will scroll past the window in an annoying fashion. While it is supposedly possible to redirect the output of `apdutool` (using `apdutool demo1.scr > demo1.output`) at the time of writing it doesn't actually work. There is also an option that claims to output to a file (`audotool -o demo1.output demo1.scr`), though this also doesn't function as it is meant to.

While there are several development environments that offer scrollable windows showing errors, things can get very difficult for the developer working from Notepad. If we are planning any serious Java Card (or, indeed, Java) development, it is worth considering investing in such an environment.

NT users can get a scrollbar on their command prompt windows by right-clicking and adjusting the screen size options.

Developing the Application

We are going to run through the development of the application we have already discussed, which will simply respond with information when presented with specific commands. We have already specified the commands we will be using, so now it's a matter of creating a Java Card application (sometimes called an applet; and sometimes a cardlet, to emphasis the difference from a J2SE Applet used in a web page) that can perform these tasks.

Java Card applications are developed in much the same way as any other Java application; they are created in a text editor and compiled with the standard Java compiler. However, once compiled, they are converted into CAP files which are then copied onto the card.

These files have several advantages over standard class files, in that they go through various transformations to make them execute more efficiently; steps that would normally be performed by the JVM at run-time. The conversion routine also confirms that the code is not using APIs not available to the **JCRE** (**Java Card Run-time Environment**) and creates an export file, and an optional JCA (Java Card Assembly file).

The export file (with the extension `.exp`) is used for applets that wish to allow other applets to access them; which would only be the case in complex applications. The JCA file contains a pseudo-code representation of our application, which can be useful for debugging or documentation.

Applet Lifecycle

A Java Card application is considered to be always active; once it has been installed, all variables will exist forever, even if power is lost to the card during a process (a remarkably common event). Power loss in the middle of a transaction can result in data loss, if values are in the process of being written, and it is always a good idea to consider what might happen during the execution of our applet. It is useful to imagine that the applet is always running, with methods being called at random times, but always in memory.

The first stage therefore, is for the application to register itself with the JCRE as being available, and then it has to be installed. Once the applet is installed, it becomes available to process commands. An application will need to select the applet, and once selected, all incoming APDUs will be delivered to the `process()` method.

Therefore, the most basic Java Card application will have a constructor, an `install()` method, and a `process` method. In addition, it is possible to have methods responding to other events, such as `select()`, which is called when the applet is selected, and `deselect()`, which is called when a different applet or file is selected. Full details of the methods available are shown in the Java Card API available from http://www.javasoft.com/products/Java Card/.

The data itself is in bytes representing ASCII values of the information needed. Smart cards (and by extension Java Card) generally use 8 byte processors, and work with ASCII, not UNICODE. In this case, the information is the username, 'bill1' and is converted as follows:

b	i	1	1	1
0x62	0x69	0x6C	0x6C	0x32

Example Java Card Applet

All Java Card applets have to be part of a package. A single package may contain many Java Card applications, but ours will contain only one:

```
package InfoAnywhere;
```

Obviously, this file will need to be in a directory called `InfoAnywhere`, in the normal Java fashion. We then need to import the classes that are specific to the Java Card API:

```
import Java Card.framework.Applet;
import Java Card.framework.APDU;
import Java Card.framework.ISO7816;
import Java Card.framework.ISOException;
```

This package contains all the Java Card classes, and is located in the file `api21.jar` located within the `<CARD_HOME>\jc211`, which we set our `CLASSPATH` to point to earlier.
Our class will extend `Applet`, as all Java Card applets do (this is not the same `Applet` used for J2SE Applets for embedding in web pages), and needs a constructor:

```
public class EmailSettings extends Applet {
```

Our constructor will be a very basic affair, just running the `register()` method of the super class `Applet`, to register the applet with the JCRE. It is perfectly possible to do far more in a constructor, and often a good idea. Given the very limited memory available within the card (though the amount of RAM varies, it's normally something in the area of 256 bytes, with cards offering around 4Kb of FLASH memory for storage); it is good programming practice to populate any variables being used here, and not to vary their size elsewhere in the applet:

```
  protected EmailSettings() {
    register();
  }
```

The `install()` method is run when the applet is first run. Remember that Java Card applets are only expected to ever run once, when deployed, and they should be considered to be running forever from that point onwards. Normally, the only thing an installer method will do is instantiate a new instance of the applet.

`bArray` is the array containing installation parameters, `bOffset` is the starting offset in `bArray`, and `bLength` is the length in bytes of the parameter data in `bArray`:

```
  public static void install(byte[] bArray, short bOffset, byte bLength) {
    new EmailSettings();
  }
```

Sending the bytes is a three stage process and is accomplished through the use of the same APDU object that was passed into the `process()` method. First, we run `setOutgoing()` to make the APDU into an output stream, then, we set the length with `setOutgoingLength()`. This setting of the length is only necessary if we are planning to output more than the normal status bytes, as we are in this case.

Finally, we `sendBytesLong()`, specifying the array of bytes to be sent, the start marker, and the number of bytes to be sent. As we shall see, JCRE will add the success status bytes (0x90, 0x00) to the end of the stream as we haven't indicated that anything went wrong. If we only wanted to send those two bytes, (for example, if we had a method which had set one of these values) then we would not need to specify any return code at all, we would just have to complete the `process()` method without generating an error:

```
  private void sendServer(APDU apdu) {
    byte result[] = new byte[13];
    apdu.setOutgoing();
    apdu.setOutgoingLength((byte) 13);
    result[0] = (byte) 0x70;
    result[1] = (byte) 0x6F;
    result[2] = (byte) 0x70;
    result[3] = (byte) 0x33;
    result[4] = (byte) 0x2E;
    result[5] = (byte) 0x77;
    result[6] = (byte) 0x72;
    result[7] = (byte) 0x6F;
    result[8] = (byte) 0x78;
    result[9] = (byte) 0x2E;
    result[10] = (byte) 0x63;
    result[11] = (byte) 0x6F;
    result[12] = (byte) 0x6D;
    apdu.sendBytesLong(result, (short) 0, (short) 13);
  }

  private void sendUser(APDU apdu) {
    byte result[] = new byte[5];
    apdu.setOutgoing();
    apdu.setOutgoingLength((byte) 5);
    result[0] = (byte) 0x62;
    result[1] = (byte) 0x69;
    result[2] = (byte) 0x6C;
    result[3] = (byte) 0x6C;
    result[4] = (byte) 0x32;
    apdu.sendBytesLong(result, (short) 0, (short) 5);
  }

  private void sendPassword(APDU apdu) {
    byte result[] = new byte[7];
    apdu.setOutgoing();
    apdu.setOutgoingLength((byte) 7);
    result[0] = (byte) 0x64;
    result[1] = (byte) 0x72;
    result[2] = (byte) 0x73;
    result[3] = (byte) 0x38;
    result[4] = (byte) 0x38;
    result[5] = (byte) 0x30;
    result[6] = (byte) 0x30;
    apdu.sendBytesLong(result, (short) 0, (short) 7);
  }
```

It is the `process()` method that will actually do all the work of the applet, responding to APDUs sent from outside. In our example, this has been kept very simple, and our example is able to respond to any of the three instructions we have defined, or generate an error if any other instruction is sent.

The method checks to see what instruction has been received by copying the input stream into the array `buffer`, then examining a particular character of that array to see what was requested. As we already know, the instruction byte will be the second byte received, but we use the constant `ISO7816.OFFSET_INS` for flexibility:

```
    public void process(APDU apdu) throws ISOException {
      byte buffer[] = apdu.getBuffer();
      byte result[];
      switch (buffer[ISO7816.OFFSET_INS]) {
      case 0x10:
        sendServer(apdu);
        return;
      case 0x20:
        sendUser(apdu);
        return;
      case 0x30:
        sendPassword(apdu);
        return;
      default:
        ISOException.throwIt(ISO7816.SW_INS_NOT_SUPPORTED);
        return;
      }
    }
  }
```

Obviously, this example is a very simple application. There are several much more complex ones included in the development kit, and we can find those in the `samples` directory.

Compiling the Example

The above example will need to be in a file called `EmailSettings`, which has to be in a directory called `InfoAnywhere`. The applet is compiled in the normal way, though the `-g` tag is used to create full debugging information, and this is required for Java Card files. The `-o` option should not be used as this optimizes for speed, not memory size.

```
javac -g InfoAnywhere\EmailSettings.java
```

Once the file is compiled, we need to convert it into a format appropriate for using with the JCRE. The converter tool will do this.

Converting the Example

The converter tool requires several parameters, and these can be specified on the command line, or in a configuration file. While there are many possible options, we'll only be looking at what's needed to get an applet working. The other options are fully documented in the user's guide, which can be found in the `doc` directory.

The three parameters we will be using, specify what kind of conversion we want to do, where the files to be converted can be found, and the AIDs of the applets and package. We start by specifying what kind of output we would like.

The following line means we want an EXP file, and a CAP file. The CAP file is the actual program which will be executed on the card, while the EXP is an Export file used when applets wish to share methods. We could also specify a JCA file, if we wanted a textual representation of our applet:

```
-out EXP CAP
```

This is the root directory in which our package directory can be found:

```
-classdir c:\projavamobile\chapter10
```

The next parameter consists of several sections in the following format:

```
-applet appletAID appletName packageName packageAID version
```

The applet name is fully qualified (starts with the package name), as that is the full name of the applet. It would be appropriate to expand this line if we have several applets in the same package:

```
-applet  0xA0:0x00:0x00:0x00:0x00:0x01 InfoAnywhere.EmailSettings InfoAnywhere
0xa0:0x0:0x0:0x0:0x00:0x00:0x1:0xc:0x1 1.0
```

As entering all of the above, every time we wanted to test a modification, would be a pain, it is possible to use a configuration file, containing all that information, which can be supplied as an option to the converter. Therefore, if we create the following file, and call it `InfoAnywhere.opt`, we can use that for our conversion:

```
-out EXP CAP
-classdir c:\projavamobile\chapter10
-applet  0xA0:0x00:0x00:0x00:0x00:0x01 InfoAnywhere.EmailSettings
InfoAnywhere
0xa0:0x0:0x0:0x0:0x00:0x00:0x1:0xc:0x1 1.0
```

This file can then be used with the following command line:

```
converter -config InfoAnywhere.opt
```

We may well find that some adjustments to the `CLASSPATH` are necessary to get this working, depending on how many libraries we have installed. This command line could be tried out if there are any problems with the conversion:

```
set
CLASSPATH=c:\jdk1.3\jre\lib\rt.jar\c:\jc212\lib\api21.jar;.;c:\jc212\lib\converter
.jar;c:\jc212\api21
```

This should produce the following output:

```
C:\WINNT\System32\cmd.exe

C:\javamobile\chapter10>converter -config InfoAnywhere.opt

Java Card 2.1.2 Class File Converter (version 1.2)
Copyright (c) 2001 Sun Microsystems, Inc. All rights reserved.

conversion completed with 0 errors and 0 warnings.

C:\javamobile\chapter10>_
```

This will create a new directory within our package directory, called `Java Card\`, and containing `InfoAnywhere.cap` and `InfoAnywhere.exp`. It's worth noting that these are the Export and CAP file we intended to create.

Running the Example

Now we want to run our example using the Java Card Development Kit emulator. We have to start by creating a couple of files. The first will indicate to the emulator what applications we want to load on to the card, and the second will provide a list of APDUs that we want sent to the card.

Regarding the programs we want to install, we have to get the installer onto the card, as this will be used to load on other applications. This installer may well be preinstalled on real cards, but we have to specify that we intend to use it in the emulator. We also need to specify that we will be using our example application:

```
com.sun.Java Card.installer.InstallerApplet 0xa0:0x0:0x0:0x0:0x62:0x3:0x1:0x8:0x1
InfoAnywhere.EmailSettings                            0xA0:0x00:0x00:0x00:0x00:0x1
```

This file is made up of the name of the applet and its AID, in that order. The AID of the installer applet is predefined. This file will be called `AppList.app`.

Creating the file to feed APDUs into the emulator is a little more complicated; we create a file containing lines of APDU code, and then use the `apdutool` to send them to the emulator. In our case we will only be sending instructions for the three commands we have defined, but we still need to start by selecting the installer applet and then asking it to install our application.

Our file will start with a `powerup` line, which will power up the emulated card. Then we send a Select command to select the installer, start it, install our applet, and tell the installer applet that we've finished. These steps are always necessary when working with the emulator, though they might be different, or nonexistent, depending on the card being used. The command sequence looks like this:

```
powerup;

echo "Selecting Installer";
0x00 0xA4 0x04 0x00 0x09 0xa0 0x00 0x00 0x00 0x62 0x03 0x01 0x08 0x01 0x7F;

echo "Start Installer";
0x80 0xB0 0x00 0x00 0x00 0x7F;
```

```
echo "Create Applet";
0x80 0xB8 0x00 0x00 0x08 0x06 0xA0 0x00 0x00 0x00 0x00 0x01 0x00 0x7F;

echo "End Installer";
0x80 0xBA 0x00 0x00 0x00 0x7F;
```

The 'echo' lines will echo that text out to the screen, so we can follow what's happening. The apdutool will also output the command sent, and the bytes received from the emulator. Once the above has been executed, we need to select our applet, send some commands, and finish with a powerdown command:

```
echo "Selecting EmailSettings";
0x00 0xa4 0x04 0x00 0x06 0xA0 0x00 0x00 0x00 0x00 0x01 127;

echo "Requesting Server Name";
0x00 0x10 0x00 0x00 0x02 0xFF 0xEE 0x7F;

echo "Requesting User Name";
0x00 0x20 0x00 0x00 0x02 0xFF 0xEE 0x7F;

echo "Requesting Password";
0x00 0x30 0x00 0x00 0x02 0xFF 0xEE 0x7F;

powerdown;
```

This file will be called test.scr, and will be provided to apdutool as a parameter.

Once the files are all prepared, it is possible to actually run the emulator, load the applet, and run the tests. In the first command window, we will need to enter the following:

```
jcwde -p 9025 AppList.app
```

This should produce the following output:

```
C:\WINNT\System32\cmd.exe - jcwde -p 9025 AppList.app

C:\javamobile\chapter10>jcwde -p 9025 AppList.app
Java Card 2.1.2 Workstation Development Environment (version 0.18).
Copyright (c) 2001 Sun Microsystems, Inc. All rights reserved.
jcwde is listening for T=0 Apdu's on TCP/IP port 9,025.
```

This shows us that the emulator is loaded and running, it is also listening on socket 9025 (which as we shall see, is the default socket for apdutool) for incoming APDU commands. We now move to our other command window and enter the following:

```
apdutool test.scr
```

This will feed the APDU commands in the file test.scr to the emulator, and show us the responses from it. By using the files described above, we should get the following output:

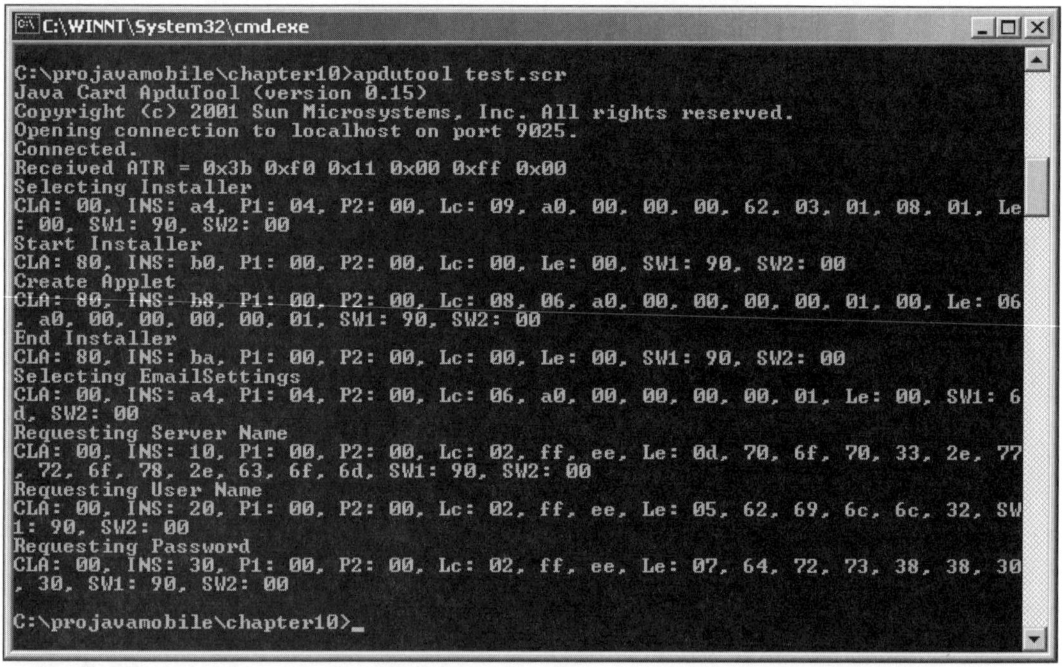

```
C:\WINNT\System32\cmd.exe

C:\projavamobile\chapter10>apdutool test.scr
Java Card ApduTool (version 0.15)
Copyright (c) 2001 Sun Microsystems, Inc. All rights reserved.
Opening connection to localhost on port 9025.
Connected.
Received ATR = 0x3b 0xf0 0x11 0x00 0xff 0x00
Selecting Installer
CLA: 00, INS: a4, P1: 04, P2: 00, Lc: 09, a0, 00, 00, 00, 62, 03, 01, 08, 01, Le
: 00, SW1: 90, SW2: 00
Start Installer
CLA: 80, INS: b0, P1: 00, P2: 00, Lc: 00, Le: 00, SW1: 90, SW2: 00
Create Applet
CLA: 80, INS: b8, P1: 00, P2: 00, Lc: 08, 06, a0, 00, 00, 00, 00, 01, 00, Le: 06
, a0, 00, 00, 00, 00, 01, SW1: 90, SW2: 00
End Installer
CLA: 80, INS: ba, P1: 00, P2: 00, Lc: 00, Le: 00, SW1: 90, SW2: 00
Selecting EmailSettings
CLA: 00, INS: a4, P1: 04, P2: 00, Lc: 06, a0, 00, 00, 00, 00, 01, Le: 00, SW1: 6
d, SW2: 00
Requesting Server Name
CLA: 00, INS: 10, P1: 00, P2: 00, Lc: 02, ff, ee, Le: 0d, 70, 6f, 70, 33, 2e, 77
, 72, 6f, 78, 2e, 63, 6f, 6d, SW1: 90, SW2: 00
Requesting User Name
CLA: 00, INS: 20, P1: 00, P2: 00, Lc: 02, ff, ee, Le: 05, 62, 69, 6c, 6c, 32, SW
1: 90, SW2: 00
Requesting Password
CLA: 00, INS: 30, P1: 00, P2: 00, Lc: 02, ff, ee, Le: 07, 64, 72, 73, 38, 38, 30
, 30, SW1: 90, SW2: 00

C:\projavamobile\chapter10>_
```

Each response starts with the command that was sent to the emulator, followed by the response from the emulator. If we work through these responses, we will see that they do indeed represent the expected strings relating to log on name, domain and password.

We should also notice that the emulator quit on receipt of the `powerdown` command.

Working with the Example

The first thing to do is to think about the `test.scr` file, and try making some modifications to that. Then we should start playing with the applet, and create a command that responds with whatever is sent to it. (We can take a look at the `HelloWorld` example included with the development kit.)

The sample applications, included with the kit, provide examples of some very important processes, such as PIN verification and applet-to-applet communications. Therefore, it is worth becoming familiar with these example applets.

Working with a Real smart card

Getting our applet onto a real smart card, and testing compatibility, will depend on the card we have chosen, and who provides it. Every manufacturer of Java Card-compatible smart cards has their own development environment and processes, not to mention a few additional features, which can provide much needed functionality. It's worth us shopping around for a card to work with, and a company we can get on with.Many companies have licensed Java Card for their own products, including the largest and most important, smart card producers; Schlumberger (with the CyberFlex), and GemPlus (with their GemXpresso product). We can obtain a full list at http://java.sun.com/products/Java Card/allies.html.

With Java Card compatible cards, we are able to install our own applications, as long as we have the root key; though the process will differ depending on from whom we obtain our card (and reader). Other types of cards may also have a mechanism for installing custom applications, but we can't assume it as it isn't normal for applications to be installed onto a card after the personalization phase.

Smart cards are generally deployed in large numbers, with a couple of million being considered a medium sized order, so we need to be aware that these companies aren't used to dealing with small numbers, and may not be sympathetic to our own personal needs. Individual card development kits, for prototyping and testing applications, are available from most card manufacturers, and are generally easy to work with.

If we are intending to deploy large numbers of cards, we will find the technical and development support available from manufacturers to be superb, with responsive technical help, and project guidance, from people completely familiar with their hardware, and experienced with hundreds of smart card applications.

Deploying Applications

It should be obvious to us that smart card applications, whether developed in Java, or otherwise, are basically very simple things. The complexity of a smart card application is in the software it communicates with; with the smart card being reduced to a secure information store, or cryptographic tool. This is a very important point; smart card applications are not developed in a vacuum but they exist to serve the needs of a developed solution and deployment will often involve not only the cards, but also card readers and desktop software as well.

It is very easy to get lost in the complexity of the concept, but it is important for us to remember that it will never be the card that does the work, but always the device with which it is communicating.

Summary

When we talk about smart card applications, it's important to remember that we are rarely talking about applications that actually live or execute on the card itself. When working with a smart card, the card will only form a very small, if essential, part of the overall application, and system designers must be careful not to push too much work onto the card, which isn't an appropriate place for it.

It is the area of security where a smart card can really show its value. Storing private keys in a really safe place is always a problem (when RAM and hard discs are suspect and vulnerable), and very little offers the security a smart card can. But for general data storage and cryptography, it isn't necessary for most developers to actually create any code to be executed on a smart card; card manufacturers have already done that work, very effectively. Secure applications can easily be created using an off-the-shelf card, and communicating with it through standard ISO7816 APDU commands.

In those unusual circumstances where a custom card application is required, Java Card certainly provides a quick and cost effective way of developing such code and is certainly easier for the experienced Java developer.

PROFESSIONAL JAVA MOBILE

TOOL SELECTION

WEB TABLET
MOBILE PHONE
MP3 PLAYER
DIGITAL CAMERA
PALMTOP
LAPTOP

NOW SELECTED

PALMTOP

11

Interfacing with Servlets and Other Elements of J2EE

By this point, we have learnt to build stand-alone mobile applications, but if we're writing commercial software, then we'll probably want to link our applications to a back-end or centralized service. Assuming you have been a little in awe of Java 2 Enterprise Edition (J2EE) and Enterprise JavaBeans (EJBs)), this chapter will help dispel the impenetrable thicket of Java Enterprise APIs, and suggest some ways to connect to your distributed users.

In this chapter, we will give a swift introduction to the core technologies and APIs underpinning J2EE, describe the different ways that J2ME and J2EE can currently connect together, and work through some example applications using a freely available open source J2EE 1.2.1 compliant application server, Jboss; covering servlets in particular. Finally, we look at what the future will hold for mobile enterprise applications.

Obviously, we cannot hope to cover any great detail of J2EE in a single chapter, especially the EJB details. However, what we do cover are the elements of J2EE that are relevant to J2ME. In particular, in this chapter, and the next, we cover:

- RMI & RMI-IIOP
- JNDI
- Servlets
- JMS

This chapter will illustrate the techniques for enabling mobile device to act as a client to enterprise application and which technologies and libraries you need to explore in more depth. If you find your appetites are whetted, then "Java Server Programming J2EE Edition" (Wrox Press, ISBN 1861004656) gives comprehensive coverage of all the J2EE APIs, including excellent chapters solely on design, testing, deployment, and performance.

To get the best from this chapter, you will have read through the previous introductory J2ME chapters, both CDC and CLDC, and have understand their limitations and also have some basic understanding of what distributed applications are.

What is Java 2 Enterprise Edition?

At a very simplistic level, (from the application developer's point of view, at least), Java 2, Enterprise Edition (J2EE) is just the standard Java 2 environment (J2SE) plus additional libraries, all wrapped up with a definition of how to package and install (or 'deploy') the applications written using these libraries. The additional libraries appear to be quite diverse, but they are all aimed at enabling distributed application development.

In addition, J2EE also defines the characteristics of an application server needed to run these distributed applications; but in this chapter, we ignore this aspect, and assume our chosen application server will run our applications, without worrying about how it does so.

The overall effect of J2EE is to move the Java platform from a language, and some convenient libraries to an industrywide definition of a complete application environment that we can just drop our Java files and configuration files onto. Using our J2EE configuration files, the application server knows how to deploy our application for us. Here is a selection of J2EE application servers that will run our example applications:

- JBoss – http://www.jboss.org (open source, and excellent for testing)
- BEA WebLogic – http://www.bea.com
- IBM WebSphere – http://www.ibm.com

For a more complete list, visit: http://java.sun.com/j2ee/licensees.html.

You will be glad to find that several J2EE libraries are already familiar to you, such as Applets, servlets, JAXP and JDBC. They are all part and parcel of J2EE. They are exactly the same libraries you have previously used, but in J2EE they are mandatory, rather than being optional extras that you download and install yourself.

Distributed Applications

A distributed application is one that is distributed across more than one physical machine, rather than the complete application residing in one location. The system architecture that a distributed application runs is diverse – it can be multiple servers, one or more client machines, or maybe mobile devices in our case, or a mixture of all these.

By application, we mean a collection of individual components or programs which, combined together, perform a useful function – for example, a client program and a server program communicating together, or a frontend component talking to a remote database component.

The rationale behind this distribution of code and data (that is, objects in Java's case) is that a centralized service performing all the processing may cause bottlenecks – CPU, I/O, disk, network, etc. – especially if our system is trying to support millions of concurrent users.

Splitting the application into distributed components means processing of data can occur where it is most convenient. For example, instead of making one central server do all the work, multiple servers can share the load of the processing, taking advantage of local resources (CPU, memory, storage). If the application needs to perform some intensive database work – maybe updating user's records – it's far better to have that code run directly on the server holding the database, rather than a central server continually reading and writing to a database across a LAN or WAN.

Similarly, but on the client-side, when a user interface needs to be displayed, it's usually better to have the client device render the pixels than some remote server generate a bit-map (or some other method of delivering the final graphic) and pass it to the client.

In summary, the simple idea behind distributed applications is to run the code and resources closest to where it's needed. Another major benefit is that there is, potentially, no single point of failure with multiple servers in a cluster sharing requests in a load-balancing manner. If a server fails, then another in its cluster can take over and handle the requests, with minimal degradation in the quality of service.

The theory behind distributed applications may be simple, but the execution is complex. Instead of just having a single server with all the resources to hand, we suddenly have to consider many other things such as:

- How do the distributed pieces of application communicate, and pass data/objects to each other?
- How do we guarantee that all our remote requests are competed successfully (a transaction), even if there's an application failure or a break in the communications link?
- How do different parts of the application find each other?
- What controls the load balancing across the distributed devices?
- How do we minimize the total cost of a transaction, be it in user-perceived-time, actual bytes, or CPU time?

The additional J2EE libraries address all these complex needs. Let's cover just the basic functions needed to create a distributed application, which are:

- **Communication**
  Data flowing between the various parts of the application
- **Location**
  How to find a remote object or resource
- **Frontend Logic and Presentation**
  A webbased thin client, or bespoke thick client UI
- **Backend Logic and Data**
  Business rules, database and transactions

We'll then have a brief look at the more specific APIs.

Communication

The different components of a distributed application need to communicate with each other. As expected, the lower levels of this communication use established protocols like TCP/IP. However, we are interested in the higherlevel application layer communication.

Since J2EE is objectoriented, we need to know how an application or component can call the methods of another object that is remote to it. J2EE specifies that objects should use the Remote Method Invocation over the Internet Inter-ORB Protocol (RMI-IIOP). It also enables Java objects to communicate with any remote CORBA (Common Object Request Broker Architecture) objects, which is useful for integration of nonJava components and applications into an enterprise application.

RMI and RMI-IIOP

RMI's single purpose is to allow us to access remote objects as if they were local to our component, with minimal code changes. In fact, the only difference should be that instead of accessing a local object, we obtain a reference to the remote object we need. Once we have that reference, we can treat the object as though it is running in our local environment.

Realistically, of course, we cannot just totally ignore the fact that the object is remote, since there may be considerable delays in access compared to a local object, which deserves consideration during the design phase of our system. Delays can result from the connectivity and also the time taken to create the remote connection, wrapping the request, and finally unwrapping the reply.

RMI is based upon the 'proxy' design pattern, in other words, it works on our behalf to deliver requests and accept replies from the remote object. When we call a method on the remote object, we are actually calling a method on the local proxy object known as the **stub** object. The stub communicates, transparently to us, to a **skeleton** object located in the same place as the remote object we want to access:

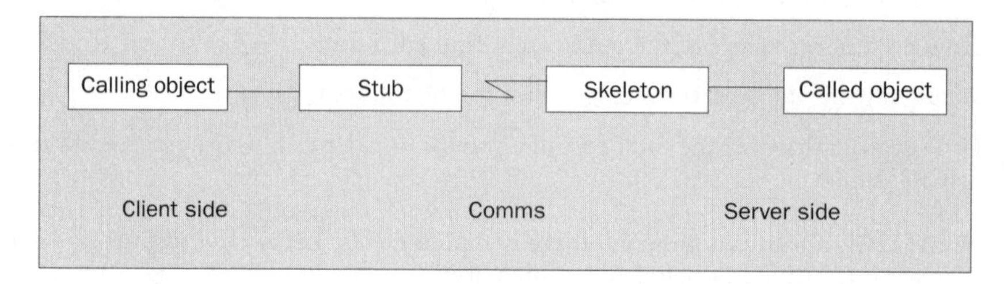

Note that the client and server descriptions only mean that the 'client' is requesting a service from the 'server' for this transaction – there is no reason why RMI cannot be peer to peer.

The job of the stub-skeleton object pair is to hide the complications of the transport and session layers from us. This includes how to convert method parameters, which may be objects themselves, into a byte stream, and convert them back into full Java objects when they reach the other end of the wire, ready for the remote object to use. This object-to-byte stream conversion is known as **marshalling**.

To give a better idea of what a very simple stub and skeleton look like, the following example should give us an idea of the work being done by these two objects.

Pseudo RMI Stub and Skeleton Example

Note that these are **not** an RMI stub and skeleton, merely a handcrafted stub and skeleton. The RMI equivalents are rather more complex, in terms of marshalling, error handling, and so on, but of course are far more robust. However, this will give us some insight into how the stub-skeleton mechanism might work in RMI. It also illustrates the point about all the extra processing involved, compared to just accessing a local `Order` object.

Suppose we have an order server that allows remote components to access its `Order` objects. Here is the definition of `Order`:

```
public interface Order {
  public String getProductName();
  public int getQuantity();
}
```

Note that it is an interface, not a class. This is because we need to abstract out an interface for both the client and server to share between them. The client-side needs the `Order` interface because it needs to know how to get the product and quantity, in this case by calling `getProductName()` and `getQuantity()`.

The real server-side object (called `OrderImpl`, below) needs to use the `Order` interface because rather than expose its inner workings, allowing remote objects to directly read its class variables, it hides the implementation details behind its two public methods. This gives the server-side the luxury of changing the way that name and quantity are stored, maybe in a database, properties file, or in our case just as private class variables, as we shall see later.

As well as defining two methods, we must also define the host where the server will reside, the port that the server and client will communicate over, and also two integer identifiers for the methods. We will place these in our `Comms` class. Both client and server will need to use the static variables in `Comms`:

```
public interface Comms {
  public static final String HOST = "localhost";
  public static final int PORT = 3456;

  public static final int GET_QUANTITY = 0;
  public static final int GET_PRODUCT_NAME = 1;
}
```

On the client-side, we need a stub class, `OrderStub`, which is an implementation of the `Order` interface. From the user's point of view it acts just like an `Order`, but under the covers, it actually connects to a remote skeleton elsewhere, to reach a remote implementation of an `Order`:

```
import java.net.Socket;
import java.io.ObjectOutputStream;
import java.io.ObjectInputStream;
import java.io.IOException;
import Comms;

public class OrderStub implements Order {

  public int getQuantity() {
    return Integer.parseInt(sendRequest(Comms.GET_QUANTITY));
  }

  public String getProductName() {
    return sendRequest(Comms.GET_PRODUCT_NAME);
  }

  private String sendRequest(int request) {
    String response = null;
    try {
```

We have used a socket for communication and also an `ObjectInputStream` and `ObjectOutputStream` to send and receive Java objects through that socket. (`ObjectInputStream` and `ObjectOutputStream` use serialization to convert objects into streams of byte data. They are performing our marshalling for us.) We send a single integer as a request to the skeleton class (described below), and receive a String in response. Obviously, a real stub class would need a far more complex interface to be usable, but for our demonstration this will suffice:

```
      Socket sock = new Socket(Comms.HOST, Comms.PORT);
      ObjectOutputStream oos =
        new ObjectOutputStream(sock.getOutputStream());
      oos.writeInt(request);
      oos.flush();

      ObjectInputStream ois = new ObjectInputStream(sock.getInputStream());
      response = (String) ois.readObject();
      oos.close();
      ois.close();
      sock.close();
    } catch (IOException ioe) {
      ioe.printStackTrace();
    } catch (ClassNotFoundException cnfe) {
      cnfe.printStackTrace();
    }
    return response;
  }
}
```

Secondly, for the client-side, we need a small piece of code to access the remote `Order` object:

```
public class Client {

  public static void main(String[] args) {
    Order localOrder = new OrderImpl("reclining chairs", 12);
    System.out.println("The local order is for " + localOrder.getQuantity()
                       + " lots of " + localOrder.getProductName());

    Order remoteOrder = new OrderStub();
    System.out.println("The remote order is for "
                       + remoteOrder.getQuantity() + " lots of "
                       + remoteOrder.getProductName());
  }
}
```

As can be seen from looking at the two `println()` statements, it is not possible to tell which `Order` is the local object, and which is the remote one, in the source code. The only difference is in the referencing of the `Order`; one uses a new statement, the other uses the `OrderStub` class to get its `Order` reference.

On the server-side we need to define an implementation of the `Order` interface, which we have called `OrderImpl`. This is the component holding the business logic for the `Order` class and contains simple `getProductName()` and `getQuantity()` methods. It is the 'real' order:

```
public class OrderImpl implements Order {

  String productName;
  int quantity;

  public OrderImpl(String name, int qty) {
    productName = name;
    quantity = qty;
  }

  public int getQuantity() {
    return quantity;
  }

  public String getProductName() {
    return productName;
  }
}
```

Also on the server-side, we need a skeleton class to handle the request made by `OrderStub`. Therefore, we need to listen for requests on the same socket (that is, the same port of the same server), and we need to translate the integer request into a real method call on the actual `ObjectImpl`, returning a String result to the `ObjectStub`. This class continually listens on the designated port for any client requests, and so uses the `ServerSocket` class:

```
import java.net.Socket;
import java.net.ServerSocket;
import java.io.ObjectOutputStream;
import java.io.ObjectInputStream;
import java.io.IOException;

public class OrderSkeleton {

  private OrderImpl theRealOrder;
  private String response;

  public OrderSkeleton(OrderImpl order) {
    theRealOrder = order;
    try {
      ServerSocket ss = new ServerSocket(Comms.PORT);

      while (true) {
        Socket sock = ss.accept();
        System.out.println("got a request");
        ObjectInputStream ois =
          new ObjectInputStream(sock.getInputStream());
        int request = ois.readInt();
        System.out.println("request id is " + request);

        switch (request) {
        case Comms.GET_QUANTITY:
          response = Integer.toString(theRealOrder.getQuantity());
          break;
        case Comms.GET_PRODUCT_NAME:
          response = theRealOrder.getProductName();
```

```
            break;
          default:
            response = null;
            break;
          }
          ObjectOutputStream oos =
            new ObjectOutputStream(sock.getOutputStream());
          oos.writeObject(response);
          oos.close();
          ois.close();
          sock.close();
        }
      } catch (IOException ioe) {
        ioe.printStackTrace();
      }
    }
  }
```

Finally, on the server-side, we have a simple executable class, `Server`, to create a remote `Order` object, and instantiate the listening skeleton:

```
public class Server {

  public static void main(String[] args) {
    OrderImpl aRemoteOrder = new OrderImpl("hot desks", 25);
    OrderSkeleton skeleton = new OrderSkeleton(aRemoteOrder);
  }
}
```

To test the example, we first start a single instance of `Server`, and then start one or more `Client` instances. You will see the clients receive and display the remote order's data. The `Server` will sit listening forever for client requests.

Our skeleton actually only serves a single order, but you can see this could be easily extended to request information about multiple orders, maybe with the client requesting orders by ID, date, and so on.

One point that might have occurred to you is that the explicit naming of the server name and port (`localhost` and `3456` in our case) for communication is inflexible when it comes to deploying the client and server. It would be far better to move these deployment details out of the code into some configuration file, or database. As we see in next example, this is exactly the sort of service that the RMI registry and JNDI offers – the location of distributed resources.

RMI Example

Now we understand the uses of a stub and skeleton in connecting clients and servers together, let's port our pseudoRMI example to the real RMI. There are several steps we need to do, but they are simple:

- Our `Order` interface must extend `java.rmi.Remote`, and its methods must all throw a `java.rmi.RemoteException`
- Our `Client` and `Server` classes need to handle the potential exceptions to match the revised `Order` interface

- Our `OrderImpl` must extend `java.rmi.server.UnicastRemoteObject`
- In the `Server` class, we need to register our server object (the instance of `OrderImpl`) with a naming service

Here are the revised source files. First, the `Order` interface:

```
import java.rmi.Remote;
import java.rmi.RemoteException;

public interface Order extends Remote {

  public String getProductName() throws RemoteException;
  public int getQuantity() throws RemoteException;
}
```

The client that creates a local order and also accesses a remote one:

```
import java.rmi.RemoteException;
import java.rmi.NotBoundException;
import java.rmi.Naming;
import java.net.MalformedURLException;

public class Client {

  public static void main(String[] args) {
    try {
      Order localOrder = new OrderImpl("reclining chairs", 12);
      System.out.println("The local order is for "
                         + localOrder.getQuantity() + " lots of "
                         + localOrder.getProductName());

      Order remoteOrder =
        (Order) Naming.lookup("rmi://localhost/aRemoteOrder");
      System.out.println("The remote order is for "
                         + remoteOrder.getQuantity() + " lots of "
                         + remoteOrder.getProductName());
    } catch (RemoteException re) {
      System.out.println("Error: problem looking up remote order");
      re.printStackTrace();
    } catch (MalformedURLException mue) {
      System.out.println("Error: incorrect URL format");
      mue.printStackTrace();
    } catch (NotBoundException nbe) {
      System.out
        .println("Error: resource not bound in naming lookup service");
      nbe.printStackTrace();
    }
  }
}
```

The actual implementation of `Order` on the server-side:

```
import java.rmi.server.UnicastRemoteObject;
import java.rmi.RemoteException;

public class OrderImpl extends UnicastRemoteObject implements Order {

  String productName;
  int quantity;

  public OrderImpl(String name, int qty) throws RemoteException {
    productName = name;
    quantity = qty;
  }

  public int getQuantity() throws RemoteException {
    return quantity;
  }

  public String getProductName() throws RemoteException {
    return productName;
  }
}
```

Finally, we have the class representing the server:

```
import java.rmi.RemoteException;
import java.rmi.Naming;
import java.net.MalformedURLException;

public class Server {

  public static void main(String[] args) {
    try {
      OrderImpl order = new OrderImpl("hot desks", 25);

      // register our order in the RMI registry so client can find it
      Naming.rebind("aRemoteOrder", order);
      System.out.println("\"aRemoteOrder\" is now in registry");
    } catch (RemoteException re) {
      System.out.println("Error creating remote order");
      re.printStackTrace();
    } catch (MalformedURLException mue) {
      System.out.println("Error: incorrect URL format");
      mue.printStackTrace();
    }
  }
}
```

Note that we no longer need our `Comms`, `OrderSkeleton`, or `OrderStub` classes. The underlying RMI handles the details we handcrafted in `Comms`, and the RMI equivalents to `OrderSkeleton` and `OrderStub` are automatically generated for us, by running the RMI compiler:

```
rmic -d classes OrderImpl
```

In the `classes\` directory, we will find two files created, `OrderImpl_Skel.class` and `OrderImpl_Stub.class`. (If you want to see the generated source, then we use the `-keep` option.)

Before starting the server, we must start a RMI registry service, which `Server` registers our `OrderImpl` with, so that the client application can find it:

```
cd classes
%JAVA_HOME%\bin\rmiregistry
```

Note that `rmiregistry` needs to find the generated classes, which is why we move to the directory holding them before starting `rmiregistry` (with the knowledge that the Java classpath always includes the current directory). We could have explicitly set the classpath, of course.

The RMI naming service uses port 1099 by default, but this may be overridden if it clashes with other services. There is a more detailed explanation of registry/naming services later in this chapter. Once the registry is running, start the server, and then one or more clients, and we will get the same result as before.

RMI is a synchronous communication API. When a request is made of a remote object, the calling thread waits for a reply from the remote end, via the stub object. It isn't a 'fire and forget' asynchronous API where we can come back later and pick up the reply at our leisure. We can code our application to do this, but it's probably better to use a standard asynchronous API rather than reinvent the wheel – see the Java Message Service (JMS) in the next chapter.

RMI-IIOP

The equivalent RMI-IIOP code differs slightly from the 'standard' RMI, in that JNDI (discussed below) is used for locating and registering objects, instead of the RMI registry, and `ObjectImpl` extends `javax.rmi.PortableRemoteObject` instead of `java.rmi.server.UnicastRemoteObject`.

The underlying runtime is quite different though, in that it is based upon a CORBA ORB, which allows our objects to communicate with nonJava components, and vice versa. We go no further in describing RMI-IIOP, but an RMI-IIOP example, with build and runtime instructions, can be found in the source download from http://www.wrox.com. This example is also very useful when reading the JNDI section in this chapter.

The final point to make is that RMI-IIOP is not available for J2ME clients, and RMI is an option only for CDC; not CDLC-based platforms. Probably the best way to bridge this gap, for CDC at least, is to write an RMI solution that connects to some proxy on the server-side. The proxy could do all the accessing of the remote objects using EJBs on behalf of the RMI client, and return results to the client accordingly. This proxy's API could be extended to serve not only RMI clients, but CLDC clients, WAP phones, and HTML devices too, assuming all the devices were wanting to reach the same server business objects.

Location

Once we can communicate between objects, whether it is by RMI, or other alternative methods described later, we need to be able to somehow specify what remote resource we want to access; be it a database, an EJB, or a message queue.

JNDI

The Java Naming and Directory Interface (JNDI), along with RMI, could arguably be considered as one of the bedrocks of J2EE. Its main purpose, from an application developer's point of view, is to allow us to access a remote resource without actually knowing its whereabouts.

This abstraction from the actual resource to a name in the JNDI directory (or database, or registry) means we avoid coding a specific service and machine IP address or name into our code, and instead, we look up the resource in the JNDI directory. It also allows the developer who wrote the back-end service to install it on a different server (or different hardware, language, etc.) and all he has to do is to change the JNDI entry details, leaving the identifying resource name the same.

If we were to request the EJB called 'Fred' then that's what we would get (or more accurately, a stub connecting us to the remote EJB 'Fred'), regardless of where it was located. This feature is related to the more familiar Domain Name Service (DNS) and Lightweight Directory Access Protocol (LDAP). JNDI is an API to access services such as these, rather than a naming service in its own right. The API defines universal access methods to naming services, rather than actually implementing a naming service itself.

The one resource we do have to know the name and location of is the JNDI directory itself, but information about all other resources should be retrieved from this for the maximum deployment flexibility.

Using JNDI is very simple. First, we need to obtain the reference to the JNDI service itself, which is known as the `InitialContext`. There are two ways of doing this, the first being to specifically define the JNDI properties required to reach the JNDI service, that is, the naming service's class, and its host and port number. In our example, we show the properties for JBoss:

```
import javax.naming.*;
...
  Properties prop = new Properties();
  prop.put(Context.INITIAL_CONTEXT_FACTORY,
                               "org.jnp.interfaces.NamingContextFactory");
  prop.put(Context.PROVIDER_URL, "localhost:1099");
  Context jndiContext = new javax.naming.InitialContext(prop);
```

Here are the equivalent properties for Sun's naming utility supplied with the JDK, `tnameserv`:

```
  Properties prop = new Properties();
  prop.put(Context.INITIAL_CONTEXT_FACTORY,
                                   "com.sun.jndi.cosnaming.CNCtxFactory");
  prop.put(Context.PROVIDER_URL, "iiop://localhost:900");
  Context jndi = new javax.naming.InitialContext(prop);
```

Your J2EE application server's parameters will no doubt differ, so you should read the server's documentation to find their specific properties.

The second way is to use some default JNDI properties. This has the advantage that the `InitialContext` code is now completely generic. It is simply one line:

```
InitialContext jndi = new InitialContext();
```

When using this method, the default JNDI properties are defined in a `jndi.properties` file placed in the path environment of your application (not the classpath, note). An example `jndi.properties` for our J2EE server, JBoss, would contain:

```
java.naming.factory.initial=org.jnp.interfaces.NamingContextFactory
java.naming.provider.url=localhost
java.naming.factory.url.pkgs=org.jboss.naming
```

Once we have the `InitialContext`, we can request a resource – an `Order`, say – with two lines of code:

```
Object obj = jndi.lookup("aRemoteOrder");
Order remoteOrder = (Order)PortableRemoteObject.narrow(obj, Order.class);
```

Note: do not directly cast the object returned by the `lookup()` method, always use `PortableRemoteObject.narrow`. Direct casting will probably work when the client and server are located on the same machine, but will not work across true distributed systems.

Here "`aRemoteOrder`" is an entry in the JNDI directory, placed there by an administrator, or automatically by a server application. Here is an example of placing an object into JNDI:

```
OrderImpl order = new OrderImpl("hot desks", 25);
// register our order in JNDI so client can find it
Properties prop = new Properties();
prop.put(Context.INITIAL_CONTEXT_FACTORY,
                                  "com.sun.jndi.cosnaming.CNCtxFactory");
prop.put(Context.PROVIDER_URL, "iiop://localhost:900");
Context jndi = new javax.naming.InitialContext(prop);
jndi.rebind("aRemoteOrder", order);
System.out.println("\"aRemoteOrder\" is now in registry");
```

Here, we see `rebind()` automatically overwrites any existing JNDI entry of the same name. In production, we would probably use `bind()`, which will throw an exception if a name is overwritten accidentally.

JNDI is very easy to use, but it can also be the source of grief, because when the JNDI `InitialContext` isn't found, everything falls around our ears. We recommend using the properties version of getting the `InitialContext` first, and then when that works, we can move to the generic single-line version, which is implementation independent.

There are plenty of tutorials on JNDI on the Web, covering how it compares to LDAP, and so on. A good place for us to start is with the JNDI FAQ at
http://java.sun.com/products/jndi/docs.html#TUTORIAL.

Now we have covered how to locate and communicate with remote objects, we can cover the more conventional areas of front-end presentation and back-end logic and data.

Front-end Logic and Presentation

For front-end request handling and presentation, J2EE holds no surprises for many Java developers. We can communicate directly with a client using sockets, RMI, or HTTP and HTTPS usually by using servlets and optionally JavaServer Pages (JSP).

Server Connectivity

In this section, we digress slightly from the pure J2EE angle and concentrate on how to connect to any back-end server, J2EE or otherwise. Our choices, as usual, are dependent on the flavor of J2ME for which we choose to develop. We have chosen to illustrate MIDP connectivity since it is the most restrictive profile currently available. However, note that MIDP classes are merely implementations of CLDC interfaces, so any profile based on CLDC will also be able to run these examples. It may be able to do rather more too, of course. The connectivity code will also run on CDC platforms.

As this is a Java book, we have chosen to demonstrate server connectivity by using servlets which are part of the J2EE suite. Apache's Jakarta Tomcat is the reference implementation for servlet enabled web servers, and is available from http://jakarta.apache.org/tomcat/.

There are four examples in this section:

- Reading content from a file
- Reading content from a file via a servlet, using the servlet to filter the content
- Writing content to a file via servlet
- Reading and writing to a JDBC-enabled database via a servlet

The first three examples can be run with any servlet-enabled web server. The fourth example needs a JDBC-enabled database. We recommend JBoss + Tomcat (available in a single download from http://www.jboss.org) as the easiest and least expensive solution for development and experimentation for J2EE, as well as these particular examples. It comes with Hypersonic as the default database. The User Manual (currently only on the Web.) describes how to configure JBoss for other commercial databases.

Although our examples are MIDP-based, all the connectivity-related code can be run using CLDC on a Palm OS compatible device, but with a different user interface in place of the MIDP elements. However, don't forget that 'plain' Palms (that is, anything but the wireless-enabled VII or VIIx) need to use 'testhttp', and need to have the `-networking` switch set when running `MakePalmApp`. Also, if we use the POSE rather than a physical Palm, then select the Redirect NetLib calls to host TCP/IP option in the Settings | Properties menu.

The lowest common denominator for J2ME connectivity, as we might expect, is the Web de facto standard protocol, HTTP. How the HTTP is carried over the air to our target device is not covered here – we just assume it gets there and back, whether by TCP/IP, WAP or other means. This means that we can take for granted that our HTTP requests will be serviced, regardless of the underlying transport.

MIDP 1.0 supports no other protocol, and so we will develop two examples that will run on MIDP, plus any other CLDC-based profile. Note that other CLDC profiles (including vendors' extensions to MIDP) are likely to support other protocols, such as HTTPS, ftp, datagram, and socket. In fact, as we were going to press, Motorola announced the availability of MIDP phones with HTTPS and socket support, thus adding 'value' to the standard MIDP 1.0. Specification.

Before we move onto the first server connected example, we refer you back to the CLDC basics (Chapter 5) to review the supported protocols other than HTTP, and the related `Connection` classes.

Basic J2ME Communication Example – Reading a File from a Web Server

We have chosen to build on an example application introduced in an earlier chapter – the Contact database (Chapter 6). The Contact database is made up of simple data types that we want to upload and download to our mobile device. A contact is made up of 5 fields:

- Name
- Phone
- E-mail
- Web URL
- Address

The format of the file is simple comma separated variables (CSV). Here are the contents of an example contacts file:

```
Richard Taylor,+447971000000,rct@poqit.com,www.poqit.com,1 Mow Meadows
Grace Charlesworth,01613390000,,,2 Mow Meadows
Joe Public,02071230000,joep@nowhere.com,www.nowhere.com,12 Acacia Avenue
```

Our first J2ME example simply reads a plain file of text from a remote server, such as the one above. It demonstrates how to write a mobile client with J2ME. As we're simply requesting a plain text file, there is no need for any bespoke server-side application – Java or otherwise. Our client gets the file contents by creating an HTTP request from the client and reading data in from the incoming byte stream.

To implement our simplest example, we need two components – an application at the mobile device end to request and display the content of a requested file, and an application at the server-end to supply the file data on request. Our server-end component is simply a web server.

Here is our J2ME application:

```
package com.poqit.j2me.client;

import java.io.InputStream;
import java.io.InputStreamReader;
import java.io.IOException;

import javax.microedition.midlet.MIDlet;
import javax.microedition.lcdui.TextBox;
import javax.microedition.lcdui.Command;
import javax.microedition.lcdui.CommandListener;
import javax.microedition.lcdui.Display;
import javax.microedition.lcdui.Displayable;
import javax.microedition.lcdui.TextField;
import javax.microedition.io.HttpConnection;
import javax.microedition.io.Connector;

public class FileRead extends MIDlet implements CommandListener {
```

```
private final static String URL =
  "http://localhost:8080/projavamobile/contacts.csv";
private static final int TEXTBOX_LENGTH = 256;
private Command exitCommand;
private Command okCommand;
private Display display;
private String file = "";

public FileRead() {
  display = Display.getDisplay(this);
  exitCommand = new Command("Exit", Command.EXIT, 1);
  okCommand = new Command("OK", Command.OK, 2);
}
```

The `startApp()` method allows the user to enter the URL of the file to be retrieved. A default URL is supplied in the `TextBox` on creation:

```
public void startApp() {
  TextBox t = new TextBox("Read file:", URL, TEXTBOX_LENGTH,
                          TextField.ANY);
  t.addCommand(exitCommand);
  t.addCommand(okCommand);
  t.setCommandListener(this);
  display.setCurrent(t);
}
```

`commandAction()` checks for two possible events; 'Exit' which stops the application, or 'OK' which fires off the request to the web server by calling `read()`. `read()` returns a result String which is then placed back in the `TextBox` to show to the user. The result can be the content of the file, or possibly an error message:

```
public void commandAction(Command c, Displayable s) {
  if (c.getCommandType() == Command.EXIT) {
    notifyDestroyed();
    return;
  }

  if (c.getCommandType() == Command.OK) {
    TextBox t = (TextBox) s;
    String result = read(t.getString());
    if (result.length() > TEXTBOX_LENGTH) {
      result = result.substring(0, TEXTBOX_LENGTH - 1);
    }
    t.setString(result);
    t.removeCommand(okCommand);
  }
}
```

The meat of the example is held in the `read()` method. Here, an `HttpConnection` is opened, using the URL typed into our `TextBox`. Note that there is no simple way to check the validity of the URL. Normally the `java.net.URL` class would be used, but it is not available in the CLDC or MIDP:

```
  public final static String read(String url) {
    String result = null;
    HttpConnection hc = null;
    InputStream is = null;
    try {
      hc = (HttpConnection) Connector.open(url);
```

We then set the request type to 'GET' (the other choices are 'POST' for sending data, and 'HEAD'). We also set some HTTP headers. Strictly, we do not have to do this, but this additional information could be of use to a servlet. For example, examining the platform header might allow a servlet to give a tailored response to the device (use of color information, a longer response message, etc.).

In our servlet, we merely print out the headers, to show that they reached the server. Note that the locale and platform system properties are not guaranteed to have a value, so we have to check that they aren't null before sending them:

```
hc.setRequestMethod(HttpConnection.GET);
String profiles = System.getProperty("microedition.profiles");
String configuration =
System.getProperty("microedition.configuration");
hc.setRequestProperty("User-Agent",
                      "Profile/" + profiles + " Configuration/"
                      + configuration);

String locale = System.getProperty("microedition.locale");
if (locale != null) {
  hc.setRequestProperty("Content-Language", locale);
}

String platform = System.getProperty("microedition.platform");
if (platform != null) {
  hc.setRequestProperty("J2ME-Platform", platform);
}
```

Once we have created the request, we just need to check the response code returned by the servlet to see if it is set to OK (a value of 200, defined in the HTTP specification), and then open the input stream and read the result sent back from the server. Note that we don't explicitly have to send the request because calling `getResponseCode()` causes the request to be sent (or in the parlance of the CLDC API, the state of the `HttpConnection` is 'Connected'):

```
      try {
        if (hc.getResponseCode() == HttpConnection.HTTP_OK) {
          is = hc.openInputStream();
```

Several other `HttpConnection` methods will also cause a waiting request to be sent, such as the `openInputStream()` method, `getType()`, and so on. The MIDP API documentation details the full list.

Having opened the `InputStream`, because we're only dealing with character data, we obtain an `InputStreamReader`, and read the characters in, line by line:

```
    InputStreamReader isReader = new InputStreamReader(is);
    int len = (int) hc.getLength();
    if (len > 0) {
      char[] chars = new char[len];
      int actual = isReader.read(chars);
      result = new String(chars);
    } else {
```

The code also allows for reading the response character by character, if the length of the response cannot be read in using `getLength()`. This can happen when the returned content is being created dynamically by the servlet:

```
          StringBuffer sb = new StringBuffer();
          int ch;
          while ((ch = is.read()) != -1) {
          sb.append((char) ch);
          }
          result = sb.toString();
        }
      } else {
        result = "Error: return code was " + hc.getResponseCode();
      }
    } catch (IOException e) {
      result = "Error: is the web server running?";
```

Finally, having got a result back, we close all the resources we used. Some exception handling has been added should our application fail. Error-handling tends to be pretty unreadable, but finding the errors is often even more painful:

```
          } finally {
            if (is != null) {
              is.close();
            }
            if (hc != null) {
              hc.close();
            }
          }
        } catch (IOException ioe) {
          result = "Error: IO exception.";
        }
        return result;
      }

      public void pauseApp() {}

      public void destroyApp(boolean unconditional) {}

      public void exit() {
        destroyApp(true);
        notifyDestroyed();
      }
    }
```

Deployment

Our first example can be run in conjunction with any web server, it doesn't even need to be a servlet-enabled one. However, before we go any further, we will look at how to set up Tomcat and the JBoss J2EE server here because we can use it for more sophisticated examples later, and in the Java Message Service chapter.

JBoss comes in two flavors – plain, and with Tomcat integrated. For our purposes, we need servlet capabilities, so we use the version that includes Tomcat. Much of the description below is for the Tomcat side of the server, and so can be applied to a plain Tomcat server if you do not want to use JBoss.

Jboss installation is simple. There are only two steps:

- Download and unzip the latest JBoss+Tomcat file from http://www.jboss.org
- Set the environment variable `%JBOSS_HOME%` to point to the JBoss directory inside the install directory, for example: `set JBOSS_HOME=C:\JBoss-2.2.2_Tomcat-3.2.2\jboss`

If JBoss doesn't start up correctly, you may need to change the following line in the `<JBOSS_HOME>\bin\run.bat`, from:

```
set JBOSS_CLASSPATH=%JBOSS_CLASSPATH%;run.jar
```

to:

```
set JBOSS_CLASSPATH=run.jar
```

That is, make JBoss ignore the current classpath.

When JBoss runs it, pumps out a lot of logging information into the console, which goes past far too fast to read. The majority of the output can be found in `<JBOSS_HOME>\log\server.log`, so we look there for any problems. Note that JBoss runs many services, which is why the log file is so complex.

Web Application Structure – the WAR file

The J2EE specification defines a structure to which web applications must adhere, and it also defines a deployment descriptor in XML format.

The web application structure has five sections:

- The public, or root, directory (for this application only)
- `WEB-INF\` directory
- `WEB-INF\web.xml` file
- `WEB-INF\classes` directory
- `WEB-INF\lib` directory

Here it is in graphical form:

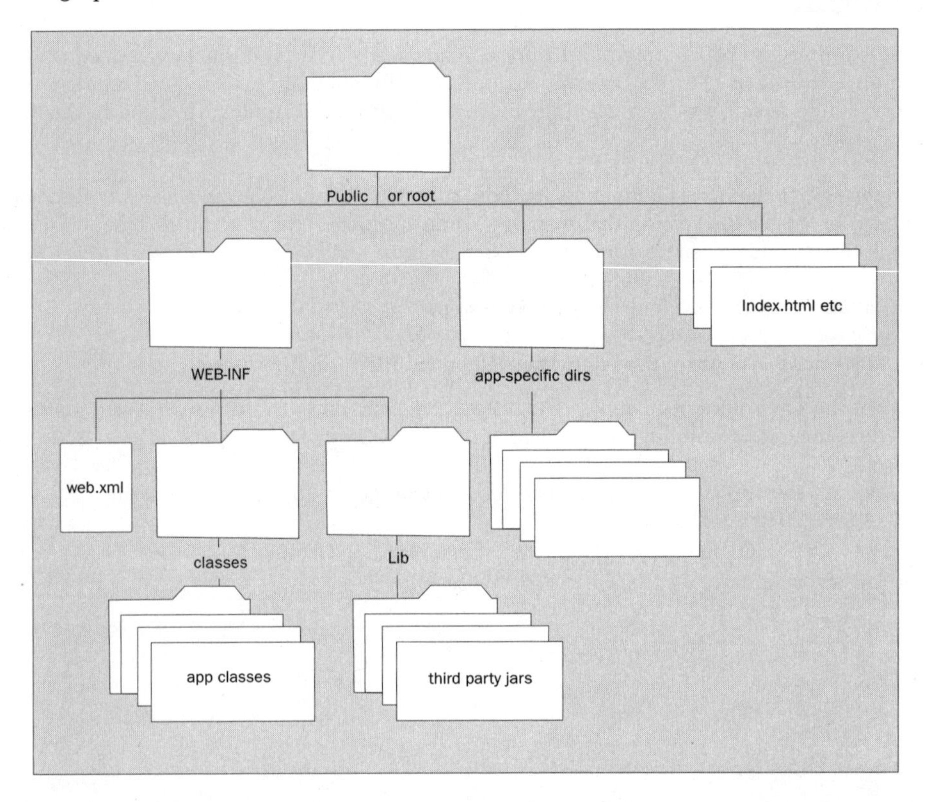

This is the structure that a web container (such as Tomcat) expects. Note that the `WEB-INF\` directory must be upper case. Obviously, delivering many individual files at once to an application server is not convenient, so the J2EE specification suggests that the files should be delivered as a JAR file, with the extension `.war` for Web ARchive. To create a WAR file, we simply move to the root directory (where the `index.html` and all the other files the application needs reside) and type:

```
jar -cvf mywarfile.war *
```

The resulting WAR file can then be handed over to a web container, such as Tomcat, for deployment. All we need to do is to place the WAR file in the directory `<TOMCAT_HOME>\webapps`, where `<TOMCAT_HOME>` is the directory where Tomcat is installed (this is within the JBoss directory structure if we are using JBoss+Tomcat). Now we start up JBoss (or stand-alone Tomcat), and the WAR file will be automatically detected, and unpacked ready to run.

The web.xml Deployment Descriptor File

The vital file within this potentially confusing structure is the `web.xml` file in the `WEB-INF\` directory. It contains the definition of our servlets, how to map those servlets to URLs, and other additional information. We do not have space to go into great detail about this file, but we show the very basic servlet setup here, which are our servlet examples in this chapter:

```
<?xml version="1.0" encoding="ISO-8859-1"?>

<!DOCTYPE web-app
    PUBLIC "-//Sun Microsystems, Inc.//DTD Web Application 2.2//EN"
    "http://java.sun.com/j2ee/dtds/web-app_2.2.dtd">

<web-app>

 <servlet>
  <servlet-name>ServletRead</servlet-name>
  <servlet-class>com.poqit.j2me.server.ServletRead</servlet-class>
 </servlet>

 <servlet>
  <servlet-name>ServletWrite</servlet-name>
  <servlet-class>com.poqit.j2me.server.ServletWrite</servlet-class>
 </servlet>

 <servlet>
  <servlet-name>ServletDB</servlet-name>
  <servlet-class>com.poqit.j2me.server.ServletDB</servlet-class>
 </servlet>

 <servlet-mapping>
  <servlet-name>ServletRead</servlet-name>
  <url-pattern>/ServletRead</url-pattern>
 </servlet-mapping>

 <servlet-mapping>
  <servlet-name>ServletWrite</servlet-name>
  <url-pattern>/ServletWrite</url-pattern>
 </servlet-mapping>

 <servlet-mapping>
  <servlet-name>ServletDB</servlet-name>
  <url-pattern>/ServletDB</url-pattern>
 </servlet-mapping>

 <mime-mapping>
  <extension>csv</extension>
  <mime-type>text/plain</mime-type>
 </mime-mapping>
</web-app>
```

The first three sections within the `<web-app>` element (the `<servlet>` tags) define which servlets are in our application, and map each name to a specific class. We could have called `ServletRead` `FredsMegaReader` – there's no limit, it's just a unique name that we use later in the file.

The second three sections (the `<servlet-mapping>` tags) map our servlets to specific URLs that the user must enter if they are to run our servlets. In our case, we just chose the same name, but with a leading forward slash. The `<url-pattern>` is added to the WAR name to make a legal URL. Here's an example for a WAR called `fred.war`, with this servlet and mapping definition:

```
<servlet>
  <servlet-name>FredsMegaReader</servlet-name>
  <servlet-class>com.poqit.j2me.server.ServletDB</servlet-class>
 </servlet>

 <servlet-mapping>
  <servlet-name>FredsMegaReader</servlet-name>
  <url-pattern>/MegaRead</url-pattern>
 </servlet-mapping>
```

To run the `ServletDB` servlet, the user would need to type in the following URL, assuming it is running on their local PC, `http://localhost/fred/MegaRead`.

A WAR file can be 'hot deployed' on some application servers, such as JBoss. Hot deploying means that the application server does not need to be restarted every time we adjust our code and build a new WAR – a useful trick when developing. We can just drop the WAR into JBoss's `deploy\` directory. JBoss spots the newer version of the WAR in the deploy directory, undeploys the old version, and replaces it with the new web application This hot deploy also works for EARs, or Enterprise ARchives, containing J2EE components, such as EJBs, etc.

On successful deployment, `<JBOSS_HOME>\log\server.log` will include something along these lines:

```
[Auto deploy] Auto deploy of file:/D:/jboss-tomcat-2.2/jboss-2.2/deploy/fred.war
[J2EE Deployer Default] Deploy J2EE application: file:/D:/jboss-tomcat-2.2/jboss-
2.2/deploy/fred.war
[J2EE Deployer Default] Create application fred.war
[J2EE Deployer Default] inflate and install module fred.war
[J2EE Deployer Default] Starting module fred.war
```

The servlet is now ready.

Servlet Example I – Filtered Reading From a File

In our second example, we will use a servlet as our server component. The assumption is that you have had some exposure to servlets at this point, since the aim of this book is to explain J2ME, not server-side technologies.

The Client side

To make the example more realistic, we will read from a file line by line, and use the servlet to perform some work on that data, rather than just deliver the raw content of the file to the J2ME device. We filter each line depending on the user's input search field:

```
package com.poqit.j2me.client;

import javax.microedition.midlet.MIDlet;
import javax.microedition.lcdui.Command;
import javax.microedition.lcdui.CommandListener;
import javax.microedition.lcdui.Display;
import javax.microedition.lcdui.Displayable;
import javax.microedition.lcdui.TextBox;
```

```
import javax.microedition.lcdui.TextField;

public class ServletRead extends MIDlet implements CommandListener {

  private final static String SERVLETURL =
                    "http://localhost:8080/projavamobile/ServletRead";
  private final static int TEXTBOX_LENGTH = 256;
  private final static int FILENAME_ENTRY = 0;
  private final static int SEARCHSTRING_ENTRY = 1;

  private int state;
  private Command exitCommand;
  private Command okCommand;
  private Display display;
  private String filename = "contacts.csv";
  private String searchString = "fred";
  private TextBox textBox;
  boolean filenameEntered;
```

This MIDlet requests specific portions of a file pointed to by the URL entered by the user. A servlet should receive this request, and do some work on behalf of the user, that is, return only the lines of the file that have the requested String in them. The application makes no attempt to format the returned file data:

```
  public ServletRead() {
    display = Display.getDisplay(this);
    exitCommand = new Command("Exit", Command.EXIT, 1);
    okCommand = new Command("OK", Command.OK, 2);
  }

  public void startApp() {
    state = FILENAME_ENTRY;
    textBox = new TextBox("Search in :", filename,
                          TEXTBOX_LENGTH, TextField.ANY);
    textBox.addCommand(exitCommand);
    textBox.addCommand(okCommand);
    textBox.setCommandListener(this);
    display.setCurrent(textBox);
  }

  public void commandAction(Command c, Displayable s) {

    // if 'Exit' key is pressed, then stop the MIDlet
    if (c.getCommandType() == Command.EXIT) {
      notifyDestroyed();
      return;
    }

    if (c.getCommandType() == Command.OK) {

      // has filename been entered by user?
      if (state == FILENAME_ENTRY) {
        state = SEARCHSTRING_ENTRY;
        filename = textBox.getString();
        textBox.setString(searchString);
        textBox.setTitle("For :");
```

```
        } else {
          searchString = textBox.getString();
          textBox.setTitle("Result :");
          doSearch();
        }
      }
    }
```

We will use the `FileRead` MIDlet's `read()` method, to save memory, since it is in the same MIDlet suite. The URL is built with the basic servlet URL, plus two parameters, `filename` and `search`:

```
    private void doSearch() {
      String result = FileRead.read(SERVLETURL + "?filename=" + filename
                                    + "&search=" + searchString);

      // Limit result length to the TextBox's capacity.
      if (result.length() > TEXTBOX_LENGTH) {
        result = result.substring(0, TEXTBOX_LENGTH - 1);
      }

      // place the result back in the TextBox, for displaying
      textBox.setString(result);

      // A run-once application, just let the user exit.
      textBox.removeCommand(okCommand);
    }

    public void pauseApp() {}

    public void destroyApp(boolean unconditional) {}

    public void exit() {
      destroyApp(true);
      notifyDestroyed();
    }
  }
```

The only difference in this application when compared to the first one is that the user has two lots of data to input, the filename, and the String to search for in each line of the file. We need to adjust the URL sent to the web server, to include the servlet's name (or rather the name mapped to our `ReadServlet` servlet), and also to include the two parameters entered by the user:
http://localhost:8080/projavamobile/ServletRead?filename=contacts.csv&search=fred

The Server side

Here is our servlet code that handles the requests received from the J2ME client (or any other source for that matter):

```
  package com.poqit.j2me.server;

  import javax.servlet.http.HttpServlet;
  import javax.servlet.http.HttpServletRequest;
  import javax.servlet.http.HttpServletResponse;
```

```
import java.io.BufferedReader;
import java.io.InputStream;
import java.io.InputStreamReader;
import java.io.PrintWriter;
import java.io.IOException;

public class ServletRead extends HttpServlet {

  protected void doGet(HttpServletRequest request,
                       HttpServletResponse response) throws IOException {
```

First, we open the `PrintWriter`, down which we return our response to the J2ME device. Then we check that the two parameters, `filename` and `search`, have been passed to us. If not, then we cannot complete the request sensibly:

```
    // get the Writer that we write our response to
    PrintWriter out = response.getWriter();

    // Read the parameters specified in the URL
    String filename = request.getParameter("filename");
    String searchString = request.getParameter("search").toLowerCase();

    if (filename == null) {
      out.println("You need to specify a file to search through.");
      return;
    }
    if (searchString == null) {
      out.println("You need to specify something to search for.");
      return;
    }
```

We then set the content type, by examining the MIME type mappings (for `.csv` in our case). At the J2ME end we can call `HttpConnection.getType()` and check that we really can display the resulting data:

```
response.setContentType(getServletContext().getMimeType(filename));

// work through the file, a line at a time, searching for the
// requested string
BufferedReader file = null;
try {
```

Having set up the response information, we can find the requested file. We want to be able to run our servlet from within a WAR, so rather than rely on the absolute URL requested, we must translate it using `getServletContext().getResourceAsStream(filename)`. (If you're using JBoss, then start it up and look in the `<JBOSS_HOME>\tmp\deploy\Default\j2me.ear\webxxxx` directory, and you will see the actual location of your deployed files.):

```
InputStream is = getServletContext().getResourceAsStream(filename);
if (is == null) {
  throw new IOException();
}
```

Once we have got the requested file, we can read it, line by line, using a `BufferedReader`, checking each line for the search String. We convert both strings to lower case before comparing, since users of real MIDP devices will not thank us for making our searches case-sensitive. Input on such devices is hard enough already:

```
        file = new BufferedReader(new InputStreamReader(is));
        String line;
        boolean foundEntry = false;

        // read the file in, a line at a time, checking for the end
        // of the file
        while ((line = file.readLine()) != null) {
          if (line.toLowerCase().indexOf(searchString) >= 0) {

            // found the requested string, so add this line to the response
            out.println(line);
            foundEntry = true;
          }
        }
        if (!foundEntry) {
          out.println("No matches were found.");
        }
      } catch (IOException ioe) {
        out.println("Couldn't find, or open, the file");
      }
      finally {

        // free up all our resources
        if (file != null) {
          file.close();
        }
        out.close();
      }
    }

    // Get Servlet information
    public String getServletInfo() {
      return "com.poqit.j2me.server.ServletRead Information";
    }
  }
```

Servlet Example II - Writing to a File

Now we have seen how to read from a file in two different ways, let's see how to write data from the mobile device to a file, again using servlets.

The Client side

The MIDP user input is very similar to the previous `ServletRead` example, but the user enters two lots of data, `filename` and `content`, which is the content to be saved to the named file. The major change is that we use `write()` in place of the previous example's `read()` method.

The first part of `write()` is the same as `read()`, because we still need an `HttpConnection`, and we still need to set the headers; but this time the request method is 'POST' instead of 'GET':

```
public final static String write(String url, String content) {
  String result = null;
  try {
    boolean status = false;
    HttpConnection hc = null;
    InputStream is = null;
    OutputStream os = null;

    hc = (HttpConnection) Connector.open(url);
    hc.setRequestMethod(HttpConnection.POST);
    String profiles = System.getProperty("microedition.profiles");
    String configuration =
                      System.getProperty("microedition.configuration");
    hc.setRequestProperty("User-Agent",
                      "Profile/" + profiles + " Configuration/"
                      + configuration);
    String locale = System.getProperty("microedition.locale");
    if (locale != null) {
      hc.setRequestProperty("Content-Language", locale);
    }

    String platform = System.getProperty("microedition.platform");
    if (platform != null) {
      hc.setRequestProperty("J2ME-Platform", platform);
    }
    try {
```

In place of `openInputStream()`, we want to write data, so we open an `OutputStream` instead, with `openOutputStream()`, and then obtain an `OutputStreamWriter`. We have only one line to write, so we only need call `writeln()` once, for a multiline content; we just loop until we have written the last line:

```
os = hc.openOutputStream();
OutputStreamWriter osWriter = new OutputStreamWriter(os);

osWriter.write(content);
osWriter.write('\r');
osWriter.write('\n');
osWriter.flush();   // Optional, openInputStream will flush
```

Having written the headers and body of our 'POST' request, we can check the response code, and if it's OK, open the `InputStream` to get the result back from the servlet. This is just the same code as the `ServletRead` example. Again, we don't need to explicitly `flush()` the 'POST' request, because calling `getResponseCode()` does this implicitly for us:

```
      // now read the result back from the servlet
      if (hc.getResponseCode() == HttpConnection.HTTP_OK) {
        is = hc.openInputStream();
        InputStreamReader isReader = new InputStreamReader(is);
        int len = (int) hc.getLength();
        if (len > 0) {
```

```
          char[] chars = new char[len];
          int actual = isReader.read(chars);
          result = new String(chars);
        } else {
          StringBuffer sb = new StringBuffer();
          int ch;
          while ((ch = is.read()) != -1) {
            sb.append((char) ch);
          }
          result = sb.toString();
        }
      } else {
        result = "Error: return code was " + hc.getResponseCode();
      }
    } catch (IOException e) {
      result = "Error: IOException.";
    }
    finally {
      if (is != null) {
        is.close();
      }
      if (os != null) {
        os.close();
      }
      if (hc != null) {
        hc.close();
      }
    }
  } catch (IOException ioe) {
    result = "Error: is the web server running, and SERVLETURL valid?";
  }
  return result;
}
```

The Server side

Here is the matching servlet:

```
package com.poqit.j2me.server;

import javax.servlet.ServletException;

import javax.servlet.http.HttpServletRequest;
import javax.servlet.http.HttpServletResponse;
import javax.servlet.http.HttpServlet;

import java.io.PrintWriter;
import java.io.FileWriter;
import java.io.BufferedReader;
import java.io.IOException;

public class ServletWrite extends HttpServlet {
```

This time, we cannot specify an absolute file address, so the servlet must know where to write the file to, in our case, this is into the directory named `WRITE_DIRECTORY`. The `WRITE_DIRECTORY` must exist when the servlet is run (although there is no reason why the servlet could not be written to auto-create the `WRITE_DIRECTORY` if it didn't). Also note it is not possible to save the file relative to the WAR. However, this is not a bad thing, since it will probably get deleted whenever the servlet WAR is undeployed; thus making us somewhat unpopular with users:

```
private final static String WRITE_DIRECTORY = "C:\\contacts\\";

public void doPost(HttpServletRequest request,
                   HttpServletResponse response) throws ServletException,
                   IOException {
```

These initial lines are just for interest. We don't use these, but they could be useful in another app. They could be useful for backward compatibility in the future, or to help decide the format and length of a response:

```
System.out.println("User-Agent is " + request.getHeader("User-Agent"));
System.out.println("Content-Language is " +
                   request.getHeader("Content-Language"));
System.out.println("J2ME-Platform is " +
                   request.getHeader("J2ME-Platform"));

// get the Writer that we write our response to
PrintWriter out = response.getWriter();

// Read the parameter specified in the URL
String filename = request.getParameter("filename");
if (filename == null) {
  out.println("You need to specify a file to save to.");
  return;
}
```

Before attempting to save any data, we first make sure we've been passed some data in the body of the HTTP request by calling `request.getContentLength()`.To write the data to a file, we create a new `FileWriter` passing in the absolute name of the file, that is, `WRITE_DIRECTORY` + `filename`. Our example will create a file if it doesn't exist, but it will not create any directory in the absolute pathname – this has to be done beforehand. Writing each line of read input is then done by calling `write(line)`:

```
// find the length of the data to be saved.
int len = request.getContentLength();
if (len < 1) {
  out.println("You need to enter some text to be saved.");
}
```

Once we've established there is data, we can obtain a `BufferedReader` using `request.getReader()`, and then read a line in at a time with `readLine()`, until there are no lines left:

```
    BufferedReader input = null;
    try {
      input = request.getReader();
      String line;
```

```
        FileWriter fw = new FileWriter(WRITE_DIRECTORY + filename);

        while ((line = input.readLine()) != null) {
          // write it away to the file
          fw.write(line);
          fw.close();
        }
        out.println("File saved successfully.");
      } catch (IOException ioe) {
        out
          .println("Couldn't create the file. Does WRITE_DIRECTORY exist?" +
                   " Is the file read-only?");
      }
      finally {
        // free up all our resources
        if (input != null) {
          input.close();
        }
        out.close();
      }
    }

    // Get Servlet information
    public String getServletInfo() {
      return "com.poqit.j2me.server.ServletWrite Information";
    }
  }
```

Database Access via a Servlet

This last example demonstrates database access via a servlet. Each contact is held in a single row of a database table, called `Contacts`. We use JBoss's default database, which we can find easily using JNDI. (Hey! – that naming/location stuff's getting useful suddenly.) We then use standard JDBC to connect to the database and create, read, and write to it as necessary.

The Client side

Our J2ME client is a little more intelligent than before, since it now offers three possible actions:

- Create the database table
- Write to the database table
- Read the entire database table

Note that we reuse the `read()` and `write()` methods defined in previous MIDlets. This is only possible if the MIDlets all belong to the same MIDlet suite, that is, that they all get loaded onto the device at once. This saves memory. If we had many more shared methods, then it would make sense to create a utility class (`ConnectorUtils.class`?) to hold them all in, and this class could then be included in any MIDlet suite that required it.

Here is the J2ME client code:

```
package com.poqit.j2me.client;

import javax.microedition.midlet.MIDlet;

import javax.microedition.lcdui.Command;
import javax.microedition.lcdui.CommandListener;
import javax.microedition.lcdui.Display;
import javax.microedition.lcdui.Displayable;
import javax.microedition.lcdui.TextBox;
import javax.microedition.lcdui.TextField;
import javax.microedition.lcdui.List;

public class ServletDB extends MIDlet implements CommandListener {
  private final static String SERVLETURL =
    "http://localhost:8080/projavamobile/servlet/ServletDB";
  private final static int TEXTBOX_LENGTH = 256;
  private final static String CREATE_ACTION = "create";
  private final static String WRITE_ACTION = "write";
  private final static String READ_ACTION = "read";
  private final static int REQUEST_NOT_SENT = 0;
  private final static int RESULT_RECEIVED = 1;

  private int state;
  private Command exitCommand;
  private Command okCommand;
  private Display display;
```

In our artificial example, the content must be in a specific format, 5 comma separated variables, with any VARCHAR data (Strings) being enclosed by single quotes:

```
  private String content =
    "'William Pear','01213543564','bill@pearshaped.com','www.pearshaped.com','2
The Grove'";
  private String action;
  private TextBox textBox;
  private List actionList;

  /**
   * POSTs data to a Servlet, for writing to a database.
   */
  public ServletDB() {
    display = Display.getDisplay(this);
    exitCommand = new Command("Exit", Command.EXIT, 1);
    okCommand = new Command("OK", Command.OK, 2);
  }

  public void startApp() {
    state = REQUEST_NOT_SENT;
    textBox = new TextBox("", "", TEXTBOX_LENGTH, TextField.ANY);
    textBox.addCommand(exitCommand);
    textBox.addCommand(okCommand);
    textBox.setCommandListener(this);
```

We use a `List` to display the three choices to the user. Only the 'write' action has any other user input – the content to save to the database:

```
    actionList = new List("Actions :", List.IMPLICIT);
    actionList.append(CREATE_ACTION, null);
    actionList.append(WRITE_ACTION, null);
    actionList.append(READ_ACTION, null);
    actionList.addCommand(exitCommand);
    actionList.setCommandListener(this);
    display.setCurrent(actionList);
  }

  public void commandAction(Command c, Displayable d) {

    // if 'Exit' key is pressed, then stop the MIDlet
    if (c.getCommandType() == Command.EXIT) {
      notifyDestroyed();
      return;
    }

    // OK pressed? (only needed for writing to database)
    // If so, send the data to the servlet
    if (c.getCommandType() == Command.OK) {
      if (state == RESULT_RECEIVED) {
        // Let user have another go...
        state = REQUEST_NOT_SENT;
        display.setCurrent(actionList);
      } else {
```

As before, once the user has entered the data, a request is fired off to the servlet, with the action being sent as an HTTP parameter named 'action'. For the 'write' action, the content above is placed in the body of the request:

```
        content = textBox.getString();

        // Use the write() method in ServletWrite.java
        String result = ServletWrite.write(SERVLETURL +
                                           "?action=write", content);
        if (result.length() > TEXTBOX_LENGTH) {
          result = result.substring(0, TEXTBOX_LENGTH - 1);
        }

        textBox.setTitle("Result :");
        textBox.setString(result);
        state = RESULT_RECEIVED;
      }
    }

    // check if the event was from a List object
    if (c == List.SELECT_COMMAND) {
      // extract the action string
      List l = (List) d;
      String action = l.getString(l.getSelectedIndex());
      if (action.equals(CREATE_ACTION) || action.equals(READ_ACTION)) {
```

Since there is no data to be passed, we can just use the `read()` method in `FileRead` to send our request, and get the result back:

```
          String result = FileRead.read(SERVLETURL + "?action=" + action);

          if (result.length() > TEXTBOX_LENGTH) {
            result = result.substring(0, TEXTBOX_LENGTH - 1);
          }
          textBox.setTitle("Result :");
          textBox.setString(result);
          display.setCurrent(textBox);
          state = RESULT_RECEIVED;
        }
        if (action.equals(WRITE_ACTION)) {
          textBox.setTitle("Data to save :");
          textBox.setString(content);
          display.setCurrent(textBox);
        }
      }
    }

    public void pauseApp() {}

    public void destroyApp(boolean unconditional) {}

    public void exit() {
      destroyApp(true);
      notifyDestroyed();
    }
  }
```

The Server-side

Our servlet, below, is significantly larger, to cope with the three options. Note that 'read' and 'create' are 'GET' requests, whereas 'write' is a 'POST' request, in other words, it contains data in the body of the HTTP request:

```
package com.poqit.j2me.server;

import javax.servlet.ServletException;

import javax.servlet.http.HttpServlet;
import javax.servlet.http.HttpServletRequest;
import javax.servlet.http.HttpServletResponse;

import java.io.PrintWriter;
import java.io.BufferedReader;
import java.io.IOException;

import java.sql.ResultSet;
import java.sql.Connection;
import java.sql.Statement;
import java.sql.SQLException;

import javax.sql.DataSource;
```

```
import javax.naming.InitialContext;
import javax.naming.NamingException;

public class ServletDB extends HttpServlet {

  // GET handles the 'create' and 'read' actions
  protected void doGet(HttpServletRequest request,
                       HttpServletResponse response) throws IOException {
    PrintWriter out = response.getWriter();

    String action = request.getParameter("action");
    if (action == null) {
      out
        .println("You need to specify 'read' or 'create' as the action parameter
in GET.");
      return;
    }

    if (action.equals("create")) {
      createTable(request, response, out);
    } else if (action.equals("read")) {
      readDB(request, response, out);
    } else {
      out.println(action + " is not a valid action parameter.");
    }
    return;
  }

  // POST handles the 'write' action
  public void doPost(HttpServletRequest request,
                     HttpServletResponse response) throws ServletException,
                     IOException {
    PrintWriter out = response.getWriter();

    String action = request.getParameter("action");
    if (action == null) {
      out.println("You need to specify 'write' as the action parameter.");
      return;
    }

    if (action.equals("write")) {
      writeDB(request, response, out);
    } else {
      out.println(action + " is not a valid action parameter.");
    }
    return;
  }
```

Each option, or action (create, write, read), has a method that performs the majority of the work of connecting to a database, creating a SQL statement to execute, and creating a reply String; be it the database content, or a status message:

```
  private void createTable(HttpServletRequest request,
                           HttpServletResponse response,
                           PrintWriter out) throws IOException {
```

```
    Statement stmt = null;
    ResultSet rs = null;
    Connection con = getConnection();

    try {
      stmt = con.createStatement();
      stmt.executeUpdate("CREATE TABLE contacts (" + "name VARCHAR (32),"
                         + "phone VARCHAR (15)," + "email VARCHAR (32),"
                         + "web VARCHAR (32)," + "address VARCHAR (64)"
                         + ")");
      out.println("contacts table created");
      stmt.close();
      con.close();
    } catch (SQLException sqle) {
      out.println("Table not created. Check the stack trace!");
      sqle.printStackTrace();
    }
  }

  private void readDB(HttpServletRequest request,
                      HttpServletResponse response,
                      PrintWriter out) throws IOException {
    Statement stmt = null;
    ResultSet rs = null;
    Connection con = getConnection();

    try {
      stmt = con.createStatement();
      rs = stmt.executeQuery("SELECT * FROM contacts ORDER BY name");

      // Read the data, a line at a time
      boolean hasEntries = false;
      while (rs.next()) {
        hasEntries = true;
        out.println(rs.getString("name") + "," + rs.getString("phone")
                    + "," + rs.getString("email") + ","
                    + rs.getString("web") + "," + rs.getString("address"));
      }
      if (!hasEntries) {
        out.println("database is empty");
      }
      stmt.close();
      con.close();
    } catch (SQLException sqle) {
      out.println("Data not fetched. Check the stack trace!");
      sqle.printStackTrace();
    }
  }

  private void writeDB(HttpServletRequest request,
                       HttpServletResponse response,
                       PrintWriter out) throws IOException {
    Statement stmt = null;
    ResultSet rs = null;
    Connection con = getConnection();
```

```
      // find the length of the data to be saved.
      int len = request.getContentLength();
      if (len < 1) {
        out.println("You need to enter some text to be saved.");
      }

      BufferedReader input = null;
      try {
        input = request.getReader();
        String line;
        try {
          stmt = con.createStatement();
```

Here we read the input in, a line at a time, checking for the end of the data. In our case, there is only one line:

```
con.setAutoCommit(false);
while ((line = input.readLine()) != null) {
```

We assume the data is in CSV (comma separated values) format, with any String values surrounded by single quotes. We then add a Mickey Mouse check to make sure there are 4 commas in the line. If not, the SQL `INSERT` will fail:

```
            if (commaCount(line) == 4) {
              stmt.executeUpdate("INSERT INTO contacts VALUES (" + line
                                            + ")");
              out.println("Data saved OK.");
            } else {
              out.println("The data must be 5 items, separated by 4 commas." +
                          " Only" + commaCount(line) + " were found");
            }
          }
          con.commit();
          // free up the resources we've used.
          stmt.close();
          con.close();
          input.close();
        } catch (SQLException sqle) {
          out.println("Data not saved. Check the stack trace!");
          sqle.printStackTrace();
        }
      } catch (IOException ioe) {
        out.println("Couldn't create the file. Does WRITE_DIRECTORY exist? " +
                    " Is the file read-only?");
      }
    }
```

For the server-side code, we need access to a JDBC-enabled database from within the servlet:

```
  private Connection getConnection() {
    try {
      Connection con = null;
      try {
```

As this example is written to run on a JNDI-enabled J2EE application server (JBoss), we can obtain a default database `Connection` using JDNI to discover a default database, in three simple lines. If you use another application server, then just use the standard method of obtaining a database `Connection` to your specific database:

```
        InitialContext jndiCntx = new InitialContext();
        DataSource ds = (DataSource) jndiCntx.lookup("java:/DefaultDS");
        con = ds.getConnection();
        return con;
      } catch (NamingException ne) {
        System.out.println("NamingException: Couldn't find the"
                           + " default database");
      }
    } catch (SQLException e) {}
    return null;
  }

  // Simple error-checking of data being saved.
  private int commaCount(String line) {
    int count = 0;
    int pos = line.indexOf(',');
    while (pos != -1) {
      count++;
      pos = line.indexOf(',', pos + 1);
    }
    return count;
  }

  // Get Servlet information
  public String getServletInfo() {
    return "com.poqit.j2me.server.ServletDB Information";
  }
}
```

It is worth noting that the error handling in this example is minimal so there are a lot of things to go wrong. However, in this case, the code really does become difficult to read. The sort of things we need to code for are closing resources regardless of the sort of errors that occurred. This is best done within a `finally` clause, since any code within `finally` is guaranteed to be run.

The other thing for us to consider at this point is that if we're really going to get serious about database access, then rather than writing reams of JDBC and error-checking code, we might want to consider using Enterprise JavaBeans (EJBs). It's vital that these applications are easy to use – our users will not thank us for sending them funky debug messages when they're in dire need of information/services. Using EJBs lets us separate the raw database access code from the servlet, and we can create a true three-tier (or 'n' tier) application with greater stability and reliability, probably along the lines of the model-view-controller design pattern.

EJB Access via a Servlet

As we saw at the beginning of this chapter, space does not allow us to cover EJB in any detail that would do justice to the complexity of the subject. However, we can say that EJB access via servlets is straightforward (assuming we know EJBs, that is) and has no effect on the J2ME client end.

We can still use the same techniques here to request and deliver data to the target device. For example, for a simple application that retrieves EJB objects (named `Contact` in our case), rather than call `executeQuery()` to fetch data from the database, we would replace the reading of a `ResultSet` code, below:

```
while (rs.next()) {
  hasEntries = true;
  out.println(rs.getString("name") + ","
              + rs.getString("phone") + ","
                   + rs.getString("email") + ","
                   + rs.getString("web") + ","
                   + rs.getString("address") + ","
                   + rs.getString("notes"));
}
```

with a call to `ContactHome.findAll()` on the EJB object's home interface (called `Contact`), and iterate through the `Enumeration` of `Contact` objects that it returns, like this:

```
// obtain the Home interface of ContactBean
Context jndiContext = new javax.naming.InitialContext();
      Object contactRef = jndiContext.lookup("ContactBean");
      ContactHome contactHome = (ContactHome)
PortableRemoteObject.narrow(contactRef,ContactHome.class);

// Now iterate through all the contacts
Enumeration contacts = contactHome.findAll();
      while (contacts.hasMoreElements())
      {
        Contact contact =
(Contact)PortableRemoteObject.narrow(contacts.nextElement(), Contact.class);
        // list contact details
     out.println(contact.getName + ","
                   + contact.getPhone() + ","
                   + contact.getEmail() + ","
                   + contact.getweb() + ","
                   + contact.getAddress() + ","
                   + contact.getNotes());
      }
```

The above code requires us to define the `Contact` and `ContactHome` classes, define the attribute mapping in the `ejb-jar.xml` file, and to have a database integrated with our application server. The setting up of EJBs is tedious, but once we have them working, our code becomes much easier to write and maintain, plus we get all the benefits from a true distributed framework supporting us.

Advanced Connectivity Topics

We now swiftly visit a few areas which we must consider when writing commercial quality applications. These items may not necessarily win us customers, but if we don't consider them, we may well lose some.

User Feedback

Mobile users like fast response times – 40 seconds can feel like a lifetime when you just want to know your bank balance. We should try and keep the user informed of progress, so they don't think the application has crashed. One way, for our examples, would be to display a progress bar (using the low-level graphics API) and to run the upload/download connectivity code as a separate thread. The threading model in CLDC and CDC is the same as J2SE, so there should be no surprises for us.

Be aware however, that not all devices are guaranteed support true threading underneath the Java facade, so we should be sure to test on the real target devices before releasing our service. We should be able to identify the device our code is running on by calling: `System.getProperty("microedition.platform")`, and deciding whether to thread or not from that information.

Predictive Loading of Data

Another touch that users appreciate is the ability to preload data in the background. Again, as above we need to use a separate thread for the connectivity code. We could, for example, preload some data or images to the J2ME client whilst they are busy typing in their username and password, which for the device, is effectively dead time, since user input is not particularly CPUintensive.

However, before nipping off to rewrite your application to perform background downloads whenever idle, don't forget that the up-and-coming 2.5G and 3G networks will most probably charge by bytes downloaded rather than the time on-line. We should give our users the option of downloading in the background, but shouldn't force it upon them.

Resuming Uploads and Downloads

There is no direct support for this in HTTP, so if we want to resume an interrupted upload or download, we will need to code our receiving application, be it the J2ME client or the servlet, to:

- Save the received data, chunk by chunk
- Remember the position in the download that it has saved so far

In our case, we could save lines one at a time, and keep a count of the lines processed. Of course, the transmitting end also needs to know about how far the download got before being interrupted, so we might pass the line number to start at an HTTP parameter, for example.

With communication time being probably the most valuable and slowest resource in the current 2G networks and the user's time being of even higher priority, this sort of defensive coding will really set our application apart from the rest. It's definitely worth spending a lot of time getting the client and server-side as efficient as possible, thus saving users a lot of their time, not to mention money.

Note that ftp supports resumption of data transfers, so if our device supports this protocol, we should use it in preference to HTTP wherever possible.

Data Formats

There is no limit to the type of data that we can send and receive via a servlet, although we should use the `Accepts` header and `Content-Type` to tell the servlet what our client can accept. For example, XML can be passed to and from the J2ME client, and several XML parsers are available for us to use. The resources towards the end of Chapter 14 show the relevant URLs.

JSP

JavaServer Pages are an enhancement of servlets, and offer similar features to Microsoft's Active Server Pages (ASP). We can embed the servlet code, and custom JSP tags, in HTML (or WML, XML, etc.) template pages. When that page is requested, the tags are replaced by dynamic data/code, the embedded code is extracted, and a servlet is created dynamically from that code (assuming there is no previous instance of that servlet).

Following on from the above example, here is a simple log-in response, where a JSP tag is used to get the user's name from the request URL, and embed it into the response page:

```
<html>
  <h1>Welcome <%= request.getParameter("user") %></h1>
  <a href="balance.html">Balance</a><br/>
  <a href="transactions.html">Transaction</a><br/>
  <a href="index.html">Log out</a>
</html>
```

Servlets can communicate with remote objects; (EJBs for example,) to access data or services, and in large systems typically do not contain business logic themselves.

Although servlets are normally used to deliver markup language over HTTP/HTTPS, there is no reason why we cannot deliver plain data over HTTP, especially if HTTP is the only `Connection` supported, as is true for MIDP 1.0. For more details on JSP, see "Professional JSP, 2nd Edition" (Wrox Press, ISBN 1861004958).

Alternative Communications

In addition to the basic RMI-IIOP protocol for remote object access, a J2EE-compliant platform must also support other communication protocols:

- Java Message Service (JMS) for message oriented middleware
- JavaMail for mail applications

JMS

The Java Message Service is an asynchronous message based communication API, as opposed to RMI's synchronous communication. Messages sent to recipients are sent without waiting for a reply, and they are not necessarily received immediately – they can be stored on a server until the recipient is ready to receive them and deal with the contents.

The sender and receiver are applications, not individual users. Also note that the applications don't have to be Java at both ends. This is because the underlying architecture, the message server or 'provider', is a third party application, which typically supports other language interfaces. This arrangement of JMS being an interface to some third party software is very similar to the way JDBC operates – it gives a Java interface into a general resource/server.

Since we cover JMS in much greater detail in the following chapter, we refer you to it for more detail, and code examples.

JavaMail

JavaMail is a specification that abstracts out the different communication protocols (POP3, IMAP, SMTP, etc.) into a neat and tidy Java API. This allows us to potentially interact with any messaging system through this single API. JavaMail is only available for pJava (and by extension the Personal Profile), and not CLDC based platforms. All that needs saying is that JavaMail development for CDC is no different compared to a standard J2SE platform. See Wrox's "Java Server Programming, J2EE Edition" for more details.

The Future

As you can tell from this chapter, J2ME-J2EE connectivity is somewhat patchy at the time of writing. J2EE expects its clients to have significant Java resources available (libraries, memory, VM, etc.) which limits us to what we can use right now. The good news is that HTTP communication is easy on all J2ME platforms, and we can expect more protocols, such as sockets and ftp, to follow soon – whether as the vendor's own components or as new, or revised, profiles.

Probably the major missing component is RMI-IIOP, which stops us directly connecting to Enterprise Java Beans. Maybe Sun will come up with a lightweight equivalent. Let's hope so – transparent connectivity to enterprise systems would not only save much manual coding, but it would also increase the reliability of the whole application, end to end.

Summary

We have described the underlying basics of J2EE – RMI and JNDI, and have seen a pseudoRMI example that could be used on J2ME devices.

Our parting thought is that true J2EE to J2ME connectivity is not generally available at the time of writing (Q2 2001), but that certain facets of distributed application development are possible, and that faster devices and more comprehensive profiles will help speed up this area of development in the very near future. The next chapter will expand on the J2EE theme, and will explore JMS and J2ME in more detail.

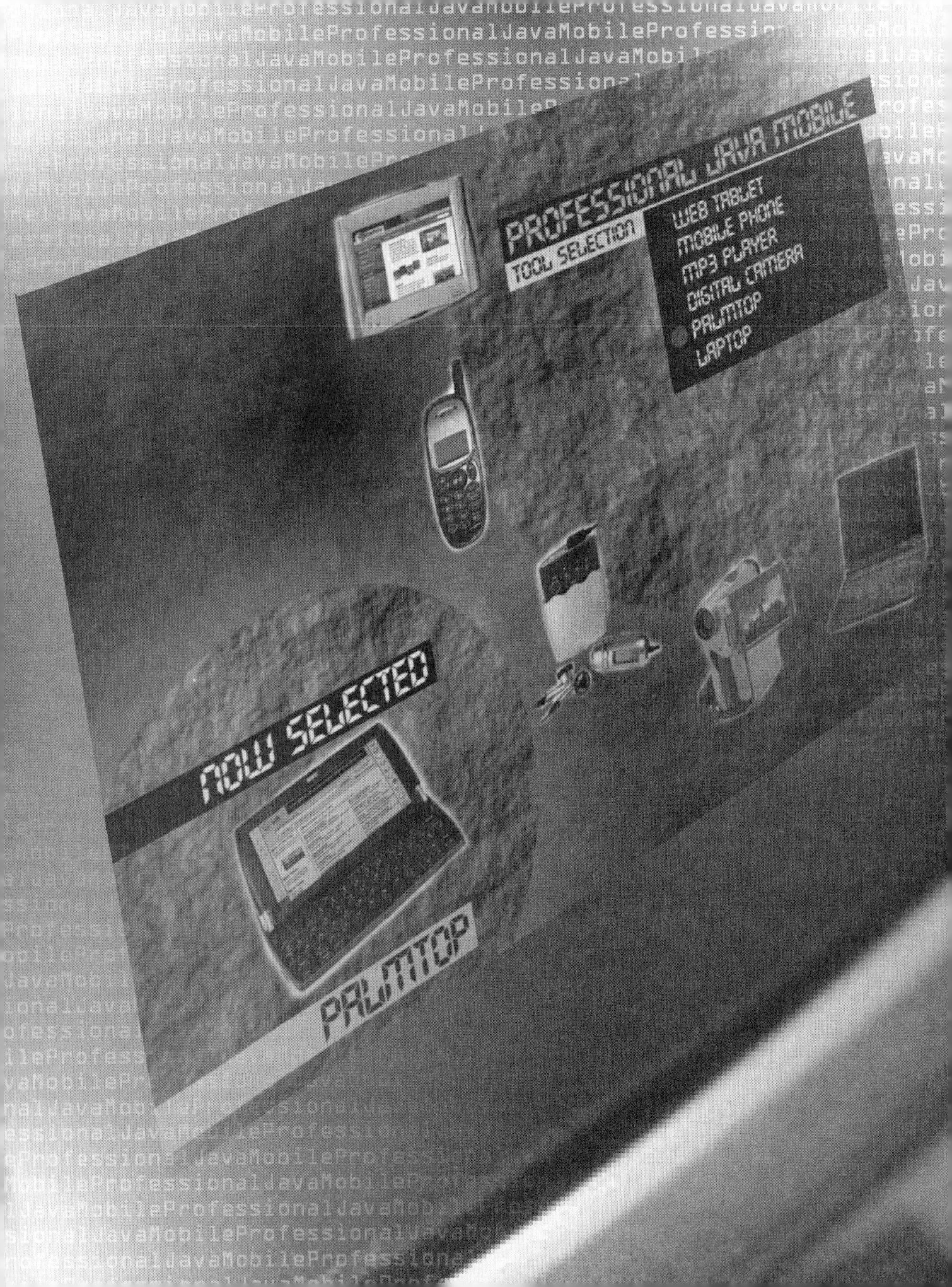

PROFESSIONAL JAVA MOBILE
TOOL SELECTION
WEB TABLET
MOBILE PHONE
MP3 PLAYER
DIGITAL CAMERA
PALMTOP
LAPTOP
NOW SELECTED
PALMTOP

12

Messaging and JMS

Having covered the majority of J2EE components relevant to mobile devices, we now concentrate on a specific component, the Java Message Service (JMS). First, we will introduce the concepts behind 'messaging' and cover the basics of JMS. In particular we will be looking at the concept of an asynchronous message. We will then go on to create a generic (J2SE) example application, which we then port to a CLDC-based mobile platform (a Palm) using a third-party JMS solution, iBus//Mobile from Softwired Inc.

Having created two versions of the same application, we will consider what limitations the J2ME platform has placed on the iBus//Mobile authors, and how that affects our J2ME applications. Finally we will see what the future of Java messaging holds.

The sample application we are going to build is an auction house bidding system. The auction house can send out new and updated lots to a multitude of potential bidders. Individual bidders can then place their own bid to usurp the current one. This is then broadcast out to the other bidders, and so on. To compile and run the two application variants described here, you need the following downloads and tools. All but the JMS-specific tools have been introduced in previous chapters:

- MIDP – www.java.sun.com/products/midp/.
- KAWT – http://www.kawt.de.
- A Palm device or Palm emulator. If you choose to use the emulator, you must also obtain a ROM image, either from a physical Palm, or by requesting one from Palm: http://www.palm.com. You will need PalmOS 3.5 or higher because 3.0 doesn't have enough heap space to run our second example. (If you're using a Palm simulator, don't forget to enable network access – right-click anywhere in the simulator and select Settings | Properties and select the Redirect NetLib calls to host TCP/IP checkbox.)

- A generic messaging server. We recommend JBoss, which includes the JBossMQ server – http://www.jboss.org.
- iBus Standard Message Server – http://www.softwired-inc.com.
- iBus//Mobile Gateway and CLDC Libraries – http://www.softwired-inc.com.

Note: Sun has yet not specified a J2ME Configuration Profile that includes JMS capabilities. Softwired (www.softwired-inc.com) have filled the gap with their iBus//Mobile solution, which runs on both CDC and CLDC platforms. Although the second example presented here uses Softwired's proprietary libraries (iBus//Mobile has a few restrictions when compared to the JMS specifications, but as versions progress it is intended that these restrictions will disappear), messaging is likely to be an important cornerstone in future mobile applications, and we feel it is well worth understanding what it can offer.

What is an Asynchronous Message?

When we talk about an asynchronous message we are talking about one where the sender and recipient are not by necessity connected to the network at the same time. In circumstances where both devices are always connected to the same network, and their locations are known, there is no great reason for considering asynchronous communications. When devices move between networks, connecting, disconnecting, and changing their locations (as mobile devices are prone to do) then it is often better to consider disconnected communication passing through a third party, particularly if the mobile device is not the one initiating the communication.

One of the main problems with sending out messages, rather than establishing a connection, is confirming their receipt. Various radio networks, including pager networks, have discovered to their cost that customers prefer a system that offers guaranteed delivery as opposed to device size or cost. The replacement of pager networks in Europe with GSM mobile phones is testament to this.

The Short Message Service

SMS was not considered central to the GSM protocol, originally being developed for the sending of handset configuration information and was more of a footnote than a true part of the protocol. The addition of the ability for users to send their own messages to each other was only an afterthought. However, it has proved spectacularly successful, with over a billion messages sent every month in Europe alone (http://www.gsm.org). This has proved a great revenue generator for the networks, who charge for every message sent. Its popularity can be attributed to several factors:

- **Fixed Cost**
  While the cost of calls may vary, the cost of a single SMS is fixed. This particularly impacts on prepaid phone owners, who only have a limited credit stored on the phone.
- **Privacy, or Politeness**
  The sending of an SMS is a private affair, which may be carried out in public. While using a mobile phone on a bus or train is considered impolite, and offers no privacy at all, the sending and receipt of SMS messages is both private and non intrusive.
- **Non Invasive to Recipient**
  Much of the value of e-mail, as a form of communication, is that it is non-invasive. The recipient can deal with the mail when they wish, and won't be distracted from whatever they are doing. SMS offers similar functionality in that the recipient can ignore the signal until they are happy to deal with it.

- **Reliability**
  The GSM network is, generally, very reliable, and SMS inherits much of that reliability. When an SMS message is sent it is stored at a gateway, before being forwarded to the receiving handset. As the handset can confirm receipt (unlike a pager message) it is possible for the gateway to wait until the handset is available before sending the message. Obviously the gateway can't hold on to messages forever, and most will retain them for a couple of days before giving up, normally enough time to reach the handset.

These features have led to the incredible popularity of SMS, despite the primitive text entry systems available to SMS users. This popularity has now extended to the point where we are starting to see devices being marketed as SMS terminals, using the GSM protocol, but supplying no voice capability!

It is interesting to note that SMS is the only example application we will be discussing that genuinely offers asynchronous messaging. The sender of a message may disconnect from the network, in the confidence that the network itself has now taken responsibility for the message delivery. In this way, it is similar to e-mail, and is sometimes termed a "**Store and Forward**" protocol in that the message is sent to a server that stores it, and forwards it when the necessary connectivity is available. Of course, when an Internet e-mail message fails to be delivered a message is sent back to the sender reporting this fact, something still lacking from the SMS networks.

Types of Asynchronous Messages

It is important to understand the two distinct ways in which asynchronous messages can operate. The definition of an asynchronous message is one where the sending and receipt are not simultaneous, and while datagram communication obviously fits this, there is another way to transmit messages to a device without opening a connection directly to that device. These Store and Forward networks are exemplified by the e-mail system we all use every day.

Store and Forward message systems are certainly the most popular, though datagram messages have their place in certain environments.

Store and Forward Networks

The concept of a Store and Forward messaging system is very easy to understand, though often more difficult to deploy. When considering Internet e-mail, the process is controlled through a protocol called **Simple Mail Transport Protocol**, or **SMTP**. An originating machine creates a message, and addresses it. It then passes the message on to either its destination or another machine, which will then take responsibility for the delivery of the message.

This process of passing messages from server to server on their way to their destination is easy to see in the headers of most e-mail messages, as the example below illustrates.

Message now available for downloading from Demon Internet:

```
Received: from punt-2.mail.demon.net by mailstore for bill@drs8800.demon.co.uk
      id 977488839:20:14746:1; Fri, 22 Dec 2000 12:40:39 GMT
```

Messaging received by Demon's incoming SMTP server:

```
Received: from anarchy.io.com ([199.170.88.101]) by punt-2.mail.demon.net
        id aa2106584; 22 Dec 2000 12:40 GMT
```

Message received by Anarchy, the outgoing SMTP server of io.com:

```
Received: from deliverator.io.com (root@deliverator.io.com [199.170.88.17])
        by anarchy.io.com (8.9.3/8.9.3) with ESMTP id GAA05423
        for <bill@drs8800.demon.co.uk>; Fri, 22 Dec 2000 06:40:03 -0600
```

Message received by io.com incoming SMTP server from wrox3 outgoing server:

```
Received: from wrox3.mail.wrox.co.uk (mail.wrox.co.uk [212.250.238.65])
        by deliverator.io.com (8.9.3/8.9.3) with ESMTP id GAA13779
        for <bill1@io.com>; Fri, 22 Dec 2000 06:40:02 -0600
```

Message received by Wrox internal SMTP server, this mail was sent by a Wrox employee for delivery outside the company:

```
Received: by WROX3 with Internet Mail Service (5.5.2448.0)
        id <WSANMX06>; Fri, 22 Dec 2000 12:43:58 -0000
```

This example is slightly confused by the fact that the times don't match for delivery. While io.com headers are marked with `-0600` to indicate they are 6 hours behind GMT, the lack of properly set clocks on servers can make it very hard to track how long a message took to be delivered. This message actually took about 8 minutes from sending to receipt, and points out a very important aspect of Store and Forward networks, delivery times are generally not guaranteed (at least, not on TCP/IP networks) which makes them less useful in mission-critical applications.

It would be possible to create a time-sensitive Store and Forward mechanism, though as IP has no guaranteed Quality of Service (delivery times) such an exercise would be of dubious value over the Internet and other networks may provide features to make such a service possible.

Of course, the big advantage of Store and Forward communications is that the recipient does not have to actually be connected to the network to receive the message; it should be waiting for them next time they connect. This means that the sender can send at their convenience and the message will be delivered at the recipient's convenience. This can be particularly useful with regard to mobile devices.

Example

Sending a message to a Store and Forward system is just like any other network programming, as the connection we are making to the server is a standard networking connection. The JavaMail API provides a comprehensive system for managing e-mail accounts, but just sending a mail is very simple (as stated by the protocol title, **Simple** Mail Transfer Protocol).

The following code shows how an SMTP mail message can easily be sent, though it ignores the responses from the server for the sake of clarity. This example uses the `getResponse()` method that is also used in Chapter 8:

```
Socket sock = new Socket("192.168.1.11",25);
InputStream in = sock.getInputStream();
OutputStream out = sock.getOutputStream();
out.write(("HELLO drs8800.demon.co.uk" + CRLF).getBytes());
System.out.println(getResponse(in));
```

```
out.write(("MAIL FROM: bill1@io.com" + CRLF).getBytes());
System.out.println(getResponse(in));
out.write(("RCPT TO: bill1@io.com" + CRLF).getBytes());
System.out.println(getResponse(in));
out.write(("DATA" + CRLF).getBytes());
System.out.println(getResponse(in));
out.write(("Here is my message" + CRLF).getBytes());
out.write(("Sent from my Java program" + CRLF).getBytes());
out.write(("." + CRLF).getBytes());
System.out.println(getResponse(in));
out.write(("QUIT" + CRLF).getBytes());
System.out.println(getResponse(in));
```

Clearly, if you planning your own system you will be defining your own protocol, though the SMTP network can provide a very useful deployed system for many applications. Using multiple servers may well be excessive, unless you are planning a worldwide network.

It should be noted that e-mail is addressed to an SMTP server that is assumed to be connected to the network at all times. While SMTP can cope with outages, and was designed to work with servers that were only connected daily, on the current Internet we generally use an e-mail client to collect the messages from the local SMTP server to our own devices using POP3 (Post Office Protocol version 3) or similar.

It is perfectly possible to run an SMTP server on a desktop machine, resulting in mail being pushed to the machine instead of pulled down by a mail client. It would also be perfectly possible to run such a server on a mobile device, providing push functionality while conforming to one of the oldest Internet standards. The only problem is knowing the IP address of the device you want to send your message to.

Tracking IP Addresses

Every computer connected to the Internet has an **IP**, or **Internet Protocol**, address, often known as a "**dotted-quad**". While the details of IP are well beyond the scope of this book, it is important to understand that every device has one, and it is only by knowing this address that we can establish any kind of communication with a device using the Internet.

Finding the IP address of a device is generally very easy. From the command line under Microsoft Windows, you can either use `ipconfig` or `winipcfg`, depending on the version:

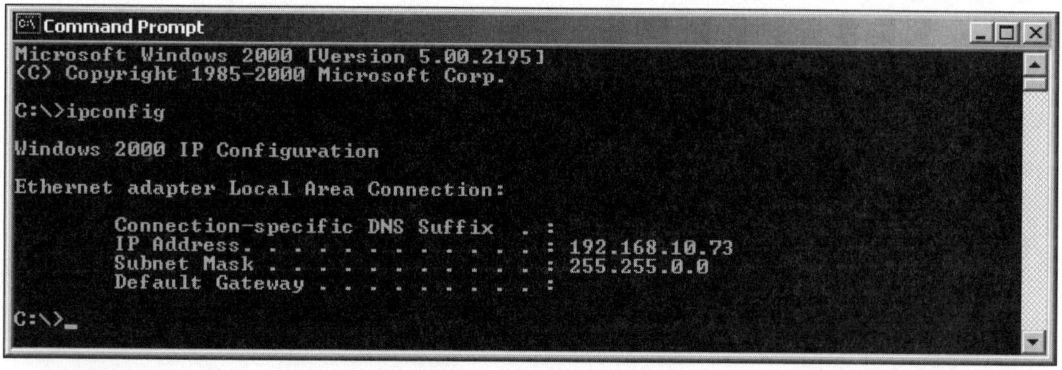

```
Command Prompt
Microsoft Windows 2000 [Version 5.00.2195]
(C) Copyright 1985-2000 Microsoft Corp.

C:\>ipconfig

Windows 2000 IP Configuration

Ethernet adapter Local Area Connection:

        Connection-specific DNS Suffix  . :
        IP Address. . . . . . . . . . . . : 192.168.10.73
        Subnet Mask . . . . . . . . . . . : 255.255.0.0
        Default Gateway . . . . . . . . . :

C:\>_
```

In Java we can use the `getLocalHost()` method of `java.net.InetAddress`:

```
InetAddress i;
i = InetAddress.getLocalHost();
```

In the above example the IP address is 192.168.10.73. It is to this address that every packet is sent, and every connection made. However, while servers on the Internet generally have IP addresses that never change, the same cannot be said for devices, particularly mobile ones.

If you are using a modem to dial into an ISP, then you will probably find your IP address is different each time you connect, if only in the last digit. This is because there is often a lack of addresses available, and ISPs keep a pool of addresses and allocate them as users connect.

Once you've been allocated an address you will keep it for the length of your connection, but next time you connect, you probably will be allocated a different address. This can be beneficial from a privacy point of view, but makes sending a message to a device using such a **dynamic IP** addressing system very difficult. This process has become so common that many offices now also use dynamic IP systems, even if the computers aren't generally disconnected at any time, just to make them easier to set-up and move around.

Contacting Mobile Devices

Applications like Internet Instant Messaging and Napster deal with dynamic IP addresses by having a central register. When we decide to use the service we run the application and it contacts the server and lets it know our IP address, devices wishing to contact we then log on to the central register and ask for our IP address, before making their connection to we. Obviously, such a system won't work if we are behind a proxy server (sharing a single IP address between the users in an office, for example).

The client would report its IP address, and the register would only see the IP address of the proxy. To get systems such as this working through a proxy it has to specifically support such applications. This system works well with desktop devices, and may work on mobiles for some applications, however, some circumstances are unique to mobiles and can cause problems.

It is generally the case that connections to desktop devices are user initiated, but that is unlikely to be true of mobile devices. If we take the example of the **GPRS** (**General Packet Radio Service**) already deployed in the UK by BTCellnet, this uses dynamic IP addressing to GSM devices, while offering an always-on Internet connection.

What this means is that IP networking is always available as long as the phone is operational in the GSM network, which should be almost all the time. Therefore an IP address will be allocated when the phone is switched on, and may persist for months (as long as the phone is switched on); however, a driver with a mobile device entering a tunnel may be disconnected from the GSM network. On emerging from the tunnel the GSM connection is re-established and another IP address allocated, which may not be the same as it had earlier.

This set of circumstances makes it particularly difficult for a message to be directed to such a mobile device. There is no way for the client software to know that it should reregister with the central server, and no way for that server to know not to attempt to contact the device. While it may be possible to use native methods to be alerted when such a connection is broken and re-established, it is not possible in pure Java.

The only workable solution, for the moment, is to use a thread that watches the local IP address, and notices when it changes, taking that as a signal to reregister with the server.

Of course, the network providers know the IP address allocated to every user, and it may be possible to use that database, if they will let you. Such a task may well be possible in pure Java, but will depend entirely on the environment used by the network provider.

Datagram-Based Messages

Below TCP lies the IP layer, which is entirely asynchronous. While TCP normally provides nice sockets for connection-based communication, datagrams can be used to send out packets with no confirmation of delivery or connection needed. This protocol is called **Undirected Datagram Protocol**, or **UDP** for short.

When to use Datagram Communications

There aren't many obvious circumstances where datagrams provide a better form of communication than connections, though they do offer a considerable processing advantage over TCP connections and for particular applications they are ideal. One example of this advantage is when using radio communications where the quality of connectivity is very poor.

In such an environment the connectivity may be so bad that socket communications are impossible, with TCP reporting that communication is impossible (while TCP will cope with lost and out of order packets, it has its limits). In such a situation it is possible to use datagrams and write your own confirmation protocol which may be much more error tolerant than TCP.

Sending datagrams is not a blocking action, so it's perfectly possible to use two threads, and have one sending repeatedly, while another waits for a confirmation response, (the receiving application would, of course, send conformations repeatedly). In this way communication can be achieved over the worst of connections.

Example Datagram Code

The following example application shows how a simple datagram sender and receiver can work together; the receiver was developed for a CDC device, though it makes no difference in this context.

The BitSender Class

First we import the appropriate libraries:

```
import java.net.DatagramSocket;
import java.net.DatagramPacket;
import java.net.InetAddress;
import java.awt.Frame;
import java.awt.TextArea;
import java.awt.Button;
import java.awt.BorderLayout;
import java.awt.event.MouseListener;
import java.awt.event.MouseEvent;
```

All that's needed is one button and an area for entering the text to be sent:

```
class BitSender implements MouseListener {
  Button send;
  TextArea message;
```

We instantiate an instance of ourselves (the `BitSender` class) so we aren't working with a static object:

```
public static void main(String args[]) {
  BitSender b = new BitSender();
  b.go();
```

This method just builds the GUI and displays it on the screen:

```
public void go() {
  Frame f = new Frame();
  f.setLayout(new BorderLayout());
  message = new TextArea();
  send = new Button("Send Message");
  send.addMouseListener(this);
  f.add(message, BorderLayout.CENTER);
  f.add(send, BorderLayout.SOUTH);
  f.setSize(200, 100);
  f.setVisible(true);
}
```

All our networking code has to be in a `try...catch` block as normal. All messages will be 20 characters long, as the length of a datagram has to be known before it can be sent (or received), so we either truncate it or pad with spaces until it's 20 characters exactly:

```
public void mouseClicked(MouseEvent e) {
  try {
    byte[] buf;
    String text = message.getText();
    while (text.length() < 20) {
      text = text + " ";
    }
    if (text.length() > 20) {
      text = text.substring(0, 20);
    }
```

To send a datagram, first a `DatagramPacket` is created and then sent through a `DatagramSocket`. Note that the length of the packet must be specified.

In this example the IP address used is 127.0.0.1, which is known as the loopback address, and refers to the local machine. The datagram is also addressed to a specific socket, in this case 1710:

```
buf = text.getBytes();
DatagramSocket d = new DatagramSocket();
DatagramPacket p =
new DatagramPacket(buf,
                   buf.length,
                   InetAddress.getByName("127.0.0.1"),
                   1710);
d.send(p);
System.out.println("Packet Length " + buf.length + " Sent");
```

Our catch block simply prints out the error:

```
    } catch (Exception e2) {
      System.out.println("Something Went Wrong");
      System.out.println(e2);
    }
  }
```

The following methods are required by all classes that implement the MouseListener interface:

```
    public void mouseEntered(MouseEvent e) {}
    public void mouseExited(MouseEvent e) {}
    public void mousePressed(MouseEvent e) {}
    public void mouseReleased(MouseEvent e) {}
  }
```

The BitTicker Class

These are the import statements for the BitTicker class, which is our receiver:

```
  import java.net.DatagramSocket;
  import java.net.DatagramPacket;
  import java.awt.Frame;
  import java.awt.Label;
  import java.awt.BorderLayout;
```

We only have a single GUI component, showing the message received:

```
  class BitTicker {
    Label show;
```

We instantiate an instance of ourselves (the BitTicker class) so we don't have to deal with a static object. Strictly speaking this isn't necessary in this example, but is common in Java classes:

```
    public static void main(String args[]) {
      BitTicker b = new BitTicker();
      b.go();
    }
```

This code creates and displays our, very basic, GUI:

```
    public void go() {
      Frame f = new Frame();
      show = new Label();
      f.setLayout(new BorderLayout());
      f.add(show, BorderLayout.CENTER);
      f.setSize(200, 100);
      f.setVisible(true);
```

We create our DatagramSocket that takes only a socket parameter. The DatagramPacket has only the byte array and an expected size as parameters. If the datagram is larger than this it will be truncated, if it's smaller then the byte array will contain null values in the remaining spaces:

```
try {
  byte[] buf = new byte[20];
  DatagramSocket d = new DatagramSocket(1710);
  DatagramPacket p = new DatagramPacket(buf, 20);
```

This loop continues forever, blocking at the `d.receive()` line, which will wait for an incoming message before proceeding to display the message. Receiving a datagram is a blocking action (the application will pause until a datagram is received). Once the message is displayed we start waiting for the next one. The `receive()` method will transfer the contents of the datagram into the array that was specified in its constructor:

```
while (true) {
  d.receive(p);
  show.setText("");
  show.setText(new String(buf));
}
```

Our `catch` block simply prints out the error:

```
      } catch (Exception e2) {
        System.out.println("Something Went Wrong");
        System.out.println(e2);
      }
    }
}
```

This example shows what would normally only be a very small part of an application: the communications section. Such an application could be integrated with other functions or even form the basis for a continuously updated information stream showing updated stock prices or similar data. While such functionality could certainly be achieved in other ways, (using socket connections for example) it would be more complex and slower to execute, as well as causing the server to hang if anything happened to the client application.

By being able to send out messages regardless of the availability of the client such a server can be extremely reliable, as well as fast. Clients can connect to the stream as they wish, though disconnected clients would still represent a load on the server.

Now that we have covered asynchronous messaging in detail, we can go on to look at JMS, and see how it makes use of asynchronous messaging.

The Auction House

Before digging into JMS itself, we'll first describe our example application, which will serve as an indicator of the types of application that JMS can be used for.

We wish to build an auction house system, where the auction lots are 'published' to all potential 'subscribed' bidders. If the bidders want to bid, they must send a message back to the auction house. The bid must identify the lot to the auction house, and also the bidder themselves, of course.

We have two applications to write – the `AuctionHouse` client and the `Bidder` client (or more accurately, the potential bidder). There is only one auction house, but multiple bidders. Here is how the different components interact. The terms topic, and queue are explained later in the chapter:

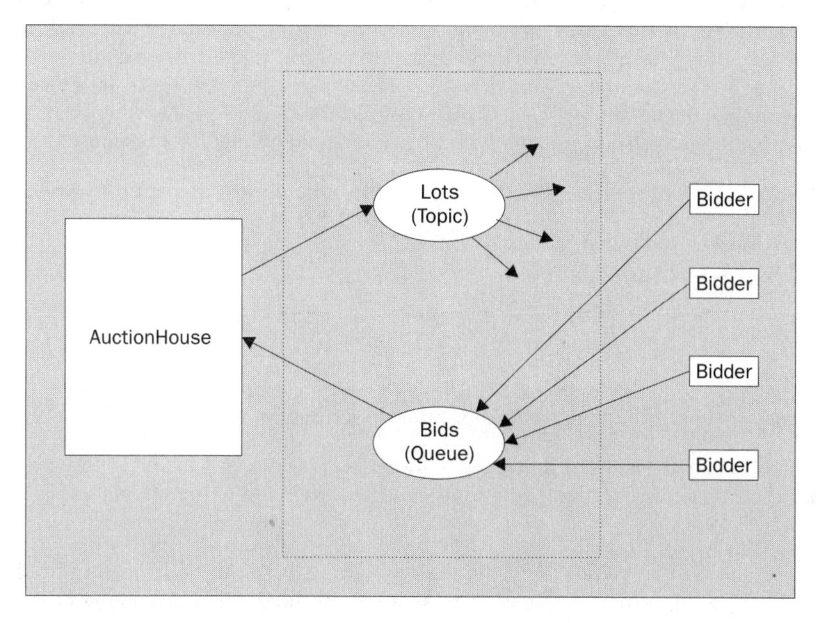

Bids go to the auction house, not direct to the other bidders. This gives the auction house a chance to filter faulty bids, exclude rogue users, allow some bidders the option of anonymity, and so on.

Bidders need to be able to join and leave the auction at will, and of course, they do not want to have to keep checking with the auction house to see if there are new, or revised, lots. They want to be told.

The auction house is necessarily simple to fit into this chapter, but it would still be quite time-consuming to write it from scratch. These are the types of problems we would have to solve:

- How do we connect the two types of client together?
- How do we send many lots to multiple clients efficiently?
- How do we guarantee a bidder's bid reaches the auction house (hand-shaking and retries would be needed)?
- What happens if the bidder becomes temporarily disconnected (assuming they are using a wireless device)?
- Can the user filter their lots (I'm only interested Gibson guitars)?

Now let's see how JMS can help us solve this distributed application problem.

MOM Basics

Messaging architecture is a good match for mobile devices, because such devices cannot generally guarantee a network connection. In such an environment synchronous communications will break down whenever the connection is lost for any length of time, because the connection is likely to timeout. Asynchronous communications, on the other hand, are suited to bursts of use, where messages are downloaded/uploaded whenever the mobile device can connect. Compare this to RMI and servlets, where the communication is synchronous, that is, a request is sent and the sender must wait, online, for a response.

To give you some idea of what messaging is useful for, here are some example applications:

- Stock market and banking alerts
- News feeds
- Auction houses
- Chat applications
- Legacy system interfacing (batched, not time-critical)

One of the chief characteristics of these applications is that data is 'pushed' to the user (they needed to register in the first place, of course), and another is, that the data being exchanged is independent of the platforms at either end.

If the messages are passed asynchronously, then they may, by definition, need to be stored somewhere before being passed on to their destination(s), as we mentioned when discussing Store and Forward networks. The intermediary that handles and potentially stores the messages is the message server, sometimes known as the 'broker' or 'provider'. Typically a third-party application, the messaging implementation is normally referred to as Message-Oriented Middleware (MOM). This MOM broker/provider is not part of JMS, but a vendor-specific implementation such as IBM's MQSeries.

Note that the participating components/applications don't have to be Java, because the MOM typically supports other language interfaces. So how does Java fit into this MOM architecture?

JMS

The Java Message Service (JMS) is a vendor-independent API for accessing the message-oriented middleware. It is not the MOM implementation itself, just the interface to MOM. This arrangement of JMS being an interface to third-party software is very similar to the way JDBC operates – it gives a Java interface into a general resource/server.

On top of the basic message delivery service, MOM systems offer the usual enterprise features, such as scalability, reliability, security, and also transactional capabilities. JMS does not actually define these features – it expects the vendors to add these enhancements, to differentiate their products. Most have done so.

MOM systems have been around much longer than Java, and the technology is mature. What JMS offers in addition is a universal set of APIs that any Java program can use regardless of the underlying MOM server implementation. Here are the building blocks for a typical messaging solution, including JMS interfaces. There are other possible architectures, but we'll consider the centralized server architecture first:

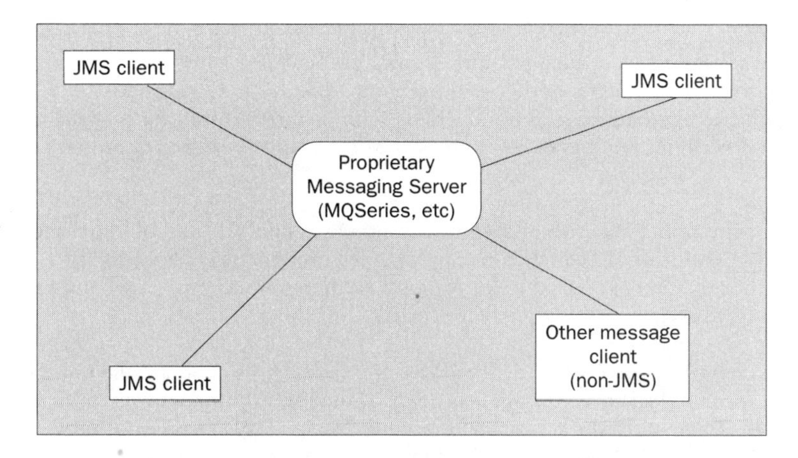

All JMS clients (and other non-JMS clients) connect only to the messaging server, never directly to each other. The server is responsible for forwarding (and storing if necessary) the messages it receives. Note that the server does not have to be a single CPU or physical system – it may be distributed for higher performance or reliability. (For more information please check *Professional JMS Programming* by Wrox Press Ltd ISBN 1-861004-93-1.) Clients can both send and receive messages, and a message can be directed to more than one receiving client as we see later.

It is worth noting that the JMS API does **not** enable different vendors of MOM servers to exchange messages. To do this a MOM 'bridge' of some form is required, and many vendors supply these as part of their offering. The bridge's job is to translate between the vendor's proprietary protocols and message formats, giving transparent access across the two vendor's systems. The following system architecture demonstrates this:

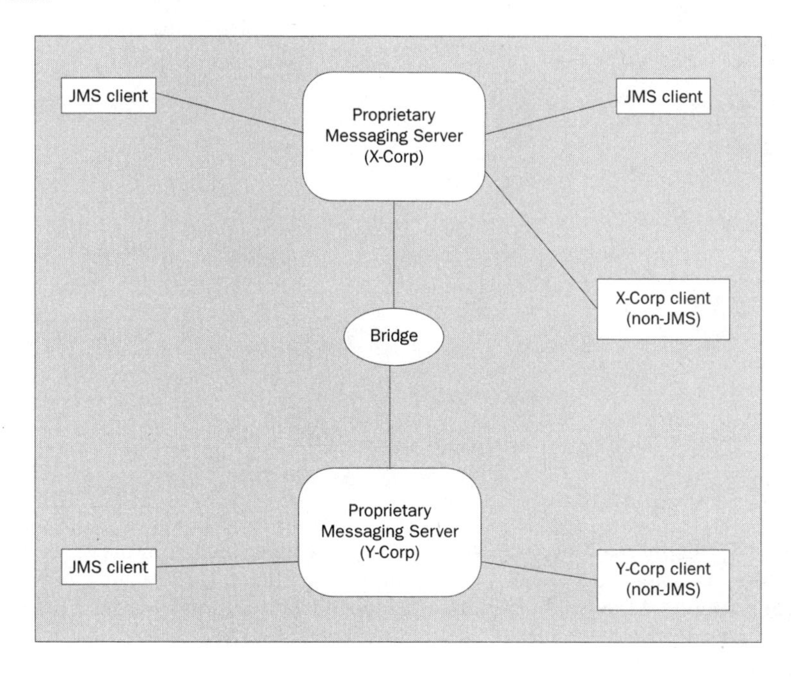

As this architecture suggests, MOM is a very useful solution for enterprise application integration (EAI) where disparate systems need to communicate, and yet they don't have any other common interfaces. With a bridge the two can talk together and exchange data (or maybe the MOM vendor has a solution for both systems, so even a bridge is not needed). This is an ideal way for interfacing to legacy mainframes and medium sized-systems such as IBM's AS/400. Vendors also offer bridges to e-mail and FTP servers.

As well as the central message server architecture, there is another way of delivering messages to more than one client – multicast IP, which is a decentralized architecture. Multicast IP is a broadcast technology rather than the standard point-to-point IP that you use every day and certain designated groups of IP addresses are reserved for multicast IP.

A messaging system can specify which IP address(es) to listen on, and all machines listening on those addresses will receive the same message, courtesy of the underlying network routers that replicates the IP packets to all listeners. In the multicast IP architecture, there is no central server, so each client is responsible for storing and forwarding their messages to other listeners.

JMS Messaging Models

JMS has two messaging models (or styles as the API refers to them), Point-to-Point messaging and Publish-and-Subscribe messaging, but both use exactly the same message format.

Point-to-Point Messaging

The Point-to-Point (P2P or PTP) messaging model is where there is only one sender and one receiver, so it's a one-to-one connection. The sender creates a message and places it into a **queue** from which the recipient extracts it. A queue is a virtual channel for communication:

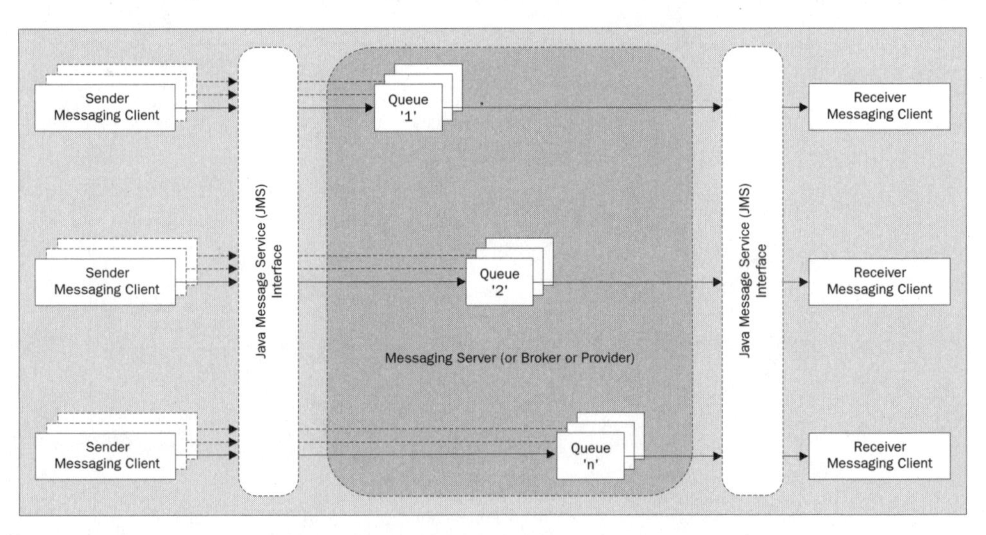

Although the PTP model may suggest a single receiver reading incoming messages, there is no reason why there cannot be multiple receivers, with each message being read by **only one** of those potential receivers. An example of this might be a load-balanced PTP solution where the first receiver that is free reads the next message in the queue.

Note that the PTP model does actually support synchronous communications, after all our tub-thumping about how messaging is based upon asynchronous communications. Another PTP-specific feature is the ability to view the queue contents messages prior to consuming the messages, which can be useful for selecting higher priority messages.

Publish-and-Subscribe Messaging

The Publish-and-Subscribe (Pub/Sub) messaging model is where there is only one sender and one or more receivers, similar to a newsgroup system or list server. Rather than a queue for sending and receiving messages, there is a **topic**, which is equivalent to an individual newsgroup. As with a queue, a topic is a virtual channel of communication. Interested clients subscribe to receive all messages sent to that topic, and can publish to that topic if they wish:

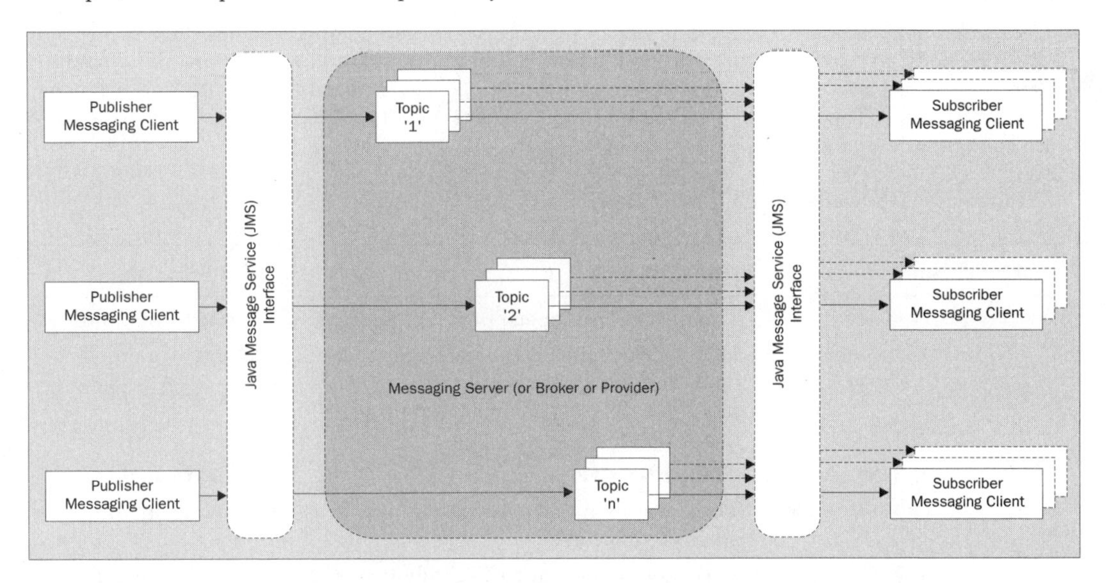

Both `Topic` and `Queue` classes extend the generic JMS `Destination` class. This is a slightly confusing name since `Destination` objects are also a source of messages, depending on whether we are sending or receiving. A less confusing name would have been `Channel`, which doesn't suggest any particular direction.

Senders and Receivers

For both models, the JMS API defines the senders and receivers of messages as being implementations of the `MessageProducer` and `MessageConsumer` interfaces respectively. For queues the sender and receiver classes are:

- `QueueSender`
- `QueueReceiver`

For topics, they are:

- `TopicPublisher`
- `TopicSubscriber`

The JMS Message

A JMS message comprises:

- Header information
- Message Properties
- Message Body (the data)

We will now discuss each of these three elements in more detail.

Header Information

The header information is standard across all JMS messages, and comprises items such as the destination, delivery mode, a reply-to destination, a timestamp, and so on. They are all defined in the `javax.jms` package, and have the naming convention `JMSxxxx` (examples are `JMSReplyTo` and `JMSTimestamp`). The messaging server automatically sets most headers, but a few, like `JMSReplyTo`, are configurable by the application.

Here are the JMS headers, with a short explanation:

Header Name	Description	Set by...
`JMSDestination`	The destination (queue or topic name).	`send()` method
`JMSDeliveryMode`	`PERSISTENT` means the message must be delivered once and only once. `NON_PERSISTENT` means the message may be lost.	`send()` method
`JMSMessageID`	Unique ID.	`send()` method
`JMSExpiration`	The time when this message expires. 0 means live for ever.	`send()` method
`JMSTimestamp`	The time the message was ready for sending to the provider.	`send()` method
`JMSPriority`	Priority level of 0-9, which is a hint to the provider about which messages to deliver first.	`send()` method
`JMSRedelivered`	Indicator saying the receiving client has been sent this message before (implicitly suggesting that no acknowledgement was received last time, from the client).	Provider
`JMSReplyTo`	The destination to reply to, if a response is expected.	Client
`JMSCorrelationID`	Unique ID to help correlate messages together. Useful for communicating with non-JMS clients, when JMS headers cannot be used.	Client
`JMSType`	The message type, usually defined by the JMS vendor, for communicating with non-JMS clients.	Client

The provider can choose to override JMSDeliveryMode, JMSExpiration, and JMSPriority, if necessary.

Message Properties

JMS message properties are optional message headers, in effect, that can be set by the application (or JMS vendor) per message. The property values that can be specified are of types int, byte, short, boolean, long, double, float, and String. The property accessor methods are defined as setXXXProperty() and getXXXProperty(). For example, to send an integer property with a message, we would call:

```
public void setIntProperty(String name, int value);
```

On receiving the message we would call the matching getXXXProperty() to read the property:

```
public int getIntProperty(String name);
```

Message properties allow clients to select messages based on some criteria applied to the properties. For example, we could design a financial application that sent stock values in the message body (discussed below), plus a message header named 'company', holding the name of the company. Each JMS client receiving the messages could then filter out just the companies they are interested in.

On the client side, to fetch all the message properties, use:

```
public Enumeration getPropertyNames();
```

We can then step through the enumerated list, one by one using Enumeration.nextElement().

We can check for a specific property with:

```
public boolean propertyExists(String name);
```

Message properties can all be cleared using:

```
public void clearProperties();
```

Note that we cannot clear individual properties – it's all or nothing.

The JMS specification suggests that carrying data in the message body is likely to be more efficient than using message properties as a delivery mechanism. The one time we may wish to use them (apart from for filtering messages) is if our application is an intermediary. The message properties can be replaced, using clearProperties(), when the message is passed on, whereas the message body cannot be, as we now see.

Message Body

The actual message body of a JMS message can be one of 6 types, defined in the `javax.jms` package:

Message Name	Description
`Message`	The super-interface for the other 5 message types below. It has no content. A message with this body still delivers headers and properties, and can be useful for event notification.
`TextMessage`	Carries a single `String`. Can carry an XML payload, etc.
`BytesMessage`	Carries a stream of bytes. Used to encode a message to match an existing vendor format.
`StreamMessage`	Carries a stream of primitive Java types, in a sequential order. Supported types are `boolean`, `byte`, `byte[]`, `char`, `short`, `int`, `long`, `float`, `double`, `String`, and `Object`.
`MapMessage`	Carries a set of name-value pairs accessible sequentially or by name. Names must be `Strings`, and values must be a primitive Java type, as per `StreamMessage`.
`ObjectMessage`	Carries a single `Object`, which must implement the `java.io.Serializable` interface. Many objects can be sent in one message by using a `Collection`.

The contents of the message body are set by using `Message.setXXX()`. The message types are all defined as interfaces rather than actual classes, since it is up to each vendor to supply the implementations.

Here is a code snippet showing a text message being constructed and sent. We also set a `String` message property, which we have called `'user'`. The `TopicSession` and `TopicPublisher` classes are
described later:

```
TopicSession session;
TopicPublisher publisher;
...
TextMessage myMessage = session.createTextMessage();
MyMessage.setText("Hello world");
myMessage.setStringProperty("user","Fred");
publisher.publish(myMessage);
```

The receiving client cannot alter the content of the message – if it tries to an exception will be thrown.

Sending a Message

Having seen the contents of a JMS message, let's step through how to send and receive one. These are the steps we take to send a message, followed by the steps to receive the same message in another client. We define the terminology later.

To send a message:

- Obtain a `ConnectionFactory`
- Create a `Connection` from the `ConnectionFactory`
- Create a `Session` from the `Connection` (for example, `TopicSession`, above)
- Obtain a `Destination` (or create a temporary one)
- Create a `MessageProducer` (for example, `TopicPublisher`, above)
- Create our message
- Send the message using the `MessageProducer`

To receive a message:

- Obtain a `ConnectionFactory`
- Create a `Connection` from the `ConnectionFactory`
- Create a `Session` from the `Connection`
- Obtain a `Destination` (or create a temporary one)
- Create a `MessageConsumer`
- Register a `MessageListener` from the `MessageConsumer`, and listen for incoming messages
- Receive the message using `MessageListener.onMessage()`

All clients must connect to the messaging server before they can send or receive any messages. This is done using a `ConnectionFactory` to create a `Connection` (either a `TopicConnection` or a `QueueConnection`). However, to communicate with the message server, we must first find it, using a JNDI lookup.

Administered Objects

To get a reference to some initial JMS objects (like a topic or queue) we need to look them up using a JNDI lookup service. JMS defines two types of objects as **administered** objects. They are resources that hold JMS configuration, and are:

- `ConnectionFactory`
- `Destination` (queues and topics)

The messaging server (provider) manages these objects, and the client cannot change them, it just uses them to gain access to the messaging server.

Connecting to a Messaging Server

Here is some example code using a ConnectionFactory to create a TopicConnection, which is discovered using JBoss's JNDI service (for details on JBoss, see the previous chapter):

```
TopicConnectionFactory tFactory =
(TopicConnectionFactory)jndi.lookup("TopicConnectionFactory");
TopicConnection tConnection = tFactory.createTopicConnection(username, password);
```

First we look up the ConnectionFactory in the JNDI directory, and then we create the Connection using the ConnectionFactory.

A Connection is typically a TCP/IP socket that can be shared by multiple Session objects (see below). On Connection creation, the client is authenticated, either by the username and password supplied or by the thread's credentials. However, if no username and password are supplied, then, according to the JMS specification, "the JDK does not define the concept of a thread's default credentials; however, it is likely this will be defined in the near future. For now, the identity of the user under which the JMS client is running should be used."

A Connection is typically a 'heavyweight' object, that is, it is potentially resource-hungry, and thus it is suggested that only one instance should be shared throughout an application. A close() method is provided to free up resources used by a Connection, and it should be called whenever the Connection is finished with.

A Connection has start() and stop() methods, which control the flow of messages into the client, but interestingly a client can send messages out even though its Connection is stopped. A Connection is in stop mode on creation, to ensure no messages are received unexpectedly.

Sessions

Once a Connection is created, one or more Session objects (a TopicSession or QueueSession) can be created using that Connection object. A Session object is a lightweight object that creates MessageProducer and MessageConsumer objects:

```
QueueSession qSession = qConnection.createQueueSession(false,
Session.AUTO_ACKNOWLEDGE);
```

In addition, Session objects can create Destination objects dynamically, TemporaryTopic and TemporarySession respectively. As the name suggests, the lifetime of these objects is limited to the lifetime of the Connection (not the Session, as you might expect), since the client, and not the messaging provider manages them. These temporary topics and queues can be used by other clients, but only if the creating client informs them of the names using a JMSReplyTo header. These are the only types of destinations that cannot be discovered using a JNDI lookup.

Sending (Point-to-Point Model)

Once a client has created a `QueueConnection`, and created a `QueueSession`, they need to do the following to send a message:

- Get a reference to the queue using JNDI (this need only be done once in an application, and could be done earlier in the application, maybe during construction)
- Create a `QueueSender`
- Call the `QueueSender.send()` method passing in the message

There are actually two ways to obtain a queue, by JNDI as mentioned above, and also from the header of a previously received message.

Using JNDI to Create a Queue

Here is the code to set up a queue to send a bid (see the auction house figure at the beginning of the chapter). The queue is a permanent one managed by the messaging provider, so we use JNDI to get its reference:

```
bidQ = (Queue)jndi.lookup("queue/Bid");
bidQSender = qSession.createSender(bidQ);
StreamMessage sm = qSession.createStreamMessage();
...
bidQSender.send(sm);
```

The naming convention (`queue/`) is specific to a JBossMQ server. Your messaging server or JNDI setup may be different since there is no naming convention defined in JMS for administered objects. Read your provider's documentation carefully to discover the correct naming scheme because JNDI gives no hints when it cannot find an object – it will bounce back with an exception.

Using the JMSReplyTo Header to Create a Queue

Our queue could have possibly been a `TemporaryQueue` obtained from a previous message's `JMSReplyTo` header, rather than looked up using JNDI. Here's how:

```
TemporaryQueue tmpQ = receivedMessage.getJMSReplyTo();
bidQSender = qSession.createSender(tmpQ);
StreamMessage sm = qSession.createStreamMessage();
...
bidQSender.send(sm);
```

Receiving (Point-to-Point Model)

Assuming we have obtained a `QueueSession` object, to receive point-to-point, we:

- Create a `QueueReceiver`
- Specify a `MessageListener` object to handle all the incoming messages
- Start the messages flowing by calling `QueueConnection.start()`

Here is an extract from our AuctionHouse class, covered in detail later, that demonstrates the steps taken to get ready for receiving bid messages – in this case from a queue:

```
private static final String BID_QUEUE = "queue/Bid";
// Queue to receive Bid request from a user
private Queue bidQ;
private QueueReceiver bidQReceiver;
...
public AuctionHouse(String username, String password) {
  ...
  bidQ = (Queue) jndi.lookup(BID_QUEUE);
  ...
  // queue to receive bids
  bidQReceiver = qSession.createReceiver(bidQ);
  bidQReceiver.setMessageListener(this);

  // only need to start connections that have incoming messages
  qConnection.start();
```

Publishing (Pub/Sub Model)

There are some small differences when using Pub/Sub, mostly just the class and method names. Once a client has created a TopicConnection, and created a TopicSession, they do the following to publish a message:

- Get a reference to the topic using JNDI, or dynamically create a TemporaryTopic by calling TopicSession.createTemporaryTopic() or Message.getJMSReplyTo[0]()
- Create a TopicPublisher
- Call the TopicPublisher.publish() method passing in the message:

```
private static final String LOT_TOPIC = "topic/Lot";
lotTopic = (Topic) jndi.lookup(LOT_TOPIC);
lotPub = tSession.createPublisher(lotTopic);
...
StreamMessage sm = tSession.createStreamMessage();
...
lotPub.publish(sm);
```

Subscribing (Pub/Sub model)

Again, the Pub/Sub steps and code are very similar to those in PTP, assuming we have obtained a TopicSession object. The steps are:

- Create a TopicSubscriber
- Specify a MessageListener object to handle all the incoming messages
- Start the messages flowing by calling TopicConnection.start()

```
public class Bidder implements MessageListener {
...
private static final String LOT_TOPIC = "topic/Lot";
lotTopic = (Topic) jndi.lookup(LOT_TOPIC);
...
tConnection = tFactory.createTopicConnection();
...
lotSub = tSession.createSubscriber(lotTopic);
lotSub.setMessageListener(this);
tConnection.start();
```

The Auction House Revisited

Now we understand the basics of JMS, we can flesh out our auction house example. If you look back to the figure at the start of this chapter, you will see that lots are published by the auction house client to many bidder clients using a topic. Contrast this to the individual bid messages, which come back by using a point-to-point queue, since there is only one target receiver/reader, the `AuctionHouse`.

Let's flesh out our design with some business classes, `Lot` and `Catalog`.

The Lot Class

A `Lot` holds the:

- Lot name
- Description of the lot
- Current price (the starting price or latest bid price)
- A suggested increment to the current bid (can be overridden by the user)
- Current bidder for that lot

The `Lot` class is comprised mostly of accessor methods. There are two interesting methods though – we define a constructor where a `javax.jms.StreamMessage` is passed in, and also a `writeStreamMessage()` method that places the contents of the `Lot` into a `StreamMessage`. We are converting the `Lot` object into a stream of basic Java types, ready for sending to a queue or topic.

We could have used the `ObjectMessage` method in JMS, and an `ObjectMessage` requires any object sent down it must be serializable, but J2ME implementations do not generally support object serialization. Therefore, we serialize it ourselves. This is trivial for our simple application, but this may be a problem you come across for more complex applications:

```
import javax.jms.StreamMessage;
import javax.jms.JMSException;

public class Lot {
  private String name;
  private String description;
  private int price;
  private int bidIncrement;
  private String currentBidder;

  public Lot(String lotName, String lotDesc, int lotPrice, int increment) {
```

```
    name = lotName;
    description = lotDesc;
    price = lotPrice;
    bidIncrement = increment;   // the suggested amount to increment a bid by
    currentBidder = "";   // initially no bidder
  }

  public Lot(Lot lot) {
    name = lot.getName();
    description = lot.getDescription();
    price = lot.getPrice();
    currentBidder = lot.getCurrentBidder();
  }

  public String getName() {
    return name;
  }

  public int getPrice() {
    return price;
  }

  public void setPrice(int newPrice) {
    price = newPrice;
  }

  public int getBidIncrement() {
    return bidIncrement;
  }

  public void setBidIncrement(int newBidIncrement) {
    bidIncrement = newBidIncrement;
  }

  public String getCurrentBidder() {
    return currentBidder;
  }

  public void setCurrentBidder(String bidder) {
    currentBidder = bidder;
  }

  public String getDescription() {
    return description;
  }

  public Lot(StreamMessage sm) {
    try {
      name = sm.readString();
      description = sm.readString();
      price = sm.readInt();
      currentBidder = sm.readString();
    } catch (JMSException jmse) {
      System.out.println("JMSException in Lot constructor");
      jmse.printStackTrace();
    }
```

```
  }

  public void writeStreamMessage(StreamMessage sm)  {
    try {
      sm.writeString(name);
      sm.writeString(description);
      sm.writeInt(price);
      sm.writeString(currentBidder);
    } catch (JMSException jmse)  {
      System.out.println("JMSException in Lot.writeStreamMessage()");
      jmse.printStackTrace();
    }
  }
}
```

This is where we create a Lot, direct from a StreamMessage:

```
public Lot(StreamMessage sm){
  try {
    name = sm.readString();
    description = sm.readString();
    price = sm.readInt();
    currentBidder = sm.readString();
  } catch(JMSException jmse){
    System.out.println("JMSException in Lot constructor");
    jmse.printStackTrace();
  }
}
```

The writeStreamMessage() method builds a Lot in the StreamMessage passed in:

```
  public void writeStreamMessage(StreamMessage sm){
    try{
      sm.writeString(name);
      sm.writeString(description);
      sm.writeInt(price);
      sm.writeString(currentBidder);
    } catch(JMSException jmse){
        System.out.println("JMSException in Lot.writeStreamMessage()");
        jmse.printStackTrace();
    }
  }
}
```

The Catalog Class

Catalog is implemented as a Vector of Lot objects. We have used the basic Vector class rather than a Collection because CLDC does not support Collection, and we will want to port this to a CLDC platform later in the chapter. Note that you can only add lots. Commercial applications would obviously need to remove lots when the bidding finishes, or if lots were withdrawn by the seller:

```
import java.util.Vector;
import java.util.Enumeration;

public class Catalog {

  // Have to use simple stuff like Vector because of CLDC limitations
  Vector lots;

  public Catalog() {
    lots = new Vector();
  }

  public void addLot(Lot l) {
    lots.addElement(l);
  }

  public Lot getLot(String name) {
    Lot lot;
    Enumeration enum = lots.elements();
    while (enum.hasMoreElements()) {
      lot = (Lot) enum.nextElement();
      if (lot.getName().equals(name)) {
        return lot;
      }
    }

    // not found
    return null;
  }

  public boolean exists(String name) {
    Lot lot;
    Enumeration enum = lots.elements();
    while (enum.hasMoreElements()) {
      lot = (Lot) enum.nextElement();
      if (lot.getName().equals(name)) {
        return true;
      }
    }

    // not found
    return false;
  }
```

The `updateLot()` method replaces an existing `Lot` with the one passed in:

```
  public void updateLot(Lot l) {
    Lot tmpLot = getLot(l.getName());
    lots.removeElement(tmpLot);
    addLot(l);
  }

  public int size() {
    return lots.size();
  }
```

```
  public Enumeration elements() {
    return lots.elements();
  }
}
```

AuctionHouse JMS Client

AuctionHouse is the class that holds all the JMS-specific code for the auction house JMS client:

```
// standard Java libraries to support naming and JMS
import javax.naming.InitialContext;
import javax.naming.Context;
import javax.naming.NamingException;

import java.util.Properties;

import javax.jms.TopicConnectionFactory;
import javax.jms.TopicConnection;
import javax.jms.TopicSession;
import javax.jms.Topic;
import javax.jms.TopicPublisher;
import javax.jms.QueueConnectionFactory;
import javax.jms.QueueConnection;
import javax.jms.QueueSession;
import javax.jms.Queue;
import javax.jms.QueueReceiver;
import javax.jms.Session;
import javax.jms.Message;
import javax.jms.StreamMessage;
import javax.jms.MessageListener;
import javax.jms.JMSException;

public class AuctionHouse implements MessageListener {

  private Catalog cat;

  // Topic for telling all users about new Lots
  private TopicConnection tConnection;
  private TopicSession tSession;
  private Topic lotTopic;
  private TopicPublisher lotPub;

  // Queue to receive Bid requests from users
  private QueueConnection qConnection;
  private QueueSession qSession;
  private Queue bidQ;
  private QueueReceiver bidQReceiver;
```

A username and password are required to create the AuctionHouse, which is an option when creating Connection objects. These must have been defined by the messaging server administration as a user, for security purposes.

The constructor is where the JMS initialization is called. An empty `Catalog` is also created to hold the lots that the auction house creates during its lifetime:

```
public AuctionHouse(String username, String password) {
  // create a private catalog to hold all our lots
  cat = new Catalog();
  initJMS(username, password);
  System.out.println("AuctionHouse started.");
}
```

`initJMS()` is where the `Connection` and `Destination` objects, the `Lot` topic and `Bid` queue, are referenced using JNDI lookup as described previously. Here is the `initJMS()` method:

```
public void initJMS(String username, String password) {
  // jBossMQ-specific topic and queue names
  String TOPIC_CONNECTION_FACTORY = "TopicConnectionFactory";
  String QUEUE_CONNECTION_FACTORY = "QueueConnectionFactory";
  String LOT_TOPIC = "topic/Lot";
  String BID_QUEUE = "queue/Bid";

  try {
    // all the JNDI-specific code to lookup Connections and Destinations
    Properties prop = new Properties();
    prop.put(Context.INITIAL_CONTEXT_FACTORY,
             "org.jnp.interfaces.NamingContextFactory");
    prop.put(Context.PROVIDER_URL, "localhost:1099");
    Context jndi = new javax.naming.InitialContext(prop);

    TopicConnectionFactory tFactory =
      (TopicConnectionFactory) jndi.lookup(TOPIC_CONNECTION_FACTORY);
    QueueConnectionFactory qFactory =
      (QueueConnectionFactory) jndi.lookup(QUEUE_CONNECTION_FACTORY);
    lotTopic = (Topic) jndi.lookup(LOT_TOPIC);
    bidQ = (Queue) jndi.lookup(BID_QUEUE);

    // we only want authenticated users running AuctionHouse
    tConnection = tFactory.createTopicConnection(username, password);
    tSession = tConnection.createTopicSession(false,
                                              Session.AUTO_ACKNOWLEDGE);

    qConnection = qFactory.createQueueConnection(username, password);
    qSession = qConnection.createQueueSession(false,
                                              Session.AUTO_ACKNOWLEDGE);

    // topic to publish any lot information
    lotPub = tSession.createPublisher(lotTopic);
```

The final step in `initJMS()` is to register `AuctionHouse` as the `MessageListener` for receiving messages on the `Bid` queue, and to start listening for those bids:

```
    // queue to receive bids
    bidQReceiver = qSession.createReceiver(bidQ);
    bidQReceiver.setMessageListener(this);
```

```
      // only need to start connections that have incoming messages
      qConnection.start();
    } catch (JMSException jmse) {
      System.out.println("JMSException in AuctionHouse constructor");
      jmse.printStackTrace();
      System.exit(-1);
    } catch (NamingException jne) {
      System.out.println("NamingException in AuctionHouse constructor");
      jne.printStackTrace();
      System.exit(-1);
    }
  }
```

`onMessage()` is the method that handles any messages received, as defined in the `javax.jms.MessageListener` interface that `AuctionHouse` implements. In our design there is only one source of incoming messages – bids from the `Bid` queue:

```
  public void onMessage(Message message) {
    try {
      if (message instanceof StreamMessage) {
        StreamMessage sm = (StreamMessage) message;
```

`onMessage()` demonstrates, somewhat artificially, the use of a JMS property. We receive the bidder's user name as a String property rather than extract it from the `Lot` with `Lot.getCurrentBidder()`. In this example application we do not attempt to filter on the "`user`" message property:

```
        String bidderName = message.getStringProperty("user");
        System.out.println("Bid received from " + bidderName);
        Lot receivedLot = new Lot(sm);
```

When a bid comes in the bid price is compared with the existing price in our `Catalog`. If the bid is not high enough, then it's rejected, and nothing else happens:

```
        // check if the bid was higher than the existing bid for that lot
        boolean bidSuccessful = false;
        Lot lot = cat.getLot(receivedLot.getName());
        if (receivedLot.getPrice() > lot.getPrice()) {
          lot.setPrice(receivedLot.getPrice());
          lot.setCurrentBidder(bidderName);
          bidSuccessful = true;
        }
```

If the bid is successful, then that bid, as a `Lot`, is published on the `Lot` topic, so that all `Bidder` clients also receive the latest bid on that lot. Note that we use a `Lot` object to represent a bid for simplicity. For a commercial application we would create a `Bid` object to minimize the data being passed in messages. On a successful bid we would also send out a specific dated confirmation directly to the successful bidder, using a queue:

```
        if (bidSuccessful) {

          // build the bid message
          sm = tSession.createStreamMessage();
```

```
            sm.setStringProperty("user", bidderName);
            lot.writeStreamMessage(sm);
            lotPub.publish(sm);
            System.out.println("Successful bid published.");
          }
        }
      } catch (JMSException jmse) {
        System.out.println("JMSException in AuctionHouse.onMessage()");
        jmse.printStackTrace();
      }
    }
```

It's worth noting that the bid is only accepted when it reaches the `AuctionHouse`. This will favor better-connected devices over those with a slow or intermittent connection. The `JMSTimestamp` header could be used to sort the messages in a fairer time order, but this is the time when the message reaches the provider, not the time that the `Bidder` client created the bid.

> *Basically any `Bidder` timestamp cannot be trusted, as it may be fraudulent. Real-world auction applications would need to take these facts in to consideration, but for simplicity we accept the bids in the order they reach the `AuctionHouse`.*

`publishNewLot()` creates the new lot from the parameters passed in, publishes the newly-created lot to the `Lot` topic by filling a `StreamMessage` using the method `Lot.writeStreamMessage()`, and finally saves the new lot in its `Catalog`:

```
    public void publishNewLot(String name, String desc, int price,
                              int bidIncrement) {
      Lot lot = new Lot(name, desc, price, bidIncrement);
      cat.addLot(lot);
      try {
        StreamMessage sm = tSession.createStreamMessage();
        lot.writeStreamMessage(sm);
        lotPub.publish(sm);
        System.out.println("New lot published:" + lot.getName() + " Price: "
                           + lot.getPrice());
      } catch (JMSException jmse) {
        System.out.println("JMSException in AuctionHouse.publishNewLot()");
        jmse.printStackTrace();
      }
    }
```

`exit()` is just a method to actively free up the resources used, that is, the `Connection` socket or similar communication route to the messaging server. Closing a `Connection` also frees up all its associated `Session` and `Destination` objects, so these do not need to be specifically closed:

```
    public void exit() {
      try {
        tConnection.close();
        qConnection.stop();
        qConnection.close();
      } catch (JMSException jmse) {
        System.out.println("JMSException in AuctionHouse.exit()");
        jmse.printStackTrace();
      }
      System.exit(0);
    }
  }
```

The AuctionHouseCmdLine Class

To run the auction house as a standalone application, we just have a simple command-line interface, `AuctionHouseCmdLine`, which creates an instance of `AuctionHouse`, and allows the user to input new lots via `java.lang.System.in`:

```
import java.io.InputStreamReader;
import java.io.BufferedReader;
import java.io.IOException;
import java.util.StringTokenizer;

public class AuctionHouseCmdLine {
  public static void main(String args[]) {
    if (args.length != 2) {
      System.out
        .println("Usage: java AuctionHouseCmdLine username password");
      return;
    }
    AuctionHouse auctionHouse = new AuctionHouse(args[0], args[1]);

    try {

      // Read all standard input and send it as a message to any Biddder
      // clients
      BufferedReader stdin =
          new java.io.BufferedReader(new InputStreamReader(System.in));
      System.out.println("Enter new lot: name, description, start price,
                                                        bid increment\n");
      System.out.println("(Example: Palm MP100,A small PDA,50,10");

      auctionHouse.publishNewLot("Palm", "A small PDA", 50, 10);
      auctionHouse.publishNewLot("Psion", "A bigger PDA", 100, 15);
      auctionHouse.publishNewLot("iPaq", "A shiny PDA", 150, 20);

      while (true) {
        String lotDesc = stdin.readLine();
        if (lotDesc != null && lotDesc.length() > 0) {
          StringTokenizer tokenizer = new StringTokenizer(lotDesc, ",");
          String name = tokenizer.nextToken();
          String desc = tokenizer.nextToken();
          int price = Integer.parseInt(tokenizer.nextToken().trim());
          int bidIncrement = Integer.parseInt(tokenizer.nextToken().trim());
          auctionHouse.publishNewLot(name, desc, price, bidIncrement);
        } else {
          auctionHouse.exit();
        }
      }
    } catch (IOException ioe) {
      System.out.println("IOException in AuctionHouseCmLine main()");
      ioe.printStackTrace();
    }
  }
}
```

Bidder JMS Client

The Bidder class is our second JMS client. It subscribes and listens to the Lot topic, and allows the user to send bids down the Bid queue to the AuctionHouse:

```
// standard Java libraries to support naming and JMS
import javax.naming.InitialContext;
import javax.naming.Context;
import javax.naming.NamingException;

import java.util.Properties;

import javax.jms.TopicConnectionFactory;
import javax.jms.TopicConnection;
import javax.jms.TopicSession;
import javax.jms.Topic;
import javax.jms.TopicSubscriber;
import javax.jms.QueueConnectionFactory;
import javax.jms.QueueConnection;
import javax.jms.QueueSession;
import javax.jms.Queue;
import javax.jms.QueueSender;
import javax.jms.Session;
import javax.jms.Message;
import javax.jms.StreamMessage;
import javax.jms.MessageListener;
import javax.jms.JMSException;

public class Bidder implements MessageListener {
  private String user;
  private BidderUI ui;
  private Catalog cat;

  // Topic variables for receiving lots
  private TopicConnection tConnection;
  private TopicSession tSession;
  private Topic lotTopic;
  private TopicSubscriber lotSub;

  // Queue variables for sending bids
  private QueueConnection qConnection;
  private QueueSession qSession;
  private Queue bidQ;
  private QueueSender bidQSender;
```

The Bidder constructor creates a new catalog, to hold any lots it receives, but this does not necessarily have exactly the same content as the AuctionHouse catalog. Why? If the AuctionHouse publishes lots before the Bidder is created, then the Bidder never receives those lots. In a commercial application where we want to tell users of lots already active and being bid upon, we might implement a Catalog queue down which each client could request a full catalog to make sure they were up to date, and also make Lot topic subscriptions 'durable' (durable subscriptions are discussed towards the end of this chapter):

```
public Bidder(String username, String password, BidderUI ui) {
  this.ui = ui;
  user = username;
  cat = new Catalog();
  initJMS(username, password);
  System.out.println("Bidder started.");
}
```

The `initJMS()` method is quite similar to its namesake in `AuctionHouse` in that it also looks up `Connection` and `Destination` objects via JNDI, but instead of listening for bids, it subscribes to the `Lot` topic, and listens for new lots:

```
public void initJMS(String username, String password) {
  // jBossMQ-specific topic and queue names
  String TOPIC_CONNECTION_FACTORY = "TopicConnectionFactory";
  String QUEUE_CONNECTION_FACTORY = "QueueConnectionFactory";
  String LOT_TOPIC = "topic/Lot";
  String BID_QUEUE = "queue/Bid";

  try {
    // all the JNDI-specific code to lookup Connections and Destinations
    Properties prop = new Properties();
    prop.put(Context.INITIAL_CONTEXT_FACTORY,
            "org.jnp.interfaces.NamingContextFactory");
    prop.put(Context.PROVIDER_URL, "localhost:1099");
    Context jndi = new javax.naming.InitialContext(prop);

    TopicConnectionFactory tFactory =
            (TopicConnectionFactory)jndi.lookup(TOPIC_CONNECTION_FACTORY);
    QueueConnectionFactory qFactory =
            (QueueConnectionFactory) jndi.lookup(QUEUE_CONNECTION_FACTORY);
    lotTopic = (Topic) jndi.lookup(LOT_TOPIC);
    bidQ = (Queue) jndi.lookup(BID_QUEUE);

    // we can, optionally, choose to authenticate the user
    qConnection = qFactory.createQueueConnection(username, password);
    qSession = qConnection.createQueueSession(false,
                                          Session.AUTO_ACKNOWLEDGE);
    bidQSender = qSession.createSender(bidQ);

    tConnection = tFactory.createTopicConnection(username, password);
    tSession = tConnection.createTopicSession(false,
                                          Session.AUTO_ACKNOWLEDGE);
    lotSub = tSession.createSubscriber(lotTopic);
    lotSub.setMessageListener(this);

    // only need to start connections that have incoming messages
    tConnection.start();
  } catch (JMSException jmse) {
    System.out.println("JMSException in Bidder constructor");
    jmse.printStackTrace();
    System.exit(-1);
  } catch (NamingException jne) {
    System.out.println("NamingException in Bidder constructor");
    jne.printStackTrace();
    System.exit(-1);
  }
}
```

onMessage() listens for incoming lots on the Lot topic sent by the AuctionHouse. If a Lot is received, the catalog of the Bidder is updated accordingly, either by adding a new lot, or by replacing an old lot with the updated one received:

```
public void onMessage(Message message) {
  if (message instanceof StreamMessage) {
    StreamMessage sm = (StreamMessage) message;
    Lot lot = new Lot(sm);
    System.out.println("Lot received: " + lot.getName() + ", Price: "
                       + lot.getPrice());
```

If this is not in our local copy of the catalog, then add it, otherwise update our current Lot:

```
    if (cat.exists(lot.getName())) {
      cat.updateLot(lot);
    } else {

      // add the new lot to the local catalog, keeping it up to date
      cat.addLot(lot);
    }
    ui.displayCatalog(cat);
    return;
  }
  System.out.println("unknown non-StreamMessage received");
}
```

makeBid() sends a Lot, encoded by using Lot.writeStreamMessage(), as a StreamMessage to the Bid queue, from which the AuctionHouse will read. The bidder's name is sent as a String property, for the AuctionHouse to extract:

```
public void makeBid(Lot lot) {
  lot.setCurrentBidder(user);
  try {
    StreamMessage sm = qSession.createStreamMessage();
    lot.writeStreamMessage(sm);
    sm.setStringProperty("user", user);
    bidQSender.send(sm);
    System.out.println("Made a bid: " + lot.getPrice() + " for "
                       + lot.getName());
  } catch (JMSException jmse) {
    System.out.println("JMSException in Bidder.makeBid()");
    jmse.printStackTrace();
  }
}
```

getCatalog() and getUserName() are accessor methods for our user interface code to use (see the BidderUI class later):

```
public Catalog getCatalog() {
  return cat;
}

public String getUserName() {
  return user;
}
```

`exit()` frees up resources before stopping the application:

```
    public void exit() {
      try {
        tConnection.close();
        qConnection.close();
      } catch (JMSException jmse) {
        System.out.println("JMSException in Bidder.exit()");
        jmse.printStackTrace();
      }
      System.exit(0);
    }
  }
```

The Bidder User Interface

For the bidder user interface, rather than having a command line, we will use a simple graphical user interface. This lets us develop a UI on the desktop, and later, place the same code onto a mobile device. We chose the standard Abstract Windowing Toolkit (AWT) that has been in existence since Java 1.0.

Although the majority of AWT has long been superceded by the Swing UI, neither CLDC nor CDC supports Swing. In fact, only the CDC supports a profile containing a subset of AWT, with CLDC currently only having MIDP support at the time of writing. However, there is a very useable CLDC implementation of AWT called kAWT (see Chapter 4). Also, it is very likely that the forthcoming PDA Profile (PDAP) will support a subset of AWT, so this has influenced our choice. We could also use KJava on the CLDC platform, but not (easily) on a desktop machine.

Our bidder user interface has three classes:

- `LoginUI`
- `BidderUI`
- `BidUI`

`LoginUI` is a modal dialog to capture the username and password when starting the application. We optionally require a valid username and password to access a `Connection`. `BidderUI` is a single `Frame` that displays the current contents of a bidder's personal catalog. Alongside each catalog entry there is a button that pops up the `BidUI` dialog to allow the user to bid for that particular lot:

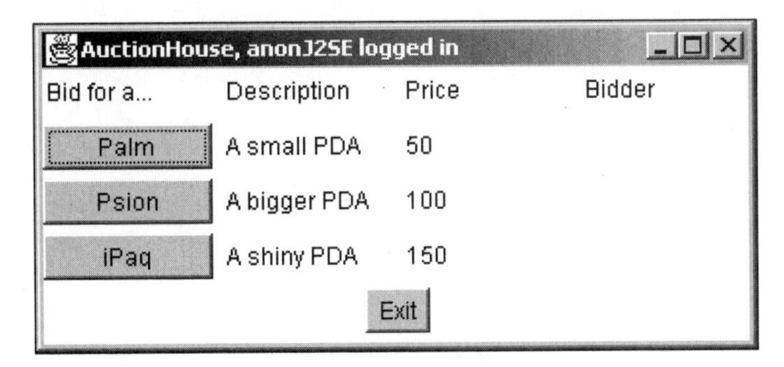

When a lot is successfully submitted, each instance of the `Bidder` class receives the updated lot, and calls `BidderUI.displayCatalog()` causing the contents of `BidderUI` to be refreshed.

The LoginUI Class

```
import java.awt.Dialog;
import java.awt.TextField;
import java.awt.Label;
import java.awt.Panel;
import java.awt.Button;
import java.awt.GridLayout;
import java.awt.FlowLayout;
import java.awt.event.ActionListener;
import java.awt.event.ActionEvent;

public class LoginUI extends Dialog implements ActionListener {
  private static final String EXIT_BUTTON_LABEL = "Exit";
  private static final String LOGIN_BUTTON_LABEL = "Login";

  private BidderUI parent;
  private TextField username;
  private TextField password;

  public LoginUI(BidderUI parent) {
    super(parent, "Auction Login", true);    // modal dialog
    this.parent = parent;
  }

  public void display() {
    Panel dialogButtonBar = new Panel(new FlowLayout());
    add("South", dialogButtonBar);

    Panel tmpPanel = new Panel(new GridLayout(2, 2, 5, 5));
    add("Center", tmpPanel);

    Label usernameLabel = new Label("Username: ");
    tmpPanel.add(usernameLabel);

    username = new TextField("anonJ2SE");
    username.setEditable(true);
    tmpPanel.add(username);

    Label passwordLabel = new Label("Password: ");
    tmpPanel.add(passwordLabel);
```

```
    password = new TextField("");
    password.setEditable(true);
    tmpPanel.add(password);

    Button exit = new Button(EXIT_BUTTON_LABEL);
    dialogButtonBar.add(exit);
    exit.addActionListener(this);

    // set up button bar along bottom
    Button login = new Button(LOGIN_BUTTON_LABEL);
    dialogButtonBar.add(login);
    login.addActionListener(this);

    // auto-resize dialog, and display it
    pack();
    show();
  }

  // button event handler
  public void actionPerformed(ActionEvent e) {
    String action = e.getActionCommand();
    if (action.equals(LOGIN_BUTTON_LABEL)) {
      parent.user = username.getText();
      parent.password = password.getText();
      dispose();
    } else if (action.equals(EXIT_BUTTON_LABEL)) {
      System.exit(0);
    }
  }
}
```

The BidderUI class

```
import java.awt.Dialog;
import java.awt.TextArea;
import java.awt.Label;
import java.awt.Panel;
import java.awt.Frame;
import java.awt.Button;
import java.awt.GridLayout;
import java.awt.FlowLayout;
import java.awt.BorderLayout;
import java.awt.event.ActionListener;
import java.awt.event.ActionEvent;
import java.util.Enumeration;

public class BidderUI extends Frame implements ActionListener {
  private static final String EXIT_BUTTON_LABEL = "Exit";

  private static Panel mainPanel;
  private static Panel buttonBar;
  private Bidder bidder;
  protected String user;
```

```
    protected String password;

    public BidderUI() {
      LoginUI loginUI = new LoginUI(this);
      loginUI.display();

      bidder = new Bidder(user, password, this);

      // title for the catalog window
      setTitle("AuctionHouse, " + bidder.getUserName() + " logged in");
```

This class defines a simple UI with a main panel, and a button bar below it:

```
      mainPanel = new Panel(new BorderLayout());
      add("Center", mainPanel);
      buttonBar = new Panel(new FlowLayout());

      // set up button bar along bottom
      Button exit = new Button(EXIT_BUTTON_LABEL);
      buttonBar.add(exit);
      exit.addActionListener(this);
      add("South", buttonBar);

      // display the catalog, if there is one
      if (bidder.getCatalog() == null) {
        displayCatalog(new Catalog());
      } else {
        displayCatalog(bidder.getCatalog());
      }
      show();
    }

    // button event handler for catalog window
    public void actionPerformed(ActionEvent e) {
      String action = e.getActionCommand();
      if (action.equals(EXIT_BUTTON_LABEL)) {
        bidder.exit();
      } else {
        // must be a bid button that's been pressed
        BidUI bidDialog = new BidUI(this);

        // bidDialog.display(bidder,action);
        Lot lot = bidder.getCatalog().getLot(action);
        bidDialog.display(bidder, lot);
      }
    }
```

The `displayCatalog()` method simply creates an `Enumeration` out of the catalog and prints each lot to the screen:

```
    public void displayCatalog(Catalog cat) {
      // clear away any existing components
      mainPanel.removeAll();
```

```
    Enumeration enum = cat.elements();
    if (enum.hasMoreElements()) {
      int rows = cat.size() + 1;
      int cols = 4;
      int hgap = 5;
      int vgap = 5;
      Panel tmpPanel = new Panel(new GridLayout(rows, cols, hgap, vgap));
      mainPanel.add("Center", tmpPanel);

      Label col1Title = new Label("Bid for a...");
      tmpPanel.add(col1Title);
      Label col2Title = new Label("Description");
      tmpPanel.add(col2Title);
      Label col3Title = new Label("Price");
      tmpPanel.add(col3Title);
      Label col4Title = new Label("Bidder");
      tmpPanel.add(col4Title);

      Lot lot;
      enum = cat.elements();
      while (enum.hasMoreElements()) {
        lot = (Lot) enum.nextElement();

        Button bidButton = new Button(lot.getName());
        tmpPanel.add(bidButton);
        bidButton.addActionListener(this);

        Label desc = new Label(lot.getDescription());
        tmpPanel.add(desc);

        Label price = new Label(Integer.toString(lot.getPrice()));
        tmpPanel.add(price);

        Label bidderName = new Label(lot.getCurrentBidder());
        tmpPanel.add(bidderName);
      }
    } else {
      TextArea emptyCatalog = new TextArea();
      emptyCatalog.setText("There are currently no items in your Catalog." +
                           "Please wait for them to arrive!");
      emptyCatalog.setEditable(false);
      mainPanel.add("Center", emptyCatalog);
    }

    // auto-resize app window in case new lots have arrived
    pack();
  }

  public static void main(String[] args) {
    BidderUI bidderUI = new BidderUI();
  }
}
```

The BidUI Class

The `BidUI` class is a `Dialog` to capture bid input:

```
import java.awt.Dialog;
import java.awt.TextField;
import java.awt.Panel;
import java.awt.Button;
import java.awt.FlowLayout;
import java.awt.event.ActionListener;
import java.awt.event.ActionEvent;

public class BidUI extends Dialog implements ActionListener {
  private static final String CANCEL_BUTTON_LABEL = "Cancel";
  private static final String BID_BUTTON_LABEL = "Place Bid";

  private BidderUI parent;
  private Lot lot;
  private Bidder bidder;
  private TextField bidPrice;

  public BidUI(BidderUI parent) {
    super(parent);
    this.parent = parent;
  }

  public void display(Bidder bidder, Lot lot) {
    this.bidder = bidder;
    this.lot = lot;

    Panel dialogButtonBar = new Panel(new FlowLayout());
    add("South", dialogButtonBar);

    TextField name = new TextField();
    name.setEditable(false);
    name.setText("Lot: " + lot.getName());
    add("North", name);

    bidPrice = new TextField(Integer.toString(lot.getPrice()
                                              + lot.getBidIncrement()));
    bidPrice.setEditable(true);
    add("Center", bidPrice);

    Button cancel = new Button(CANCEL_BUTTON_LABEL);
    dialogButtonBar.add(cancel);
    cancel.addActionListener(this);

    // set up button bar along bottom
    Button placeBid = new Button(BID_BUTTON_LABEL);
    dialogButtonBar.add(placeBid);
    placeBid.addActionListener(this);

    // auto-resize dialog, and display it
    pack();
    show();
  }
```

```
    // button event handler
    public void actionPerformed(ActionEvent e) {
      String action = e.getActionCommand();
      if (action.equals(BID_BUTTON_LABEL)) {
```

Here we make a copy of the current lot, so we don't change the actual price in our catalog. We want the AuctionHouse to do this for us, as a confirmation that it received our bid successfully (note that we cannot use Cloneable in CLDC):

```
        Lot lotCopy = new Lot(lot);
        lotCopy.setPrice(Integer.parseInt(bidPrice.getText()));
        bidder.makeBid(lotCopy);
        parent.displayCatalog(bidder.getCatalog());
        dispose();
      } else if (action.equals(CANCEL_BUTTON_LABEL)) {
        dispose();
      }
    }
  }
```

The most interesting points here are that we have purposefully used a simple layout (GridLayout) because the more comprehensive GridBagLayout is not supported by kAWT. Also, we could have placed the check for an invalid bid in this class, but we wanted an excuse to demonstrate the use of queues.

Running the Auction House Application

Running this example should be straightforward, regardless of the vendor you choose. We chose to run it on the open source J2EE JBoss application server, which conveniently includes, among many other goodies, a JMS provider, JBossMQ (originally named SpyderMQ). It is available from www.jboss.org.

If you choose not to use JBossMQ, there are four small code changes to make, and they are all JNDI-related. Change all four definitions of the topic and queue JNDI lookup names to match your vendor's naming scheme. They are defined as Strings at the top of initJMS() in AuctionHouse.java and Bidder.java.

Also, don't forget to make sure that the application can find the default JNDI settings (defined in jndi.properties in the environment path (not the classpath)), or change the new InitialContext() to be loaded with specific parameters by using the new InitialContext(params) version of the call, where params is a Properties object, as described in Chapter 11.

These are the steps to configure your messaging server, and run the two types of client:

- Create the Lot topic and Bid queue using your vendor's administration tool.
- Add some example bidders ' usernames and passwords using your vendor's administration tool.
- Start your messaging server
- Start one or more BidderUI instances. They will have no entries in their catalogs.
- Start one AuctionHouseCmdLine instance. The bidder consoles will all receive the initial three lots created automatically by the AuctionHouse.

Note: JBossMQ on Win98 seems not to support more than three simultaneous sockets for some unknown reason, so if you find the same limitation on your own development environment, limit yourself to two Bidder clients, and one AuctionHouse. It appears to work fine on Linux.

Here are the JBossMQ specific steps. Edit conf\jbossmq.xml, and add the following lines anywhere within the root <Server> element:

```
<Topic><Name>Lot</Name></Topic>
<Queue><Name>Bid</Name></Queue>
```

In the same file, add the bidder usernames and passwords to the <UserManager> element, in the following format. Here we choose not to have a password defined for our user, 'anonJ2SE':

```
<User>
  <Name>anonJ2SE</Name>
  <Password></Password>
</User>
```

Start JBoss. You need make no other changes to the default JBoss configuration – JBossMQ is started by default.

Start the Bidder client. Here is an example script where JBOSS_HOME is an environment variable that points to your JBoss home directory (the script is all one line):

```
java  -cp
.;%JBOSS_HOME%\lib\ext\jta-spec1_0_1.jar;%JBOSS_HOME%\lib\ext\jms.jar;%JBOSS_HOME%
\lib\ext\jnpserver.jar;%JBOSS_HOME%\lib\ext\jbossmq.jar BidderUI
```

Having started the AuctionHouseCmdLine (passing in a valid username and password on the command line) you should see all Bidder windows update to show the example lots that AuctionHouseCmdLine creates.

The lots have no bidder associated with them since they're all new. Click on one of the bid buttons, and place a bid greater than the start price. All Bidder clients will again update to show the new price for that lot, plus the name of the bidder that placed the successful bid. The AuctionHouseCmdLine console window also holds status information about bids received.

Example Palm iBus//Mobile Application

Our Bidder JMS client is suitable for converting to run on a Palm platform using the CLDC Configuration plus kAWT and Softwired's iBus//Mobile libraries. Let's see what iBus//Mobile offers, and then see what code changes we must make to make our Bidder run on a Palm.

iBus//Mobile is a combination of CLDC/CDC libraries plus a gateway. The libraries sit on the mobile device, and give us the ability to create JMS classes, albeit using a ch.softwired package instead of javax.jms. (In future version the developer will have the choice of importing the javax.jms package at the cost of having a lightly larger JAR file – approximately 20K)

The Softwired JMS library is a subset of the full JMS. The limitations imposed by the libraries being a subset are relatively small, as we find in the rest of this chapter. The libraries also offer an alternative naming system to JNDI, since JNDI is not available on CLDC or CDC. The classes take very little space – 60Kb or so for the library, plus as low as 50Kb run-time RAM.

The iBus//Mobile Gateway is needed to translate between the different protocols used by mobile clients (such as WAP) and the TCP/IP used by Softwired's iBus Messaging server. iBus//Mobile also offers alternative ways of delivering messages to thin clients such as SMS phones, but this is outside the scope of this chapter.

> *Note that iBus//Mobile can only communicate with Softwired's iBus messaging server so that is our JMS provider for this example. In future releases, the WebLogic provider will be supported.*

We do need to make a small change to our `AuctionHouse`, as well as the iBus//Mobile-related changes to the `Bidder` client. As we are changing the underlying messaging provider, the naming service-related code in `AuctionHouse` must change to match the iBus messaging server.

Code Changes

The number of changes to port our example to a Palm using kAWT and iBus//Mobile is remarkably small, and is a credit to those two vendors. The changes are limited to the manner in which iBus//Mobile replaces JNDI with its own lookup methods.

The majority of the steps to convert the code are within the `initJMS()` methods of `AuctionHouse` and `Bidder`:

- Change the import statements from `java.jms.*` and `javax.naming.*` to their `ch.softwired.*` equivalent libraries and classes.
- Change the code to discover `ConnectionFactory` objects, because iBus//Mobile has to use its own naming lookup methods (JNDI doesn't exist on a CLDC platform).
- You may need to change the code to discover topics and queues because iBus//Mobile has to use its own lookup methods. JNDI is also supported in iBus//Mobile so this isn't strictly necessary.
- Remove the catch statement block for `NamingException` in the `AuctionHouse` and `Bidder` constructors, since this is only thrown by the JNDI methods.
- Optional: modify `BidderUI` to suit a Palm-sized screen. We chose to omit the `Lot` description from the catalog display on the Palm client, and also display the catalog in a scrollable window, since space is tight.

Here is the source code. A full listing can be downloaded from http://www.wrox.com.
First, the `AuctionHouse` class. The import statements have changed from `javax.jms.*` and `javax.naming.*` to the `ch.softwired` equivalent libraries and classes:

```
// iBus//Mobile-specific libraries to support naming and JMS
import ch.softwired.mobilejms.TopicConnectionFactory;
import ch.softwired.mobilejms.TopicConnection;
import ch.softwired.mobilejms.TopicSession;
import ch.softwired.mobilejms.Topic;
```

```
import ch.softwired.mobilejms.TopicPublisher;
import ch.softwired.mobilejms.QueueConnectionFactory;
import ch.softwired.mobilejms.QueueConnection;
import ch.softwired.mobilejms.QueueSession;
import ch.softwired.mobilejms.Queue;
import ch.softwired.mobilejms.QueueReceiver;
import ch.softwired.mobilejms.Session;
import ch.softwired.mobilejms.Message;
import ch.softwired.mobilejms.StreamMessage;
import ch.softwired.mobilejms.MessageListener;
import ch.softwired.mobilejms.JMSException;
import ch.softwired.mobileibus.naming.NameServer;
import ch.softwired.mobileibus.qos.ProtocolStackRegistry;
import ch.softwired.mobileibus.qos.QoSConst;
```

The `AuctionHouse.initJMS()` method has the iBus//Mobile naming code in place of the standard JNDI calls:

```
public void initJMS(String username, String password) {
  String LOT_TOPIC = "topic/Lot";
  String BID_QUEUE = "queue/Bid";

  try {
    // Note the different protocol stack to that of the J2ME Bidder client
    ProtocolStackRegistry.registerTcpQos("qos-tcp", "AuctionHouse",
                                         "localhost");
    TopicConnectionFactory tFactory =
      NameServer.lookupTopicConnectionFactory("qos-tcp");
    QueueConnectionFactory qFactory =
      NameServer.lookupQueueConnectionFactory("qos-tcp");

    // we only want authenticated users running AuctionHouse.
    // However, iBus Standard Edition does not authenticate.
    tConnection = tFactory.createTopicConnection();
    tSession = tConnection.createTopicSession(false,
                                          Session.AUTO_ACKNOWLEDGE);

    qConnection = qFactory.createQueueConnection();
    qSession = qConnection.createQueueSession(false,
                                          Session.AUTO_ACKNOWLEDGE);

    lotTopic = tSession.createTopic(LO);
    lotPub = tSession.createPublisher(lotTopic);

    bidQ = qSession.createQueue(BID_QUEUE);
    bidQReceiver = qSession.createReceiver(bidQ);
    bidQReceiver.setMessageListener(this);

    // only need to start connections that have incoming messages
    qConnection.start();
  } catch (JMSException jmse) {
    // iBus//Mobile-specific libraries to support naming and JMS
    System.out.println("JMSException in AuctionHouse constructor");
    jmse.printStackTrace();
    System.exit(-1);
  }
}
```

The two ConnectionFactory objects, tFactory and qFactory, are now located using Softwired's NameServer object, rather than JNDI. Also, the Softwired protocol stack needs to be specified, for the connection to the iBus//Mobile gateway.

The protocol stack methods for a PC and a Palm are different – ProtocolStackRegistry.registerTcpQos() for the PC, and ProtocolStackRegistry.registerQos() for the Palm. Softwired defines several protocol stacks, but you can create your own if one does not exist for your device. We will not go into further detail here, but the iBus//Mobile documentation covers protocol stacks comprehensively.

The two Destination objects for AuctionHouse, lotTopic and bidQ, are created dynamically, rather than located via a JNDI lookup. According to the JMS specification, when using the createTopic() and createQueue() methods, we are not actually creating topics and queues, but obtaining references to ones that are already in existence using a (potentially) vendor-specific String. This is not true of the iBus//Mobile API, which really does create the topic or queue for us. Normally, creation of such Destination objects is the job of the JMS provider, rather than the JMS code itself.

The final change is that the NamingException catch is no longer needed, since its JNDI-specific.

For Bidder, the modified initJMS() method is very similar to the one for AuctionHouse above:

```
public void initJMS(String username, String password) {

  // iBus-specific topic and queue names
  String LOT_TOPIC = "topic/Lot";
  String BID_QUEUE = "queue/Bid";

  // Name used for iBus logging
  String NAME = "kAWTBidder";

  try {
    String qosName = "qos-" + NAME;
    Hashtable params = new Hashtable();
    params.put(QoSConst.KEY_URL, "socket://localhost:8738");
    params.put(QoSConst.KEY_DOWN_POLL_INTERVAL, "5000");
    params.put(QoSConst.KEY_DO_RECONNECT, "true");
    ProtocolStackRegistry.registerQos(qosName, params, username,
                                      false,   /* doStoreFwd */
                                      false,   /* doReliability */
                                      false);  /* doSecurity */
    TopicConnectionFactory tFactory =
                    NameServer.lookupTopicConnectionFactory(qosName);
    QueueConnectionFactory qFactory =
                    NameServer.lookupQueueConnectionFactory(qosName);

    // We can, optionally, choose to authenticate the user here.
    // However, iBus Standard Edition does not authenticate.
    qConnection = qFactory.createQueueConnection();
    qSession = qConnection.createQueueSession(false,
                                          Session.AUTO_ACKNOWLEDGE);
    bidQ = qSession.createQueue(BID_QUEUE);
    bidQSender = qSession.createSender(bidQ);

    tConnection = tFactory.createTopicConnection();
```

```
        tSession = tConnection.createTopicSession(false,
                                                   Session.AUTO_ACKNOWLEDGE);
        lotTopic = tSession.createTopic(LOT_TOPIC);
        lotSub = tSession.createSubscriber(lotTopic);
        lotSub.setMessageListener(this);

        // only need to start connections that have incoming messages
        tConnection.start();
      } catch (JMSException jmse) {
        System.out.println("JMSException in Bidder constructor");
        System.out.println("JMSException error code is: "
                           + jmse.getErrorCode());
        jmse.printStackTrace();
        System.exit(-1);
      }
    }
```

Bidder's `onMessage()` method has one slightly obscure alteration, which is required because kAWT is not thread-safe, and yet we need to call it within a thread. In `onMessage()`, we cannot simply display the catalog straight after we have received a new/update lot as we did in the J2SE version, like so:

```
ui.displayCatalog(cat);
```

Instead, we must make sure the kAWT catalog displaying work happens after the `onMessage()` thread finishes. We do this by creating a temporary class instance, `doDisplayCatalog`, that implements `Runnable`, and we then place an event on the AWT system event queue. This causes the `doDisplayCatalog.run()` method to execute some time later, when it's safe to do so, thus displaying the catalog:

```
    Runnable doDisplayCatalog = new Runnable() {
      public void run() {
      ui.displayCatalog(cat);
    }
  };
  Toolkit.getDefaultToolkit().getSystemEventQueue()
         .invokeLater(doDisplayCatalog);
```

The only other change we make is to `BidderUI.displayCatalog()`, where we do not display the lot description on a Palm because of lack of screen space. Our `GridLayout` now only needs 3 columns, one each for the bid button, price, and bidder. We also place the `GridLayout` in a `ScrollPane`, since the screen can only display approximately 6 lots.

To test this mobile version of the JMS client, we need to install and test the iBus and iBus//Mobile components.

Obtaining and Installing the iBus Server

Go to www.softwired-inc.com, register, and download the standard iBus server. A 30-day license file will be e-mailed to you. Copy the e-mailed `ibuslicense.bin` to the `<iBus_Home>\server\lib\` directory, according to the installation instructions. Start the server by running the `startserver` batch file in `<iBus_Home>\server\bin\`. The server can be stopped by using the `stopserver` batch file in the same directory.

Obtaining and Installing the iBus//Mobile Gateway and Libraries

Go to www.softwired-inc.com, register, and download. A URL will be e-mailed to you linking you to the download instructions and the download itself. Follow the instructions, and install the gateway and corresponding license.

Reboot, start the iBus server, and then start the gateway using the `startgateway` batch file found in the `<iBusMobile_install>\gateway\bin\` directory. (Note that the scripts do not work on Windows 98 without minor adjustments.) It should connect to the already-running iBus server successfully. As with the server, there is a corresponding shut down script included for the mobile gateway.

Testing the Basic iBus Installation

Note: The following instructions apply to iBus//Mobile 2.0.1. Future versions may change. Read the accompanying Softwired documentation if in doubt.

Softwired supply some basic example JMS clients for us to test our iBus and iBus//Mobile installation. They are called `sproducer` and `sconsumer`. Even better, they supply Palm `.prc` versions too, and the source is available if you wish to modify them yourself.

The directory structure of both the iBus and iBus//Mobile installations is slightly confusing, and the compiled PC example clients are not actually supplied in plain class format. They are in a JAR, `ibusmob-jms-j2se.jar`, located in `<iBusMobile_install>\client\j2se\lib\`.

You can run them (on Windows/NT at least) from the **Start** menu, or by running the `startclient.bat` or `startclient.sh` script in `<iBusMobile_install>\client\j2se\bin\`, passing in the appropriate client class, `ch.softwired.mobileibus.tool.sproducer -i`, or `ch.softwired.mobileibus.tool.sconsumer -i`.

The example source is in `<iBusMobile_install>\client\j2se\src\ch\softwired\mobileibus\tool\`.

The library for compiling the `AuctionHouse` source is in `<iBusMobile_install>\client\j2se\lib\ibusmob-jms-j2se.jar`.

The pre-built Palm `.prc` clients are in `<iBusMobile_install>\client\palm\prc\`.

The source is in `<iBusMobile_install>\client\palm\src\ch\softwired\mobileibus\tool\`.

You will also find the preverified Palm classes necessary for compilation, and bundling into your own `Bidder.prc`, in `<iBusMobile_install>\client\palm\lib\preverify\`.

AuctionHouse-specific iBus Configuration

No specific configuration is needed to iBus, unless you are using the Enterprise Edition, and wish to add users/passwords. However, we still need to configure our `Lot` topic and our `Bid` queue.

iBus provides an administrative console for us to create topics and queues, which can be downloaded from the same source as the server and gateway. Once this is installed, start the server as described above. The admin console is started in a similar manner using the `startAdmin` script found in the `<iBusAdmin_Home>\adminClient\bin\` directory. Log on using the defaults and click on the **Messages** tab. Now click on either the **Topics** or **Queues** folder in the sidebar. Create a new queue called `queue/Bid` and a new topic called `topic/Lot`. This is all we need to do to create `Destination` objects in iBus.

Palm Simulators

The Palm simulator, POSE, is available from http://www.palmos.com, as described in Chapter 4. Note that you need to have OS3.5 or higher, because iBus//Mobile needs the extra Java heap space which 3.5 offers (223Kb as opposed to the 95Kb for OS3.0).

Also note that iBus//Mobile can be run on other mobile devices, not just a Palm. In particular, iBus//Mobile comes complete with Symbian examples, and it also has extensive support for 'dumb' non-J2ME clients such as WAP and SMS phones and pagers. This means none of your potential mobile users are excluded from your messaging application. See the Softwired iBus//Mobile documentation for greater detail on how to implement these non-J2ME mobile clients.

kAWT Installation

kAWT is downloadable from http://www.kawt.de, as described in Chapter 4.

Building the JMS Clients

Our Palm `Bidder` client can be built in the same manner as described in Chapter 4. Be sure to include both iBus//Mobile and kAWT libraries when using `MakePalmApp` to create your Palm executable. Here are some example Windows 98 scripts. You will need to modify them to suit your environment.

One environment variable is pre-defined outside the scripts. `IBUSMOB_HOME` is the installation directory of iBus//Mobile. This is set during installation of iBus//Mobile, so you shouldn't need to do this step yourself.

Our directory structure is:

- `.`
  Scripts and destination for `BidderUI.prc` Palm executable
- `src`
  Source files
- `tmp`
  Compiled Palm classes (not preverified)
- `preverifiy`
  Preverified Palm classes
- `classes`
  J2SE classes, for `AuctionHouse`

The first script, `makeBidder.bat`, compiles the individual classes for the bidder client. Note that there are line wraps in this listing due to the book's page width. Download the script source from http://www.wrox.com if you are unsure of where the breaks occur:

```
REM mkdir tmp

REM The directory holding jme_cldc and kAWT directories
set KVM_ROOT=C:

set
KVM_CLASSES=%KVM_ROOT%\j2me_cldc1.0.2\bin\common\api\classes;%KVM_ROOT%\j2me_cldc1
.0.2\tools\palm\src\classes.jar
```

```
set KAWT_CLASSES=%KVM_ROOT%\kAWT\kawt.jar

REM The directory holding the iBus preverified libraries
set IBUS_PALM_CLASSPATH=%IBUSMOB_HOME%\client\palm\lib\preverify

set CLASSPATH=%KVM_CLASSES%;%KAWT_CLASSES%;%IBUS_PALM_CLASSPATH%

javac -g:none -d tmp -classpath tmp;%CLASSPATH% -bootclasspath
%KVM_ROOT%\j2me_cldc1.0.2\bin\kjava\api\classes src\Lot.java
javac -g:none -d tmp -classpath tmp;%CLASSPATH% -bootclasspath
%KVM_ROOT%\j2me_cldc1.0.2\bin\kjava\api\classes src\Catalog.java
javac -g:none -d tmp -classpath tmp;%CLASSPATH%;src -bootclasspath
%KVM_ROOT%\j2me_cldc1.0.2\bin\kjava\api\classes src\Bidder.java
javac -g:none -d tmp -classpath tmp;%CLASSPATH% -bootclasspath
%KVM_ROOT%\j2me_cldc1.0.2\bin\kjava\api\classes src\LoginUI.java
javac -g:none -d tmp -classpath tmp;%CLASSPATH% -bootclasspath
%KVM_ROOT%\j2me_cldc1.0.2\bin\kjava\api\classes src\BidUI.java
javac -g:none -d tmp -classpath tmp;%CLASSPATH% -bootclasspath
%KVM_ROOT%\j2me_cldc1.0.2\bin\kjava\api\classes src\BidderUI.java

cls

call makeBidderPRC.bat
```

The second script preverifies the classes, and builds the Palm application. The `PRE_OUTPUT` directory needs to exist before running this script:

```
REM The directory holding jme_cldc and kAWT directories
set KVM_ROOT=C:

REM The directory holding your compiled classes,i.e. ch\softwired\mobileibus\tool
for the example source
set COMPILED_CLASSES=tmp

REM The directory holding the iBus preverified libraries
set IBUS_PALM_CLASSPATH=%IBUSMOB_HOME%\client\palm\lib\preverify

set
KVM_CLASSES=%KVM_ROOT%\j2me_cldc1.0.2\bin\common\api\classes;%KVM_ROOT%\j2me_cldc1
.0.2\tools\palm\src\classes.jar
set KAWT_CLASSES=%KVM_ROOT%\kAWT\kawt.jar
set CLASSPATH=%KVM_CLASSES%;%KAWT_CLASSES%;%CLASSPATH%
set
PRE_CLASSPATH=%COMPILED_CLASSES%;%KVM_CLASSES%;%IBUS_PALM_CLASSPATH%;%KAWT_CLASSES%

set PRE_OUTPUT=preverify
REM mkdir %PRE_OUTPUT%

REM first PREVERIFY
%KVM_ROOT%\j2me_cldc1.0.2\bin\win32\preverify -classpath %PRE_CLASSPATH% -d
%PRE_OUTPUT% %COMPILED_CLASSES%

REM Remove unverified classes, to avoid confusion on subsequent builds.
del tmp\*.class
```

```
REM roll them all into a jar
cd %PRE_OUTPUT%
jar cf bidder.jar *
cd ..

REM Make the Palm app
REM  makePalmApp variables
set MAKEPALMAPP_CP=%PRE_OUTPUT%\bidder.jar;%KAWT_CLASSES%;%IBUS_PALM_CLASSPATH%
set MAKEPALMAPP_BCP=%KVM_ROOT%\j2me_cldc1.0.2\bin\common\api\classes
java palm.database.MakePalmApp -networking -v -version "1.0" -bootclasspath
%MAKEPALMAPP_BCP% -classpath %MAKEPALMAPP_CP% -o Bidder.prc BidderUI

REM Lastly, remove preverified classes, to avoid confusion on subsequent builds.
del preverify\*.class
```

The last script, for building the desktop `AuctionHouse` client uses a different iBus//Mobile Java library, the J2SE one, and of course we don't need to preverify or use `MakePalmApp`:

```
javac -d classes -classpath %IBUSMOB_HOME%\client\j2se\lib\ibusmob-jms-j2se.jar
src\Lot.java
javac -d classes -classpath
classes;%IBUSMOB_HOME%\client\j2se\lib\ibusmob-jms-j2se.jar src\Catalog.java
javac -d classes -classpath
classes;%IBUSMOB_HOME%\client\j2se\lib\ibusmob-jms-j2se.jar src\AuctionHouse.java
javac -d classes -classpath
classes;%IBUSMOB_HOME%\client\j2se\lib\ibusmob-jms-j2se.jar
src\AuctionHouseCmdLine.java
```

You may also wish to consider using the 'ant' make utility that is available from http://www.apache.org rather than using unintelligent scripts. It is a Java application so you don't need any additional executables or resources.

Running the Palm JMS Client

To run this example:

- Start iBus server.
- Start the iBus//Mobile gateway.
- Start Palm emulator, and load the `BidderUI.prc`.
- Start one or more Palm `BidderUI` clients, and log in. You can use the default username with no password because the standard iBus server does not authenticate the user. Note that connecting to the provider can take 10-20 seconds.
- Start the `AuctionHouseCmdLine` (the iBus//Mobile version, not the JBoss version).

Note that you can start `AuctionHouse` first, but the Palm clients will miss the default lots that `AuctionHouse` creates. If you do start the Palm clients first, then just enter some more lots on the `AuctionHouse` command line, and they will appear on the Palms. Now try entering a bid on one of the Palms, and make sure that bid is received by the `AuctionHouse`, and sent back out to all Palms. Their catalogs will refresh, showing the new bid price, and the bidder's name.

> **Note: versions of iBus//Mobile before 2.0.2 deliver queue messages intermittently, so if you use these versions you may find some of your bids do not reach the `AuctionHouse` client until some time later, usually after restarting the iBus server.**

Here is the result:

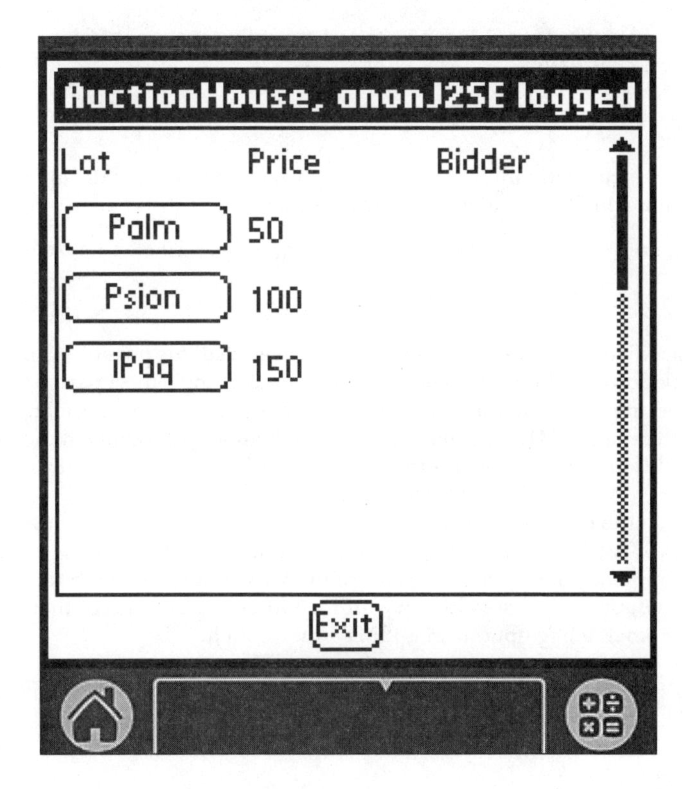

The Compromises

Now that we've got the mobile JMS client running, let's look back and review what we had to compromise on to make it run. There are several areas:

CLDC compromises:

- Use of a `Vector` rather than a more sophisticated `Collection` for our catalog of lots.
- Encoding/decoding our `Lot` objects for placing in a `StreamMessage` rather than using an `ObjectMessage`. No CLDC support for serialization is the underlying reason that this is not supported by iBus//Mobile.

JMS (iBus//Mobile) compromises:

- iBus//Mobile does not yet support the MIDP RecordStore database, so we could not run it on a MIDP platform.

UI compromises:

- kAWT was used in place of the non-existent PDAP
- GridLayout rather than GridBagLayout used for layout control, because kAWT does not support GridBagLayout

Device compromises:

- Palms cannot currently multi-task so it wasn't possible to allow the application to run in the background, which is the ideal for many messaging clients
- We could not use Palm OS 3.0, because of lack of heap space required for iBus//Mobile

Is It a Success?

The answer is a qualified 'yes'. Yes because we have proved that it is possible to port a standard J2EE client to a J2ME platform with minimal code changes, albeit using a third party UI library in place of any official AWT-supporting profile at the time of writing. However, the combination of J2ME's immaturity, and the limited nature of a J2ME environment make such applications quite difficult to set up, because we have used several third party components that are both pushing the limits of the technology.

This is not to belittle the third-party libraries in any way. It isn't currently a straightforward case of downloading a J2ME-J2EE toolkit, writing the application code, and pressing a magic button. But then again, J2EE development is never easy because of the very nature of distributed applications. Not only are they difficult to deploy, they are also very difficult to debug because of their distributed nature, combined with minimal debug options in a J2ME environment.

If you need a messaging solution, and want to second-guess the PDAP Profile, then getting to grips with iBus//Mobile and kAWT will give you an excellent head start in understanding the necessary environment.

The MIDP Client

Softwired also supplies a MIDP version of JMS, so we will create a MIDP client too. We can run MIDP and kAWT together against the same AuctionHouse, and watch them interact.

The changes to the MIDP implementation compared to the kAWT version are mostly related to the user interface, and the method of building the client.

For our MIDP client, in place of the three UI classes (LoginUI, BidderUI, and BidUI), we have a single class JMSMIDlet instead:

```
import javax.microedition.midlet.MIDlet;
import javax.microedition.lcdui.Displayable;
import javax.microedition.lcdui.Display;
import javax.microedition.lcdui.Form;
import javax.microedition.lcdui.TextField;
import javax.microedition.lcdui.TextBox;
import javax.microedition.lcdui.List;
```

```
import javax.microedition.lcdui.Choice;
import javax.microedition.lcdui.StringItem;
import javax.microedition.lcdui.Command;
import javax.microedition.lcdui.CommandListener;

import java.util.Enumeration;

import ch.softwired.mobilejms.MessageListener;

public class JMSMIDlet extends MIDlet implements CommandListener {

  // button/menu labels
  private static final String CANCEL_BUTTON_LABEL = "Cancel";
  private static final String BID_BUTTON_LABEL = "Place Bid";
  private static final String EXIT_BUTTON_LABEL = "Exit";
  private static final String OK_BUTTON_LABEL = "OK";
  private static final String LOGIN_BUTTON_LABEL = "Login";

  private static final String LOTNAME_DELIMITER = ":";

  // state, to identify which screen the 'OK' button has been pressed on
  private static final int USERNAME_SCREEN = 0;
  private static final int PASSWORD_SCREEN = 1;
  private static final int BID_SCREEN = 3;

  // order of items in bid Form
  private static final int DESC_ITEM = 0;
  private static final int PRICE_ITEM = 1;
  private static final int BID_ITEM = 2;
  private static final int BIDDER_ITEM = 3;

  private int screen;
  private Command exitCommand;
  private Command okCommand;
  private Command cancelCommand;

  private Bidder bidder;
  private String user;
  private String password;

  // the current lot
  private Lot lot;

  public Display display;

  public JMSMIDlet() {
    display = Display.getDisplay(this);
    exitCommand = new Command(EXIT_BUTTON_LABEL, Command.EXIT, 1);
    okCommand = new Command(OK_BUTTON_LABEL, Command.OK, 1);
    cancelCommand = new Command(CANCEL_BUTTON_LABEL, Command.CANCEL, 1);
  }
```

The default username is added in the `startApp()` method:

```
public void startApp() {
  if (bidder == null) {
    TextBox t = new TextBox("Username:", "anonMIDP", 32, TextField.ANY);
    t.addCommand(exitCommand);
    t.addCommand(okCommand);
    t.setCommandListener(this);
    screen = USERNAME_SCREEN;
    display.setCurrent(t);
  } else {
    displayCatalog(bidder.getCatalog());
  }
}
```

As we are using MIDP, the screen space is potentially even more limited than in kAWT, so a new `Form` is used for each bid placement:

```
public void displayBid(Bidder bidder, String lotName) {
  lot = bidder.getCatalog().getLot(lotName);

  Form form = new Form(lotName + ":");
  form.insert(DESC_ITEM,
              new StringItem("", "Desc:" + lot.getDescription() + "\n"));
  form.insert(PRICE_ITEM, new StringItem("", "Price:" + lot.getPrice()));

  TextField textField =
    new TextField("Your bid:", Integer.toString(lot.getPrice()
          + lot.getBidIncrement()), 32, TextField.NUMERIC);
  form.insert(BID_ITEM, textField);

  form.insert(BIDDER_ITEM,
              new StringItem("",
                             "Current Bidder:" + lot.getCurrentBidder()
                             + "\n"));

  okCommand = new Command(BID_BUTTON_LABEL, Command.OK, 1);
  form.addCommand(cancelCommand);
  form.addCommand(okCommand);
  form.setCommandListener(this);
  screen = BID_SCREEN;
  display.setCurrent(form);
}
```

A new `List` is created each time the catalog is displayed anew:

```
public void displayCatalog(Catalog cat) {
  List list;
  Enumeration enum = cat.elements();
  if (enum.hasMoreElements()) {
    list = new List("Lot name: price", Choice.IMPLICIT);
    while (enum.hasMoreElements()) {
      lot = (Lot) enum.nextElement();
      list.append(lot.getName() + LOTNAME_DELIMITER + " "
                  + lot.getPrice(), null);
```

```
      }
      list.addCommand(exitCommand);
      list.setCommandListener(this);
      display.setCurrent(list);
    } else {
      Form form = new Form("Catalog empty");
      form
        .append("There are currently no items in your Catalog." +
                " Please wait for them to arrive!");
      form.addCommand(exitCommand);
      form.setCommandListener(this);
      display.setCurrent(form);
    }
  }

  public void commandAction(Command c, Displayable s) {
    // check if the event was from a List object
    if (c == List.SELECT_COMMAND) {
      List l = (List) s;
      String choice = l.getString(l.getSelectedIndex());

      // pick out just the first part of the List string
      String lotName = choice.substring(0,
                                        choice.indexOf(LOTNAME_DELIMITER));
      displayBid(bidder, lotName);
    } else if (c.getCommandType() == Command.EXIT) {
      destroyApp(true);
    } else if (c.getCommandType() == Command.CANCEL) {
      startApp();
    } else if (c.getCommandType() == Command.OK) {
      if (screen == USERNAME_SCREEN) {
        TextBox t = (TextBox) s;
        user = t.getString();

        // now display password entry field
        t = new TextBox("Password:", "", 32, TextField.ANY);
        t.addCommand(cancelCommand);
        t.addCommand(new Command("Login", Command.OK, 1));
        t.setCommandListener(this);
        screen = PASSWORD_SCREEN;
        display.setCurrent(t);
      } else if (screen == PASSWORD_SCREEN) {
        TextBox t = (TextBox) s;
        password = t.getString();

        // create a Bidder instance, and connect to the JMS provider
        bidder = new Bidder(user, password, this);
        displayCatalog(bidder.getCatalog());
      } else if (screen == BID_SCREEN) {
        Form form = (Form) s;
        String priceStr = ((TextField) form.get(BID_ITEM)).getString();

        // Make copy of the current lot, so we don't change the actual
        // price in
        // our catalog. We want the AuctionHouse to do this for us, as a
        // confirmation that it received our bid successfully.
```

```
          // (Note cannot use Cloneable in CLDC...)
          Lot lotCopy = new Lot(lot);
          lotCopy.setPrice(Integer.parseInt(priceStr));
          bidder.makeBid(lotCopy);
          displayCatalog(bidder.getCatalog());
        }
      }
    }

    public void pauseApp() {}

    public void destroyApp(boolean unconditional) {
      if (bidder != null) {
        bidder.exit();
      }
      notifyDestroyed();
    }
  }
```

The Bidder class has minor changes because the class passed into the constructor is now a MIDlet and not a Lot, and Catalog classes stay identical. Also, the initJMS() method needs to call a slightly different version of the registerQos() method, where the gateway connection URL is passed as a parameter of its own, as well as within the params Hashtable:

```
String url = "socket://localhost:8738";
ProtocolStackRegistry.registerQos(url,
                                  qosName,
                                  params,
                                  username,
                                  false, /* doStoreFwd    */
                                  false, /* doReliability */
                                  false);/* doSecurity    */
```

To build and run the MIDP client, we must somehow get the iBus//Mobile classes into our MIDlet. Currently, the only way to do this is to unzip the iBus//Mobile libraries, and then zip our classes up with these unzipped files. (A method of including JARs in MIDlet suites is promised for a future MIDP release.) Here is the script to do this:

```
REM Environment
REM ----------------
set J2MEWTK_ROOT=D:\J2MEWTK

REM compile
REM ----------
cd src
javac -g:none -d ..\tmpclasses -classpath
..\tmpclasses;..\classes;%IBUSMOB_HOME%\client\midp\lib\ibusmob-jms-midp.jar
-bootclasspath %J2MEWTK_ROOT%\lib\midpapi.zip Lot.java
javac -g:none -d ..\tmpclasses -classpath
..\tmpclasses;..\classes;%IBUSMOB_HOME%\client\midp\lib\ibusmob-jms-midp.jar
-bootclasspath %J2MEWTK_ROOT%\lib\midpapi.zip Catalog.java
javac -g:none -d ..\tmpclasses -classpath
.;..\tmpclasses;..\classes;%IBUSMOB_HOME%\client\midp\lib\ibusmob-jms-midp.jar
-bootclasspath %J2MEWTK_ROOT%\lib\midpapi.zip Bidder.java
```

```
javac -g:none -d ..\tmpclasses -classpath
.;..\tmpclasses;..\classes;%IBUSMOB_HOME%\client\midp\lib\ibusmob-jms-midp.jar
-bootclasspath %J2MEWTK_ROOT%\lib\midpapi.zip JMSMIDlet.java

REM preverifiy
REM --------------
cd ..
%J2MEWTK_ROOT%\bin\preverify -classpath
%J2MEWTK_ROOT%\lib\midpapi.zip;tmpclasses;%IBUSMOB_HOME%\client\midp\lib\ibusmob-j
ms-midp-nonpreverified.jar -d classes tmpclasses

REM unjar the iBus//Mobile classes into our classes directory. (Only needs to be
run once!)
REM We need to include them in our MIDlet jar.
REM
--------------------------------------------------------------------------------
--------------------
copy %IBUSMOB_HOME%\client\midp\lib\ibusmob-jms-midp.jar classes
cd classes
jar xf ibusmob-jms-midp.jar
cd ..

REM Create application jar, in bin directory
REM ----------------------------------------------------
cd bin
jar cmf MANIFEST.MF JMSMIDlet.jar -C ..\classes .
```

You can run the resulting client from the J2MEWTK or a MIDlet runner utility. Here is the resulting client:

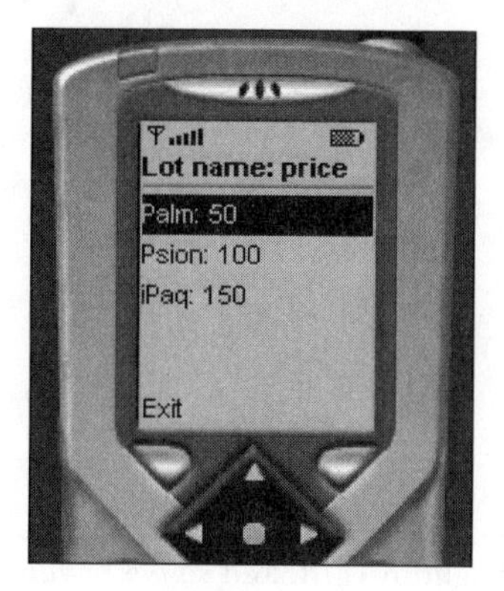

Enhancements

The application is ripe for improvement, as you've no doubt noticed. Here are some ideas, if you choose to experiment further:

- Allow catalog requests, so that every bidder can ask for an up-to-date view of the current lots, rather than just having its own snapshot.
- Allow bidders to specifically track particular lots. One way this can be done is by allocating each `Lot` its own temporary bid queue. Bidders then subscribe to that queue and only get to see price changes for the lots they choose to track. (Another way would be to send a `LotCategory` – 'guitars', for example – as a message header, and allow bidders to select the categories they were interested in. This would be slightly less flexible, but the messaging provider would probably require less processing power and resources if there are many lots.)

- Pass bids and lots as objects rather than primitive types. This requires RMI support on the J2ME client, which is available on PersonalJava, and will be in the forthcoming RMI Profile.
- Display the lot details on the J2ME device using a `Dialog`, leaving just the name and price in `displayCatalog()`.
- Add content checking of entry fields: i.e. are bids numeric, are usernames and passwords in a valid format?
- Make bids transactional (see below).
- Set an expiration time for each lot, after which bids cannot be taken, and the auction finishes for that lot. Using a `JMSExpiration` message header is also useful if a client has not connected for a long time. Expired lots will need deleting from the catalog.

Here is an improved architecture, with a 'tracking' queue:

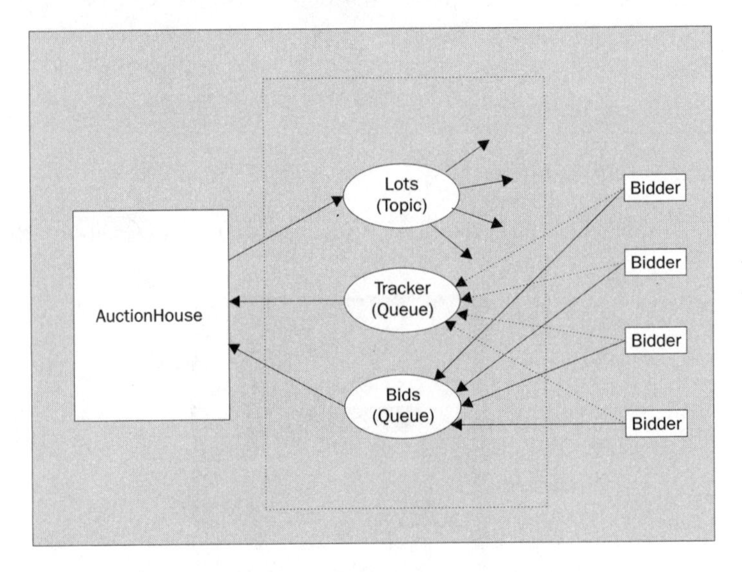

Other JMS Features

Our example was necessarily straightforward, and so we haven't yet covered many JMS features. We refer you to *Professional JMS*, Wrox Press ISBN 1-861004-93-1 for greater detail.

The Future

So what does the future hold for messaging as an architecture? Currently it is quite an under-used J2EE architecture, but it has a good future.

There are several developments that will drive messaging:

- Firstly the existence of JMS itself, which distances the developer from the peculiarities of each vendor's messaging system. This is good news – developers can create applications without worrying about picking the 'wrong' platform, and operators must differentiate on quality of product rather than obscure technical advantages.

- The advent of high-speed packet-based third-generation mobile communications (3G). This will give us much quicker access, and with its 'always-on' mode of working, it means that the 'push' mechanism of messaging will be very attractive. In fact, the messaging architecture would make an ideal bi-directional successor to the current Short Message Service (SMS) that is so prevalent in Europe.

- The new message-driven bean in J2EE, versions 1.3 and higher, will help pull messaging into the mainstream of J2EE. This bean allows J2EE applications to process JMS messages asynchronously. Ordinary enterprise beans allow you to send messages and to receive them synchronously, but not asynchronously. Like a message listener in a standalone JMS client, a message-driven bean contains an `onMessage()` method that is called automatically when a message arrives. The EJB container also automatically performs several setup tasks that a standalone client would have had to do, such as:

    - Creating a message consumer (a `QueueReceiver` or `TopicSubscriber`) to receive the messages. The message-driven bean is associated with a destination and connection factory at deployment time. Durable subscription and message selectors are also specified at deployment time.

    - Registering the message listener.

    - Specifying a message acknowledgment mode.

  A message-driven bean is similar in some ways to a stateless session bean – it's relatively short-lived and has no state for a specific client.

- Multi-tasking mobile devices. The ideal messaging target device needs to multi-task, with typical messaging clients sitting in the background listening for incoming messages, and popping up accordingly to inform the user of the new bid, stock price, and so on. We need a multi-tasking Palm, mobile phone, or whatever device we wish to use that day.

Summary

In this chapter we examined asynchronous messaging and investigated the basic elements of JMS, and why it is a platform particularly suited to mobile devices. We then wrote a desktop-based application, and ported that application to a Palm, noting the differences, and also the missing pieces that had to be plugged by third parties – in particular the JMS implementation for J2ME and the AWT implementation for CLDC. Lastly we covered the more advanced features of JMS. You should now know enough to create your own JMS-based applications, whether they be mobile solutions, desktop solutions, or a mixture of the two.

The main points to learn from this chapter are:

- JMS is particularly suited to asynchronous connectivity, such as 'push' services.

- JMS's ability to save and resend messages is vital with intermittently connected mobile devices.

- JMS abstracts out the underlying messaging system, allowing you to swap message servers with minimal (or zero) code changes.

- JMS is a candidate for commercial mobile applications right now, but you must use third-party solutions. However, this is a small loss compared to the gain in programming ease, and the code will easily port to other JMS solutions in the future.

PROFESSIONAL JAVA MOBILE

TOOL SELECTION

WEB TABLET
MOBILE PHONE
MP3 PLAYER
DIGITAL CAMERA
PALMTOP
LAPTOP

NOW SELECTED

PALMTOP

13

Developing a J2ME Application for the MIDP and the Palm

In this chapter we shall look into developing a J2ME application for use on the Palm powered handheld platform and the Mobile Information Device Profile (MIDP). We will begin by describing design guidelines and then include a case study of a J2ME game application called the Towers of Hanoi. The case study describes the appearance and behavior of the application as well as the user interface design, the application design, and implementation for both platforms. We will also be considering design and implementation porting issues.

We shall limit the design and implementation of our application to the Spotless API for the Palm powered handheld device and the MIDP API for the mobile device. Both of these utilize the Connected Limited Device Configuration API. The Spotless system was a result of a research project carried out by Sun Microsystems and was one of the first attempts to get Java to run on a small device. This product was to evolve into the KVM, the virtual machine for the J2ME configuration. The MIDP system is the first Profile released to complement the CLDC for mobile devices.

In this chapter we shall be using a small slice of the Spotless API called the Palm overlay for the CLDC. We shall also be using the MIDP MIDlet user interface package.

Designing for Portability

Portability brings many design limitations for the software designer and thus requires disciplined, well-informed design decisions. The MID Profile has certain features like high-level APIs that are designed to encourage programming for portability on small devices (see Chapter 5 for more details). Another feature that can help in designing for portability is Object-Oriented abstraction. In Java this can be achieved with interfaces and abstract classes.

Designing for Small Devices

Writing for small devices brings about even more design limitations in wireless technology such as:

- **Small screen sizes and memory or processor power limits.**
  Certain devices will only permit text items of a certain length because of the limited screen sizes.
- **Users' skills and their expectations.**
  Some mobile device users are not familiar with computer applications. They will expect to see easy-to-use resilient applications. However, Palm OS users can be expected to be more familiar with computerized devices.
- **Variety of usability aspects available on the target devices.**
  Some devices may allow screen presses like mouse events whilst other devices may only allow keyboard input.

The rest of this chapter is focused on a detailed case study for developing a J2ME application.

The Towers of Hanoi

The subject of our case study is the Towers of Hanoi game. The objective of the Towers of Hanoi game is to move the rings from the 1st tower on the left hand side to the 3rd tower on the right. You can only move one ring at a time, but you cannot place a larger ring on top of a smaller ring. For a 5 ring tower the minimum theoretical number of moves is 31.

For the purpose of this case study we shall use the J2ME Wireless Toolkit and the Palm OS emulator to run the application. The Wireless toolkit can be downloaded from the following address http://java.sun.com/products/j2mewtoolkit/download.html. The Palm OS emulator has to be downloaded separately from http://www.palmos.com/dev/tech/tools/emulator/. We also need to obtain a Palm ROM image to be able to run the Palm OS emulator. To get hold of a ROM image you can download from the following address http://www.palmos.com/dev/tech/tools/emulator/#roms/.

For more details on both these environments see Chapter 4.

The J2ME Wireless Toolkit

Here we can see the Wireless Toolkit emulator screen, which runs on a PC desktop:

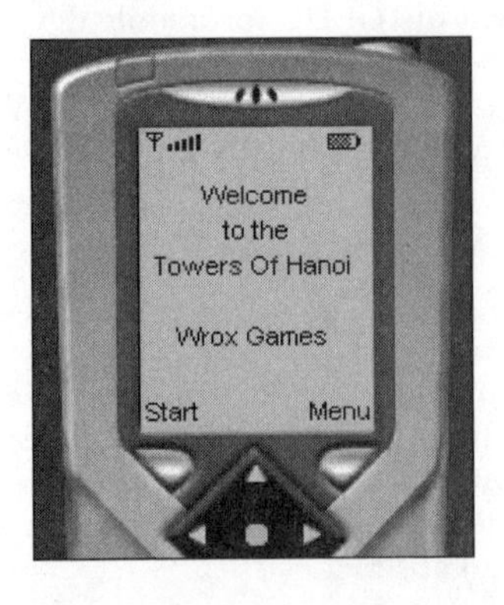

The Palm OS Emulator

The emulator software does not include ROM images. It is like a computer without an operating system. There are two sources of ROM images: ROM image files downloaded from the Resource Pavilion or ROM images uploaded from an actual device:

The User Interface Design

Usability is one of the key issues defining software quality as it is concerned with the effectiveness and efficiency of the user interface and the user's reaction to that interface. A good user interface is user-friendly, consistent, and simple.

There are, however, issues of usability like refresh speed of the screen, screen layout, and also keeping consistency between the different screen layouts and the two applications for the two devices. We have considered keeping consistency between the different screen layouts of the same application to be more important than keeping consistency between the two applications on two devices. This is quite simply because it is unlikely that the same user will own two devices.

Unlike the Spotless system, the Mobile Information Device Profile has no windowing system; instead it has screen and command abstractions. The MIDP has high-level and low-level APIs, as discussed in Chapter 5. The high-level APIs are highly abstracted, while the low-level APIs require more design work to be portable as they are device-specific commands. For more information about these concepts consult the white paper "*Applications for Mobile Information Devices, Helpful Hints for Application Developers and User Interface Designers using the Mobile Information Device Profile*" by Sun Microsystems, which is available at http://java.sun.com.

The Screen Navigation

The Unified Modeling Language (UML) state diagram below shows the screen navigation as transitions between states, which represent different screens of the application. The option the user chooses to take is used as the name of the event. The Welcome View will lead into the Main Game View. This in turn will lead to the Game Over View. The user can also access the Help View and the About View from the Main Game View. There is an Error page that can be displayed at any point within the application:

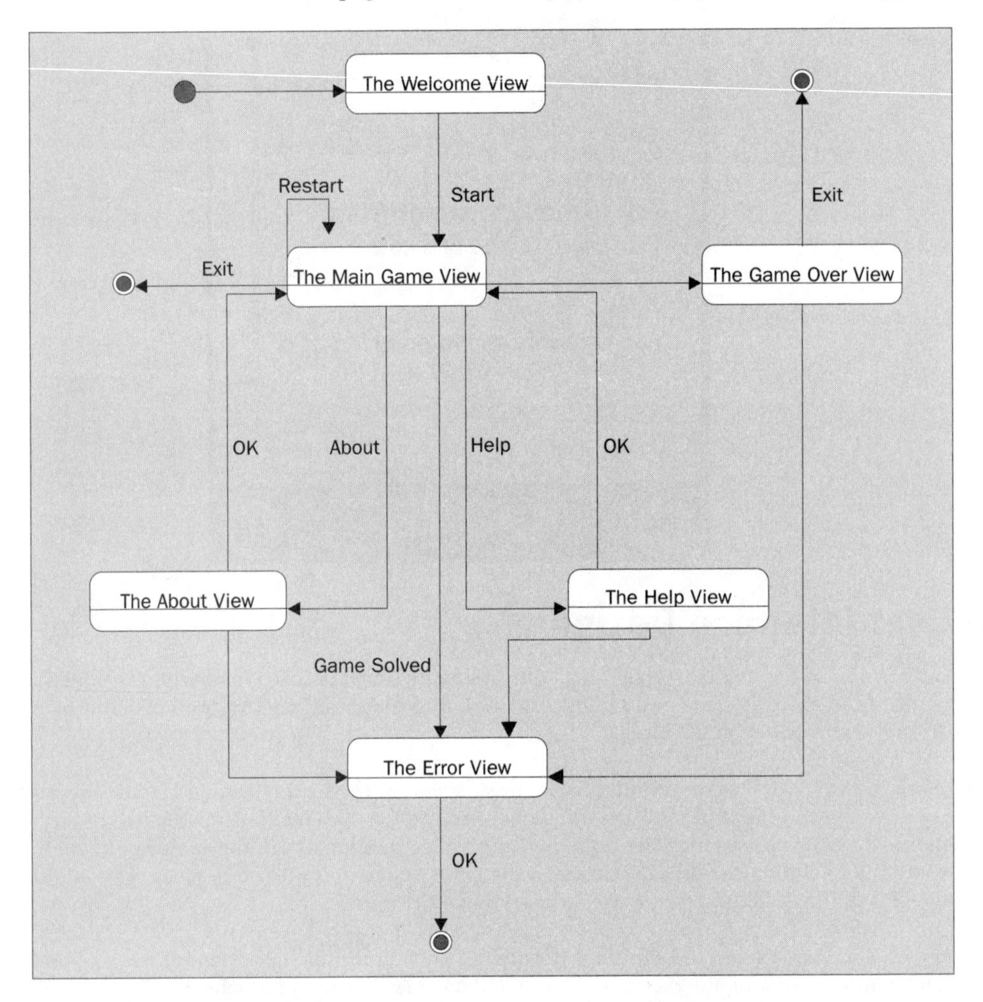

The Screen Design

For consistency the Exit and Start buttons always appear on the same side of the screen on a device. The dimensions for each tower have been adjusted in proportion with the screen dimensions for each device.

There is an element of symmetry in the screen design. The user interface objects are arranged around an invisible central vertical axis:

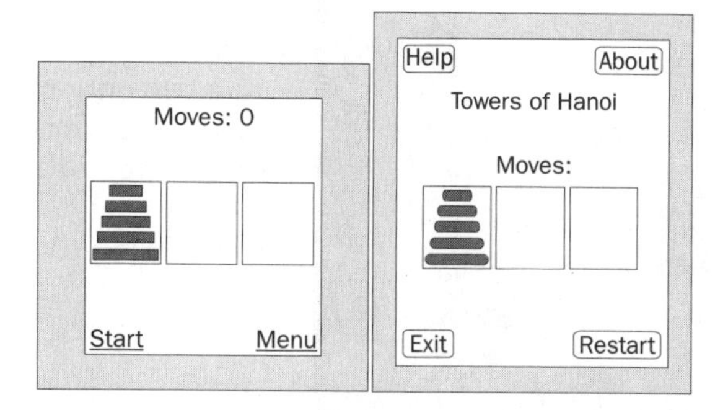

The Welcome View

This is the first View, which the user is presented with after starting the applicatio,n containing the title of the game:

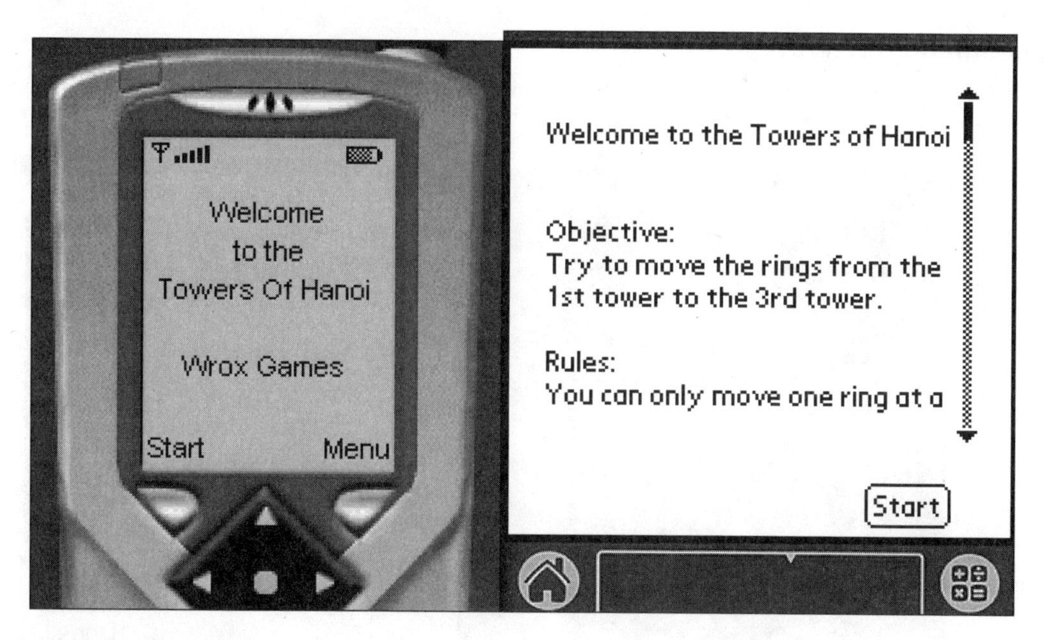

The Main Game View

The Main Game View shows three towers. The tower to the left has five rings arranged on top of each other. By selecting a source tower that has a ring and then selecting a target tower, the ring gets moved from the first tower to the target tower. Each time a move is made the number of moves is updated. You can press the Exit button to leave the application and the Start button to restart the application:

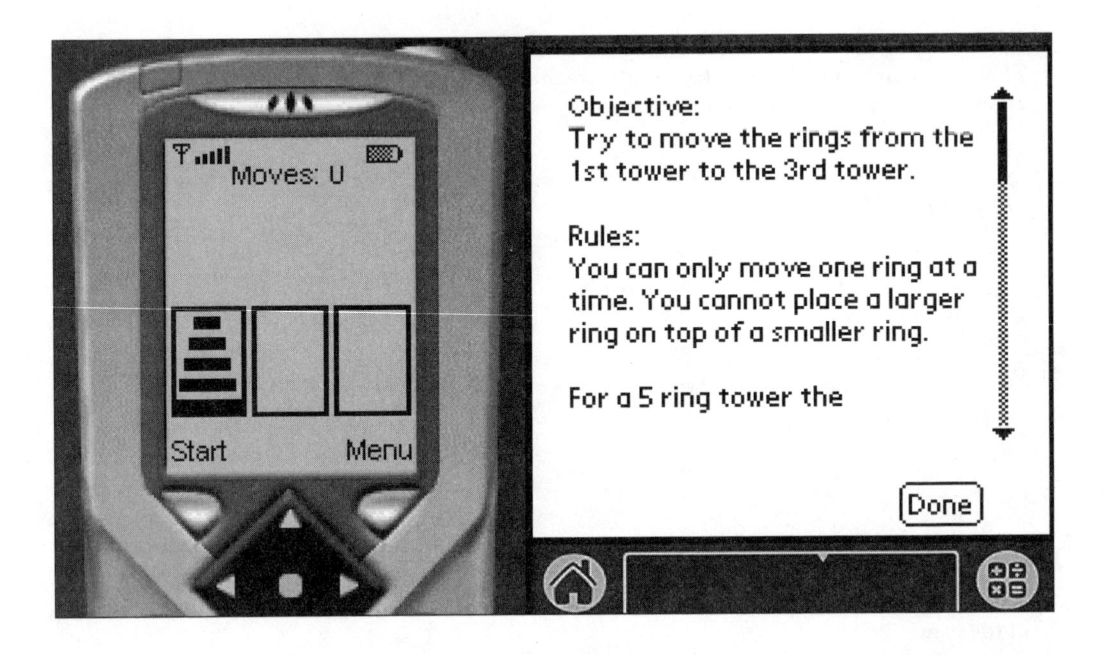

The Game Over View

The Game Over View is shown when the game ends and has a few lines of text that state the number of moves it took to solve the game:

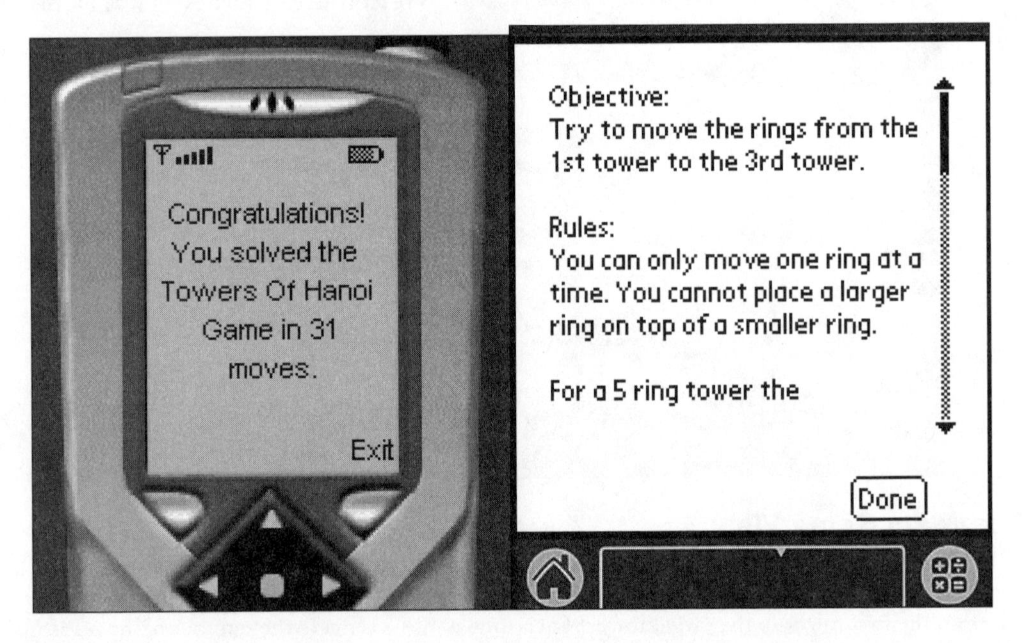

The Help View

At any time during the application lifecycle the user can select the Help View. This View consists of a few lines of text that describe how the game is played. The following text is included in the Help View:

Objective:
Try to move the rings from the 1st tower to the 3rd tower.
Rules:
You can only move one ring at a time. You cannot place a larger ring on top of a smaller ring.
For a 5 ring tower the theoretical minimum number of moves is 31.
Good luck!

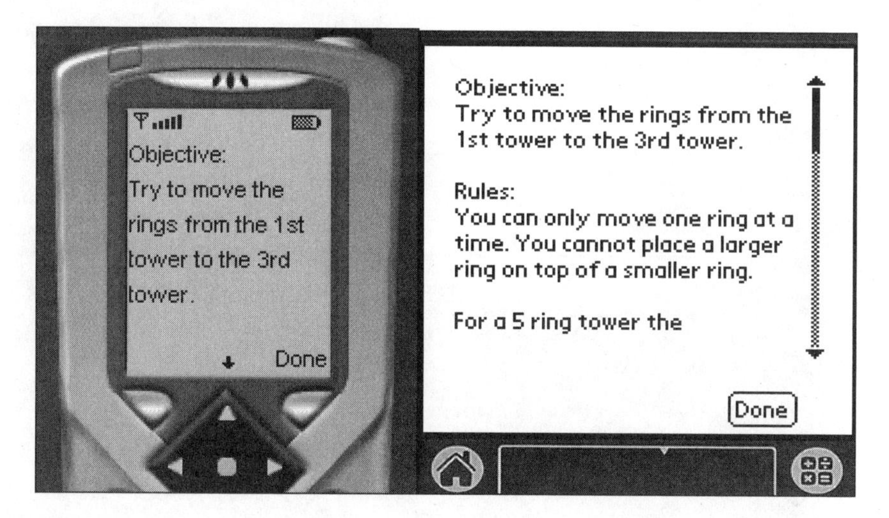

The About View

The About View can also be selected at any time. This View contains the name of the application and the version number of the API used:

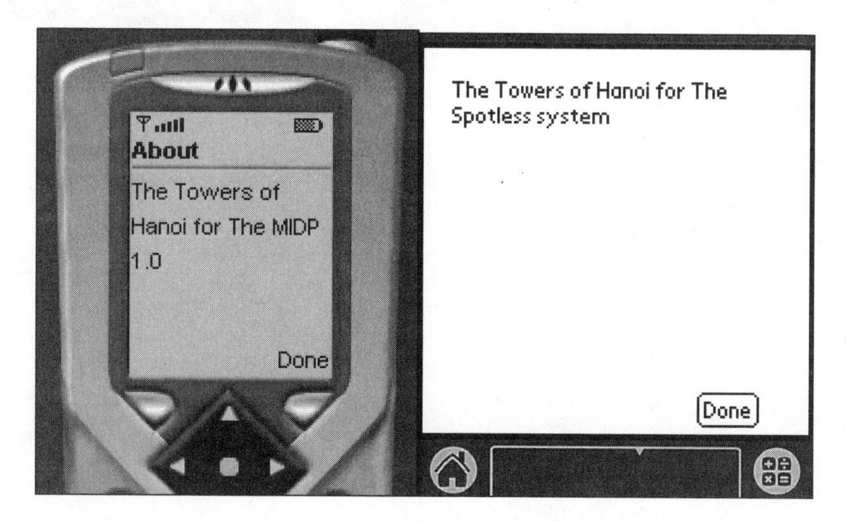

Commands and Action Keys

Soft keys resemble menu commands or buttons on a windowing application whereas hard keys are actual device keys like a telephone key pad. The MIDP implementation relies on hard and soft keys and no screen input. The Palm implementation relies on soft keys and screen input. All soft keys on the Palm handheld device are in the form of round buttons:

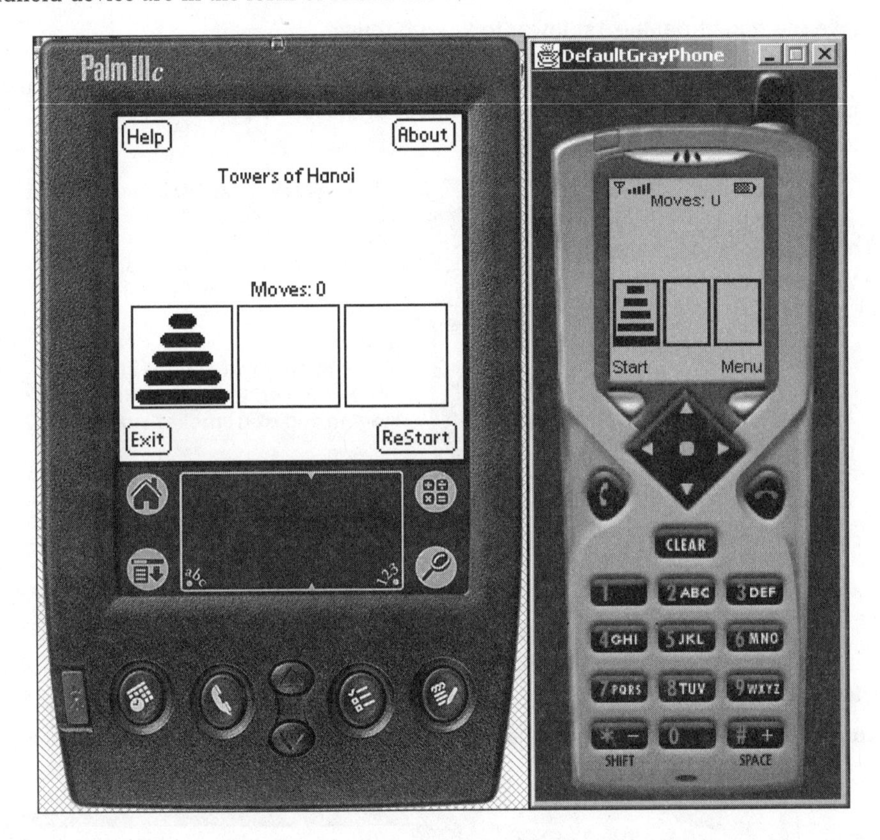

The Application Design

So far we have looked at the user interface design for the Towers of Hanoi game. In the following few sections we are going to take a closer look at the application design. By application design we mean the logical and physical structures of the application and how these all fit together.

High-Level Design

The application design is abstracted out into three layers: the Games and Hanoi shared layers and the Hanoi-MIDlet/Hanoi-Spotless layer. The shared layers are CLDC-specific. Hanoi-MIDlet is MIDP-specific and Hanoi-Spotless is Spotless-specific, as shown in the following diagram:

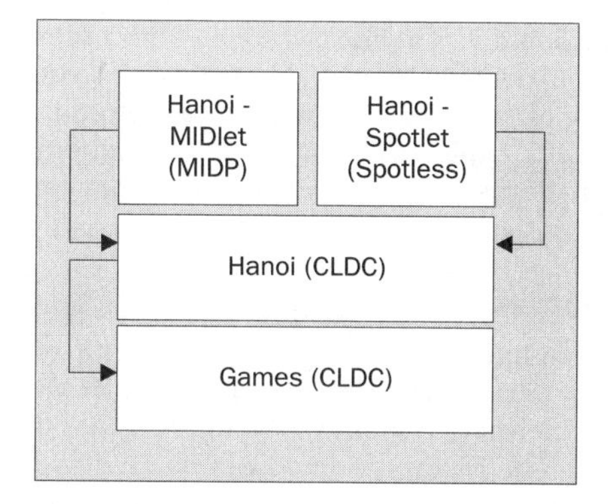

The abstract logic is captured in the shared layer, where most of the classes live while the user interface logic is captured in the Hanoi-MIDlet and Hanoi-Spotless layers. There is only one-way dependency, from top to bottom. The shared layer can be reused unchanged to support a new Connected Limited Device Profile.

A more detailed breakdown of the above diagram is as follows:

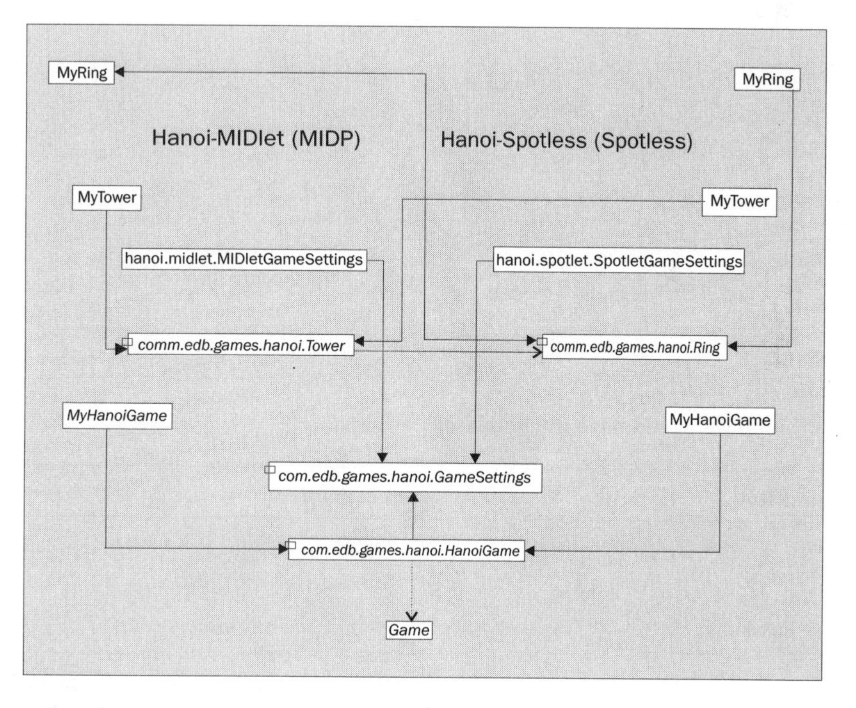

Detailed Design

Designing to a fine-grain level, with small methods and modular, loosely coupled classes makes code more portable. Parts of the code can be overridden to suit different platforms. Being a pure object-oriented language Java inherently makes this easier.

In 1994, Gamma, Helm, Johnson and Vlissides wrote an outstanding book called "Design Patterns Elements of Reusable Object-Oriented Software". In this book a design pattern is defined as a description of communicating objects and classes that are customized to solve a general design problem in a particular context. The book explores a set of 23 design patterns. In the Hanoi game application we will use two of these design patterns: the Factory Method design pattern and the Template Method design pattern.

The Factory Method Design Pattern

The Factory Method defines an interface for creating an object, but lets sub-classes decide which class to instantiate. In this context the `HanoiGame` class has two Factory Methods: `createRing()` and `createTower()`. It wants to delegate responsibility and localizes the knowledge of `Ring` and `Tower` creation to its derived classes. The Factory Method can be used to connect parallel class hierarchies. In this case we have three parallel hierarchies: the `Game` hierarchy the `Tower` hierarchy and the `Ring` hierarchy.

By using this pattern, device specific implementations for `Ring` and `Tower` have the same interface and are used uniformly by the abstract class `HanoiGame`. The logic is abstracted out in the abstract class `HanoiGame` and the `Tower` and `Ring` classes. Normally the Factory Method has four participants as follows:

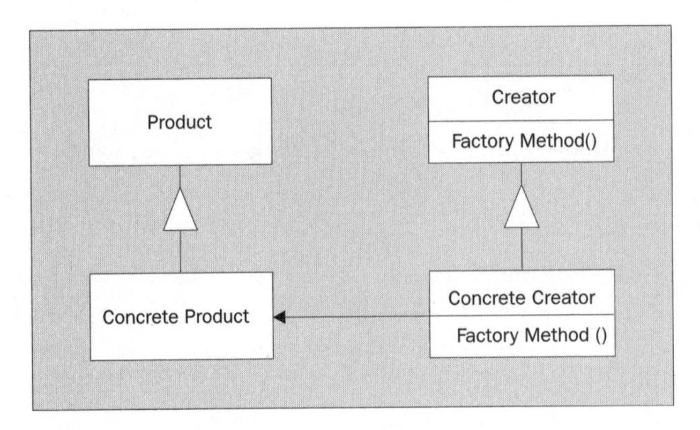

In our application we would have the following participants:

Participant Role	Participant
Products	`Ring` `Tower`
Concrete Products	`HanoiSpotless.MyRing`, `HanoiSpotless.MyTower`, `HanoiMIDlet.MyRing`, `HanoiMIDlet.MyTower`
Creator	`HanoiGame` class
Concrete Creators	`HanoiSpotless.MyHanoiGame`, `HanoiMIDlet.MyHanoiGame`

The Template Method Design Pattern

The Template Method design pattern defines the skeleton of an algorithm in an operation, deferring some steps to subclasses. A Template Method lets subclasses redefine certain steps of an algorithm without changing the algorithm structure. This pattern is used in the `GameSettings` hierarchy of the Hanoi application. The `GameSettings` abstract class has a number of abstract methods and final methods. The diagram below shows the `GameSettings` hierarchy:

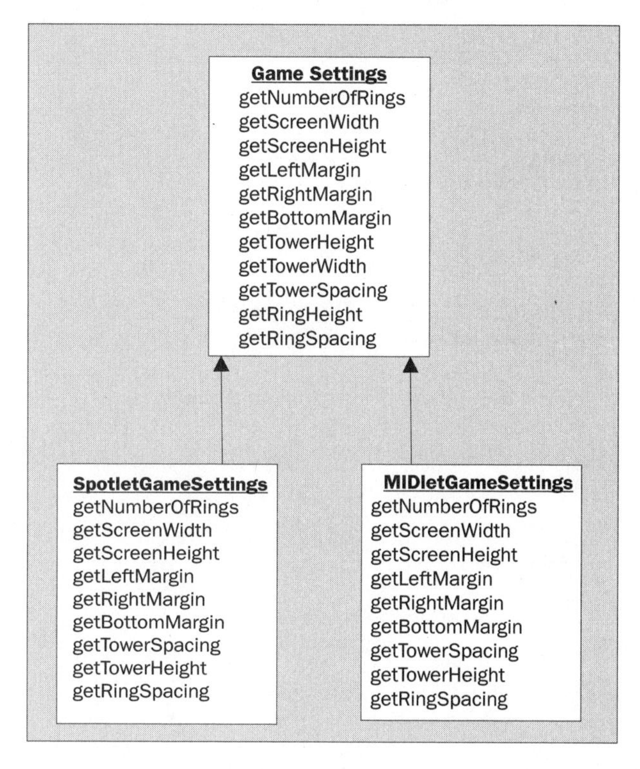

Each device has a different screen size that varies in proportion. The Wireless Toolkit has screen dimensions of 96x100 and the Palm OS emulator has screen dimensions of 160x160. In effect the dimensions of GUI objects for each device are different. The Template Method design pattern allows these values to be calculated independently without affecting the game logic implementation.

Participants:

- Abstract Class (`GameSettings`)
- Concrete Class (`MIDletGameSettings`, `SpotlessGameSettings`)

The Implementation

The Hanoi Game implementation is installed in three parts, as described in the **High Level Design** section. These are the shared implementation, the MIDP implementation, and the Palm implementation.

For detailed instructions on building MIDP and Palm applications see Chapter 4. We shall assume a working knowledge of these processes for the rest of this chapter.

Shared Implementation

The Game interface provides a game abstraction definition and it can be initialized, started and exited. A Game can have an Introduction View, a Main View, a Finish View, a Help View, an About View and finally an Error View:

```
package com.edb.games;

// Simple game application interface

public interface Game {

  // Initializes the game
  public abstract Game initialize();

  // Starts the game
  public abstract Game start();

  // Exits the game
  public abstract Game exit();
```

The following code returns the specific Views for the game:

```
  // Returns the Intro View for this game
  public abstract Object getIntroView();

  // Returns Main View for this game
  public abstract Object getMainView();

  // Returns the last View for this game
  public abstract Object getGameOverView();

  // Returns the Help View for this game
  public abstract Object getHelpView();

  // Returns the About View for this game
  public abstract Object getAboutView();

  // Returns Error View for this game
  public abstract Object getErrorView(String error);

  // Returns true if the game is rolling other wise returns false
  public abstract boolean isTheGameRolling();
}
```

Any game application can implement this interface. While this interface can be compiled under any Java platform, because it actually simplifies a game abstraction it will be more suitable for small implementations.

The HanoiGame Abstract Class

This class implements the `Game` interface. It is responsible for creating three towers and the five rings:

```
package com.edb.games.hanoi;

import com.edb.games.Game;
```

This class represents the Hanoi base class for the Hanoi game. The design of this class and its derived classes is based on the Factory Method design pattern:

```
public abstract class HanoiGame implements Game {

  static final boolean SELECTED = true;

  private Tower[] _towers;
  private Tower _lastSelectedTower = null;
  private int _number_of_moves = 0;
  private GameSettings _settings = null;

  // Creates instance of me
  public HanoiGame(GameSettings settings) {
    _towers = new Tower[3];
    _settings = settings;
    _towers[0] = createTower(settings.getNumberOfRings(),
                             settings.getLeftMargin(),
                             settings.getScreenHeight()
                             - settings.getBottomMargin()
                             - settings.getTowerHeight(), _settings);
    _towers[1] = createTower(settings.getNumberOfRings(),
                             settings.getLeftMargin()
                             + settings.getTowerWidth()
                             + settings.getTowerSpacing(),
                             settings.getScreenHeight()
                             - settings.getBottomMargin()
                             - settings.getTowerHeight(), _settings);
    _towers[2] = createTower(settings.getNumberOfRings(),
                             settings.getLeftMargin()
                             + settings.getTowerWidth() * 2
                             + settings.getTowerSpacing() * 2,
                             settings.getScreenHeight()
                             - settings.getBottomMargin()
                             - settings.getTowerHeight(), _settings);
  }
```

The following Factory Method creates a tower:

```
  protected abstract Tower createTower(int totalNumberOfRings,
                                       int xposition, int yposition,
                                       GameSettings settings);
```

This Factory Method creates a ring:

```
protected abstract Ring createRing(int size, GameSettings settings);

protected abstract void notifyGameSolved();

// paints current view for this game
public void paint(Object paintData) {
  this._towers[0].paint(paintData);
  this._towers[1].paint(paintData);
  this._towers[2].paint(paintData);
}

// starts me
public Game start() {
  this._lastSelectedTower = null;
  this._number_of_moves = 0;
  for (int i = 0; i < _settings.getNumberOfRings(); i++) {
    _towers[0].removeRing();
    _towers[1].removeRing();
    _towers[2].removeRing();
  }
  for (int i = _settings.getNumberOfRings(); i >= 1; i--) {
    Ring tempRing = createRing(i, _settings);
    _towers[0].addRing(tempRing);
  }
  return this;
}
```

This code implements the game's logic:

```
// selects a tower
public void select(int towerSelected) {
  // Find newly selected tower
  Tower newSelectedTower = _towers[towerSelected];
  if (_lastSelectedTower == null) {
    if (!newSelectedTower.isEmpty()) {
      _lastSelectedTower = newSelectedTower;
      newSelectedTower.setSelected(true);
    }
  }

  // same tower.pressed?
  else {
    if (_lastSelectedTower == newSelectedTower) {
      _lastSelectedTower = null;
      newSelectedTower.setSelected(false);
    }

    // different.tower pressed?
    else {
      // if (_lastSelectedTower != newSelectedTower)
      Ring tempRing = _lastSelectedTower.removeRing();
      if (!newSelectedTower.addRing(tempRing)) {
        _lastSelectedTower.addRing(tempRing);
```

```
            _lastSelectedTower.setSelected(false);
            _lastSelectedTower = null;
          } else {
            _lastSelectedTower.setSelected(false);
            newSelectedTower.setSelected(false);
            _lastSelectedTower = null;
            _number_of_moves++;
          }
        }
      }
      if (isGameSolved()) {
        notifyGameSolved();
      }
    }
```

The following method returns `true` if the game is solved and `false` if it is not:

```
    public final boolean isGameSolved() {
      return _towers[2].isSolved();
    }

    // Return my set of towers
    public final Tower[] getTowers() {
      return _towers;
    }

    // Returns my current set of moves
    public final int getNumberOfMoves() {
      return _number_of_moves;
    }
  }
```

The important thing to note about the above class is that it abstracts the logic of the game implementation away from the user interface implementation. In this way it can support more than one user interface API. The `createTower()` and `createRing()` abstract methods are Factory Methods as described in the *Detailed Design* section.

The Tower Abstract Class

The following source code represents the `Tower` object in the Hanoi game. This is the abstract class that has to be extended to support different platforms. The `paint()`, `unpaint()` and `paintBorder()` methods are UI specific and have to be implemented by derived classes:

```
  package com.edb.games.hanoi;

  // The Tower class knows its height and position.
  // Rings can be added to and removed from a Tower.
  // The Tower knows if it isEmpty, if it isSolved, or if has been
  //pressed.
  // The Tower can paint itself and instructs the rings to do likewise.

  public abstract class Tower {

    protected int _totalNumberOfRings;
```

```
protected int _currentNumberOfRings;
protected int _xposition;
protected int _yposition;
protected Ring[] _rings;
protected boolean _selected;
private GameSettings _settings;

// Creates instance of me
public Tower(int totalNumberOfRings, int xposition, int yposition,
             GameSettings settings) {
  _totalNumberOfRings = totalNumberOfRings;
  _xposition = xposition;
  _yposition = yposition;
  _rings = new Ring[totalNumberOfRings];
  _settings = settings;
}
```

This method identifies whether the tower is full of rings or not:

```
public final boolean isSolved() {
  return (_totalNumberOfRings == _currentNumberOfRings);
}

// Returns my game settings
protected final GameSettings getSettings() {
  return _settings;
}

// Returns true if this tower is empty other wise returns false
public final boolean isEmpty() {
  return (_currentNumberOfRings == 0);
}

// Returns true if this tower has been pressed other wise returns
// false
public final boolean pressed(int x, int y) {
  if ((x >= _xposition && x <= _xposition + _settings.getTowerWidth())
          && (y >= _yposition
              && y <= _yposition + _settings.getTowerHeight())) {
    return true;
  } else {
    return false;
  }
}
```

This method sets the tower in a selected or unselected state:

```
Tower setSelected(boolean selected) {
  _selected = selected;
  return this;
}

public final boolean isSelected() {
  return _selected;
```

```
    }

    // Adds a ring to the me.
    public boolean addRing(Ring newRing) {
      // if no _rings then addRing
      if (_currentNumberOfRings == 0
              || _rings[_currentNumberOfRings - 1].getSize()
                 > newRing.getSize()) {
        _rings[_currentNumberOfRings] = newRing;
        _currentNumberOfRings++;
        return true;
      } else {
        return false;
      }
    }
```

This method removes the topmost ring from the tower. If this tower has no rings this method does nothing:

```
    public Ring removeRing() {
      if (_currentNumberOfRings == 0) {
        return null;
      } else {
        _currentNumberOfRings--;
        Ring tempRing = _rings[_currentNumberOfRings];
        _rings[_currentNumberOfRings] = null;
        return tempRing;
      }
    }

    // Paints me
    public abstract void paint(Object paintData);

    // Erases me
    public abstract void unpaint(Object paintData, int x, int y);

    // Paints my border
    public abstract void paintBorder(Object paintData, int x, int y);
  }
```

The Ring Abstract Class

The fourth piece of the shared layer is the Ring abstract class:

```
  package com.edb.games.hanoi;

  // The Ring class knows it's size and how to paint itself.

  public abstract class Ring {

    private int _size;
    private GameSettings _settings = null;

    // Creates instance of me
    public Ring(int Size, GameSettings settings) {
```

```
    this._size = Size;
    this._settings = settings;
  }

  // Returns my size
  public final int getSize() {
    return this._size;

  }

  // Returns my settings
  protected final GameSettings getSettings() {
    return _settings;
  }

  // Paints me
  public abstract void paint(Object paintData, int x, int y);

  // Erases me
  public abstract void unpaint(Object paintData, int x, int y);
}
```

The GameSettings Abstract Class

The `GameSettings` class is the last piece in the shared layer:

```
package com.edb.games.hanoi;
```

This class represents the game settings for the Hanoi game. The design of this class and its potential derived classes are based on the Template Method design pattern:

```
public abstract class GameSettings {

  public static final int LEFT_TOWER = 0;
  public static final int CENTER_TOWER = 1;
  public static final int RIGHT_TOWER = 2;

  // Returns Number of rings
  public abstract int getNumberOfRings();

  // Returns Screen width
  public abstract int getScreenWidth();

  // Returns Screen height
  public abstract int getScreenHeight();

  // Returns left margin
  public abstract int getLeftMargin();

  // Returns right margin
  public abstract int getRightMargin();

  // Returns bottom margin
  public abstract int getBottomMargin();
```

```
    // Returns tower height
    public abstract int getTowerHeight();

    // Returns tower width
    public final int getTowerWidth() {
      return (getScreenWidth() - getLeftMargin() - getRightMargin() -
                                                (getTowerSpacing() * 2))/3;
    }

    // Returns tower spacing
    public abstract int getTowerSpacing();

    // Returns ring height
    public final int getRingHeight() {
      return (getTowerHeight() - (getRingSpacing() * (getNumberOfRings() -
                                                    1)))/getNumberOfRings();
    }

    // Returns ring spacing
    public abstract int getRingSpacing();
  }
```

MIDP Implementation

The MIDP implementation has one public class, the `HanoiMIDlet` class, and two package classes: the `HanoiCanvas` and the `MIDletGameSettings` classes.

The HanoiMIDlet Class

The `HanoiMIDlet` class extends the MIDP `MIDlet` class. The package name is:

```
package com.edb.games.hanoi.midlet;
```

To extend the `MIDlet` class you need to import the following class:

```
import javax.microedition.midlet.MIDlet;
```

Other imports by this class are the following:

```
import javax.microedition.lcdui.Canvas;
import javax.microedition.lcdui.Graphics;
import javax.microedition.lcdui.Alert;
import javax.microedition.lcdui.Displayable;
import javax.microedition.lcdui.Display;
import javax.microedition.lcdui.CommandListener;
import javax.microedition.lcdui.Command;
import javax.microedition.lcdui.Screen;
import com.edb.games.Game;
import com.edb.games.hanoi.HanoiGame;
import com.edb.games:hanoi.Tower;
import com.edb.games.hanoi.Ring;
import com.edb.games.hanoi.GameSettings;
```

The HanoiMIDlet class also includes the implementation of three inner classes. They are the MyHanoiGame, MyRing, and MyTower classes. The class is final because it is not meant to be extended:

```
public final class HanoiMIDlet extends MIDlet {

  private Game _game = null;
  private CommandListener _commandListener = null;
```

The startApp() method initializes this class. This is one of the standard MIDlet methods that should be implemented by derived classes:

```
    public void startApp() {
      try {
        _game = new MyHanoiGame(new MIDletGameSettings());
        setCommandListener();
        setIntroView();
        while (!_game.isTheGameRolling()) {
          Thread.yield();
        }
      } catch (Exception e) {
        if (_game != null) {
          setErrorView("Error: " + e.toString());
        } else {
          System.out.println("Error: " + e.toString());
        }
      }
    }
```

Another two methods that should be implemented by derived classes are the pauseApp() and destroyApp() methods. In our example they have empty implementation:

```
    public void pauseApp() {}

    public void destroyApp(boolean unconditional) {}
```

The HanoiGame class also acts as a CommandListener. In the following private method we are setting the commands for our application:

```
    private void setCommandListener() {
      _commandListener = new CommandListener() {
        public void commandAction(Command command, Displayable d) {
          if (command.getLabel().equals("Start")) {
            _game.start();
            setMainView();
            ((Canvas) _game.getMainView()).repaint();
          }
          if (command.getLabel().equals("Help")) {
            setHelpView();
          }
          if (command.getLabel().equals("About")) {
            setAboutView();
          }
```

```
        if (command.getLabel().equals("Exit")) {
          _game.exit();
          destroyApp(false);
          notifyDestroyed();
        }
      }
    };
  }
```

The following private methods set the various Views for the game. Each method delegates to the game inner class instance to get the required View and this is set as the current View. This behavior is implemented in a standard way:

```
  // Sets my Intro View
  private void setIntroView() {
    Display.getDisplay(this).setCurrent((Displayable) _game.getIntroView());
  }

  // Sets my Main View
  private void setMainView() {
    Display.getDisplay(this).setCurrent((Displayable) _game.getMainView());
  }

  // Sets my last View
  private void setGameOverView() {
    Display.getDisplay(this)
      .setCurrent((Displayable) _game.getGameOverView());
  }

  // Sets my Help View
  private void setHelpView() {
    Display.getDisplay(this).setCurrent((Displayable) _game.getHelpView());
  }

  // Sets my About View
  private void setAboutView() {
    Display.getDisplay(this).setCurrent((Displayable) _game.getAboutView());
  }

  // Sets my Error View
  private void setErrorView(String error) {
    Display.getDisplay(this)
      .setCurrent((Displayable) _game.getErrorView(error));
  }
```

In this section we shall implement the three inner classes: the MyHanoiGame inner class, the MyRing inner class, and the MyTower inner class. The first is the MyHanoiGame, the MIDP implementation of the abstract HanoiGame class, which we have reviewed in the shared layer implementation:

```
  private class MyHanoiGame extends HanoiGame {

    private Displayable _introScreen;
    private Displayable _gameCanvas;
    private Displayable _gameOverScreen;
```

```
private Alert _helpView;
private Alert _aboutView;
private Alert _errorView;
```

All the private variables of this class are references to the six Views of the Hanoi game. The constructor initializes the game instance with a specialized MIDP `GameSettings` implementation:

```
// Creates instance of me
MyHanoiGame(GameSettings settings) {
  super(settings);
}

// Initializes the game
public Game initialize() {
  return this;
}

// Exits the game
public Game exit() {
  return this;
}
```

The following methods provide `getXXX()` implementation for all the Views in this game. As Java objects are quite expensive, the new operator is only used once. The next time the same View is requested, the original View gets returned. This saves on memory used on the heap:

```
// Returns the intro View for this game
public Object getIntroView() {
  if (_introScreen == null) {
    _introScreen = new Canvas() {
      public void paint(Graphics graphics) {
        setGraphics(graphics, this);
        graphics.drawString("Welcome", getWidth() / 2,
                            (getHeight() / 8) + 15,
                            Graphics.HCENTER | Graphics.BOTTOM);
        graphics.drawString("to the", getWidth() / 2,
                            (getHeight() / 8) + 30,
                            Graphics.HCENTER | Graphics.BOTTOM);
        graphics.drawString("Towers Of Hanoi ", getWidth() / 2,
                            (getHeight() / 8) + 45,
                            Graphics.HCENTER | Graphics.BOTTOM);
        graphics.drawString("Wrox Games ", getWidth() / 2,
                            (getHeight() / 8) + 75,
                            Graphics.HCENTER | Graphics.BOTTOM);
      }
    };
    _introScreen.addCommand(new Command("Start", Command.SCREEN, 1));
    _introScreen.addCommand(new Command("Help", Command.SCREEN, 2001));
    _introScreen.addCommand(new Command("About", Command.SCREEN, 2002));
    _introScreen.addCommand(new Command("Exit", Command.SCREEN, 2003));
    _introScreen.setCommandListener(_commandListener);
  }
  return _introScreen;
}
```

This method returns Main View for this game:

```
public Object getMainView() {
  if (_gameCanvas == null) {
    _gameCanvas = new HanoiCanvas(this);
    _gameCanvas.addCommand(new Command("Start", Command.SCREEN, 1));
    _gameCanvas.addCommand(new Command("Help", Command.SCREEN, 2001));
    _gameCanvas.addCommand(new Command("About", Command.SCREEN, 2002));
    _gameCanvas.addCommand(new Command("Exit", Command.SCREEN, 2003));
    _gameCanvas.setCommandListener(_commandListener);
  }
  return _gameCanvas;
}
```

The following method returns the last View for this game informing the player the number of moves they took to solve the game:

```
// Returns last View for this game
public Object getGameOverView() {
  if (_gameOverScreen == null) {
    _gameOverScreen = new Canvas() {
      public void paint(Graphics graphics) {
        setGraphics(graphics, this);
        graphics.drawString("Congratulations! ", getWidth() / 2,
                            (getHeight() / 8) + 15,
                            Graphics.HCENTER | Graphics.BOTTOM);
        graphics.drawString("You solved the  ", getWidth() / 2,
                            (getHeight() / 8) + 30,
                            Graphics.HCENTER | Graphics.BOTTOM);
        graphics.drawString("Towers Of Hanoi ", getWidth() / 2,
                            (getHeight() / 8) + 45,
                            Graphics.HCENTER | Graphics.BOTTOM);
        graphics.drawString("Game in " + getNumberOfMoves(),
                            getWidth() / 2, (getHeight() / 8) + 60,
                            Graphics.HCENTER | Graphics.BOTTOM);
        graphics.drawString("moves.", getWidth() / 2,
                            (getHeight() / 8) + 75,
                            Graphics.HCENTER | Graphics.BOTTOM);
      }
    };
    _gameOverScreen.addCommand(new Command("Exit", Command.SCREEN, 0));
    _gameOverScreen.setCommandListener(_commandListener);
  }
  return _gameOverScreen;
}
```

Here the Help, About, and Error Views for this game are returned:

```
public Object getHelpView() {
  if (_helpView == null) {
    _helpView = new Alert("Help");
    _helpView.setTimeout(Alert.FOREVER);
    _helpView
      .setString("Objective:\nTry to move the rings from the 1st tower to the
```

```
3rd tower.\n\nRules:\nYou can only move one ring at a time. You cannot place a
larger ring on top of a smaller ring.\n\nFor a 5 ring tower the theoretical
minimum number of moves is 31.\n\nGood luck!");
      }
      return _helpView;
    }

    // Returns About View for this game
    public Object getAboutView() {
      if (_aboutView == null) {
        _aboutView = new Alert("About");
        _aboutView.setTimeout(Alert.FOREVER);
        _aboutView.setString("The Towers of Hanoi for The MIDP 1.0");
      }
      return _aboutView;
    }

    // Returns Error View for this game
    public Object getErrorView(String error) {
      if (_errorView == null) {
        _errorView = new Alert("Error");
        _errorView.setTimeout(Alert.FOREVER);
        _errorView.setString(error);
      }
      return _errorView;
    }
```

This next method is designed to return `true` or `false` if the game Main View runs on a separate thread. As all Views run on the same thread, this method always return `true`:

```
    // Returns true if the game is rolling other wise returns false
    public boolean isTheGameRolling() {
      return true;
    }
```

The following two methods are the Factory Methods we described earlier in the *Detailed Design* section. Here a special MIDP implementations of `Ring` and `Tower` get created and returned:

```
    // Factory Method, creates a tower
    protected Tower createTower(int totalNumberOfRings, int xposition,
                                int yposition, GameSettings settings) {
      return new MyTower(totalNumberOfRings, xposition, yposition,
                         settings);
    }

    // Factory Method, creates a ring
    protected Ring createRing(int size, GameSettings settings) {
      return new MyRing(size, settings);
    }

    // Callback, notifies that this game has been solved
    protected void notifyGameSolved() {
      setGameOverView();
    }
```

This method sets the graphics for this game:

```
    private void setGraphics(Graphics graphics, Canvas canvas) {
      graphics.setColor(0xffffff);
      graphics.fillRect(0, 0, canvas.getWidth(), canvas.getHeight());
      graphics.setColor(0);
    }
  }
```

The `MyRing` inner class is the MIDP implementation of the `Ring` abstract class that we have reviewed earlier. It provides MIDP-specific implementations for the abstract methods `paint()` and `unpaint ()`:

```
  // Inner class represents my ring implementation
  private class MyRing extends Ring {

    // Creates me
    MyRing(int size, GameSettings settings) {
      super(size, settings);
    }

    // Refreshes the display of this ring
    public void paint(Object graphics, int x, int y) {
      int ringHeight = getSettings().getRingHeight();
      int step = (getSettings().getTowerWidth() - 4)
                  / getSettings().getNumberOfRings();
      step = step - step % 2;    // Make step even!
      int ringWidth = getSettings().getTowerWidth()
                      - ((getSettings().getNumberOfRings() - getSize())
                        * step);
      int ringX = ((getSettings().getNumberOfRings() - getSize()) * step)
                  / 2;
      ((Graphics) graphics).fillRect(x + ringX, y, ringWidth,
                                     getSettings().getRingHeight());
    }

    // Removes this ring from the display
    public void unpaint(Object graphics, int x, int y) {
      ((Graphics) graphics).drawRect(x, y,
                                     getSettings().getNumberOfRings(),
                                     getSettings().getRingHeight());
    }
  }
```

The `MyTower` inner class is the MIDP implementation of the `Tower` abstract class that we have reviewed earlier. It provides MIDP specific implementation for the abstract methods `paint()`, `unpaint()` and `paintBorder()`:

```
  // Inner class represents my tower implementation
  private class MyTower extends Tower {

    // Creates instance of me
    MyTower(int totalNumberOfRings, int xposition, int yposition,
            GameSettings settings) {
      super(totalNumberOfRings, xposition, yposition, settings);
```

```
    }

    // Refreshes the display of this tower.
    public void paint(Object graphics) {
      int y = 0;
      paintBorder((Graphics) graphics, _xposition, _yposition);
      unpaint((Graphics) graphics, _xposition, _yposition);
      y = _yposition + getSettings().getTowerHeight()
          - getSettings().getRingHeight();
      for (int i = 0; i <= _currentNumberOfRings - 1; i++) {
        _rings[i].paint((Graphics) graphics, _xposition, y);
        y = y - getSettings().getRingSpacing()
            - getSettings().getRingHeight();
      }
    }
```

The following method removes this tower from the display:

```
    public void unpaint(Object graphics, int x, int y) {
      ((Graphics) graphics).drawRect(x, y, getSettings().getTowerWidth(),
                                     getSettings().getTowerHeight());
    }

    // Paints the border for this tower
    public void paintBorder(Object graphics, int x, int y) {
      if (_selected) {
        ((Graphics) graphics).drawRect(x - 1, y - 1,
                                       getSettings().getTowerWidth() + 2,
                                       getSettings().getTowerHeight() + 2);
        ((Graphics) graphics).drawRect(x - 1, y - 1,
                                       getSettings().getTowerWidth() + 2,
                                       getSettings().getTowerHeight() + 2);
      } else {
        ((Graphics) graphics).drawRect(x - 1, y - 1,
                                       getSettings().getTowerWidth() + 2,
                                       getSettings().getTowerHeight() + 2);
        ((Graphics) graphics).drawRect(x - 1, y - 1,
                                       getSettings().getTowerWidth() + 2,
                                       getSettings().getTowerHeight() + 2);
      }
    }
  }
}
```

The MIDletGameSettings Class

The `MIDletGameSettings` class implements the abstract class called `GameSettings`. The main function of this class is to provide device-specific settings:

```
package com.edb.games.hanoi.midlet;

import com.edb.games.hanoi.GameSettings;

final class MIDletGameSettings extends GameSettings {
```

```
  private static int NUMBER_OF_RINGS = 5;
  private static final int SCREEN_WIDTH = 96;
  private static final int SCREEN_HEIGHT = 100;
  private static final int LEFT_MARGIN = 2;
  private static final int RIGHT_MARGIN = 2;
  private static final int BOTTOM_MARGIN = 5;
  private static final int TOWER_SPACING = 5;
  private static final int TOWER_HEIGHT = 40;
  private static final int RING_SPACING = 3;

  // Returns Number of rings
  public int getNumberOfRings() {
    return NUMBER_OF_RINGS;
  }

  // Returns Screen width
  public int getScreenWidth() {
    return SCREEN_WIDTH;
  }

  // Returns Screen height
  public int getScreenHeight() {
    return SCREEN_HEIGHT;
  }

  // Returns left margin
  public int getLeftMargin() {
    return LEFT_MARGIN;
  }

  // Returns right margin
  public int getRightMargin() {
    return RIGHT_MARGIN;
  }

  // Returns bottom margin
  public int getBottomMargin() {
    return BOTTOM_MARGIN;
  }

  // Returns tower spacing
  public int getTowerSpacing() {
    return TOWER_SPACING;
  }

  // Returns tower height
  public int getTowerHeight() {
    return TOWER_HEIGHT;
  }

  // Returns ring spacing
  public int getRingSpacing() {
    return RING_SPACING;
  }
}
```

The HanoiCanvas Class

The final class of the MIDP implementation is the `HanoiCanvas`. This is a special MIDP `Canvas` implementation that represents the Main View of the MIDP Hanoi implementation.

This class is internal to this package so therefore it has package access only:

```
package com.edb.games.hanoi.midlet;

import javax.microedition.lcdui.Canvas;
import javax.microedition.lcdui.Graphics;
import javax.microedition.lcdui.Alert;
import javax.microedition.lcdui.Displayable;
import javax.microedition.lcdui.Display;
import javax.microedition.lcdui.CommandListener;
import javax.microedition.lcdui.Command;
import javax.microedition.lcdui.Screen;
import com.edb.games.Game;
import com.edb.games.hanoi.HanoiGame;
import com.edb.games.hanoi.GameSettings;

// Represents the canvas for this game

final class HanoiCanvas extends Canvas {

  private HanoiGame _game = null;

  // Creates an instance of me
  public HanoiCanvas(HanoiGame game) {
    this._game = game;
    _game.start();
  }

  // Paints me
  protected void paint(Graphics graphics) {
    graphics.setColor(0xffffff);
    graphics.fillRect(0, 0, getWidth(), getHeight());
    graphics.setColor(0);
    graphics.drawString("Moves: " + _game.getNumberOfMoves(),
                        getWidth() / 2, (getHeight() / 8),
                        Graphics.BOTTOM | Graphics.HCENTER);
    _game.paint(graphics);
  }
```

This is a callback method that is used when a key on the MID is pressed:

```
  protected void keyPressed(int key) {
    int action = getGameAction(key);
    int tower = 0;
    switch (action) {
    case RIGHT:
      tower = GameSettings.RIGHT_TOWER;
      break;
    case LEFT:
      tower = GameSettings.LEFT_TOWER;
```

```
          break;
        case FIRE:
          tower = GameSettings.CENTER_TOWER;
          break;
        default:
          return;    // nothing we recognize, exit
        }
        _game.select(tower);
        repaint();
      }
    }
```

Building the Classes

As we are using the J2ME Wireless Toolkit, building the classes becomes very easy. First of all, create a new project called Hanoi, and specify the full name of the main class for this application:

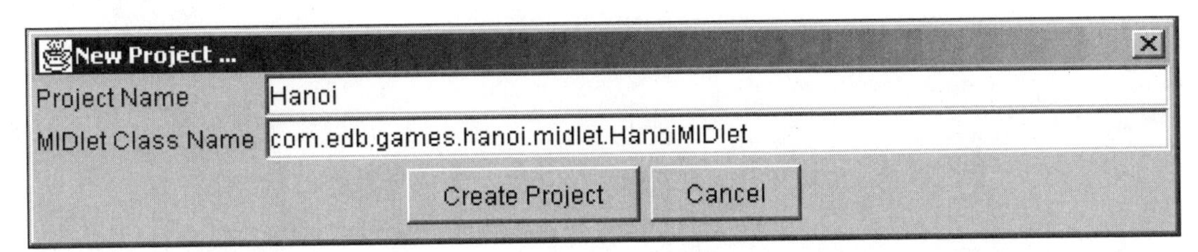

After accepting the settings, simply move the source files for our application into the `J2MEWTK\apps\Hanoi\src\` directory and press Build. The Toolkit will compile and preverify all the files, as well as creating a MIDlet JAR file and JAD. Finally, press Run to start the MIDlet suite.

Palm OS Implementation

The Palm OS Spotless implementation has exactly the same structures as the MIDP implementation but except that all the user interface APIs used belong to the Spotless system APIs.

The HanoiSpotless Class

First of all we shall look into the `HanoiSpotless` implementation that extends the `Spotless` class. Naturally these classes are packaged under a separate package that is called spotless as follows:

```
package com.edb.games.hanoi.spotless;
```

This class imports the following classes:

```
import com.sun.kjava.Button;
import com.sun.kjava.Dialog;
import com.sun.kjava.DialogOwner;
import com.sun.kjava.Graphics;
import com.sun.kjava.Spotless;
```

In addition it imports the shared layer interfaces and abstract classes that we have discussed earlier:

```
import com.edb.games.Game;
import com.edb.games.hanoi.HanoiGame;
```

```
import com.edb.games.hanoi.Tower;
import com.edb.games.hanoi.Ring;
import com.edb.games.hanoi.GameSettings;

// main class for a Hanoi Spotless game
public final class HanoiSpotless extends Spotless {

  private MyHanoiGame _game = null;
  static Graphics _graphics = Graphics.getGraphics();

  static Button _exitButton;
  static Button _startButton;
  static Button _helpButton;
  static Button _aboutButton;
```

When instantiated this class creates an instance of the `MyHanoiGame` class that is a `HanoiGame` Spotless implementation, and sets the Introduction View on the screen:

```
  // Creates instance of me
  HanoiSpotless() {
    _game = new MyHanoiGame(new SpotlessGameSettings());
    setIntroView();
  }
```

Unlike a MIDP implementation a Spotless implementation has to have a `main()` method which in this case is as follows:

```
  // Starts up the Hanoi game
  public static void main(String[] args) {
    HanoiSpotless hanoi = null;
    try {
      hanoi = new HanoiSpotless();
    } catch (Exception e) {
      if (hanoi != null && hanoi._game != null) {
        hanoi.setErrorView("Error: " + e.toString());
      } else {
        System.out.println("Error: " + e.toString());
      }
    }
  }
```

The `penDown()` method is a callback method. Each time a pen event is received this method is called:

```
  // callback called when on a pen down event
  public void penDown(int x, int y) {
    if (_exitButton.pressed(x, y)) {
      _game.exit();
    } else if (_startButton.pressed(x, y)) {
      _game.start();
    } else if (_helpButton.pressed(x, y)) {
      setHelpView();
    } else if (_aboutButton.pressed(x, y)) {
      setAboutView();
```

```
    } else {

      // Find newly selected tower
      int selectedTower = 0;
      for (int i = 0; i < 3; i++) {
        if (_game.getTowers()[i].pressed(x, y)) {
          selectedTower = i;
          break;
        }
      }
      _game.select(selectedTower);
    }
  }

  // Refreshes the screen
  public void refreshScreen() {
    _game.paintTowers();
    _game.drawStrings();
  }
```

The following group of methods set the different Views for the game, by delegating to the `game` instance associated with this `Spotless` implementation:

```
  // Sets my Intro View
  private void setIntroView() {
    ((Dialog) _game.getIntroView()).showDialog();
    ;
  }

  // Sets my last View
  private void setGameOverView() {
    ((Dialog) _game.getGameOverView()).showDialog();
    ;
  }

  // Sets my Help View
  private void setHelpView() {
    ((Dialog) _game.getHelpView()).showDialog();
    ;
  }

  // Sets my About View
  private void setAboutView() {
    ((Dialog) _game.getAboutView()).showDialog();
    ;
  }

  // Sets my Error View
  private void setErrorView(String error) {
    ((Dialog) _game.getErrorView(error)).showDialog();
    ;
  }
```

As in the HanoiMIDlet implementation, the HanoiSpotless implementation has three inner classes: the MyHanoiGame Spotless system implementation, (of the HanoiGame abstract class), the MyTower implementation (of the Tower abstract class) and the MyRing implementation (of the Ring abstract class):

```
// Inner class represents my game implementation
private class MyHanoiGame extends HanoiGame implements DialogOwner {
```

Similar to the MIDP implementation, the MyHanoiGame class has a constructor, an initialize(), start(), and exit() method:

```
// Creates instance of me
MyHanoiGame(GameSettings settings) {
  super(settings);
  _exitButton = new Button("Exit", 2, 145);
  _startButton = new Button("ReStart", 122, 145);
  _helpButton = new Button("Help", 1, 1);
  _aboutButton = new Button("About", 130, 1);
  _graphics.drawString(" Moves: " + this.getNumberOfMoves(), 60, 75,
                       _graphics.PLAIN);
}

// Initializes the game
public Game initialize() {
  return this;
}

// Starts the game
public Game start() {
  super.start();
  _graphics.clearScreen();
  paintButtons();
  paintTowers();
  drawStrings();
  return this;
}

// Exits the game
public Game exit() {
  System.exit(0);
  return this;
}
```

The following three methods are private fine-grain methods used by other coarse-grain methods in the same class:

```
// Paints my buttons
private void paintButtons() {
  _exitButton.paint();
  _startButton.paint();
  _helpButton.paint();
  _aboutButton.paint();
}
```

```
    // Refreshes the display of my towers
    private void paintTowers() {
      getTowers()[0].paint(null);
      getTowers()[1].paint(null);
      getTowers()[2].paint(null);
    }

    // Draws title and number of moves
    private void drawStrings() {
      _graphics.drawString("Towers of Hanoi", 46, 21);
      _graphics.drawString(" Moves: " + _game.getNumberOfMoves(), 60, 75,
                           _graphics.PLAIN);
    }
```

As in the MIDP implementation the following private methods set the various Views for the game. Each method delegates with the game inner class instance in order to get the required View and this is set as the current View. This action is implemented in a standard way. Obviously the actual GUI objects returned by these methods have different explicit types:

```
    // Returns the intro View for this game
    public Object getIntroView() {
      String introText =
        "\nWelcome to the Towers of Hanoi\n\nObjective:\nTry to move the rings
 from the 1st tower to the 3rd tower.\n\nRules:\nYou can only move one ring at a
 time. You cannot place a larger ring on top of a smaller ring.\n\nFor a 5 ring
 tower the theoretical minimum number of moves is 31.\n\nGood luck!";
      Dialog dlg = new Dialog(this, "Towers of Hanoi", introText, "Start");
      return dlg;
    }

    // Returns main View for this game
    public Object getMainView() {
      return null;
    }
```

Here the final View of the game is returned that details the number of moves it took to complete the game:

```
    public Object getGameOverView() {
      String solvedText =
        "Congratulations!\nYou solved the\nTowers of Hanoi\n  in "
        + getNumberOfMoves() + " moves!";
      Dialog dlg = new Dialog(this, "Congrats", solvedText, "Exit");
      return dlg;
    }
```

The following methods return the Help and About Views respectively for this game:

```
    public Object getHelpView() {
      String introText =
        "Objective:\nTry to move the rings from the 1st tower to the 3rd
 tower.\n\nRules:\nYou can only move one ring at a time. You cannot place a larger
 ring on top of a smaller ring.\n\nFor a 5 ring tower the theoretical minimum
```

```
number of moves is 31.\n\nGood luck!";
      Dialog dlg = new Dialog(this, "Help", introText, "Done");
      return dlg;
    }

    // Returns about View for this game
    public Object getAboutView() {
      String introText = "The Towers of Hanoi for The Spotless system";
      Dialog dlg = new Dialog(this, "About", introText, "Done");
      return dlg;
    }

    // Returns about View for this game
    public Object getErrorView(String error) {
      String introText = "Error: " + error;
      Dialog dlg = new Dialog(this, "Error", introText, "Done");
      return dlg;
    }
```

The `isTheGameRolling()` method is designed to return `true` or `false` if the game Main View runs on a separate thread. As all Views run on the same thread, this method always return `true`:

```
    public boolean isTheGameRolling() {
      return true;
    }
```

The following two methods are the Factory Methods we described earlier in the **Detailed Design** section. Here special Spotless system implementations of `Ring` and `Tower` are created and returned:

```
    // Factory Method, creates a tower
    protected Tower createTower(int totalNumberOfRings, int xposition,
                                int yposition, GameSettings settings) {
      return new MyTower(totalNumberOfRings, xposition, yposition,
                         settings);
    }

    // Factory Method, creates a ring
    protected Ring createRing(int size, GameSettings settings) {
      return new MyRing(size, settings);
    }
```

This callback method informs us that the game has been solved:

```
    protected void notifyGameSolved() {
      setGameOverView();
    }
```

The `dialogDismissed()` method is a `dialogOwner()` callback method called when a dialog gets dismissed:

```
    // callback, called when a dialog is dismissed
    public void dialogDismissed(String title) {
```

```
      if (title.equals("Towers of Hanoi")) {
        register(NO_EVENT_OPTIONS);
        start();
      } else if (title.equals("About") || title.equals("Help")) {
        register(NO_EVENT_OPTIONS);
        _graphics.clearScreen();
        paintButtons();
        paintTowers();
        drawStrings();
      } else {
        register(NO_EVENT_OPTIONS);
        exit();
      }
    }
```

This class overrides the `select()` method implementation parent class to do a refresh:

```
    // selects a tower
    public void select(int towerSelected) {
      super.select(towerSelected);
      if (!isGameSolved()) {
        refreshScreen();
      }
    }
  }
```

The second inner class is the `MyRing` implementation of the `Ring` class. Like all the other inner classes of this class this private member inner class cannot be instantiated from outside this class:

```
  // Inner class represents my ring implementation
  private class MyRing extends Ring {

    // Creates me
    MyRing(int size, GameSettings settings) {
      super(size, settings);
    }
```

Like the MIDP implementation of the `Ring` abstract class, this class has standard implementation of the `paint()` and `unpaint()` methods:

```
    // Refreshes the display of this ring
    public void paint(Object paintData, int x, int y) {
      int ringHeight = getSettings().getRingHeight();
      int step = (getSettings().getTowerWidth() - 4)
                 / getSettings().getNumberOfRings();
      step = step - step % 2;    // Make step even!
      int ringWidth = getSettings().getTowerWidth()
                      - ((getSettings().getNumberOfRings() - getSize())
                         * step);
      int ringX = ((getSettings().getNumberOfRings() - getSize()) * step)
                  / 2;
      Graphics.drawRectangle(x + ringX, y, ringWidth,
                             getSettings().getRingHeight(), Graphics.PLAIN,
                             4);
```

```
    }

    // Removes this ring from the display
    public void unpaint(Object paintData, int x, int y) {
      Graphics.drawRectangle(x, y, getSettings().getTowerWidth(),
                             getSettings().getRingHeight(), Graphics.ERASE,
                             4);
    }
  }
```

The last inner class is the `MyTower` inner class. Again the `MyTower` class is a private member inner class that cannot be instantiated from outside this class:

```
  // Inner class represents my tower implementation
  private class MyTower extends Tower {

    // Creates instance of me
    MyTower(int totalNumberOfRings, int xposition, int yposition,
            GameSettings settings) {
      super(totalNumberOfRings, xposition, yposition, settings);
    }
```

Like the MIDP implementation of the `Tower` abstract class it has standard implementation for the three methods `paint()`,`unpaint()` and `paintBorder()`:

```
    // Refreshes the display of this tower.
    public void paint(Object paintData) {
      int y = 0;
      paintBorder(paintData, _xposition,
                  _yposition);                          // Draw the border of the
tower
      unpaint(paintData, _xposition, _yposition);   // Unpaint the _rings
      y = _yposition + getSettings().getTowerHeight()
          - getSettings().getRingHeight();

      // Paint the _rings in the tower
      for (int i = 0; i <= _currentNumberOfRings - 1; i++) {
        _rings[i].paint(paintData, _xposition, y);
        y = y - getSettings().getRingSpacing()
            - getSettings().getRingHeight();
      }
    }
```

The `unpaint()` method removes this tower from the display:

```
    public void unpaint(Object paintData, int x, int y) {
      Graphics.drawRectangle(x, y, getSettings().getTowerWidth(),
                             getSettings().getTowerHeight(),
                             Graphics.ERASE, 0);
    }
```

The `paintBorder()` method draws the border for this tower:

```
    public void paintBorder(Object paintData, int x, int y) {
      if (_selected) {
        Graphics.drawBorder(x - 1, y - 1,
                            getSettings().getTowerWidth() + 2,
                            getSettings().getTowerHeight() + 2,
                            Graphics.ERASE, Graphics.SIMPLE);
        Graphics.drawBorder(x - 1, y - 1,
                            getSettings().getTowerWidth() + 2,
                            getSettings().getTowerHeight() + 2,
                            Graphics.PLAIN, Graphics.GRAY);
      } else {
        Graphics.drawBorder(x - 1, y - 1,
                            getSettings().getTowerWidth() + 2,
                            getSettings().getTowerHeight() + 2,
                            Graphics.ERASE, Graphics.GRAY);
        Graphics.drawBorder(x - 1, y - 1,
                            getSettings().getTowerWidth() + 2,
                            getSettings().getTowerHeight() + 2,
                            Graphics.PLAIN, Graphics.SIMPLE);
      }
    }
  }
}
```

The SpotlessGameSettings Class

The final class in this package is the `SpotlessGameSettings` class, the implementation of the abstract class called `GameSettings`. The settings values contained in this class are Palm OS-specific:

```
package com.edb.games.hanoi.spotless;

import com.edb.games.hanoi.GameSettings;

final class SpotlessGameSettings extends GameSettings {

  private static int NUMBER_OF_RINGS = 5;

  // Change this to set the difficuly level
  private static final int SCREEN_WIDTH = 160;
  private static final int SCREEN_HEIGHT = 160;
  private static final int LEFT_MARGIN = 6;
  private static final int RIGHT_MARGIN = 6;
  private static final int BOTTOM_MARGIN = 25;
  private static final int TOWER_SPACING = 6;
  private static final int TOWER_HEIGHT = 45;
  private static final int RING_SPACING = 2;

  public int getNumberOfRings() {
    return NUMBER_OF_RINGS;
  }
  public int getScreenWidth() {
    return SCREEN_WIDTH;
  }
```

```
    public int getScreenHeight() {
      return SCREEN_HEIGHT;
    }
    public int getLeftMargin() {
      return LEFT_MARGIN;
    }
    public int getRightMargin() {
      return RIGHT_MARGIN;
    }
    public int getBottomMargin() {
      return BOTTOM_MARGIN;
    }
    public int getTowerSpacing() {
      return TOWER_SPACING;
    }
    public int getTowerHeight() {
      return TOWER_HEIGHT;
    }
    public int getRingSpacing() {
      return RING_SPACING;
    }
  }
```

Building the Classes

To build the classes we have put together a batch file that has the following content:

```
set CLDC_HOME=c:\j2me_cldc1.0.2

mkdir temp
mkdir classes
mkdir bin

javac -g:none -bootclasspath %CLDC_HOME%\bin\kjava\api\classes -classpath

%CLDC_HOME%\bin\kjava\api\classes;. -d % temp com\edb\games\*.java
javac -g:none -bootclasspath %CLDC_HOME%\bin\kjava\api\classes -classpath
%CLDC_HOME%\bin\kjava\api\classes;. -d temp com\edb\games\hanoi\*.java
javac -g:none -bootclasspath %CLDC_HOME%\bin\kjava\api\classes -classpath
%CLDC_HOME%\bin\kjava\api\classes;. -d temp com\edb\games\hanoi\spotless\*.java

%CLDC_HOME%\bin\win32\preverify -d classes -classpath
%CLDC_HOME%\bin\kjava\api\classes;. temp

java -classpath %CLDC_HOME%\tools\palm\src;. palm.database.MakePalmApp -v -version
"1.0"

-bootclasspath %CLDC_HOME%\bin\kjava\api\classes -classpath classes -o
bin\HanoiSpotless.prc com.edb.games.hanoi.spotless.HanoiSpotless
```

The final line builds the `HanoiSpotless.prc` file. For a detailed look at each of these commands, please see Chapter 4.

Running the Application

To run the application on the Palm emulator we will first need to install the KVM files onto the emulator. The two necessary files (`KVM.prc` and `KVMUtils.prc`) are found in the `<CLDC_HOME>\bin\kjava\palm\` directory. We load these as we would any other application. We should also install the `HanoiSpotless.prc` using the same procedure. To start the application all we do is double click on the Hanoi icon.

We can also test our application using the `kvmkjava.exe`, which is a test tool that comes with the CLDC download. This will save time installing our application on the Palm OS emulator when we are developing. As can be seen below, the tool has a primitive looking Palm emulator user interface:

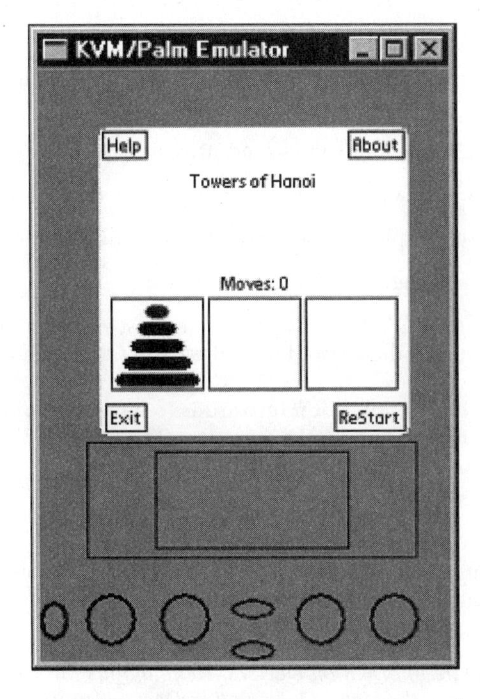

To run the `kvmkjava.exe` test tool from the command prompt, run the following command (where `<OS>` is the OS that you are working with):

```
<CLDC_HOME>\bin\kjava\<OS>\kvmkjava com.edb.games.hanoi.spotless.HanoiSpotless
```

Summary

In this chapter we discussed designing and implementing the Hanoi game application and porting it on two platforms, the Mobile Information Device Profile and the Spotless System for the Palm OS. When developing applications for these two platforms we must adhere the following:

- When writing code, move as much of the application logic, into the CLDC layer.
- Write loosely, coupled code, with as little dependency as possible, to promote reuse.
- Keep the user interface simple.
- Make the navigation clear and keep it shallow with the least number of screens possible.
- Make button locations, label names, and titles consistent. Buttons should always appear on the same location.
- Limit text input, pre-fill default values, and provide selection rather than text input.
- Display user, centric messages and avoid ambiguity in text.
- Do not have too much display on one screen. It is not desirable to force the user to scroll.
- Think about error handling.
- Remember to use the CLDC test tools to run and test your application. Once you are happy with the software you can then install it on the Palm OS emulator.

Ideally the way to take the Hanoi application forward is to have multiple implementations on different profiles, sharing the same configuration implementation. The follow up to the MID profile is the PDA profile that can make, great porting ground for the Hanoi application.

References

- Gamma Erich, Helm Richard, Johnson Ralph and Vlissides John, **Design Patterns ISBN 0-201-63361-2: Elements of Reusable Object-Oriented Software**.: Addison-Wesley Publishing Company, Inc, 1994.
- **Applications for** *Mobile Information Devices, Helpful hints for Application Developers and User Interface Designers using the Mobile Information Device Profile*. A Sun white paper: http://java.sun.com/.
- Hix, Deborah & Hartson, H Rex *Developing User Interface ISBN 0-471578134* : John Wiley & Sons, Inc, 1993.

PROFESSIONAL JAVA MOBILE

TOOL SELECTION

WEB TABLET
MOBILE PHONE
MP3 PLAYER
DIGITAL CAMERA
PALMTOP
LAPTOP

NOW SELECTED

PALMTOP

14

Case Study: A Portable Expert

The prototype system described in this study is a basic Expert System designed to run on the Palm. Expert Systems came about through research in AI, and are a way of abstracting logic so that inferences can be made according to given rules to perform diagnostic functions. It should be suitable for many classification and diagnostic tasks, though it was prompted by a request for a (botanical) tree recognition application. In terms of performance, the end result leaves a lot to be desired, but it demonstrates the feasibility of the basic concept.

Whether or not the expert system side of this study is of interest to other developers, the way in which (minimal) XML is used for data persistence and transfer could easily be adapted for many other portable applications, and is the main focus of this chapter. We will also discuss the advisability of using XML in mobile applications later in the chapter; however, the ability to manipulate XML on the desktop or server before loading it onto the device opens the door to a wide range of data sources, and of course there is no need to develop another data format for the device.

As will be shown here, XML (or rather a reduced variety of it) is certainly an option for some applications.

Initial Analysis

The virtual machine we will be using is the KVM, so we shall dub the application **Kex**, for 'KVM Expert System'. We will also be using the MVC design pattern and will begin by dividing the application functionality into Model, View, and Controller. Since this style of design is prevalent, we should have no problems finding examples to base the system on.

An Expert System is fundamentally just a set of questions, and answers that depend on the replies given for identification. Each question answered narrows the possible answers down until only one answer is possible or no more questions are left, in which case a number of potential answers exist. An example of this is a system for identifying trees according to given criteria. The questions will be along the lines of: do the tree's leaves fall during winter? Does the tree have thin needle-like leaves? Does the tree have leaves in bundles? Does the tree have leaves or needles in bundles of five? Does the tree have cones? We will use this example for illustration wherever necessary in the remainder of the chapter.

Production Rules

It is apparent that as we will be modeling characteristics of real-world objects in Java (trees), then the OO paradigm will be useful when it comes to representing our information (in fact, this paradigm was largely developed from the AI representation of objects using frames and slots). In terms of reasoning, within existing expert systems there is a prominent paradigm we can use known as the **production system**, comprising **production rules** and methods to manipulate them.

The production rule is nothing more than an `IF ... THEN ...` statement. These rules always express a conditional statement with an antecedent clause (`IF`) and a consequent clause (`THEN`). These clauses will contain **facts**, which represent items in the system, for example `IF facta THEN factb`. A simple representation system (or knowledge base) can be built using just facts and rules.

For our tree-recognition system we have our target or **goal** facts such as 'Oak' and 'Elm' and intermediate facts that the user may not be interested in such as 'Deciduous' or 'Conifer'. An '`IS-A`' relationship is implied in this kind of fact, for example 'This tree **is a** sycamore'. Note that the actual contents of a fact, for example 'Oak' is irrelevant to the system (it has no concept of 'Oak' above its relation to other facts, which we will define in rules; the abstraction allows us to substitute any knowledge base without modification to the system's logic) so the word 'Oak' is acting as nothing more than an identifier or **token**.

Within this system the information is defined in terms of relationships between facts, and an inference mechanism (algorithm) then operates on these relationships. A practical consequence of this is that usually the order in which the statements appear in such a declarative system doesn't matter because the algorithm will decide the order of evaluation – or to put it another way, the order in which we ask the user questions is irrelevant.

The relationships in production systems are the rules. An example of a rule would be '`IF tree loses its leaves in winter THEN tree is deciduous`'. Of course, we can incorporate Boolean operators, for example '`IF deciduous AND NOT leaves have green undersides THEN Holm Oak`'.

Looking at these two examples, a structural link can be seen – the consequent clause of the first rule (the consequence) is an antecedent (a condition) to the second rule. When conditions of the first rule are satisfied, the first antecedent of the second is asserted. Clearly, to identify a particular tree we are somehow going to have to make use of the structure of these links, and this is the job of our inference engine.

Our Requirements for the Model-View-Controller System

We have now pretty much determined the general form of the Model, that is, it will contain a representation of a knowledge domain, comprising of facts and rules, and an inference mechanism. The easiest way for the inference engine to operate is directly on the knowledge domain data, so that at each step of the information discovery (obtained from the user), and reasoning, the knowledge domain contains the factual knowledge we have determined so far. We will see later that there is a downside to this approach, but the benefits outweigh the costs.

The user interface (UI) will have to allow the user to load in the knowledge base to initialize the system, and as in any software project give appropriate notification of progress, errors, etc. Due to the nature of the target device, the loading of data will not be quite as straightforward as, for example, hooking into a database with JDBC, but from the point of view of the UI this is of no consequence.

The core of the UI will be concerned with getting rule antecedents from the user – that is asking questions that will satisfy the `IF` part of rules. The least complicated way of doing this will be to formulate the antecedents as questions that provide us with Boolean `true/false` data. In more human terms this translates into 'Yes'/'No' questions, with a 'Don't Know' option so that the user won't get stuck. Finally, the UI must supply the user with the end results. The View will then comprise the visual components of the UI, and the Controller will handle transitions between different stages – domain loading, questioning, results, and dealing with errors based on user interactions and the behavior of the Model.

System Overview

To help explain the role of different parts of the system, here we have a diagram of the final application:

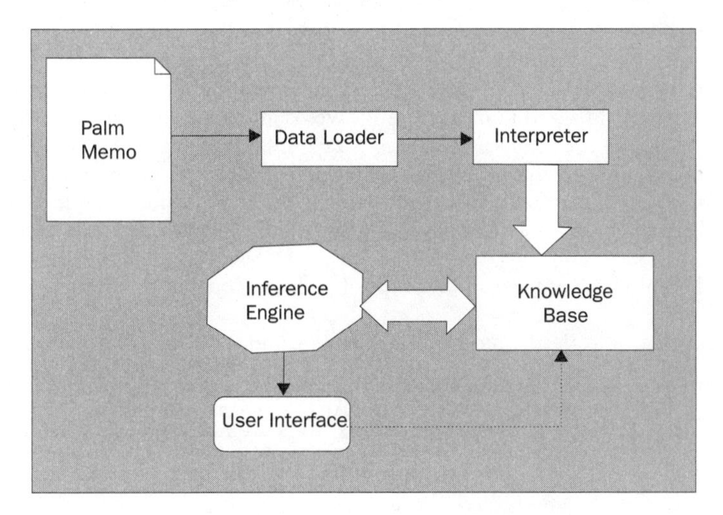

As you can see, the knowledge base data will be provided as a Palm Memo entry, which will be in the form of an XML document. The selected data will be loaded into the system and interpreted to create the knowledge base. The inference engine will choose questions to ask the user, and modify the domain knowledge based on the user's responses until either a goal is satisfied or no further questions are left.

After the results have been displayed the user may reinitialize the knowledge base for another run, in which case an in-memory copy of the domain knowledge will be used to avoid having to reload and interpret the data again. Memory **is** limited, but for this application the loading time was such a major bottleneck (due to limitations of KJava) that it was judged better to take this approach. The program flow will be determined by the class containing the user interface and will be based on user interactions and the behavior of the inference engine.

From Abstract to Concrete

At this point in the book, you've probably already got everything we will using:

- Palm handheld
- POSE emulator with ROM image and KVM installed
- Personal computer

- J2ME CLDC/KVM Palm Release (CLDC version 1.0 & KJava)
- Java compiler
- Text editor/IDE
- Documentation for all the above

There is an emulator supplied with J2ME (`kvm.exe`), but the Palm OS Emulator (POSE) from Palm provides much more accurate hardware-level emulation and so should be used in preference.

Sample Knowledge

It's probably been a while since you last heard the phrase "Garbage In, Garbage Out", but it's certainly worth bearing in mind when dealing with expert systems. For the system to be useful, the knowledge on which it operates must be accurate. Acquisition of domain knowledge is a complex field, but the knowledge we aim to use will be taken directly from classification trees in a book. This should be considerably more straightforward than the alternative of interrogating a human expert.

The size of this knowledge base will be quite unwieldy when all we wish to do is define simple operations. So to test the operation of our system we will imagine a rather primitive expert, who more or less knows how to tell the difference between four things. We have interviewed this expert and elicited their knowledge.

There are four different kinds of trees in this expert's world. You can tell them apart because:

- An oak has entire leaves (smooth straight edges) and its twigs are not aromatic when crushed.
- A smoketree has leaves that are entire and its twigs are aromatic when crushed
- A poplar does not have entire leaves and its twigs are not aromatic when crushed
- A sassafras does not have entire leaves but its twigs are aromatic when crushed

Notice that eliminating all of the other possibilities implicitly identifies the last tree type. We can now determine the facts and rules, and state them formally.

For facts, we have the targets we are trying to identify: Oak, Smoketree, Poplar, and Sassafras. To these we can also add 'leaves are entire' and 'twigs are aromatic when crushed' which act as intermediate facts. We can now list the facts, each comprising a token and two basic attributes:

Token	Goal	Question
Oak	Yes	
Poplar	Yes	
Sassafras	Yes	
Smoketree	Yes	
'Are twigs aromatic when crushed?'		Yes
'Are leaves entire?'		Yes
Leaves are entire		

In this case, "Are leaves entire" acts as an intermediate fact. Conceptually, we establish whether the tree has leaves that are entire first, and then combine this fact with a check for aromatic twigs when crushed. This produces the following production rules:

- `IF 'Are leaves entire?' THEN Leaves are entire`
- `IF NOT 'Are twigs aromatic when crushed?' AND Leaves are entire THEN Oak`
- `IF 'Are twigs aromatic when crushed?' AND Leaves are entire THEN Smoketree`
- `IF NOT 'Are twigs aromatic when crushed?' AND NOT Leaves are entire THEN Poplar`

Knowledge Base Classes

We need a class to represent a fact one to represent, a rule, and one to wrap up the combined facts and rules into the domain knowledge base. We will be asking our user questions, and each of these we can represent as a fact. We can easily set an attribute to denote that the fact is a question. Similarly we can record whether or not a fact is a goal with a Boolean attribute `true` or `false`. The logical `NOT` that can appear in a rule also needs representation. We can now build a block diagram of the data structure of a knowledge base:

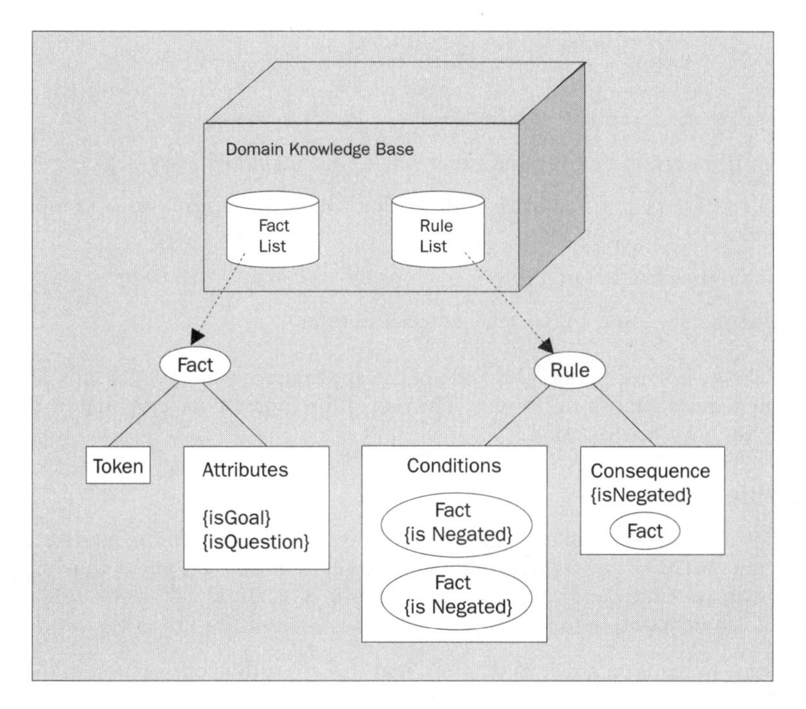

A rapid prototyping exercise yielded some other ideas that will make the next stage of development quite straightforward. For instance, take a rule:

```
IF condition THEN fact
```

Where `condition` would be the result of the state of one or more facts, for example:

```
IF factA AND factB THEN factC
```

We can see that irrespective of the details of the inference engine, the facts that make up the antecedents (`factA` and `factB`) will be handled slightly differently from that of the consequent (`factC`), although they are essentially the same, having a token and attributes. In object-oriented terms, this suggests one parent class with two descendants. Note also that in the diagram above, the facts that make up the conditions have their 'negated' attribute held internally, whereas this attribute of the consequence will be contained in the enclosing rule, as this attribute is more closely associated with the rule than the fact.

We decided earlier to allow the inference engine to operate directly on a domain, as it tries to establish rules and prove facts. This means that some form of run-time variable is needed to reflect the state of proven (or disproven) facts and established rules. This can be implemented by simply adding an attribute to the fact object to show whether or not we know the state of the fact (`isKnown`) and an attribute to show for a fact with a known state whether the fact is `true` or `false` (`isAffirmed`). The rule object will have an attribute to show if all its conditions have been established (`isFired`). As the inference engine progresses, some rules may have one or more conditions satisfied but still have some yet to be determined. We can simply remove conditions from each rule, changing the rule itself at run time.

We can now start coding the knowledge base classes. At this point we'd better start keeping an eye on our target class libraries, which will be the CLDC (Connected, Limited Device Configuration) and the KJava API for the Palm.

An appropriate set of classes is as follows, which we will package into `com.kex.model`:

- `FactBase` – the fact element base class
- `Fact` – an extension of `FactBase` that will be our main fact class
- `Condition` – an extension of `FactBase` that will be specifically used to represent facts in the antecedant
- `Rule` – a representation of a rule
- `KnowledgeBase` – our initial sets of facts and rules

As the focus of this case study is the XML side of the application, we will only look at those classes that directly manipulate the XML on the device. The rest of the source code can be downloaded from the Wrox web site: http://www.wrox.com.

The KHashTable Class

The first piece of code we will look at is for a utility class, used widely in the model. Our facts have attributes to denote whether they represent questions, goals, or neither, and a convenient way of storing collections of attribute-value pairs is as keys and values in a `Hashtable`, which unlike the `Collection` classes of Java 2 can be found in the CLDC.

This is a good time to mention a design decision that will affect the construction of the whole knowledge base. We will be loading in our data from an external source, and as we will see later this is very time-consuming, so once loaded it will be desirable to return to the initial state of the domain for subsequent runs without having to reload.

Our inference engine will operate on the domain as information becomes available, facts become known, and rules are fired, thus modifying it. This will mean having a working copy of the domain and maintaining a copy intact, before processing.

The normal Java handling of data structures such as the `Hashtable` and `Vector` is by reference, which would normally have meant that our original copy of the domain would have been modified in the same manner as the working copy. What we need is a way of making a 'deep' copy of the objects, a facility our model classes need to allow us to maintain and restore a copy of the initial state. This facility is very useful, allowing one to transfer information around freely without affecting the original data.

The FactBase Class

This will be the fact element base class, from which we will extend classes to represent (antecedent) conditions and (consequent) facts. We have already seen two attributes common to both, for example whether or not the fact/condition is a goal, and whether or not the fact/condition is a question. These characteristics we will store as attribute-value pairs in an instance of `KHashtable`.

The Hashtable can store any number of these attribute-value pairs, so we can express the goal and question attributes within a single `KHashtable` object for the `FactBase` class. We will name this object `avpairs`. We will also need to store the 'contents' of a fact/condition, such as its token value, ('Oak', Poplar etc.) for which we can simply use a String, which we will name `value`. Although `value` identifies the fact/condition, it will be convenient to define a numeric identifier, `id`.

The Fact Class

This will represent the fact object as initially loaded into the knowledge base and in its role as the consequent of an `IF...THEN...` statement. To make things easier later in our inference engine, a list of the rules in which a particular fact occurs will also be stored within the fact. The whole purpose of this expert system is to establish the truth or otherwise of facts, based on the rules in the knowledge base and the user's input. We will not only need a flag for each fact to say if it is true or false, but also a flag to say whether or not this has been established. We will thus have the class attributes `stateknown` for our state of knowledge about the fact, and `affirmed` for its truth or falsehood.

`Fact` also defines a `toString()` method that provides relevant information.

The Condition Class

This is another extension of `FactBase`, with an extra attribute which indicates whether or not the condition is negated, for example in the statement `IF a AND NOT b THEN c`, the negated attribute of condition `a` is `false`, and the negated attribute of fact `b` is `true`. Unlike the `Fact` objects a `Condition` object is 'fixed' in a rule and does not change its value, so has no working attributes such as `stateknown` or `affirmed`.

The Rule Class

The list of facts the rule refers to will be contained in an instance of the `KnowledgeBase` class (which will be described shortly), as will the rule itself, so we will give `Rule` a `KnowledgeBase` class attribute. The rule needs to have a set of conditions. Rather than have a redundant array of `Condition` objects, we can represent these as a list of references to the corresponding `Fact` objects, each reference having an associated value corresponding to the 'negated' flag of the condition.

We have defined a structure that is ideal for storing this kind of data, the `KHashtable`. We have to find a way of internally representing the Boolean modifier of the consequent of the rule, that is to say, be able to tell the difference between `IF a THEN b` and `IF a THEN NOT b`. This information appears to be associated with the consequent fact, suggesting that we use an approach like the 'negated' attribute of `Condition`, though it is far easier to associate it with the rule, as a class attribute.

We will again be providing a verbose `toString()` method, and this time it is complicated enough to warrant caching the return String. When the first call is made on `toString()`, the return value will be placed in `outstring`, and a flag `dirty` will be set to `false`. If another call is made to the method before any changes have been made to the rule, `outstring` will be returned again. However, if removing a condition has modified the rule, the `dirty` flag will be set to `true` and the return string will be rebuilt.

The `trigger()` method handles the internal processing of the rule. The method is sent a `Fact`, and this is checked against the conditions of the rule. The required behavior is best shown by example:

- trigger fact: `a, affirmed = true`
- rule: `IF a AND b THEN c`

In this case, fact `a` is matched, causing a partial satisfaction of the rule, leaving it as follows:

```
IF b THEN c
```

If we continue with the same rule and assert fact `b`:

- trigger fact: `b, affirmed = true`
- rule: `IF b THEN c`

All conditions are satisfied, and the rule can be fired and the `affirmed` flag of `c` set to `true`.

However, if we return to the first rule, and assert that `a` is `false`:

- trigger fact: `a, affirmed = false`
- rule: `IF a AND b THEN c`

Following Boolean logic, if `a` is `false`, `a AND b` must also be `false`, so the conditions are satisfied, and the rule can be fired and the `affirmed` flag of fact `c` set to `false`.

When the attribute of the rule is `false (NOT c)`:

- trigger fact: `a, affirmed = true`
- rule: `IF a AND b THEN NOT c`

The result is the modified rule:

```
IF b THEN NOT c
```

For coding this kind of logic, a **lot** of time can be saved by working things out with pencil and paper **first**.

The method begins in a straightforward fashion:

```
public Fact trigger(Fact triggerfact) {
    boolean match;
    Condition condition;
```

```
    int conid;
    boolean connegated;
    int triggerid = triggerfact.getId();
    boolean conditionstate = false;
    boolean resultstate = false;

    for (Enumeration ec = conditions.keys(); ec.hasMoreElements(); ) {
      Object key = ec.nextElement();
      conid = ((Integer) key).intValue();
      connegated = conditions.getBoolean(key);
      match = (conid == triggerid);
      if (match) {
        try {
```

We will now see why a pencil and paper were called for – `conditionstate` is a combination of the 'affirmed' state of `triggerfact` and the 'negated' value of the condition. If the two are the same (`true/true` or `false/false`) then condition state will be `false`, if they are different (`true/false` or `false/true`) it becomes `true`. This is the exclusive-or logical operation which is available in Java using the ^ symbol. This line is inside a `try...catch` block as `triggerfact.isAffirmed()` will throw an exception if the fact is marked as not known:

```
          conditionstate = triggerfact.isAffirmed() ^ connegated;
        } catch (Exception e) {
          System.err.println(e);
        }
        if (conditionstate) {
          conditions.remove(key);
          dirty = true;
          knowBase.update(this);
          KLog.println("Updating : " + this);
          if (conditions.size() == 0) {
            fired = true;
          }
        } else {
          fired = true;
        }

        if (fired) {
          resultstate = isNegated() ^ conditionstate;
          Fact rulefact = getFact();
          rulefact.setAffirmed(resultstate);
          return rulefact;
        }
      }
    }
    return null;
  }
```

The KnowledgeBase Class

This class will perform two functions: it will act as a repository for our initial facts and rules, and as a 'working memory' on which the inference engine can operate. To allow changes to be made to the data, we will have two `update()` methods, one taking a fact as its argument and the other taking a rule (the update methods will also be used in the data loading process, as more detail becomes available about an item). There will also be a lookup table to quickly access a fact by its token.

The Engine Class

The inference engine will operate by selecting the question that occurs in most rules, which will be passed back to the user interface, which will at some point return the information whether or not the answer to the question is true. This will hopefully produce more known facts, which can be fed back into the system until all the consequences of the user's response have been resolved. It may be that during this process of resolution a goal is proven true, so our inference engine needs to check for this and notify other parts of the system.

Similarly, the engine will need to give an appropriate response if it runs out of questions before a goal has been reached. When a known fact causes a rule to fire, this will produce another known fact that will need to be processed and so we have a `Vector` to hold a queue of these 'discovered' facts. To indicate when the inference engine has either run out of questions or has proven a goal we have the flag `done`. The `count` variable is something of a luxury, to indicate in the log the number of facts that have been asserted.

Using XML for Knowledge Persistence

We will come to the practicalities of loading data on the Palm shortly, but first let us first look at how we can express our knowledge base in XML. We will express the facts as simple elements:

```
<fact>Helicopter</fact>
```

The rules take advantage of the nesting available in XML, and use the XML attributes to reflect the attributes of the particular element, so we can express the rule `IF Are crushed twigs aromatic AND NOT Leaves are entire THEN Sassafras`

```
<rule>
  <fact>Sassafras</fact>
  <condition>Are crushed twigs aromatic</condition>
  <condition negated="true">Leaves are entire</condition>
</rule>
```

The XML document should start with a prolog describing the document. Here we can give the version of XML that will be used (1.0, the current standard, see http://www.w3.org/TR/REC-xml/) and the character encoding (UTF-8, an ASCII-compatible Unicode used by Java, among other things, see RFC 2279). In addition, if the document is to be read directly by humans then comments are advisable. We can also add an attribute to the parent element `<domain>` to provide its name.

Putting all this together, we can succinctly describe our test knowledge base as an XML document:

```
Kex "Example"
<?xml version='1.0' encoding='UTF-8'?>
<!DOCTYPE domain SYSTEM "kex.dtd">

<domain name="Test">

  <!--  *** facts ***  -->

  <fact goal="true">Oak</fact>
  <fact goal="true">Poplar</fact>
  <fact goal="true">Sassafras</fact>
  <fact goal="true">Smoketree</fact>
  <fact question="true">Are leaves entire</fact>
  <fact question="true">Are twigs aromatic when crushed</fact>
  <fact>alive</fact>
```

```
  <!--  *** rules ***  -->

  <rule>
    <fact>Oak</fact>
    <condition> Are twigs aromatic when crushed </condition>
    <condition>Leaves are entire</condition>
  </rule>

  <rule>
    <fact>Poplar</fact>
    <condition negated="true">Are twigs aromatic when crushed</condition>
    <condition negated="true">Leaves are entire</condition>
  </rule>

  <rule>
    <fact>Sassafras</fact>
    <condition>Are twigs aromatic when crushed</condition>
    <condition negated="true">Leaves are entire</condition>
  </rule>

  <rule>
    <fact>Smoketree</fact>
    <condition>Leaves are entire</condition>
  </rule>

</domain>
```

The first line is not XML, but there is a reason for including this, which will be explained shortly.

You may have noticed that a Document Type Definition (DTD) is specified in the `DOCTYPE` tag. By defining this, we can use a standard XML editor (on the desktop) to write and check the validity of our knowledge domain files. This is very useful, but it must be remembered that a document that conforms to this DTD may not necessarily contain valid information for our expert system – there may be logical flaws.

The DTD (`kex.dtd`) is defined as follows:

```
<?xml encoding="UTF-8"?>

<!-- DTD for KVM Expert System Knowledge Base Domains -->

<!ELEMENT domain (fact+,rule+)>
<!ATTLIST domain name CDATA #REQUIRED>
<!ELEMENT fact (#PCDATA)>
<!ATTLIST fact goal (true | false) "false">
<!ATTLIST fact question (true | false) "false">
<!ATTLIST fact negated (true | false) "false">
<!ELEMENT rule (fact,condition+)>
<!ELEMENT condition (#PCDATA)>
<!ATTLIST condition negated (true | false) "false">
```

We will impose on our XML data files an additional, non-standard requirement, that all facts must be defined before the rules. This will enable our system to check one important part of the logical validity of the data, that all facts used in the rules have already been defined. A bonus of this is that it makes interpretation of the data a little easier.

Reading a Palm Memo

XML consists of plain text; so to get our knowledge base data into the application we need to find a good way of representing text on the Palm. Here we could build our own low-level format, but again there is an existing technique we can use, which will require the minimum of coding. An application supplied with the Palm is 'Memo', which is essentially a text editor, which stores its data in a standard Palm database (PDB). So, to us, the Memo can act as an off-the-shelf persistence mechanism.

There are two key advantages with using the Palm Memo for data storage. First, it will allow us to manually enter and edit expert knowledge directly on the Palm using a standard application. Secondly, and probably more importantly, it allows us to use an existing data transfer medium, in the form of the conduit used to move Memo data from the desktop onto the Palm. We can prepare our knowledge base files on the desktop using a text or XML editor, then copy and paste the text into either the Palm Desktop or Outlook notes. A short Hotsync later, we have our data on board, ready to use.

I/O Classes

Reading the contents of a Memo is relatively straightforward, though in this area the KJava API is seriously weak, and the result is that the Memo reader will be considerably slower than it could be.

The PdbReader Class

The purpose of this class is to retrieve a named document from the Palm Memo database. The key to this operation is the KJava `Database` class. Though it is possible to create new PDB databases using this class, all we are concerned with in this project is opening an existing Memo. In these days of namespaces and resource locators, it is somehow refreshing to see that the Palm database only has a name (roughly corresponding to a file name) and two identifiers, each a representation of a 4-character string.

Only the identifiers are needed to open a database. The first of these is the `CreatorId`, which is a unique company/developer ID registered with Palm. The second is the `typeId`, which refers to the type of data stored in the database, that is the application that will use it. We will be reading from the Memo applications database, which is called `MemoDB` and has the `CreatorId` `memo` and `typeId` `DATA`.

There is only one constructor in the `Database` class:

```
public Database(int typeID, int creatorID, int mode)
```

This will create a `Database` object and open the database. The mode is one of three constants, `READONLY`, `WRITEONLY`, and `READWRITE`. Note that the identifier arguments are of type `int`. This value is obtained by interpreting the IDs as a kind of base-256 number. A utility method `getId()` is provided in our reader class to make these conversions:

```
package com.kex.io;

import com.sun.kjava.Database;

public class PdbReader {

  private Database db;

  public PdbReader(String typeIdString, String creatorIdString,
                   int mode) throws Exception {
    this(getId(typeIdString), getId(creatorIdString), mode);
```

```
    }

    public PdbReader(int typeId, int creatorId, int mode) throws Exception {
      db = new Database(typeId, creatorId, mode);
      if (!db.isOpen()) {
        throw new Exception("Database unavailable.");
      }
    }

    private PdbReader() {}

    public long getNumberOfRecords() {
      return db.getNumberOfRecords();
    }
```

The data from the Memo database is retrieved as a series of single-field records, each presented in its entirety as a byte array (partial reads are not implemented in the KJava release):

```
    public byte[] readRecord(int recordNumber) {
      byte[] record = null;
      try {
        record = db.getRecord(recordNumber);
      } catch (Exception e) {
        return null;
      }
      return record;
    }

    public void close() {
      db.close();
      db = null;
    }

    public static int getId(String string) {
      if (string.length() != 4) {
        return -1;
      }
      int total = 0;
      char[] ch = new char[4];
      string.getChars(0, 4, ch, 0);
      for (int i = 0; i < 4; i++) {
        total += (int) ch[i] << ((3 - i) * 8);
      }
      return total;
    }
  }
```

The KMemoReader Class

We will check the records retrieved against the name of the knowledge base document we are looking for until we find the document required or run out of records. Though it would be possible to keep our documents on the Palm exactly as in the sample XML listing we saw earlier, this leads to the listed name of in the Memo application being the fairly uninformative `"<?xml version=..."` line. To make it more user friendly we will add a title to document, on the first line of the Memo, in the format:

```
Kex "documentname"
```

We can now use the string `Kex` to identify Kex data, and the document name to find the particular document we request. Note the use of quotes to delimit the document name. The reading will be carried out by an instance of `KMemoReader`, a class that gets its memo-reading capability by extending the class above:

```java
package com.kex.io;

import com.sun.kjava.Database;
import com.kex.util.KLog;
import com.kex.control.Status;

public class KMemoReader extends PdbReader {

  public KMemoReader() throws Exception {
    super("DATA", "memo", Database.READONLY);
  }

  public byte[] getDocument(String name) {
    name = "\"" + name + "\"";
    int namelength = name.length();
    byte[] record;

    String head = new String();
    for (int recordNumber = 0; recordNumber < getNumberOfRecords();
         recordNumber++) {
      KLog.println("Reading record " + recordNumber);

      record = super.readRecord(recordNumber);

      if (record.length > 20) {
        head = new String(record, 0, 20);
        if (head.regionMatches(true, 0, "Kex", 0, 3)) {
          KLog.println("\nFound a Kex KB :\n" + head + "...\n");
          if (head.regionMatches(true, 4, name, 0, namelength)) {

            return record;
          }
        }
      }
```

Reading through the data can be a lengthy process, as each record must be read in full before we can check it, so we show the user a simple progress indicator, adding an asterisk to a message box for every record we examine:

```java
        Status.message += "*";
    }
     return null;
  }
}
```

The above class provides us with a byte array containing our knowledge base in XML format.

Extracting the Data

As you might expect, a lot of work has been done on reading XML into Java, and the standard tool for this is SAX, the Simple API for XML. This operates by serially reading in a character stream (the XML file) and triggering events when it encounters XML tags. These events will be handled by a custom class that creates Java objects as required by the application.

There are many implementations of SAX available, a standard package being `org.xml.sax`. Most implementations of SAX are too large to consider for our limited target device, and may well use classes not available in the CLDC API. Looking at our application's requirements we can see that we only need to implement the interpretation of simple element tags and their attributes.

We can draw on SAX to get a good pattern for building our interpreter, and this approach will make our classes more amenable to integration with standard SAX systems later if we so wish. We will be making quite a few simplifications and customizations, and we will build our knowledge base objects directly from the output of the interpreter.

SAX

The first release of SAX is gradually being replaced by SAX2, and the original classes are now deprecated but maintained for compatibility. For our purposes, we only need to look at the earlier, more rudimentary classes. The classes in the SAX API of interest to us are:

- `DocumentHandler`
  This is the main event-handling interface of SAX, and it defines methods that will be called when particular parts of the XML document are encountered. These methods (shown without arguments) are as follows:

  - `characters()`
  - `endDocument()`
  - `endElement()`
  - `ignorableWhiteSpace()`
  - `processingInstruction()`
  - `setDocumentLocator()`
  - `startDocument()`
  - `startElement()`

  Most of these methods are self-explanatory, `characters()` receives notification of character data (for example the text of an element) and `processingInstruction()` is used to generate objects that will fulfill a processing task. The `setDocumentLocator()` method will receive an object for locating the origin of SAX document events.

- `HandlerBase`
  This is a basic implementation of the `DocumentHandler` interface, together with three other SAX interfaces. It is provided as a convenience so the developer can extend these interfaces without having to implement every method.

- `Parser`
  This is the interface that will supply the events that will be received by a `DocumentHandler`. First the method `setDocumentHandler()` is called, then `parse()` will cause the parser to run through the XML document and send events to the `DocumentHandler`.

- `InputSource`
  This tells SAX where to obtain the XML data it is required to process.

The parser in SAX operates on a serial stream of data, triggering events as characteristic sequences of characters are recognized. These sequences correspond to XML tags, which ultimately represent object models. The serial approach is however quite logical, as in text-based formats the information representation is serial in nature. Of course, until the elements have been interpreted no hierarchy can be built.

A significant aspect of SAX is that it doesn't need to hold large amounts of data in memory, and in principal, there is no limit to the document length. The data we obtain from a Palm Memo will be a whole record, so we lose this advantage, but we still need to analyze this character by character, which is trivial from a byte array. The parser will trigger events; these events will be handled by an implementation of the `DocumentHandler` interface.

XML Classes

We will use a stripped-down, customized subset of the SAX classes as follows:

- `com.kex.DocumentHandler`
  A cut-down version of the `DocumentHandler` interface
- `com.kex.KexDocumentHandler`
  An implementation of `XDocumentHandler` which will populate a `KDomain` with facts and rules (we have included the full '`Kex`' name as this class is much more application-specific than most of the other classes)
- `com.kex.ThinParser`
  A cut-down version of `Parser` that will generate events corresponding to XML elements and their attributes

We have eliminated the `HandlerBase` class, as we have minimized the number of methods we need to implement in our handler from the interface. The `InputSource` class is redundant, as we can directly pass the byte array read by `KMemoReader`.

The DocumentHandler Interface

This is quite similar to SAX's `DocumentHandler` interface, except we have removed methods we won't be needing (if fact we don't really need `startDocument()` and `endDocument()` in our application, but these are retained as they are good markers for debugging):

```
package com.kex.xml;

import java.util.Hashtable;

public interface DocumentHandler {

  public abstract void startDocument();
  public abstract void endDocument();
  public abstract void startElement(String name, Hashtable avpairs);
  public abstract void endElement(String name);
  public abstract void characters(char[] ch, int start, int length);
}
```

Both the parser and the document handler are essentially state machines. The parser operates at a character level, keeping track of the current position in a text document, for instance whether or not it is inside a tag. The document handler carries out the same kind of role at a higher level, interpreting the layout that is, the nesting hierarchy of XML elements and building (domain, fact, and rule) objects for our application from this.

The parser that will be calling an implementation of `DocumentHandler` as described below.

The ThinParser Class

This class scans through the character array source, calling the methods of the document handler when it encounters the appropriate sequence of characters. It includes a little error checking – the `testChar()` method checks to see if a character is that expected in valid XML. If this isn't the case then an exception is thrown, providing information as to the location of the error. In addition, stack is maintained to keep a track of the level of element nesting, so that if this is invalid then an exception is also thrown:

```
package com.kex.xml;

import java.util.Hashtable;
import java.util.Stack;

public class ThinParser {

  private DocumentHandler handler;
  private char currentChar;
  private char[] source;

  private int charCount;
  private int firstChar;
  private int lineCount;

  private Stack tagStack = null;

  public final static char CHAR_A = 65;
  public final static char CHAR_Z = 90;
  public final static char CHAR_a = 97;
  public final static char CHAR_z = 122;
  public final static char CHAR_0 = 48;
  public final static char CHAR_9 = 57;

  public final static char CHAR_BANG = 33;

  public final static Hashtable emptyHashtable = new Hashtable();

  public ThinParser() {}
```

The source character array and the document handler are set by the following methods:

```
  public void setSource(char[] source) {
    this.source = source;
  }

  public void setDocumentHandler(DocumentHandler handler) {
    this.handler = handler;
  }
```

When parse is called some variables are initialized and the method that will actually scan the character array is called, `readDoc()`:

```
  public void parse() {
    currentChar = 0;
    tagStack = new Stack();
    charCount = 0;
    lineCount = 1;
    readDoc();
  }
```

Most of the methods are simple checks for particular kinds of character:

```
private boolean isWhitespace(char c) {
  return (c < CHAR_BANG) || (c > CHAR_z);
}

private boolean isLetterOrDigit(char c) {
  return (c >= CHAR_A && c <= CHAR_Z) || (c >= CHAR_a && c <= CHAR_z)
        || (c >= CHAR_0 && c <= CHAR_9);
}

private int getLineNumber() {
  return lineCount;
}

private String getCharString() {
  return new String(source, firstChar - 1, charCount - firstChar);
}

private void readChar() throws ParseException {
  try {
    currentChar = source[charCount++];
  } catch (ArrayIndexOutOfBoundsException e) {
    throw new ParseException("unexpected end of file");
  }
  if (currentChar == '\n') {
    lineCount++;
  }
}

private void eatTag() throws ParseException {
  while (currentChar != '>') {
    readChar();
  }
}

private void testChar(char c) throws ParseException {
  if (currentChar != c) {
    throw new ParseException("read : \"" + currentChar
                           + "\", expected : \"" + c + "\"");
  }
}
```

The `readDoc()` method joins up the individual `readStartElement()`, `readContent()`, and `readEndElement()` methods with a conditional structure that reflects that found in serialized XML:

```
private void readDoc() {
  boolean inXml = false;
  handler.startDocument();
  try {
    while (charCount < source.length) {
      eatWhitespace();
      if (currentChar == '<') {
        inXml = true;
        readChar();
        if (isLetterOrDigit(currentChar)) {
          readStartElement();
          readContent();
        } else {
          if (currentChar == '/') {
```

```
              readEndElement();
            }
            eatTag();
          }
        } else {

          if (inXml) {
            testChar('>');
          }
          readChar();
        }

      }

      handler.endDocument();
    } catch (ParseException e) {
      System.err.println(e);
    }
  }
```

The next method starts reading immediately after the '<' of an element opening tag, extracting the name of the element followed by any XML attributes:

```
  private void readStartElement() throws ParseException {
    String name = readWord();
    eatWhitespace();
    Hashtable avPairs;
    if (currentChar != '>') {
      avPairs = readAttributes();
    } else {
      avPairs = emptyHashtable;
    }
    tagStack.push(name);
    handler.startElement(name, avPairs);
    testChar('>');
    readChar();

  }
```

The next method reads the attributes within a start element tag:

```
  private Hashtable readAttributes() throws ParseException {
    Hashtable avPairs = new Hashtable();
    String attribute;
    String value;
    while (currentChar != '>') {
      attribute = readWord();
      eatWhitespace();
      testChar('=');
      eatWhitespace();
      readChar();
      testChar('\"');
      readChar();
      value = readString();
      testChar('\"');
      readChar();
      avPairs.put(attribute, value);
      eatWhitespace();
    }
    testChar('>');
    return avPairs;
  }
```

The next method pulls out the characters found between tag pairs, so `<tag>this stuff</tag>`:

```
    private void readContent() throws ParseException {
      if (isLetterOrDigit(currentChar)) {
        readString();
        handler.characters(source, firstChar - 1, charCount - firstChar);
      } else {
        readChar();
      }
    }

    private void readEndElement() throws ParseException {
      if (tagStack.empty()) {
        throw new ParseException("tag mismatch");
      }
      readChar();
      eatWhitespace();
      String name = readWord();
      if (!name.equals((String) tagStack.pop())) {
        throw new ParseException("tag mismatch");
      }
      handler.endElement(name);
      eatWhitespace();
      testChar('>');
    }

    private void eatWhitespace() throws ParseException {
      while (isWhitespace(currentChar)) {
        readChar();
      }
    }

    private String readString() throws ParseException {
      return readString(true);
    }

    private String readWord() throws ParseException {
      return readString(false);
    }

    private String readString(boolean includeWhitespace)
          throws ParseException {
      firstChar = charCount;
      while (isLetterOrDigit(currentChar)
            || (includeWhitespace && isWhitespace(currentChar))) {
        readChar();
      }
      return getCharString();
    }
```

An inner class is used to provide a custom exception to be thrown when errors with the parser class are encountered:

```
    public class ParseException extends Exception {
      public ParseException(String msg) {
        System.err.println("\nXML parse error - " + msg
                          + " on or before line " + getLineNumber());
        System.exit(1);
      }
    }
  }
```

The KexDocumentHandler Class

The parser and document handler above can be used wherever a basic XML parser is required. To use them it is necessary to provide a means for the parser to get at the XML characters, and an implementation of the handler to catch the events. The class that will implement a document handler here to populate the knowledge base is as follows. The current state, that is, what kind of a tag we are looking at is held in the variable `tag`:

```
package com.kex.xml;

import java.util.Hashtable;
import com.kex.model.Fact;
import com.kex.model.Rule;
import com.kex.model.Condition;
import com.kex.model.KnowledgeBase;
import com.kex.util.KLog;
import com.kex.util.KHashtable;
import com.kex.util.KException;

public class KexDocumentHandler implements DocumentHandler {

  private KnowledgeBase knowBase = new KnowledgeBase();
  private int nesting;
  private int tag = NONE;
  private Fact fact;
  private Rule rule;
  private Fact rulefact;
  private Condition condition;
  private KHashtable avpairs = new KHashtable();

  final static int NONE = 0;
  final static int FACT = 1;
  final static int RULEFACT = 2;
  final static int CONDITION = 3;

  public KexDocumentHandler() {}

  public KnowledgeBase getKnowledgeBase() {
    return knowBase;
  }

  public void startDocument() {
    KLog.println("Start Document");
  }

  public void endDocument() {
    KLog.println("\nEnd Document");
  }
```

The most significant aspects of the data are found around the starting tag, with the XML attributes being passed along with the `startElement()` call. The cases below correspond to the levels of nesting within the XML document:

```
  public void startElement(String name, Hashtable pairs) {

    // KHashtable
    avpairs = new KHashtable(pairs);
    nesting++;
    switch (nesting) {
    case 1:
```

```
      // knowBase
      Object nameo = avpairs.get("name");
      if (nameo != null) {
        String domname = (String) nameo;
        knowBase.setName(domname);
      }
      break;
    case 2:

      // facts and rules
      if (name.equals("fact")) {
        tag = FACT;
        fact = new Fact();
        fact.setAVPairs(avpairs);
      }
      if (name.equals("rule")) {
        rule = new Rule(knowBase);
        knowBase.addRule(rule);
      }
      break;
    case 3:

      // within rules
      tag = NONE;
      if (name.equals("fact")) {
        tag = RULEFACT;
      }
      if (name.equals("condition")) {
        tag = CONDITION;
        condition = new Condition();
        condition.setAVPairs(avpairs);
      }
      break;
    }
  }

  public void characters(char[] ch, int start, int length) {
    String s = new String(ch, start, length);
    Fact cfact = null;

    switch (tag) {
    case FACT:
      fact.setValue(s);
      break;
    case RULEFACT:
      try {
        rulefact = new Fact(knowBase.getFact(s));
        if (avpairs.getBoolean("negated")) {
          rule.setNegated(true);
        } else {
          rule.setNegated(false);
        }
      } catch (Exception e) {
        System.err.println(e);
      }
      break;
    case CONDITION:
      try {
        cfact = knowBase.getFact(s);
      } catch (Exception e) {
        System.err.println(e);
      }
```

```
        cfact.addRuleRef(rule);
        knowBase.update(cfact);
        condition.setValue(s);
        condition.setId(cfact.getId());
        rule.addCondition(condition);
        break;
      default:
        try {
          throw new KException("Interpretation error.");
        } catch (Exception e) {
          System.err.println(e);
        }
        break;
      }

    }
```

When the `endElement()` method is called this is the signal that a whole rule or fact has been read, so the details are placed in the knowledge base:

```
    public void endElement(String name) {
      nesting--;
      if (name.equals("fact") && tag != RULEFACT) {
        knowBase.addFact(fact);
      }
      if (name.equals("rule")) {
        try {
          rule.setFact(rulefact);
        } catch (Exception e) {
          System.err.println(e);
        }
        knowBase.update(rule);
      }
    }
  }
```

The KbLoader Class

The class that will pass the data loaded from the Palm Memo into the parser is as follows:

```
  package com.kex.io;

  import com.kex.model.KnowledgeBase;
  import com.kex.control.Status;
  import com.kex.xml.ThinParser;
  import com.kex.xml.DocumentHandler;
  import com.kex.xml.KexDocumentHandler;
  import com.kex.util.KLog;

  public class KbLoader {

    public static KnowledgeBase load(String name) {
      KMemoReader memo = null;
      try {
        memo = new KMemoReader();
      } catch (Exception e) {
        System.err.println(e);
      }
      Status.message = "Searching for \"" + name + "\"\n";
      byte[] document = memo.getDocument(name);
```

```
      memo.close();
      if (document == null) {
        return null;
      }
      Status.message = "Loading " + name + "...\n";

      char[] charData = new char[document.length];
      for (int i = 0; i < document.length; i++) {
        charData[i] = (char) document[i];
      }
      ThinParser parser = new ThinParser();
      parser.setSource(charData);

      document = null;
      DocumentHandler handler = new KexDocumentHandler();
      parser.setDocumentHandler(handler);
      KLog.resetTimer();
      parser.parse();
      KLog.logTime();
      Status.message = null;
      return ((KexDocumentHandler) handler).getKnowledgeBase();
    }
  }
```

User Interface

As well as needing device-specific code for the I/O class `KMemoReader`, we will also have to use specialized classes for graphics and event handling on the Palm. The dozen or so visual components of the KJava API are used in a similar manner to those of the more familiar AWT.

The user interface is based around six states, shown in the diagram below. Five of these represent states where the application is waiting for user input, and the sixth is a transitory state, where an informational, Working... message is displayed on screen while the application is carrying out processing in the background. The first five states correspond directly to separate sections of the code, though the message state will be called from various points in the program where a delay in the application's response is likely:

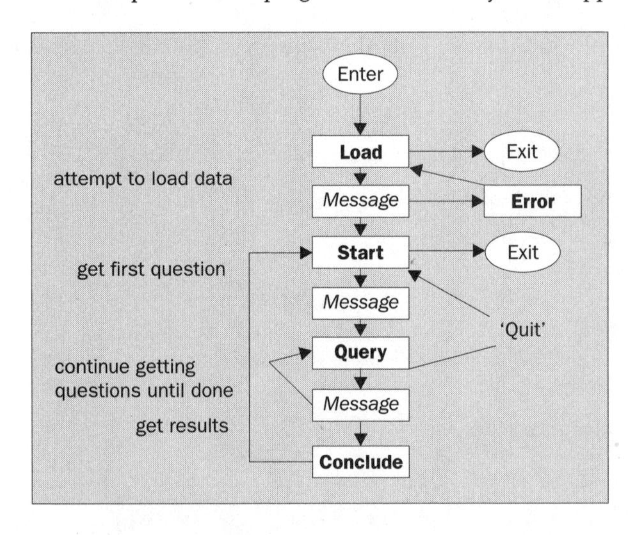

Trouble-Shooting the Knowledge Bases

Here are a few tips on getting the full application running:

- A validating XML editor on the desktop is currently the easiest way of inputting data.
- If a particular knowledge base isn't working as expected, check the KVMutil log, and double-check that the entries in the memo are as intended, as duplicate facts etc. will give spurious results. Remember the Boolean logic is strict, so it may be advisable to draw a truth table for any rule that seems to be misbehaving.
- Note that this version of the application has not been tested on large data sets (more than about 30 facts and rules). It is anticipated that larger sets will have a disproportionate hit on performance, as the function of comparisons is not linear with the number of items. It is also uncertain at which point the application will cease to operate altogether, due to the finite amount of stack available to the KVM.

Summary

This case study has implemented an expert system that uses XML to introduce its knowledge base onto a Palm hand held computer. This is no mean feat, as XML is not generally recommended for use on mobile computing devices as its sheer size can bring expensive penalties to an application. If we add to that the size of the Java XML parsers, using XML doesn't look like a viable prospect. However, we have seen in this chapter that it is perfectly possible to write cut down versions of SAX classes that can be used on a CLDC-specified device.

By using the Palm platform's native Memo file format to transport the XML onto the device we achieved two things. First we minimized the amount of coding which was required, as the Memo format can act as an off-the-shelf persistence mechanism. No new, low-level format was needed. Second, we could take advantage of an existing data transfer system, which means that we don't have to write any new protocols or custom storage structures.

Hopefully we have convinced you that XML is not such a bad proposition on a mobile device, and that with a bit of work we can leverage it into our applications.

Further Reading

See the following books for more information about SAX and other XML parsers in Java:

- Professional-Java-XML (Wrox Press, ISBN 186100401X)
- Java-XML-Programmers Reference (Wrox Press, ISBN 1861005202)

XML

The following is a list of web sites that contain useful XML information:

- World Wide Web Consortium – http://www.w3.org
- XML Software – http://www.xmlsoftware.com
- XML.org – http://www.xml.org
- There is lots of information regarding SAX at the W3C, and another good site is maintained by David Megginson, who led the initiative that created SAX – http://www.megginson.com

- TinyXML – http://www.gibaradunn.srac.org/tiny/
- NanoXML – http://nanoxml.sourceforge.net
- kXML (and more) – http://www.sourceforge.net
- Xerces (the XML parser from the Apache team) – http://xml.apache.org/xerces-j/index.html
- XMLTree – http://www.xmltree.com

Expert Systems

"*Expert Systems, Principles and Programming PWS ISBN 0-534950-53-1*" by Giarratano and Riley is a classic text, the first half of which describes Expert Systems in general and the second half is devoted to the CLIPS expert system shell (the authors were among the developers of the shell, and Gary Riley regularly contributes to the CLIPS mailing list):

- CLIPS – http://www.ghgcorp.com/clips/CLIPS.html
- Jess (Java Expert System Shell) – http://herzberg1.ca.sandia.gov/jess/
- SOAR Project – http://bigfoot.eecs.umich.edu/~soar/

There are good public FAQs on expert systems and AI in general, one of the sources of which is http://www.landfield.com/faqs/.

PROFESSIONAL JAVA MOBILE

TOOL SELECTION

WEB TABLET
MOBILE PHONE
MP3 PLAYER
DIGITAL CAMERA
PALMTOP
LAPTOP

NOW SELECTED

PALMTOP

15

Case Study: Mobile Positioning

The ability to determine the geographic position of a client opens up many possibilities for the WAP service developer. According to its supporters "mobile location solutions", which combine mobile locations systems and location specific services represent the next killer application of the mobile industry.

This chapter covers using a WAP phone as a thin client to a location sensitive J2EE driven hotel search engine. The idea of the case study application is to give the user the ability to quickly find the closest hotel in relation to the positioning of the client phone.

The chapter is based around a simple but complete and fully functional application, written in JSP (generating both HTML and WML) and EJB. A comprehensive code run-through as well as deployment instructions will be provided. The application relies on an emulation of a Mobile Positioning Center provided by Ericsson through the MPS-SDK 3.0.

Mobile Location Solutions

We all have experienced the information overflow on the Internet today. The amount of data and services can be quite overwhelming, especially for the new user. In the wireless world we are more or less required to limit information heavily and add ability to access only relevant data to a particular situation. The user will not be satisfied with just "browsing the web" over the phone – it is too slow, cumbersome, and expensive for small devices.

This challenge should be taken as a new possibility for adapting information to position and user context, creating a new paradigm of service design. We should therefore limit the information we present to what is relevant to a specific user at a specific time and at a specific location.

This chapter deals with the location part of the context.

Location

On the conventional Internet there is no built-in concept of location. Mobile phones, on the other hand, are intrinsically bound to the location of the client just by the nature of the cell network. This can be used to create customized information based on the geographic context of the user.

Location is also implicit in many business practices and relevant to enterprise computing. The physical layout of an office, for instance, may represent organizational structure. Any business that is spatially distributed may benefit from location sensitivity. This pertains not only to the wireless systems, but to any system – whether it's sales support, inventory control or tracking and logistics.

The table below shows some potential location-based services:

Service	Example
Position	Where am I? Where is X?
Events	Medical alert. Public announcements.
Assets	Fleet management. Security.
Routes	How do I get to X?
Context/Overview	What building am I standing in front of?
Directories	Looking for nearest X. Where can I buy X?

Location-sensitive services may well be the next big innovation in human-computer interaction. Consider the fact that we are all brought up to think in three dimensions. The concept that objects relate to each other in space with respective distance may be used to create a more natural way of organizing computer systems in general than the various schemes currently employed – be they hierarchical or relational. But before we go too far astray trying to recreate the Internet in the shape of science-fiction novels, let's focus on the matter at hand.

Mobile Positioning

Mobile Positioning is the ability to determine the location of a wireless enabled device, such as a mobile phone – and thus the user of that device. The technology and services for this exist today, but expect a substantial growth in the field as progress is made in utilizing this. In the US, regulation is driving positioning technology, as operators are required to provide location information for emergency calls. By the end of 2001 this system will be in place, and that should ignite a whole business of creating localized services for the US market and the rest of the world.

There are currently two main strategies for mobile positioning:

- Network-oriented
- Handset-oriented

Network-Oriented Positioning

The basic unit of the cell network is the cell. A mobile phone is connected to a specific cell at any given time. Since knowledge of the location of the cell base station is available the phone location can be approximated. This – the simplest of location schemes – is called cell of origin (**COO**):

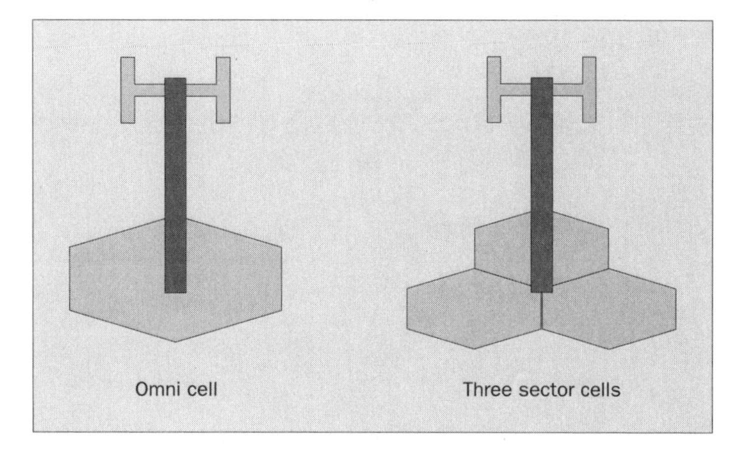

This gives an accuracy of a couple of hundred meters in urban areas where antennas are plentiful, but only a very rough approximation in rural areas – it is quite possible that accuracy will be reduced to within 35 kilometers.

Finer grained location is being made possible through the concept of timing advance. Cell phones are regularly 'pinged' by their base stations. We can get an idea of how far away the phone is from the base station by measuring the time delay it takes for this signal to reach the station. Sometimes we can also use the fact that some base stations cover their cell with several antennas covering each angle. Using these two schemes we are now at the level of mobile positioning system currently employed in GSM networks:

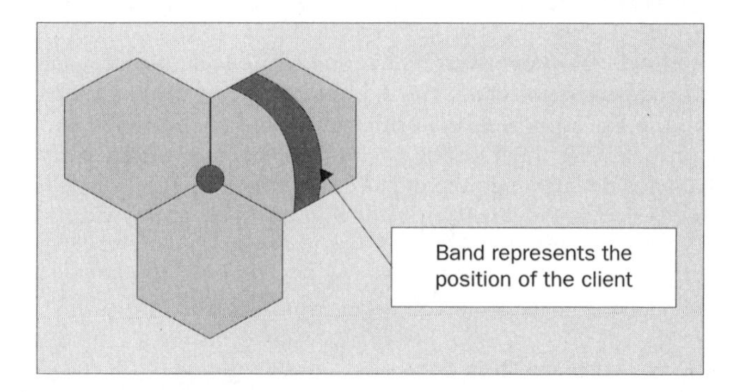

Future Developments of Network-Oriented Positioning

An even greater level of accuracy can theoretically be achieved by triangulating the handset from several base stations. This is commonly known as **TOA** – Time of Arrival. By measuring the time difference it takes for the signal to reach the various stations an accurate location can be achieved for the handset. The great advantage of this technology is that is requires no special additions to the handset, however a big limitation is the fact that the base stations need to be synchronized with a very high accuracy (nano-second range) – which is currently not possible.

Another technology that is being pursued is **AOA** – Angle of Arrival. This idea here is to create an array of antennas at various base stations throughout the network (see diagram below). By knowing the angle accurately from several antennas the handset can be pinpointed. The weakness of this approach is cost, since the calibrated arrays are expensive to employ and maintain.

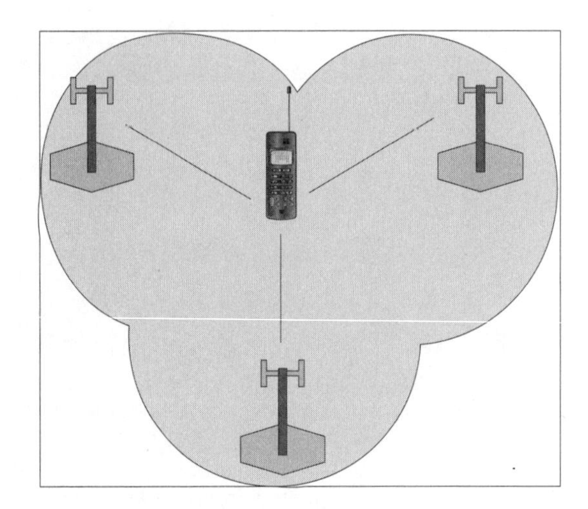

Finally, various systems for signal attenuation and location pattern matching are being discussed and implemented. In specific areas the signature of the wireless RF signal is unique to a specific location, due to so-called multi-path phenomena. By measuring the radio "fingerprint" in different sites and adding a powerful pattern matching system to the network accurate location pinpointing may be achieved. One problem here of course is that the fingerprint changes when the topography does with new buildings, etc. and also with weather and solar activity.

Developing Mobile Positioning Solutions

As developers we need a development system to convey the location information to our application. Unfortunately there is currently no set standard, but a couple of vendor-specific solutions exist. In this chapter we have used the Ericsson Mobile Positioning Solutions Software Development kit (MPS-SDK 3.0). This is not the only development kit for positioning, but one that is fairly easy to use with Java, and it includes an emulator for developing prototypes and demos. In addition, it is also currently employed in real-life positioning systems (The Swedish Yellow Pages service, www.gulasidorna.se, is an example of this).

Some of the current players in location ibformation provision are:

- TruePosition: www.trueposition.com
- CellPoint Systems AB: www.cellpt.com
- US Wireless: www.uswcorp.com
- Ericsson Mobile Positioning System (MPS): www.ericsson.se
- SnapTrack: www.snaptrack.com
- BT Cellnet: www.btcellnet.co.uk
- CPS Cursor: www.cursor-system.com
- SignalSoft Corporation: www.signalsoftcorp.com

Handset-Oriented Positioning

The other main strategy for establishing mobile location is the handset-driven approach, primarily using the Global Positioning System (**GPS**). GPS is a satellite-based system, which gives good outdoors accuracy of 10m resolution or less. One drawback is that it doesn't work well indoors or in urban areas. Also adding GPS functionality to the handset would naturally increase the price of that handset. This, together with increased demands on batteries and need for a special GPS aerial, presents a challenge for device providers. This chapter therefore focuses on network-oriented services.

Ericsson MPS

The Ericsson Mobile Positioning System is one vendor specific, production-ready solution that exists today. Its main component is a server-based Mobile Positioning Center that must be installed on the telco's network. The system includes software extensions for the mobile network that must also be installed.

The MPS is GSM-based and relies on COO positioning enhanced by time advance and angle for three sector cells. The API itself is independent of positioning method and thus ready for integration with GPS or network-based solutions. The protocol to communicate with the MPC is request/response based and implemented in XML. However the MPS-SDK Java classes provide a wrapper for this, so that you do not have to specifically create and parse XML to use the API.

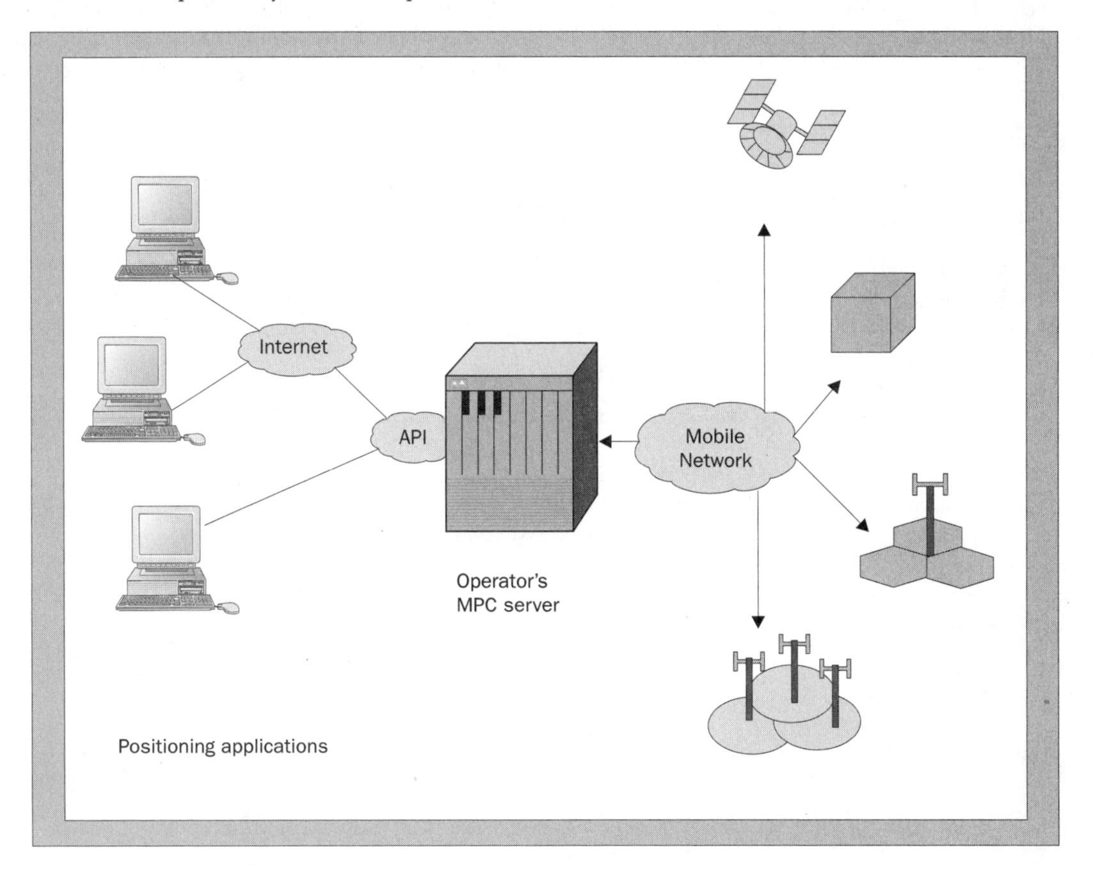

The responsibility for installing the positioning system lies with the network provider/telco. Thus you as a service provider need to start negotiating promptly if you want to deploy a real-life system using this API. This unfortunately is the failing point for services being developed today. Dealing with the providers can be a bit daunting, as is the lack of standards in this area. Expect a standard to emerge as the demand for location-based services increases. The up side of this is that when total positioning control is given to the mobile network operator, possible revenue streams increase. This means that better and more numerous services can be created in the long run.

The hotel-search application runs an emulation of an MPC provided with the MPS-SDK that can be used for proof-of-concept development. Also included in the SDK are sample clients to the emulated MPC that you can use to test your data.

Mobile Positioning Protocol, version 3.0

The Mobile Positioning Protocol (MPP) is used to interface with the Mobile Positioning Center (MPC). It is an application-level protocol for positioning that uses standard HTTP 1.0 to transfer specially formatted requests and responses. The protocol is proprietary to Ericsson and is subject to change as the work with developing standards for positioning progresses. Notably the ETSI (European Telecommunication Standards Institute) are working on a standard and ANSI are also committed to producing standards in this area.

The client in the MPP system, referred to as the Location Service Client (LCS), must create a special request message (or Location Immediate Request (LIR)) and post to a certain URL to receive the location data – the XML formatted Location Immediate Answer (LIA). The URL used for submitting is always http://host:port/newRequest/, here host and port are configurable by the bearer company who installs the MPC:

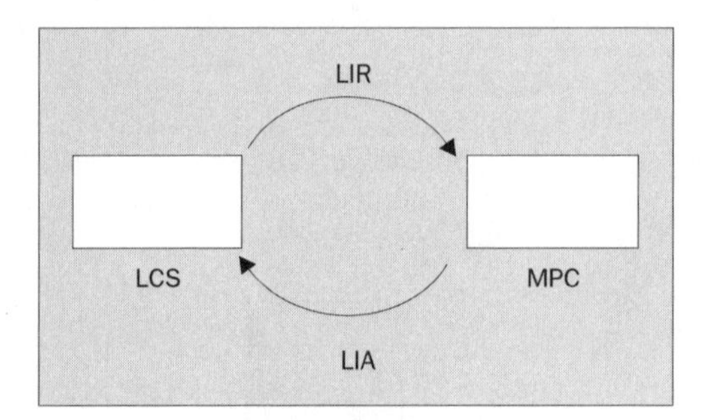

The Request

The LIR message (request) will contain information about the phones to position giving the Mobile Station ID, essentially the phone number, to the MPC. Here is a sample LIR:

```
<?xml version= '1.0'?>
<!DOCTYPE REQ>
<REQ ver="3.00">
  <CLIENT>
    <ID>aUser</ID>
    <PWD>aPwd</PWD>
  </CLIENT>
  <LIR>
    <GEO_INFO>
      <COORD_SYS>LL</COORD_SYS>
```

```
      <DATUM>WGS-84</DATUM>
      <FORMAT>IDMS0</FORMAT>
    </GEO_INFO>
    <MSIDS>
      <MSID>461011334411</MSID>
      <MSID>461011334414</MSID>
      <MSID_RANGE>
        <START_MSID>461011334500</START_MSID>
        <STOP_MSID>461011334599</STOP_MSID>
      </MSID_RANGE>
    </MSIDS>
  </LIR>
</REQ>
```

The above message will identify the LCS with the given ID and password and attempt a positioning of a group of phone numbers in the latitude/longitude coordinate system (see below).

The Response

If a properly formatted request has been received correctly the MPC will respond with a LIA message. Note that the successful arrival of a response doesn't necessarily mean that the requested phone will be successfully positioned (it could be switched off for instance). It just means that the request arrived properly to the server.

Here is a sample LIA in response to the above request:

```
<?xml version= '1.0'?>
<!DOCTYPE ANS>
<ANS ver="3.00">
  <LIA>
    <GMT_OFF> +0100 </GMT_OFF>
    <POS msid="1234512345">
      <PD>
        <TIME> 20000626171825 </TIME>
        <ARC>
          <LL_POINT>
            <LAT> N301628.3 </LAT>
            <LONG> W974425.2 </LONG>
          </LL_POINT>
          <IN_RAD> 1100 </IN_RAD>
          <OUT_RAD> 1650 </OUT_RAD>
          <START_ANGLE> 120 </START_ANGLE>
          <STOP_ANGLE> 240 </STOP_ANGLE>
        </ARC>
      </PD>
    </POS>
    <POS msid="1234512346">
      <PD>
        <TIME> 20000626171825 </TIME>
        <ARC>
          <LL_POINT>
            <LAT> N301630.3 </LAT>
            <LONG> W974450.2 </LONG>
          </LL_POINT>
          <IN_RAD> 0 </IN_RAD>
```

```
              <OUT_RAD> 1650 </OUT_RAD>
              <START_ANGLE> 120 </START_ANGLE>
              <STOP_ANGLE> 240 </STOP_ANGLE>
            </ARC>
          </PD>
        </POS>
      </LIA>
    </ANS>
```

This response gives the positioning of two of the requested phones. Note that the contents of the response are a time marker (of the format YYYYMMDDHHMMSS) and an arc giving the area in which the phone has been located. The coordinate given is the point of origin of the arc, the location of the base station.

For more information on the XML format of the request and response please refer to the official documentation from which the above has been adapted. The specification for the MPP is included in the MPC-SDK as a PDF with the tantalizing name "Mobile Positioning Protocol – Version 3.0" document. Fortunately for the Java developer we do not really have to worry about this XML format as the classes in the MPS-SDK encapsulates this information for our benefit.

The Coordinate System

There are many reference systems for describing location on the earth. Such a system is generally referred to as a geodetic datum. An example system is the World Geodetic System 1984 (WGS-84). This is the default system, and the only datum supported by the MPC Emulator supplied with the MPS-SDK. It is also the datum used with GPS. The coordinate system that goes with WGS-84 is the geodetic latitude/longitude system.

Latitude of a given location or point is the angle from the equatorial plane as shown here:

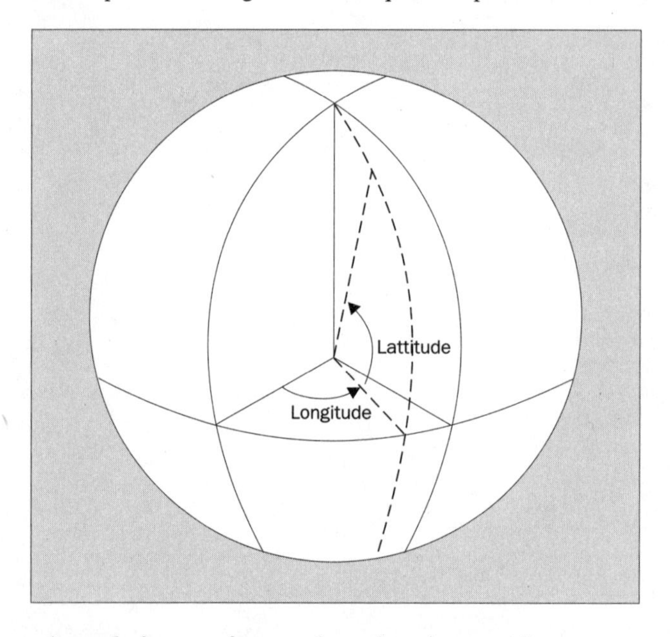

The latitude is as shown the angle from a reference plane, the prime meridian.

The coordinates that are received can have different formats. The only one available in the current release gives the latitude/longitude as:

`N301628` (latitude)
`W974425` (longitude)

The different parts of the format `N 30 16 28` breaks down to N – Angle North or South of equator, `30` – number of degrees, `16` - number of minutes of a degree and `28` – number of seconds of a degree.

Using the SDK

The SDK contains:

- An MPC emulator, running as a servlet
- Two demonstration clients to the emulator, written in Java
- A map tool for creating demo data (not shown here)
- Documentation
- Supporting Java classes for creating clients to the emulator

The Emulator

The emulator is implemented as a servlet and uses a number of configurable files to simulate a couple of pre-defined mobile clients as they move around. Once the emulator is started it will start reading the instructions in those files, calculating the positions of all the phones and updating them over time. When the end of the test data has been reached it will start over. The idea is to configure the data in the files to suit your demonstration purposes. The files are:

- `Authfile.txt` – this file lists the **Mobile Stations** (MS) for the system, each on its own individual line. Each mobile station is a phone that has been registered with the MPC. The user and password given is the authentication of the LCS, not of the mobile station. Remember that many applications could be benefiting from the same MPC service.
- `Routfile.txt` – describes how the stations move between the cells. Each entry in this file describes the mobile stations as they move through their route. The first column is the ID of the Mobile Station. The second is an offset time after which it will go to the next entry. The third column is the cell ID defined in the `Celldata.txt` file and finally the distance and mode of the phone is given:

```
46708123456789      10      100      5000     busy
```

- `Celldata.txt` – defines the cells of the system. The first column is the cell ID, the second is the longitude and the third is the latitude. The fourth column is the cell type (either `Omni` or `CircleSector`), the fifth is the radius and the sixth is the direction, in degrees:

```
100       E153806    N603758  CircleSector   5500       100
```

The API

The MPS-SDK package included has the following Java classes for facilitating communication with the MPC:

- `Coordinate` – encapsulates a point, for example a longitude/latitude location
- `LocationRequest` – the class for creating the request and sending it to the MPC
- `LocationResult` – holds the response
- `MPSException` – is thrown in case of error.

Sample client code is shown below:

```
LocationRequest posRequest =
                    new LocationRequest("localhost", 80, "aUser", "aPwd");
posRequest.addMSISDN("46777100009");
posRequest.send();

LocationResult posResult;
while ((posResult = posRequest.getLocationResult()) != null) {
  int theMSISDN = posResult.getPositionItem();
  Coordinate theCoord = posResult.getCoord();

  String theLongitude = theCoord.getLatitude());
  String theLongitude = theCoord.getLongitude());

  // Do something with it...
}
```

Hotel-Search

The main part of this chapter is devoted to presenting a full application allowing the mobile user to retrieve a list of the closest hotels to their current location: the hotel-search application:

Hotel-search is a dual WAP/HTML application. Although the application itself has a WML front-end it relies on a web interface to register the user.

Use Cases

The hotel-search application has the following use case scenario:

- **Web User**
  Register – need to enter profile information about the user, specifically register a phone number
- **Web Administrator**
  Display and edit contents of the database for both hotels and users
- **Wireless User**
  Login – connect the current user to a stored profile
  Search for hotels – main use case of the application
  Get info on hotel – user may want the phone number of a hotel, etc.

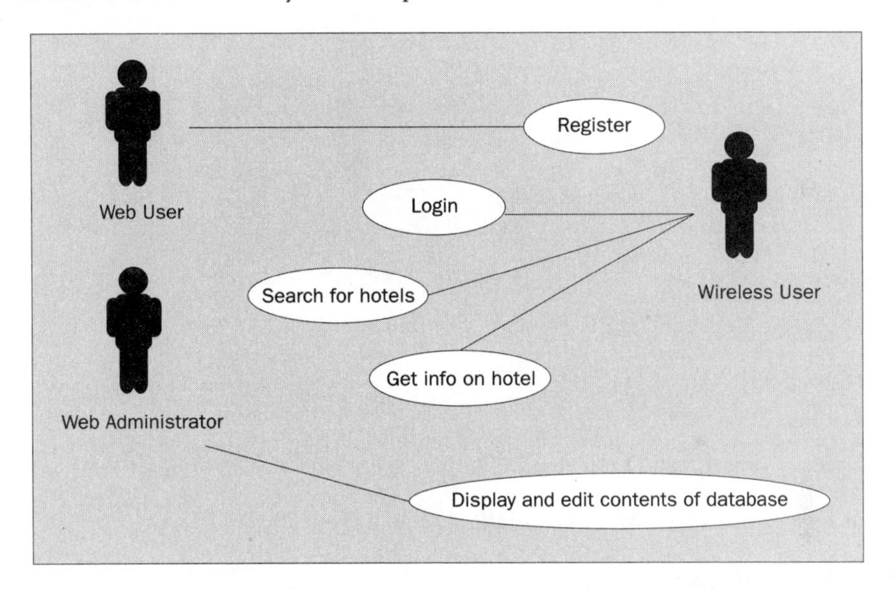

Architecture

The application is deployed on an Orion application server (www.orionserver.com). The Mobile Positioning Center (MPC) Emulator is deployed separately on a JRun Server. The design reason for keeping the MPC Emulator separate from the client application is simply that this more closely corresponds to how we would communicate with the MPC in real life. It would not reside in the same JVM as the application but would rather be distributed somewhere on the bearer's network so it makes more sense to have it served by different servers. The technical reason for choosing JRun, rather that twin Orion servers is that there seems to be a slight bug in the emulator that Orion picks up but doesn't seem to cause any problems for JRun (It also works nicely with Apache and JServ).

Persistence in the application is implemented using CMP – Container Managed Persistence and the default handler for Orion will be an in-memory Hypersonic DB running in the same process as Orion. The application is auto-deployed without using an `.ear` archive. See the **Installation and Deployment** section later in the chapter:

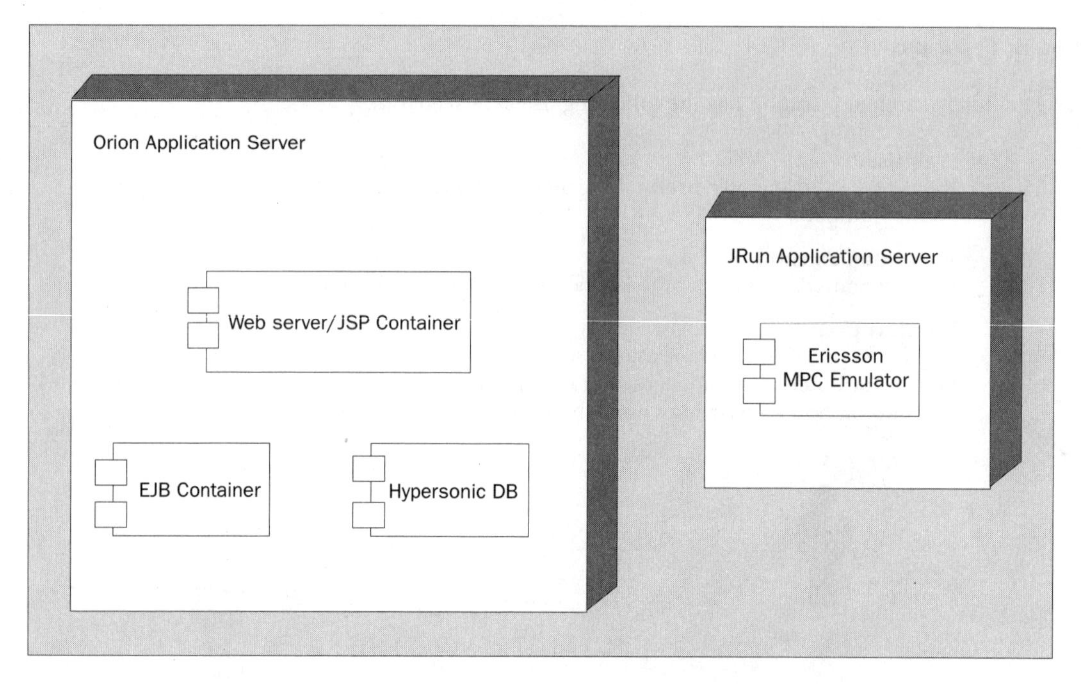

Software used:

- Orion Application Server 1.5.2
- JRun Server 3.1
- Ericsson MPS-SDK 3.0
- YoSpace Smartphone Emulator 2.0, http://www.yospace.com or other WAP emulator

We will now go through the process of installing each of these.

Orion Application Server

- Download files from http://www.orionserver.com. Orion is free for development.
- Follow the steps in http://www.orionserver.com/docs/install.html.
- Sample installation directory `C:\orion\`.

JRun

- Download from http://www.jrun.com/Products/JRun/. JRun is free for development.
- Follow the installation wizard.
- Sample installation directory `C:\JRun\`.

MPS SDK 3.0

- Download MPS-SDK 3.0 rev B from http://www.ericsson.com/developerzone/. Note that you will have to register for the Ericsson developer program, which is free.
- Unzip into `C:\mps-sdk3.0\`.
- Copy files from `C:\mps-sdk3.0\Emulator\newRequest\` to `C:\JRun\servers\default\default-app\WEB-INF\classes\`
- Copy `C:\mps-sdk3.0\ClassLibrary\lib\mpssdk.jar` to `C:\JRun\servers\default\default-app\WEB-INF\lib\` and to `C:\orion\lib\`.
- Add the following mapping to JRun in the `C:\Jrun\servers\default\default-app\WEB-INF\web.xml`:

```
<servlet>
    <servlet-name>newRequest</servlet-name>
    <init-param>
        <param-name>configpath</param-name>
        <param-value>
          c:\JRun\servers\default\default-app\WEB-INF\MPP
        </param-value>
    </init-param>
    <servlet-class>newRequest</servlet-class>
</servlet>

<servlet-mapping>
    <url-pattern>/newRequest</url-pattern>
    <servlet-name>newRequest</servlet-name>
</servlet-mapping>

<servlet>
    <servlet-name>Version</servlet-name>
    <servlet-class>Version</servlet-class>
</servlet>

<servlet-mapping>
    <url-pattern>/Version</url-pattern>
    <servlet-name>Version</servlet-name>
</servlet-mapping>
```

This is done because the MPP prescribes that the MPC is called by the special URL http://host:port/newRequest/. In the emulator a servlet is mapped to this URL; the `newRequest` servlet:

- Copy content of `C:\mps-sdk3.0\Emulator\MPP\` directory to `C:\JRun\servers\default\default-app\WEB-INF\MPP\`. This directory contains the configuration files for the emulator. Theses files can be configured and adapted for your use. See previous section and Ericsson documentation for this:
    - `Authfile.txt`
    - `Celldata.txt`
    - `Confdata.txt`
    - `Routfile.txt`

- Start or restart JRun.
- You can test the installation of the MPC Emulator by running the test apps supplied in the SDK (just run the batch file in C:\mps-sdk3.0\ClassLibrary\examples\<EXAMPLE_NAME>\) using the following parameters:
  - Host: localhost
  - Port: 8100 (JRun webserver)
  - User: aUser (as defined in Authfile.txt)
  - Password: aPwd
  - MS: 46708123456789 for example (see Authfile.txt)

The Input Output Demo

This demo allows us to add a phone number to the list to the left (as shown below) and other parameters for the MPC. Pressing Send will compile the request and send it to the MPC. The Result tab will contain the results from the server:

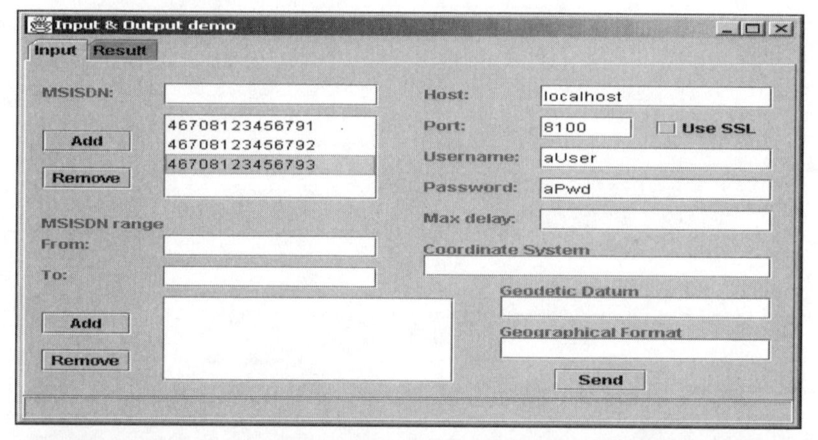

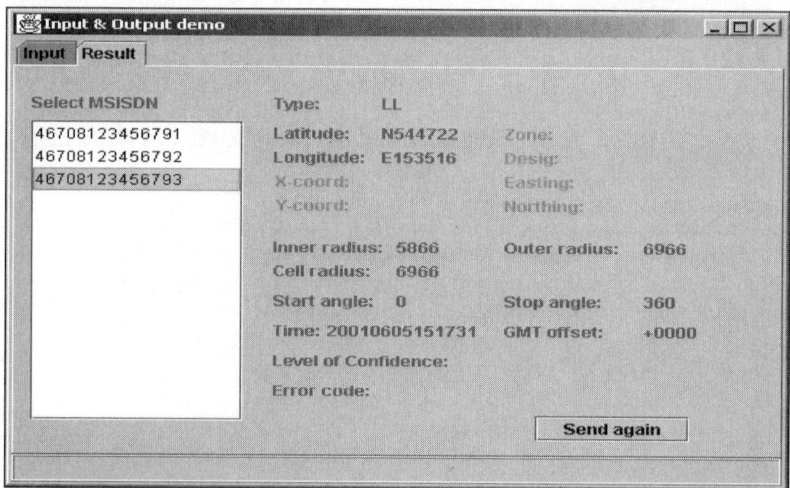

The Position Areas Visualization Client

There is also a more visually appealing demo of essentially the same thing, i.e. a sample interaction with the MPC Emulator. Give it the required data and press Get Position:

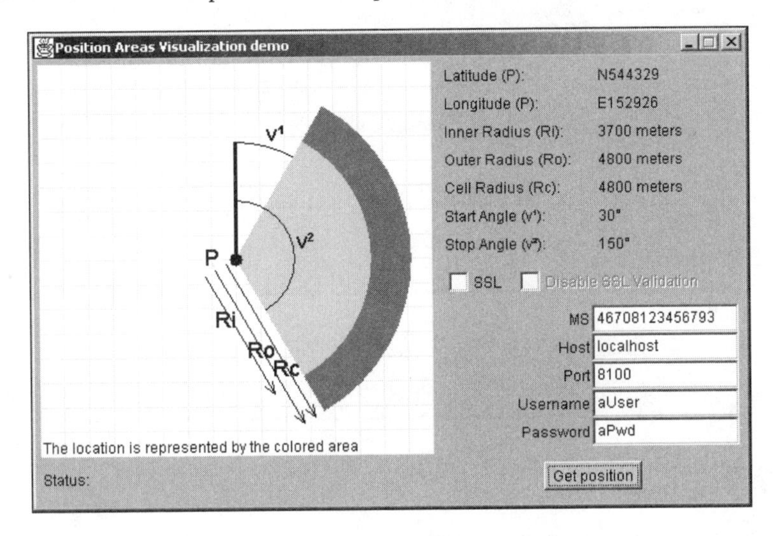

Application Tier

The application database has two tables: `Users` and `Hotels`. The database is deployed automatically by Orion server and is for all practical purposes not the most important aspect of this application. We will assume that this is a resource we have access to. All interaction with the database occurs through Entity EJBs. The `Hotels` table stores relevant hotel information including latitude and longitude position. The `Users` table notably contains the phone number needed for positioning.

The view of the database shown below has been created using the utility JSPs presented in the **Presentation Tier** section. The positions of the hotels have been chosen based on the default configuration files of the MPC Emulator.

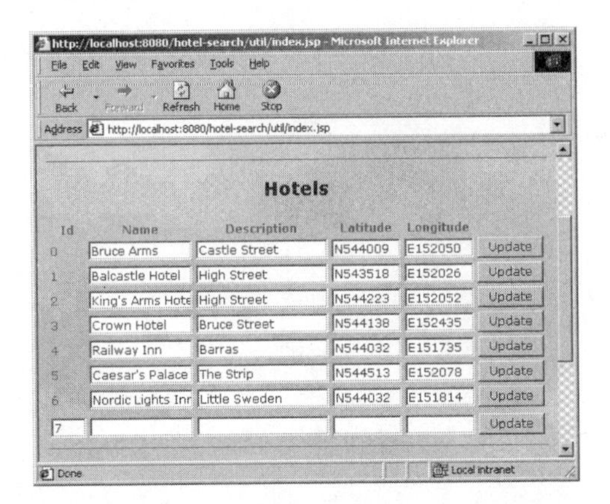
http://localhost:8080/hotel-search/util/index.jsp - Microsoft Internet Explorer

Hotels

Id	Name	Description	Latitude	Longitude	
0	Bruce Arms	Castle Street	N544009	E152050	Update
1	Balcastle Hotel	High Street	N543518	E152026	Update
2	King's Arms Hote	High Street	N544223	E152052	Update
3	Crown Hotel	Bruce Street	N544138	E152435	Update
4	Railway Inn	Barras	N544032	E151735	Update
5	Caesar's Palace	The Strip	N544513	E152078	Update
6	Nordic Lights Inr	Little Sweden	N544032	E151814	Update
7					Update

Quoting 46708123456789 will take the user on a path passing close to all the hotels:

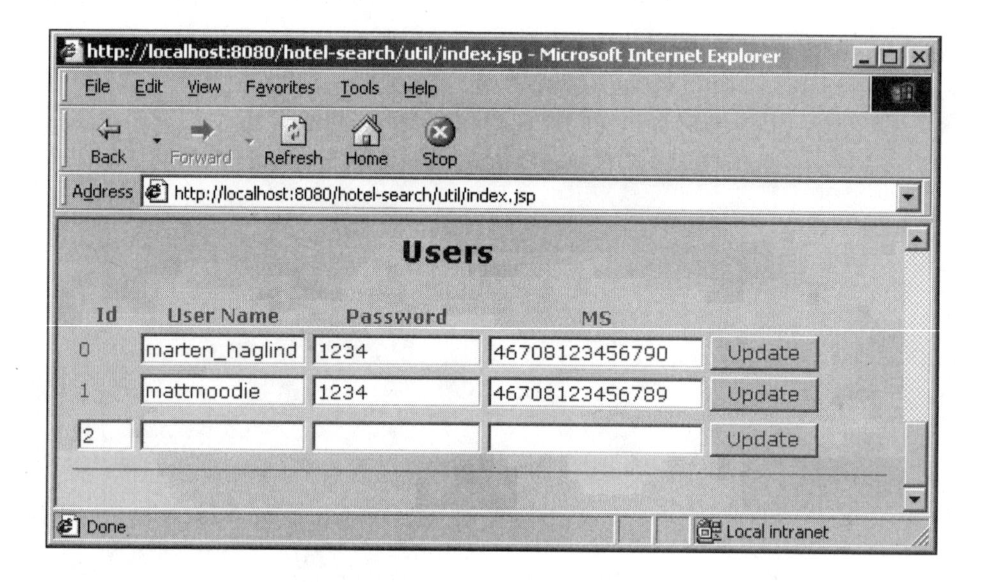

Enterprise JavaBeans

In the following section we create the application logic for the application, implemented in EJB, with two entity beans and one session bean. Knowledge of the J2EE model is assumed. Also note that some of the EJB code has been left out, as it is not directly relevant, or is too trivial. Full code can be downloaded from http://www.wrox.com.

The Home Interfaces

Home interfaces are used to locate beans. `LocationHandler` is the session bean, which implements the positioning functionality. It is created as appropriate by the presentation tier using its home interface:

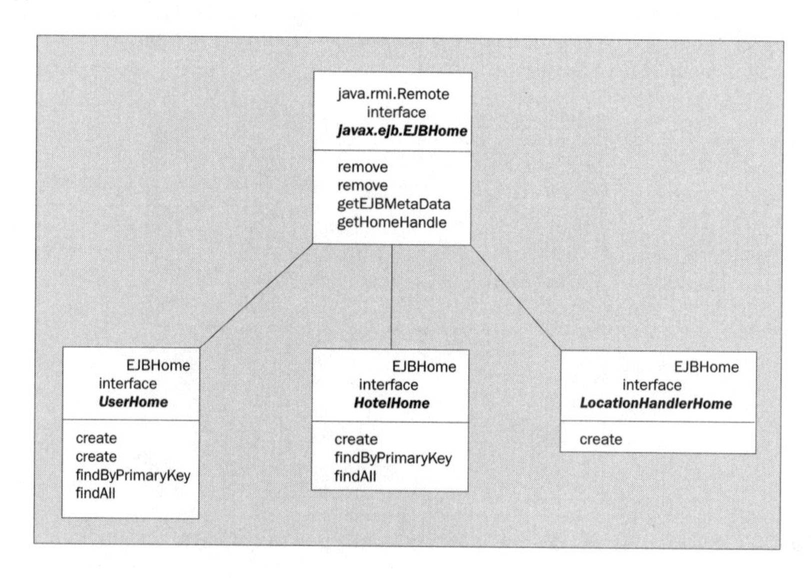

The LocationHandler Interface

Here is the `LocationHandler` interface for creating a `LocationHandler` session:

```
public interface LocationHandlerHome extends EJBHome {
```

The constructor will create a new `LocationHandler`:

```
    public LocationHandler create() throws CreateException, RemoteException;
}
```

The UserHome Interface

The `UserHome` interface creates or finds `User` entities. For the entity beans, create means that we are adding a row to the database, whereas the finder methods will retrieve existing data. The create method without parameters will create a new `User` with an index of the highest available index+1:

```
public interface UserHome extends EJBHome {
```

The `create(long id)` method will create a new user by specifying a user ID. This will throw a `CreateException` should the ID be already in use:

```
    public User create(long id) throws CreateException, RemoteException;
```

The `create()` method will create a new user with the next available index:

```
    public User create() throws CreateException, RemoteException;
```

The `findByPrimaryKey()` method will retrieve the user by primary key (user ID):

```
    public User findByPrimaryKey(long key)
            throws RemoteException, FinderException;
```

The `findAll()` method retrieves a list of all current users:

```
    public Collection findAll() throws RemoteException, FinderException;
}
```

The HotelHome Interface

The `HotelHome` Interface for creating or finding `Hotel` entities:

```
public interface HotelHome extends EJBHome {
```

The `create()` method will create a new hotel by specifying a hotel ID. This will throw an exception if the ID is already in use:

```
    public Hotel create(long id) throws CreateException, RemoteException;
```

The `findByPrimaryKey()` method will retrieve a hotel by primary key (hotel ID):

```
    public Hotel findByPrimaryKey(long key)
            throws RemoteException, FinderException;
```

The `findAll()` method retrieves a list of all stored hotels:

```
    public Collection findAll() throws RemoteException, FinderException;
  }
```

The Remote Interfaces

The remote interfaces define the functionality of the EJBs as available for clients of the application tier, i.e. the JSPs/presentation tier:

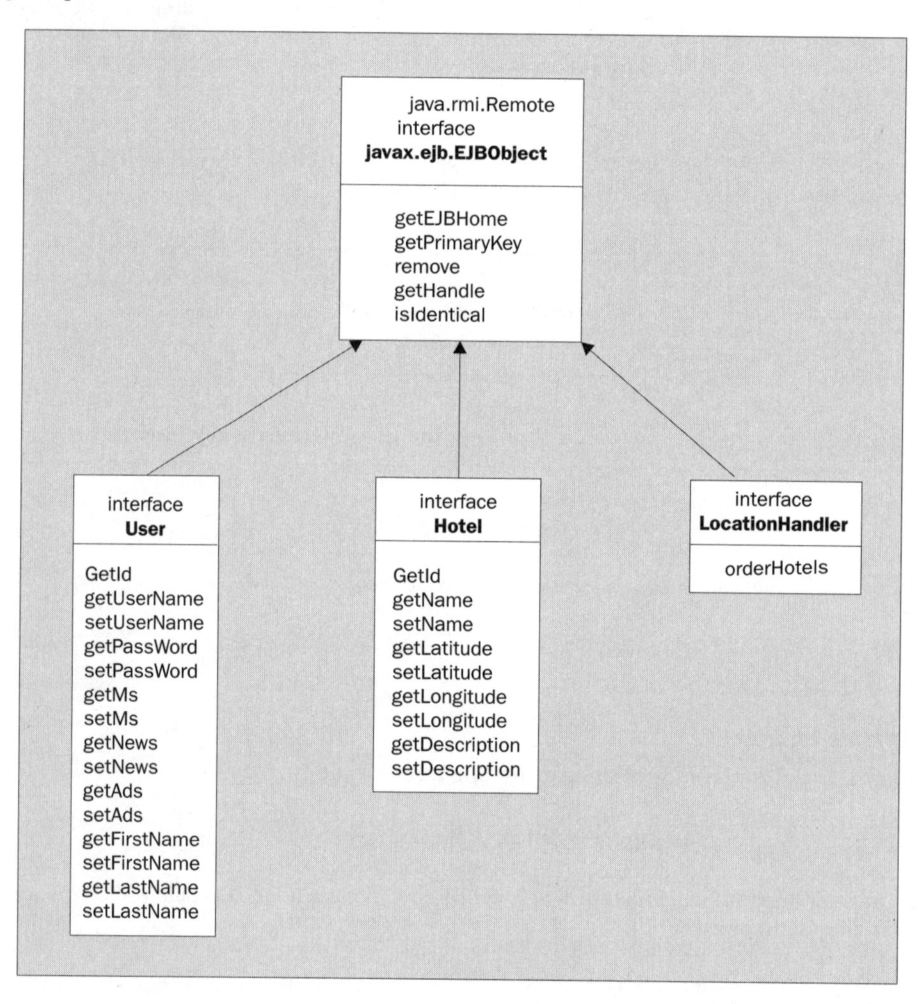

The LocationHandler Interface

The remote interface for the `LocationHandler` defines the only available functionality for accessing location data. The `orderHotels()` method will return a `Collection` of `HotelAccessor` objects.

The `LocationHandler` interface defines the `LocationHandler` session. `LocationHandler` is used to determine which hotels are closest to a given client, and is responsible for interacting with the Mobile Positioning Center:

```
public interface LocationHandler extends EJBObject {
```

Gives the three closest hotels to the given user ordered in ascending order based on distance from user:

```
  public Collection orderHotels(User user,
                                Collection hotels) throws RemoteException;
}
```

The User Interface

The two entities are defined below. They correspond to the tables in the database. Individual bean objects will map to individual rows in the database. Notice the accessors and mutators for the persistent fields:

```
public interface User extends EJBObject {

  public long getId() throws RemoteException;

  public String getUserName() throws RemoteException;
  public void setUserName(String value) throws RemoteException;

  public String getPassWord() throws RemoteException;
  public void setPassWord(String value) throws RemoteException;
```

The following is the accessor for obtaining the `Ms` field that stores the phone number of the client:

```
  public String getMs() throws RemoteException;
  public void setMs(String value) throws RemoteException;

  public String getFirstName() throws RemoteException;
  public void setFirstName(String value) throws RemoteException;

  public String getLastName() throws RemoteException;
  public void setLastName(String value) throws RemoteException;
}
```

The Hotel Interface

The Hotel Interface defines the Hotel entity:

```
public interface Hotel extends EJBObject {
```

getID() is the accessor for the ID field. An ID uniquely identifies each Hotel:

```
    public long getId() throws RemoteException;

    public String getName() throws RemoteException;
    public void setName(String value) throws RemoteException;
```

getLatitude() is the accessor method for the latitude field. Latitude is the angular distance North or South of the equator:

```
    public String getLatitude() throws RemoteException;
    public void setLatitude(String value) throws RemoteException;
```

getLongitude() is the accessor method for the longitude field. Longitude is the angular distance East or West of the prime meridian:

```
    public String getLongitude() throws RemoteException;
    public void setLongitude(String value) throws RemoteException;
```

getDescription() is the accessor method for the description field. Contains info to be displayed to the user for the hotel, typically the address. Can contain WML formatted links and information:

```
    public String getDescription() throws RemoteException;
    public void setDescription(String value) throws RemoteException;
}
```

The Bean Implementation

The implementation for the entity beans is almost trivial, as Container Managed Persistency (CMP) is used. The implementation for LocationHandler is very interesting, however. Here is where the communication to the MPC happens. Note the HotelAccessor class that the LocationHandler returns to the presentation layer:

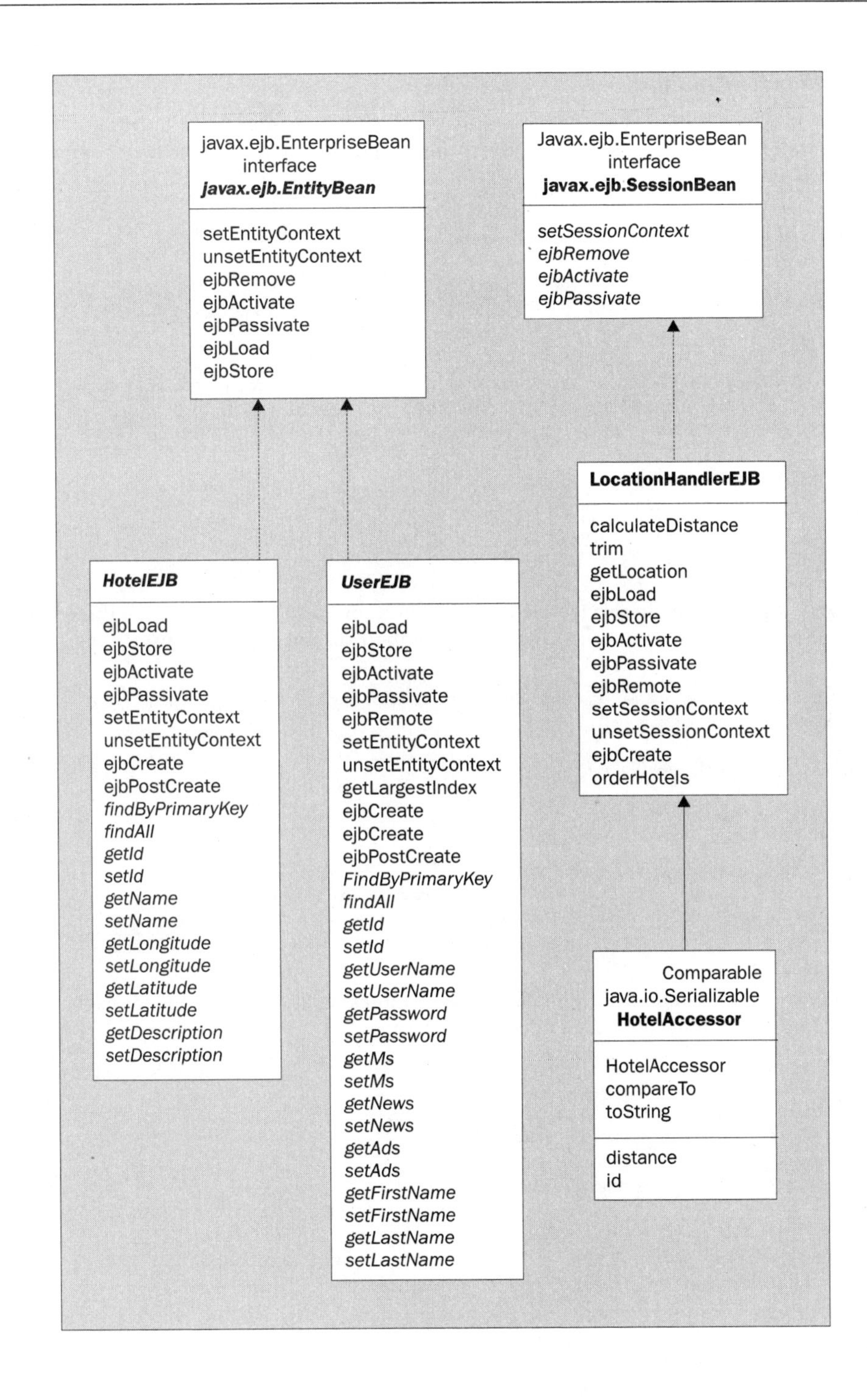
javax.ejb.EnterpriseBean
interface
javax.ejb.EntityBean
setEntityContext
unsetEntityContext
ejbRemove
ejbActivate
ejbPassivate
ejbLoad
ejbStore
Javax.ejb.EnterpriseBean
interface
javax.ejb.SessionBean
setSessionContext
ejbRemove
ejbActivate
ejbPassivate
LocationHandlerEJB
calculateDistance
trim
getLocation
ejbLoad
ejbStore
ejbActivate
ejbPassivate
ejbRemote
setSessionContext
unsetSessionContext
ejbCreate
orderHotels
HotelEJB
ejbLoad
ejbStore
ejbActivate
ejbPassivate
setEntityContext
unsetEntityContext
ejbCreate
ejbPostCreate
findByPrimaryKey
findAll
getId
setId
getName
setName
getLongitude
setLongitude
getLatitude
setLatitude
getDescription
setDescription
UserEJB
ejbLoad
ejbStore
ejbActivate
ejbPassivate
ejbRemote
setEntityContext
unsetEntityContext
getLargestIndex
ejbCreate
ejbCreate
ejbPostCreate
FindByPrimaryKey
findAll
getId
setId
getUserName
setUserName
getPassword
setPassword
getMs
setMs
getNews
setNews
getAds
setAds
getFirstName
setFirstName
getLastName
setLastName
Comparable
java.io.Serializable
HotelAccessor
HotelAccessor
compareTo
toString
distance
id

The LocationHandlerEJB Class

Here is the implementation class for the `LocationHandler` session bean. `LocationHandler` determines which hotels are closest to a given client, and is responsible for interacting with the Mobile Positioning Center:

```
public class LocationHandlerEJB implements SessionBean {

  // Hostname to MPC Emulator running on JRun.
  public static final String MPC_HOST = "localhost";
  // Port for MPC Emulator running on JRun.
  public static final int MPC_PORT = 8100;
  // Username to MPC. Should correspond to entries in Authfile.txt
  public static final String MPC_USERNAME = "aUser";
  // Password to MPC. Should correspond to entries in Authfile.txt
  public static final String MPC_PASSWORD = "aPwd";
  // Emulator does not support SSL
  public static final boolean MPC_USE_SSL = false;

  protected SessionContext context;
```

`orderHotels()` is the main service method of `LocationHandler`. It gives the three closest hotels to the given user ordered in ascending order based on their distance from the user:

```
public Collection orderHotels(User user, Collection hotels)
  throws RemoteException {

  ArrayList list = new ArrayList();

  try {

    // Logging to facilitate debugging
    System.out.println("Attempting to retrieve hotels for user "
                       + user.getUserName() + ". (Phone number "
                       + user.getMs() + ")");
```

We need to loop through the list of hotels supplied and calculate the distance for each, finally placing the `HotelAccessor` object for each in an `ArrayList`:

```
    Iterator iter = hotels.iterator();
    while (iter.hasNext()) {
      Hotel hotel = (Hotel) iter.next();
      HotelAccessor ha =
        new HotelAccessor(hotel.getId(),
                 calculateDistance(getLocation(user.getMs()),
                                   hotel.getLatitude(),
                                   hotel.getLongitude()));
      list.add(ha);

      // Logging to facilitate debugging
      System.out.println("Adding a hotel" + ha);

    }
```

Refer to the `HotelAccessor` class for `compareTo()`, to see the sort order:

```
      Collections.sort(list);
      Collections.reverse(list);

      // Logging to facilitate debugging
      System.out.println("Done ordering. Returning list: " + list);

    } catch (Exception ex) {
      ex.printStackTrace();
      throw new RemoteException("Communication with MPC failed: " + ex);
    }

    // Only return the three closest hotels
    return new ArrayList(list.subList(0, 3));
  }
```

The `calculateDistance()` helper method calculates the distance from a `mpssdk.Coordinate` to a latitude and longitude given as Strings. The formula used is the well-known Pythagoras' theorem ($a^2+b^2=c^2$):

```
  private double calculateDistance(Coordinate from, String toLatitude,
                                   String toLongitude) {
    double result =
      Math.sqrt(Math.pow((trim(from.getLatitude()) - trim(toLatitude)), 2)
                + Math.pow((trim(from.getLongitude()) - trim(toLongitude)),
                           2));
    return result;
  }
```

The `trim()` helper method parses a longitude or latitude as total seconds of a degree:

```
  private long trim(String in) {
    int len = in.length();
    long minutes = Long.parseLong(in.substring(len - 4, len - 2));
    long seconds = Long.parseLong(in.substring(len - 2, len));
    return (minutes * 60 + seconds);
  }
```

`getLocation()` is a helper method that abstracts the communication to the MPC. `ms` is the phone number of the client being located. The `Coordinate` returned contains the location of the client's cell:

```
  private mpssdk.Coordinate getLocation(String ms) {
    mpssdk.Coordinate result = null;
    try {
```

First we create a request object to be sent to the MPC:

```
  LocationRequest posRequest = new LocationRequest(MPC_HOST, MPC_PORT,
          MPC_USERNAME, MPC_PASSWORD, MPC_USE_SSL);
```

We must add phone numbers to the request before sending. We could add several numbers in one go:

```
posRequest.addMSISDN(ms);
```

The request is parsed into XML format before being sent to the MPC:

```
posRequest.send();
LocationResult posResult;
```

We then retrieve the results. There will be only one coordinate, since we only added one phone number:

```
      if ((posResult = posRequest.getLocationResult()) != null) {
        result = posResult.getCoord();
      }
    } catch (Exception e) {
      System.out.println("getLocation failed: " + e);
      e.printStackTrace();
    }
    return result;
  }
```

Note that there are a number of methods required for bean implementations, but as they are entirely standard for this application, we have left them out.

The HotelAccessor Class

The `HotelAccessor` is a helper class used to return hotel info from the `orderHotels()` method of the `LocationHandler`:

```
public class HotelAccessor implements Comparable, java.io.Serializable {

  // Hotel id
  long id;

  // Distance from current user.
  double distance;

  public double getDistance() {
    return distance;
  }

  public long getId() {
    return id;
  }

  public HotelAccessor(long id, double distance) {
    this.id = id;
    this.distance = distance;
  }
```

This is a comparator to order the results based on distance:

```
    public int compareTo(Object o) {
      if (((HotelAccessor) o).distance > this.distance) {
        return +1;
      } else if (((HotelAccessor) o).distance < this.distance) {
        return -1;
      } else {
        return 0;
      }
    }

    public String toString() {
      return super.toString() + " (Id: " + id + ", Distance:" + distance
             + ")";
    }
  }
```

The UserEJB Class

`UserEJB` defines the `User` implementation. This is quite trivial, since we are using CMP – Container Managed Persistency, so we have left out all the getters and setters, along with all the other trivial methods required by a bean implementation:

```
  public abstract class UserEJB implements EntityBean {

    protected EntityContext context;
```

The `getLargestIndex()` method finds the largest available index. This is not the most efficient algorithm:

```
    private long getLargestIndex() throws CreateException {
      long id = 0;
      long tmp = 0;
      try {
        Iterator iter =
          (((UserHome) context.getEJBHome()).findAll()).iterator();
        while (iter.hasNext()) {
          tmp = ((User) iter.next()).getId();
          if (tmp > id) {
            id = tmp;
          }
        }
      } catch (Exception ex) {
        System.out.println(ex);
        throw new CreateException(ex.toString());
      }
      return id;
    }
  }
```

The HotelEJB Class

HotelEJB defines the Hotel implementation. This is again quite trivial because we are using CMP:

```
public abstract class HotelEJB implements EntityBean {

  protected EntityContext context;
}
```

Presentation Tier

The presentation tier of the application is implemented as JSPs using custom tags to access the application tier (EJBs). The presentation tier is tasked with the generation of both HTML and WML.

HTML

The user is welcomed by an index page when accessing the URL for the application, http://(host)/hotel-search/. Although this page currently is pure HTML and should be implemented as such for performance it is for now defined as a JSP for simplicity:

```
<%@ page errorPage="../oops.jsp" %>

<html>

<head>
    <link rel="stylesheet" href="hotel.css">
    <title>Hotel-Search</title>
</head>

<body background="background.jpg">

<h3 align="center">Welcome to</h3>
<div align="center"><img src="hotel-search.gif"></div>
<br><hr><br>
<a href="reg/index.jsp">Register for hotel-search</a>

</body>

</html>
```

The Registration Index JSP

When following the link to 'Register for hotel-search' the user will end up in a registration page as displayed below. Here user information is entered and posted back to the application. Notice the numerical password consisting of 4 characters to simplify login from a WAP user agent. Also notice that a select list has been already filled in with existing phone numbers. This is to simplify the demonstration of the application:

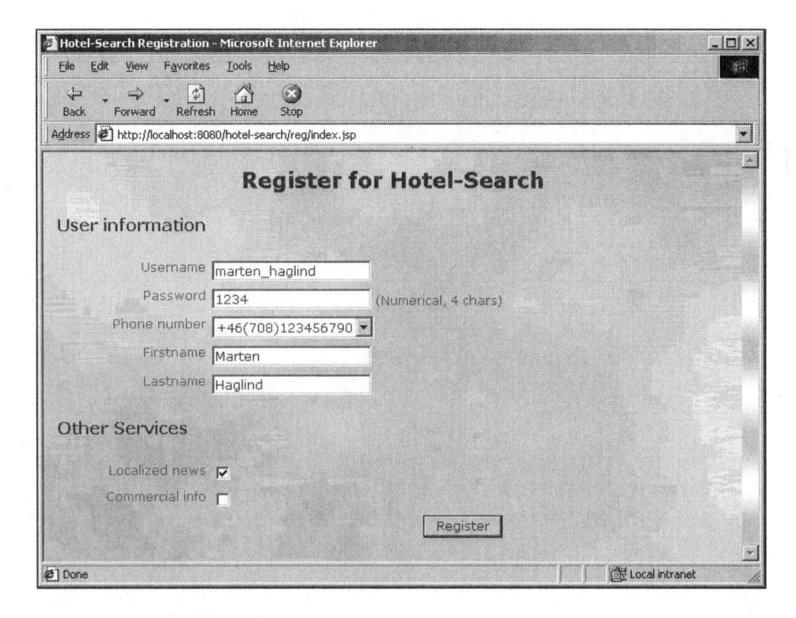

The code for the registration page is shown below:

```
<%@ page errorPage="../oops.jsp" %>

<html>
<head>
  <link rel="stylesheet" href="../hotel.css">
  <title>Hotel-Search Registration</title>
</head>

<body background="../background.jpg">

  <h3 align="center">Register for Hotel-Search</h3>
  <form action="register.jsp" method="POST">
    <table cellpadding="2" cellspacing="2">
      <tr>
        <td valign="bottom"><h5>User information</h5></td>
      </tr>
      <tr>
        <td valign="top" align="right">Username</td>
        <td valign="top" align="left">
          <input type="text" name="userName" size="23">
        </td>
      </tr>
      <tr>
        <td valign="top" align="right">Password</td>
        <td valign="top" align="left">
          <input type="text" name="passWord" size="23">
            (Numerical, 4 chars)
        </td>
      </tr>
      <tr>
        <td valign="top" align="right">Phone number</td>
```

```
          <td valign="top" align="left">
            <select name="ms" size="1">
              <option value="46708123456789">+46(708)123456789</option>
              <option value="46708123456790">+46(708)123456790</option>
              <option value="46708123456791">+46(708)123456791</option>
              <option value="46708123456792">+46(708)123456792</option>
              <option value="46708123456793">+46(708)123456793</option>
            </select>
          </td>
        </tr>
        <tr>
          <td valign="top" align="right">Firstname</td>
          <td valign="top" align="left">
            <input type="text" name="firstName" size="23">
          </td>
        </tr>
        <tr>
          <td valign="top" align="right">Lastname</td>
          <td valign="top" align="left">
            <input type="text" name="lastName" size="23">
          </td>
        </tr>
        <tr>
          <td valign="bottom"><br><h5>Other Services</h5></td>
        </tr>
        <tr>
          <td valign="top" align="right">Localized news</td>
          <td>
            <input type="checkbox" name="news" value="ON">
          </td>
        </tr>
        <tr>
          <td valign="top" align="right">Commercial info</td>
          <td><input type="checkbox" name="ads" value="ON"></td>
        </tr>
        <tr>
          <td align="right" colspan="2">
            <input type="submit" value="Register" name="submit">
          </td>
        </tr>
      </table>
    </form>
  </body>
  </html>
```

The Registration JSP

The registration form is submitted to `register.jsp`. This will display a confirmation message to the user after storing the new user in the database:

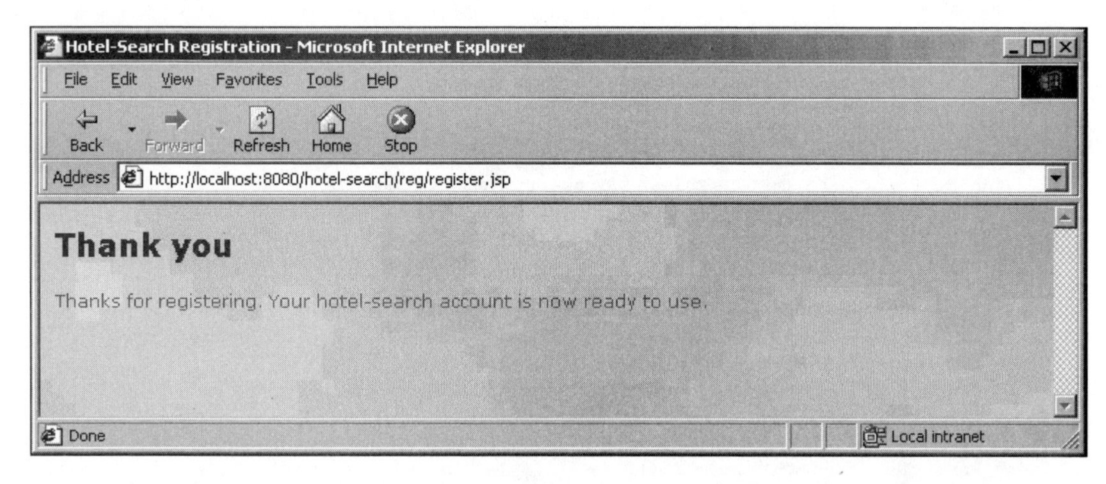

Here is the implementation for the registration. Notice the use of Orion custom tags for creating `User` beans and setting properties. The Orion `ejbtags` have nothing to do with positioning as such, but are an excellent and easy way to get to EJBs from JSP.

The taglib in use is available from http://www.orionserver.com/tags/ejbtags/ including instructions on download, installation and usage.

The main custom tags defined are (note: to use these you must include a taglib directive: `<%@ taglib uri="ejbtags" prefix="ejb"%>`):

- `useHome` – to locate and use a home interface:

```
<ejb:userHome id="myBarHome" type="foo.BarHome" location= "java:comp/env/ejb/Bar">
```

- `useBean` – to locate and narrow an EJB (used together with create below):

```
<ejb:useBean id="myApa" type="foo.Apa" scope="session"/>
```

- `createBean` – for creating or looking up beans. Notice that we need to use a home for this – created with the `useHome` tag:

```
<ejb:useBean id="myApa" type="foo.Apa" scope="session">
  <ejb:createBean instance="<%=aHome.create()%>">
</ejb:useBean>
```

Here is the register page with the above tags included:

```
<%@ page errorPage="../oops.jsp" %>
<%@ taglib uri="ejbtags" prefix="ejb" %>

<ejb:useHome
```

```
    id="userHome"
    type="mobilejava.UserHome"
    location="java:comp/env/ejb/User" />


<ejb:useBean id="userBean" type="mobilejava.User" scope="request">
    <ejb:createBean instance="<%=userHome.create()%>" />
</ejb:useBean>

<jsp:setProperty name="userBean" property="userName" param="userName"/>
<jsp:setProperty name="userBean" property="ms" param="ms"/>
<jsp:setProperty name="userBean" property="passWord" param="passWord"/>
<jsp:setProperty name="userBean" property="firstName" param="firstName"/>
<jsp:setProperty name="userBean" property="lastName" param="lastName"/>

<% if (request.getParameter("news") != null) { %>
    <jsp:setProperty name="userBean" property="news" value="true"/>
<% } if (request.getParameter("ads") != null) { %>
    <jsp:setProperty name="userBean" property="ads" value="true"/>
<% } %>

<% long userId=userBean.getId(); %>

<html>

<head>
    <link rel="stylesheet" href="../hotel.css">
    <title>Hotel-Search Registration</title>
</head>

<body background="../background.jpg">
    <h3>Thank you</h3>
    <p>Thanks for registering. Your hotel-search account is now ready to use.</p>
</body>
</html>
```

The Error JSP

In case something should go wrong with the registration, we have defined the following error page, oops.jsp:

```
<%@ page isErrorPage="true" %>

<html>
<head>
    <link rel="stylesheet" href="hotel.css">
    <title>Hotel-Search</title>
</head>

<body background="background.jpg">
    <h3 align="center">An exception has occured</h3><hr><br>
    <p><%=exception%></p>
</body>
</html>
```

The Utilities Index JSP

For the purpose of maintaining the database there is an administration interface to the entities available (as displayed earlier in the chapter). It is activated by accessing http://(host)/hotel-search/util/:

```
<%@ page errorPage="../oops.jsp" %>
<%@ taglib uri="ejbtags" prefix="ejb" %>

<ejb:useHome
    id="hotelHome"
    type="mobilejava.HotelHome"
    location="java:comp/env/ejb/Hotel" />

<ejb:useHome
    id="userHome"
    type="mobilejava.UserHome"
    location="java:comp/env/ejb/User" />


<html>
<head><link rel="stylesheet" href="../hotel.css"></head>

<body background="../background.jpg">

<h2>Administer Hotel-Search</h2>
<hr>

<form action="http://localhost:8100/Version" method="POST">
    <input type="submit" value="MPC Version">
</form>

<hr>
```

Here we display all the hotels that have been registered:

```
<table>
  <tr>
    <td align="center" colspan="6"><h4>Hotels</h4></td>
  </tr>
  <tr>
    <th>Id</th>
    <th>Name</th>
    <th>Description</th>
    <th>Latitude</th>
    <th>Longitude</th>
  </tr>

  <% int i=0; %>

  <ejb:iterate id="hotel" type="mobilejava.Hotel"
               collection="<%=hotelHome.findAll()%>" max="100">

    <tr>
      <form action="edit_hotel.jsp" method="GET">
        <input type="hidden" value="<%=hotel.getId()%>" name="id"/>
          <td><%=hotel.getId()%></td>
```

```
          <td>
            <input type="text" value="<%=hotel.getName()%>"
                   name="name" size="15"/>
          </td>
          <td>
            <input type="text" value="<%=hotel.getDescription()%>"
                   name="description" size="20"/>
          </td>
          <td>
            <input type="text" value="<%=hotel.getLatitude()%>"
                   name="latitude" size="9"/>
          </td>
          <td>
            <input type="text" value="<%=hotel.getLongitude()%>"
                   name="longitude" size="9"/>
          </td>
          <td>
            <input type="submit" value="Update">
          </td>
        </form>
    </tr>
    <% i++; %>
  </ejb:iterate>

  <tr></tr>
  <tr>
    <form action="add_hotel.jsp" method="GET">
      <td><input type="text" value="<%=i%>" name="id" size="3"/></td>
      <td><input type="text" value="" name="name" size="15"/></td>
      <td><input type="text" value="" name="description" size="20"/></td>
      <td><input type="text" value="" name="latitude" size="9"/></td>
      <td><input type="text" value="" name="longitude" size="9"/></td>
      <td><input type="submit" value="Update"></td>
    </form>
  </tr>
</table>

<hr/>
```

Now we display all our registered users:

```
<table>
  <tr>
    <td align="center" colspan="5"><h4>Users</h4></td>
  </tr>
  <tr>
    <th>Id</th><th>User Name</th><th>Password</th><th>MS</th>
  </tr>

  <% i=0; %>

  <ejb:iterate
    id="user" type="mobilejava.User"
    collection="<%=userHome.findAll()%>" max="100">
```

```
      <tr>
        <form action="edit_user.jsp" method="GET">
          <input type="hidden" value="<%=user.getId()%>" name="id"/>
          <td><%=user.getId()%></td>
          <td><input type="text" value="<%=user.getUserName()%>"
                     name="userName" size="15"/></td>
          <td><input type="text" value="<%=user.getPassWord()%>"
                     name="passWord" size="15"/></td>
          <td><input type="text" value="<%=user.getMs()%>"
                     name="ms" size="20"/>   </td>
          <td><input type="submit" value="Update"></td>
        </form>
      </tr>
      <% i++; %>
    </ejb:iterate>

    <tr></tr>
    <tr>
      <form action="add_user.jsp" method="GET">
        <td><input type="text" value="<%=i%>" name="id" size="3"/></td>
        <td><input type="text" value="" name="userName" size="15"/></td>
        <td><input type="text" value="" name="passWord" size="15"/></td>
        <td><input type="text" value="" name="ms" size="20"/></td>
        <td><input type="submit" value="Update"></td>
      </form>
    </tr>
  </table>
```

and to end the page:

```
      <hr>
    </body>
  </html>
```

In the utility page there are links to:

- `Add_user.jsp`
- `Edit_user.jsp`
- `Add_hotel.jsp`
- `Edit_hotel.jsp`

These are helper pages whose functions should be obvious. They will all return the user to the main page once they have performed their function. Code is given below for `add_user` and `edit_user`. `add_hotel` and `edit_hotel` are similar and only really differ in what create/finder method is used.

The add_user JSP

```
<%@ page errorPage="../oops.jsp" %>
<%@ taglib uri="ejbtags" prefix="ejb" %>
<% long id = Long.parseLong(request.getParameter("id")); %>

<ejb:useHome
```

```
    id="hotelHome"
    type="mobilejava.UserHome"
    location="java:comp/env/ejb/User" />

<ejb:useBean id="hotelBean" type="mobilejava.User" scope="request">
    <ejb:createBean instance="<%=hotelHome.create(id)%>" />
</ejb:useBean>

<jsp:setProperty name="hotelBean" property="userName" param="userName"/>
<jsp:setProperty name="hotelBean" property="passWord" param="passWord"/>
<jsp:setProperty name="hotelBean" property="ms" param="ms"/>

<% response.sendRedirect("index.jsp"); %>
```

The edit_user JSP

```
<%@ page errorPage="../oops.jsp" %>
<%@ taglib uri="ejbtags" prefix="ejb" %>
<% long id = Long.parseLong(request.getParameter("id")); %>

<ejb:useHome
    id="hotelHome"
    type="mobilejava.UserHome"
    location="java:comp/env/ejb/User" />

<ejb:useBean id="hotelBean" type="mobilejava.User" scope="request">
    <ejb:createBean instance="<%=hotelHome.findByPrimaryKey(id)%>" />
</ejb:useBean>

<jsp:setProperty name="hotelBean" property="userName" param="userName"/>
<jsp:setProperty name="hotelBean" property="passWord" param="passWord"/>
<jsp:setProperty name="hotelBean" property="ms" param="ms"/>

<% response.sendRedirect("index.jsp"); %>
```

WML

The WML below is displayed in the Ericsson 380 as simulated by the SmartPhone Emulator. When the user first accesses the WML site though the phone they will be prompted to login.

The WML Login JSP

The site can be accessed by phone as

http://(host)/hotel-search/wml/

or

http://(host)/hotel-search/wml/login.jsp?user=(user)

The latter can be saved as a favorite on the phone to save time logging in when accessing the service:

The code for this page is shown below. Again, explaining WML is out of scope for this chapter and so knowledge is assumed. If you do not know WML, you should still get a fair idea from the code below:

```
<%@page contentType="text/vnd.wap.wml" errorPage="../oops_wml.jsp"%>
<% String user = request.getParameter("user"); %>

<?xml version="1.0"?>
<!DOCTYPE wml PUBLIC "-//WAPFORUM//DTD WML 1.1//EN"
                     "http://www.wapforum.org/DTD/wml_1.1.xml">

<!--
    Login deck for Hotel-Search
-->

<wml>
  <card title="Login" id="login">
    <p align="center">
      <img src="hotel-search.wbmp" alt="Hotel-Search" />
    </p>
    <p>Please login</p>
```

The user name can be given a parameter (`user`) and will be pre-filled in. If this is so then `password` should be all numerical:

```
    <% if (user != null) { %>
      <p>User: <%=user%>
      <br/>
      Password:
      <input format="*n" name="password" type="password"/>
      <anchor>
        Login
        <go href="do_login.jsp" method="post">
          <postfield name="user" value="<%=user%>"/>
          <postfield name="password" value="$password"/>
        </go>
      </anchor>
```

```
        </p>
        <% } else { %>
        <p>User: <input format="*m" name="user" type="text"/>
        <br/>
        Password: <input format="*n" name="password" type="password"/>
        <anchor>
          Login
          <go href="do_login.jsp" method="post">
            <postfield name="user" value="$user"/>
            <postfield name="password" value="$password"/>
          </go>
        </anchor>
      </p>
      <% } %>

  </card>

</wml>
```

The WML do_login JSP

When the user has logged in and is authenticated by the application (primarily to get access to the stored phone number, needed for the positioning) the user will be routed to the main page of the hotel-search application:

```
<%@ taglib uri="ejbtags" prefix="ejb" %>

<ejb:useHome id="userHome"
             type="mobilejava.UserHome"
             location="java:comp/env/ejb/User" />

<%
  String user = request.getParameter("user");
  String password = request.getParameter("password");
%>

<ejb:iterate id="userBean"
             type="mobilejava.User"
             collection="<%=userHome.findAll()%>"
             max="100">

<%
  if (userBean.getUserName().equals(user)) {
    if (userBean.getPassWord().equals(password)) {
      // User authenticated
      session.setAttribute("loggedIn","Yes");
      session.setAttribute("userId", "" + userBean.getId());
    } else {
      // Wrong password
      session.setAttribute("loggedIn","No");
    }
  }
%>

</ejb:iterate>

<jsp:forward page="index.jsp"/>
```

This deck illustrates one point of browser differences that can vex the WAP developer. In this case the navigational menu is displayed as recommended by Phone.Com for the UP browser but as normal links for any other browser. In this case we identify the browser by querying the user-agent header for the browser type.

A splash card is also displayed, demonstrating the use of timers in WML:

The WML Index JSP

```
<%@page contentType="text/vnd.wap.wml" errorPage="../oops_wml.jsp"%>

<%
    // If not logged in - forward to login page
    if ((session.getAttribute("loggedIn") == null)
        || (session.getAttribute("loggedIn").equals("No"))) { %>
    <jsp:forward page="login.jsp"/>
<% } %>


<%
    // Determine if user is Phone.Com
    String browser = request.getHeader("user-agent");
    if (browser.indexOf("UP") !=-1) {
        browser ="Phone.Com";
    } else {
        browser ="Other";
    }
%>

<?xml version="1.0"?>
<!DOCTYPE wml PUBLIC "-//WAPFORUM//DTD WML 1.1//EN"
                     "http://www.wapforum.org/DTD/wml_1.1.xml">
```

Here is the main deck of the hotel-search application:

```
<wml>

<!-- Splash card displayed for 2,5 seconds with logo -->
  <card title="Welcome" ontimer="#menu"  id="splash">
    <timer value="25"/>
    <p align="center">
      <img src="hs-logo.wbmp" alt="Hotel-Search Logo" />
    </p>
    <p>
      <small>Welcome to Hotel-Search</small>
    </p>
  </card>

<!-- Main menu card -->
  <card title="Main Menu" id="menu">
    <p align="center">
      <img src="hotel-search.wbmp" alt="Hotel-Search" />
    </p>

    <p>
      Welcome
    </p>
    <p>
```

The menu is rendered differently in the UP SDK because navigational selects work nicely in that browser:

```
<%
  if (browser.equals("Phone.Com")) {
%>
    <select>
      <option onpick="/hotel-search/wml/search.jsp">
        Perform Search</option>
      <option onpick="#about">About</option>
    </select>

<%
  // Else use conventional links
  } else {
%>

      <a href="search.jsp">Perform Search</a> 
      <a href="#about">About</a>

<% } %>

    </p>
  </card>

<!-- About card -->
  <card title="About"  id="about">
    <p align="center">
```

```
        <img src="hs-logo.wbmp" alt="Hotel-Search Logo" />
      </p>
      <p>
        <small>
          The hotel-search application is a demonstration
          of using a wml browser as a thin client to a
          Mobile Positioning Center. Written by Marten Haglind.<br/>
          <anchor>Back<prev/></anchor>
        </small>
      </p>
    </card>
  </wml>
```

After the user follows the link to perform a search there is a slight delay as the application queries the location center for the closest hotel. The results are displayed as below. The distance measurements within parenthesis are given here for testing as seconds of a degree, which is not very useful for the end user and but should rather be converted into something more useful:

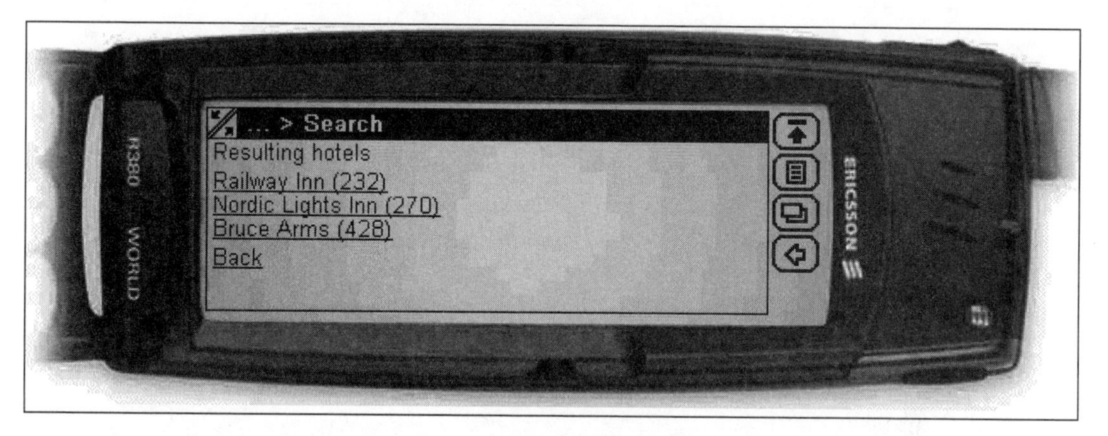

The hotels are also selectable to display more hotel-info:

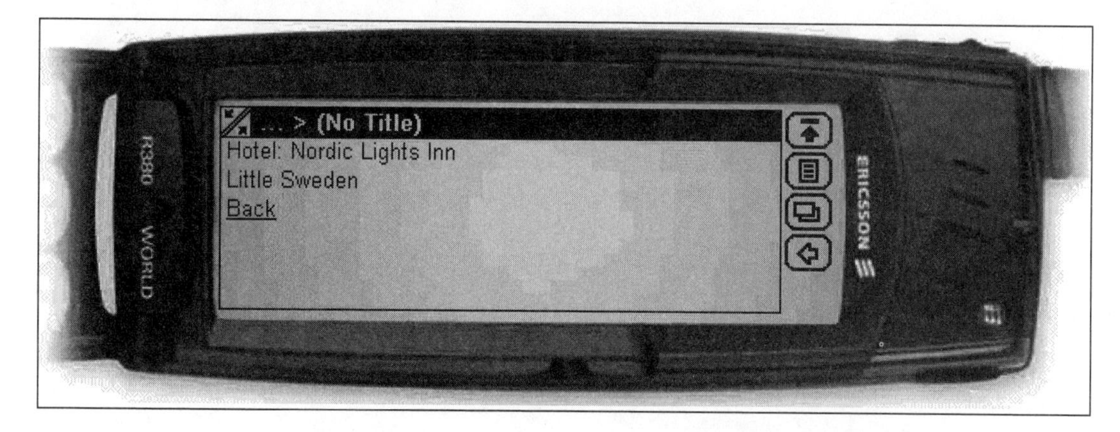

After a while the simulation has moved on and the same user will get a different result when accessing the deck again, simulating that they have moved and are now is closer to the "Bruce Arms":

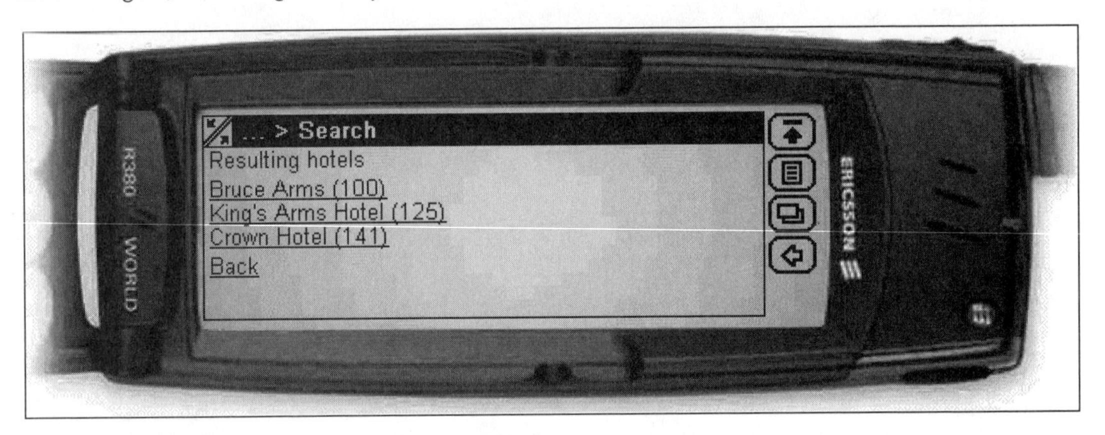

The WML Search JSP

The code for this is shown below:

```
<%@ taglib uri="ejbtags" prefix="ejb" %>
<%@ page contentType="text/vnd.wap.wml" import="java.util.*,mobilejava.*"
errorPage="../oops_wml.jsp" %>

<%
    // If not logged in forward to login page
    if ((session.getAttribute("loggedIn") == null)
        || (session.getAttribute("loggedIn").equals("No"))) { %>
    <jsp:forward page="login.jsp"/>
<% } %>

<%
    // Extract user id from session
    long id = Long.parseLong((String) session.getAttribute("userId"));
%>

<ejb:useHome
    id="locationHandlerHome"
    type="mobilejava.LocationHandlerHome"
    location="java:comp/env/ejb/LocationHandler" />

<ejb:useHome
    id="userHome"
    type="mobilejava.UserHome"
    location="java:comp/env/ejb/User" />

<ejb:useHome
    id="hotelHome"
    type="mobilejava.HotelHome"
    location="java:comp/env/ejb/Hotel" />
```

```
<ejb:useBean id="locationHandler"
    type="mobilejava.LocationHandler" scope="request">
    <ejb:createBean instance="<%=locationHandlerHome.create()%>" />
</ejb:useBean>

<ejb:useBean id="user" type="mobilejava.User" scope="request">
    <ejb:createBean instance="<%=userHome.findByPrimaryKey(id)%>" />
</ejb:useBean>

<?xml version="1.0"?>
<!DOCTYPE wml PUBLIC "-//WAPFORUM//DTD WML 1.1//EN"
                     "http://www.wapforum.org/DTD/wml_1.1.xml">

<wml>
```

The following code asks the user-agent not to cache the deck. According to current specifications, the user-agent may treat this request as a hint and is free to ignore it:

```
  <head>
    <meta http-equiv="Cache-Control" content="no-cache" forua="true"/>
  </head>

  <card title="Search" id="login">
    <p>
      Resulting hotels
    </p>
    <p>
<%
```

Here is where we call the LocationHandler. This is where the work gets done:

```
Collection col = locationHandler.orderHotels(user, hotelHome.findAll());
```

When the results arrive, we loop through them and create links. The hotel name and distance (in seconds of degree) is displayed:

```
  Iterator iter = col.iterator();
  while (iter.hasNext()) {
    mobilejava.HotelAccessor acc = (mobilejava.HotelAccessor) iter.next();
    mobilejava.Hotel hotel = hotelHome.findByPrimaryKey(acc.getId());
%>

      <a href="#hotel<%=hotel.getId()%>">
        <%=hotel.getName()%>
        (<%=(int)acc.getDistance()%>)
      </a>
      <br/>

<% } // end of while loop %>

    </p>
    <p>
      <anchor>Back<prev/></anchor>
```

```
    </p>
  </card>

<%
  // Loop through again to create separate cards for hotel descriptions
  iter = col.iterator();
  while (iter.hasNext()) {
    mobilejava.HotelAccessor acc = (mobilejava.HotelAccessor) iter.next();
    mobilejava.Hotel hotel = hotelHome.findByPrimaryKey(acc.getId());
%>

  <card id="hotel<%=hotel.getId()%>">
    <p>Hotel: <%=hotel.getName()%></p>
    <p><%=hotel.getDescription()%></p>
    <p><anchor>Back<prev/></anchor></p>
  </card>

<% } // end of while loop %>

</wml>
```

The WML Error JSP

An error page for WML is also provided:

```
<%@ page contentType = "text/vnd.wap.wml" isErrorPage="true" %>
<?xml version="1.0"?>
<!DOCTYPE wml PUBLIC "-//WAPFORUM//DTD WML 1.1//EN"
                     "http://www.wapforum.org/DTD/wml_1.1.xml">
<wml>
    <card title="Exception">
        <p><big>An exception has occured!<big><br/></p>
        <p><small><%=exception%></small></p>
    </card>
</wml>
```

Here are some screenshots of the application from other emulators:

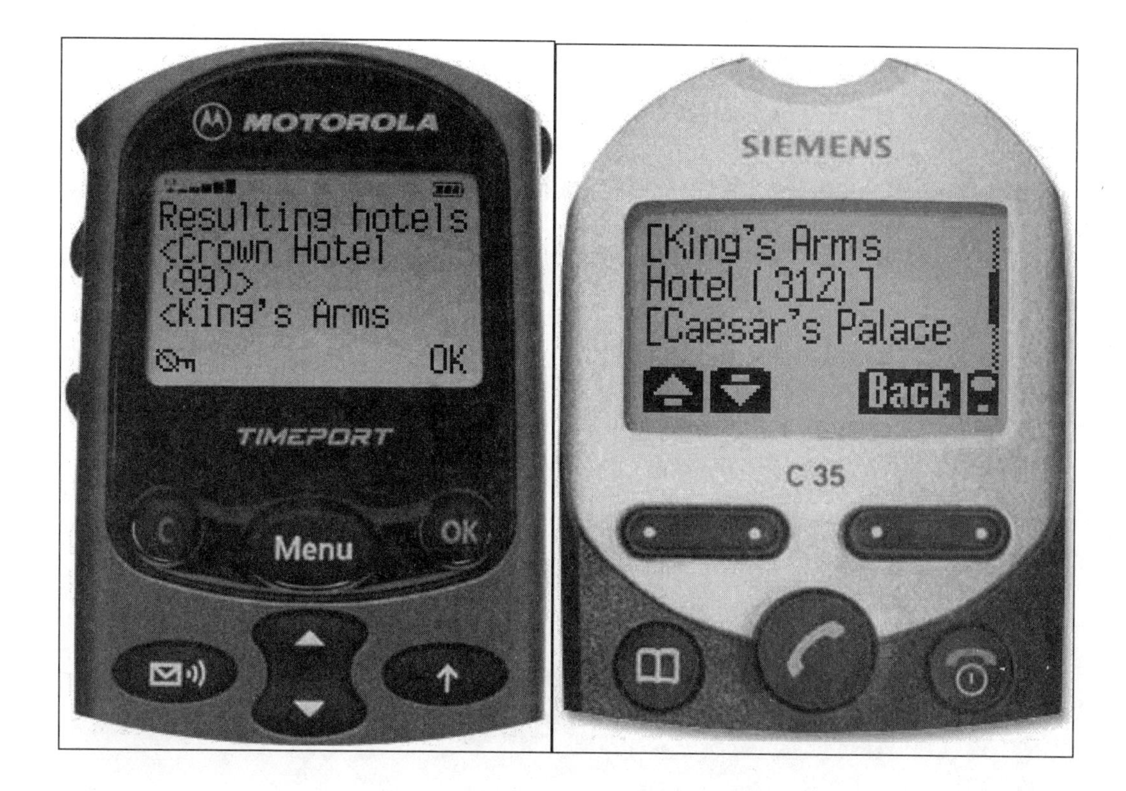

Installation & Deployment

This section gives hints and instructions for installing and deploying this application. The notes assume that you are running Windows but this could easily be adapted to another operating system such as Linux.

Hotel-search

Install the hotel-search application by extracting `hotel-search.zip` into the `orion\applications\` directory. The `hotel-search.zip` can be downloaded from the Wrox web site: `http://www.wrox.com`. Alternatively, build the following directory structure for the application:

```
\orion\
        applications\
                       hotel-search\
                                      META-INF\
                                      hotel-search-ejb\
                                                        META-INF\
                                                        mobilejava\
                                      hotel-search-web\
                                                        WEB-INF\
                                                                classes\
                                                                lib\
                                                        reg\
                                                        util\
                                                        wml\
```

We also need to update the configuration for Orion:

- Add to `orion\config\server.xml` to make the Orion server aware of the new application:

```
<application name="hotel-search" path="../applications/hotel-search/" />
```

- Add to `orion\config\default-web-site.xml`:

```
<web-app
    application="hotel-search"
    name="hotel-search-web"
    root="/hotel-search/" />
```

- Start Orion by giving the command:

```
java -showversion -Xms64m -Xmx256m -jar orion.jar
```

when in the `orion\` main directory. The application should auto-deploy itself:

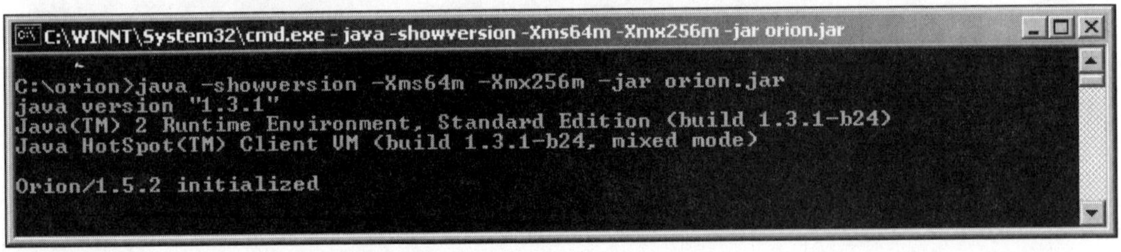

```
C:\WINNT\System32\cmd.exe - java -showversion -Xms64m -Xmx256m -jar orion.jar

C:\orion>java -showversion -Xms64m -Xmx256m -jar orion.jar
java version "1.3.1"
Java(TM) 2 Runtime Environment, Standard Edition (build 1.3.1-b24)
Java HotSpot(TM) Client VM (build 1.3.1-b24, mixed mode)

Orion/1.5.2 initialized
```

Deployment Descriptors

The application uses three main XML descriptors as specified by J2EE:

- `application.xml` for defining the Enterprise Application
- `ejb-jar.xml` for containing deployment info for the EJBs
- `web.xml` for defining the web application

The application.xml File

The following file should be placed in the `hotel-search\META-INF\` directory. It describes the hotel-search application:

```
<?xml version="1.0"?>
<!DOCTYPE application PUBLIC "-//Sun Microsystems, Inc.//DTD J2EE Application
1.2//EN" "http://java.sun.com/j2ee/dtds/application_1_2.dtd">

<application>
  <display-name>Hotel-Search Application Archive</display-name>
  <module><ejb>hotel-search-ejb</ejb></module>
  <module>
```

```
      <web>
        <web-uri>hotel-search-web</web-uri>
        <context-root>/</context-root>
      </web>
    </module>
  </application>
```

The ejb-jar.xml File

The following file should be placed in `hotel-search-ejb\META-INF\`. This file describes our EJBs:

```
<ejb-jar>
  <display-name>hotel-search-ejb</display-name>
  <enterprise-beans>
    <entity>
      <display-name>mobilejava.Hotel</display-name>
      <ejb-name>mobilejava.Hotel</ejb-name>
      <home>mobilejava.HotelHome</home>
      <remote>mobilejava.Hotel</remote>
      <ejb-class>mobilejava.HotelEJB</ejb-class>
      <persistence-type>Container</persistence-type>
      <prim-key-class>java.lang.Long</prim-key-class>
      <reentrant>False</reentrant>
      <cmp-version>2.x</cmp-version>
      <cmp-field><field-name>id</field-name></cmp-field>
      <cmp-field><field-name>name</field-name></cmp-field>
      <cmp-field><field-name>longitude</field-name></cmp-field>
      <cmp-field><field-name>latitude</field-name></cmp-field>
      <cmp-field><field-name>description</field-name></cmp-field>
      <primkey-field>id</primkey-field>
      <abstract-schema-name>Hotel</abstract-schema-name>
    </entity>
    <entity>
      <display-name>mobilejava.User</display-name>
      <ejb-name>mobilejava.User</ejb-name>
      <home>mobilejava.UserHome</home>
      <remote>mobilejava.User</remote>
      <ejb-class>mobilejava.UserEJB</ejb-class>
      <persistence-type>Container</persistence-type>
      <prim-key-class>java.lang.Long</prim-key-class>
      <reentrant>False</reentrant>
      <cmp-version>2.x</cmp-version>
      <cmp-field><field-name>id</field-name></cmp-field>
      <cmp-field><field-name>ms</field-name></cmp-field>
      <cmp-field><field-name>news</field-name></cmp-field>
      <cmp-field><field-name>ads</field-name></cmp-field>
      <cmp-field><field-name>userName</field-name></cmp-field>
      <cmp-field><field-name>passWord</field-name></cmp-field>
      <cmp-field><field-name>firstName</field-name></cmp-field>
      <cmp-field><field-name>lastName</field-name></cmp-field>
      <primkey-field>id</primkey-field>
      <abstract-schema-name>User</abstract-schema-name>
    </entity>
    <session>
      <description>For communicating with the MPC</description>
```

```
      <display-name>mobilejava.LocationHandler</display-name>
      <ejb-name>mobilejava.LocationHandler</ejb-name>
      <home>mobilejava.LocationHandlerHome</home>
      <remote>mobilejava.LocationHandler</remote>
      <ejb-class>mobilejava.LocationHandlerEJB</ejb-class>
      <session-type>Stateful</session-type>
      <transaction-type>Container</transaction-type>
      <ejb-ref>
        <ejb-ref-name>Hotel</ejb-ref-name>
        <ejb-ref-type>Entity</ejb-ref-type>
        <home>mobilejava.HotelHome</home>
        <remote>mobilejava.Hotel</remote>
      </ejb-ref>
    </session>
  </enterprise-beans>
  <assembly-descriptor>
    <container-transaction>
      <method>
        <ejb-name>Hotel</ejb-name>
        <method-name>*</method-name>
      </method>
      <trans-attribute>Supports</trans-attribute>
    </container-transaction>
    <container-transaction>
      <method>
        <ejb-name>User</ejb-name>
        <method-name>*</method-name>
      </method>
      <trans-attribute>Supports</trans-attribute>
    </container-transaction>
  </assembly-descriptor>
</ejb-jar>
```

The web.xml File

The following file should be placed in `hotel-search-web\WEB-INF\`. This file creates references to the EJBs and taglibs that we are using:

```
<?xml version="1.0"?>
<!DOCTYPE web-app PUBLIC "-//Sun Microsystems, Inc.//DTD Web Application 2.3//EN"
"http://java.sun.com/j2ee/dtds/web-app_2_3.dtd">

<web-app>
  <display-name>hotel-search-web</display-name>
  <welcome-file-list>
    <welcome-file>index.jsp</welcome-file>
  </welcome-file-list>
  <taglib>
    <taglib-uri>ejbtags</taglib-uri>
    <taglib-location>/WEB-INF/lib/ejbtags.jar</taglib-location>
  </taglib>
  <login-config>
    <auth-method>BASIC</auth-method>
  </login-config>
  <ejb-ref>
    <ejb-ref-name>ejb/Hotel</ejb-ref-name>
```

```
      <ejb-ref-type>Entity</ejb-ref-type>
      <home>mobilejava.HotelHome</home>
      <remote>mobilejava.Hotel</remote>
    </ejb-ref>
    <ejb-ref>
      <ejb-ref-name>ejb/User</ejb-ref-name>
      <ejb-ref-type>Entity</ejb-ref-type>
      <home>mobilejava.UserHome</home>
      <remote>mobilejava.User</remote>
    </ejb-ref>
    <ejb-ref>
      <ejb-ref-name>ejb/LocationHandler</ejb-ref-name>
      <ejb-ref-type>Session</ejb-ref-type>
      <home>mobilejava.LocationHandlerHome</home>
      <remote>mobilejava.LocationHandler</remote>
    </ejb-ref>
  </web-app>
```

Summary

Mobile positioning may become the killer application of the wireless Internet. This case study has demonstrated how you can develop localized services using current technologies creating demos and prototypes. We have discussed generic techniques for locating mobile devices in cell networks and some possible uses for personalizing an application based on position context. In lue of a standardized protocol we have used a vendor-specific technology – the Ericsson Mobile Positioning System with its corresponding SDK. The application was built using standard J2EE technologies with a WML front-end GUI for maximum client-type penetration.

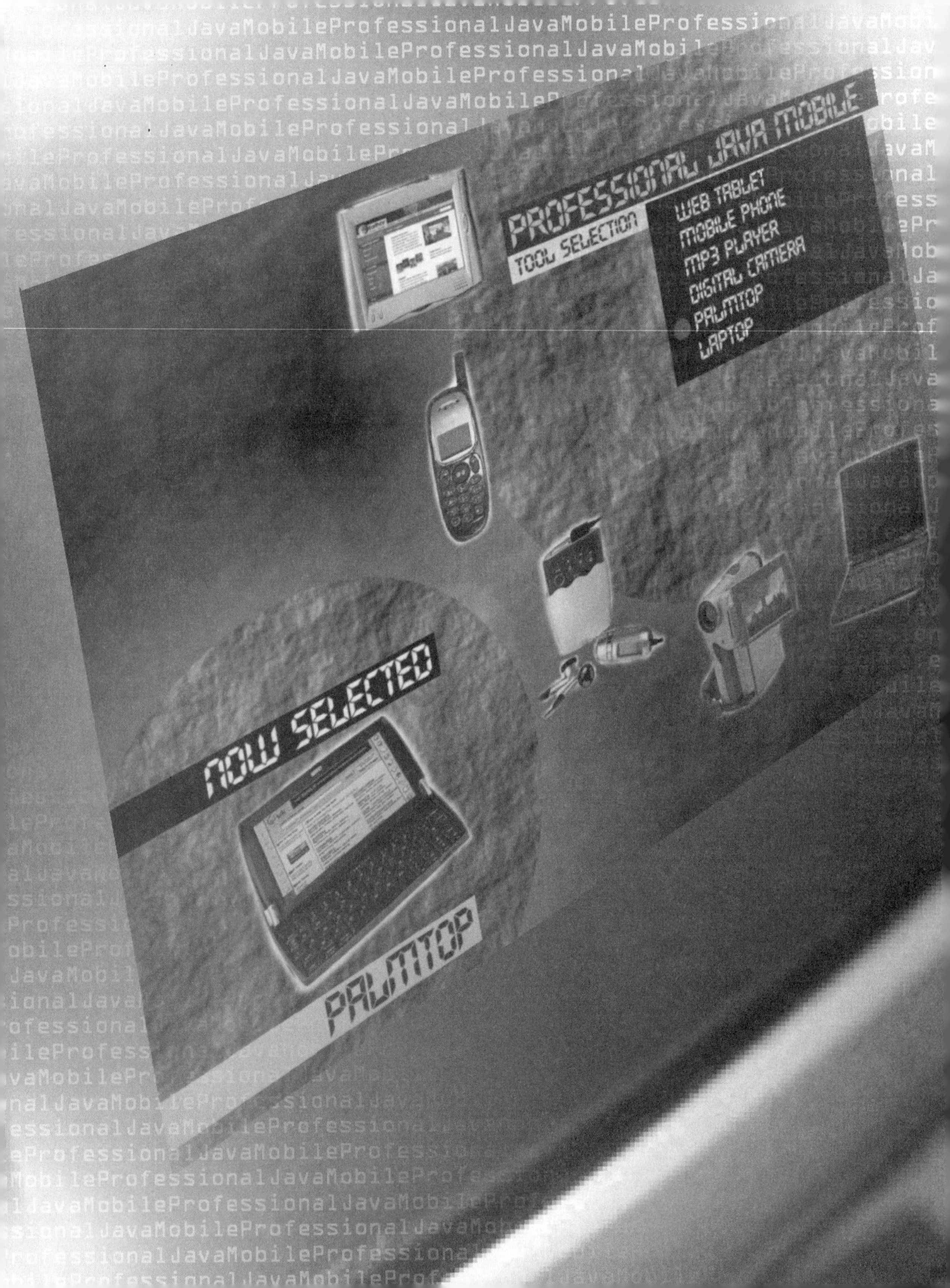

PROFESSIONAL JAVA MOBILE

TOOL SELECTION

WEB TABLET
MOBILE PHONE
MP3 PLAYER
DIGITAL CAMERA
PALMTOP
LAPTOP

NOW SELECTED

PALMTOP

16

Security

Countless seminars, workshops, and keynote talks at conferences present, explain, and promote a vision of the future where information access and e-commerce transactions take place from all sorts of device, while on the go. Near the end of such talks there is usually a slide entitled "Big Challenges" and invariably the first bullet point on that slide, in big letters, is "SECURITY". At best, that is followed by other keywords such as "authorization" and "authentication".

The importance of all these is quickly stressed and the presentation swiftly moves on to the final slide where everybody is reminded of the good things once more. The security issue, however, invariably remains open.

The reasons for avoiding the problem until the last possible moment are easy to understand. Security is a multi-faceted, complicated matter with no immediately obvious solutions. Securing a system is a war between those who wish to protect it and those who wish to crack it and that does not make very good marketing material. Nevertheless, the hard truth remains that to have a secure system you need to build a secure system, not a system that has security added to it.

How do we begin to tackle this problem? What are the concerns and what are the available solutions? More importantly for this book, how do they all fit in the context of mobile devices and mobile code? These are the questions we pose and attempt to partly answer in this chapter. It should be noted that security is such a large subject that a complete coverage would only be possible with several volumes. The aim here is to make the reader aware of the main issues. We will cover the following subjects:

- What are secure systems?
- What is computer security?
- Security in the context of mobile devices
- Security in the context of mobile code
- Security changes from the J2SE to the CDC and the CLDC
- Network security for mobile devices

Setting the Scene

In this section, we will lay the basis for the rest of the chapter. We will talk about secure systems and the issues facing any designer wishing to build a secure system in the abstract. We then transfer these issues to mobile devices and mobile code and see what the threats are and how they are accentuated by the limitation of mobile devices.

What is a Secure System?

What do we mean by security? A dictionary-like definition might be "The state of being free from danger or injury". In this case, a secure system is one that is built so that attacks on it will not cause harm, either through loss of data, compromised data or incorrect results. Furthermore, a secure system should not have any internal processes that may cause harm.

Unfortunately, such a system is near impossible to build. Even if one could be diligent and methodic enough to produce a system with no bugs so as to avoid the possibility that the system enters an insecure state due to internal processes, it is nearly impossible to foresee and cater for every possible external attack. Does this mean that the game is lost and that it is pointless to spend money securing a system since it will eventually be cracked? Well, not quite.

There is something to be gained by making a system hard to crack, even if you know that it may eventually happen. It minimizes the number of people that will try to crack it because not everyone has the necessary resources. In the old days this meant having the necessary manpower and expertise in explosives and lock picking to break a safe, in the modern world it translates into computing power and mathematical expertise to break codes. We should point out that there are cases where a "good enough" level of protection is sufficient. For example, what is the point of detonating an expensive explosive device to open a safe whose contents do no cover the cost of cracking it?

Nevertheless, as we said, given enough time, incentive, and resources every system can be cracked. Therefore, our earlier definition of a secure system may be correct but it is not realistic. What more is there to security then? If a system is compromised, it is just as important, if not more, to be aware of that, so as to stop using the same keys and procedures again.

Furthermore, it is important to know who compromised your system in order to prosecute them. Therefore, detection and prosecution need to be added to our definition of a secure system. Our definition now looks like this: a secure system is one where there are processes in place to prevent and detect **the likelihood** of attacks to the system and **minimize the impact** of the attack, as well as prosecute the offenders through the available legal and judicial channels. Security is thus divided into three tasks: **detection**, **prosecution**, and **prevention**.

It is important to stress the use of the word "processes". In this context, we do not mean merely algorithms embedded in software. Certainly, that is part of the issue but more crucially, these processes need to extend to the physical devices and the way they are secured as well as spelling out a code of practice for the people who use the software and access the physical devices. More often than not, a security breach occurs not because a cracker managed to break a cryptographic code but because an aggravated employee uses his or her own access password to enter the system and cause harm.

Secure Computer Systems

How do we understand prevention, detection and prosecution in terms of secure **computer** systems and what processes should be in place to realize these tasks?

Detection and Prosecution

To start with, it has to be said that very little is actually done on the detection and prosecution aspects of security for computer systems, especially as it refers to the networked world. For these tasks to be performed effectively it implies that a computer system should be able to realize when an intruder has compromised or is attempting to compromise the system and, having done that, identify and prosecute the intruder.

However, what most systems focus on is preventing attacks. Once you are through a certain login/authentication process, which is a prevention mechanism, there is very little to indicate that you are there unlawfully except, suspicious actions such as deleting all the files and crashing machines. A serious attacker is likely to be more subtle. Nevertheless, even if you do manage to detect an intruder, because of obvious behavior, and manage to link that behavior to a person or group of persons, there is still no guarantee that you will be able to have justice exacted upon them.

Usually detecting the identity of the intruder and achieving prosecution requires time, dedication, and impressive detective work combined with network expertise and even then the whole thing might fail if the attacker is in a country that does not have an extradition treaty with the country where the attack took place! Only a few countries (mainly in what is called the "Western" world) have any sort of legislation directed specifically at computer crime. Unfortunately, we are still a long way from establishing international treaties and the appropriate technology standards that will allow for effective prosecution.

Prevention

Sometimes, even if the attack is the most simple you could imagine and you know it is happening, it is hard to prevent it. For example, the most effective and simplistic class of attacks of recent times is the Denial of Service (DoS) attack. The basic idea behind a DoS attack is to flood a server with so many data packets that it cannot cope and therefore will crash or will have to be shut down. This is a very serious attack because it uses the feature for which the network has been designed, communication, to bring the system down.

No communication medium, from snail mail to the Internet' is secure. In fact, even mobile phones have suffered DoS attacks by being sent a huge number of SMS messages. This attack took place in Germany. Apparently, a person going by the nickname "HSE" wrote a Visual Basic 5.0 program that used a public SMS service (of which there are a lot on the Net these days)' to flood mobile phones thus rendering them incapable of receiving "real" messages since there was no more memory on the phone.

Realizing the difficulties of detection and prosecution, it is no wonder that computer security focuses almost entirely on prevention or protection. The problem is that for prevention mechanisms to work in the absence of detection and prosecution they must be perfect and this is far from being true.

Access Control

For computer security, therefore, the main concepts are **confidentiality, integrity,** and **availability**. These are all different ways of thinking about access control.

Confidentiality

Confidentially refers to controlling who can access and read what data. There has been a lot of research on this particular area due to its importance to the military. Military documents all have security levels assigned to them from "Confidential" to "Top Secret" and other variations. All this structure is in place to ensure that only authorized persons read potentially very important information. The two most important tasks at this level are identification and authentication, two issues which we will come back to repeatedly in this chapter.

Integrity

Integrity has to do with ensuring that data is in the appropriate state, in other words ensuring that it is as the last authorized person left it. As confidentiality has to do with who can see what data, integrity has to do with who can write what on which data. Nevertheless, because integrity is closely linked with authentication and authorization, and those terms are more tangible and better understood, work has focused on them. If the system can reasonably guarantee that unauthorized persons do not access data then that guarantee partly extends to the integrity of the data as well. However, this assumption allows aggravated members of an organization to cause havoc since they do have authorization.

Availability

Availability is once more a difficult term to nail down since it is about guaranteeing that access to data is possible at all times. Of course, this has security implications in terms of having data completely erased from your system and therefore no longer available but it also partially makes reference to the need of computer systems to be safely built so that they do not crash or produce bugs that make access to data impossible. After all if you are not able to access the data, even though it is there and nobody else has seen it, it is just as bad as having the data erased maliciously.

In more concrete terms however availability refers to such things as preventing DoS attacks or preventing attackers from blocking access to data physically or digitally.

Security in the Context of Mobile Devices

We have talked about security requirements for computer systems in general by introducing the notions of confidentiality, integrity, and availability of data. As we mentioned, all these issues boil down to controlling access to the data. Access control is a multi-faceted problem, which refers not only to such things as using passwords and digital signatures but also to physical access to machines.

Who can enter a building? How can we make sure that the people in the vicinity of a machine are trusted? Could they look at the screen and get enough information to break in? When thinking in terms of desktop computers such problems are somewhat banal. By now people can build relatively secure buildings that can reasonably ensure that only authorized personnel is in the building.

The USA army even has terms such as Secure Compartmented Information Facilities, (SCIFs). These facilities not only protect physical access to machines but also prevent intrusion via side channels such as reading the radiation emitted from cryptographic equipment, or the power and data lines or computer screen. It may seem far-fetched but it is possible to gain valuable information simply by measuring power emission. To put it simply the more power emitted might mean more computation and that information could lead to a hunch about the cryptographic algorithms used, etc.etc. Yes, unfortunately people do have the time, the money, and the motivation to research such ways of attacks.

Nevertheless, when the devices are stationary physical access can be controlled more easily. The only physical protection mobile devices have, on the other hand, is the person who carries them and we humans get distracted ever so easily. In the past couple of years we have all been amused by stories of military and governmental officials losing their laptops, carrying sensitive state information in public places. Often the thief had no idea of the status of the owner of the laptop and their only motive was to make some quick money. In more sinister cases, however, the identity of the owner may have been the reason for the theft.

Access Control Problems and Solutions

All these issues create a completely new area of access control problems that portable devices need to deal with. How can we prevent access to the device by unauthorized people, even when they have the device sitting in front of them and they are allowed to pick it apart?

To tackle this problem, mobile phones have Personal Identification Numbers (PINs) that the user must enter in order to activate the phone. If the wrong PIN is entered three times then the phone is blocked. The only way for the user to unblock it is to enter another number called a Personal Unblocking Key (PUK) that is usually provided with the phone manual. The idea is that if a legitimate user forgets the PIN they may remember the PUK or they could return home and find it. The only thing this prevents, however, is running up a high phone bill. The thief can always change SIM cards and use the physical device.

For laptops, other solutions have been developed recently. There are programs that can be embedded stealthily on a laptop's hard disk that check in with a control center whenever the machine is plugged into a phone line. If the owner of the laptop calls the control center and informs them that the laptop has been stolen the next time it is plugged into a phone socket it will call the center in the background and provide information such as the phone number, IP address, etc. etc.

This can eventually lead to tracing the thief. Amazingly enough this scheme actually worked a few times! However, the problems are obvious. As soon as thieves catch up, they will quickly find a work-around such as removing the hard disk and doing a complete wipe of it. In the future solutions will probably include some form of biometric authentication via a thumb or a retina scan. For the time being, however, PINs and PUKs are the best we have. As we will see later on, even the latest recommendations for secure mobile transactions use them.

Digital Attacks on Wireless Networks

We have discussed the security problems posed by mobile devices because of the much higher probability that they might get lost. That is not the only difference from stationary devices, however. The second problem has more to do with purely digital attacks and is a result of the structure of wireless networks. Such networks are composed of cells (that's where the term cell-phone comes from).

Each cell has a receiver-transmitter antenna that covers a certain geographic area and is able to cater for a number of mobile phones by dividing the available bandwidth among the phones currently in the cell. When a mobile phone enters a cell, it registers itself with that cell and that information propagates to central posts, which will take care of routing calls to the appropriate cells.

In fact, this is the basis of a certain class of solutions to localization services. Mobile phone companies are hoping to install enough cells and come up with smart algorithms so as to be able to locate any mobile phone within a few meters and provide geographically relevant information. Now, the security problem arises when someone substitutes a legitimate antenna with one that is able to masquerade to the phone as legitimate but instead of sending information to the appropriate center diverts that information to its own center.

For phone conversations this may not be very exciting and in any case, if the intruders do not forward the call further into the network, it is immediately detected since you will not be able to talk to the person you called. If it is an e-commerce transaction, however, the devious antenna may be able to fool your phone into surrendering all credit card information without you realizing a thing. This is certainly a sophisticated attack and not everyone can pull it off, but if the motivation is there, (in other words if enough transactions take place via wireless networks) then someone is bound to try it.

What this rather extreme example illustrates is that the inherent decentralization of wireless networks makes them more susceptible to attacks since there are more points of failure along the way. By allowing the formation of adhoc networks by peer nodes we must rely on network protocols to control the co-operation of these nodes and we must be able to trust nodes to do what is required.

If one node decides to transmit wrong routing information, this might bring down a whole section of the network. If all the nodes were administered by a single, centralized position then some of the problems would be minimized (although single point-of-failure problems then arise). into however, as users move between countries or geographical areas they use a different service provider, a process known as roaming. The user must be assured that it is safe to transmit information through different service providers.

Another inherent characteristic of mobile networks is the high probability of a disconnection, either because the available bandwidth in a certain cell has been consumed or because the user is going through a tunnel or in to a building. Current e-commerce applications are not equipped to handle transactions that have a very low reliability rating.

At every re-connection the principles must be re-authenticated in order to assure the safety of the transactions. This needs to happen in a manner that is not annoying for the user so that the user does not abandon the process. When we examine the proposition of the mobile transaction group later on we will see that a lot of the focus of the work is on creating a satisfying user experience while at the same time retaining security.

Security in the Context of Mobile Code

In the preceding section we discussed the security issues that arise because of physical mobility. In this section, we will examine the problems that arise due to code mobility. The concepts of code mobility have long been studied in labs and there is extensive published and freely available research on the subject.

One of the earliest papers, comes from the Xerox Palo Alto Research Center and talks about "worm" programs "that span machine boundaries and also replicate themselves on idle machines". Although mobile code is very attractive for building software systems there has always been one great obstacle: security.

Trusting software is a big issue even when the software is bought at a shop and you are sure that it is coming from a reputable manufacturer. Imagine having code that is downloaded and run on your machine from potentially anywhere and without you being aware of it! Nevertheless, this is what we do every day when we move from one web site to another. Applets, JavaScript, ActiveX controls are all types of mobile code that execute on our machines often without us realizing it.

With the introduction of the Java Virtual Machine, a model for code mobility was firmly established as the defacto approach to mobile programs. This model places mobile code into a containment area on an executing environment and it does not allow the code to access local resources unless it has permission from the owner of the containment area. Java calls this containment area a sandbox. Within the sandbox, programs are allowed to play as much as they like, but they cannot directly access resources outside the sandbox.

The following diagram illustrates a simple reference scenario for the generalized case of mobile code. An executing environment has some software running, which is trusted and acts as the virtual machine for mobile code. The virtual machine controls access to the underlying resources by containing the mobile code in a sandbox. This way if the mobile code is malicious it cannot do any direct harm to real resources.

When a program is to move from one executing environment to the other, the virtual machine will package it in some way (for example Java serialization) and send it to the other machine. There the code is unpacked (de-serialized), its credentials verified, and then allowed to execute in the sandbox.

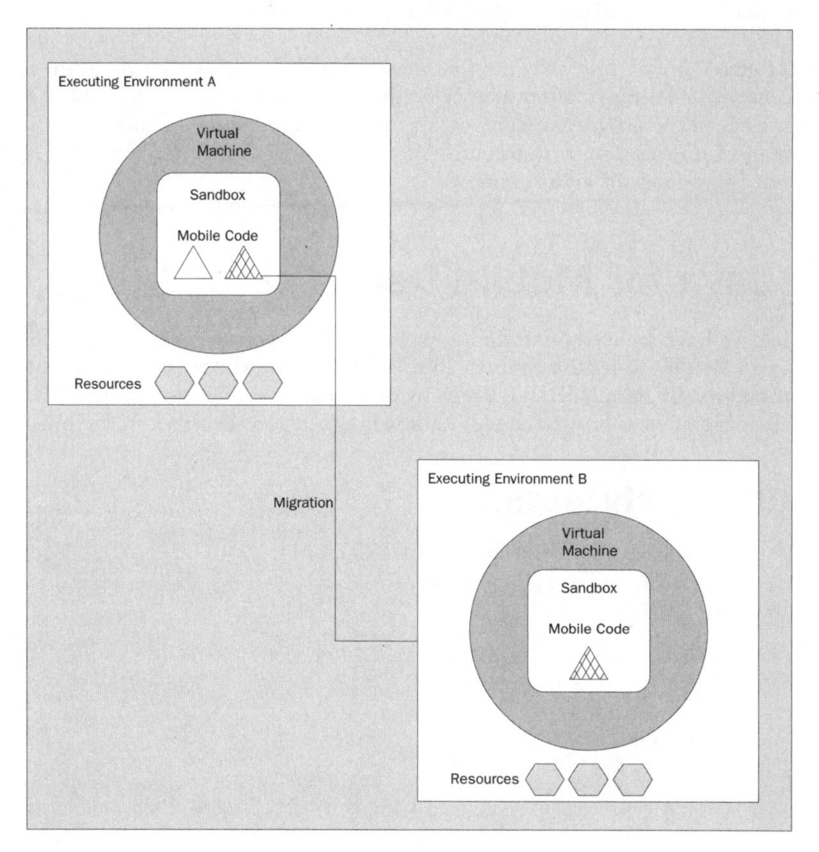

The first type of program to gain widespread acceptance with such a model were JavaApplets. Applets use the virtual machine provided by the web browser and can add such features to a web site as navigation, games, advertisements, etc. etc. Since applets only execute within the sandbox and cannot directly access resources, the user can be sure that even if the applet is malicious it will not cause great harm.

Security holes have been discovered that actually allowed Applets to gain access to the file system, but this problem was due to an implementation issue and this has now been fixed. Unfortunately even if the model works well, the implementation will always have problems. As the saying goes: "In theory, there is no difference between practice and theory. In practice there is."

Applets are a one-way type of mobile code, since they only move from a server to a client and not the other way round. As software evolves, we will see programs that move from one client to another searching out ideal executing environments to perform their task. A good example of these types of programs is the IBM-developed and now open source Aglets, a mobile agent system. The security risks in those cases will increase since the code may have started out from a trusted base but' it has since passed through unreliable clients.

Furthermore, even if the virtual machine and sandbox are bug free, contained mobile code may still be able to cause harm by hogging memory resources. For example, imagine an applet that enters a continuous loop that simply fills the memory with objects. After a while, if the virtual machine does not impose upper limits to memory usage, the machine may crash. Even if a limit is imposed some of the memory is permanently unavailable.

A more sinister attack is a mobile program that masquerades as some other program and convinces the user to surrender sensitive information such as credit card details. There is very little any sandbox can do to stop that since at the programmatic level the code is perfectly legitimate. What is required is an authentication and authorization infrastructure for every program that may come on to your device. We will discuss such issues later on in the chapter.

Securing Java for Mobile Devices

Up to this point, we have focused on outlining problems and security risks posed by mobile devices and mobile code in an attempt to set the context. In this section, we will see how some of the issues we outlined are addressed by the J2ME. We begin by comparing the security of the J2SE to the J2ME to get a clear understanding of what has changed and how it affects the security of programs.

J2SE Security Architecture

The most important parts of the J2SE security architecture are:

- The bytecode verifier
- The class loader
- The security manager
- The access controller

The following figure illustrates how the components fit together:

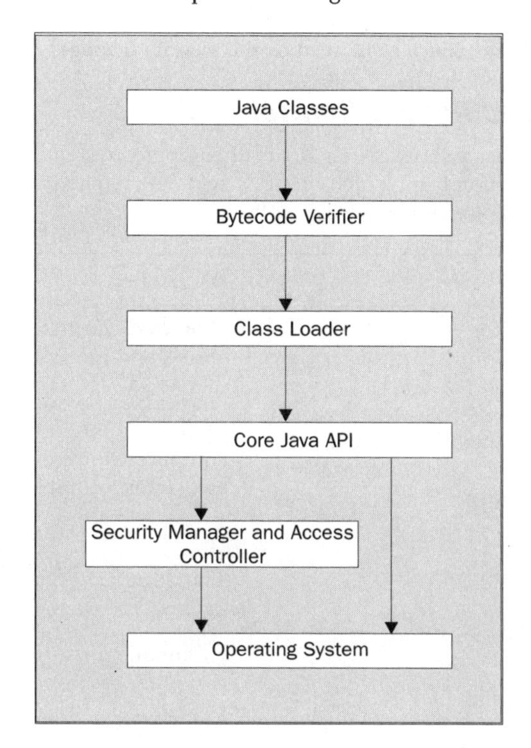

Bytecode Verifier

Java code is compiled to `class` files that are collections of bytecodes, instructions for the Java Virtual Machine. When a `class` file is to be loaded into the JVM the bytecode verifier is responsible for checking that the structure of the file is a valid one that the JVM can accept and execute. This verification process is built into the JVM and it is not accessible to the developer.

This is an important security measure since if a `class` file is tampered with while in transit the bytecode verifier might catch some of the problems and refuse to load the file into the JVM. One of the things that the verifier checks is that there are no stack overflows and underflows, a usual hacker trick for causing a system to crash.

Class Loader

From a security point of view the job of a class loader can be simply described as making sure that malicious code does not interact with benevolent code. This is achieved by establishing namespaces within the virtual machine. Every class has a unique name within a namespace and every class in the JVM is associated with a class loader. Classes loaded by different class loaders belong to different namespaces and cannot interact with each other.

Java 2 (JDK 1.2) introduced a new class loader and access control architecture that provided significant improvements from JDK1.1. Every class loader has a parent, with the root class loader being the bootstrap class loader that loads the Java API classes.

When a class is loaded it is placed into a protection domain. Each protection domain is associated with a certain set of rights of access to the underlying system resources. These rights are described by a security policy, which is an ASCII file, and are enforced by the security manager and the access control mechanisms.

Security Manager and Access Control

Since Java 2, every class has a set of access rights to computer system resources based on the signer of the Java class and the location from which the Java class was fetched. The general elements of this access control architecture are shown on the table below:

Element	Description
Real Computer Resources	Protected computer resources such as network access and file system access.
`Permission` objects	A `Permission` object represents an access right to the computer system resources. Each computer resource has a permission object associated with it.
`ProtectionDomain`	A set of classes that have the same access rights.
`SecurityManager`	The `SecurityManager` is mostly there for backwards compatibility.
Security Policy	Essentially a contract defining rights of use of computer resources between a class and the JVM.
`AccessController`	This handles the bulk of the tasks of ensuring that the security policy is adhered to.
Security Management Tools	JDK1.2 comes with some tools that can be used to create certificates, and security policies, digitally sign Java archives and verify the signatures.

When a class attempts to make use of computer resources the resource management code will invoke the `AccessController`'s `checkPermission()` method. This method takes a `Permission` object as an argument and will throw an `AccessControlException` to indicate that the class does not have the right to access the requested computer resource.

This execution chain is illustrated in the following diagram:

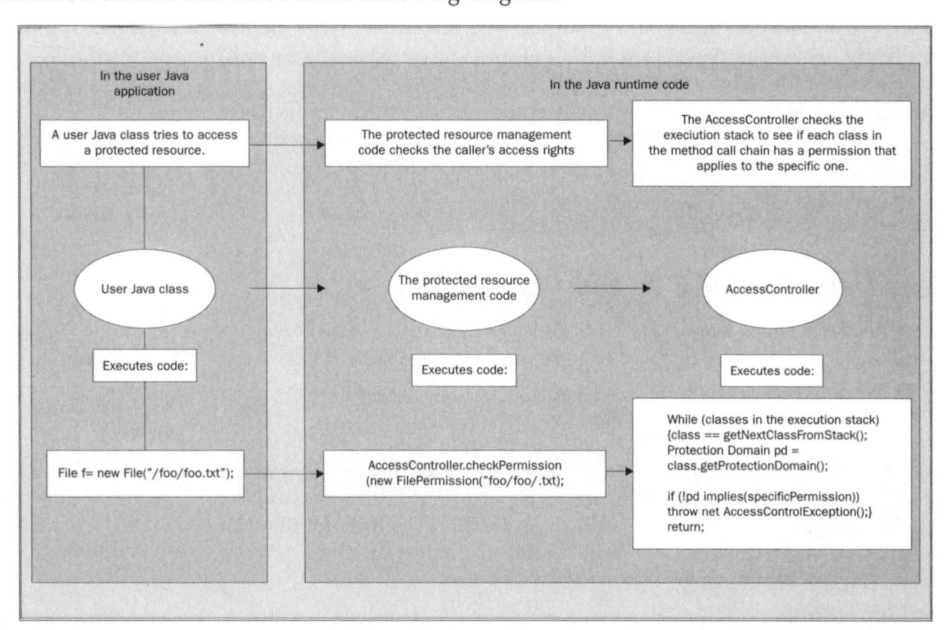

This security architecture allows fine-grained control of what programs are allowed to do when executing in a JVM and thus provides a powerful mechanisms for securing applications. The J2SE also provides encryption and authentication facilities, which allow programs to sign code and authenticate code based on public/private keys.

J2ME Security Architecture

With the J2ME and especially the CLDC Connection Limited Device configuration a lot of cutbacks had to be made in order to cater for the limited capabilities of mobile devices. We will first look at the changes from the J2SE to the CDC Connected Device Configuration and then focus on the CLDC and the MIDP.

Changes from J2SE to the CDC

The CDC requires a fully completed Java Virtual Machine and as such coul,d via the addition of profiles, cater for what is now the entire Java 2 Platform' Standard Edition API. As a configuration, the CDC defines `java.security` and `java.security.cert` packages, which are subsets of the corresponding packages in the J2SE. The CDC **does not** define:

- `java.security.acl`
  These are the classes and interfaces that have been superseded by the `java.security` package.

- `java.security.interfaces`
  This package provides interfaces for generating RSA (Rivest, Shamir and Adleman Assymetric Cipher algorithm) and DSA (Digital Signature Algorithm) keys.

- `java.security.spec`
  Classes and interfaces for key specifications and algorithm parameter specifications.

By including some of the basic security packages into the CDC, the programmer can have fine-grained control over application access to the underlying resources. What cannot, at the time of writing, be done is gain ready access to cryptographic functions as those packages have been removed.

The implication is that if you wish to write applications that might need to send credit card information over an unreliable network the only way to implement the required cryptographic functionality is to do it yourself. However it is not realistic or reliable to expect every application to implement its own cryptographic functions, so it is very probable that as profiles are developed for the CDC we will see optimized cryptographic packages coming in.

Changes from the J2SE to the CLDC

While with the CDC it was a case of limiting some of the packages included in the configuration but keeping the same security architecture for the JVM, with the CLDC everything has been significantly cut down.

The reasons for this are quite clear: on the one hand CLDC devices have severely limited capabilities and on the other hand the JVM was designed to be as simple as possible so as to avoid faulty or malicious code that could crash the whole device. It is interesting to note that while we are perfectly OK with a PC crashing we would not be very understanding if a mobile phone crashed, since our expectations are different.

Classfile Verification

We have seen how for the J2SE `.class` file verification happened at the moment when the class was loaded into a JVM. For the CLDC this process had to be optimized in order to cut down on performance costs. The process has been divided into two phases: a pre-verification phase and an in-device verification. The basic difference is that the pre-verification phase makes changes to the structure of the Java `class` file in order to make in-device verification simpler and more efficient.

In-device verification happens at the native-code level and not via class loaders as for the J2SE. This significantly reduces risks of a security breach at that level although it is still to be seen if someone could manipulate the pre-verification process to fool the in-device verification into loading a malicious class.

No Native Function Support

The CLDC does not provide for JNI (Java Native Interface) support, which means that an application developer cannot define native functions and access them via a Java program. Although at first glance this may look like a limitation of the CLDC, device manufacturers consider it a feature since it makes it even more difficult to write code that could crash the device.

Reflection

The CLDC does not support reflection, so the application programmer cannot write code that attempts to explore other classes. The reason behind this is again a combination of performance considerations along with a wish to make sure that CLDC applications are robust and cannot crash a whole device.

Security in the MID Profile

The MID profile makes no changes to the security architecture of the CLDC. It also does not provide any additional classes for secure network connection. As the specification states it relies on the underlying network support of the micro browser on the device to provide secure end-to-end connectivity.

End-to-end security refers to a secure path all the way from the sender to the receiver of the information.

Security for Network Communication

In the previous section we have seen how the J2ME with the CLDC and MIDP provide some security for Java applications by taking a very minimalist approach and not allowing applications to do anything that may harm the underlying device. In this section we will see what support there is for secure network communications and ultimately e-commerce transactions.

This support does not come from the Java programming language but from the underlying wireless communication protocols and more specifically from the Wireless Application Protocol (WAP).

There are three important tasks to be achieved in order to provide reasonable security for network transactions:

- Identify the parties at each end of the transaction
- Provide them with the appropriate authorization level to perform the transaction
- Secure the information transmitted over the communication channel

These tasks are accomplished via different cryptographic tools, which we will examine in the following sections.

Public Key Infrastructure

Public Key Infrastructure (PKI) is a system designed to support the safe authentication and authorization of documents via digital signatures and encryption. In order to understand PKI we must make a small digression into general encryption systems. There are two types of encryption systems: **single-key** and **dual-key** (or symmetric and asymmetric dual-key encryption). With single-key encryption systems the same key encrypts and decrypts a message. The sender and receiver, therefore, will have to share this key and make sure that it does not fall into the wrong hands.

Obviously, for the Internet a single-key encryption mechanism is highly insecure since the probability that the key will be intercepted while it is being sent from the creator to the user is high.

With dual-key encryption we have different keys for encrypting and decrypting a message. If a document is encrypted with one key it can be decrypted with the other and vice-versa. As an added bonus, if one key falls into the wrong hands it cannot be used to guess the other.

PKI uses dual-key cryptography. One of the keys is kept private and it is used to encrypt the message while the other key is made public and it is used to decrypt messages. The advantages of this system are obvious. If Alice wants to send a message to Bob (Alice and Bob are cryptographers' favorite people, cunningly extended from A and B), Alice needs to send Bob her public key first, or register her public key with a trusted third party such as VeriSign.

Alice can then encrypt her message and send it to Bob. Bob will use the public key to decrypt the message and if successful he can be reasonably sure that Alice encrypted the message since only Alice has the right key. As you can see, this system is useful for identifying who sent the message but not for hiding the information contained in that message since anyone with Alice's public key would be able to decrypt the message.

This is the basis for **digital signatures**. In fact there is no point in encrypting the whole message, since what we are after is not hiding the content of a message but proving the identity of the creator of the message. What is encrypted is a CRC of the message along with some information about the signer. What digital signatures solve is the problem of key distribution.

Keys can be public and can be used to gain a certain degree of confidence about who the sender of the document is. Of course, this only works as long as the private keys are kept truly private, the assumption being that since private keys do not need to move about keeping them private is easier. If someone breaks into Alice's machine and steals the private key then that person could masquerade as Alice.

In order to create trust in such an environment about the validity of digital signatures there need to be trusted third parties that will vouch for the validity of digital signatures. These third parties produce certificates that private key holders can provide as proof of their trustworthiness. These certificates are public keys that are digitally signed by the third party, which is usually called a Certificate Authority (CA).

Digital signatures and Certificate Authorities provide the building blocks for a Public Key Infrastructure. Of course, PKI is not the answer to all problems but is does provide some progress towards authenticating and authorizing users. In order to have an effective PKI a lot of effort needs to be spent to educate users in protecting their private keys, in distributing and maintaining the private keys and in distributing and maintaining public directories of private keys.

As a final note, I should point out that in some cases, symmetric and asymmetric cryptography could be combined to provide improved performance. For example, in SSL, asymmetric encryption is used to swap symmetric keys securely and from there on symmetric encyption (which is more cost effective), is used. Furthermore, while we focused on ensuring who the sender is, the same technique can be used for protection. In this case Bob could use Alice's public key to encrypt the data that can then only be decrypted by Alice's private key.

Securing the Communication Channels

PKI answers the authentication and authorization problem but only partly the problem of encrypting data so that it can pass through an unreliable network. These tasks have been delegated to the Wireless Transport Security (WLTS) Layer in the WAP stack (for mobile phones using WAP). Although it is true that certain mobile devices can use TSL1.0 (Transport Layer Security, see http://www.ietf.org/rfc/rfc2246.txt for more information)here the focus is placed on the efforts in the field of mobile phones that are essentially more limited devices.

The WTLS layer provides compression, encryption, and authentication (all optional services). The functionality of the WTLS layer is similar to TLS 1.0 used in TCP/IP networks. A number of encryption and authentication mechanisms are available, providing the developer with a certain amount of freedom as to what security mechanisms to implement.

WAP Security Problems

Perhaps the most salient criticism raised against WAP 1.1 is that there was a gap in it's security model. This was due to the fact that while the WTLS protocol does provide encryption, the "endpoint" of the encrypted data is the WAP gateway. Once data is at the gateway it has to be decrypted so as to understand what the request from the mobile device is. If the request is for a secure transaction with a web server, this means that the data needs to be encrypted once more, using SSL for TCP/IP communication, and sent to the web server. The web server's response will once more have to be decrypted and encrypted back to WTLS in order to be sent back to the mobile device.

If the gateway is a machine that you can trust, that is all very well. The problem arises when the content provider does not trust the gateway. This is a very common situation since gateways are often supplied by a third party, and as the user moves so might the gateway the user is going through. In summary this is a classic "person in the middle" problem:

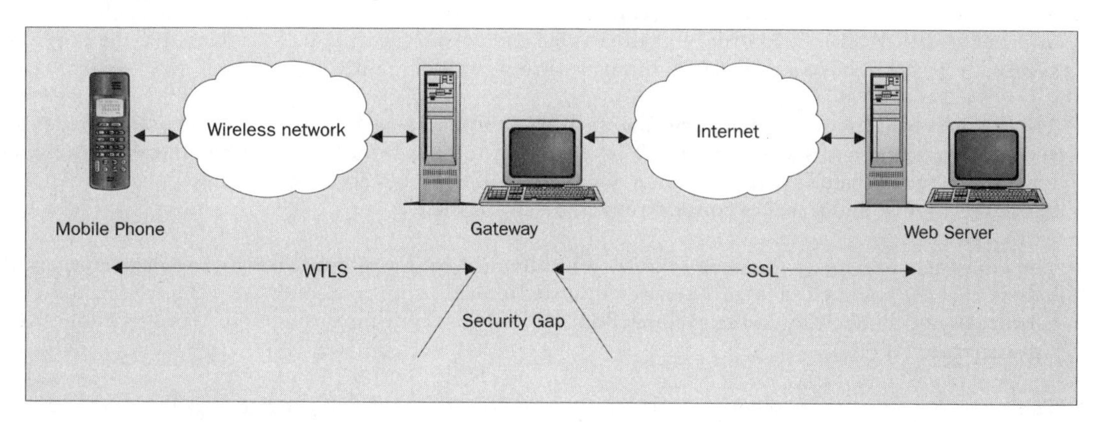

Possible Solutions

The solutions available before WAP 1.2 included:

- Making the origin (web) server the gateway so no encoding/decoding issues occur, since it all happens within the same machine or domain
- Providing content only through your own gateway

Either of theses two measures includes, among others, the following considerations:

- Making sure that the gateway never stores decrypted data on secondary storage
- Ensuring that decrypted data is removed from memory as quickly as possible
- Physical security for the gateway with access only from authorized administrators
- Limited and very secure remote access control

A third solution is:

- Buying software to be installed on the gateway of your wireless service provider to decrease the possibility of a security infringement

Such a solution usually involves making sure that the encryption/decryption processes happen within a single process, as quickly as possible (milliseconds) and ensuring that it is not possible to copy the data to persistent storage. Clearly, this can only work if the gateway owner actually agrees to install the software, something that can not always be enforced.

Security Considerations

The problem with all such solutions is that they exhibit the classic symptoms of a hack. They are not robust and they do not scale. These problems need to be solved for effective m-commerce today because there are a lot of devices and gateways out there with WAP 1.1 browsers. As we have mentioned before it is important to consider what audience a WAP application will reach in order to develop an efficient solution.

What is required is a mechanism that will allow the Wireless Application Environment (WAE) agent to encrypt data and have it delivered encrypted all the way to the content provider. Such a solution is introduced with WAP 1.3. In order to achieve this end-to-end security, it is assumed that the provider of secure services will have a gateway within its secure service domain.

The scenario is as follows: a user is surfing through an insecure gateway and at some point decides to access and make use of secure services. They will inform the gateway of this intent and will supply the URL of the secure gateway. A navigation document will be obtained from the insecure gateway pointing to the secure one, and a secure connection will be established.

The above scenario raises the issue of trust. Why should I trust a certain gateway and its owner, and believe that the transaction is truly secure? In order to deal with this issue WAP 1.2 introduced digital certificates and Public Key Infrastructure (PKI) security mechanisms through the Wireless Public Key Infrastructure (WPKI).

There are, however, other security considerations to do with other deficiencies of WTLS and with the bearer technologies. For an overview of such issues you can reference the following paper: Markku-Juhani Saarinen," *Attacks Against The WAP WTLS Protocol*" University of Jyvaskyla, 1999.

Mobile Electronic Transactions Initiative

The WAP Standard was designed in an effort to provide a set of protocols and APIs through which one could develop applications for mobile phones, including e-commerce applications. The WTLS layer was aimed to provide the required secure channel with the use of PKI providing the necessary authentication and authorization mechanisms.

Since then, however, it has become apparent that it is also necessary to standardize the procedures that users will have to go through to perform electronic transactions. Good procedures through a consistent user interface would decrease the possibility of mistakes and increase the user's trust in the device.

With such issues in mind, Ericsson, Nokia, and Motorola founded the Mobile Electronic Transactions initiative. The goal of MeT is to provide a common framework for mobile e-commerce that embraces current technologies such as WTLS, WPKI, and Bluetooth. The aim is to produce a consistent framework of procedures through which secure transactions should take place. Since its initial foundation Panasonic, Siemens and Sony have also joined the group.

The first series of specifications were published on 21st February 2001. These specification focus on standardizing the user experience with secure transactions over a series of different scenarios ranging from account-based payment, event ticketing, retail shopping, SET wallet, WAP Banking and WAP Shopping. It is hoped that through the provision of these standardized interfaces users will become as comfortable with mobile phones and electronic transactions as they are with using ATMs and credit cards.

The web site of the MeT initiative is http// www.mobiletransaction.org.

Summary

In this chapter we have briefly discussed the security issues raised by mobile devices due to physical and code mobility and have examined how Java attempts to tackle there problems. We have also discussed securing network transactions through PKI and encrypted channels based on WTLS. Finally, we discussed the latest initiative by device manufacturers aimed at providing a unique consistent network for mobile transactions.

As mentioned in the introduction, security is an ongoing issue and solutions are never easy. Even with the latest development, there still exists many security holes for the determined attacker. However, it is becoming increasingly harder and less cost effective to deploy such attacks.

PROFESSIONAL JAVA MOBILE
TOOL SELECTION
WEB TABLET
MOBILE PHONE
MP3 PLAYER
DIGITAL CAMERA
PALMTOP
LAPTOP
NOW SELECTED
PALMTOP

17

Designing J2ME Software

Java has enabled a new generation of software systems to reach a larger audience via the World Wide Web. Rebooting a server, or asking the local expert user for advice, is no longer an appropriate usage model. Issues that were easily resolved at a departmental level with client-server systems cannot be handled in this new deployment environment. J2ME extends this reach even further beyond the fixed-line Internet, to a fluid and chaotic deployed base. Our systems support and user-training models, with current indifferent design and loose architectural practices, might not scale to the new mobile Internet.

Java has already extended the reach of our software systems across the Internet. As these applications become mobile and the users operate the system from remote locations outside the immediate reach of support and training services, support, advice and training become harder to supply and obtain. Consequently, usability and robustness must be designed into mobile systems from the ground up.

What is Design?

Software design is an activity that should deliver certain basic benefits to a software system. All systems can be said to have a design, regardless of whether any formal, or conscious, design decisions were considered. However, in this chapter, design is treated as a specific and separate phase in the development process.

What is design and what is good design? One answer to this question is that a system is well designed if a user can:

- Figure out what the system does and do it easily
- Be able to tell what is going on as the system works
- Understand the output from the system
- Recover from any errors or exceptions
- Undo actions easily

Sounds simple enough, but the software designer must create systems that cater for a wide range of, often conflicting, requirements and priorities in order to satisfy the following 'customers':

- Sponsors/champions
- Project management
- Developers
- Sales staff
- The purchasing agent
- End-users
- Support and maintenance staff

Sponsors

The business sponsor (also known as the champion) behind the development project is a critical group or individual, without whom the project cannot succeed or even commence. The sponsor champions the project within the organization planning to use the software, raising the profile of the project at the right management level and generally fighting the corner of the project team in order to obtain credibility and resources.

The sponsors of the project have either direct or indirect control over the project's budget. They define the priorities of the project and its business case. Furthermore, the sponsor will have a vitally important role in determining purchasing and acceptance criteria for the final product with the purchasing agent and end-users.

The designer must work closely with the software architect, project manager, sales team (in the case of an external client), and senior developers, in order to ensure that the software's design is communicated effectively to the sponsor or sponsors and that the specific requirements of this important group, or individual, are visibly addressed.

Project Management

The project management team will typically be concerned with the following issues:

- Resource availability – of necessary skill sets and materials
- Cost justification of expenditure on any new infrastructure and skill development/training
- Risk management
- Time to market
- How easy the application will be to deploy and support

When communicating the design to this group, the focus must be to demonstrate that innovation and complexity in the design can be both successfully managed and turned into a profit and/or a value-generating activity. In this chapter, we will consider these aspects of design in greater depth.

Developers

Developers want to program using design plans that give a clear sense of purpose, both technically and functionally. In addition, the design should contain plans that are easy to implement. If this is not possible, then the design should at least be as clear as possible in areas such as technical risk.

Clarity and communication of intention are the key factors in creating design plans from which developers can work. Implementation details should be clearly explained in order to enable the programmers to investigate and implement software in accordance with the design.

The developers will be responsible for implementing the designs and for the testing of the system throughout the development process against the system definition provided by the designer.

Sales Staff

Where an application is being sold to another company to generate revenue, the sales team need to be given clear pointers on what makes the software attractive to clients. The software designer must be able to demonstrate the value of both underlying technical and user-interface decisions. Software that meets customer expectations must be both created and demonstrable, in order to keep the market happy.

The Purchasing Agent

In a perfect world the purchasing agent and end user would have motivations that are perfectly aligned. However, in reality the end user seldom has the same influence over the process as the person who has the budgetary control to authorize the purchase. The software designer should be aware of this issue and plan accordingly. Purchasing agents are likely to be highly sensitive to such factors as cost, appearance, feature list, and the prestige value of owning the software.

The designer should be conscious that certain technology choices can influence whether or not a purchase is made. For example, dot com companies are more likely to buy Java applications than those constructed in COBOL, no matter how well the COBOL software may work.

End-users

End-users are frequently given new software and expected to be instantly more productive than they were before. Using new software can be a stressful experience for many people and the designer must make the adoption and learning experience as smooth as possible.

The user group is, of course, the key 'customer' for the designer. While other 'customers' may be instrumental in ensuring the adoption of the software, it is unlikely that it will continue to be used if the end-users encounter serious problems with the final product. Usability, robustness, and speed must therefore be designed into the software, to enable the end-user to assimilate rapidly the skills and techniques they, will need to use the system.

The bulk of the designer's work in user interface design and in the non-functional aspects of the system (such as robustness, speed and scalability) are critical factors in the real success of a design.

Support and Maintenance Staff

The designer must create software designs that can be easily maintained and, if the need arises, easily extended at a later date. Useful diagnostics, ease of maintenance, and good annotation are essential and these features must be the result of conscious and transparent design choices by the software designer.

System Design versus System Architecture

Architecture and design are often confused but there are stark differences between the two. Although an architect is responsible for designing quality software, the architect must, in addition, appreciate a great deal more than a designer about the software development process and the technology environment into which the software will be deployed.

Building a House, Building a City

Software design and systems architecture differ in many respects. A design can stand alone and relates to the performance of a distinct and definable set of actions. Architecture relates a single system's design to the design and implementation of similar and related systems within a technical market or client organization.

The difference between designing a single standalone device-based application and creating a software architecture for an effective infrastructure of mobile applications, can be likened to the difference between building a single house and building a city:

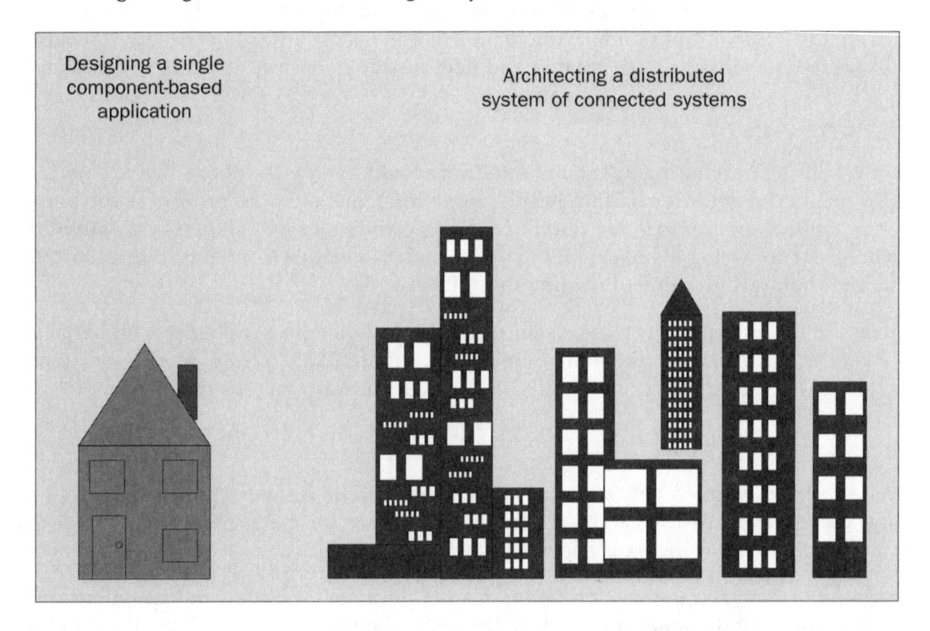

Architecture ensures that common metaphors are used between systems and that the implementation of a design fits with existing systems within the client's infrastructure. Architecture, in its relation to design, ensures that designs:

- Meet a continuity of intention
- Are designed to a greater plan
- Have implementations that can be planned and cost-estimated

Managing Complexity

Creating software is an exercise in managing complexity in both business and technical domains; the divergent needs of the 'customers' outlined above, lead to complex requirement scenarios that intertwine and may conflict. These requirements must be met and managed by the system design. Consequently, design is an **essential** activity for the production of quality software.

There are two major elements to software system design:

- User-centered design
- Infrastructure and system-level design

To understand design completely, we must therefore approach the subject from both a user-centered and a technical perspective.

User-Centered Design

No matter how good the design and its implementation, there is always a user to misunderstand things and 'mess things up'. As we have already discussed, user-centered design should:

- Increase the ease with which the user can determine the purpose and function of the system at a given instant
- Make the current state of the system apparent to the user
- Make visible the result of any user-action
- Allow graceful recovery from errors and exceptions
- Allow the user to easily undo any action

In his groundbreaking book, "The Design of Everyday Things" (The MIT Press, 1998 ISBN 0-262640-37-6), Donald Norman developed a set of principles for transforming difficult tasks into simple ones, by applying a user-centered design. His book is a must-read for anyone involved in software systems design. All of Norman's basic principles can be applied, especially in the J2ME environment, resulting in impressive improvements in the interface and usability.

Norman's seven principles for reducing complexity in design can be briefly summarized as follows:

- Use familiar, real-world, metaphors to communicate the intentions and functions of the system.
- Simplify the structure and performance of tasks.
- Make functionality visible and transparent.
- Map the system's conceptual models to easily understandable and comprehensible 'natural' models.
- Exploit constraints to limit user options and make choices obvious and clear.
- Design for error, make it easy to reverse operations and make operations and action success clear.
- If a natural metaphor for the system's functionality cannot be found, develop a single standardized and consistently used conceptual model for using the system. In other words: if it cannot be made explicitly clear, at least make it standard.
- Communicate intention and create usability.

Systems are more readily digested and utilized if the users can easily formulate and maintain their own conceptual model of the system. The user's conceptual model is developed based on their experience of successfully using the system.

It is important for designers of computing systems to realize that the design of the system is the only mechanism by which they can communicate the intentions and functionality of the system to the end users. The system must be capable of both performing intended actions and communicating the underlying intention of the system, in a single, easily understandable metaphor:

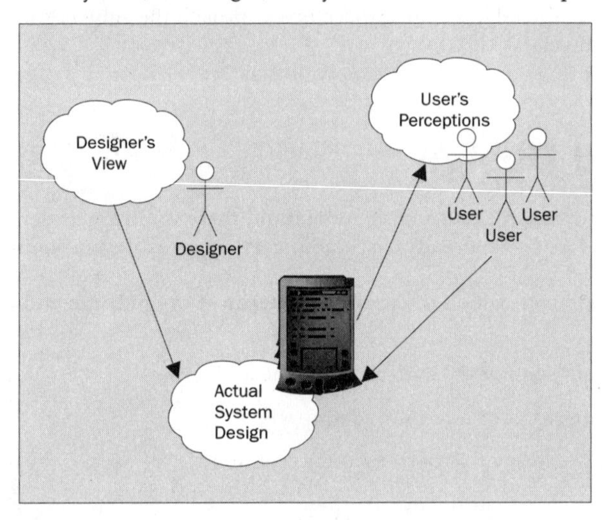

Designing systems capable of clear communication of intent that are simple and effective to use by both new and experienced users is a tall-order for the designer. It requires a significant amount of experience, intuition, and iterative refinement.

Short and Long-term Memory and Conceptual Models

The average user can maintain five unrelated options in their mind at any one time. In a complex software system, the design must consider this and break the functionality, workflow, and options into easily digestible chunks.

The role of design, for mobile technology in particular, should be to make common tasks appropriate to people on the move. This means making them quick and simple to achieve. Mobile technology can facilitate and simplify by:

- Automating simple and repetitive tasks
- Restructuring the workflow of tasks to take advantage of the technology
- Providing consistency, clues, and structure, making it easier for the user to complete their task
- Providing visibility, transparency, and reporting on the progress of particular tasks

Click Depth and Small Screens

Click depth refers to the number of user clicks and choices that must be negotiated to reach a particular function or service in the system. A common piece of advice on the fixed-line World Wide Web, is that the user should be able to navigate to any particular piece of information or functionality, within three clicks.

This tends to lead to highly flattened views of a system, with pages each containing a multitude of data items and functional options. Providing the user with an overwhelming set of options and choices at any one time can cause severe difficulty in any system – having more than five options at any one point is probably too much. The difficulty with this approach is further exacerbated by the small screen size and reduced navigation capacity and ease of input on the typical mobile or consumer device.

Smaller device-based applications must allow the user to choose the functionality they require from the business functions available; preconceptions about layout, structure, and content must be kept to a minimum. This is especially true for devices with limited bandwidth and computational resources, where redundant functionality will cost the user both unnecessary time delays and expense. This is the challenge that must be met in order to accelerate the acceptance of J2ME-based devices. It will require a new way of thinking about the presentation of information and the input of data through non-computer terminal devices.

Designing for Smaller Screens

Even a basic computer monitor can display 800x600 pixels and a single line of text can accommodate 100 or more readable characters. We also have color and a wide variety of different styles, typefaces, and graphics at our disposal in order to make the browsing experience both easy and pleasurable for the user.

The screen size of a smart phone may be in the region of 80x60 pixels, smaller by a factor of ten, and the screen size of an average PDA device, for example, the Palm V Organiser, is 160x160 pixels. The designer of applications written in J2ME therefore has far less screen real estate at their disposal.

Initially, it may be easy to design J2ME interfaces that are driven by the structure of the basic functional requirements. However, to avoid implementing rather clunky UIs, designers must realize that to design device-based applications successfully, they must rethink traditional e-commerce and web-based application design completely. It is necessary to abandon all preconceptions about layout and functionality in order to take advantage of these devices and avoid their intrinsic user interface limitations.

The key areas to target are:

- Reduction of the amount of user input – it tends to be slow and cumbersome even on a PDA
- Simplification of the menu structure
- Giving easy access to the main menu, to stop the user from feeling lost
- Consistent naming and/or graphical-positioning of user options
- Reassuring the user with visual clues (such as the synchronization of menu options, titles and purpose of screens)
- Providing the ability to undo any choice easily
- Giving the user feedback on the successful execution of their actions

Navigation is more difficult with a small screen, so the menus have to be extremely easy to navigate and the structures as shallow as possible. Clicks/selections must be kept to an absolute minimum; simplicity and brevity are the name of the game.

Making a Mobile System Explorable

Encouraging experimentation by the users can increase and improve the usage of the system dramatically. User exploration of the system can have the following benefits:

- Improved user confidence with increased product usage
- The user can experiment with potentially useful functionality, confident that they will not run into difficulties
- As the user explores the system, loyalty and user-buy-in are promoted as the user learns to utilize the system to solve they particular need or problem

Designs for mobile and device systems, particularly those with graphical user interfaces, should be explorable. The designer should strive to make the system explorable by promoting a sensible and self-referential user interface. If a natural analogy to explain the system options and functions cannot be found, it is best to standardize the method of presenting the functionality. This standardization ensures that the system is learned as quickly as possible, and conforms to expectations from similar systems where possible. It also allows new functionality to be explored in a familiar environment.

Design to Cope with Error

Software design for mobile and device-based systems must be particularly robust, capable of dealing with exceptions gracefully and recovering operations as automatically as possible.

As mentioned at the start of the chapter, whereas a helpdesk can usually be contacted when a user experiences problems on a PC, for a user of a mobile device, expert help will almost certainly not be as readily available. Robustness in device technology must consequently meet high user expectations. Failures and exceptions must be caught and the effects handled by the programmers therefore a higher degree of automatic recovery from errors is desirable. The design of device-based software must consequently incorporate safety measures, assertion, and parameter checks at a code level.

Reverse Usability

Just as particular care should be taken in achieving the peak in usability, enabling the user to comprehend and utilize complex software systems, hard design choices have to be made both to supply the necessary functionality and to reduce user errors.

For example, in the design of a contact-management software system, users may be given the ability to add, amend, and remove contact details. A busy salesperson on the move, using the mobile interface to this system via his PDA, will need to be able to add and amend contact details easily and efficiently while paying only partial attention to what he is doing, which the designer must consider when designing for this particular scenario.

While the ability to remove contact details from the database is clearly a necessary function, it also needs to be accident-friendly, providing obvious and jolting visual clues to warn the user that what he/she is attempting may not be want he/she wants to achieve. A fine balance needs to be struck: the warnings must be sufficiently clear to alert the user to the possibility that he/she might be about to undertake an action that will be regretted.

At the same time, these warnings must not hinder the actions of a user who does want to undertake that process; a busy user is unlikely to respond well to being forced to confirm an intention repeatedly. It is therefore necessary to make accidents difficult whilst still maintaining the design's assistance to deliberate intention. Most importantly, the ability to reinstate details easily is essential, if the warnings were ignored and an unintentional deletion has taken place.

The guideline is that dangerous user actions should be flagged with jolting and informative visual clues to avoid erroneous user actions, while maintaining system usability for those who actually want to perform such tasks. Potentially 'dangerous' actions such as deletions should be easily undone or reversed (even after several other actions).

Infrastructure and System-level Design

System design and constructing software using programming techniques and tools is just as important as user-centered design and requires a different set of skills. Developing applications suitable for the J2ME market place requires an intimate knowledge of the Java programming language, a feeling for efficiency, and inevitably a knowledge and understanding of network and distributed technologies.

However, J2ME does make it easier to develop applications that will port to a number of device-based platforms and Java provides a number of class libraries that provide useful and easily understandable abstractions for constructing network-aware applications.

A Programmer's Perspective on J2ME

The developer of Java applications for devices must learn three types of programming techniques:

- Construction – basic language and library utilization skills
- Techniques and tricks – an advanced understanding of the environment
- Elaboration – the usage of patterns of design and code refactoring techniques

Construction of J2ME-based Applications

J2ME allows for elegant object-oriented software design. Using Java to program on the device enables access to all the traditional benefits of Java; we explored some of these benefits in Chapter 2 but in case anyone needs a reminder, they are as follows:

- Portability
- Ability to write component-based software to achieve software reuse
- Simplicity of language
- Safety and security
- Availability of developers

Portability

J2ME allows for extensive portability within the constraints of the configurations and profiles for which the code was designed. For example, while applications written for the Personal Profile of the CDC will be easily capable of deployment on the desktop under Java 2 Standard Edition, this will be of limited practical use since the user interface will be too confined to take full advantage of a desktop machine. This dilemma is made even more apparent when comparing porting software between more limited devices.

Portability with J2ME means appropriate portability; in other words, portability between devices conforming to the same configurations and profiles. See Chapter 7 for more details on porting applications.

Techniques and Tricks

Designing and developing software successfully for J2ME requires the knowledge of certain tricks and techniques. These approaches take optimal advantage of the benefits of mobile technology. For example:

- Follow good design patterns and principles when constructing mobile applications; essentially, keep the mobile application's structure flat, navigable and simple, but do invite exploration of the system. Keep the design clean and simple.
- Focus on applications where the use of J2ME and mobile technology provides a clear advantage.
- Treat the mobile channel as one delivery mechanism among many. Isolate the delivery mechanism from the business logic in the application code.
- Look at the overall business plan, consider how the mobile channel will interact, or cause interaction with other client-facing channels and mechanisms (for example the web site, the call center, etc.).

Elaboration – Patterns and Metaphors for Software Design

Designs are rarely unique; even if the particular business problem or mobile channel is being developed for the first time, designers can draw upon a well of experience and learning from their predecessors and their own previous experience.

Beginning component-based development for J2ME-based systems should involve a review of existing patterns of development and design. A strong movement in the last decade towards the formulation of industry and design-domain experience into standard templates of design, called patterns, has advanced the way in which we design software.

Objects provide us with a way to model the artefacts and entities in the business domain. J2ME gives us the ability to create software components that map these objects into more coarsely-grained deployable units of software. Patterns allow us to share the structures within these object and components, in order to engineer timeless and mature system designs successfully, even in new situations.

Introduction to Component-based Software Design

"From a sequence of these individual patterns, whole buildings with the character of nature will form themselves within your thoughts, as easily as sentences."

Christopher Alexander, "The Timeless Way of Building" (Oxford University Press, 1979).

The idea behind component-based software design is a simple one; by combining the objects within your software into coarser-grained entities, which follow the good practice of encapsulating functionality into a separate deployable unit of code, you can achieve better software design in terms of:

- Testing and deploying components separately
- Obtaining better reuse from the component than is possible with finer-grained objects
- Creating practices of design that allow for the rapid integration of components to construct robust software systems

Component-based development grew out of the traditions and methods of object-oriented design as a way of achieving higher levels of functional separation and reuse of functional components of the system. These concepts come out of the 'loose coupling, high cohesion' approach to software construction developed in the 1970s.

For example an 'employee' class may be used separately within the coding of several business systems, but an 'employee selection' component can be inserted without change into a number of projects each using the same interface to retrieve an employee without any knowledge of the underlying storage or retrieval mechanisms.

J2ME systems that must be developed for a large number of potential platforms can benefit from this type of componentization. By separating the software system into a number of complementary but separate units, you can increase the likelihood of reuse as the system and new systems, are developed for new devices and business domains.

Components versus Objects

Software components and objects within a software system are not interchangeable terms. An object represents a single artifact within a software system; it has a unique identity and an associated persistent state. A software component, on the other hand, represents a more coarsely grained system artifact; it encapsulates an entire business process. It is a separate and independent deployable unit of software and has no persistent state.

Perceived Problems with Component-based Systems

It is often argued that the increase in generalization and the additional code and effort needed to achieve modularity and the separation of component responsibility, leads to problems in the following areas for component-based systems:

- Performance
- Security
- Reliability an additional area of concern with regard to component-based systems for mobile Java devices

However, as we shall see, with the benefit of good design and prudent utilization of the strengths of Java on the device and the server, many of these perceived problems can be overcome.

Performance

Performance measurements for a mobile application indicate how the application can make best use of the underlying device technology to extract the maximum in speed, responsiveness and efficiency, with the minimum of power consumption and memory usage. A correctly written piece of Java code will run slightly slower than the same piece of code written in natively compiled languages, such as C or C++, on any specific platform. However, Java has many other advantages; for example, it allows an ease of code construction and portability that reduce the importance of the virtual machine performance.

Performance issues for enterprise systems are more usually solved by high-level design choices (for example the use of load balancing in a distributed computing environment) than by the use of native code to provide performance enhancements. Nevertheless, technologies emerging from Sun and IBM, among others, such as Java **hot-spot** technology and so-called **Just-in-Time** compilers obviate many Java code performance problems on the server, creating close to native compilation execution speeds. Improvements for the mobile Java platform, especially as the device technology develops to become more powerful, will bring similar benefits, in time, for J2ME.

The primary question of performance should not be "how slow is Java compared to C or C++?" but rather "how fast does this system need to run and how much are we willing to sacrifice in the area of system performance, in order to gain in the areas of development productivity and portability?"

Security

The security of systems wholly composed of components is only ever as strong as the weakest link. Finding and securing this weakest link can be problematic, especially if the source code for the components is either not available, or is not familiar to the developer in a complex system. Even worse, applying security fixes to one element in the system, may actually increase the overall security weakness, since other components may be affected in unpredictable ways.

Using component frameworks from known vendors is becoming a common way of alleviating this concern, but it remains a problem that must be assessed for all component-based development projects, especially those making use of third party products.

As J2ME is an industry standard, security problems will be found and fixed at a slightly unsettling, but effective, pace. As open standards come to be used for construction and especially network connectivity of device-oriented systems, the development community will benefit from the existing maturity of security solutions in this area.

Reliability

J2ME as a device-level computing platform lacks the maturity and deployed base of many native tools. However, J2ME is attracting overwhelming industry support, principally from the device manufacturers themselves. J2ME has therefore leveraged many of the lessons learned from the development of device technology and can be expected to mature rapidly as the best minds in the industry baptize it in a myriad of technical situations, extending and developing the platform. This will occur as rapidly as we have seen Java mature on the server-side.

We can also expect J2ME to benefit from its industry wide backing and community-based development programme; innovations and developments are coming to the J2ME platform at a rate that far exceeds the support and development of the competing device-based development platforms.

Device-based Programming

An excellent resource for programmers developing object-oriented component-based systems for devices is "*Doing Hard Time ISBN 0201489375*" (Addison-Wesley, 1999), in which Bruce Powel Douglass outlines the basic properties of a real-time system aimed at device-oriented systems. Many of the principles for embedded and real-time systems that he explores are directly applicable to the world of J2ME.

Considerations for Device-based Systems

The systems targeted by mobile Java software will be primarily aimed at devices, but what is a device-based system? A device-based system will typically be used in more extreme environments, with less backup and support than that present for most wired-network PCs, as discussed earlier in this chapter.

A device-based system will typically have more limitations to overcome than a PC-based system in terms of available computational and network resource. Furthermore, the system may have some real-time elements and the expectations and demands in responsiveness and robustness will almost certainly be higher.

For these reasons, the design is going to be difficult and an essential aspect of designing for device-based systems will be to manage the inherent complexity and avoid introducing additional levels of complication; in other words, the 'Keep it Simple, Stupid' (KISS) principle applies.

The major benefit of Java in a device-programming context is that it facilitates the rapid development of device-based systems, in a way that neither C nor Assembler can match. This ease of coding and rapid development in the construction of device-based systems is the major reason for switching to Java when designing a device-based application. Java allows for a compelling development environment, far in advance of the other tools currently available in this technology space.

Whereas before only those skilled in real-time C and Assembly programming could develop on these device-centered platforms, J2ME opens it up to a huge number of Java developers. Although Java programmers can quickly begin to use the J2ME environment, they should aim to emulate the best practices and lessons learned by their über-geek predecessors.

The essential properties of a device-based application are:

- Timeliness
- Responsiveness
- Predictability, correctness, and robustness
- Fault tolerance and safety

The designer of J2ME software must consider the factors listed above, if they are to create services and applications that are successful in the device market. We shall discuss each of these properties below.

Timeliness

The handling of events and task processing within set timeframes is the essential characteristic of any system with pretensions of having a real-time nature.

A system can be considered 'real-time' if it is one in which the deadlines for responding to events and carrying out actions, meet the performance criteria. Applications must be designed to strictly meet the requirements of performing certain actions at specific times and respond to certain events within specified timescales. Failure to meet these criteria must be considered a serious system error in real-time software.

Feedback to the User and System Responsiveness

Allowing the user to continue using the application while data is saved, or the network is accessed, and providing feedback to the user if the application is genuinely inaccessible, can dramatically improve the user's attitude towards the system's performance. Performing appropriate tasks in the background and preventing the user from experiencing unnecessary delays, is one of the easiest ways to improve both performance and usability.

The speed of applications is as much a matter of perceived responsiveness as actual execution time; keeping the user informed and engaged should be as much a focus of development effort as reducing microseconds in the processing time of complex calculations. This feedback is particularly important in a mobile and device-application context, where users expect rapid responses to their actions and will have little tolerance for unexplained delays and pauses in the software's operation.

Predictability, Correctness, and Robustness

Predictability, correctness, and robustness are key aspects of device-based systems. Users will not tolerate systems that exhibit behavior that is erratic, are prone to failures, damage data, or are perceived to behave in an irrational manner.

The latency in network connectivity, problems with wireless coverage and the current restrictions in bandwidth in the mobile wireless networks could easily cause applications to behave in a way that would appear to be illogical, or at best, inconsistent to the user.

The designer of mobile systems must cater for these environmental issues and provide smooth, consistent coverage and system performance for the user. This is essential to build confidence and satisfaction with the software in use on the device.

Fault Tolerance and Safety

Fault tolerance and safety refer to the ability of device-based software to behave in abnormal and supernormal conditions. Quite apart from the correct behavior that must be exhibited by software on a device in normal conditions, the same software must respond safely without permanent or irrevocable harm to the system's data even in the worst of conditions.

A fault-tolerant system is able to identify and report faults; whether the fault is in the device software, in the network, or even in the hardware, the system should be able to:

- Correct the fault without (or with minimal) user interaction and continue processing
- Restore itself to a stable state and avoid damaging its data
- Advise the user of its state and the actions taken
- Enter a fail-safe handling routine, or in a truly hopeless situation, fail as gracefully as possible

Error-handling and recovery must be built into the underlying system design and thought must be given to how the programmers of the system will handle exceptions. While people will tolerate PC crashes and failures, they are far less likely to in mobile applications, especially as these systems become more commonly used in safety critical environments, such as automobiles.

The design then, should assist programmers in identifying possible error conditions. It should address the question of how to incorporate ideas such as component-heartbeats (in order to monitor network availability, or for the network to determine a device or other service's availability) and how the developers can best log and report failures and errors.

Optimization and Correctness in Design

There are a number of key lessons that can be usefully applied in the design of J2ME-based systems. Many of these have been derived over the years from software design in other areas. They are just as valuable, or even more so, in the J2ME environment:

- Get the infrastructure right early on
- Use the techniques of code refactoring throughout the coding process

- Optimize at a code-level late in the development cycle
- Analyze to determine where you really need to optimize
- Optimize the object creation process
- Assert yourself
- Refactor with a plan
- Code defensively: try and catch

Get the Infrastructure Right Early On

The best way to create an optimal, well designed and performing system is to get the system design right. The exact meaning of 'right' is defined in terms of performance and good-design, so the last sentence does represent a bit of a tautology. However, 'right' for a designer must be defined in terms of the software's eventual acceptance by and success with the various user groups that we discussed at the beginning of this chapter.

On a technical and infrastructure level, the successful design of mobile systems rests on a number of critical factors:

- Does the business process or function lend itself naturally to the chosen delivery mechanism?
- Where is the split in the workload between the mobile client and the server?
- How will the mobile client and server communicate (choosing between the most optimal or clearest communication strategies, from XML, RMI, direct JDBC calls etc.)?
- How will the user-interface metaphor be constructed?
- What client, server hardware, and carrier network will the application initially be designed for?

Use the Techniques of Code Refactoring Throughout the Coding Process

Good coding practices and techniques making use of design and refactoring patterns, can greatly enhance the maintainability and correctness of your code. Refactoring techniques allow programmers to review and safely change code to ensure that design problems and metaphors can be reflected throughout the codebase. They also ensure that the benefits of changes and insights can be successfully applied and re-engineered throughout the application. However, it is particularly tempting in Java projects, given the usual rapid development and time-pressured nature of projects, to refrain from following through with these good practices.

Design changes to particular elements should be extended to other areas of the project with a common design metaphor or construct; code-refactoring patterns enable programmers and designers to identify and safely follow through with any necessary modifications to elements of the software infrastructure.

> **The best way to achieve performant code and meet project deadlines, is to maintain a consistent and well-written code base.**

If neglected, these non-visible factors associated with a quality code-base will eventually lead to degradation in the system's overall quality and extensibility. Learn how to refactor code and apply common patterns; it is an essential skill that all senior developers and technical project managers should develop.

Optimize at a Code-level Late in the Development Cycle

It is essential to refactor code throughout the project, starting at day one of the development, in order to maintain consistency, efficiency and to identify and eliminate bugs. However, writing and editing code specifically to add wizardry and performance-tuning measures too early in the development cycle is something to beware of.

Optimization tricks and techniques can hinder the continuing design and refactoring process that should be natural to a development process. The essential message is that hacking code too early to improve performance can get in the way of good design. Edit with care: optimization often introduces errors and conflicts with the clarity and maintainability of the system's code.

This is not to say that you should write non-optimal code and ignore good practice, but rather that the careful optimization of code is better performed once the system nears completion and its performance has been analyzed for problems. Optimization techniques can be focused on specific problem areas: for example, reduction in network calls, or particular program loops and functions. Design the software, develop to the plan, measure its performance, and then optimize.

Analyze to Determine Where You Really Need to Optimize

Performance tuning while maintaining code correctness and quality, revolves around identifying and concentrating on those areas that actually require optimization. It is extremely important to utilize code profiling techniques to ensure that work to optimize code occurs only where substantial benefits are possible and not in areas where the code is seldom executed or where efficiency saving has little impact on the overall application. Code optimization to improve speed, memory usage, or network efficiency – as opposed to refactoring – can require a number of tradeoffs.

Improving Startup Time versus Execution Speed

It is generally straightforward to improve the startup time of an application by deferring a task until it is actually required. This is particularly beneficial if the deferred task is not always required in the execution of the program.

Improve Network Performance versus Result Processing

Reducing the number of network calls and messages can dramatically improve the response time of an application and so a common performance-enhancing technique is to reduce the number of network calls by batching up many separate requests for information into a single network request. The response from such a request does, however, require additional processing to uncover the individual answers and so on a smaller device with more limited processing power, this is a tradeoff that must be carefully evaluated.

Improving Performance versus Reducing Memory Usage

Many performance optimizations require extra coding to achieve the desired enhancement. For example, caching is a common technique used to avoid unnecessary requests and for improving performance. However, along with many other techniques, caching can lead to increased memory utilization and as such can lead to performance problems and out-of-memory errors on memory-constrained devices such as those targeted by J2ME.

Don't Over-optimize or Optimize Unnecessarily

Refactoring code is an important part of the development process, but optimization to suit a particular deployment platform can limit the future portability of the codebase and therefore, excessive optimization – particularly those optimizations aimed at a particular device – should be approached with extreme caution. An additional 'gotcha' to many optimizations is that, by enhancing code performance with tricks, you place your codebase outside the remit of future compiler technology enhancements, thus damaging the potential performance of your code at a later date.

Optimize the Object Creation Process

Creating new objects on the fly is a resource-hungry process, especially if these objects are numerous and short-lived. Reducing this expenditure in object-creation can lead to significant performance gains. There are a number of techniques that can be easily used to avoid excessive object-creation and the associated garbage collection that will occur.

Reuse Existing Objects

Processing objects in a loop, for instance, reusing the same object instance that is reinitialized at the start of the cycle, is significantly more efficient than creating a new instance for each cycle of the loop. A typical example of this type of performance problem is the concatenation of strings in a loop:

```
String strNames = "";
for (int i = 0; to i < names.numberOfNames(), ++i) {
    strNames += names.getNameAtPosition(i);
    strNames += "\n";
}
```

This tends to create many short-lived `String` objects that can damage the performance of code and consume valuable resources.

A more efficient mechanism would be to use a single instance of a `StringBuffer` object, as shown:

```
StringBuffer strbufNames = new StringBuffer();
for (int i = 0; to i < names.numberOfNames(), ++i) {
    strbufNames.append(names.getNameAtPosition(i));
    strbufNames.append("\n");
}
```

Make Use of Object Pools

Object pools and caches can reduce the overhead in creating and accessing commonly used resources in the system. The use of object pools can dramatically improve the performance for certain tasks and is an essential, if sometimes complex, technique for creating applications in a distributed environment. For example, database connection objects and other types of server connection, such as sockets, can be stored in a simple pool and reinitialized as and when necessary rather than created each time they are used.

Caching Information on the Device

The judicious use of an object cache can produce significant improvements in application performance. Caches are commonly used at both an application and a system level in client-server applications: for example, all popular browsers cache web pages in order to make surfing bearable. All major databases rely on caching techniques to improve access to frequently accessed data and the commonly used technique of utilizing virtual memory is a form of caching.

Caching is thus an essential technique for device-based applications. This is doubly true for applications that need to operate both on and off the network. However, there is something of a black-art in designing an efficient object-cache; caching too much, or allowing inconsistencies in the objectcache, can lead to performance problems, data loss and other serious application-level bugs.

Issues with Designing a Client-side Cache

The design of a client-side cache is a problem that was faced in many client-server architectures and the lessons learnt in that period can be applied to the J2ME universe. Several questions must be asked and answered by the designer of a device cache:

What Should be Cached?

Objects should be placed in a client-side cache when the overhead saved by storing them in this way more than pays for the overhead of retrieving them. Such objects are typically those that are more frequently accessed than they are changed and are not too large in memory.

How Much Data Should be Cached on the Device?

How much should be cached depends on a number of issues:

- How much resource is available for the cache?
- Will the application need to cache information for off-line usage?
- Which objects lend themselves to caching?

As you can imagine there is no fixed answer to these questions. In general terms, enough should be cached to improve performance, while not consuming unacceptable amounts of resource – this is a fine balance. As with many balances in system development, small enhancements can lead to significant system performance improvements if crafted in the correct way. In this way, even the caching of small numbers of frequently used objects in the system, can lead to significant enhancements to the overall performance of the software.

How Should the Cache be Updated?

The updating of the cache is made considerably easier in device-based applications if only a single application on the client will be updating the cache. If this is not the case, then an 'Observer' pattern-based framework will be needed to monitor changes to the cached item in the server-side and to update and notify the client-side cache with any external changes. Furthermore, on all J2ME-based systems, resource constraints are a serious issue and the cache must be maintained within strict size limits.

A cache manager can remove objects from the cache, as it becomes full, in a number of ways. Typical strategies include:

- First In, First Out (FIFO) stacks
- Tracking access to the objects and removing the least recently used objects
- Associating a time-out to the objects and sweeping through the cache periodically to remove 'dead' objects

Cache management techniques designed and developed in other software environments are just as applicable to J2ME.

How Can Cache Integrity with the Server Datastore be Maintained?

Data integrity is a major issue for caching. When caching information locally that can be updated in a multi-user server-side environment, certain issues must be considered:

- When will cached information be preferred to re-retrieving information?
- How will conflicts between client device and server be resolved?
- How will external and server-side updates to items stored in the cache be handled?

We have already considered how, in an environment with a permanent connection to the network, design patterns such as the 'Observer' pattern from the 'Gang of Four' – see "Design Patterns" by Erich Gamma, Richard Helm, Ralph Johnson, and John Vlissides (Addison-Wesley, 1994 0201633612) – can be usefully employed to update items actively as they change on the server or client.

This client-server cache synchronization will itself involve performance tradeoffs and concerns such as how often does the client checks the cache and by what mechanism does the 'cache observer' receives notification from the server to inform it that changes have occurred to specific cached objects

Judicious choices must also be made in order to handle write conflicts and this is especially true in any application that needs to maintain a level of offline usage. Database record-locking strategies, time-stamping of objects in the main data store, and object attribute comparisons, are all possible options to avoid potential inconsistencies and each strategy has its relative merits and problems. For example, for heavily updated objects in the system, pessimistic database record locking (locking an object's rows in the database when it is read), while safe (since no-one else can update the rows while you 'hold' the object for update in a locked state), can severely damage system performance, since everyone else is kept waiting while this occurs.

As with the answer to the question 'What should be cached?', the answer to the problem of maintaining cache to server-side integrity is highly subjective. This is an issue for both the cache developer and the business user. All assumptions in this area should always be double-checked against the business requirements and road-tested with user case scenarios.

Assert Yourself

Maintain all your debugging code alongside your production code and use assertion statements to test the validity of pre and post conditions for all-important processes:

```
boolean _ASSERT = true;

if(_ASSERT) {
    System.out.println("Debug information");
    // Assertion test statements
}
```

If the `_ASSERT` flag is set to `false`, the Java compiler will cut out this "dead" code in the compilation process. However, the Java compiler will only remove the code if it is unreachable, so do remember to recompile with the `_ASSERT` flag set to `false` before you ship the software.

Assertion statements can aid debugging and, since they can be contained in conditional blocks to allow removal from the final production compilation, they do not affect the size or performance of the final release code. A Java expert group is currently investigating creating a standard language facility for generating and maintaining assertion statements under the auspices of JSR 41.

Refactor with a Plan

We have already mentioned the practice of code refactoring and the use of these coding techniques to avoid introducing unnecessary errors while optimizing code.

Refactoring was introduced by the Smalltalk community and entails a series of good practices and patterns for changing and transforming code. The consistency with which code changes are made is sometimes erratic and changes to the design of both objects and components can have an unpredictable effect on the rest of the system if proper caution is not exercised. For example, good configuration management tools and/or procedures are a prerequisite for code refactoring to be viable in a team environment.

The patterns of code refactoring ensure that potential pitfalls of change are highlighted and that seemingly small changes to methods, objects, and components, are conducted in such a way as to shield the rest of the development effort from error while bringing the maximum benefit to both the programmers and project.

For more information on code refactoring and refactoring patterns, you can consult the book "*Refactoring* ISBN 0201485672" by Martin Fowler (Addison-Wesley, 1999).

Code Defensively

Defensive coding techniques represent an established and mature approach to avoiding and handling errors and exceptions within the system. As with all sensible strategies, it is made up of entirely common sense approaches.

The use of `try...catch` blocks to handle anticipated errors and erroneous fringe scenarios helps device-based systems to catch and handle unexpected events in order to handle failure gracefully.

Out of memory errors on the creation of complex and memory hungry objects are a typical problem for J2ME-based software. While the handling of releasing memory is the job of the garbage collector (which should ensure that unreferenced memory is freed as quickly as possible), it is the duty of the programmer to use good defensive programming practices to ensure that potential memory exceptions are checked.

On the creation of new objects in J2ME-based applications that may require significant allocation of resources, programmers should ensure that a `try...catch` block is set up to ensure that such `java.lang.OutOfMemoryError` error conditions are checked.

Once an exception is caught the programmer should endeavor to fix the problem and allow the user to continue using the system as well as possible, or to fail as gracefully as possible. However, in many device-based systems, failure is not an option and the system must be able to recover from all common exceptions. We must always remember that while people have come to accept the occasional failure of PC-based software, embedded systems often run in situations without support-desks and on devices that demand high-levels of reliability and therefore cannot be allowed to crash.

J2ME Coding Standards

Coding conventions are as important as design and architecture standards and disregard for proper care in the crafting of code can lead to difficulty in reuse and maintenance in the future:

- 80% of the lifetime cost of a software system goes to maintenance
- Software is rarely maintained for its whole life by the original programmer
- Code conventions improve the readability of the code, allowing programmers to understand new code more quickly and thoroughly
- Bad code style, or the use of unsupported coding tricks, may negate the benefits of and even create difficulties for, future compiler enhancements

Sun provides a document outlining its suggested coding conventions, which have been adopted by many Java developers. For further details see: http://java.sun.com/docs/codeconv/html/CodeConvTOC.doc.html. For a more readable guide book to Java coding style, see: "The Elements of Java Style" by Allan Vermeulen et al. of Rogue Wave Software, Inc. (Cambridge University Press, 2000).

Availability of Developers

As a relatively young area of technology, Java does suffer from a constraining shortage of developers, but is now rapidly attracting talented people. If the availability of Java skills is compared to the relatively narrow current world of embedded and device programming, it can be seen that Java has a substantially larger and faster growing skills base from which to draw talent. The availability of this talent should substantially reduce the development and maintenance costs of projects, since replacement programmers and those with domain experience should be easier to find.

Java and J2ME can provide a pool of trained people to take J2ME code and redevelop it to suit new developments in device technology and new business situations, in a way that, due to the lack of trained programmers, has not been possible with traditional device technology.

Factors for Small Connected Systems

A key decision in the design of small mobile systems is where, and how, to balance the loading of server-side and client portions of the application. We took a closer look at the models for client-server and collaborative distribution models in Chapter 2. For now we simply need to consider how the device is to be used and what the requirements of this usage dictate.

The Continuing Need for Client-side Computing

While it is vitally important to take advantage of both the device's and the server's computing power, the benefit of using device technology centers on its portable and immediate nature, which requires two things:

- Immediacy in response – which is at odds with the intermittent and slow communication of wireless networking technologies
- Specific functionality – usage tailored for the task

Wireless networks tend to exhibit behavior that is not conducive to thin-client solutions, in that they:

- Tend to have limited bandwidth even when compared to a home PC and a modem
- Have a higher degree of latency in the network response
- Have lower connection stability than fixed-line networks
- Suffer from less than 100% coverage and as a result have less predictable availability

Disconnected Access

Being able to construct software that can run locally on the device ensures the benefits of mobility can be used to the greatest advantage and the negative effects of this poor network coverage can be minimized. This can be achieved through the prudent use of object caching to enable local execution and to maximize the benefits of a network connection once it is available.

There are two main models for connecting to the network from a mobile device:

- Periodic synchronization via a cable or cradle
- Wireless network connection

The benefit of wireless network access is simply mobility. Synchronization with a server-side system can be done while the device is within the range of a wireless network cell, or micro-cell and there is no need for the user to place the device into a desk cradle or to attach any cables.

What Sort of Applications Benefit from J2ME?

There are essentially two types of applications that can be developed using J2ME:

- Applications that would have traditionally been used on PCs or laptops, but can now be developed for sophisticated smart-phones or PDAs
- Applications that are peculiar to a device or embedded system (such as bar-code reader software, or a fridge temperature control application)

In the first case, it is important to decide which applications to develop for the mobile platform on the basis of how the users of the software will benefit from the ability to perform the business function away from a PC or laptop.

In the latter case, it is necessary to consider carefully how the use of Java technology will add benefit to the software and the software development process. Applications that can benefit from the mobility provided by using Java 2 Micro Edition to design and code the software, typically display the following characteristics:

- Provide relevant content and/or services to the mobile user, but require a more sophisticated user-interface than can be developed using a browser-based interface. Possible reasons not to use a browser include:
  - Need for complex client-side input validation
  - Need for sophisticated user-interface components (such as grid-boxes, tabs that need to be independently refreshed)
  - Need for off-line usage

- Provide little in the way of user input, or if input is required, it should be separated into a mobile client that can be used 'on the road' and a desktop component that can be used for any heavy user input
- Show clear and compelling benefits from the ability to operate remotely or on the move

An example of a business process that could specifically benefit from the mobile channel might be an application to accept the entry of timesheets from consultants who need to keep a precise record of their utilization, but who are infrequently at a PC that is connected to their company's network.

As discussed in Chapter 2, a mobile application enabling the simple input of information detailing their time spent on different activities would be an ideal application for a PDA, or a sophisticated smart-phone, wirelessly networked mobile platform.

Summary

J2ME provides programmers with configurations and profiles that attack and solve some of the design and development issues for constructing applications for mobile devices. However, there is still much for the designer of device-based software to do.

Software design is concerned with planning and the communication of plans to those involved in the software project. In addition, at the start of the chapter, we investigated how the design ultimately communicates the intent of the system to the software's users.

We have examined in this chapter how user-centered design needs to be considered and how this usability is a particularly important issue for the designer of J2ME applications. This is because the typical user will be working with a little screen, either a very small or no keyboard, and might not have access to helpdesks, but nonetheless needs services and applications that are both reliable and easy to use.

Usability is only part of the story. Good software engineering design practice and effective code-level implementation are as relevant as ever to J2ME programmers. Combining J2ME with these coding good practices should enable the development of efficient software using an environment tailored to the device.

Software design is an activity that should deliver certain basic benefits to a software system. The cost of change at this early stage is relatively low, so the more development and implementation issues that can be solved in the design phase, the lower the cost of the project and the more likely it will be to succeed.

Good design practices must be learned from experience, since J2ME is a new technology, much of the experience needed must be leveraged from other areas of software development into this space. The lessons learned in the development of client-server, device-based, and real-time systems could all be applied to J2ME, while of course paying close attention to the user-centered and system-level design issues, which are specific to the mobile world.

PROFESSIONAL JAVA MOBILE
TOOL SELECTION
WEB TABLET
MOBILE PHONE
MP3 PLAYER
DIGITAL CAMERA
PALMTOP
LAPTOP
NOW SELECTED
PALMTOP

Appendix A - API Comparison

Packages that contain supporting classes are listed alphabetically. All packages list interfaces first, then classes, followed by exceptions and finally errors.

Package java.awt

J2SE Classes	CDC	CLDC	Foundation Profile	MID Profile
AlphaComposite	No	No	No	No
AWTEvent	No	No	No	No
AWTEventMulticaster	No	No	No	No
AWTPermission	No	No	No	No
BasicStroke	No	No	No	No
BorderLayout	No	No	No	No
Button	No	No	No	No
Canvas	No	No	No	Yes
CardLayout	No	No	No	No
Checkbox	No	No	No	No
CheckboxGroup	No	No	No	No
CheckboxMenuItem	No	No	No	No
Choice	No	No	No	Yes

J2SE Classes	CDC	CLDC	Foundation Profile	MID Profile
Color	No	No	No	No
Component	No	No	No	No
ComponentOrientation	No	No	No	No
Container	No	No	No	No
Cursor	No	No	No	No
Dialog	No	No	No	No
Dimension	No	No	No	No
Event	No	No	No	No
EventQueue	No	No	No	No
FileDialog	No	No	No	No
FlowLayout	No	No	No	No
Font	No	No	No	Yes
FontMetrics	No	No	No	No
Frame	No	No	No	No
GradientPaint	No	No	No	No
Graphics	No	No	No	Yes
Graphics2D	No	No	No	No
GraphicsConfigTemplate	No	No	No	No
GraphicsConfiguration	No	No	No	No
GraphicsDevice	No	No	No	No
GraphicsEnvironment	No	No	No	No
GridBagConstraints	No	No	No	No
GridBagLayout	No	No	No	No
GridLayout	No	No	No	No
Image	No	No	No	Yes
Insets	No	No	No	No
JobAttributes	No	No	No	No
JobAttributes.Default SelectionType	No	No	No	No
JobAttributes.DestinationType	No	No	No	No
JobAttributes.DialogType	No	No	No	No

J2SE Classes	CDC	CLDC	Foundation Profile	MID Profile
JobAttributes.Multiple DocumentHandlingType	No	No	No	No
JobAttributes.SidesType	No	No	No	No
Label	No	No	No	No
List	Yes	No	Yes	Yes
MediaTracker	No	No	No	No
Menu	No	No	No	No
MenuBar	No	No	No	No
MenuComponent	No	No	No	No
MenuItem	No	No	No	No
MenuShortcut	No	No	No	No
PageAttributes	No	No	No	No
PageAttributes.ColorType	No	No	No	No
PageAttributes.MediaType	No	No	No	No
PageAttributes.Orientation RequestedType	No	No	No	No
PageAttributes.OriginType	No	No	No	No
PageAttributes.PrintQuality Type	No	No	No	No
Panel	No	No	No	No
Point	No	No	No	No
Polygon	No	No	No	No
PopupMenu	No	No	No	No
PrintJob	No	No	No	No
Rectangle	No	No	No	No
RenderingHints	No	No	No	No
RenderingHints.Key	No	No	No	No
Robot	No	No	No	No
Scrollbar	No	No	No	No
ScrollPane	No	No	No	No
SystemColor	No	No	No	No

Table continued on following page

J2SE Classes	CDC	CLDC	Foundation Profile	MID Profile
TextArea	No	No	No	No
TextComponent	No	No	No	No
TextField	No	No	No	No
TexturePaint	No	No	No	No
Toolkit	No	No	No	No
Window	No	No	No	No
AWTException	No	No	No	No
FontFormatException	No	No	No	No
IllegalComponentState Exception	No	No	No	No
AWTError	No	No	No	No

Package java.io

J2SE Classes	CDC	CLDC	Foundation Profile	MID Profile
DataInput	Yes	Yes	Yes	Yes
DataOutput	Yes	Yes	Yes	Yes
Externalizable	Yes	No	Yes	No
FileFilter	Yes	No	Yes	No
FilenameFilter	Yes	No	Yes	No
ObjectInput	Yes	No	Yes	No
ObjectInputValidation	Yes	No	Yes	No
ObjectOutput	Yes	No	Yes	No
ObjectStreamConstants	Yes	No	Yes	No
Serializable	Yes	No	Yes	No
BufferedInputStream	Yes	No	Yes	No
BufferedOutputStream	Yes	No	Yes	No
BufferedReader	Yes	No	Yes	No
BufferedWriter	Yes	No	Yes	No
ByteArrayInputStream	Yes	Yes	Yes	Yes
ByteArrayOutputStream	Yes	Yes	Yes	Yes
CharArrayReader	No	No	Yes	No
CharArrayWriter	No	No	Yes	No

J2SE Classes	CDC	CLDC	Foundation Profile	MID Profile
DataInputStream	Yes	Yes	Yes	Yes
DataOutputStream	Yes	Yes	Yes	Yes
File	Yes	No	Yes	No
FileDescriptor	Yes	No	Yes	No
FileInputStream	Yes	No	Yes	No
FileOutputStream	Yes	No	Yes	No
FilePermission	Yes	No	Yes	No
FileReader	Yes	No	Yes	No
FileWriter	Yes	No	Yes	No
FilterInputStream	Yes	No	Yes	No
FilterOutputStream	Yes	No	Yes	No
FilterReader	No	No	Yes	No
FilterWriter	No	No	No	No
InputStream	Yes	Yes	Yes	Yes
InputStreamReader	Yes	Yes	Yes	Yes
LineNumberInputStream	No	No	No	No
LineNumberReader	No	No	Yes	No
ObjectInputStream	Yes	No	Yes	No
ObjectInputStream.GetField	Yes	No	Yes	No
ObjectOutputStream	Yes	No	Yes	No
ObjectOutputStream.PutField	Yes	No	Yes	No
ObjectStreamClass	Yes	No	Yes	No
ObjectStreamField	Yes	No	Yes	No
OutputStream	Yes	Yes	Yes	Yes
OutputStreamWriter	Yes	Yes	Yes	Yes
PipedInputStream	Yes	No	Yes	No
PipedOutputStream	Yes	No	Yes	No
PipedReader	No	No	Yes	No
PipedWriter	No	No	Yes	No
PrintStream	Yes	Yes	Yes	Yes
PrintWriter	Yes	No	Yes	No

Table continued on following page

J2SE Classes	CDC	CLDC	Foundation Profile	MID Profile
PushbackInputStream	Yes	No	Yes	No
PushbackReader	No	No	Yes	No
RandomAccessFile	No	No	Yes	No
Reader	Yes	Yes	Yes	Yes
SequenceInputStream	No	No	Yes	No
SerializablePermission	Yes	No	Yes	No
StreamTokenizer	Yes	No	Yes	No
StringBufferInputStream	No	No		No
StringReader	No	No	Yes	No
StringWriter	No	No	Yes	No
Writer	Yes	Yes	Yes	Yes
CharConversionException	Yes	No	Yes	No
EOFException	Yes	Yes	Yes	Yes
FileNotFoundException	Yes	No	Yes	No
InterruptedIOException	Yes	Yes	Yes	Yes
InvalidClassException	Yes	No	Yes	No
InvalidObjectException	Yes	No	Yes	No
IOException	Yes	Yes	Yes	Yes
NotActiveException	Yes	No	Yes	No
NotSerializableException	Yes	No	Yes	No
ObjectStreamException	Yes	No	Yes	No
OptionalDataException	Yes	No	Yes	No
StreamCorruptedException	Yes	No	Yes	No
SyncFailedException	Yes	No	Yes	No
UnsupportedEncodingException	Yes	Yes	Yes	Yes
UTFDataFormatException	Yes	Yes	Yes	Yes
WriteAbortedException	Yes	No	Yes	No

Package java.lang

J2SE Classes	CDC	CLDC	Foundation Profile	MID Profile
Cloneable	Yes	No	Yes	No
Comparable	Yes	No	Yes	No

J2SE Classes	CDC	CLDC	Foundation Profile	MID Profile
Runnable	Yes	Yes	Yes	Yes
Boolean	Yes	Yes	Yes	Yes
Byte	Yes	Yes	Yes	Yes
Character	Yes	Yes	Yes	Yes
Character.Subset	Yes	No	Yes	No
Character.UnicodeBlock	Yes	No	Yes	No
Class	Yes	Yes	Yes	Yes
ClassLoader	Yes	No	Yes	No
Compiler	No	No	Yes	No
Double	Yes	No	Yes	No
Float	Yes	No	Yes	No
InheritableThreadLocal	Yes	No	Yes	No
Integer	Yes	Yes	Yes	Yes
Long	Yes	Yes	Yes	Yes
Math	Yes	Yes	Yes	Yes
Number	Yes	No	Yes	No
Object	Yes	Yes	Yes	Yes
Package	Yes	No	Yes	No
Process	Yes	No	Yes	No
Runtime	Yes	Yes	Yes	Yes
RuntimePermission	Yes	No	Yes	No
SecurityManager	Yes	No	Yes	No
Short	Yes	Yes	Yes	Yes
StrictMath	Yes	No	Yes	No
String	Yes	Yes	Yes	Yes
StringBuffer	Yes	Yes	Yes	Yes
System	Yes	Yes	Yes	Yes
Thread	Yes	Yes	Yes	Yes
ThreadGroup	Yes	No	Yes	No
ThreadLocal	Yes	No	Yes	No
Throwable	Yes	Yes	Yes	Yes

Table continued on following page

J2SE Classes	CDC	CLDC	Foundation Profile	MID Profile
Void	Yes	No	Yes	No
ArithmeticException	Yes	Yes	Yes	Yes
ArrayIndexOutOfBounds Exception	Yes	Yes	Yes	Yes
ArrayStoreException	Yes	Yes	Yes	Yes
ClassCastException	Yes	Yes	Yes	Yes
ClassNotFoundException	Yes	Yes	Yes	Yes
CloneNotSupportedException	Yes		Yes	
Exception	Yes	Yes	Yes	Yes
IllegalAccessException	Yes	Yes	Yes	Yes
IllegalArgumentException	Yes	Yes	Yes	Yes
IllegalMonitorStateException	Yes	Yes	Yes	Yes
IllegalStateException	Yes		Yes	Yes
IllegalThreadStateException	Yes	Yes	Yes	Yes
IndexOutOfBoundsException	Yes	Yes	Yes	Yes
InstantiationException	Yes	Yes	Yes	Yes
InterruptedException	Yes	Yes	Yes	Yes
NegativeArraySizeException	Yes	Yes	Yes	Yes
NoSuchFieldException	Yes	No	Yes	No
NoSuchMethodException		No		No
NullPointerException	Yes	Yes	Yes	Yes
NumberFormatException	Yes	Yes	Yes	Yes
RuntimeException	Yes	Yes	Yes	Yes
SecurityException	Yes	Yes	Yes	Yes
StringIndexOutOfBounds Exception	Yes	Yes	Yes	Yes
UnsupportedOperation Exception	Yes	No	Yes	No
AbstractMethodError	Yes	No	Yes	No
ClassCircularityError	Yes	No	Yes	No
ClassFormatError	Yes	No	Yes	No
Error	Yes	Yes	Yes	Yes

J2SE Classes	CDC	CLDC	Foundation Profile	MID Profile
ExceptionInInitializerError	Yes	No	Yes	No
IllegalAccessError	Yes	No	Yes	No
IncompatibleClassChange Error	Yes	No	Yes	No
InstantiationError	Yes	No	Yes	No
InternalError	Yes	No	Yes	No
LinkageError	Yes	No	Yes	No
NoClassDefFoundError	Yes	No	Yes	No
NoSuchFieldError	Yes	No	Yes	No
NoSuchMethodError	Yes	No	Yes	No
OutOfMemoryError	Yes	Yes	Yes	Yes
StackOverflowError	Yes	No	Yes	No
ThreadDeath	Yes	No	Yes	No
UnknownError		No	Yes	No
UnsatisfiedLinkError	Yes	No	Yes	No
UnsupportedClassVersion Error	Yes	No	Yes	No
VerifyError	Yes	No	Yes	No
VirtualMachineError	Yes	Yes	Yes	Yes

Package java.lang.ref

J2SE Classes	CDC	CLDC	Foundation Profile	MID Profile
PhantomReference	Yes	No	Yes	No
Reference	Yes	No	Yes	No
ReferenceQueue	Yes	No	Yes	No
SoftReference	Yes	No	Yes	No
WeakReference	Yes	No	Yes	No

Package java.lang.reflect

J2SE Classes	CDC	CLDC	Foundation Profile	MID Profile
InvocationHandler	Yes	No	Yes	No
Member	Yes	No	Yes	No

Table continued on following page

J2SE Classes	CDC	CLDC	Foundation Profile	MID Profile
AccessibleObject	Yes	No	Yes	No
Array	Yes	No	Yes	No
Constructor	Yes	No	Yes	No
Field	Yes	No	Yes	No
Method	Yes	No	Yes	No
Modifier	Yes	No	Yes	No
Proxy		No		No
ReflectPermission	Yes	No	Yes	No
InvocationTargetException	Yes	No	Yes	No
UndeclaredThrowable Exception	Yes	No	Yes	No

Package java.math

J2SE Classes	CDC	CLDC	Foundation Profile	MID Profile
BigDecimal	No	No	No	No
BigInteger	Yes	No	Yes	No

Package java.net

J2SE Classes	CDC	CLDC	Foundation Profile	MID Profile
ContentHandlerFactory	Yes	No	Yes	No
DatagramSocketImplFactory	Yes	No	Yes	No
FileNameMap	Yes	No	Yes	No
SocketImplFactory		No	Yes	No
SocketOptions	Yes	No	Yes	No
URLStreamHandlerFactory	Yes	No	Yes	No
Authenticator		No	Yes	No
ContentHandler	Yes	No	Yes	No
DatagramPacket	Yes	No	Yes	No
DatagramSocket	Yes	No	Yes	No
DatagramSocketImpl	Yes	No	Yes	No

J2SE Classes	CDC	CLDC	Foundation Profile	MID Profile
HttpURLConnection	No	No	Yes	No
InetAddress	Yes	No	Yes	No
JarURLConnection	Yes	No	Yes	No
MulticastSocket	No	No	Yes	No
NetPermission	Yes	No	Yes	No
PasswordAuthentication	No	No	Yes	No
ServerSocket	No	No	Yes	No
Socket	No	No	Yes	No
SocketImpl	No	No	Yes	No
SocketPermission	Yes	No	Yes	No
URL	Yes	No	Yes	No
URLClassLoader	Yes	No	Yes	No
URLConnection	Yes	No	Yes	No
URLDecoder	No	No	Yes	No
URLEncoder	No	No	Yes	No
URLStreamHandler	Yes	No	Yes	No
BindException	Yes	No	Yes	No
ConnectException	No	No	Yes	No
MalformedURLException	Yes	No	Yes	No
NoRouteToHostException	No	No	Yes	No
ProtocolException	Yes	No	Yes	No
SocketException	Yes	No	Yes	No
UnknownHostException	Yes	No	Yes	No
UnknownServiceException	Yes	No	Yes	No

Package java.rmi

J2SE Classes	CDC	CLDC	Foundation Profile	MID Profile
Remote	No	No	No	No
MarshalledObject	No	No	No	No
Naming	No	No	No	No

Table continued on following page

J2SE Classes	CDC	CLDC	Foundation Profile	MID Profile
RMISecurityManager	No	No	No	No
AccessException	No	No	No	No
AlreadyBoundException	No	No	No	No
ConnectException	No	No	Yes	No
ConnectIOException	No	No	No	No
MarshalException	No	No	No	No
NoSuchObjectException	Yes	No	Yes	No
NotBoundException	No	No	No	No
RemoteException	No	No	No	No
RMISecurityException	No	No	No	No
ServerError	No	No	No	No
ServerException	No	No	No	No
ServerRuntimeException	No	No	No	No
StubNotFoundException	No	No	No	No
UnexpectedException	No	No	No	No
UnknownHostException	Yes	No	Yes	No
UnmarshalException	No	No	No	No

Package java.security

J2SE Classes	CDC	CLDC	Foundation Profile	MID Profile
Certificate	Yes	No	Yes	No
DomainCombiner	Yes	No	Yes	No
Guard	Yes	No	Yes	No
Key	Yes	No	Yes	No
Principal	No	No	Yes	No
PrivateKey	No	No	Yes	No
PrivilegedAction	Yes	No	Yes	No
PrivilegedExceptionAction	Yes	No	Yes	No
PublicKey	Yes	No	Yes	No
AccessControlContext	Yes	No	Yes	No

J2SE Classes	CDC	CLDC	Foundation Profile	MID Profile
AccessController	Yes	No	Yes	No
AlgorithmParameterGenerator	No	No	Yes	No
AlgorithmParameter GeneratorSpi	No	No	Yes	No
AlgorithmParameters	No	No	Yes	No
AlgorithmParametersSpi	No	No	Yes	No
AllPermission	Yes	No	Yes	No
BasicPermission	Yes	No	Yes	No
CodeSource	Yes	No	Yes	No
DigestInputStream	No	No	Yes	No
DigestOutputStream	Yes	No	Yes	No
GuardedObject	Yes	No	Yes	No
Identity	No	No	Yes	No
IdentityScope	No	No	Yes	No
KeyFactory	No	No	Yes	No
KeyFactorySpi	No	No	Yes	No
KeyPair	No	No	Yes	No
KeyPairGenerator	No	No	Yes	No
KeyPairGeneratorSpi	No	No	Yes	No
KeyStore	No	No	Yes	No
KeyStoreSpi	No	No	Yes	No
MessageDigest	Yes	No	Yes	No
MessageDigestSpi	Yes	No	Yes	No
Permission	Yes	No	Yes	No
PermissionCollection	Yes	No	Yes	No
Permissions	Yes	No	Yes	No
Policy	Yes	No	Yes	No
ProtectionDomain	Yes	No	Yes	No
Provider	Yes	No	Yes	No
SecureClassLoader	Yes	No	Yes	No
SecureRandom	No	No	Yes	No

Table continued on following page

J2SE Classes	CDC	CLDC	Foundation Profile	MID Profile
SecureRandomSpi	No	No	Yes	No
Security	Yes	No	Yes	No
SecurityPermission	Yes	No	Yes	No
Signature	No	No	Yes	No
SignatureSpi	No	No	Yes	No
SignedObject	No	No	Yes	No
Signer	No	No	Yes	No
UnresolvedPermission	Yes	No	Yes	No
AccessControlException	Yes	No	Yes	No
DigestException	Yes	No	Yes	No
GeneralSecurityException	Yes	No	Yes	No
InvalidAlgorithmParameter Exception	No	No	Yes	No
InvalidKeyException	Yes	No	Yes	No
InvalidParameterException	Yes	No	Yes	No
KeyException	Yes	No	Yes	No
KeyManagementException	No	No	Yes	No
KeyStoreException	No	No	Yes	No
NoSuchAlgorithmException	Yes	No	Yes	No
NoSuchProviderException	Yes	No	Yes	No
PrivilegedActionException	Yes	No	Yes	No
ProviderException	Yes	No	Yes	No
SignatureException	Yes	No	Yes	No
UnrecoverableKeyException	No	No	Yes	No

Package java.security.acl

J2SE Classes	CDC	CLDC	Foundation Profile	MID Profile
Acl	No	No	Yes	No
AclEntry	No	No	Yes	No
Group	No	No	Yes	No
Owner	No	No	Yes	No

J2SE Classes	CDC	CLDC	Foundation Profile	MID Profile
Permission	Yes	No	Yes	No
AclNotFoundException	No	No	Yes	No
LastOwnerException	No	No	Yes	No
NotOwnerException	No	No	Yes	No

Package java.security.cert

J2SE Classes	CDC	CLDC	Foundation Profile	MID Profile
X509Extension	No	No	Yes	No
Certificate	Yes	No	Yes	No
Certificate.CertificateRep	Yes	No	Yes	No
CertificateFactory	No	No	Yes	No
CertificateFactorySpi	No	No	Yes	No
CRL	No	No	Yes	No
X509Certificate	No	No	Yes	No
X509CRL	No	No	Yes	No
X509CRLEntry	No	No	Yes	No
CertificateEncodingException	Yes	No	Yes	No
CertificateException	Yes	No	Yes	No
CertificateExpiredException	No	No	Yes	No
CertificateNotYetValidException	No	No	Yes	No
CertificateParsingException	No	No	Yes	No
CRLException	No	No	Yes	No

Package java.security.interfaces

J2SE Classes	CDC	CLDC	Foundation Profile	MID Profile
DSAKey	No	No	Yes	No
DSAKeyPairGenerator	No	No	Yes	No
DSAParams	No	No	Yes	No
DSAPrivateKey	No	No	Yes	No

Table continued on following page

J2SE Classes	CDC	CLDC	Foundation Profile	MID Profile
DSAPublicKey	No	No	Yes	No
RSAKey	No	No	Yes	No
RSAPrivateCrtKey	No	No	Yes	No
RSAPrivateKey	No	No	Yes	No
RSAPublicKey	No	No	Yes	No

Package java.security.spec

J2SE Classes	CDC	CLDC	Foundation Profile	MID Profile
AlgorithmParameterSpec	No	No	Yes	No
KeySpec	No	No	Yes	No
DSAParameterSpec	No	No	Yes	No
DSAPrivateKeySpec	No	No	Yes	No
DSAPublicKeySpec	No	No	Yes	No
EncodedKeySpec	No	No	Yes	No
PKCS8EncodedKeySpec	No	No	Yes	No
RSAKeyGenParameterSpec	No	No	Yes	No
RSAPrivateCrtKeySpec	No	No	Yes	No
RSAPrivateKeySpec	No	No	Yes	No
RSAPublicKeySpec	No	No	Yes	No
X509EncodedKeySpec	No	No	Yes	No
InvalidKeySpecException	No	No	Yes	No
InvalidParameterSpec Exception	No	No	Yes	No

Package java.sql

J2SE Classes	CDC	CLDC	Foundation Profile	MID Profile
Array	Yes	No	Yes	No
Blob	No	No	No	No
CallableStatement	No	No	No	No
Clob	No	No	No	No

J2SE Classes	CDC	CLDC	Foundation Profile	MID Profile
Connection	Yes	Yes	Yes	Yes
DatabaseMetaData	No	No	No	No
Driver	No	No	No	No
PreparedStatement	No	No	No	No
Ref	No	No	No	No
ResultSet	No	No	No	No
ResultSetMetaData	No	No	No	No
SQLData	No	No	No	No
SQLInput	No	No	No	No
SQLOutput	No	No	No	No
Statement	No	No	No	No
Struct	No	No	No	No
Date	Yes	Yes	Yes	Yes
DriverManager	No	No	No	No
DriverPropertyInfo	No	No	No	No
SQLPermission	No	No	No	No
Time	No	No	No	No
Timestamp	No	No	No	No
Types	No	No	No	No
BatchUpdateException	No	No	No	No
DataTruncation	No	No	No	No
SQLException	No	No	No	No
SQLWarning	No	No	No	No

Package java.text

J2SE Classes	CDC	CLDC	Foundation Profile	MID Profile
AttributedCharacterIterator	No	No	Yes	No
CharacterIterator	No	No	Yes	No
Annotation	No	No	Yes	No
AttributedCharacterIterator. Attribute	No	No	Yes	No

Table continued on following page

J2SE Classes	CDC	CLDC	Foundation Profile	MID Profile
AttributedString	No	No	Yes	No
BreakIterator	No	No	Yes	No
ChoiceFormat	Yes	No	Yes	No
CollationElementIterator	No	No	Yes	No
CollationKey	No	No	Yes	No
Collator	No	No	Yes	No
DateFormat	Yes	No	Yes	No
DateFormatSymbols	Yes	No	Yes	No
DecimalFormat	Yes	No	Yes	No
DecimalFormatSymbols	Yes	No	Yes	No
FieldPosition	Yes	No	Yes	No
Format	Yes	No	Yes	No
MessageFormat	Yes	No	Yes	No
NumberFormat	Yes	No	Yes	No
ParsePosition	Yes	No	Yes	No
RuleBasedCollator	No	No	Yes	No
SimpleDateFormat	Yes	No	Yes	No
StringCharacterIterator	No	No	Yes	No
ParseException	Yes	No	Yes	No

Package java.util

J2SE Classes	CDC	CLDC	Foundation Profile	MID Profile
Collection	Yes	No	Yes	No
Comparator	Yes	No	Yes	No
Enumeration	Yes	Yes	Yes	Yes
EventListener	No	No	Yes	No
Iterator	Yes	No	Yes	No
List	No	No	No	No
ListIterator	Yes	No	Yes	No
Map	Yes		Yes	No

J2SE Classes	CDC	CLDC	Foundation Profile	MID Profile
Map.Entry	Yes	C	Yes	No
Observer		No	Yes	No
Set	Yes	No	Yes	No
SortedMap	Yes	No	Yes	No
SortedSet	Yes	No	Yes	No
AbstractCollection	Yes	No	Yes	No
AbstractList	Yes	No	Yes	No
AbstractMap	Yes	No	Yes	No
AbstractSequentialList	Yes	No	Yes	No
AbstractSet	Yes	No	Yes	No
ArrayList	Yes	No	Yes	No
Arrays	Yes	No	Yes	No
BitSet	Yes	No	Yes	No
Calendar	Yes	No	Yes	Yes
Collections	Yes	No	Yes	No
Date		No		No
Dictionary	Yes	No	Yes	No
EventObject		No	Yes	No
GregorianCalendar	Yes	No	Yes	No
HashMap	Yes	No	Yes	No
HashSet	Yes	No	Yes	No
Hashtable	Yes	Yes	Yes	Yes
LinkedList	Yes	No	Yes	No
ListResourceBundle	Yes	No	Yes	No
Locale	Yes	No	Yes	No
Observable		No	Yes	No
Properties	Yes	No	Yes	No
PropertyPermission	Yes	No	Yes	No
PropertyResourceBundle	Yes	No	Yes	No
Random	Yes	Yes	Yes	Yes
ResourceBundle	Yes	No	Yes	No

J2SE Classes	CDC	CLDC	Foundation Profile	MID Profile
SimpleTimeZone	Yes	No	Yes	No
Stack	Yes	Yes	Yes	Yes
StringTokenizer	Yes	No	Yes	No
Timer	No	No	Yes	Yes
TimerTask	No	No	Yes	Yes
TimeZone	Yes	Yes	Yes	Yes
TreeMap	Yes	No	Yes	No
TreeSet	Yes	No	Yes	No
Vector	Yes	Yes	Yes	Yes
WeakHashMap	Yes	No	Yes	No
ConcurrentModification Exception	Yes	No	Yes	No
EmptyStackException	Yes	Yes	Yes	Yes
MissingResourceException	Yes	No	Yes	Np
NoSuchElementException	Yes	Yes	Yes	Yes
TooManyListenersException	No	No	Yes	No

Package java.util.jar

J2SE Classes	CDC	CLDC	Foundation Profile	MID Profile
Attributes	Yes	No	Yes	No
Attributes.Name	Yes	No	Yes	No
JarEntry	Yes	No	Yes	No
JarFile	Yes	No	Yes	No
JarInputStream	Yes	No	Yes	No
JarOutputStream		No	Yes	No
Manifest	Yes	No	Yes	No
JarException	Yes	No	Yes	No

Package java.util.zip

J2SE Classes	CDC	CLDC	Foundation Profile	MID Profile
Checksum	Yes	No	Yes	No
Adler32	No	No	Yes	No
CheckedInputStream	No	No	Yes	No
CheckedOutputStream	No	No	Yes	No
CRC32	Yes	No	Yes	No
Deflater	No	No	Yes	No
DeflaterOutputStream	No	No	Yes	No
GZIPInputStream	No	No	Yes	No
GZIPOutputStream	No	No	Yes	No
Inflater	Yes	No	Yes	No
InflaterInputStream	Yes	No	Yes	No
ZipEntry	Yes	No	Yes	No
ZipFile	Yes	No	Yes	No
ZipInputStream	Yes	No	Yes	No
ZipOutputStream	No	No	Yes	No
DataFormatException	Yes	No	Yes	No
ZipException	Yes		Yes	No

Package javax.naming.directory

J2SE Classes	CDC	CLDC	Foundation Profile	MID Profile
Attribute	No	No	No	No
Attributes	Yes	No	Yes	No
DirContext	No	No	No	No
BasicAttribute	No	No	No	No
BasicAttributes	No	No	No	No
InitialDirContext	No	No	No	No
ModificationItem	No	No	No	No
SearchControls	No	No	No	No
SearchResult	No	No	No	No

Table continued on following page

J2SE Classes	CDC	CLDC	Foundation Profile	MID Profile
AttributeInUseException	No	No	No	No
AttributeModification Exception	No	No	No	No
InvalidAttributeIdentifier Exception	No	No	No	No
InvalidAttributesException	No	No	No	No
InvalidAttributeValue Exception	No	No	No	No
InvalidSearchControls Exception	No	No	Yes	No
InvalidSearchFilterException	No	No	No	No
NoSuchAttributeException	No	No	No	No
SchemaViolationException	No	No	No	No

Package javax.swing

J2SE Classes	CDC	CLDC	Foundation Profile	MID Profile
Action	No	No	No	No
BoundedRangeModel	No	No	No	No
ButtonModel	No	No	No	No
CellEditor	No	No	No	No
ComboBoxEditor	No	No	No	No
ComboBoxModel	No	No	No	No
DesktopManager	No	No	No	No
Icon	No	No	No	No
JComboBox.KeySelection Manager	No	No	No	No
ListCellRenderer	No	No	No	No
ListModel	No	No	No	No
ListSelectionModel	No	No	No	No
MenuElement	No	No	No	No
MutableComboBoxModel	No	No	No	No
Renderer	No	No	No	No

J2SE Classes	CDC	CLDC	Foundation Profile	MID Profile
RootPaneContainer	No	No	No	No
Scrollable	No	No	No	No
ScrollPaneConstants	No	No	No	No
SingleSelectionModel	No	No	No	No
SwingConstants	No	No	No	No
UIDefaults.ActiveValue	No	No	No	No
UIDefaults.LazyValue	No	No	No	No
WindowConstants	No	No	No	No
AbstractAction	No	No	No	No
AbstractButton	No	No	No	No
AbstractCellEditor	No	No	No	No
AbstractListModel	No	No	No	No
ActionMap	No	No	No	No
BorderFactory	No	No	No	No
Box	No	No	No	No
Box.Filler	No	No	No	No
BoxLayout	No	No	No	No
ButtonGroup	No	No	No	No
CellRendererPane	No	No	No	No
ComponentInputMap	No	No	No	No
DebugGraphics	No	No	No	No
DefaultBoundedRangeModel	No	No	No	No
DefaultButtonModel	No	No	No	No
DefaultCellEditor	No	No	No	No
DefaultComboBoxModel	No	No	No	No
DefaultDesktopManager	No	No	No	No
DefaultFocusManager	No	No	No	No
DefaultListCellRenderer	No	No	No	No
DefaultListCellRenderer. UIResource	No	No	No	No
DefaultListModel	No	No	No	No

Table continued on following page

J2SE Classes	CDC	CLDC	Foundation Profile	MID Profile
DefaultListSelectionModel	No	No	No	No
DefaultSingleSelectionModel	No	No	No	No
FocusManager	No	No	No	No
GrayFilter	No	No	No	No
ImageIcon	No	No	No	No
InputMap	No	No	No	No
InputVerifier	No	No	No	No
JApplet	No	No	No	No
JButton	No	No	No	No
JCheckBox	No	No	No	No
JCheckBoxMenuItem	No	No	No	No
JColorChooser	No	No	No	No
JComboBox	No	No	No	No
JComponent	No	No	No	No
JDesktopPane	No	No	No	No
JDialog	No	No	No	No
JEditorPane	No	No	No	No
JFileChooser	No	No	No	No
JFrame	No	No	No	No
JInternalFrame	No	No	No	No
JInternalFrame.JDesktopIcon	No	No	No	No
JLabel	No	No	No	No
JLayeredPane	No	No	No	No
JList	No	No	No	No
JMenu	No	No	No	No
JMenuBar	No	No	No	No
JMenuItem	No	No	No	No
JOptionPane	No	No	No	No
JPanel	No	No	No	No
JPasswordField	No	No	No	No
JPopupMenu	No	No	No	No

J2SE Classes	CDC	CLDC	Foundation Profile	MID Profile
JPopupMenu.Separator	No	No	No	No
JProgressBar	No	No	No	No
JRadioButton	No	No	No	No
JRadioButtonMenuItem	No	No	No	No
JRootPane	No	No	No	No
JScrollBar	No	No	No	No
JScrollPane	No	No	No	No
JSeparator	No	No	No	No
JSlider	No	No	No	No
JSplitPane	No	No	No	No
JTabbedPane	No	No	No	No
JTable	No	No	No	No
JTextArea	No	No	No	No
JTextField	No	No	No	No
JTextPane	No	No	No	No
JToggleButton	No	No	No	No
JToggleButton.ToggleButton Model	No	No	No	No
JToolBar	No	No	No	No
JToolBar.Separator	No	No	No	No
JToolTip	No	No	No	No
JTree	No	No	No	No
JTree.DynamicUtilTreeNode	No	No	No	No
JTree.EmptySelectionModel	No	No	No	No
JViewport	No	No	No	No
JWindow	No	No	No	No
KeyStroke	No	No	No	No
LookAndFeel	No	No	No	No
MenuSelectionManager	No	No	No	No
OverlayLayout	No	No	No	No
ProgressMonitor	No	No	No	No

Table continued on following page

J2SE Classes	CDC	CLDC	Foundation Profile	MID Profile
ProgressMonitorInputStream	No	No	No	No
RepaintManager	No	No	No	No
ScrollPaneLayout	No	No	No	No
ScrollPaneLayout.UIResource	No	No	No	No
SizeRequirements	No	No	No	No
SizeSequence	No	No	No	No
SwingUtilities	No	No	No	No
Timer	No	No	Yes	Yes
ToolTipManager	No	No	No	No
UIDefaults	No	No	No	No
UIDefaults.LazyInputMap	No	No	No	No
UIDefaults.ProxyLazyValue	No	No	No	No
UIManager	No	No	No	No
UIManager.LookAndFeelInfo	No	No	No	No
ViewportLayout	No	No	No	No
UnsupportedLookAndFeel Exception	No	No	No	No

Package javax.swing.filechooser

J2SE Classes	CDC	CLDC	Foundation Profile	MID Profile
FileFilter	Yes	No	Yes	No
FileSystemView	No	No	No	No
FileView	No	No	No	No

Package org.omg.CORBA

J2SE Classes	CDC	CLDC	Foundation Profile	MID Profile
ARG_IN	No	No	No	No
ARG_INOUT	No	No	No	No
ARG_OUT	No	No	No	No
BAD_POLICY	No	No	No	No

J2SE Classes	CDC	CLDC	Foundation Profile	MID Profile
BAD_POLICY_TYPE	No	No	No	No
BAD_POLICY_VALUE	No	No	No	No
CTX_RESTRICT_SCOPE	No	No	No	No
Current	No	No	No	No
CurrentOperations	No	No	No	No
CustomMarshal	No	No	No	No
DataInputStream	Yes	Yes	Yes	Yes
DataOutputStream	Yes	Yes	Yes	Yes
DomainManager	No	No	No	No
DomainManagerOperations	No	No	No	No
DynAny	No	No	No	No
DynArray	No	No	No	No
DynEnum	No	No	No	No
DynFixed	No	No	No	No
DynSequence	No	No	No	No
DynStruct	No	No	No	No
DynUnion	No	No	No	No
DynValue	No	No	No	No
IDLType	No	No	No	No
IDLTypeOperations	No	No	No	No
IRObject	No	No	No	No
IRObjectOperations	No	No	No	No
Object	Yes	Yes	Yes	Yes
OMGVMCID	No	No	No	No
Policy	Yes	No	Yes	No
PolicyOperations	No	No	No	No
PRIVATE_MEMBER	No	No	No	No
PUBLIC_MEMBER	No	No	No	No
Repository	No	No	No	No
UNSUPPORTED_POLICY	No	No	No	No

Table continued on following page

J2SE Classes	CDC	CLDC	Foundation Profile	MID Profile
UNSUPPORTED_POLICY_VALUE	No	No	No	No
VM_ABSTRACT	No	No	No	No
VM_CUSTOM	No	No	No	No
VM_NONE	No	No	No	No
VM_TRUNCATABLE	No	No	No	No
_IDLTypeStub	No	No	No	No
_PolicyStub	No	No	No	No
Any	No	No	No	No
AnyHolder	No	No	No	No
AnySeqHelper	No	No	No	No
AnySeqHolder	No	No	No	No
BooleanHolder	No	No	No	No
BooleanSeqHelper	No	No	No	No
BooleanSeqHolder	No	No	No	No
ByteHolder	No	No	No	No
CharHolder	No	No	No	No
CharSeqHelper	No	No	No	No
CharSeqHolder	No	No	No	No
CompletionStatus	No	No	No	No
CompletionStatusHelper	No	No	No	No
Context	No	No	No	No
ContextList	No	No	No	No
CurrentHelper	No	No	No	No
CurrentHolder	No	No	No	No
DefinitionKind	No	No	No	No
DefinitionKindHelper	No	No	No	No
DoubleHolder	No	No	No	No
DoubleSeqHelper	No	No	No	No
DoubleSeqHolder	No	No	No	No
DynamicImplementation	No	No	No	No
Environment	No	No	No	No

J2SE Classes	CDC	CLDC	Foundation Profile	MID Profile
ExceptionList	No	No	No	No
FieldNameHelper	No	No	No	No
FixedHolder	No	No	No	No
FloatHolder	No	No	No	No
FloatSeqHelper	No	No	No	No
FloatSeqHolder	No	No	No	No
IdentifierHelper	No	No	No	No
IDLTypeHelper	No	No	No	No
Initializer	No	No	No	No
IntHolder	No	No	No	No
LongHolder	No	No	No	No
LongLongSeqHelper	No	No	No	No
LongLongSeqHolder	No	No	No	No
LongSeqHelper	No	No	No	No
LongSeqHolder	No	No	No	No
NamedValue	No	No	No	No
NameValuePair	No	No	No	No
NameValuePairHelper	No	No	No	No
NVList	No	No	No	No
ObjectHelper	No	No	No	No
ObjectHolder	No	No	No	No
OctetSeqHelper	No	No	No	No
OctetSeqHolder	No	No	No	No
ORB	No	No	No	No
PolicyHelper	No	No	No	No
PolicyHolder	No	No	No	No
PolicyListHelper	No	No	No	No
PolicyListHolder	No	No	No	No
PolicyTypeHelper	No	No	No	No
Principal	No	No	No	No
PrincipalHolder	No	No	No	No

Table continued on following page

J2SE Classes	CDC	CLDC	Foundation Profile	MID Profile
RepositoryIdHelper	No	No	No	No
Request	No	No	No	No
ServerRequest	No	No	No	No
ServiceDetail	No	No	No	No
ServiceDetailHelper	No	No	No	No
ServiceInformation	No	No	No	No
ServiceInformationHelper	No	No	No	No
ServiceInformationHolder	No	No	No	No
SetOverrideType	No	No	No	No
SetOverrideTypeHelper	No	No	No	No
ShortHolder	No	No	No	No
ShortSeqHelper	No	No	No	No
ShortSeqHolder	No	No	No	No
StringHolder	No	No	No	No
StringValueHelper	No	No	No	No
StructMember	No	No	No	No
StructMemberHelper	No	No	No	No
TCKind	No	No	No	No
TypeCode	No	No	No	No
TypeCodeHolder	No	No	No	No
ULongLongSeqHelper	No	No	No	No
ULongLongSeqHolder	No	No	No	No
ULongSeqHelper	No	No	No	No
ULongSeqHolder	No	No	No	No
UnionMember	No	No	No	No
UnionMemberHelper	No	No	No	No
UShortSeqHelper	No	No	No	No
UShortSeqHolder	No	No	No	No
ValueBaseHelper	No	No	No	No
ValueBaseHolder	No	No	No	No
ValueMember	No	No	No	No

J2SE Classes	CDC	CLDC	Foundation Profile	MID Profile
ValueMemberHelper	No	No	No	No
VersionSpecHelper	No	No	No	No
VisibilityHelper	No	No	No	No
WCharSeqHelper	No	No	No	No
WCharSeqHolder	No	No	No	No
WStringValueHelper	No	No	No	No
BAD_CONTEXT	No	No	No	No
BAD_INV_ORDER	No	No	No	No
BAD_OPERATION	No	No	No	No
BAD_PARAM	No	No	No	No
BAD_TYPECODE	No	No	No	No
Bounds	No	No	No	No
COMM_FAILURE	No	No	No	No
DATA_CONVERSION	No	No	No	No
FREE_MEM	No	No	No	No
IMP_LIMIT	No	No	No	No
INITIALIZE	No	No	No	No
INTERNAL	No	No	No	No
INTF_REPOS	No	No	No	No
INV_FLAG	No	No	No	No
INV_IDENT	No	No	No	No
INV_OBJREF	No	No	No	No
INV_POLICY	No	No	No	No
INVALID_TRANSACTION	No	No	No	No
MARSHAL	No	No	No	No
NO_IMPLEMENT	No	No	No	No
NO_MEMORY	No	No	No	No
NO_PERMISSION	No	No	No	No
NO_RESOURCES	No	No	No	No
NO_RESPONSE	No	No	No	No
OBJ_ADAPTER	No	No	No	No

Table continued on following page

J2SE Classes	CDC	CLDC	Foundation Profile	MID Profile
OBJECT_NOT_EXIST	No	No	No	No
PERSIST_STORE	No	No	No	No
PolicyError	No	No	No	No
SystemException	No	No	No	No
TRANSACTION_REQUIRED	No	No	No	No
TRANSACTION_ ROLLEDBACK	No	No	No	No
TRANSIENT	No	No	No	No
UNKNOWN	No	No	No	No
UnknownUserException	No	No	No	No
UserException	No	No	No	No
WrongTransaction	No	No	No	No

Package org.omg.CORBA_2_3.portable

J2SE Classes	CDC	CLDC	Foundation Profile	MID Profile
Delegate	No	No	No	No
InputStream	Yes	Yes	Yes	Yes
ObjectImpl	No	No	No	No
OutputStream	Yes	Yes	Yes	Yes

Package org.omg.CORBA.portable

J2SE Classes	CDC	CLDC	Foundation Profile	MID Profile
BoxedValueHelper	No	No	No	No
CustomValue	No	No	No	No
IDLEntity	No	No	No	No
InvokeHandler	No	No	No	No
ResponseHandler	No	No	No	No
Streamable	No	No	No	No
StreamableValue	No	No	No	No
ValueBase	No	No	No	No

J2SE Classes	CDC	CLDC	Foundation Profile	MID Profile
ValueFactory	No	No	No	No
Delegate	No	No	No	No
InputStream	Yes	Yes	Yes	Yes
ObjectImpl	No	No	No	No
OutputStream	Yes	Yes	Yes	Yes
ServantObject	No	No	No	No
ApplicationException	No	No	No	No
IndirectionException	No	No	No	No
RemarshalException	No	No	No	No
UnknownException	No	No	No	No

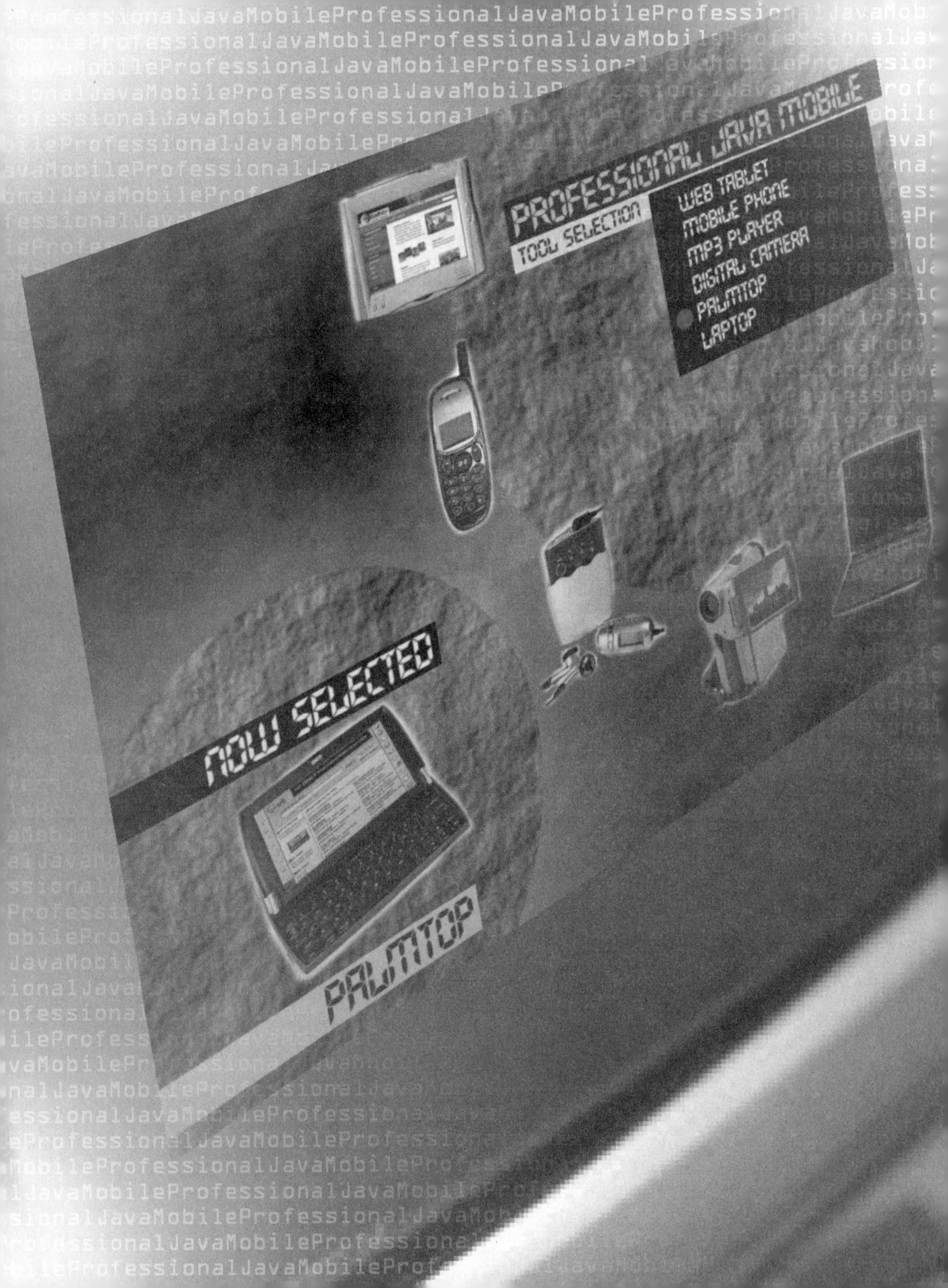

PROFESSIONAL JAVA MOBILE
TOOL SELECTION
WEB TABLET
MOBILE PHONE
MP3 PLAYER
DIGITAL CAMERA
PALMTOP
LAPTOP
NOW SELECTED
PALMTOP

B

The CLDC Classes

The following appendix lists the classes available in the CLDC. The classes represent a subset of the classes in the Java 2 Standard Edition, and if you are familiar with those you should have no problem with them as, their usage is identical although is some cases the source code has been optimized for limited devices.

Due to the general familiarity with the basic classes we have kept description of the various classes to a minimum. If you have any trouble with using a class, please refer to the official java documentation available through http://java.sun.com

The new IO classes can be found in Appendix D. Although you can use the classes listed below for standard IO operations, the recommended method for maximum robustness and cross platform compatibility is using the `javax.microedition.io` classes as listed in the aforementioned appendix.

java.io Package

This package provides system input and output through data streams, serialization and the file system.

Interfaces

DataInput

```
public interface DataInput
```

The `DataInput` interface provides for reading bytes from a binary stream and reconstructing from them data in any of the Java primitive types.

DataOutput

```
public interface DataOutput
```

The `DataOutput` interface provides for converting data from any of the Java primitive types to a series of bytes and writing these bytes to a binary stream.

Classes

InputStream

```
public abstract class InputStream extends Object
```

This abstract class is the superclass of all classes representing an input stream of bytes. Applications that need to define a subclass of `InputStream` must always provide a method that returns the next byte of input.

DataInputStream

```
public class DataInputStream extends InputStream implements DataInput
```

A data input stream lets an application read primitive Java data types from an underlying input stream in a machine-independent way. An application uses a data output stream to write data that can later be read by a data input stream.

Reader

```
public abstract class Reader extends Object
```

This is an abstract class for reading character streams.

Writer

```
public abstract class Writer extends Object
```

This is an abstract class for writing to character streams.

OutputStream

```
public abstract class OutputStream extends Object
```

This abstract class is the superclass of all classes representing an output stream of bytes.

DataOutputStream

```
public class DataOutputStream extends OutputStream implements DataOutput
```

A data input stream lets an application write primitive Java data types to an output stream in a portable way.

InputStreamReader

```
public class InputStreamReader extends Reader
```

An `InputStreamReader` is a bridge from byte streams to character streams: It reads bytes and translates them into characters.

PrintStream

```
public class PrintStream extends OutputStream
```

A `PrintStream` adds functionality to another output stream, namely the ability to print representations of various data values conveniently.

OutputStreamWriter

```
public class OutputStreamWriter extends Writer
```

An `OutputStreamWriter` is a bridge from character streams to byte streams: Characters written to it are translated into bytes. The encoding that it uses may be specified by name, or the platform's default encoding may be accepted.

ByteArrayInputStream

```
public class ByteArrayInputStream extends InputStream
```

A `ByteArrayInputStream` contains an internal buffer that contains bytes that may be read from the stream. An internal counter keeps track of the next byte to be supplied by the `read` method.

ByteArrayOutputStream

```
public class ByteArrayOutputStream extends OutputStream
```

This class implements an output stream in which the data is written into a byte array. The buffer automatically grows as data is written to it. The data can be retrieved using `toByteArray()` and `toString()`.

IOException

```
public class IOException extends Exception
```

Signals that an I/O exception of some sort has occurred. This class is the general class of exceptions produced by failed or interrupted I/O operations.

InterruptedIOException

```
public class InterruptedIOException extends IOException
```

Signals that an I/O operation has been interrupted. An `InterruptedIOException` is thrown to indicate that an input or output transfer has been terminated because the thread performing it was terminated.

EOFException

```
public class EOFException extends IOException
```

Signals that an end of file or end of stream has been reached unexpectedly during input.

UTFDataFormatException

```
public class UTFDataFormatException extends IOException
```

Signals that a malformed UTF-8 string has been read in a data input stream or by any class that implements the `DataInput` interface.

UnsupportedEncodingException

```
public class UnsupportedEncodingException extends IOException
```

This signals that character encoding is not supported.

java.util Package

This package contains the collections framework, legacy collection classes, date and time facilities and miscellaneous utility classes.

Interfaces

Enumeration

```
public interface Enumeration
```

An object that implements the `Enumeration` interface generates a series of elements, one at a time. Successive calls to the `nextElement` method return successive elements of the series.

Classes

Date

```
public class Date extends Object
```

The class `Date` represents a specific instant in time, with millisecond precision. This class has been made into a subset for the MID Profile based on JDK 1.3. In the full API, the class `Date` had two additional functions. It allowed the interpretation of dates as year, month, day, hour, minute, and second values.

Calendar

```
public abstract class Calendar extends Object
```

`Calendar` is an abstract class for getting and setting dates using a set of integer fields such as `YEAR`, `MONTH`, `DAY`, and so on.

Random

```
public class Random extends Object
```

An instance of this class is used to generate a stream of pseudo-random numbers.

Hashtable

```
public class Hashtable extends Object
```

This class implements a hashtable, which maps keys to values. Any non-`null` object can be used as a key or as a value.

This is a hashtable enumerator class. This class should remain opaque to the client. It will use the `Enumeration` interface.

Vector

```
public class Vector extends Object
```

The `Vector` class implements a growable array of objects. Like an array, it contains components that can be accessed using an integer index.

Stack

```
public class Stack extends Vector
```

The `Stack` class represents a last-in-first-out (LIFO) stack of objects.

TimeZone

```
public abstract class TimeZone extends Object
```

`TimeZone` represents a time zone offset and it also figures out daylight savings.

EmptyStackException

```
public class EmptyStackException extends RuntimeException
```

Thrown by methods in the `Stack` class to indicate that the stack is empty.

NoSuchElementException

```
public class NoSuchElementException extends RuntimeException
```

Thrown by the `nextElement` method of an `Enumeration` to indicate that there are no more elements in the enumeration.

java.lang Package

This package provides classes that are fundamental to the design of the Java programming language.

Interfaces

Runnable

```
public interface Runnable
```

The `Runnable` interface should be implemented by any class whose instances are intended to be executed by a thread. The class must define a method of no arguments called `run`.

Classes

Object

```
public class Object
```

Class `Object` is the root of the class hierarchy. Every class has `Object` as a superclass. All objects, including arrays, implement the methods of this class.

Throwable

```
public class Throwable extends Object
```

The `Throwable` class is the superclass of all errors and exceptions in the Java language.

Character

```
public final class Character extends Object
```

The `Character` class wraps a value of the primitive type `char` in an object. An object of type `Character` contains a single field whose type is `char`.

Byte

```
public final class Byte extends Object
```

The `Byte` class is the standard wrapper for byte values.

Integer

```
public final class Integer extends Object
```

The `Integer` class wraps a value of the primitive type `int` in an object. An object of type `Integer` contains a single field whose type is `int`.

String

```
public final class String extends Object
```

The `String` class represents character strings.

Thread

```
public class Thread extends Object implements Runnable
```

This class represents a thread of execution in a program. The Java Virtual Machine allows an application to have multiple threads of execution running concurrently.

Class

```
public final class Class extends Object
```

Instances of the class `Class` represent classes and interfaces in a running Java application.

StringBuffer

```
public final class StringBuffer extends Object
```

A string buffer implements a mutable sequence of characters.

Math

```
public final class Math extends Object
```

The class `Math` contains methods for performing basic numeric operations.

Boolean

```
public final class Boolean extends Object
```

The `Boolean` class wraps a value of the primitive type `boolean` in an object.

Short

```
public final class Short extends Object
```

The `Short` class is the standard wrapper for short values.

System

```
public final class System extends Object
```

The `System` class contains several useful class fields and methods. It cannot be instantiated.

Long

```
public final class Long extends Object
```

The Long class wraps a value of the primitive type long in an object.

Runtime

```
public class Runtime extends Object
```

Every Java application has a single instance of class Runtime that allows the application to interface with the environment in which the application is running.

Exception

```
public class Exception extends Throwable
```

The class Exception and its subclasses are a form of Throwable that indicates conditions that a reasonable application might want to catch.

RuntimeException

```
public class RuntimeException extends Exception
```

RuntimeException is the superclass of those exceptions that can be thrown during the normal operation of the Java Virtual Machine.

InterruptedException

```
public class InterruptedException extends Exception
```

Thrown when a thread is waiting, sleeping, or otherwise paused for a long time and another thread interrupts it using the interrupt method in class Thread.

IllegalArgumentException

```
public class IllegalArgumentException extends RuntimeException
```

Thrown to indicate that a method has been passed an illegal or inappropriate argument.

IllegalThreadStateException

```
public class IllegalThreadStateException extends IllegalArgumentException
```

Thrown to indicate that a thread is not in an appropriate state for the requested operation.

NumberFormatException

```
public class NumberFormatException extends IllegalArgumentException
```

Thrown to indicate that the application has attempted to convert a string to one of the numeric types, but that the string does not have the appropriate format.

InstantiationException

```
public class InstantiationException extends Exception
```

Thrown when an application tries to create an instance of a class using the `newInstance` method in class `Class`, but the specified class object cannot be instantiated because it is an interface or is an abstract class.

SecurityException

```
public class SecurityException extends RuntimeException
```

Thrown by the security manager to indicate a security violation.

NegativeArraySizeException

```
public class NegativeArraySizeException extends RuntimeException
```

Thrown if an application tries to create an array with negative size.

IndexOutOfBoundsException

```
public class IndexOutOfBoundsException extends RuntimeException
```

Thrown to indicate that an index of some sort (such as to an array, to a string, or to a vector) is out of range. Applications can subclass this class to indicate similar exceptions.

ArrayIndexOutOfBoundsException

```
public class ArrayIndexOutOfBoundsException extends IndexOutOfBoundsException
```

Thrown to indicate that an array has been accessed with an illegal index. The index is either negative or greater than or equal to the size of the array.

ClassNotFoundException

```
public class ClassNotFoundException extends Exception
```

Thrown when an application tries to load in a class through its string name (using the `forName` method in class `Class`) but no definition for the class with the specified name could be found.

IllegalAccessException

```
public class IllegalAccessException extends Exception
```

Thrown when an application tries to load in a class, but the currently executing method does not have access to the definition of the specified class, because the class is not public and in another package.

ClassCastException

```
public class ClassCastException extends RuntimeException
```

Thrown to indicate that the code has attempted to cast an object to a subclass of which it is not an instance.

ArrayStoreException

```
public class ArrayStoreException extends RuntimeException
```

Thrown to indicate that an attempt has been made to store the wrong type of object into an array of objects.

StringIndexOutOfBoundsException

```
public class StringIndexOutOfBoundsException extends IndexOutOfBoundsException
```

Thrown by the `charAt` method in class `String` and by other `String` methods to indicate that an index is either negative or greater than or equal to the size of the string.

NullPointerException

```
public class NullPointerException extends RuntimeException
```

Thrown when an application attempts to use `null` in a case where an object is required.

IllegalMonitorStateException

```
public class IllegalMonitorStateException extends RuntimeException
```

Thrown to indicate that a thread has attempted to wait on an object's monitor or to notify other threads waiting on an object's monitor without owning the specified monitor.

ArithmeticException

```
public class ArithmeticException extends RuntimeException
```

Thrown when an exceptional arithmetic condition has occurred. For example, an integer "divide by zero" throws an instance of this class.

Error

```
public class Error extends Throwable
```

An `Error` is a subclass of `Throwable` that indicates serious problems that a reasonable application should not try to catch.

VirtualMachineError

```
public abstract class VirtualMachineError extends Error
```

Thrown to indicate that the Java Virtual Machine is broken or has run out of resources necessary for it to continue operating.

OutOfMemoryError

```
public class OutOfMemoryError extends VirtualMachineError
```

Thrown when the Java Virtual Machine cannot allocate an object because it is out of memory and no more memory could be made available by the garbage collector.

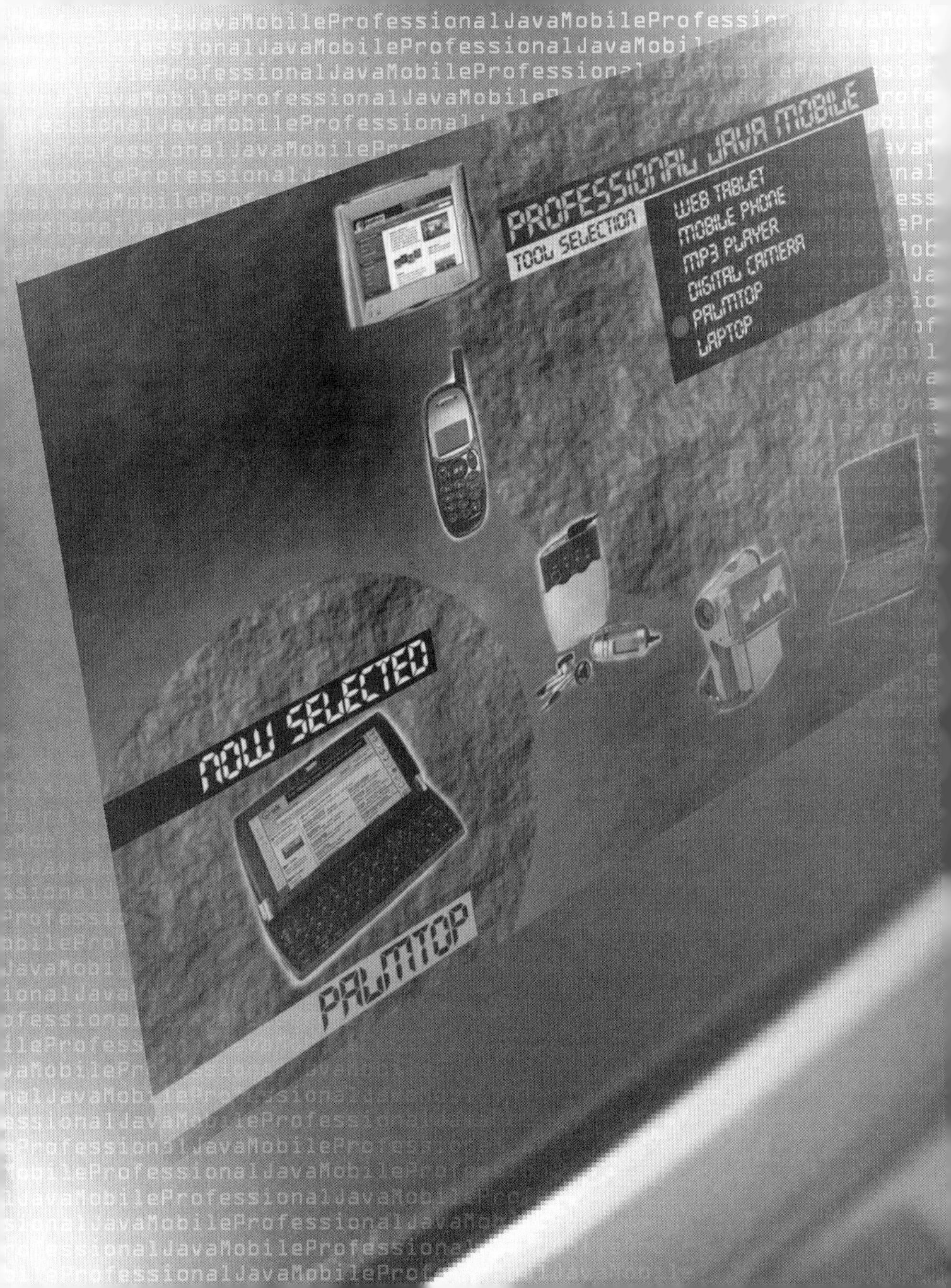
PROFESSIONAL JAVA MOBILE
TOOL SELECTION
WEB TABLET
MOBILE PHONE
MP3 PLAYER
DIGITAL CAMERA
PALMTOP
LAPTOP
NOW SELECTED
PALMTOP

C

The MIDP classes

The following classes are those found in addition to the CDC classes as listed in the previous appendix. These classes define additional functionality for devices that match the MIDP Profile and define graphical functionality, access to the `Midlet` class and its associated classes. (The `Midlet` class defines the lifecycle for MIDP applications - see Chapters 4 and 5 for further information). Finally, data access classes as defined in the `javax.microedition.rms` package are listed and explained in brief.

For further information on the usage of all of these classes please refer to the aforementioned chapters on the MIDP and the CLDC.

javax.microedition.lcdui Package

Classes

Image

```
public class Image
```

The `Image` class is used to hold graphical image data.

Methods

Signature	Notes
`public static Image createImage(final int width, final int height) throws IllegalArgumentException`	Creates a new, mutable image for off-screen drawing. The width and height of the image must both be greater than zero: ❑ `width`: the width of the new image, in pixels ❑ `height`: the height of the new image, in pixels
`public static Image createImage(Image image)`	Creates an immutable image from a source image: ❑ `source`: the source image to be copied
`public static Image createImage(String name) throws java.io.IOException`	Creates an immutable image from decoded image data obtained from the named resource: ❑ `name`: the name of the resource containing the image data in one of the supported image formats
`public static Image createImage(final byte imagedata, final int imageoffset, final int imagelength) throws ArrayIndexOutOfBoundsException, IllegalArgumentException`	Creates an immutable image that is decoded from the data stored in the specified byte array at the specified offset and length. The data must be in a self-identifying image file format supported by the implementation, such as PNG: ❑ `imageData`: the array of image data in a supported image format ❑ `imageOffset`: the offset of the start of the data in the array ❑ `imageLength`: the length of the data in the array
`public Graphics getGraphics() throws IllegalStateException`	Creates a new `Graphics` object that renders to this image. This image must be mutable; it is illegal to call this method on an immutable image.
`public int getWidth()`	Returns: width of the image.
`public int getHeight()`	Returns: height of the image.
`public boolean isMutable()`	Check if this image is mutable. Mutable images can be modified by rendering to them through a `Graphics` object obtained from the `getGraphics()` method of this object.

Displayable

```
public abstract class Displayable
```

An object that has the capability of being placed on the display. A `Displayable` object may have commands and listeners associated with it. The contents displayed and their interaction with the user are defined by subclasses.

Methods

Signature	Notes
`public boolean isShown()`	Checks if the `Displayable` is actually visible on the display.
`public void addCommand(Command cmd)`	Adds a command to the `Displayable`. The implementation may choose, for example, to add the command to any of the available soft buttons or place it in a menu: ❑ `cmd`: the command to be added
`public void removeCommand (Command cmd)`	Removes a command from the `Displayable`: ❑ `cmd`: the command to be removed
`public void setCommandListener (CommandListener l)`	Sets a listener for command to this `Displayable`, replacing any previous `CommandListener`. A `null` reference is allowed and has the effect of removing any existing listener: ❑ `l`: the new listener, or null

AlertType

```
public class AlertType
```

The `AlertType` provides an indication of the nature of an `Alert`. `Alert` objects are used by an application to present information to the user and wait before proceeding.

Constructors

Signature	Notes
`protected AlertType()`	Protected constructor for subclasses

Fields

Signature	Notes
`public static final AlertType INFO`	Alert for information
`public static final AlertType WARNING`	Alert for warning

Table continued on following page

Signature	Notes
`public static final AlertType ERROR`	Alert for error
`public static final AlertType ALARM`	Alert for alarm
`public static final AlertType CONFIRMATION`	Alert for confirmation

Methods

Signature	Notes
`public boolean playSound(Display display)`	Alerts the user by playing the sound for this `AlertType`. The AlertType instance is used as a hint by the device to generate an appropriate sound.

Screen

```
public abstract class Screen extends Displayable
```

The common superclass of all high-level user interface classes. The contents displayed and their interaction with the user are defined by subclasses.

Using subclass-defined methods, the application may change the contents of a `Screen` object while it is shown to the user. If this occurs, and the `Screen` object is visible, the display will be updated automatically. That is, the implementation will refresh the display in a timely fashion without waiting for any further action by the application.

Constructors

Signature	Notes
`Screen()`	Creates a new `Screen` object with no title and no ticker
`Screen(String title)`	Creates a new `Screen` object with the given title and with no ticker: ❑ `title`: the title of the `Screen`, or null for no title

Methods

Signature	Notes
`public String getTitle()`	Gets the title of the `Screen`. Returns `null` if there is no title.

Signature	Notes
`public void setTitle(String s)`	Sets the title of the `Screen`. If `null` is given, removes the title: ❑ `s`: the new title, or `null` for no title
`public void setTicker(Ticker ticker)`	Set a ticker for use with this `Screen`, replacing any previous ticker: ❑ `ticker`: the ticker object used on this screen
`public Ticker getTicker()`	Returns the `Ticker` object used, or null if no ticker is present

TextBox

```
public class TextBox extends Screen
```

A `TextBox` has a maximum size, which is the maximum number of characters that can be stored in the object at any time (its capacity). This limit is enforced when the user is editing text within the `TextBox`, as well as when the application program calls methods on the TextBox that modify its contents.

Constructors

Signature	Notes
`public TextBox(String title, String text, final int maxSize, final int constraints) throws IllegalArgumentException, IllegalArgumentException, IllegalArgumentException, IllegalArgumentException`	Creates a new `TextBox` object with the given title String, initial contents, maximum size in characters, and constraints: ❑ `title`: the title text to be shown with the display ❑ `text`: the initial contents of the text editing area, `null` may be used to indicate no initial content ❑ `maxSize`: the maximum capacity in characters ❑ `constraints`: see the input constraints section of the `TextField` class

Methods

Signature	Notes
`public String getString()`	Returns the contents of the `TextBox` as a String value.
`public void setString(String text) throws IllegalArgumentException, IllegalArgumentException`	Sets the contents of the `TextBox` as a String value, replacing the previous contents: ❑ `text`: the new value of the `TextBox`

Table continued on following page

Signature	Notes
`public int getChars(final char data) throws ArrayIndexOutOfBoundsException`	Copies the contents of the `TextBox` into a character array starting at index zero. Array elements beyond the characters copied are left unchanged: ❑ `data`: the character array to receive the value
`public void setChars(final char data, final int offset, final int length) throws ArrayIndexOutOfBoundsException, IllegalArgumentException, IllegalArgumentException`	Sets the contents of the `TextBox` from a character array, replacing the previous contents: ❑ `data`: the source of the character data ❑ `offset`: the beginning of the region of characters to copy ❑ `length`: the number of characters to copy
`public void insert(String src, final int position) throws IllegalArgumentException, IllegalArgumentException`	Inserts a string into the contents of the `TextBox`: ❑ `src`: the String to be inserted ❑ `position`: the position at which insertion is to occur
`public void insert(final char data, final int offset, final int length, final int position) throws ArrayIndexOutOfBoundsException, IllegalArgumentException, IllegalArgumentException`	Inserts a sub-range of an array of characters into the contents of the `TextBox`: ❑ `data`: the source of the character data ❑ `offset`: the beginning of the region of characters to copy ❑ `length`: the number of characters to copy ❑ `position`: the position at which insertion is to occur
`public void delete(final int offset, final int length) throws StringIndexOutOfBoundsException, IllegalArgumentException`	Deletes characters from the `TextBox`: ❑ `offset`: the beginning of the region to be deleted ❑ `length`: the number of characters to be deleted
`public int getMaxSize()`	Returns the maximum size (number of characters) that can be stored in this `TextBox`.

Signature	Notes
`public int setMaxSize(final int maxSize) throws IllegalArgumentException`	Sets the maximum size (number of characters) that can be contained in this `TextBox`: ❑ `maxSize`: the new maximum size
`public int size()`	Returns the number of characters that are currently stored in this `TextBox`.
`public int getCaretPosition()`	Returns the current input position.
`public void setConstraints(final int constraints) throws IllegalArgumentException`	Sets the input constraints of the `TextBox`: ❑ `constraints`: see the input constraints section of the `TextField` class
`public int getConstraints()`	Returns: the current constraints value

Item

```
public abstract class Item
```

A superclass for components that can be added to a `Form` or an `Alert`. All `Item` objects have a label field, which is a string that is attached to the item. The label is typically displayed near the component when it is displayed within a screen.

Constructors

Signature	Notes
`Item(String label)`	Creates a new item with a given label: ❑ `label`: the label string; `null` is allowed
`Item()`	Creates a new item with a `null` label

Methods

Signature	Notes
`public void setLabel(String label)`	Sets the label of the `Item`: ❑ `label`: the label string
`public String getLabel()`	Returns the label of this `Item` object

ChoiceGroup

```
public class ChoiceGroup extends Item implements Choice
```

A `ChoiceGroup` is a group of selectable elements intended to be placed within a `Form`. The group may be created with a mode that requires a single choice to be made or that allows multiple choices.

Constructors

Signature	Notes
`public ChoiceGroup(String label, final int choiceType) throws IllegalArgumentException`	Creates a new, empty `ChoiceGroup`, specifying its title and its type: ❑ `label`: the item's label ❑ `choiceType`: either `EXCLUSIVE` or `MULTIPLE`
`public ChoiceGroup(String label, final int choiceType, String stringElements, Image imageElements) throws NullPointerException, NullPointerException, IllegalArgumentException, IllegalArgumentException`	Creates a new `ChoiceGroup`, specifying its title, the type of the `ChoiceGroup`, and an array of `String` and `Image` objects to be used as its initial contents: ❑ `label`: the item's label ❑ choiceType: `EXCLUSIVE` or `MULTIPLE` ❑ `stringElements`: set of strings specifying the string parts of the `ChoiceGroup` elements ❑ `imageElements`: set of images specifying the image parts of the `ChoiceGroup` elements

Methods

Signature	Notes
`public int size()`	Returns the number of elements in the `ChoiceGroup`
`public String getString(final int i) throws IndexOutOfBoundsException`	Returns the String part of the element: ❑ `elementNum`: the index of the element to be queried
`public Image getImage(final int i) throws IndexOutOfBoundsException`	Returns: the image part of the element, or `null` if there is no image: ❑ `elementNum`: the number of the element to be queried
`public int append(String stringElement, Image imageElement) throws IllegalArgumentException, NullPointerException`	Returns: the assigned index of the element: ❑ `stringPart`: the String part of the element to be added ❑ `imagePart`: the image part of the element to be added, or `null` if there is no image part

Signature	Notes
`public void insert(final int index, String stringElement, Image imageElement) throws IndexOutOfBoundsException, IllegalArgumentException, NullPointerException`	❑ `elementNum`: the index of the element where insertion is to occur ❑ `stringPart`: the string part of the element to be inserted ❑ `imagePart`: the image part of the element to be inserted, or `null` if there is no image part
`public void delete(final int index) throws IndexOutOfBoundsException`	❑ elementNum: the index of the element to be deleted
`public void set(final int index, String stringElement, Image imageElement) throws IndexOutOfBoundsException, IllegalArgumentException, NullPointerException`	❑ `elementNum`: the index of the element to be set ❑ `stringPart`: the string part of the new element ❑ `imagePart`: the image part of the element, or null if there is no image part
`public boolean isSelected(final int index) throws IndexOutOfBoundsException`	Returns the selection state of the element: ❑ `elementNum`: the index of the element to be queried
`public int getSelectedIndex()`	Returns the index number of an element in the `ChoiceGroup` that is selected, or `-1` if none.
`public int getSelectedFlags(final boolean selectedArray_return) throws IllegalArgumentException, NullPointerException`	Queries the state of a `ChoiceGroup` and returns the state of all elements in the boolean array `selectedArray_return`: ❑ `selectedArray_return`: array to contain the results.
`public void setSelectedIndex(final int index, final boolean selected) throws IndexOutOfBoundsException`	For `ChoiceGroup` objects of type `MULTIPLE`, this simply sets an individual element's selected state. For `ChoiceGroup` objects of type `EXCLUSIVE`, this can be used only to select an element: ❑ `elementNum`: the number of the element. Indexing of the elements is zero-based. ❑ `selected`: the new state of the element. `true`=selected, `false`=not selected.

Table continued on following page

Signature	Notes
`public void setSelectedFlags(final boolean selectedArray) throws IllegalArgumentException, NullPointerException`	❑ Attempts to set the selected state of every element in the `ChoiceGroup`. The array must be at least as long as the size of the `ChoiceGroup`: `selectedArray`: an array in which the method collect the selection status

List

```
public class List extends Screen implements Choice
```

The `List` class is a `Screen` containing a list of choices. Most of the behavior is common with class `ChoiceGroup` and the common API is defined in interface `Choice`.

Constructors

Signature	Notes
`public List(String title, final int listType) throws IllegalArgumentException`	Creates a new, empty `List`, specifying its title and the type of the list: ❑ `title`: the screen's title ❑ `listType`: one of `IMPLICIT`, `EXCLUSIVE`, or `MULTIPLE`
`public List(String title, final int listType, String stringElements, Image imageElements) throws NullPointerException, NullPointerException, IllegalArgumentException, IllegalArgumentException`	Creates a new `List`, specifying its title, the type of the `List`, and an array of strings and images to be used as its initial contents: ❑ `title`: the screen's title ❑ `listType`: one of `IMPLICIT`, `EXCLUSIVE`, or `MULTIPLE` ❑ `stringElements`: set of strings specifying the string parts of the List elements ❑ `imageElements`: set of images specifying the image parts of the List elements

Fields

Signature	Notes
`public static final Command SELECT_COMMAND`	SELECT_COMMAND is a special command that `commandAction()` can use to recognize if the user did the select operation on a `IMPLICIT List`.

Methods

Signature	Notes
`public int size()`	Returns the number of elements in the `List`
`public String getString(final int index) throws IndexOutOfBoundsException`	Returns the string part of the element: ❑ `elementNum`: the index of the element to be queried
`public Image getImage(final int index) throws IndexOutOfBoundsException`	Returns: the image part of the element, or `null` if there is no image: ❑ `elementNum`: the number of the element to be queried
`public int append(String stringElement, Image imageElement) throws IllegalArgumentException, NullPointerException`	Returns the assigned index of the element: ❑ `stringPart`: the string part of the element to be added ❑ `imagePart`: the image part of the element to be added, or `null` if there is no image part
`public void insert(final int index, String stringElement, Image imageElement) throws IndexOutOfBoundsException, IllegalArgumentException, NullPointerException`	❑ `elementNum`: the index of the element where insertion is to occur ❑ `stringPart`: the string part of the element to be inserted ❑ `imagePart`: the image part of the element to be inserted, or null if there is no image part
`public void delete(final int index) throws IndexOutOfBoundsException`	❑ `elementNum`: the index of the element to be deleted
`public void set(final int index, String stringElement, Image imageElement) throws IndexOutOfBoundsException, IllegalArgumentException, NullPointerException`	Sets the element referenced by `elementNum` to the specified element, replacing the previous contents of the element
`public boolean isSelected(final int index) throws IndexOutOfBoundsException`	Returns the selection state of the element: ❑ `elementNum`: index to element to be queried
`public int getSelectedIndex()`	Returns the index of selected element, or -1 if none
`public int getSelectedFlags(final boolean selectedArray_return) throws IllegalArgumentException, NullPointerException`	Returns the number of selected elements in the `Choice`: ❑ `selectedArray_return`: array to contain the results

Table continued on following page

Signature	Notes
`public void setSelectedIndex(final int index, final boolean selected) throws IndexOutOfBoundsException`	❑ `elementNum`: the index of the element, starting from zero ❑ `selected`: the state of the element, where `true` means selected and `false` means not selected
`public void setSelectedFlags(final boolean selectedArray) throws IllegalArgumentException, NullPointerException`	❑ `selectedArray`: an array in which the method collect the selection status

TextField

```
public class TextField extends Item
```

A `TextField` is an editable text component that may be placed into a `Form`. It can be given a piece of text that is used as the initial value.

The `TextField` shares the concept of **input constraints** with the `TextBox` object. The different constraints allow the application to request that the user's input be restricted in a variety of ways.

Constructors

Signature	Notes
`public TextField(String label, String text, final int maxSize, final int constraints) throws IllegalArgumentException, IllegalArgumentException, IllegalArgumentException, IllegalArgumentException`	Creates a new `TextField` object with the given label, initial contents, maximum size in characters, and constraints: ❑ `label`: item label ❑ `text`: the initial contents, or `null` if the TextField is to be empty ❑ `maxSize`: the maximum capacity in characters ❑ `constraints`: the input constraints

Fields

Signature	Notes
`public static final int ANY`	The user is allowed to enter any text.
`public static final int EMAILADDR`	The user is allowed to enter an e-mail address.
`public static final int NUMERIC`	The user is allowed to enter only an integer value.

Signature	Notes
`public static final int PHONENUMBER`	The user is allowed to enter a phone number. The phone number is a special case, since a phone-based implementation may be linked to the native phone dialing application.
`public static final int URL`	The user is allowed to enter a URL.
`public static final int PASSWORD`	The text entered must be masked so that the characters typed are not visible. The actual contents of the text field are not affected, but each character is displayed using a mask character such as "*". The `PASSWORD` modifier can be combined with other input constraints by using the logical OR operator (|).
`public static final int CONSTRAINT_MASK`	The mask value for determining the constraint mode. The application should use the logical `AND` operation with a value returned by `getConstraints()` and `CONSTRAINT_MASK` in order to retrieve the current constraint mode, in order to remove any modifier flags such as the `PASSWORD` flag.

Methods

Signature	Notes
`public String getString()`	Returns the contents of the `TextField` as a string value.
`public void setString(String text) throws IllegalArgumentException, IllegalArgumentException`	Sets the contents of the `TextField` as a string value, replacing the previous contents. ❑ `text`: the new value of the `TextField`
`public int getChars(final char data) throws ArrayIndexOutOfBoundsException`	Copies the contents of the `TextField` into a character array starting at index zero. Array elements beyond the characters copied are left unchanged. ❑ `data`: the character array to receive the value
`public void setChars(final char data, final int offset, final int length) throws ArrayIndexOutOfBoundsException, IllegalArgumentException, IllegalArgumentException`	Sets the contents of the `TextField` from a character array, replacing the previous contents: ❑ `data`: the source of the character data ❑ `offset`: the beginning of the region of characters to copy ❑ `length`: the number of characters to copy

Table continued on following page

Signature	Notes
`public void insert(String src, final int position) throws IllegalArgumentException, IllegalArgumentException`	Inserts a string into the contents of the `TextField`: ❑ `src`: the string to be inserted ❑ `position`: the position at which insertion is to occur
`public void insert(final char data, final int offset, final int length, final int position) throws ArrayIndexOutOfBoundsException, IllegalArgumentException, IllegalArgumentException`	Inserts a subrange of an array of characters into the contents of the `TextField`: ❑ `data`: the source of the character data ❑ `offset`: the beginning of the region of characters to copy ❑ `length`: the number of characters to copy ❑ `position`: the position at which insertion is to occur
`public void delete(final int offset, final int length) throws StringIndexOutOfBoundsException, IllegalArgumentException`	Deletes characters from the `TextField`: ❑ `offset`: the beginning of the region to be deleted ❑ `length`: the number of characters to be deleted
`public int getMaxSize()`	Returns the maximum size (number of characters) that can be stored in this `TextField`.
`public int setMaxSize(final int maxSize) throws IllegalArgumentException`	Sets the maximum size (number of characters) that can be contained in this `TextField`: ❑ `maxSize`: the new maximum size
`public int size()`	Returns the number of characters that are currently stored in this `TextField`.
`public int getCaretPosition()`	Return the current input position. For some UIs this may block some time and ask the user about the intended caret position, on some UIs may just return the caret position.
`public void setConstraints(final int constraints) throws IllegalArgumentException`	Sets the input constraints of the `TextField`. If the current contents of the `TextField` do not match the new constraints, the contents are set to empty: ❑ `constraints`: Sets the input constraints of the `TextField`.
`public int getConstraints()`	Returns the current input constraints of the `TextField`.

Canvas

```
public abstract class Canvas extends Displayable
```

The `Canvas` class is a base class for writing applications that need to handle low-level events and to issue graphics calls for drawing to the display. Game applications will likely make heavy use of the `Canvas` class. From an application development perspective, the `Canvas` class is interchangeable with standard `Screen` classes, so an application may mix and match `Canvas` with high-level screens as needed.

Constructors

Signature	Notes
`protected Canvas()`	

Fields

Signature	Notes
`public static final int UP`	Constant for the `UP` game action.
`public static final int DOWN`	Constant for the `DOWN` game action.
`public static final int LEFT`	Constant for the `LEFT` game action.
`public static final int RIGHT`	Constant for the `RIGHT` game action.
`public static final int FIRE`	Constant for the `FIRE` game action.
`public static final int GAME_A`	Constant for the general purpose "A" game action.
`public static final int GAME_B`	Constant for the general purpose "B" game action.
`public static final int GAME_C`	Constant for the general purpose "C" game action.
`public static final int GAME_D`	Constant for the general purpose "D" game action.
`public static final int KEY_NUM0`	keyCode for ITU-T key 0.
`public static final int KEY_NUM1`	keyCode for ITU-T key 1.
`public static final int KEY_NUM2`	keyCode for ITU-T key 2.
`public static final int KEY_NUM3`	keyCode for ITU-T key 3.
`public static final int KEY_NUM4`	keyCode for ITU-T key 4.
`public static final int KEY_NUM5`	keyCode for ITU-T key 5.
`public static final int KEY_NUM6`	keyCode for ITU-T key 6.

Table continued on following page

Signature	Notes
`public static final int KEY_NUM7`	keyCode for ITU-T key 7.
`public static final int KEY_NUM8`	keyCode for ITU-T key 8.
`public static final int KEY_NUM9`	keyCode for ITU-T key 9.
`public static final int KEY_STAR`	keyCode for ITU-T key "star" (*).
`public static final int KEY_POUND`	keyCode for ITU-T key "pound" (#).

Methods

Signature	Notes
`public int getWidth()`	Returns the width of the displayable area in pixels.
`public int getHeight()`	Returns the height of the displayable area in pixels.
`public boolean isDoubleBuffered()`	Checks if the Graphics is double buffered by the implementation.
`public boolean hasPointerEvents()`	Checks if the platform supports pointer press and release events.
`public boolean hasPointerMotionEvents()`	Checks if the platform supports pointer motion events (pointer dragged).
`public boolean hasRepeatEvents()`	Checks if the platform can generate repeat events when key is kept down.
`public int getKeyCode(final int gameAction) throws IllegalArgumentException`	Returns a key code that corresponds to the specified game action on the device: ❑ `gameAction`: the game action
`public String getKeyName(final int keyCode) throws IllegalArgumentException`	Returns an informative key string for a key. The String returned will resemble the text physically printed on the key: ❑ `keyCode`: the key code being requested
`public int getGameAction(final int keyCode) throws IllegalArgumentException`	Returns the game action associated with the given key code of the device: ❑ `keyCode`: the key code
`protected void keyPressed(final int keyCode)`	Called when a key is pressed. The `getGameAction()` method can be called to determine what game action, if any, is mapped to the key: ❑ `keyCode`: The key code of the key that was pressed.

Signature	Notes
`protected void keyRepeated(final int keyCode)`	Called when a key is repeated (held down): ❑ `keyCode`: The key code of the key that was repeated
`protected void keyReleased(final int keyCode)`	❑ Called when a key is released: `keyCode`: The key code of the key that was released
`protected void pointerPressed(final int x, final int y)`	Called when the pointer is pressed: ❑ `x`: The horizontal location where the pointer was pressed (relative to the `Canvas`) ❑ `y`: The vertical location where the pointer was pressed (relative to the `Canvas`)
`protected void pointerReleased(final int x, final int y)`	Called when the pointer is released: ❑ `x`: The horizontal location where the pointer was released (relative to the `Canvas`) ❑ `y`: The vertical location where the pointer was released (relative to the `Canvas`)
`protected void pointerDragged(final int x, final int y)`	Called when the pointer is dragged: ❑ `x`: The horizontal location where the pointer was dragged (relative to the `Canvas`) ❑ `y`: The vertical location where the pointer was dragged (relative to the `Canvas`)
`public final void repaint(final int x, final int y, final int width, final int height)`	Requests a repaint for the specified region of the `Screen`. Calling this method may result in subsequent calls to `paint()`, where the passed Graphics object's clip region will include at least the specified region: ❑ `x`: the x coordinate of the rectangle to be repainted ❑ `y`: the y coordinate of the rectangle to be repainted ❑ `width`: the width of the rectangle to be repainted ❑ `height`: the height of the rectangle to be repainted

Table continued on following page

Signature	Notes
`public final void repaint()`	Requests a repaint for the entire `Canvas`. The effect is identical to `repaint(0, 0, getWidth(), getHeight());`
`public final void serviceRepaints()`	Forces any pending repaint requests to be serviced immediately. This method blocks until the pending requests have been serviced.
`protected void showNotify()`	The implementation calls `showNotify()` immediately prior to this canvas being made visible on the display. `Canvas` subclasses may override this method to perform tasks before being shown, such as setting up animations, starting timers, etc.
`protected void hideNotify()`	The implementation calls `hideNotify()` shortly after the canvas has been removed from the display. `Canvas` subclasses may override this method in order to pause animations, revoke timers, etc.
`protected abstract void paint(Graphics g)`	Renders the canvas. The application must implement this method in order to paint any graphics: ❑ `g`: the `Graphics` object to be used for rendering the canvas

StringItem

```
public class StringItem extends Item
```

An item that can contain a String. A `StringItem` is display-only; the user cannot edit the contents. Both the label and the textual content of a `StringItem` may be modified by the application. The visual representation of the label may differ from that of the textual contents.

Constructors

Signature	Notes
`public StringItem(String label, String text)`	Creates a new `StringItem` object with the given label and textual content: ❑ `label`: the Item label ❑ `text`: the text contents

Methods

Signature	Notes
`public String getText()`	Returns the text contents of the `StringItem`, or null if the `StringItem` is empty.
`public void setText(String text)`	Sets the text contents of the `StringItem`: ❑ `text`: the new content

Form

```
public class Form extends Screen
```

A `Form` is a `Screen` that contains an arbitrary mixture of items: images, read-only text fields, editable text fields, editable date fields, gauges, and choice groups. In general, any subclass of the `Item` class may be contained within a form. The implementation handles layout, traversal, and scrolling. None of the components contained within has any internal scrolling; the entire contents scrolls together. Note that this differs from the behavior of other classes, the `List` for example, where only the interior scrolls.

Constructors

Signature	Notes
`public Form(String title)`	Creates a new, empty `Form`: ❑ `title`: the title of the `Form`, or `null` for no title
`public Form(String title, Item items) throws IllegalStateException, NullPointerException`	Creates a new `Form` with the specified contents. This is identical to creating an empty `Form` and then using a set of `appendItem()` methods: ❑ `title`: the title String ❑ `items`: the array of items to be placed in the `Form`, or `null` if there are no items

Methods

Signature	Notes
`public int append(Item item) throws IllegalStateException, NullPointerException`	Adds an item into the `Form`. Strings are filled so that current line is continued if possible: ❑ `item`: the `Item` to be added.
`public int append(String str) throws NullPointerException`	Adds an item consisting of one String to the form: ❑ `str`: the String to be added

Table continued on following page

Signature	Notes
`public int append(Image image) throws IllegalArgumentException, NullPointerException`	Adds an item consisting of one `Image` to the form: ❑ img: the image to be added
`public void insert(final int index, Item item) throws IndexOutOfBoundsException, IllegalStateException, NullPointerException`	Inserts an item into the `Form` just prior to the item specified. The size of the Form grows by one: ❑ `itemNum`: the index where insertion is to occur ❑ `item`: the item to be inserted
`public void delete(final int index) throws IndexOutOfBoundsException`	Deletes the Item referenced by itemNum. The size of the Form shrinks by one. It is legal to delete all items from a Form. The itemNum parameter must be within the range [0..size()-1], inclusive. ❑ `itemNum`: the index of the item to be deleted
`public void set(final int index, Item item) throws IndexOutOfBoundsException, IllegalStateException, NullPointerException`	Sets the item referenced by `itemNum` to the specified item, replacing the previous item. The previous item is removed from this `Form`: ❑ `itemNum`: the index of the item to be replaced ❑ `item`: the new item to be placed in the `Form`
`public Item get(final int index) throws IndexOutOfBoundsException`	Returns the item at a given position. The contents of the `Form` are left unchanged: ❑ `itemNum`: the index of items
`public void setItemStateListener(ItemStateListener iListener)`	Sets the `ItemStateListener` for the `Form`, replacing any previous `ItemStateListener`: ❑ `iListener`: the new listener, or `null` to remove it
`public int size()`	Returns the number of items in the `Form`.

ImageItem

```
public class ImageItem extends Item
```

A class that provides layout control when `Image` objects are added to a `Form` or an `Alert`. Each `ImageItem` object contains a reference to an `Image` object. This image must be immutable. If the `Image` object were not required to be immutable, the application could paint into it at any time, potentially requiring the containing `Form` or `Alert` to be updated on every graphics call.

Constructors

Signature	Notes
`public ImageItem(String label, Image img, final int layout, String altText) throws IllegalArgumentException, IllegalArgumentException`	Creates a new `ImageItem` with the given label, image, layout directive, and alternate text String: ❑ `label`: the label string ❑ `img`: the image, must be immutable ❑ `layout`: a combination of layout directives ❑ `altText`: the text that may be used in place of the image

Fields

Signature	Notes
`public static final int LAYOUT_DEFAULT`	Use the default formatting of the "container" of the image.
`public static final int LAYOUT_LEFT`	Image should be close to left-edge of the drawing area.
`public static final int LAYOUT_RIGHT`	Image should be close to right-edge of the drawing area.
`public static final int LAYOUT_CENTER`	Image should be horizontally centered.
`public static final int LAYOUT_NEWLINE_BEFORE`	A new line should be started before the image is drawn.
`public static final int LAYOUT_NEWLINE_AFTER`	A new line should be started after the image is drawn.

Methods

Signature	Notes
`public Image getImage()`	Returns the `Image` contained within the `ImageItem`, or `null` if there is no contained image.
`public void setImage(Image img) throws IllegalArgumentException`	Sets the `Image` object contained within the `ImageItem`. The image must be immutable: ❑ `img`: the new image
`public String getAltText()`	Returns the text String to be used if the image exceeds the device's capacity to display it.

Table continued on following page

Signature	Notes
`public void setAltText(String altText)`	Sets the alternate text of the `ImageItem`, or `null` if no alternate text is provided: ❑ `text`: the new alternate text
`public int getLayout()`	Returns the layout directives used for placing the image.
`public void setLayout(final int layout) throws IllegalArgumentException`	Sets the layout directives: ❑ `layout`: a combination of layout directive values

Ticker

```
public class Ticker
```

Implements a "ticker-tape," a piece of text that runs continuously across the display. The direction and speed of scrolling are determined by the implementation. While animating, the ticker string scrolls continuously. That is, when the string finishes scrolling off the display, the ticker starts over at the beginning of the String.

Constructors

Signature	Notes
`public Ticker(String str) throws NullPointerException`	Constructs a new `Ticker` object, given its initial contents String: ❑ `str`: string to be set for the `Ticker`

Methods

Signature	Notes
`public void setString(String str) throws NullPointerException`	Sets the String to be displayed by this ticker: ❑ `str`: String to be set for the `Ticker`
`public String getString()`	Returns the string currently being scrolled by the ticker

Font

```
public final class Font
```

The `Font` class represents fonts and font metrics. Fonts cannot be created by applications. Instead, applications query for fonts based on font attributes and the system will attempt to provide a font that matches the requested attributes as closely as possible.

Fields

Signature	Notes
`public static final int STYLE_PLAIN`	The plain style constant. This may be combined with the other style constants for mixed styles.
`public static final int STYLE_BOLD`	The bold style constant. This may be combined with the other style constants for mixed styles.
`public static final int STYLE_ITALIC`	The italicized style constant. This may be combined with the other style constants for mixed styles.
`public static final int STYLE_UNDERLINED`	The underlined style constant. This may be combined with the other style constants for mixed styles.
`public static final int SIZE_SMALL`	The "small" system-dependent font size.
`public static final int SIZE_MEDIUM`	The "medium" system-dependent font size.
`public static final int SIZE_LARGE`	The "large" system-dependent font size.
`public static final int FACE_SYSTEM`	The "system" font face.
`public static final int FACE_MONOSPACE`	The "monospace" font face.
`public static final int FACE_PROPORTIONAL`	The "proportional" font face.

Methods

Signature	Notes
`public static Font getDefaultFont()`	Returns the default font of the system.
`public static Font getFont(final int face, final int style, final int size) throws IllegalArgumentException`	Obtains an object representing a font having the specified face, style, and size. If a matching font does not exist, the system will attempt to provide the closest match. This method **always** returns a valid font object, even if it is not a close match to the request: ❑ `face`: one of `FACE_SYSTEM`, `FACE_MONOSPACE`, or `FACE_PROPORTIONAL` ❑ `style`: `STYLE_PLAIN`, or a combination of `STYLE_BOLD`, `STYLE_ITALIC`, and `STYLE_UNDERLINED` ❑ `size`: one of `SIZE_SMALL`, `SIZE_MEDIUM`, or `SIZE_LARGE`

Table continued on following page

Signature	Notes
`public int getStyle()`	Returns the style of the font.
`public int getSize()`	Returns the size of the font.
`public int getFace()`	Returns the face of the font.
`public boolean isPlain()`	Returns `true` if the font is plain.
`public boolean isBold()`	Returns `true` if the font is bold.
`public boolean isItalic()`	Returns `true` if the font is italic.
`public boolean isUnderlined()`	Returns `true` if the font is underlined.
`public int getHeight()`	Returns the standard height of a line of text in this font. This value includes sufficient spacing to ensure that lines of text painted this distance from anchor point to anchor point are spaced as intended by the font designer and the device.
`public int getBaselinePosition()`	Returns the distance in pixels from the top of the text to the text's baseline.
`public int charWidth(final char ch)`	Returns the advance width of the specified character in this Font. The advance width is the amount by which the current point is moved from one character to the next in a line of text, and thus includes proper inter-character spacing: ❑ `ch`: the character to be measured
`public int charsWidth(final char ch, final int offset, final int length) throws ArrayIndexOutOfBoundsException, NullPointerException`	Returns the advance width of the characters in `ch`, starting at the specified offset and for the specified number of characters (length). The advance width is the amount by which the current point is moved from one character to the next in a line of text: ❑ `ch`: The array of characters ❑ `offset`: The index of the first character to measure ❑ `length`: The number of characters to measure

Signature	Notes
`public int stringWidth(String str) throws NullPointerException`	Returns the total advance width for showing the specified string in this Font. The advance width is the amount by which the current point is moved from one character to the next in a line of text: ❑ `str`: the String to be measured.
`public int substringWidth(String str, final int offset, final int len) throws StringIndexOutOfBoundsException, NullPointerException`	Returns the total advance width for showing the specified substring in this Font. The advance width is the amount by which the current point is moved from one character to the next in a line of text: ❑ `str`: the String to be measured. ❑ `offset`: zero-based index of first character in the substring ❑ `len`: length of the substring.

Graphics

```
public class Graphics
```

Provides simple 2D geometric rendering capability. Drawing primitives are provided for text, images, lines, rectangles, and arcs. Rectangles and arcs may also be filled with a solid color. Rectangles may also be specified with rounded corners.

Fields

Signature	Notes
`public static final int HCENTER`	Constant for centering text and images horizontally around the anchor point
`public static final int VCENTER`	Constant for centering images vertically around the anchor point
`public static final int LEFT`	Constant for positioning the anchor point of text and images to the left of the text or image
`public static final int RIGHT`	Constant for positioning the anchor point of text and images to the right of the text or image
`public static final int TOP`	Constant for positioning the anchor point of text and images above the text or image
`public static final int BOTTOM`	Constant for positioning the anchor point of text and images below the text or image

Table continued on following page

Signature	Notes
`public static final int BASELINE`	Constant for positioning the anchor point at the baseline of text
`public static final int SOLID`	Constant for the `SOLID` stroke style
`public static final int DOTTED`	Constant for the `DOTTED` stroke style

Methods

Signature	Notes
`public void translate(final int x, final int y)`	Translates the origin of the graphics context to the point (x, y) in the current coordinate system. All coordinates used in subsequent rendering operations on this graphics context will be relative to this new origin: ❑ `x`: the x coordinate of the new translation origin ❑ `y`: the y coordinate of the new translation origin
`public int getTranslateX()`	Returns the X coordinate of the translated origin of this graphics context.
`public int getTranslateY()`	Returns the Y coordinate of the translated origin of this graphics context.
`public int getColor()`	Returns the current color.
`public int getRedComponent()`	Returns the red component of the current color.
`public int getGreenComponent()`	Returns the green component of the current color.
`public int getBlueComponent()`	Returns the blue component of the current color.
`public int getGrayScale()`	Returns the current grayscale value of the color being used for rendering operations. If the color was set by `setGrayScale()`, that value is simply returned. If the color was set by one of the methods that allows setting of the red, green, and blue components, the value returned is computed from the RGB color components (possibly in a device-specific fashion) that best approximates the brightness of that color.

Signature	Notes
`public void setColor(final int red, final int green, final int blue) throws IllegalArgumentException`	Sets the current color to the specified RGB values. All subsequent rendering operations will use this specified color: ❑ `red`: The red component of the color being set in range 0-255. ❑ `green`: The green component of the color being set in range 0-255. ❑ `blue`: The blue component of the color being set in range 0-255.
`public void setColor(final int RGB)`	Sets the current color to the specified RGB values. All subsequent rendering operations will use this specified color: ❑ `RGB`: the color being set
`public void setGrayScale(final int value) throws IllegalArgumentException`	Sets the current grayscale to be used for all subsequent rendering operations. For monochrome displays, the behavior is clear. For color displays, this sets the color for all subsequent drawing operations to be a gray color equivalent to the value passed in: ❑ `value`: the desired grayscale value
`public Font getFont()`	Returns the current font
`public void setStrokeStyle(final int style) throws IllegalArgumentException`	Sets the stroke style used for drawing lines, arcs, rectangles, and rounded rectangles. This does not affect fill, text, and image operations: ❑ `style`: can be `SOLID` or `DOTTED`
`public int getStrokeStyle()`	Returns the stroke style used for drawing operations.
`public void setFont(Font font)`	Sets the font for all subsequent text rendering operations: ❑ `font`: the specified font
`public int getClipX()`	Returns the X offset of the current clipping area, relative to the coordinate system origin of this graphics context. Separating the `getClip` operation into two methods returning integers is more performance and memory efficient than one `getClip()` call returning an object.

Table continued on following page

Signature	Notes
`public int getClipY()`	Returns the Y offset of the current clipping area, relative to the coordinate system origin of this graphics context.
`public int getClipWidth()`	Returns the width of the current clipping area.
`public int getClipHeight()`	Returns the height of the current clipping area.
`public void clipRect(final int x, final int y, final int width, final int height)`	Intersects the current clip with the specified rectangle. The resulting clipping area is the intersection of the current clipping area and the specified rectangle. This method can only be used to make the current clip smaller. To set the current clip larger, use the `setClip()` method: ❑ `x`: the x coordinate of the rectangle to intersect the clip with ❑ `y`: the y coordinate of the rectangle to intersect the clip with ❑ `width`: the width of the rectangle to intersect the clip with ❑ `height`: the height of the rectangle to intersect the clip with
`public void setClip(final int x, final int y, final int width, final int height)`	Sets the current clip to the rectangle specified by the given coordinates. Rendering operations have no effect outside of the clipping area: ❑ `x`: the x coordinate of the new clip rectangle ❑ `y`: the y coordinate of the new clip rectangle ❑ `width`: the width of the new clip rectangle ❑ `height`: the height of the new clip rectangle
`public void drawLine(final int x1, final int y1, final int x2, final int y2)`	Draws a line between the coordinates (x1,y1) and (x2,y2) using the current color and stroke style: ❑ `x1`: the x coordinate of the start of the line ❑ `y1`: the y coordinate of the start of the line ❑ `x2`: the x coordinate of the end of the line ❑ `y2`: the y coordinate of the end of the line

Signature	Notes
`public void fillRect(final int x, final int y, final int width, final int height)`	Fills the specified rectangle with the current color: ❑ `x`: the x coordinate of the rectangle to be filled ❑ `y`: the y coordinate of the rectangle to be filled ❑ `width`: the width of the rectangle to be filled ❑ `height`: the height of the rectangle to be filled
`public void drawRect(final int x, final int y, final int width, final int height)`	Draws the outline of the specified rectangle using the current color and stroke style. The resulting rectangle will cover an area (`width` + 1) pixels wide by (`height` + 1) pixels tall: ❑ `x`: the x coordinate of the rectangle to be drawn ❑ `y`: the y coordinate of the rectangle to be drawn ❑ `width`: the width of the rectangle to be drawn ❑ `height`: the height of the rectangle to be drawn
`public void drawRoundRect(final int x, final int y, final int width, final int height, final int arcWidth, final int arcHeight)`	Draws the outline of the specified rounded corner rectangle using the current color and stroke style. The resulting rectangle will cover an area (`width` + 1) pixels wide by (`height` + 1) pixels tall: ❑ `x`: the x coordinate of the rectangle to be drawn ❑ `y`: the y coordinate of the rectangle to be drawn ❑ `width`: the width of the rectangle to be drawn ❑ `height`: the height of the rectangle to be drawn ❑ `arcWidth`: the horizontal diameter of the arc at the four corners ❑ `arcHeight`: the vertical diameter of the arc at the four corners

Signature	Notes
`public void fillRoundRect(final int x, final int y, final int width, final int height, final int arcWidth, final int arcHeight)`	Fills the specified rounded corner rectangle with the current color: ❑ `x`: the x coordinate of the rectangle to be filled ❑ `y`: the y coordinate of the rectangle to be filled ❑ `width`: the width of the rectangle to be filled ❑ `height`: the height of the rectangle to be filled ❑ `arcWidth`: the horizontal diameter of the arc at the four corners ❑ `arcHeight`: the vertical diameter of the arc at the four corners
`public void fillArc(final int x, final int y, final int width, final int height, final int startAngle, final int arcAngle)`	Fills a circular or elliptical arc covering the specified rectangle: ❑ `x`: the x coordinate of the upper-left corner of the arc to be filled ❑ `y`: the y coordinate of the upper-left corner of the arc to be filled ❑ `width`: the width of the arc to be filled ❑ `height`: the height of the arc to be filled ❑ `startAngle`: the beginning angle. ❑ `arcAngle`: the angular extent of the arc, relative to the start angle.
`public void drawArc(final int x, final int y, final int width, final int height, final int startAngle, final int arcAngle)`	Draws the outline of a circular or elliptical arc covering the specified rectangle, using the current color and stroke style: ❑ `x`: the x coordinate of the upper-left corner of the arc to be drawn. ❑ `y`: the y coordinate of the upper-left corner of the arc to be drawn. ❑ `width`: the width of the arc to be drawn ❑ `height`: the height of the arc to be drawn ❑ `startAngle`: the beginning angle ❑ `arcAngle`: the angular extent of the arc, relative to the start angle.

Signature	Notes
`public void drawString(String str, final int x, final int y, final int anchor) throws NullPointerException, IllegalArgumentException`	Draws the specified String using the current font and color. The x,y position is the position of the anchor point: ❑ `str`: the String to be drawn ❑ `x`: the x coordinate of the anchor point ❑ `y`: the y coordinate of the anchor point ❑ `anchor`: the anchor point for positioning the text
`public void drawSubstring(String str, final int offset, final int len, final int x, final int y, final int anchor) throws StringIndexOutOfBoundsException, IllegalArgumentException, NullPointerException`	Draws the specified String using the current font and color. The x,y position is the position of the anchor point: ❑ `str`: the String to be drawn ❑ `offset`: zero-based index of first character in the substring ❑ `len`: length of the substring ❑ `x`: the x coordinate of the anchor point ❑ `y`: the y coordinate of the anchor point ❑ `anchor`: the anchor point for positioning the text
`public void drawChar(final char character, final int x, final int y, final int anchor) throws IllegalArgumentException`	Draws the specified character using the current font and color: ❑ `character`: the character to be drawn ❑ `x`: the x coordinate of the anchor point ❑ `y`: the y coordinate of the anchor point ❑ `anchor`: the anchor point for positioning the text
`public void drawChars(final char data, final int offset, final int length, final int x, final int y, final int anchor) throws ArrayIndexOutOfBoundsException, IllegalArgumentException, NullPointerException`	Draws the specified characters using the current font and color: ❑ `data`: the array of characters to be drawn ❑ `offset`: the start offset in the data ❑ `length`: the number of characters to be drawn ❑ `x`: the x coordinate of the anchor point ❑ `y`: the y coordinate of the anchor point ❑ `anchor`: the anchor point for positioning the text

Table continued on following page

Signature	Notes
`public void drawImage(Image img, final int x, final int y, final int anchor) throws IllegalArgumentException, NullPointerException`	Draws the specified `Image` by using the anchor point. The image can be drawn in different positions relative to the anchor point by passing the appropriate position constants: ❑ `img`: the specified image to be drawn ❑ `x`: the x coordinate of the anchor point ❑ `y`: the y coordinate of the anchor point ❑ `anchor`: the anchor point for positioning the image

DateField

```
public class DateField extends Item
```

A `DateField` is an editable component for presenting date and time (calendar) information that may be placed into a `Form`. The value for this field can be initially set or left unset. If the value is not set then the UI for the field shows this clearly. The field value for "not initialized state" is not a valid value and `getDate()` for this state returns `null`.

Constructors

Signature	Notes
`public DateField(String label, final int mode) throws IllegalArgumentException`	Creates a `DateField` object with the specified mode: ❑ `label`: item label ❑ `mode`: the input mode, one of `DATE`, `TIME` or `DATE_TIME`
`public DateField(String label, final int mode, TimeZone timeZone) throws IllegalArgumentException`	Creates a date field in which calendar calculations are based on the specific `TimeZone` object and the default calendaring system for the current locale: ❑ `label`: item label ❑ `mode`: the input mode, one of `DATE`, `TIME` or `DATE_TIME` ❑ `timeZone`: a specific time zone, or `null` for the default time zone

Fields

Signature	Notes
`public static final int DATE`	Input mode for date information (day, month, year). With this mode this `DateField` allows the user to only modify the date value. The time information of the `date` object is ignored.
`public static final int TIME`	Input mode for time information (hours and minutes). With this mode this `DateField` allows the user only to modify the time.
`public static final int DATE_TIME`	Input mode for date (day, month, year) and time (minutes, hours) information. With this mode this `DateField` allows the user to modify both time and date information.

Methods

Signature	Notes
`public Date getDate()`	Returns date value of this field.
`public void setDate(Date date) throws IllegalArgumentException`	Sets a new value for this field: ❑ `date`: new value for this field
`public int getInputMode()`	Returns the input mode for this date field. Valid input modes are `DATE`, `TIME` and `DATE_TIME`.
`public void setInputMode(final int mode) throws IllegalArgumentException`	Set input mode for this date field: ❑ `mode`: the input mode, must be one of `DATE`, `TIME` or `DATE_TIME`

Gauge

```
public class Gauge extends Item
```

The `Gauge` class implements a bar graph display of a value intended for use in a form. `Gauge` is optionally interactive. The values accepted by the object are small integers in the range zero through a maximum value established by the application. The application is expected to normalize its values into this range. The device is expected to normalize this range into a smaller set of values for display purposes. Doing so will not change the actual value contained within the object. The range of values specified by the application may be larger than the number of distinct visual states possible on the device, so more than one value may have the same visual representation.

Constructors

Signature	Notes
`public Gauge(String label, final boolean interactive, final int maxValue, final int initialValue) throws IllegalArgumentException`	Creates a new `Gauge` object with the given label, in interactive or non-interactive mode, with the given maximum and initial values: ❑ `label`: the label of the `Gauge` ❑ `interactive`: tells whether the user can change the value ❑ `maxValue`: the maximum value ❑ `initialValue`: the initial value in the range [0..`maxValue`]

Methods

Signature	Notes
`public void setValue(final int value)`	Sets the current value of this `Gauge` object: ❑ `value`: the new value
`public int getValue()`	Returns the current value of this `Gauge` object
`public void setMaxValue(final int maxValue) throws IllegalArgumentException`	Sets the maximum value of this `Gauge` object: ❑ `maxValue`: the new maximum value
`public int getMaxValue()`	Returns the maximum value of this `Gauge` object
`public boolean isInteractive()`	Tells whether the user is allowed to change the value of the `Gauge`

Display

```
public class Display
```

`Display` represents the manager of the display and input devices of the system. It includes methods for retrieving properties of the device and for requesting that objects be displayed on the device. Other methods that deal with device attributes are primarily used with `Canvas` objects and are thus defined there instead of here.

There is exactly one instance of `Display` per MIDlet and the application can get a reference to that instance by calling the `getDisplay(MIDlet midlet)` method. The application may call the `getDisplay()` method from the beginning of the `startApp()` call until the `destroyApp()` call returns. The `Display` object returned by all calls to `getDisplay()` will remain the same during this time.

Methods

Signature	Notes
`public static Display getDisplay(MIDlet midlet) throws NullPointerException`	Returns the `Display` object that is unique to this MIDlet: ❑ `midlet`: Midlet of the application
`public boolean isColor()`	Returns information about color support of the device.
`public int numColors()`	Returns the number of colors (if `isColor()` is `true`) or gray levels (if `isColor()` is `false`) that can be represented on the device. Note that number of colors for black and white display is 2.
`public Displayable getCurrent()`	Gets the current `Displayable` object for this MIDlet. The `Displayable` object returned may not actually be visible on the display if the MIDlet is running in the background, or if the `Displayable` is obscured by a system screen. The `Displayable.isShown()` method may be called to determine whether the `Displayable` is actually visible on the display.
`public void setCurrent(Displayable next)`	Requests that a different `Displayable` object be made visible on the display. The change will typically not take effect immediately. It may be delayed so that it occurs between event delivery method calls, although it is not guaranteed to occur before the next event delivery method is called. The `setCurrent()` method returns immediately, without waiting for the change to take place. Therefore, a call to `getCurrent()` shortly after a call to `setCurrent()` is unlikely to return the value passed to `setCurrent()`. Calls to `setCurrent()` are not queued. A delayed request made by a `setCurrent()` call may be superseded by a subsequent call to `setCurrent()`: ❑ `next`: the `Displayable` requested to be made current

Table continued on following page

Signature	Notes
`public void setCurrent(Alert alert, Displayable nextDisplayable) throws NullPointerException, IllegalArgumentException`	Requests that this `Alert` be made current, and that the `nextDisplayable` be made current after the `Alert` is dismissed. This call returns immediately regardless of the timeout value of the `Alert` or whether it is a modal alert. The `nextDisplayable` must not be an `Alert`: ❑ `alert`: the alert to be shown ❑ `nextDisplayable`: the `Displayable` to be shown after this alert is dismissed
`public void callSerially (Runnable r)`	Causes the `Runnable` object `r` to have its `run()` method called later, serialized with the event stream, soon after completion of the repaint cycle: ❑ `r`: instance of interface `Runnable` to be called

Command

```
public class Command
```

The `Command` class is a construct that encapsulates the semantic information of an action. The behavior that the command activates is not encapsulated in this object. This means that the `Command` contains only information about a "command" not the actual action that happens when the command is activated. The action is defined in a `CommandListener` associated with the `Screen`. `Command` objects are presented in the user interface and the way they are presented may depend on the semantic information contained within the command.

Constructors

Signature	Notes
`public Command(String label, final int commandType, final int priority) throws IllegalArgumentException`	Creates a new `Command` object with the given label, type, and priority: ❑ `label`: the label String ❑ `commandType`: one of `BACK`, `CANCEL`, `HELP`, `OK`, `SCREEN`, or `STOP` ❑ `priority`: the priority value of the `Command`

Fields

Signature	Notes
`public static final int SCREEN`	Specifies an application-defined command that pertains to the current screen. Examples could be "Load" and "Save".
`public static final int BACK`	A navigation command that returns the user to the logically previous screen. The jump to the previous screen is not done automatically by the implementation but by the `CommandListener` provided by the application.
`public static final int CANCEL`	A `Command` that is a standard negative answer to a dialog implemented by the current screen. Nothing is cancelled automatically by the implementation; cancellation is implemented by the `CommandListener` provided by the application.
`public static final int OK`	A `Command` that is a standard positive answer to a dialog implemented by the current screen. Nothing is done automatically by the implementation; any action taken is implemented by the `CommandListener` provided by the application.
`public static final int HELP`	This `Command` specifies a request for on-line help. No help information is shown automatically by the implementation. The `CommandListener` provided by the application is responsible for showing the help information.
`public static final int STOP`	A `Command` that will stop some currently running process, operation, etc. Nothing is stopped automatically by the implementation. The cessation must be performed by the `CommandListener` provided by the application.
`public static final int EXIT`	A `Command` used for exiting from the application. When the user invokes this command, the implementation does not exit automatically. The application's `CommandListener` will be called, and it should exit the application if it is appropriate to do so.
`public static final int ITEM`	With this `Command` type the application can hint to the implementation that the command is specific to a particular item on the screen. For example, an implementation of `List` can use this information for creating context sensitive menus.

Methods

Signature	Notes
`public String getLabel()`	Returns the label of the `Command`
`public int getCommandType()`	Returns the type of the `Command`
`public int getPriority()`	Returns the priority of the `Command`

Alert

```
public class Alert extends Screen
```

An `Alert` is a screen that shows data to the user and waits for a certain period of time before proceeding to the next screen. An `Alert` is an ordinary screen that can contain text (String) and image, and which handles events like other screens.

The intended use of `Alert` is to inform the user about errors and other exceptional conditions.

Constructors

Signature	Notes
`public Alert(String title)`	Constructs a new, empty `Alert` object with the given title: ❑ `title`: the title String, or `null`
`public Alert(String title, String alertText, Image alertImage, AlertType alertType) throws IllegalArgumentException`	Constructs a new `Alert` object with the given title, content String and image, and alert type. The layout of the contents is implementation dependent: ❑ `title`: the title String, or `null` if there is no title ❑ `alertText`: the String contents, or `null` if there is no String ❑ `alertImage`: the image contents, or `null` if there is no image ❑ `alertType`: the type of the `Alert`, or `null` if the `Alert` has no specific type

Fields

Signature	Notes
`public static final int FOREVER`	`FOREVER` indicates that an `Alert` is kept visible until the user dismisses it. It is used as a value for the parameter to `setTimeout()` to indicate that the alert is modal. Instead of waiting for a specified period of time, a modal `Alert` will wait for the user to take some explicit action, such as pressing a button, before proceeding to the next screen.

Methods

Signature	Notes
`public int getDefaultTimeout()`	Returns the default time for showing an `Alert`. This is either a positive value, which indicates a time in milliseconds, or the special value `FOREVER`.
`public int getTimeout()`	Returns the time this `Alert` will be shown. This is either a positive value, which indicates a time in milliseconds, or the special value `FOREVER`.
`public void setTimeout(final int time) throws IllegalArgumentException`	Sets the time for which the `Alert` is to be shown. This must either be a positive time value in milliseconds, or the special value `FOREVER`: ❑ `time`: timeout in milliseconds, or `FOREVER`
`public AlertType getType()`	Returns the type of the `Alert`.
`public void setType(AlertType type)`	Sets the type of the `Alert`: ❑ `type`: an `AlertType`, or `null` if the `Alert` has no specific type
`public String getString()`	Returns the text string used in the `Alert`.
`public void setString(String str)`	Sets the text string used in the `Alert`: ❑ `str`: the text String of the `Alert`, or `null` if there is no text
`public Image getImage()`	Returns the `Image` used in the `Alert`.
`public void setImage(Image img) throws IllegalArgumentException`	Sets the `Image` used in the `Alert`: ❑ `img`: the image of the `Alert`, or `null` if there is no image

Table continued on following page

Signature	Notes
`public void addCommand(Command cmd) throws IllegalStateException`	A `Command` is not allowed on an `Alert`, so this method will always throw an `IllegalStateException` whenever it is called: ❑ `cmd`: the `Command`
`public void setCommandListener(CommandListener l) throws IllegalStateException`	A `Listener` is not allowed on an `Alert`, so this method will always throw an `IllegalStateException` whenever it is called: ❑ `l`: the Listener

Interfaces

CommandListener

```
public interface CommandListener
```

This interface is used by applications that need to receive high-level events from the implementation. An application will provide an implementation of a `Listener` (typically by using a nested class or an inner class) and will then provide an instance of it on a `Screen` in order to receive high-level events on that `Screen`.

Methods

Signature	Notes
`public void commandAction(Command c, Displayable d)`	Indicates that a command event has occurred on `Displayable d`: ❑ `c`: a `Command` object identifying the command. This is either one of the applications that have been added to `Displayable` with `addCommand(Command)` or is the implicit `SELECT_COMMAND` of `List`. ❑ `d`: the `Displayable` on which this event has occurred.

ItemStateListener

```
public interface ItemStateListener
```

This interface is used by applications that need to receive events that indicate changes in the internal state of the interactive items within a `Form` screen.

Methods

Signature	Notes
`public void itemStateChanged(Item item)`	Called when the internal state of an `Item` has been changed by the user: ❑ `item`: the item that was changed

Choice

```
public interface Choice
```

`Choice` defines an API for a user interface component implementing a selection from a predefined number of choices. Such UI components are `List` and `ChoiceGroup`. The contents of the `Choice` are represented with Strings and optional images. The application may provide `null` for the image if the element does not have an image part. If the application provides an image, the implementation may choose to ignore the image if it exceeds the capacity of the device to display it. If the implementation displays the image, it will be displayed adjacent to the text string and the pair will be treated as a unit.

Fields

Signature	Notes
`public static final int EXCLUSIVE`	`EXCLUSIVE` is a choice having exactly one element selected at a time.
`public static final int MULTIPLE`	`MULTIPLE` is a choice that can have arbitrary number of elements selected at a time.
`public static final int IMPLICIT`	`IMPLICIT` is a choice in which the currently focused item is selected when a `Command` is initiated. Note: `IMPLICIT` is not accepted by `ChoiceGroup`.

Methods

Signature	Notes
`public int size()`	Returns the number of elements present
`public String getString(final int elementNum) throws IndexOutOfBoundsException`	Returns the String part of the element referenced by `elementNum`: ❑ `elementNum`: the index of the element to be queried
`public Image getImage(final int elementNum) throws IndexOutOfBoundsException`	Returns the `Image` part of the element referenced by `elementNum`: ❑ `elementNum`: the index of the element to be queried

Table continued on following page

Signature	Notes
`public int append(String stringElement, Image imageElement) throws IllegalArgumentException, NullPointerException`	Appends an element to the `Choice`. The added element will be the last element of the `Choice`: ❑ `stringPart`: the String part of the element to be added ❑ `imagePart`: the image part of the element to be added, or `null` if there is no image part
`public void insert(final int elementNum, String stringElement, Image imageElement) throws IndexOutOfBoundsException, IllegalArgumentException, NullPointerException`	Inserts an element into the `Choice` just before the element specified: ❑ `elementNum`: the index of the element where insertion is to occur ❑ `stringPart`: the String part of the element to be inserted ❑ `imagePart`: the image part of the element to be inserted, or `null` if there is no image part
`public void delete(final int elementNum) throws IndexOutOfBoundsException`	Deletes the element referenced by `elementNum`. It is legal to delete all elements from a `Choice`: ❑ `elementNum`: the index of the element to be deleted
`public void set(final int elementNum, String stringElement, Image imageElement) throws IndexOutOfBoundsException, IllegalArgumentException, NullPointerException`	Sets the element referenced by `elementNum` to the specified element, replacing the previous contents of the element: ❑ `elementNum`: the index of the element to be set ❑ `stringPart`: the String part of the new element ❑ `imagePart`: the image part of the element, or `null` if there is no image part
`public boolean isSelected(final int elementNum) throws IndexOutOfBoundsException`	Returns a boolean value indicating whether this element is selected: ❑ `elementNum`: the index of the element to be queried

<table>
<tr><th>Signature</th><th>Notes</th></tr>
<tr><td><code>public int getSelectedIndex()</code></td><td>Returns the index number of an element in the <code>Choice</code> that is selected. For <code>Choice</code> types <code>EXCLUSIVE</code> and <code>IMPLICIT</code> there is at most one element selected, so this method is useful for determining the user's choice. Returns -1 if there are no elements in the <code>Choice</code> or no element has been selected.</td></tr>
<tr><td><code>public int getSelectedFlags(final boolean selectedArray_return) throws IllegalArgumentException, NullPointerException</code></td><td>Queries the state of a <code>Choice</code> and returns the state of all elements in the boolean array <code>selectedArray_return</code>. Note: this is a result parameter. It must be at least as long as the size of the <code>Choice</code> as returned by <code>size()</code>. If the array is longer, the extra elements are set to false:

❑ <code>selectedArray_return</code>: array to contain the results</td></tr>
<tr><td><code>public void setSelectedIndex(final int elementNum, final boolean selected) throws IndexOutOfBoundsException</code></td><td>For <code>MULTIPLE</code>, this simply sets an individual element's selected state.

For <code>EXCLUSIVE</code>, this can be used only to select any element, that is, the <code>selected</code> parameter must be <code>true</code>. When an element is selected, the previously selected element is deselected. If <code>selected</code> is <code>false</code>, this call is ignored. If the element was already selected, the call has no effect.

For <code>IMPLICIT</code>, this can be used only to select any element, that is, the <code>selected</code> parameter must be <code>true</code>. When an element is selected, the previously selected element is deselected. If <code>selected</code> is <code>false</code>, this call is ignored. If the element was already selected, the call has no effect:

❑ <code>elementNum</code>: the index of the element, starting from zero

❑ <code>selected</code>: the state of the element, where <code>true</code> means selected and <code>false</code> means not selected</td></tr>
<tr><td><code>public void setSelectedFlags(final boolean selectedArray) throws IllegalArgumentException, NullPointerException</code></td><td>Attempts to set the selected state of every element in the <code>Choice</code>. The array must be at least as long as the size of the <code>Choice</code>. If the array is longer, the additional values are ignored:

❑ <code>selectedArray</code>: an array in which the method collects the selection status</td></tr>
</table>

javax.microedition.midlet Package

The MIDlet package defines Mobile Information Device Profile applications and the interactions between the application and the environment in which the application runs.

The MIDlet package defines Mobile Information Device Profile applications and the interactions between the application and the environment in which the application runs. An application of the

Mobile Information Device Profile is a MIDlet.

The MIDlet lifecycle defines the protocol between a `MIDlet` and its environment through the following:

- A simple well-defined state machine
- A concise definition of the MIDlet's states
- APIs to signal changes between the states

Classes

MIDlet

```
public abstract class MIDlet
```

A MIDlet is a MID Profile application. The application must extend this class to allow the application management software to control the MIDlet and to be able to retrieve properties from the application descriptor and notify and request state changes. The methods of this class allow the application management software to create, start, pause, and destroy a MIDlet.

A MIDlet is a set of classes designed to be run and controlled by the application management software via this interface. The states allow the application management software to manage the activities of multiple MIDlets within a runtime environment. It can select which MIDlets are active at a given time by starting and pausing them individually.

The application management software maintains the state of the MIDlet and invokes methods on the MIDlet to change states. The MIDlet implements these methods to update its internal activities and resource usage as directed by the application management software. The MIDlet can initiate some state changes itself and notifies the application management software of those state changes by invoking the appropriate methods.

Constructors

Signature	Notes
`protected MIDlet()`	Protected constructor for subclasses.

Methods

Signature	Notes
`protected abstract void startApp() throws MIDletStateChangeException`	Signals the MIDlet to start and enter the Active state. In the Active state the MIDlet may hold resources. The method will only be called when the MIDlet is in the Paused state. Two kinds of failures can prevent the service from starting, transient and non-transient. For transient failures the `MIDletStateChangeException` exception should be thrown. For non-transient failures the `notifyDestroyed()` method should be called.
`protected abstract void pauseApp()`	Signals the MIDlet to stop and enter the Paused state. In the Paused state the MIDlet must release shared resources and become quiescent. This method will only be called when the MIDlet is in the Active state.
`protected abstract void destroyApp(final boolean unconditional) throws MIDletStateChangeException`	Signals the MIDlet to terminate and enter the Destroyed state. In the Destroyed state the MIDlet must release all resources and save any persistent state. This method may be called from the Paused or Active states. MIDlets should perform any operations required before being terminated, such as releasing resources or saving preferences or state: ❑ `unconditional`: If `true` when this method is called, the MIDlet must cleanup and release all resources. If `false` the MIDlet may throw a `MIDletStateChangeException` to indicate it does not want to be destroyed at this time.
`public final void notifyDestroyed()`	Used by a MIDlet to notify the application management software that it has entered into the Destroyed state. The application management software will not call the MIDlet's `destroyApp()` method, and all resources held by the MIDlet will be considered eligible for reclamation. The MIDlet must have performed the same operations (cleanup, releasing of resources etc) it would have if the `destroyApp()` had been called.

Table continued on following page

Signature	Notes
`public final void notifyPaused()`	Notifies the application management software that the MIDlet does not want to be active and has entered the Paused state. Invoking this method will have no effect if the MIDlet is destroyed, or if it has not yet been started.
`public final String getAppProperty(String key) throws NullPointerException`	Provides a MIDlet with a mechanism to retrieve named properties from the application management software. The properties are retrieved from the combination of the application descriptor file and the manifest. If the attribute in the descriptor has the same name as an attribute in the manifest the value from the descriptor is used and the value from the manifest is ignored: ❑ `key`: the name of the property
`public final void resumeRequest()`	Provides a MIDlet with a mechanism to indicate that it is interested in entering the Active state. Calls to this method can be used by the application management software to determine which applications to move to the Active state. When the application management software decides to activate this application it will call the `startApp()` method.

MIDletStateChangeException

```
public class MIDletStateChangeException extends Exception
```

Signals that a requested MIDlet state change failed. This exception is thrown by the MIDlet in response to state change calls into the application via the MIDlet interface.

Constructors

Signature	Notes
`public MIDletStateChangeException()`	Constructs an exception with no specified detail message
`public MIDletStateChangeException (String s)`	Constructs an exception with the specified detail message: ❑ `s`: the detail message

javax.microedition.rms Package

Classes

RecordStore

```
public class RecordStore
```

A class representing a record store. A record store consists of a collection of records that will remain persistent across multiple invocations of the MIDlet. The platform is responsible for making its best effort to maintain the integrity of the MIDlet's record stores throughout the normal use of the platform, including reboots, battery changes, etc.

Methods

Signature	Notes
`public static void deleteRecordStore(String recordStoreName) throws RecordStoreException, RecordStoreNotFoundException`	Deletes the named record store. MIDlet suites are only allowed to operate on their own record stores, including deletions. If the record store is currently open by a MIDlet when this method is called, or if the named record store does not exist, a `RecordStoreException` will be thrown: ❑ `recordStoreName`: The MIDlet suite unique record store to delete
`public static RecordStore openRecordStore(String recordStoreName, final boolean createIfNecessary) throws RecordStoreException, RecordStoreNotFoundException, RecordStoreFullException`	Open (and possibly create) a record store associated with the given MIDlet suite. If this method is called by a MIDlet when the record store is already open by a MIDlet in the MIDlet suite, this method returns a reference to the same `RecordStore` object: ❑ `recordStoreName`: The MIDlet suite unique name, not to exceed 32 characters, of the record store ❑ `createIfNecessary`: If `true`, the record store will be created if necessary

Table continued on following page

Signature	Notes
`public void closeRecordStore() throws RecordStoreNotOpenException, RecordStoreException`	This method is called when the MIDlet requests to have the record store closed. Note that the record store will not actually be closed until `closeRecordStore()` is called as many times as `openRecordStore()` was called. In other words, the MIDlet needs to make a balanced number of close calls as open calls before the record store is closed. When the record store is closed, all listeners are removed. If the MIDlet attempts to perform operations on the `RecordStore` object after it has been closed, the methods will throw a `RecordStoreNotOpenException`.
`public static String listRecordStores()`	Returns an array of the names of record stores owned by the MIDlet suite. Note that if the MIDlet suite does not have any record stores, this function will return `null`. The order of `RecordStore` names returned is implementation dependent.
`public String getName() throws RecordStoreNotOpenException`	Returns the name of this `RecordStore`.
`public int getVersion() throws RecordStoreNotOpenException`	Each time a record store is modified (record added, modified or deleted), it's version is incremented. This can be used by MIDlets to quickly tell if anything has been modified. The initial version number is implementation dependent. The increment is a positive integer greater than 0. The version number only increases as the `RecordStore` is updated.
`public int getNumRecords() throws RecordStoreNotOpenException`	Returns the number of records currently in the record store.
`public int getSize() throws RecordStoreNotOpenException`	Returns the amount of space, in bytes, that the record store occupies. The size returned includes any overhead associated with the implementation, such as the data structures used to hold the state of the record store, etc.
`public int getSizeAvailable() throws RecordStoreNotOpenException`	Returns the amount of additional room (in bytes) available for this record store to grow. Note that this is not necessarily the amount of extra MIDlet-level data which can be stored, as implementations may store additional data structures with each record to support integration with native applications, synchronization, etc.

Signature	Notes
`public long getLastModified() throws RecordStoreNotOpenException`	Returns the last time the record store was modified, in the format used by `System.currentTimeMillis()`.
`public void addRecordListener(RecordListener listener)`	Adds the specified `RecordListener`. If the specified listener is already registered, it will not be added a second time. When a record store is closed, all listeners are removed: ❑ `listener`: the `RecordChangedListener`
`public void removeRecordListener(RecordListener listener)`	Removes the specified `RecordListener`. If the specified listener is not registered, this method does nothing: ❑ `listener`: the `RecordChangedListener`
`public int getNextRecordID() throws RecordStoreNotOpenException, RecordStoreException`	Returns the record ID of the next record to be added to the record store. This can be useful for setting up pseudo-relational relationships. That is, if we have two or more record stores whose records need to refer to one another, we can predetermine the record IDs of the records that will be created in one record store, before populating the fields and allocating the record in another record store.
`public int addRecord(final byte data, final int offset, final int numBytes) throws RecordStoreNotOpenException, RecordStoreException, RecordStoreFullException`	Adds a new record to the record store. The record ID for this new record is returned. This is a blocking atomic operation. The record is written to persistent storage before the method returns: ❑ `data`: The data to be stored in this record. If the record is to have zero-length data (no data), this parameter may be `null`. ❑ `offset`: The index into the data buffer of the first relevant byte for this record. ❑ `numBytes`: The number of bytes of the data buffer to use for this record (may be zero).
`public void deleteRecord(final int recordId) throws RecordStoreNotOpenException, InvalidRecordIDException, RecordStoreException`	The record is deleted from the record store. The `recordId` for this record is **not** reused: ❑ `recordId`: The ID of the record to delete

Table continued on following page

Signature	Notes
`public int getRecordSize(final int recordId) throws RecordStoreNotOpenException, InvalidRecordIDException, RecordStoreException`	Returns the size (in bytes) of the MIDlet data available in the given record: ❑ `recordId`: The ID of the record to use in this operation
`public int getRecord(final int recordId, final byte buffer, final int offset) throws RecordStoreNotOpenException, InvalidRecordIDException, RecordStoreException, ArrayIndexOutOfBoundsException`	Returns the data stored in the given record: ❑ `recordId`: The ID of the record to use in this operation ❑ `buffer`: The byte array in which to copy the data ❑ `offset`: The index into the buffer in which to start copying
`public byte getRecord(final int recordId) throws RecordStoreNotOpenException, InvalidRecordIDException, RecordStoreException`	Returns a copy of the data stored in the given record: ❑ `recordId`: The ID of the record to use in this operation
`public void setRecord(final int recordId, final byte newData, final int offset, final int numBytes) throws RecordStoreNotOpenException, InvalidRecordIDException, RecordStoreException, RecordStoreFullException`	Sets the data in the given record to that passed in. After this method returns, a call to `getRecord(int recordId)` will return an array of `numBytes` size containing the data supplied here: ❑ `recordId`: The ID of the record to use in this operation ❑ `newData`: The new data to store in the record ❑ `offset`: The index into the data buffer of the first relevant byte for this record ❑ `numBytes`: The number of bytes of the data buffer to use for this record

Signature	Notes
`public RecordEnumeration enumerateRecords(RecordFilter filter, RecordComparator comparator, final boolean keepUpdated) throws RecordStoreNotOpenException`	Returns an enumeration for traversing a set of records in the record store in an optionally specified order. The filter, if not `null`, will be used to determine what subset of the record store records will be used. The comparator, if not `null`, will be used to determine the order in which the records are returned. If both the filter and comparator are `null`, the enumeration will traverse all records in the record store in an undefined order. This is the most efficient way to traverse all of the records in a record store: ❑ `filter`: if not `null`, will be used to determine what subset of the record store records will be used. ❑ `comparator`: if not `null`, will be used to determine the order in which the records are returned. ❑ `keepUpdated`: If `true`, the enumerator will keep its enumeration current with any changes in the records of the record store. Use with caution as there are possible performance consequences.

RecordStoreException

```
public class RecordStoreException extends Exception
```

Thrown to indicate a general exception occurred in a record store operation.

Constructors

Signature	Notes
`public RecordStoreException(String message)`	Constructs a new `RecordStoreException` with the specified detail message: ❑ `message`: the detail message
`public RecordStoreException()`	Constructs a new `RecordStoreException` with no detail message

InvalidRecordIDException

```
public class InvalidRecordIDException extends RecordStoreException
```

Thrown to indicate an operation could not be completed because the record ID was invalid.

Constructors

Signature	Notes
`public InvalidRecordIDException(String message)`	Constructs a new `InvalidRecordIDException` with the specified detail message: ❑ `message`: the detail message.
`public InvalidRecordIDException()`	Constructs a new `InvalidRecordIDException` with no detail message

RecordStoreNotFoundException

```
public class RecordStoreNotFoundException extends RecordStoreException
```

Thrown to indicate an operation could not be completed because the record store could not be found.

Constructors

Signature	Notes
`public RecordStoreNotFoundException(String message)`	Constructs a new `RecordStoreNotFoundException` with the specified detail message: ❑ `message`: the detail message
`public RecordStoreNotFoundException()`	Constructs a new `RecordStoreNotFoundException` with no detail message.

RecordStoreFullException

```
public class RecordStoreFullException extends RecordStoreException
```

Thrown to indicate an operation could not be completed because the record store system storage is full.

Constructors

Signature	Notes
`public RecordStoreFullException(String message)`	Constructs a new `RecordStoreFullException` with the specified detail message: ❑ `message`: the detail message
`public RecordStoreFullException()`	Constructs a new `RecordStoreFullException` with no detail message.

RecordStoreNotOpenException

```
public class RecordStoreNotOpenException extends RecordStoreException
```

Thrown to indicate that an operation was attempted on a closed record store.

Constructors

Signature	Notes
`public RecordStoreNotOpenException(String message)`	Constructs a new `RecordStoreNotOpenException` with the specified detail message: ❑ `message`: the detail message.
`public RecordStoreNotOpenException()`	Constructs a new `RecordStoreNotOpenException` with no detail message.

Interfaces

RecordEnumeration

```
public interface RecordEnumeration
```

A class representing a bidirectional record store `Record` enumerator. The `RecordEnumeration` logically maintains a sequence of the record ID's of the records in a record store. The enumerator will iterate over all (or a subset, if an optional record filter has been supplied) of the records in an order determined by an optional record comparator.

Methods

Signature	Notes
`public int numRecords()`	Returns the number of records available in this enumeration's set. That is, the number of records that have matched the filter criterion. Note that this forces the `RecordEnumeration` to fully build the enumeration by applying the filter to all records, which may take a non-trivial amount of time if there are a lot of records in the record store.
`public byte nextRecord() throws InvalidRecordIDException, RecordStoreNotOpenException, RecordStoreException`	Returns a copy of the **next** record in this enumeration, where next is defined by the comparator and/or filter supplied in the constructor of this enumerator. The byte array returned is a copy of the record. Any changes made to this array will **not** be reflected in the record store. After calling this method, the enumeration is advanced to the next available record.
`public int nextRecordId() throws InvalidRecordIDException`	Returns the `recordId` of the **next** record in this enumeration, where **next** is defined by the comparator and/or filter supplied in the constructor of this enumerator. After calling this method, the enumeration is advanced to the next available record.
`public byte previousRecord() throws InvalidRecordIDException, RecordStoreNotOpenException, RecordStoreException`	Returns a copy of the **previous** record in this enumeration, where **previous** is defined by the comparator and/or filter supplied in the constructor of this enumerator. The byte array returned is a copy of the record. Any changes made to this array will **not** be reflected in the record store. After calling this method, the enumeration is advanced to the next (previous) available record.
`public int previousRecordId() throws InvalidRecordIDException`	Returns the `recordId` of the **previous** record in this enumeration, where **previous** is defined by the comparator and/or filter supplied in the constructor of this enumerator. After calling this method, the enumeration is advanced to the next (previous) available record.
`public boolean hasNextElement()`	Returns true if more elements exist in the **next** direction.
`public boolean hasPreviousElement()`	Returns true if more elements exist in the **previous** direction.
`public void reset()`	Returns the index point of the enumeration to the beginning.

Signature	Notes
`public void rebuild()`	Request that the enumeration be updated to reflect the current record set. Useful for when a MIDlet makes a number of changes to the record store, and then wants an existing `RecordEnumeration` to enumerate the new changes.
`public void keepUpdated(final boolean keepUpdated)`	Used to set whether the enumeration will keep its internal index up to date with the record store record additions/deletions/changes. Note that this should be used carefully due to the potential performance problems associated with maintaining the enumeration with every change: ❑ `keepUpdated`: if `true`, the enumerator will keep its enumeration current with any changes in the records of the record store. Use with caution as there are possible performance consequences. If `false` the enumeration will not be kept current and may return record IDs for records that have been deleted or miss records that are added later. It may also return records out of order that have been modified after the enumeration was built.
`public boolean isKeptUpdated()`	Returns `true` if the enumeration keeps its enumeration current with any changes in the records.
`public void destroy() throws IllegalStateException`	Frees internal resources used by this `RecordEnumeration`. MIDlets should call this method when they are done using a `RecordEnumeration`. If a MIDlet tries to use a `RecordEnumeration` after this method has been called, it will throw an `IllegalStateException`.

RecordComparator

```
public interface RecordComparator
```

An interface defining a comparator that compares two records (in an implementation-defined manner) to see if they match or what their relative sort order is. The application implements this interface to compare two candidate records. The return value must indicate the ordering of the two records. The compare method is called by `RecordEnumeration` to sort and return records in an application specified order.

Fields

Signature	Notes
`public static final int EQUIVALENT`	`EQUIVALENT` means that in terms of search or sort order, the two records are the same. This does not necessarily mean that the two records are identical.
`public static final int FOLLOWS`	`FOLLOWS` means that the left (first parameter) record **follows** the right (second parameter) record in terms of search or sort order.
`public static final int PRECEDES`	`PRECEDES` means that the left (first parameter) record **precedes** the right (second parameter) record in terms of search or sort order.

Methods

Signature	Notes
`public int compare(final byte rec1, final byte rec2)`	Returns `RecordComparator.PRECEDES` if `rec1` precedes `rec2` in sort order, or `RecordComparator.FOLLOWS` if `rec1` follows `rec2` in sort order, or `RecordComparator.EQUIVALENT` if `rec1` and `rec2` are equivalent in terms of sort order: ❑ `rec1`: The first record to use for comparison. Within this method, the application must treat this parameter as read-only. ❑ `rec2`: The second record to use for comparison. Within this method, the application must treat this parameter as read-only.

RecordListener

```
public interface RecordListener
```

A listener interface for receiving `Record` changed/added/deleted events from a record store.

Methods

Signature	Notes
`public void recordAdded(RecordStore recordStore, final int recordId)`	Called when a record has been added to a record store: ❑ `recordStore`: the `RecordStore` in which the record is stored ❑ `recordId`: the record ID of the record that has been added

Signature	Notes
`public void recordChanged(RecordStore recordStore, final int recordId)`	Called after a record in a record store has been changed. If the implementation of this method retrieves the record, it will receive the changed version: ❑ `recordStore`: the `RecordStore` in which the record is stored ❑ `recordId`: the `recordId` of the record that has been changed
`public void recordDeleted(RecordStore recordStore, final int recordId)`	Called after a record has been deleted from a record store. If the implementation of this method tries to retrieve the record from the record store, an `InvalidRecordIDException` will be thrown: ❑ `recordStore`: the `RecordStore` in which the record was stored ❑ `recordId`: the `recordId` of the record that has been deleted

RecordFilter

```
public interface RecordFilter
```

An interface defining a filter that examines a record to see if it matches (based on an application-defined criteria). The application implements the `match()` method to select records to be returned by the `RecordEnumeration`. Returns true if the candidate record is selected by the `RecordFilter`. This interface is used in the record store for searching or subsetting records.

Methods

Signature	Notes
`public boolean matches(final byte candidate)`	Returns true if the candidate matches the implemented criterion: ❑ `candidate`: The record to consider. Within this method, the application must treat this parameter as read-only.

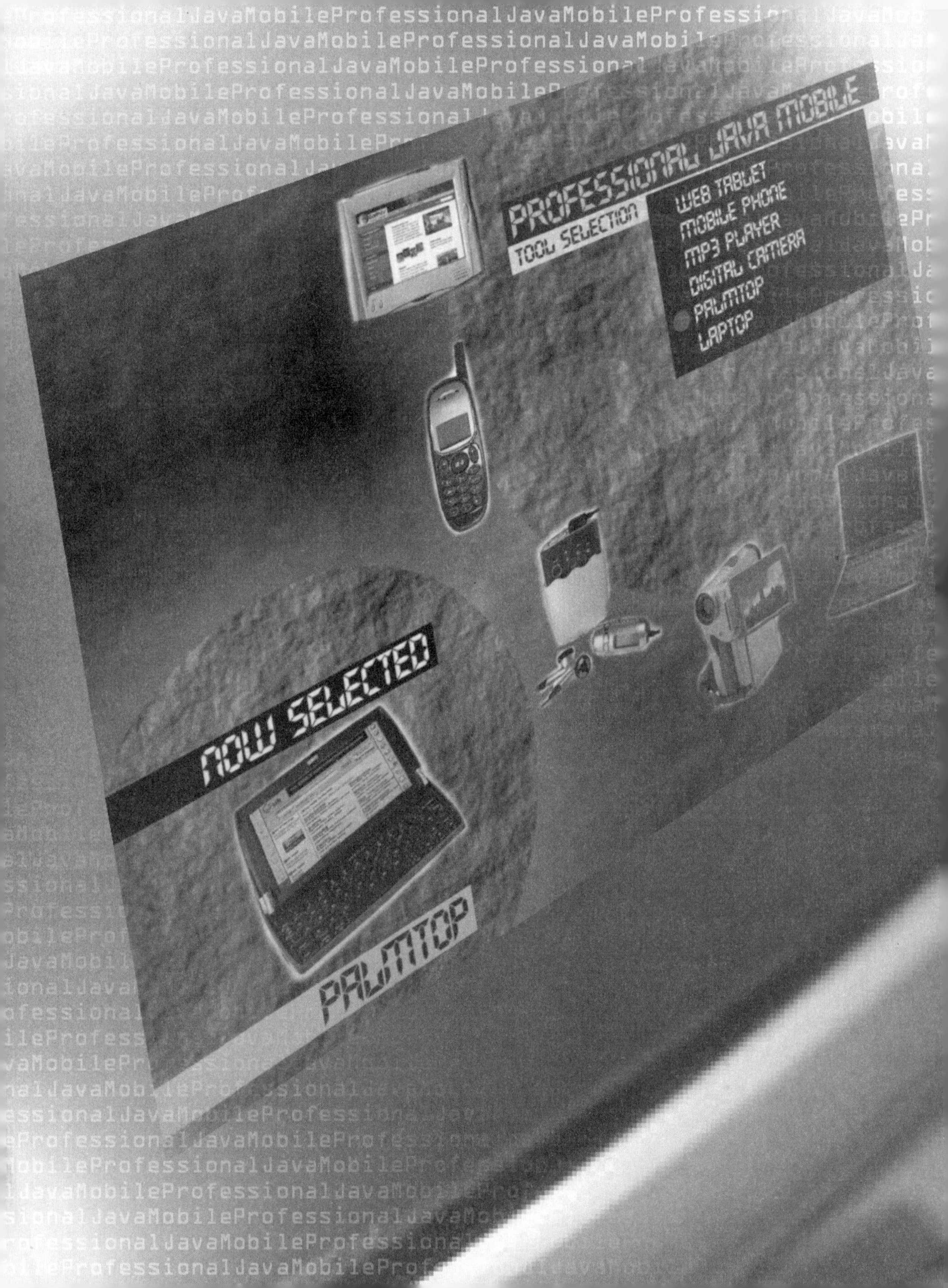

PROFESSIONAL JAVA MOBILE

TOOL SELECTION

WEB TABLET
MOBILE PHONE
MP3 PLAYER
DIGITAL CAMERA
PALMTOP
LAPTOP

NOW SELECTED

PALMTOP

D

javax.microedition.io Package

The classes listed here represent the new IO classes defined in the CLDC, and now available in the CDC. Each class is explained below but for a more detailed explanation on the usage of these classes please refer to Chapter 3.

Classes

Connector

```
public class Connector extends Object
```

This class is a placeholder for the static methods used to create all the connection objects.

This is done by dynamically looking up a class the name of which is formed from the platform name and the protocol of the requested connection. The parameter string describing the target conforms to the URL format as described in RFC 2396. This takes the general form {scheme}:[{target}][{parms}], where:

- {scheme} is the name of a protocol such as http
- The {target} is normally some kind of network address, but protocols may regard this as a fairly flexible field when the connection is not network-oriented
- Any {parms} are formed as a series of equates on the form ";x=y" such as ;type=a

An optional second parameter may be specified to the open function. This is a mode flag that indicates to the protocol handler the intentions of the calling code. The options here are to specify if the connection is going to be read (`READ`), written (`WRITE`), or both (`READ_WRITE`). The validity of these flag settings is protocol dependent. For instance, a connection for a printer would not allow read access, and would throw an `IllegalArgumentException` if this was attempted. Omitting this parameter results in `READ_WRITE` being used by default.

An optional third parameter is a Boolean flag to indicate if the calling code has been written in such a way as to handle timeout exceptions. If this is selected the protocol may throw an `InterruptedIOException` when it detects a timeout condition. This flag is only a hint to the protocol handler and it is no guarantee that such exceptions will be thrown. Omitting this parameter results in no exceptions being thrown. The timeout period is not specified in the open call because this is protocol specific. Protocol implementers can either hardwire an appropriate value or read them from an external source such as the system properties.

Four functions are provided just to gain access to an input or output stream because of the common occurrence of this task.

Constructors

Signature	Notes
`private Connector()`	Prevent instantiation

Fields

Signature	Notes
`public static final int READ`	Access mode
`public static final int WRITE`	Access mode
`public static final int READ_WRITE`	Access mode
`private static String platform`	The platform name
`private static boolean j2me`	True if we are running on a J2ME system
`private static String classRoot`	The root of the classes

Methods

Signature	Notes
`static void <clinit>()`	Class initializer
`public static Connection open(String name) throws IllegalArgumentException, ConnectionNotFoundException, java.io.IOException`	Create and open a `Connection`: ❑ `name`: The URL for the connection
`public static Connection open(String name, final int mode) throws IllegalArgumentException, ConnectionNotFoundException, java.io.IOException`	Create and open a `Connection`: ❑ `name`: The URL for the connection ❑ `mode`: The access mode

Signature	Notes
`public static Connection open(String name, final int mode, final boolean timeouts) throws IllegalArgumentException, ConnectionNotFoundException, java.io.IOException`	Create and open a `Connection`: ❑ `name`: The URL for the connection ❑ `mode`: The access mode ❑ `timeouts`: A flag to indicate that timeouts should be enabled
`private static Connection openPrim(String name, final int mode, final boolean timeouts, String platfrm) throws ClassNotFoundException, IllegalArgumentException, ConnectionNotFoundException, java.io.IOException, java.io.IOException, IllegalArgumentException`	Create and open a `Connection`: ❑ `name`: The URL for the connection ❑ `mode`: The access mode ❑ `timeouts`: A flag to indicate that timeouts should be enabled
`public static java.io.DataInputStream openDataInputStream(String name) throws IllegalArgumentException, ConnectionNotFoundException, java.io.IOException`	Create and open a connection input stream: ❑ `name`: The URL for the connection
`public static java.io.DataOutputStream openDataOutputStream(String name) throws IllegalArgumentException, ConnectionNotFoundException, java.io.IOException`	Create and open a connection output stream: ❑ `string`: The URL for the connection
`public static java.io.InputStream openInputStream(String name) throws IllegalArgumentException, ConnectionNotFoundException, java.io.IOException`	Create and open a connection input stream: ❑ `name`: The URL for the connection
`public static java.io.OutputStream openOutputStream(String name) throws IllegalArgumentException, ConnectionNotFoundException, java.io.IOException`	Create and open a connection output stream: ❑ `name`: The URL for the connection

ConnectionNotFoundException

```
public class ConnectionNotFoundException extends java.io.IOException
```

This class is used to signal that a connection target cannot be found.

Constructors

Signature	Notes
`public ConnectionNotFoundException()`	Constructs a `ConnectionNotFoundException` with no detail message. A detail message is a string that describes this particular exception.
`public ConnectionNotFoundException(String s)`	Constructs a `ConnectionNotFoundException` with the specified detail message. A detail message is a string that describes this particular exception: ❑ s: the detail message

Interfaces

StreamConnection

```
public interface StreamConnection
```

This interface defines the capabilities that a stream connection must have.

StreamConnectionNotifier

```
public interface StreamConnectionNotifier
```

This interface defines the capabilities that a connection notifier must have.

Methods

Signature	Notes
`public StreamConnection acceptAndOpen() throws java.io.IOException`	Returns a `StreamConnection` that represents a server side socket connection.

ContentConnection

```
public interface ContentConnection
```

This interface defines the stream connection over which content is passed.

Methods

Signature	Notes
`public String getType()`	Returns the type of content that the resource connected to is providing. E.g. if the connection is via HTTP, then the value of the `content-type` header field is returned.
`public String getEncoding()`	Returns a string describing the encoding of the content that the resource connected to is providing, e.g. if the connection is via HTTP, the value of the `content-encoding` header field is returned.
`public long getLength()`	Returns the length of the content that is being provided, e.g. if the connection is via HTTP, then the value of the `content-length` header field is returned.

Datagram

```
public interface Datagram
```

This is the generic datagram interface. It represents an object that will act as the holder of data to be sent or received from a datagram connection. The `DataInput` and `DataOutput` interfaces are extended by this interface to provide a simple way to read and write binary data in and out of the datagram buffer. A special function `reset()` may be called to reset the read/write point to the beginning of the buffer.

Methods

Signature	Notes
`public String getAddress()`	Returns the address in string form, or `null` if no address was set.
`public byte getData()`	Returns the buffer.
`public int getLength()`	Returns the length of the data.
`public int getOffset()`	Returns the offset into the data buffer.

Signature	Notes
`public void setAddress(String addr) throws IllegalArgumentException`	Sets datagram address. The parameter String describing the target of the datagram takes the form: {protocol}:{target}, e.g. The "target" can be "//{host}:{port}" (but is not necessarily limited to this). In this example a datagram connection for sending to a server could be addressed datagram://123.456.789.12:1234. Note that if the address of a datagram is not specified, then it defaults to that of the connection: ❑ `addr`: the new target address as a URL
`public void setAddress(Datagram reference) throws IllegalArgumentException`	Sets datagram address, copying the address from another datagram: ❑ `reference`: the datagram whose address will be copied as the new target address for this datagram.
`public void setLength(final int len) throws IllegalArgumentException`	Sets the length: ❑ `len`: the new length of the data
`public void setData(final byte buffer, final int offset, final int len) throws IllegalArgumentException`	Sets the buffer, offset and length: ❑ `buffer`: the data buffer ❑ `offset`: the offset into the data buffer ❑ `len`: the length of the data in the buffer
`public void reset()`	Resets the read/write pointer and zeros the offset and length parameters.

OutputConnection

```
public interface OutputConnection
```

This interface defines the capabilities that an output stream connection must have.

Methods

Signature	Notes
`public java.io.OutputStream openOutputStream() throws java.io.IOException`	Opens and returns an output stream for a connection
`public java.io.DataOutputStream openDataOutputStream() throws java.io.IOException`	Opens and returns a data output stream for a connection

HttpConnection

```
public interface HttpConnection
```

This interface defines the necessary methods and constants for an HTTP connection.
HTTP is a request-response protocol in which the parameters of request must be set before the request is sent. The connection exists in one of two states:

- Setup, in which the connection has not been made to the server
- Connected, in which the connection has been made, request parameters have been sent and the response is expected

The following methods may be invoked only in the Setup state:

- `setRequestMethod()`
- `setRequestProperty()`

The transition from Setup to Connected is caused by any method that requires data to be sent to or received from the server. The following methods cause the transition to the Connected state:

- `openInputStream()`
- `openOutputStream()`
- `getLength()`
- `getType()`
- `getEncoding()`
- `getHeaderField()`
- `getResponseCode()`
- `getResponseMessage()`
- `getHeaderFieldInt()`
- `getHeaderFieldDate()`

The following methods may be invoked at any time:

- close()
- getRequestMethod()
- getRequestProperty()
- getURL()
- getProtocol()
- getHost()
- getFile()
- getRef()
- getPort()
- getQuery()

Example Using StreamConnection

We will now look at an example using StreamConnection, where no HTTP specific behavior is needed or used:

```
void getViaStringConnection(String url) throws IOException {

  StreamConnection c = null;
  InputStream s = null;
  try {
```

Connector.open() is used to open a URL and a StreamConnection is returned. From the StreamConnection the InputStream is opened:

```
    c = (StreamConnection)Connector.open(url);
    s = c.openInputStream();
    int ch;
```

The InputStream is used to read every character until end of file (-1). If an exception is thrown the connection and stream are closed:

```
    while ((ch = s.read()) != -1)
    {
      ...
    }
  } finally {
    if (s != null) {
      s.close();
    }
    if (c != null) {
      c.close();
    }
  }
}
```

Example Using ContentConnection

This example does a simple read of a URL using `ContentConnection`. No HTTP specific behavior is needed or used.

```
void getViaContentConnection(String url) throws IOException {

  ContentConnection c = null;
  InputStream is = null;
  try {
```

`Connector.open()` is used to open a URL and a `ContentConnection` is returned:

```
    c = (ContentConnection)Connector.open(url);
```

The `ContentConnection` may be able to provide the length. If the length is available, it is used to read the data in bulk:

```
    int len = (int)c.getLength();
    if (len > 0) {
```

From the `StreamConnection` the `InputStream` is opened. It is used to read every character until end of file (-1):

```
      is = c.openInputStream();
      byte[] data = new byte[len];
      int actual = is.read(data);
      ...
    } else {
      int ch;
      while ((ch = is.read()) != -1) {
        ...
      }
    }
```

If an exception is thrown the connection and stream are closed:

```
  } finally {
      if (is != null) {
      is.close();
      }
      if (c != null) {
        c.close();
      }
  }
}
```

Example Using HttpConnection

Here we read the HTTP headers and the data using `HttpConnection`.

```
void getViaHttpConnection(String url) throws IOException {

  HttpConnection c = null;
  InputStream is = null;
  try {
```

Again, `Connector.open()` is used to open the URL but htis time an `HttpConnection` is returned:

```
    c = (HttpConnection)Connector.open(url);
```

Getting the `InputStream` will open the connection and read the HTTP headers. They are stored until requested:

```
    is = c.openInputStream();
```

The HTTP headers are read and processed. If the length is available, it is used to read the data in bulk:

```
    // Get the ContentType
    String type = c.getType();

    // Get the length and process the data
    int len = (int)c.getLength();
    if (len > 0) {
      byte[] data = new byte[len];
      int actual = is.read(data);
      ...
    } else {
        int ch;
        while ((ch = is.read()) != -1)
        {
          ...
        }
    }
  } finally {
      if (is != null) {
        is.close();
      }
      if (c != null) {
        c.close();
      }
   }
}
```

Example Using POST with HttpConnection

Post a request with some headers and content to the server and process the headers and content. Connector.open() is used to open the URL and a `HttpConnection` is returned. The request method is set to POST and request headers set. A simple command is written and flushed. The HTTP headers are read and processed. If the length is available, it is used to read the data in bulk. From the `StreamConnection` the `InputStream` is opened. It is used to read every character until end of file (-1). If an exception is thrown the connection and stream is closed.

```
void postViaHttpConnection(String url) throws IOException {
  HttpConnection c = null;
  InputStream is = null;
  OutputStream os = null;
  try {
    c = (HttpConnection)Connector.open(url);

    // Set the request method and headers
    c.setRequestMethod(HttpConnection.POST);
    c.setRequestProperty("If-Modified-Since", "29 Oct 1999 19:43:31 GMT");
    c.setRequestProperty("User-Agent", "Profile/MIDP-1.0 Configuration/CLDC-1.0");
    c.setRequestProperty("Content-Language", "en-US");

    // Getting the output stream may flush the headers os = c.openOutputStream();
    os.write("LIST games\n".getBytes());
    os.flush();

    // Optional, openInputStream will flush
    // Opening the InputStream will open the connection
    // and read the HTTP headers. They are stored until
    // requested.
    is = c.openInputStream();

    // Get the ContentType String type = c.getType();
    processType(type);

    // Get the length and process the data
    int len = (int)c.getLength();
    if (len > 0) { byte[] data = new byte[len];
      int actual = is.read(data);
      process(data);
    } else {
        int ch;
        while ((ch = is.read()) != -1) {
          process((byte)ch);
        }
    }
  } finally {
      if (is != null) {
        is.close();
      }
      if (os != null) {
        os.close();
      }
      if (c != null) {
        c.close();
      }
  }
}
```

Simplified Stream Methods on Connector

The `Connector` class defines the following convenience methods for retrieving an input or output stream directly for a specified URL:

- `InputStream openDataInputStream(String url)`
- `DataInputStream openDataInputStream(String url)`
- `OutputStream openOutputStream(String url)`
- `DataOutputStream openDataOutputStream(String url)`

Please be aware that using these methods implies certain restrictions. We will not get a reference to the actual connection, but rather just references to the input or output stream of the connection. Not having a reference to the connection means that we will not be able to manipulate or query the connection directly. This in turn means that we will not be able to call any of the following methods:

- `getRequestMethod()`
- `setRequestMethod()`
- `getRequestProperty()`
- `setRequestProperty()`
- `getLength()`
- `getType()`
- `getEncoding()`
- `getHeaderField()`
- `getResponseCode()`
- `getResponseMessage()`

Fields

Signature	Notes
`public static final String HEAD`	HTTP Head method.
`public static final String GET`	HTTP Get method.
`public static final String POST`	HTTP Post method.
`public static final int HTTP_OK`	200: The request has succeeded.
`public static final int HTTP_CREATED`	201: The request has been fulfilled and resulted in a new resource being created.
`public static final int HTTP_ACCEPTED`	202: The request has been accepted for processing, but the processing has not been completed.

Signature	Notes
public static final int HTTP_NOT_AUTHORITATIVE	203: The returned meta-information in the entity-header is not the definitive set as available from the origin server.
public static final int HTTP_NO_CONTENT	204: The server has fulfilled the request but does not need to return an entity-body, and might want to return updated meta-information.
public static final int HTTP_RESET	205: The server has fulfilled the request and the user agent **should** reset the document view that caused the request to be sent.
public static final int HTTP_PARTIAL	206: The server has fulfilled the partial GET request for the resource.
public static final int HTTP_MULT_CHOICE	300: The requested resource corresponds to any one of a set of representations, each with its own specific location, and agent- driven negotiation information is being provided so that the user (or user agent) can select a preferred representation and redirect its request to that location.
public static final int HTTP_MOVED_PERM	301: The requested resource has been assigned a new permanent URI and any future references to this resource **should** use one of the returned URIs.
public static final int HTTP_MOVED_TEMP	302: The requested resource resides temporarily under a different URI.
public static final int HTTP_SEE_OTHER	303: The response to the request can be found under a different URI and **should** be retrieved using a GET method on that resource.
public static final int HTTP_NOT_MODIFIED	304: If the client has performed a conditional GET request and access is allowed, but the document has not been modified, the server **should** respond with this status code.
public static final int HTTP_USE_PROXY	305: The requested resource **must** be accessed through the proxy given by the `Location` field.
public static final int HTTP_TEMP_REDIRECT	307: The requested resource resides temporarily under a different URI.
public static final int HTTP_BAD_REQUEST	400: The request could not be understood by the server due to malformed syntax.

Signature	Notes
`public static final int HTTP_UNAUTHORIZED`	401: The request requires user authentication. The response **must** include a `WWW-Authenticate` header field containing a challenge applicable to the requested resource.
`public static final int HTTP_PAYMENT_REQUIRED`	402: This code is reserved for future use.
`public static final int HTTP_FORBIDDEN`	403: The server understood the request, but is refusing to fulfil it. Authorization will not help and the request **should not** be repeated.
`public static final int HTTP_NOT_FOUND`	404: The server has not found anything matching the Request-URI. No indication is given of whether the condition is temporary or permanent.
`public static final int HTTP_BAD_METHOD`	405: The method specified in the Request-Line is not allowed for the resource identified by the Request-URI.
`public static final int HTTP_NOT_ACCEPTABLE`	406: The resource identified by the request is only capable of generating response entities which have content characteristics not acceptable according to the accept headers sent in the request.
`public static final int HTTP_PROXY_AUTH`	407: This code is similar to 401 (Unauthorized), but indicates that the client must first authenticate itself with the proxy.
`public static final int HTTP_CLIENT_TIMEOUT`	408: The client did not produce a request within the time that the server was prepared to wait. The client **may** repeat the request without modifications at any later time.
`public static final int HTTP_CONFLICT`	409: The request could not be completed due to a conflict with the current state of the resource.
`public static final int HTTP_GONE`	410: The requested resource is no longer available at the server and no forwarding address is known.
`public static final int HTTP_LENGTH_REQUIRED`	411: The server refuses to accept the request without a defined content length.
`public static final int HTTP_PRECON_FAILED`	412: The precondition given in one or more of the request-header fields evaluated to false when it was tested on the server.

Signature	Notes
public static final int HTTP_ENTITY_TOO_LARGE	413: The server is refusing to process a request because the request entity is larger than the server is willing or able to process.
public static final int HTTP_REQ_TOO_LONG	414: The server is refusing to service the request because the Request-URI is longer than the server is willing to interpret.
public static final int HTTP_UNSUPPORTED_TYPE	415: The server is refusing to service the request because the entity of the request is in a format not supported by the requested resource for the requested method.
public static final int HTTP_UNSUPPORTED_RANGE	416: A server **should** return a response with this status code if a request included a RANGE request-header field , and none of the range-specifier values in this field overlap the current extent of the selected resource, and the request did not include an If-Range request-header field.
public static final int HTTP_EXPECT_FAILED	417: The expectation given in an EXPECT request-header field could not be met by this server, or, if the server is a proxy, the server has unambiguous evidence that the request could not be met by the next-hop server.
public static final int HTTP_INTERNAL_ERROR	500: The server encountered an unexpected condition that prevented it from fulfilling the request.
public static final int HTTP_NOT_IMPLEMENTED	501: The server does not support the functionality required to fulfil the request.
public static final int HTTP_BAD_GATEWAY	502: The server, while acting as a gateway or proxy, received an invalid response from the upstream server it accessed in attempting to fulfil the request.
public static final int HTTP_UNAVAILABLE	503: The server is currently unable to handle the request due to a temporary overloading or maintenance of the server.
public static final int HTTP_GATEWAY_TIMEOUT	504: The server, while acting as a gateway or proxy, did not receive a timely response from the upstream server specified by the URI or some other auxiliary server it needed to access in attempting to complete the request.
public static final int HTTP_VERSION	505: The server does not support, or refuses to support, the HTTP protocol version that was used in the request message.

Methods

Signature	Notes
`public String getURL()`	Return a string representation of the URL for this connection.
`public String getProtocol()`	Returns the protocol name of the URL of this `HttpConnection`. e.g., HTTP or HTTPS.
`public String getHost()`	Returns the host information of the URL of this `HttpConnection`. e.g. host name or IPv4 address.
`public String getFile()`	Returns the file portion of the URL of this `HttpConnection`. `null` is returned if there is no file.
`public String getRef()`	Returns the ref portion of the URL of this `HttpConnection`. RFC2396 specifies the optional fragment identifier as the text after the crosshatch (#) character in the URL. This information may be used by the user agent as additional reference information after the resource is successfully retrieved. The format and interpretation of the fragment identifier is dependent on the media type [RFC2046] of the retrieved information.
`public String getQuery()`	Returns the query portion of the URL of this `HttpConnection`. RFC2396 defines the query component as the text after the last question-mark (?) character in the URL.
`public int getPort()`	Returns: the network port number of the URL for this `HttpConnection`. The default HTTP port number (80) is returned if there was no port number in the string passed to `Connector.open()`.
`public String getRequestMethod()`	Get the current request method. e.g. HEAD, GET, POST. The default value is GET.

Signature	Notes
`public void setRequestMethod(String method) throws java.io.IOException`	Set the method for the URL request, only the following are legal, subject to protocol restrictions: ❑ GET ❑ POST ❑ HEAD The default method is GET: ❑ `method`: the HTTP method
`public String getRequestProperty(String key)`	Returns the value of the named general request property for this connection: ❑ `key`: the keyword by which the request is known (e.g., "accept")
`public void setRequestProperty(String key, String value)`	Sets the general request property. If a property with the key already exists, overwrite its value with the new value. Note: HTTP requires all request properties that can legally have multiple instances with the same key to use a comma-separated list syntax which enables multiple properties to be appended into a single property: ❑ `key`: the keyword by which the request is known (e.g., "`accept`") ❑ `value`: the value associated with it
`public int getResponseCode() throws java.io.IOException`	Returns the HTTP response status code. It parses responses like: `HTTP/1.0 200 OK HTTP/1.0 401 Unauthorized` and extracts the integers 200 and 401 respectively from the response (i.e., the response is not valid HTTP).

Signature	Notes
`public String getResponseMessage() throws java.io.IOException`	Gets the HTTP response message, if any, returned along with the response code from a server. From responses like: ❑ `HTTP/1.0 200 OK HTTP/1.0 404 Not Found` Extracts the Strings `OK` and `Not Found` respectively. Returns `null` if none could be discerned from the responses (the result was not valid HTTP).
`public long getExpiration() throws java.io.IOException`	Returns the expiration date of the resource that this URL references, or 0 if not known. The value is the number of milliseconds since January 1, 1970 GMT.
`public long getDate() throws java.io.IOException`	Returns the sending date of the resource that the URL references, or `0` if not known. The value returned is the number of milliseconds since January 1, 1970 GMT.
`public long getLastModified() throws java.io.IOException`	Returns the value of the `last-modified` header field. The result is the number of milliseconds since January 1, 1970 GMT.
`public String getHeaderField(String name) throws java.io.IOException`	Returns the value of the named header field, or `null` if there is no such field in the header: ❑ `name`: of a header field.
`public int getHeaderFieldInt(String name, final int def) throws java.io.IOException`	Returns the value of the named field parsed as a number. This form of `getHeaderField()` exists because some connection types (e.g., `http-ng`) have pre-parsed headers. Classes for that connection type can override this method and short-circuit the parsing: ❑ `name`: the name of the header field. ❑ `def`: the default value.

Signature	Notes
`public long getHeaderFieldDate(String name, final long def) throws java.io.IOException`	Returns the value of the named field parsed as date. The result is the number of milliseconds since January 1, 1970 GMT represented by the named field. This form of `getHeaderField()` exists because some connection types (e.g., `http-ng`) have pre-parsed headers. Classes for that connection type can override this method and short-circuit the parsing: ❑ `name`: the name of the header field. ❑ `def`: a default value.
`public String getHeaderField(final int n) throws java.io.IOException`	Returns the value of the nth header field or `null` if the array index is out of range. An empty String is returned if the field does not have a value. ❑ `n`: the index of the header field
`public String getHeaderFieldKey(final int n) throws java.io.IOException`	Returns the key of the nth header field or `null` if the array index is out of range: ❑ `n`: the index of the header field

InputConnection

```
public interface InputConnection
```

This interface defines the capabilities that an input stream connection must have.

Methods

Signature	Notes
`public java.io.InputStream openInputStream() throws java.io.IOException`	Open and return an input stream for a connection
`public java.io.DataInputStream openDataInputStream() throws java.io.IOException`	Open and return a data input stream for a connection

Connection

```
public interface Connection
```

This is the most basic type of generic connection. Only the `close()` method is defined. The `open()` method defined here because opening is always done by the `Connector.open()` methods.

Methods

Signature	Notes
`public void close() throws java.io.IOException`	Closes the connection. When the connection has been closed access to all methods except this one will cause an `IOException` to be thrown. Closing an already closed connection has no effect. Streams derived from the connection may be open when method is called. Any open streams will cause the connection to be held open until they themselves are closed.

DatagramConnection

```
public interface DatagramConnection
```

This interface defines the capabilities that a datagram connection must have:

- The parameter string describing the target of the connection takes the form: {protocol}:[//{host}]:{port}
- A datagram connection can be opened in a "client" mode or a "server" mode. If the //{host} is missing then it is opened as a "server" (by "server", this means that a client application initiates communication). When the //{host} is specified the connection is opened as a client.

Some examples are:

- A datagram connection for accepting datagrams: datagram://:1234
- A datagram connection for sending to a server: datagram://123.456.789.12:1234

Note that the port number in "server mode" (unspecified host name) is that of the receiving port. The port number in "client mode" (host name specified) is that of the target port. The reply to port in both cases is never unspecified. In "server mode", the same port number is used for both receiving and sending. In "client mode", the reply-to port is always dynamically allocated.

Methods

Signature	Notes
`public int getMaximumLength()`	Get the maximum length a datagram can be.
`public int getNominalLength()`	Get the nominal length of a datagram.
`public void send(Datagram dgram) throws java.io.IOException, InterruptedIOException`	Send a datagram: ❑ `dgram`: A datagram
`public void receive(Datagram dgram) throws java.io.IOException, InterruptedIOException`	Receive a datagram: ❑ `dgram`: A datagram
`public Datagram newDatagram(final int size)`	Make a new datagram object automatically allocating a buffer: ❑ `size`: The length of the buffer to be allocated for the datagram
`public Datagram newDatagram(final int size, String addr)`	Make a new datagram object: ❑ `size`: The length of the buffer to be used ❑ `addr`: The address to which the datagram must go
`public Datagram newDatagram(final byte buf, final int size) throws IllegalArgumentException`	Make a new datagram object: ❑ `buf`: The buffer to be used in the datagram ❑ `size`: The length of the buffer to be allocated for the datagram
`public Datagram newDatagram(final byte buf, final int size, String addr) throws IllegalArgumentException`	Make a new datagram object: ❑ `buf`: The buffer to be used in the datagram ❑ `size`: The length of the buffer to be used ❑ `addr`: The address to which the datagram must go

PROFESSIONAL JAVA MOBILE

TOOL SELECTION

WEB TABLET
MOBILE PHONE
MP3 PLAYER
DIGITAL CAMERA
PALMTOP
LAPTOP

NOW SELECTED

PALMTOP

Index

A Guide to the Index

The index is arranged hierarchically, in alphabetical order, with symbols preceding the letter A. Most second-level entries and many third-level entries also occur as first-level entries. This is to ensure that users will find the information they require however they choose to search for it.

Symbols

***7 device**
Java 2 Micro Edition, 11
'Hello World' application
EPOC operating system, 32
Kilo Virtual Machine, 28
MIDlets, 25
2.5G/3G networks
charging, 431
2-tier client–server application model
Java 2 Micro Edition, 55
3G (third generation) networks
Internet screen phones, 335
Java 2 Micro Edition, 35, 40
smart phones, 335
3-tier client–server application model
Java 2 Micro Edition, 56

A

About View
Game interfaces, 506
Towers of Hanoi game, 501, 517, 527
abstract classes
Factory Method, 504
glue layer, 222
implementation, 520
paint(), 530
paintBorder(), 530
Template Method, 505
Towers of Hanoi game, 507
unpaint(), 530
UserInterface not allowed to be, 251
abstract logic
Towers of Hanoi game, 503
Abstract Window Toolkit (AWT)
application conversion, 276, 293
Connected Device Configuration, 270, 297
Connected Limited Device Configuration, 272, 297, 303
messaging, 469
accept method
ServerSocket, 94, 96
acceptAndOpen() method
server socket implementation, 94
SimpleServer class, 97
StreamConnectionNotifier, 96
Accepts header
data formats, 431
access
disconnected, 652
access controllers
elements, 622
execution chain, 623
J2SE security architecture, 620, 622
security, 616
AccessControlException, 622
accessor methods
Hotel interface, 584
messaging, 468
Action Keys
Towers of Hanoi game, 502
actionListener() method
ContactApp, 239
UIListener interface, 230
activation
smart card file structure, 373
ActiveSync
synchronization, 329
ActiveX
security, 619
add_user JSP
mobile positioning case study, 597
addActions() method
UserInterfaceMIDP class, 247, 250
addCommand() method, 160
addExceptDate() method
Repeat class, 352
addField() method
AggregateField class, 344
addLine() method
application conversion, 282
addParameter() method
ItemField class, 345
addRecord() method, 193
KJava, 255
addRecordListener() method, 201
addRepeatDate() method
Repeat class, 352

Address Book package
JavaPhone API, architecture, 338
javax.pim.addressbook, 344
sample application, 345
inserting contacts, 345
running examples, 350
updating contacts, 347
Symbian implementation of JavaPhone API, 344
address books. See contact databases
Address class
javax.net.datagram, 355
administered objects, 453
ADPU (Application Data Programming Units). See Application Data Programming Units (ADPU)
AggregateField class
addField() method, 344
Aglets
mobile agent system, 620
AID (Application Identifiers). See Application Identifiers (AID)
AIF builder tool
Symbian implementation of JavaPhone API, 340
ALARM
Alert, 168
alert warnings
design considerations, 638
Alerts
method, 168
Screen, 167
analysis workshop
time-sheet data entry
Java 2 Micro Edition architecture, 64, 75
anchor points
text, 188
Angle of Arrival (AOA)
mobile positioning, 567
ant make utility, 484
ANY
TextBox, 162
AOA (Angle of Arrival). See Angle of Arrival (AOA)
Apache
Jakarta Tomcat, 406
APDU (Application Protocol Data Units). See Application Protocol Data Units (APDU)
apdutool application, 382, 388
API (Application Programmer Interface). See Application Programmer Interface (API)
APOP command
POP3, 101
append method, 165
applets. See also Java Card applications
Application Identifiers, 378
Java 2 Micro Edition evolution, 13
Java Card
PIN, 374
Java Card example, 384
lifecycles, 383
security, 619, 620
application commands
Towers of Hanoi game, 514
application conversion
Connected Device Configuration, 269, 297, 301
Connected Limited Device Configuration, 269, 272, 295
Line class, 297
Slide class, 297
devices, 293
Graphical User Interface design, 291
JavaCheck, 272, 287
Pocket PC device, 293
Slide class, 284
slide editors, 274
Confirmation class, 285
Insertion class, 280, 292
Line class, 285
MouseListener interface, 282
Status class, 287
SlideEditor class, 276, 282
SlideViewer class, 282, 292
Touchable interface, 271, 274, 293
Application Data Programming Units (ADPU)
apdutool, 382, 388
ISO7816 commands, 374
Java Card Development Kit emulator, 388
smart card communication, 371
application descriptor file
MIDlets, 142
application design
Towers of Hanoi game, 502
Application Identifiers (AID)
smart cards, 378
Application Programmer Interface (API)
CLDC, 495
high-level, 161
Java 2 Enterprise Edition, 59
Java Card, 370, 383
JavaPhone, 335
low-level, 176
MIDP, 133, 495
MPS-SDK package, 573
Application Protocol Data Units (APDU)
Java Card, 19
application running
Towers of Hanoi game, 533
application tier
mobile positioning case study, 579
application.xml file
mobile positioning case study, 608
applications
apdutool, 382, 388
AuctionHouse client, 445
Bidder client, 445
complete mobile, 217
construction, 639
decreasing size, 209
definition, 394
delivery, 209
development, 495
distributed, 394
e-mail client, 102
hotel-search. See hotel-search application
J2ME, 652
Java, 96
Java Card, 367
kAWT, 152
KJava, 149
KJava Palm, 150
management, 209
MIDP
starting, 251
networking, 96
POP3, 102
sockets, 96
starting KJava, 265
arcAngle, 183

architecture
compared to design, 633
Java 2 Enterprise Edition, 58
Java 2 Micro Edition, 51, 60
mobile devices, 78
service-oriented, 61
messaging, 446
Software Architecture Analysis Method, 75
time-sheet data entry, 63
architecture security
J2ME, 623
J2SE, 620
arcs, 183
ASSERT
design considerations, 649
assertion design considerations, 649
asymmetric encryption systems, 625
asynchronous communications, 446
asynchronous messages, 436, 437
attributes
MIDlet and MIDlet suites, 144
auction house example
code conversions, 477
JMS, 444, 457
AuctionHouse
JMS client, 461
AuctionHouse-Specific iBus configuration, 481
Authfile.txt
MPC emulator, 573
availability
security, 616
AWT (Abstract Window Toolkit). See Abstract Window Toolkit (AWT).

B

BACK
buttons, 165
commands, 166
events, 165
background downloads, 431
back-up synchronization, 310
bandwidths
synchronization, 313
base profiles
Foundation Profile, 84
base stations
mobile positioning, 566, 572
Base user interface class
Displayable, 159
baselines
anchor ponts, 190
images, 190
beans
entity, 581, 584
implementation, 584
location with Home interfaces in hotel-search application, 580
Bidder class
JMS client, 466
bidder user interface, 469
BidderUI class, 469, 470, 471
BidUI class, 469, 472, 474
bind() method
JNDI, 405
BitSender class
datagram code, 441
BitTicker
datagram code, 443
Bluetooth network
mobile devices, 36, 39
Bluetooth technology
security, 628
synchronization, 332
Boolean values
synchronization, 320
synchronization protocols, 311
bridges
MOM, 447
browsers
Java 2 Micro Edition, 13, 41
WAP development, 601
BufferedReader
servlet file reading, 418
building classes
Towers of Hanoi game, 523, 532
buildList() method
application conversion, 277
business classes
ContactApp, 233
JMS, 457
business logic
ContactApp, 234
separating from low-level implementation, 219
business sponsors
software design, 632
business systems
Java 2 Micro Edition architecture, 53
object classes
time-sheet data entry case study, 69
button bar
KJava screen layout, 257
Button object
application conversion, 298
BUTTONHEIGHT, 258
buttons
ContactApp, 234
labels, 228
multi-edit screens, 262
pen events, 263
bytecode verifiers
J2SE security architecture, 620, 621

C

C language
Virtual Machines, 84
C++
mobile devices, 12
CA (Certificate Authority). See Certificate Authority (CA)
caches
client-side design, 648
data integrity, 649
design considerations, 648
objects, 647
calculateDistance() method
mobile positioning case study, 587

Calendar package
accessing schedule information, 352
adding schedule information, 353
creating to-do lists, 353
JavaPhone API, architecture, 338
javax.pim.calendar, 352
Symbian implementation of JavaPhone API, 352
CalendarDatabase class
javax.pim.calendar, 352
CalendarEntry class
javax.pim.calendar, 352
CalendarToDo class
javax.pim.calendar, 352
callback
Spotless implementation, 528
Towers of Hanoi game, 522
Canvas
coordinate system, 179
Displayable, 159
graphics, 178
Hanoiclass, 522
images, 190
low-level graphics, 178
MIDP implementation, 522
Canvas.getWidth/Height(), 179
Canvas.repaint(), 178
cardItems() method
ContactDatabase class, 348
cardlets. See applets;Java Card applications
cards. See also smart cards
credit, 369
SIM, 368
carets
flashing, 264
carrier technology
Java 2 Micro Edition, 34
case sensitive searching
MIDP, 418
catalog class
JMS, 459
catch blocks
messaging, 443, 444
CDC. See Connected Device Configuration
CDC (Connected Device Configuration). See Connected Device Configuration (CDC).
CDLC-based platforms
RMI, 403
cell of origin (COO), 566
mobile positioning, 569
Celldata.txt
MPC emulator, 573
cells
mobile positioning, 566
network security, 617
centralized server architecture
messaging, 446
Certificate Authority (CA), 626
champions
software design, 632
changeOfSlide() method
application conversion, 278
characters() method
XML expert systems, 558
charging
bytes versus time, 431
CHARHEIGHT
title space, 260
check methods
unsupported low-level interface methods, 192
check Permission(), 622
Choice
Lists, 164
ChoiceGroup
Forms, 169
circles
drawing, 191
circular arcs, 183
class building
Towers of Hanoi game, 532
class files
bytecode verifier, 621
class loaders
eliminated KVM functionality, 127
J2SE security architecture, 620, 621
class preverification, 483
class variables
ContactApp, 235
UserInterfaceMIDP class, 246
classes
abstract, 222, 251
Towers of Hanoi game, 507
Address, 355
AggregateField, 344
Base User Interfaces, 159
bidder user interface, 469
BitSender, 441
BitTicker, 443
CalendarDatabase, 352
CalendarEntry, 352
CalendarToDo, 352
Canvas, 522
CmdLine
AuctionHouse, 465
Command, 160
Comms, 231
RMI, 397
Condition
expert systems, 543
Confirmation
slide editors, 285
Connected Limited Device Configuration, 131
ConnectionBaseInterface, 92
Connector, 87, 92
ConnectorUtils.MIDlets, 422
Contact, 221
contact database user interface, 228
ContactApp
business and control class, 233
separation of business logic from low-level implementation, 220
source, 234
UIListener interface, 230
ContactCard, 344, 350
ContactDatabase, 344, 348, 356
ContactServlet, 240
Coordinate
MPS-SDK package, 573
Datagram, 355
DatagramNameService, 355
DatagramService, 355
DatagramServiceFactory, 355
DB, 227

classes (continued)
DBKJava, 253
DBMIDP, 242
Displayable, 159
DocumentHandler
Simple API for XML, 551
Engine
expert systems, 546
Fact
expert systems, 543
FactBase
expert systems, 543
Factory, 105
Factory Method, 504
Field, 222
filters, 199
font, 190
GamesSettings, 531
glue layer, 222
Graphics, 179
HandlerBase
Simple API for XML, 551
HanoiCanvas, 522
HanoiSpotlet, 523
HotelAccessor, 584, 588
HotelEJB, 590
HttpConnection, 203
I/O
expert systems, 548
InputSource
Simple API for XML, 551
Insertion
slide editors, 280, 292
ItemField, 345
J2SE platform, 85
JtapiPeerFactory, 343
KbLoader
XML, 559
KexDocumentHandler
XML, 557
KHashTable
expert systems, 542
KJava-specific, 253
KMemoReader
expert systems, 549
KnowledgeBase
expert systems, 541, 545
Line
slide editors, 285, 297
List
application conversion, 276
LocaleData, 86
LocationHandlerEJB
mobile positioning case study, 586
LocationRequest
MPS-SDK package, 573
LocationResult
MPS-SDK package, 573
MailClient, 102, 109
MailStore, 102, 104
MainKJava, 265
MainMIDP, 251
Message, 103
MIDlet, 135, 142
MIDletGameSettings, 520
MIDP implemetation, 513
MPSException
MPS-SDK package, 573
Parser
Simple API for XML, 551
PdbReader
expert systems, 548
PopProxy, 120
PopReader, 102, 105
PopServer, 114
PowerMonitor, 361
PowerWarningType, 361
pre-verification, 149
Protocol, 87
pseudo-business, 233
Record, 223
Repeat, 352
RepeatRule, 352
Rule
expert systems, 543
SelectScrollTextBox, 259
SimpleClient, 98
SimpleServer, 97
Slide
application conversion, 284
Connected Limited Device Configuration, 297
SlideEditor
application conversion, 276, 282
SlideViewer
application conversion, 282, 292
Spotlet
Towers of Hanoi game, 523
Spotlet, 258
SpotletGamesSettings, 531
Status
slide editors, 287
StreamConnectionNotifier, 97
StringTokenizer, 108
Template Method, 505
ThinParser
XML, 553
TimerTask, 206
UserEJB, 589
UserInterfaceKJava, 257
UserInterfaceMIDP, 246
UserProfile, 351
XML
expert systems, 552
classfile pre-verifier, 128
classfile verification
(CLDC), 624
CLDC. See Connected Limited Device Configuration
clear() method
contact database design, 227
DBMIDP class, 243
click depth
small screens, 636
client building
JMS, 482
client code
synchronization, 325
client sockets
creation, 90
opening, 87
client tier
Java 2 Micro Edition architecture, 56
clients
exchange synchronization, 318
JMS, 461
JMS bidder, 466

clients (continued)
Mobile Positioning Protocol, 570
synchronization, 315
synchronization code, 325
synchronization exchange, 318
synchronization flow charts, 314
client-server application models
Java 2 Micro Edition, 55
client-server caches, 649
client–server deployment
Java 2 Micro Edition, 55
client-side computing
need for, 651
client-side design
caches, 648
clipping, 185
clipRect(), 186
close() method
Connection interface, 87
contact database design, 227
UserInterface supporting method, 230
closeRecordStore(), 193
closeRecordStore() method
DBMIDP class, 243
CmdLine class
AuctionHouse, 465
CMP (Container Managed Persistency). See Container Managed Persistency (CMP)
code
client synchronization, 325
game logic implementation, 508
server synchronization, 320
synchronization, 320
Views
Towers of Hanoi game, 506
code conversions
auction house, 477
messaging system, 475
code refactoring
design, 645
code techniques
defensiveness, 650
code-levels
optimization, 646
coding processes
optimal design, 645
coding standards
J2ME, 650
collaborative–dynamic–distributed application model
Java 2 Micro Edition, 57
color, 180
com.wrox.cdc.io.j2se.serversocket.Protocol, 94
com.wrox.cdc.j2se.socket.Protocol, 93
Command buttons
Forms, 173
Command class, 160
command keys
Towers of Hanoi game, 502
command()
smart cards, 377
commandAction() method
MIDlets, 136, 208
rectangles, 182
UserInterfaceMIDP class, 250
Web server communication, 408
CommandListener interface, 161
Spotlet class, 263
Towers of Hanoi game, 514
UserInterfaceMIDP class, 250
commands
creation for Java Card applications, 376
ISO7816, 374
MIDlet constructor, 135
POP3, 100, 101, 107, 108, 113
smart card, 371
comma-separated (CSV) text files, 220
Common Object Request Broker Architecture (CORBA), 396
Java 2 Enterprise Edition, 59
Comms class
RMI, 397
user interface (UI) design, 231
communication
distributed applications
J2EE, 395
J2ME example, 407
smart card example, 372
with smart cards, 370
Communication API package
JavaPhone API, architecture, 338
communication channels
security, 626
communication model
Bluetooth network, 37
Wireless LAN, 38
communications networks
security, 625
Community Draft
Java Community Process, 46
compatibility
CLDC and CDC compared, 132
CLDC, Javacard and EmbeddedJava, 132
Java Card applications, 390
compilers
individual class scripts, 482
compiling
Java Card applets, 386
MIDlet example, 137
complexity management
software design, 634
component-based software design
introduction, 640
component-based systems
percieved problems, 641
performance, 641
reliability, 642
security, 642
components
architecture
devices, 53
Java 2 Micro Edition, 61
compared to objects, 641
diagrams
time-sheet data entry case study, 73
conceptual methods
long term memory, 636
short term memory, 636
concrete classes
Template Method, 505
Condition class
expert systems, 543

conduits
  synchronization, 331
confidentiality, 616
configuration, 14
  Java 2 Micro Edition, 15, 16
  messaging servers, 475
CONFIRMATION
  Alert, 168
Confirmation class
  slide editors
    application conversion, 285
connect() method
  PopReader class, 107
Connected Device Configuration (CDC)
  application conversion, 269, 270, 272, 297, 301
    PersonalJava, 270
  CLDC comparison, 132
  Foundation Profile, 81
  installation, 82
  Java 2 Micro Edition, 16, 30
  Linux platform, 82
  networking, 294
  porting, 293
  security, 83, 623
  synchronization, 326
  VxWorks platform, 82
Connected Limited Device Configuration (CLDC), 125, 130, 157
  API, 495
  application conversion, 269, 272, 295
  CDC comparison, 132
  classes, 131
  development environment, 147
  downloading, 148, 217
  environment testing, 148
  expert systems, 542, 551
  Foundation Profile, 83
  Java 2 Micro Edition, 16, 20
  JavaCard and EmbeddedJava comparison, 132
  JMS, 459
  KVM, 126
  messaging, 476, 485
  mobile devices, 23
  Mobile Information Device Profile, 22
  networking, 295
  porting, 294
  security, 623, 624
  server connectivity, 406
  synchronization code, 325
  system properties, 201
  Towers of Hanoi game, 502
Connected Limited Devices
  definition, 126
connected systems
  design factors, 651
Connection interface, 87
ConnectionBaseInterface, 93
  code, 92
ConnectionFactory
  messaging, 453, 454, 479
ConnectionNotFoundException, 92
connections
  messaging servers, 454
  server socket Protocol, 94
  sockets and serversockets, 91
  socket Protocol, 93
connectivity
  advanced topics, 430
  Palm, 151
Connector class
  open() method, 87
  Protocol class instantiation, 87
  socket connections, 92
Connector object
  networking, 295
CONSTRAINT MASK
  TextBox, 162
construction
  J2ME applications, 639
constructors
  ContactApp, 236
  DBKJava class, 253
  Java Card applets, 384
  Java Card application development, 383
  MainMIDP class, 251
  MIDlet class, 135
  MyHanoiGame class, 526
  protocol synchronization, 311
  Record class, 223
  Spotlet, 259
Consumer Application Server, 84
Contact class
  contact databases, 221
contact databases, 217
  design, 222
  format, 220
  Palm/KJava port, 252
  reading files from web servers, 407
  requirements, 218
ContactApp class
  business and control class, 233
  class variables, 235
  generic data passed to UIListener, 250
  separation of business logic from low-level implementation, 220
  source, 234
  UIListener interface, 230
ContactCard class
  getFields() method, 350
  vCard standard, 344
ContactDatabase class
  cardItems() method, 348
  javax.pim.addressbook, 344
  openDatabase() method, 348, 356
  updateItem() method, 349
ContactServlet class
  ContactApp, 240
Container Managed Persistency (CMP)
  bean implementation, 584, 589
  mobile positioning case study, 575
ContentConnection interface, 87
  Palm CLDC, 231
Content-Type
  data formats, 431
control classes
  ContactApp, 233
controller
  Model–View–Controller system, 52
    expert systems, 538
conversion
  Java Card applets, 386
COO (cell of origin). See cell of origin (COO)

Coordinate class
MPS-SDK package, 573
coordinate system
mobile positioning, 572
coordinates
Canvas, 179
CORBA (Common Object Request Broker Architecture). See Common Object Request Broker Architecture (CORBA)
correctness in design, 644
costs
SMS, 436
country.code
microedition.locale, 202
CRC message encryption, 626
create() method
ContactApp, 237
HotelHome interface, 581
UserHome interface, 581
create(long id) method
UserHome interface, 581
createBean
Orion custom tag, 593
createImage()
image downloads, 208
createNewsFlash()
timers, 207
createRecords(), 194
createRing()
Towers of Hanoi game, 504
createTower
Towers of Hanoi game, 504
creating protocol synchronization, 311
creationTime
synchronization, 313, 322, 326
creatorID
Palm databases, 253
credit cards
compared with smart cards, 369
cryptography
Java Card, 370
Crystal devices
Symbian implementation of JavaPhone API, 339
CSV (comma-separated) text files. See comma-separated (CSV) text files
currentScreen class variable
ContactApp, 235
custom tags
User bean creation/properties, 593
CVM. See C language, Virtual Machines
CyberFlex, 390
cyclic files
smart cards, 373

D

d.receive()
messaging, 444
daemon threads
eliminated KVM functionality, 127
data
caches, 648
integrity, 649
data extraction
expert systems, 550
data formats
servlets, 431
data loss
Java Card applications, 383
data packets
UDP, 88
database access
servlets, 422
database tier
Java 2 Micro Edition architecture, 56
database.merge() method
Comms class, 232
databases
contact, 217
design, 222
format, 220
Palm/KJava ports, 252
requirements, 218
interaction through Entity EJBs, 579
KJava, 254
merging, 266
Palm
creatorID, 253
sharing, 266
Datagram class
javax.net.datagram, 355
Datagram interface
J2ME, 89
UDP data packets, 88
Datagram package
datagram addressing, 355
JavaPhone API, architecture, 338
javax.net.datagram, 355
sending and receiving datagrams, 357
Symbian implementation of JavaPhone API, 355
DatagramConnection
UDP datapackets, 88
UDP programming, 89
DatagramNameService class
javax.net.datagram, 355
lookup() method, 356
lookupAll() method, 356
DatagramPacket, 442, 443
datagrams
code example, 441
messaging, 441
sending and receiving, 89
DatagramService class
getMaximumLength() method, 359
getServerService() method, 358
getService() method, 355
javax.net.datagram, 355
parseAddress() method, 355
receive() method, 355
send() method, 355
DatagramServiceFactory class
javax.net.datagram, 355
DatagramSocket, 442, 443
DataInputStream
image downloads, 208
DateField
Forms, 170
DB class
contact database design, 227
DBKJava class
Palm/KJava port, 253

DBMIDP class
MIDP specific database class, 242
debugging
Java 2 Micro Edition, 54
Mobile INformation Device Profile, 211
decentralized architecture
multicast IP, 448
decryption systems, 625
default database connection
JNDI, 429
defensive coding techniques, 650
DELE command
POP3, 101, 113, 118
delete() method
ContactApp, 238
deleteMessage() method
readMessage(), 109
deleteRecord() method
contact database design, 227
KJava, 257
deleteRecordStore(), 193, 198
deleteSlide() method
application conversion, 279
deletion
Palm/KJava records, 254, 256
records
DBMIDP class, 246
delivery
applications, 209
Denial of Service (DoS) attacks, 615
deployment
client–server
Java 2 Micro Edition, 55
deployment descriptor file
web.xml, 412
descriptors
XML, 608
deselect() method
Java Card application development, 383
design
applications
Towers of Hanoi game, 502
compared to architecture, 633
correctness, 644
databases, 222
error handling, 638
Factory Method, 504
fine grain level
Towers of Hanoi game, 503
Graphical User Interface
application conversion, 291
high-level
Towers of Hanoi game, 502
J2ME software, 631
limitations in wireless technology, 496
metaphors, 640
optimization, 644
patterns, 640
portability, 495
reverse usability, 638
screens, 498
small devices, 496
Template Method, 505
user interfaces, 497
contact databases, 228
desktop
AuctionHouse, 484
Destination objects
AuctionHouse, 479
destroyApp() method, 161
MainMIDP class, 252
MIDlets, 24, 137
rectangles, 182
Towers of Hanoi game, 514
details() method
ContactApp, 236
detection
security, 615
developers
software design, 632
development
J2ME applications, 495
Java 2 Micro Edition, 51, 53
time-sheet data entry case study, 69
development cycles
optimal code-levels, 646
development environment
CLDC, 147
development platforms
Mobile Information Device Profile, 211
device compromises
messaging, 486
device identification
threading, 431
device-based programming, 642
correctness, 644
fault tolerance, 644
predictability, 644
robustness, 644
safety, 644
system responsiveness, 643
timeliness, 643
user feedback, 643
devices
application conversion, 293
Connected Limited Device Configuration, 295
component-based systems, 53
design for small, 496
MIDP
case sensitive searching, 418
mobile, 7
*7, 11
3G (third generation) networks, 35, 40
Bluetooth network, 36, 39
Connected Limited Device Configuration, 20
EmbeddedJava, 19
General Packet Radio Service, 35
Global System for Mobile Communication, 34
Infrared network, 40
infrastructure, 78
Java 2 Micro Edition, 7, 14
Java Card, 17
Kilo Virtual Machine, 21
Wireless LAN, 38
time-sheet data entry
Java 2 Micro Edition architecture, 77
dialogDismissed()
Towers of Hanoi game, 528
dialogOwner()
Towers of Hanoi game, 528
digital attacks
security, 617
wireless networks, 617
digital signatures, 626

direct casting
JNDI, 405
Direct Sequence Spread Spectrum (DSSS)
wireless communication, 37
directory structure
messaging, 482
DirectoryConnection, 90
disconnected access, 652
display
contact fields, 260
Display class
color, 180
greyscale, 180
Display.isColor(), 180
Display.numColors(), 180
Display.setCurrent()
textbox, 163
Displayable
Canvas, *159*
class, *159*
Screen, 159
timers, 207
Displayables
addActions() method, 250
displayCatalog(), 472
displayFields() method
UserInterface display methods, 229
UserInterfaceMIDP class, 247
displayList() method
UserInterface display methods, 229
distributed applications
communication
J2EE, 395
definition, 394
location
J2EE, 403
distributed systems
Java 2 Micro Edition, 62
Document Type Definition (DTD)
expert systems, 547
DocumentHandler class
Simple API for XML, 551
DocumentHandler interface
XML expert systems, 552
DoS (Denial of Service). See Denial of Service (DoS)
DOTTED lines, 186
dotted-quad, 439
double data types
eliminated KVM functionality, 128
double-buffering
graphics, 180
download() method
ContactApp, 238
DOWNLOAD_PATH
ContactApp, 235
downloadAndMerge() method
Comms class, 231
downloadOK() method
ContactApp, 238
downloads
background, 431
CLDC, 148, 217
images, 208
Java Card API, 383
Java Card Development Kit, 381
JBoss + Tomcat, 406
JRun server, 576
kAWT, 151
KJava, 148
MIDP, 217
MPS SDK 3.0, 577
Orion application server, 576
resuming, 431
ROM images, 218
taglibs, 593
drawArc(), 187
drawChar() text rendering, 190
drawChars() text rendering, 190
drawImage(), 190
drawRect(), 187
drawRectangle(), 181, 185
drawRoundRect(), 187
drawString() text rendering, 190
drawSubstring() text rendering, 190
DSSS (Direct Sequence Spread Spectrum). See Direct Sequence Spread Spectrum (DSSS)
DTD (Document Type Definition). *See* Document Type Definition (DTD)
dual-key encryption systems, 625
dynamic IP addressing, 440

E

EAI (enterprise application integration). See enterprise application integration (EAI)
EAR (Enterprise ARchive files)
hot deployment, 414
eatWhitespace() method
XML expert systems, 556
edit() method
ContactApp, 237
edit_user JSP
mobile positioning case study, 598
editField() method
UserInterface display methods, 229
UserInterfaceMIDP class, 248
editFields() method
UserInterface display methods, 229
editions
Java 2, 14
editListener() method
ContactApp, 238, 239
UIListener interface, 230
ejb-jar.xml file
mobile positioning case study, 609
EJBs (Enterprise JavaBeans). See Enterprise JavaBeans (EJBs)
ejbtags
EJBs from JSP, 593
ellipse
creation, 184
elliptical arcs, 183
e-mail
client construction, 102
SMTP, 439
EMAILADDR
TextBox, 162
EmbeddedJava
CLDC and JavaCard, 132
Java 2 Micro Edition, 19
JavaPhone API, 336

emulation environment
Mobile Information Device Profile, 26
PersonalJava
application conversion, 274, 292, 293
emulators
emulator file system
Symbian implementation of JavaPhone API, 339
Mobile OInformation Device Profile, 211
MPC, 575
description, 573
input output demo, 578
position areas visualization client, 579
supported datum, 572
testing, 578
Palm, 218
Palm Operating System Emulator
expert systems, 539
Palm OS, 496, 497
SmartPhone, 598
testing JTAPI with emulator, 363
Wireless Toolkit, 496
enablers
Java 2 Micro Edition, 9
encryption systems, 625
endElement() method
XML expert systems, 559
end-users
software design, 633
Engine class
expert systems, 546
enterprise application integration (EAI), 448
Enterprise ARchive files (EAR). See EAR (Enterprise ARchive files)
Enterprise JavaBeans (EJBs)
database access, 429
ejbtags, 593
Java 2 Enterprise Edition, 58, 60, 69
mobile positioning case study, 580
servlet access, 429
enterprise systems
performance, 641
entity beans
implementation, 584
mobile positioning case study, 581
enumeration
messaging, 472
Enumeration.next Element(), 197
enumurators
record retrieval, 196
environments
CLDC development, 147
EPOC operating system
Personal Java, 32
running Java aplications, 340
Symbian implementation of JavaPhone API, 340
EQUIVALENT
records, 198
Ericsson
mobile electronic transactions initiative, 628
Ericsson Mobile Positioning Solutions (MPS-SDK 3.0), 568, 569
mobile positioning case study, 577
ERROR
Alert, 168
error classes
eliminated KVM functionality, 128
error handling
database access, 429
design, 638
device-based programming, 644
Error JSP
mobile positioning case study, 594
Error View
Game interfaces, 506
Towers of Hanoi game, 517
ETSI (European Telecommunications Standards Institute). See European Telecommunications Standards Institute (ETSI)
European Telecommunications Standards Institute (ETSI)
mobile positioning standards, 570
event handlers
ContactApp, 239
listeners, 341
Meta events, 341
observers, 341
penDown(), 259, 263
Spotlets, 263
Symbian implementation of JavaPhone API, 341
events
BACK, 165
EXIT, 165
high-level user interface, 160
Lists, 165
low-level, 176
RecordListener, 201
exception handling
J2ME server connectivity, 410
exchanges
client synchronization, 318
server synchronization, 318
EXCLUSIVE Lists, 164
executables
Palm, 150
execution speed
compared to start-up time, 646
exit, MyHanoiGame class, 526
EXIT event, 165
EXIT command, 166
MIDlet constructor, 135
exit() method
Main interface, 231
MainMIDP class, 252
expert systems
Palm Operating System Emulator, 539
portable devices, 537
explorability design in mobile systems, 637
extraction, data
expert systems, 550

F

face
font attributes, 188
Fact class
expert systems, 543
FactBase class
expert systems, 543
Factory classes, 105
Connector, 87

Factory Method design pattern
HanoiGame abstract class, 507
MIDP implementation, 518
Spotless implementations, 528
Towers of Hanoi game, 504
fault tolerance
device-based programming, 644
feedback
user, 430
FHSS (Frequency-Hopping Spread Spectrum). See Frequency-Hopping Spread Spectrum (FHSS)
Field class
databases, 222
Field objects
multiple edit, 249
fields
contact database format, 221
input
pen events, 263
KJava editing, 261
synchronization protocols, 311
file system, emulator
Symbian implementation of JavaPhone API, 339
file writing
servlets, 418
filename
servlet instruction, 235
files
smart cards, 372, 373
fillArc(), 183
filled circles
creation, 184
fillRect(), 181
fillRoundRect(), 185
Fills, 181
Filter class, 199
filtered file reading
servlets, 414
filtering
contact database, 266
filters
records, 198
final View
Towers of Hanoi game, 527
finalization
eliminated KVM functionality, 127
finally clause
error handling, 429
findAll() method
HotelHome interface, 582
UserHome interface, 581
findByPrimaryKey() method
HotelHome interface, 582
UserHome interface, 581
fine grain level design
Towers of Hanoi game, 503, 526
Finish View
Game interfaces, 506
FirstPerson Inc., 13
fixed costs
SMS, 436
fixed infrastructure topology
wireless communication, 38
Flash
timers, 207
float data types
eliminated KVM functionality, 128
flow charts
client synchronization, 314
server synchronization, 315
synchronization, 315
client, 314
server, 315
focus
UserInterfaceKJava class, 262
FOLLOWS
records, 198
Font class, 190
Font.getFace(), 188
Font.getFont(), 188
Font.getSize(), 188
Font.getStyle(), 188
fonts
attributes, 188
low-level graphics, 188
obtaining, 188
Forms
interactive, 173
methods, 172
Screen, 169
timers, 206
FormTest() method, 174
Forte
J2MEWTK build, 140
Forward networks
messaging, 437, 438
Foundation Profile
base profiles, 84
Connected Device Configuration (CDC), 81
installation, 82
Java 2 Micro Edition, 32
Java Applications, 96
java.text.resources, 86
javax.microedition.io, 86
RMI functionality, 121
supported classes, 85
frames
application conversion, 291
Frequency-Hopping Spread Spectrum (FHSS)
wireless communication, 37
front-end logic
J2EE, 405
ftp
resuming interrupted data transfers, 431

G

Game hierarchies
Factory Method, 504
Game interface
shared implementation, 506
game logic implementation
code, 508
Game Over View
Towers of Hanoi game, 500
Game views
Towers of Hanoi game, 506
games
instance initialization, 516
Towers of Hanoi, 496

GameSettings
abstract class, 512
class, 531
hierarchies, 505
implementation, 520
gateways
iBus//Mobile, 477, 481
security, 627
Gauge
Forms, 170
interactive Forms, 173
GCF. See Generic Connection Framework
GemPlus
GemXpresso, 390
General Packet Radio Service (GPRS)
Java 2 Micro Edition, 35
General Packet Radio Service (GPRS), 440
Generic Connection Framework (GCF), 90
javax.microedition.io, 86
geodetic datum system
mobile positioning, 572
GET
networking, 203
Get Password command
smart card programming, 380
Get response command
smart cards, 376
Get Server Name command
smart card programming, 379
Get User Name command
smart card programming, 380
getAddresses() method
Provider interface, 343
getAvailableLocales()
java.text.resources, 86
getBaselinePosition(), 190
getBatteryLevel() method
PowerMonitor class, 361
getBytes() method
contact database design, 225
getClip(), 185
getClipHeight/Width(), 185
getCurrentUser() method
UserProfile class, 351
getDB() method
MainMIDP class, 252
getDescription() method, 584
getEstimatedSecondsRemaining() method
PowerMonitor class, 361
getFields() method
contact database design, 226
ContactCard class, 350
getFieldValue()
contact database design, 225
getGameAction(), 177
getId() method, 584
expert systems, 548
getInputStream() method, 91
server sockets, 96
getJtapiPeer() method
JtapiPeerFactory class, 343
getKeyCode(), 177
getLargestIndex() method
UserEJB class, 589
getLastModified()
records, 201
getLatitude() method, 584
getLength()
HttpConnection, 204
getLocation() method
mobile positioning case study, 587
getLongitude() method, 584
getMaximumLength() method
DatagramService class, 359
getMaxSize(), 163
getNextRecordID(), 196
getNumberOfRecords()
DBKJava class, 254
getOutputStream() method
server sockets, 96
getPK() method
contact database design, 225
getProductName() method
RMI, 397
getProvider() method
JtapiPeer interface, 343, 354
getQuantity() method
RMI, 397
getResourceAsStream(filename) method
WAR files, 417
getResponse() method
message systems, 438
networking, 294
synchronization, 321, 326
getResponseCode() method
POST requests, 419
getServerService() method
DatagramService class, 358
getService() method
DatagramService class, 355
getServices() method
JtapiPeer interface, 343
getServletContext() method
WAR files, 417
getSize() method, 109
MailClient, 111
getStrokeStyle(), 186
getTranslateX(), 192
getTranslateY(), 192
getVersion ()
records, 201
getXXX() implementation
Towers of Hanoi game, 516
global graphics
synchronization, 325
Global Positioning System (GPS)
mobile positioning, 569, 572
Global System for Mobile Communication (GSM)
Java 2 Micro Edition, 34
global variables
CRLF, 326
synchronization, 325
glue layer class, 222
glue layers, 220
go() method
application conversion, 277
goal facts
expert systems, 538
GPRS (General Packet Radio Service). See General Packet Radio Service (GPRS).
GPS (Global Positioning System). See Global Positioning System (GPS)

Graphical User Interface (GUI)
application conversion
design, 291
go() method, 277
Swing API, 270
EmbeddedJava, 19
messaging, 469
graphics
low-level, 178
Towers of Hanoi game, 519
Graphics class, 179
Graphics object
application conversion, 297
Graphics.getFont(), 188
Graphics.setFont(), 188
grayscale, 180
Green project
Java 2 Micro Edition, 11
GridLayout
messaging, 475, 480
GSM (Global System for Mobile Communication. See Global System for Mobile Communication (GSM)
GSM Subscriber Identity Module
Java Card compatible SIMs, 367
Multos operating system, 370
GUI (Graphical User Interface). See Graphical User Interface (GUI)

H

hackers, 628
half duplex communication
Java Card, 18
handleKeyDown() method, 265
HandlerBase class
Simple API for XML, 551
handset-oriented positioning
mobile phone location, 569
HanoiCanvas class, 522
HanoiGame abstract class, 507
HanoiMIDlet Class, 513
HanoiSpotlet class, 523
hard keys
Towers of Hanoi game, 502
hard masks
smart card file structure, 373
Hashtables
expert systems, 542
hasNextElement(), 197
hasPointerMotionEvents(), 177
hasPreviousElement(), 197
hasRepeatEvents(), 177
HEAD
networking, 203
header information
JMS, 450
Help View
Game interfaces, 506
Towers of Hanoi game, 501, 517, 527
hideNotify() events, 177
hierarchies
Factory Method, 504
Template Method, 505
high-level APIs, 161
Alert, 167
Form, 169
Lists, 164
Textbox, 162
high-level design
Towers of Hanoi game, 502
high-level user interface events, 160
Home Gateways
Java Embedded Server, 44
Home interfaces
mobile positioning case study, 580
hot-deployment
WAR/EAR files, 414
Hotel interface
mobile positioning case study, 583
HotelAccessor class
mobile positioning case study, 584, 588
HotelEJB class
mobile positioning case study, 590
HotelHome interface
mobile positioning case study, 581
hotel-search application
application tier, 579
database tables, 579
deployment, 575, 607
description, 570
Home interfaces, 580
input output demo, 578
installation, 607
logic, 580
presentation tier, 579, 590
Remote interfaces, 582
use cases, 575
HotJava browser
Java 2 Micro Edition, 13
Hotspot application, 15, 16
HotSync Technology, 330
HTML
generation by presentation tier, 590
HTTP
resuming interrupted data transfers, 431
HttpConnection class, 203
ContactServlet class, 241
ContentConnection replacement, 231

I

I/O classes
expert systems, 548
I18N
text, 190
iBus
installation tests, 481
iBus configurations
AuctionHouse-Specific, 481
iBus server
installation, 480
iBus//Mobile
gateways, 477, 481
libraries, 481
messaging, 476
MIDP client, 490
ID variables
synchronization variables, 311

IDs
record retrieval, 196
IllegalArgumentException, 93
IllegalStateException
Alerts, 167
Image.getGraphics(), 180, 190
ImageItem
Forms, 170
images
downloads, 208
low-level graphics, 190
iMode Technology (NTT DoCoMo)
Java 2 Micro Edition, 41
implementation
Hanoi game, 505
Java 2 Micro Edition, 17
IMPLICIT Lists, 164
IMPLICIT mode
UserInterfaceMIDP class, 247
incremental files
smart cards, 374
individual class script compilation, 482
inference engine
expert systems, 538, 539, 542, 545
INFO
Alert, 168
Infrared network
mobile devices, 40
infrastructure
optimal design, 645
system-level design, 639
initial work
time-sheet data entry
Java 2 Micro Edition, 69
InitialContext
JNDI, 404
initialized()
MyHanoiGame class, 526
Initiation
Java Community Process, 46
initJMS(), 467, 477
inner classes
HanoiMIDlet class, 514, 515
HanoiSpotless implementation, 526
MyRing, 529
MyTower, 530
Spotless implementation, 526
InputConnection interface, 87, 88
InputSource class
Simple API for XML, 551
InputStream
HttpConnection, 203
InputStreamReader, 409
Insert button
slide editors
application conversion, 282
Insertion class
slide editors
application conversion, 280, 292
Insertion panel
application conversion, 292
slide editors, 292
insertSlide() method
application conversion, 279, 282, 292
Install package
JavaPhone API, architecture, 338
install() method
Java Card applets, 384
Java Card application development, 383
installation
iBus server, 480
iBus//Mobile gateways, 481
iBus//Mobile libraries, 481
instance variables
Slide class
application conversion, 284
integrity, 616
interactive Forms, 173
interactiveTest(), 174
interface definitions
contact databases, 228
interfaces. See also user interfaces
CommandListener, 161
Home, 580
Hotel, 583
HotelHome, 581
ItemStateListener, 173
Java Native Interface (JNI), 624
JtapiPeer, 343, 354
LocationHandler, 581, 583
low-level API, 176
Main, 230
MobileProvider, 354
MobileRadio, 354
NetworkSelection, 354
PowerMonitorListener, 361
Provider, 343
Remote, 582
RMI, 397
shared implementation, 506
UIListener, 230, 234
User, 583
UserHome, 581
UserProfileImpl, 351
International Standards Organization
ISO7816, 368
internationalization
text, 190
Internet Instant Messaging
dynamic IP addressing, 440
Internet Protocol addresses
tracking, 439
Internet screen phones, 335
JavaPhone API, phone profile, 337
interrupted data transfers
resuming, 431
InterruptedIOException
timeouts, 88
Introduction View
Game interfaces, 506
IntroMIDlet1, 135
ints
slide editors
application conversion, 282
IOException
socket programming, 91
isDoubleBuffered()
graphics, 180
ISO7816
command APDUs, 374
smart cards, 368, 370
isShown() method, 160

isTheGameRolling()
Towers of Hanoi game, 528
ItemField class
addParameter() method, 345
itemStateChanged() method, 173
ItemStateListener interface, 173
iteration
MailClient, 110

J

J2EE (Java 2 Enterprise Edition). See Java 2 Enterprise Edition (J2EE).
J2EE JBoss application server
auction house, 475
J2ME (Java 2 Micro Edition). See Java 2 Micro Edition (J2ME).
J2MEWTK (J2ME Wireless Toolkit)
compiling and running MIDlets, 137, 138
J2SE (Java 2 Standard Edition). See Java 2 Standard Edition (J2SE).
J2SE platform
Foundation Profile, 83
supported classes, 85
JAD (Java application descriptor). See Java application descriptor (JAD)
Jakarta Tomcat. See also Tomcat
servlet-enabled web servers, 406
JAM (Java Application Manager). See Java Application Manager (JAM).
JAR files
manifest file, 143
MIDlet suites, 134
WAR files, 412
Java
IP address tracking, 440
Symbian Platform and, 336
using smart cards with, 377
Java 2 Enterprise Edition (J2EE), 14
Application Programmer Interface, 59
architecture, 58, 60
communication
distributed applications, 395
protocols, 432
definition, 394
front-end logic, 405
interfacing APIs, 393
JavaBeans, 58, 60, 69
libraries, 394
limitations, 433
location, 403
server linking, 54
virtual machine, 16
web application structure, 411
Java 2 Micro Edition (J2ME)
2-tier client–server application model, 55
3G (third generation) networks, 35, 40
3-tier client–server application model, 56
application benefits, 652
application develpoment, 495
applications construction, 639
architecture, 51, 60
component-based systems, 53
time-sheet data entry case study, 63
architecture security, 623
Bluetooth network, 36, 39
browsers, 41
carrier technology, 34
client–server deployment, 55
coding standards, 650
communication example, 407
Connected Device Configuration, 30
application conversion, 269
Connected Limited Device Configuration, 20
Datagram interface, 89
debugging, 54
development, 51, 53
time-sheet data entry case study, 69
EmbeddedJava, 19
evolution, 11
expert systems, 540
Foundation Profile, 32
General Packet Radio Service, 35
Global System for Mobile Communication, 34
implementations, 17
Infrared network, 40
J2MEWTK (Wireless Toolkit), 137, 138
Java Card, 17, 21
Java Community Process, 46
Java Embedded Server, 44
Java Virtual Machine, 21
JavaTV API, 31
Jini network, 43
layers, 15
limited device components, 130
mobile devices, 7, 14
networking, 62
n-tier client–server application model, 57
Open Services Gateway Initiative, 45
Personal Digital Assistant, 27
Personal Profile, 30
programming techniques, 639
Remote Method Invocation Profile, 32
SDK, 17
Symbian EPOC, 32
security, 128
security architecture, 623
server linking, 54
service-oriented architecture, 61
software design, 631
SPIN® questioning, 65
Spotless Palm, 27
techniques, 639
tricks, 639
Wireless LAN, 38
Wireless Markup Language, 41
Wireless Network Technology, 36, 41
Wireless Toolkit, 496
Java 2 Standard Edition (J2SE), 14
architecture security, 620
Connected Device Configuration
application conversion, 269
Connected Limited Device Configuration
application conversion, 272
security architecture, 620
virtual machine, 16, 21
Java application descriptor (JAD)
MIDlets, 143
Java Application Manager (JAM)
KVM utility, 130
mobile devices, 23
Java application server
Java 2 Micro Edition, 60
time-sheet data entry case study, 77

Java applications
Foundation Profile, 96
Java Card
API, 370, 383
applets, 384
applications, 367
Java 2 Micro Edition, 17, 21
networking, 18
Java Card applications
APDU identification, 376
compatibility, 390
creating, 378
development, 383
power loss, 383
Java Card Development Kit, 381, 388
Java Card Runtime Environment (JCRE), 383
Java Code Compact (JCC)
KVM utility, 130
Java Community Process (JCP)
Java 2 Micro Edition, 46
Java Embedded Server (JES)
Java 2 Micro Edition, 44
Java Message Service (JMS), 435, 446
auction house example, 444, 457
Bidder class client, 466
building clients, 482
components, 450
Converting to run on Palm platforms, 476
iBus//Mobile compromises, 486
Java 2 Enterprise Edition, 60, 432
MIDP version, 486
models, 448
Palm platform conversions, 476
running Palm, 484
Timestamp header, 464
Java Naming and Directory Interface API (JNDI)
Java 2 Enterprise Edition, 59, 404
Java Native Interface (JNI)
eliminated KVM functionality, 127
security, 624
Java Remote Method Invocation (RMI) Profile
Java 2 Micro Edition, 32
Java security
mobile devices, 620
Java Transaction API
Java 2 Enterprise Edition, 60
Java Virtual Machine (JVM). See also virtual machine
Java 2 Micro Edition, 16, 21
KVM subset, 126
security, 619
java.rmi.RemoteException, 400
java.rmi.server.UnicastRemoteObject, 401
java.security, 623
java.security.cert, 623
java.text.resources
classes, 86
JavaBeans
Java 2 Enterprise Edition, 58, 60, 69
JavaCard
CLDC and EmbeddedJava, 132
JavaCheck
application conversion, 272, 287, 289, 290
JavaMail API
Java 2 Enterprise Edition, 59, 432
JavaPhone API, 335
architecture, 336
Address Book package, 338
Calendar package, 338
Communication API package, 338
Datagram package, 338
Install package, 338
Internet Screen Phone Profile, 337
JTAPI Core package, 337
JTAPI Mobile package, 337
Power Management package, 338
Power Monitor package, 338
Smart Phone Profile, 337
SSL package, 338
User Profile package, 338
Personal Profile, 31
Symbian Platform and, 336
Symbian implementation of JavaPhone API, 339
Address Book package, 344
Calendar package, 352
Datagram package, 355
JTAPI Core package, 341
JTAPI Mobile package, 354
Power Monitor package, 360
User Profile package, 351
Symbian SDK, 362
JTAPI implementation, 362
testing JTAPI with emulator, 363
JavaScript
security, 619
JavaServer Pages (JSP)
add_user, 597
dynamic servlet creation, 431
edit_user, 598
Error page, 594
registration, 593
registration index, 590
utilities index, 595
WML do_login, 600
WML error, 606
WML index, 601
WML login, 598
WML search, 604
JavaTV API
Personal Profile, 31
javax.microedition.io
classes, 86
javax.net.datagram
Address class, 355
Datagram class, 355
Datagram package, 355
DatagramNameService class, 355
DatagramService class, 355
DatagramServiceFactory class, 355
javax.pim.addressbook
Address Book package, 344
ContactDatabase class, 344
extends javax.pim.database.Database, 344
javax.pim.calendar
Calendar package, 352
CalendarDatabase class, 352
CalendarEntry class, 352
CalendarToDo class, 352
Repeat class, 352
RepeatRule class, 352
javax.pim.database.Database
extended by javax.pim.addressbook, 344
javax.power.monitor
Power Monitor package, 360
PowerMonitor class, 360
PowerMonitorListener interface, 360
PowerWarningType class, 360

javax.telephony
JTAPI Core package, 341
javax.telephony.capabilities
JTAPI Core package, 341
javax.telephony.events
JTAPI Core package, 341
javax.telephony.mobile
JTAPI Mobile package, 354
MobileProvider interface, 354
MobileRadio interface, 354
NetworkSelection interface, 354
JBoss
archive file hot-deployment, 414
setup, 411
Tomcat, 406, 411
JBossMQ, 475, 476
JCC (Java Code Compact). See Java Code Compact (JCC)
JCP (Java Community Process). See Java Community Process (JCP)
JCRE (Java Card Runtime Environment). See Java Card Runtime Environment (JCRE)
JDBC API
Java 2 Enterprise Edition, 59
JDBC-enabled databases
JBoss + Tomcat, 406
JES (Java Embedded Server). See Java Embedded Server (JES)
Jini
RMI profile, 121
Jini network
Java 2 Micro Edition, 43
JMS (Java Message Service). See Java Message Service (JMS).
JNDI (Java Naming and Directory Interface). See Java Naming and Directory Interface (JNDI).
JNI (Java Native Interface). See Java Native Interface (JNI).
JRun server
mobile positioning case study, 575, 576
JSP (JavaServer Pages). See JavaServer Pages (JSP)
JSR 066. See RMI profile
JTAPI Core package
JavaPhone API, architecture, 337
javax.telephony, 341
javax.telephony.capabilities, 341
javax.telephony.events, 341
Symbian implementation of JavaPhone API, 341
JTAPI Mobile package
JavaPhone API, architecture, 337
javax.telephony.mobile, 354
Symbian implementation of JavaPhone API, 354
JtapiPeer interface
getProvider() method, 343, 354
getServices() method, 343
JtapiPeerFactory class
getJtapiPeer() method, 343
JVM (Java Virtual Machine). See Java Virtual Machine (JVM).

K

kAWT (KVM Abstract Windowing Toolkit)
CLDC development environment, 147
download, 151
sample application, 152
testing Palm application, 154
kAWT installation, 482
KbLoader class
XML expert systems, 559
Kex (KVM Expert System). ***See*** **KVM Expert System (Kex)**
KexDocumentHandler class
XML expert systems, 557
key controls
Towers of Hanoi game, 502
key events
KJava user interface, 257
keyDown() method
application conversion, 300
Spotlets, 263, 265
keyPressed() events, 177
keyReleased() events, 177
keyRepeated() events, 177
keys, primary
KJava record deletion, 256
readRecord(), 245
Record class, 223, 224
KHashTable class
expert systems, 542
Kilo Virtual Machine (KVM), 125, 126
See also virtual machine
'Hello World' application, 28
alternatives, 212
eliminated functionality, 127
expert systems, 537, 561
implementations, 129
mobile devices, 21, 23, 27
porting, 129
resources, 213
utilities, 129
KJava
application building, 149
application testing, 149
CLDC development environment, 147
Database, 254
displaying lists, 259
download, 148
field editing, 261
key events, 257
list size, 260
Palm application testing, 150
pen events, 257
record deletion, 254, 256
screen layout, 257
specific classes, 253
starting applications, 265
user interface and database classes, 252
KMemoReader class
expert systems, 549
knowledge base
expert systems, 538, 540, 541, 542
trouble shooting, 561
XML, 546
KnowledgeBase class
expert systems, 545
KVM (Kilo Virtual Machine). ***See*** **Kilo Virtual Machine (KVM).**
KVM Abstract Windowing Environment (kAWT). See kAWT (KVM Abstract Windowing Toolkit)
KVM Expert System (Kex), 537, 549

kvmkjava.exe
test tool, 533

L

labels
TextField, 259
language
microedition.locale, 202
languages
C++, 12
Oak, 12, 16
last View
Towers of Hanoi game, 517
latitude
mobile positioning, 572
layers
high-level design, 502
Java 2 Micro Edition
time-sheet data entry case study, 74
LCS (Location Service Clients). See Location Service Clients (LCS)
LIA (Location Immediate Answer). See Location Immediate Answer (LIA)
libraries
iBus//Mobile, 476, 481
third party, 212
limited devices
characteristics, 126
components of J2ME, 130
Line class
application conversion, 285
Connected Limited Device Configuration, 297
linear fixed files
smart cards, 373
lines
low-level graphic, 186
Linux
Connected Device Configuration (CDC)
implementation, 82
LIR (Location Immediate Requests). See Location Immediate Requests (LIR)
LIST
user interface design, 229
List class
application conversion, 276
LIST command
POP3, 101
list() method
ContactApp, 234, 236
listen() method
PopServer class, 115, 116
listeners
JavaPhone API events, 341
UIListener interface, 230, 234
listListener() method
ContactApp, 239
UIListener interface, 230
lists
events, 165
high-level API, 164
KJava, 260
pen events, 263
UserInterfaceMIDP class, 247
loading data
predictive, 431
Local Area Wireless Network Technology
Java 2 Micro Edition, 36, 41
LocaleData
java.text.resources, 86
location
concept, 566
coordinate system, 572
current players, 568
distributed applications
J2EE, 403
services, 566
location hiding
Jini network, 43
Location Immediate Answer (LIA)
mobile positioning, 570, 571
Location Immediate Requests (LIR)
mobile positioning, 570
Location Service Clients (LCS)
Mobile Positioning Protocol, 570
LocationHandler interface
implementation, 584
mobile positioning case study, 581, 583
session bean in hotel-search application, 580
WML search JSP, 605
LocationHandlerEJB class
mobile positioning case study, 586
LocationRequest class
MPS-SDK package, 573
LocationResult class
MPS-SDK package, 573
logic
business
ContactApp, 234
separation from low-level implementation, 219
front-end, 405
mobile positioning case study, 580
Towers of Hanoi game, 503
LoginUI class, 469, 470
long term memory
conceptual models, 636
longitude
mobile positioning, 572
lookup() method
DatagramNameService class, 356
distributed applications, 405
lookupAll() method
DatagramNameService class, 356
Lot class
JMS, 457
low-level events, 176
low-level graphics, 178
translations, 192
low-level implementation
separation from business logic, 219
low-level interface methods
unsupported, 192
low-level user interface API, 176

M

magnetic strips
credit cards, 369
MailClient class
e-mail client, 102, 109

MailStore class
e-mail client, 102, 104
main area
KJava screen layout, 257
main class variable
ContactApp, 235
Main interface
contact databases, 230
Main View
Game interfaces, 506
Towers of Hanoi game, 499, 517
main() method
application conversion, 276, 299
kAWT application, 152
MailClient, 111
Spotless implementation, 524
MAINHEIGHT, 258
MainKJava class, 265
MainMIDP class
starting MIDP applications, 251
Maintenance
Java Community Process, 47
maintenance staff
software design, 633
MakePalmApp, 150
management
applications, 209
manifest files
MIDlet suites, 143
marshalling
RMI, 396
masks
smart card file structure, 373
matches()
records, 198
MAXHEIGHT
list size, 260
MAXX
list size, 260
Memo
Palm expert systems, 548, 552, 559, 561
memory
contact databases, 266
memory cards, 369
memory usage reductions
compared to performance, 646
Menu object
Connected Device Configuration
application conversion, 271
menus. See lists
merge() method
contact database design, 227, 228
merging
databases, 266
message body
JMS, 450, 452
Message class
e-mail client, 103
message encryption/decryption systems, 625
message properties
JMS, 450, 451
MessageConsumer, 454
JMS, 449
MessageListener, 462
Message-oriented middleware (MOM), 446, 448
bridge, 447
Java Message Service, 432
MessageProducer
JMS, 449, 454
messaging
application examples, 446
architecture, 446
asynchronous, 436, 437
code conversions, 475
datagrams, 441
enhancement methods, 491
Forward networks, 437, 438
MIDP, 486
models, 448
receiving, 453
sending, 452
Store networks, 437, 438
messaging servers
administration, 461
configurations, 475
connections, 454
Met (Mobile Electronic Transcations). See Mobile Electronic Transcations (MeT)
Meta events, JavaPhone API events, 341
methods
About View
Towers of Hanoi game, 517
accept(), 94, 96
acceptAndOpen(), 88, 94, 96, 97
accessor, 584
actionListener(), 230, 239
addActions(), 247, 250
addCommand(), 160
addExceptDate(), 352
addField(), 344
addLine()
application conversion, 282
addParameter(), 345
addRecord(), 255
addRepeatDate(), 352
Alert, 168
basic MIDlet creation, 135
bind()
JNDI, 405
buildList()
application conversion, 277
calculateDistance(), 587
cardItems(), 348
changeOfSlide()
application conversion, 278
characters()
XML, 558
clear(), 227, 243
close(), 87, 227
UserInterface, 230
closeRecordStore(), 243
commandAction(), 136, 250
Web server communication, 408
connect(), 107
create(), 237, 581
create(long id), 581
database.merge(), 232
delete(), 238
deleteMessage(), 109
deleteRecord(), 227, 257

methods (continued)
deleteSlide()
application conversion, 279
deselect(), 383
destroyApp(), 137, 161, 252
MIDlets, 24
details(), 236
displayFields(), 229, 247
displayList(), 229
download(), 238
downloadAndMerge(), 231
downloadOK(), 238
eatWhitespace()
XML, 556
edit(), 237
editField(), 229, 248
editFields(), 229
editListener(), 230, 238, 239
endElement()
XML, 559
Error View
Towers of Hanoi game, 517
event handling, 259
exit(), 231, 252
Factory (design pattern), 504
findAll(), 581, 582
findByPrimaryKey(), 581, 582
Forms, 172
FormTest, 174
getAddresses(), 343
getAvailableLocales(), 86
getBatteryLevel(), 361
getBytes(), 225
getCurrentUser(), 351
getDB(), 252
getDescription(), 584
getEstimatedSecondsRemaining(), 361
getFields(), 226, 350
getFieldValue(), 225
getId(), 584
expert systems, 548
getInputStream(), 91, 96
getJtapiPeer(), 343
getLargestIndex(), 589
getLatitude(), 584
getLocation(), 587
getLongitude(), 584
getMaximumLength(), 359
getNumberOfRecords(), 254
getOutputStream(), 96
getPK(), 225
getProductName()
RMI, 397
getProvider(), 343, 354
getQuantity()
RMI, 397
getResourceAsStream(filename), 417
getResponse()
networking, 294
getResponseCode(), 419
getServerService(), 358
getService(), 355
getServices(), 343
getSize(), 109, 111
go()
application conversion, 277
handleKeyDown(), 265
Help View
Towers of Hanoi game, 517
insertSlide()
application conversion, 279, 282, 292
install(), 383, 384
isShown, 160
itemStateChanged, 173
keyDown(), 263, 265
application conversion, 300
last View
Towers of Hanoi game, 517
list(), 234, 236
listen(), 115, 116
listener, 234
listListener(), 230, 239
lookup(), 356
distributed applications, 405
lookupAll(), 356
Main View
Towers of Hanoi game, 517
main(), 111, 152
application conversion, 276, 299
merge(), 227, 228
multiEditListener(), 230, 237
newDatagram(), 89
notifyDestroyed(), 136, 142, 252
numRecords(), 227, 254, 255
open(), 87, 88, 105, 227, 253
DBMIDP class, 242
openDatabase(), 348, 356
openOutputStream(), 419
openPrim(), 93, 95
openXXXStream() helper, 88
orderHotels(), 583, 586, 588
paint(), 259
paintInsertion()
application conversion, 302
paintSlides()
application conversion, 303
parseAddress(), 355
pauseApp(), 137, 147, 252
MIDlets, 24
penDown(), 263
application conversion, 301
pressed()
application conversion, 301
private fine grain, 526
process(), 383, 385
processRequest(), 115
protocol stacks, 479
read(), 416
database access, 425
Web server communication, 408
readAllMail(), 110
readAttributes()
XML, 555
readBoolean(), 89
readContent()
XML, 554
readDoc()
XML, 553, 554
readEndElement()
XML, 554
readLine(), 98, 107, 108, 115, 421
readMail(), 110
readMessage(), 108
readRecord(), 227, 245
readRecords(), 227, 236, 244
readStartElement()
XML, 554
rebind()
JNDI, 405

methods (continued)
receive(), 89, 355
record-related, 227
register(), 384
application conversion, 299
request.getContentLength(), 421
select(), 383
send(), 355
serviceRepaints, 176
setCurrentUser(), 351
setImplementation(), 351
setPK(), 225
setRecord(), 255
setUIListener(), 230
setVisible()
application conversion, 281, 286
startApp(), 136, 147, 251
MIDlets, 24
Web server communication, 408
startElement()
XML, 557
synchronization protocol creation, 312, 313
Template (design pattern), 505
testChar()
XML, 553
tf.loseFocus(), 264
toFields(), 226
toString()
expert systems, 543, 544
trigger()
expert systems, 544
trim(), 587
unsupported low-level interfaces, 192
update()
expert systems, 545
updateItem(), 349
upload(), 231
uploadOK(), 238
UserInterface display, 229
usingExternalPowerSource(), 361
viewSlide()
application conversion, 283
write(), 205, 418
writeChars(), 89
writeln(), 419
writeRecord(), 227, 245
microedition
configuration, 202
microedition.encoding, 202
microedition.io, 86
microedition.locale, 202
microedition.platform, 202
microedition.profiles, 202
microedition.protocolpath, 87
MicroEdition-Configuration, 145
MicroEdition-Profile, 145
Microsoft Windows
IP address
tracking, 439
MID (Mobile Information Devices). See Mobile Information Devices (MID)
middle tier
Java 2 Micro Edition architecture, 56
MIDlet Classes
MIDP implemetation, 513
MIDlet connections, 208
MIDlet layers
Towers of Hanoi game, 503
MIDlet suites, 142
attributes, 144
JAR files, 134
manifest file, 143
MIDlet to MIDlet connections, 208
MIDlet user interface package
MIDP, 495
MIDlet-Data-Size, 145
MIDlet-Description, 145
MIDletGameSettings class, 520
MIDlet-Icon, 145
MIDlet-Info-URL, 145
MIDlet-Jar-Size, 145
MIDlet-Jar-URL, 145
MIDlet-Name, 144
MIDlets
application descriptor file, 142
attributes, 144
compiling and running, 137
definition, 134
file reading, 415
launching and controlling, 146
MIDlet suites, 142
Mobile Information Device Profile, 24
shared methods, 422
simple, 135
TextBox, 136
MIDlet-Vendor, 144
MIDlet-Version, 144
MIDP (Mobile Information Device Profile). See Mobile Information Device Profile (MIDP).
MIME type mapping
servlets, 417
MINX
list size, 260
MINY
list size, 260
mirror synchronization, 308, 310, 312, 320
mobile agent system
Aglets, 620
mobile applications
performance, 641
mobile code
security, 618
mobile devices
*7, 11
3G (third generation) networks, 35, 40
Bluetooth network, 36, 39
Connected Limited Device Configuration, 20, 23
contacting, 440
EmbeddedJava, 19
future, 213
General Packet Radio Service, 35
Global System for Mobile Communication, 34
Infrared network, 40
infrastructure
Java 2 Micro Edition, 78
Java 2 Micro Edition, 7, 14
Java Application Manager, 23
Java Card, 17
JAVA security, 620
Kilo Virtual Machine, 21, 23, 27
messaging architecture, 446
security, 616, 620
Wireless LAN, 38
mobile electronic transactions (MeT) initiative, 628

Mobile Information Device Profile (MIDP), 125, 132, 157
API, 495
application development, 495
applications
starting, 251
Canvas, 522
case sensitive searching, 418
classes, 513
Connected Limited Device Configuration, 22, 23
device examples, 138
downloading, 217
Factory Method, 518
Hanoi game, 505
J2ME application development, 495
JMS, 486
messaging, 486
MIDlet user interface package, 495
MIDlets, 24
port
contact databases, 242
Record Management System, 193
security, 624
server connectivity, 406
system properties, 202
User Interfaces, 157
Wireless Toolkit, 26
Mobile Information Devices (MID)
hardware requirements, 133
software requirements, 134
Symbian Platform, 335
mobile location solutions, 565
mobile phones
cells, 566
location concept, 566
mobile positioning, 565. See also hotel-search application
case study. See hotel-search application
coordinate system, 572
definition, 566
Enterprise JavaBeans, 580
GPS, 569
handset-oriented, 569
network-oriented, 566
Mobile Positioning Centers (MPC)
emulators, 573, 575, 578
input output demo, 578
position areas visualization client, 579
Ericsson Mobile Positioning System, 569
Mobile Positioning Protocol, 570
Mobile Positioning Protocol (MPP)
description, 570
mobile positioning solutions
development, 568
mobile systems
explorability design, 637
MobileProvider interface
javax.telephony.mobile, 354
MobileRadio interface
javax.telephony.mobile, 354
models
JMS, 448
Model–View–Controller system
expert systems, 538
Java 2 Micro Edition, 52
three-tier applications, 429
MOM (Message-oriented middleware). See Message-oriented middleware (MOM)
Mondex
Multos operating system, 369
Motorola
mobile electronic transactions initiative, 628
MouseListener interface
application conversion, 279, 282
messaging, 443
MPC (Mobile Positioning Centers). See Mobile Positioning Centers (MPC)
MPP (Mobile Positioning Protocol). See Mobile Positioning Protocol (MPP)
MPSException class
MPS-SDK package, 573
MPS-SDK 3.0 (Ericsson Mobile Positioning Solutions). See Ericsson Mobile Positioning Solutions (MPS-SDK 3.0)
multicast IP
decentralized architecture, 448
MULTIEDIT
user interface design, 229
multiEditListener() method
ContactApp, 237
UIListener interface, 230
multi-path phenomena
mobile positioning, 568
multiple edit Field objects
UserInterfaceMIDP class, 249
multiple fields
KJava editing, 261
MULTIPLE Lists, 164
Multos
smart card operating system, 369
mutator methods
Slide class
application conversion, 284
MyHanoiGame class, 526
HanoiMIDlet class, 514
inner HanoiMIDlet class, 515
MyRing class
HanoiMIDlet class, 514
inner class, 519, 529
inner HanoiMIDlet class, 515
MyTower class
HanoiMIDlet class, 514
inner class, 519, 530
inner HanoiMIDlet class, 515

N

NAMEEND_ST, 228, 247, 260
naming services
registering server objects, 401
Napster
dynamic IP addressing, 440
National Registered Application Provider (RID), 378
native function support
CLDC, 624
navigation
Forms, 173
screens, 498
small screens, 637
Netscape browser
Java 2 Micro Edition evolution, 13
network communications security, 625
network performance
compared to result processing, 646

networking
Connected Device Configuration, 294
Connected Limited Device Configuration, 295
Java 2 Micro Edition, 62
Java Card, 18
Mobile Information Device Profile, 203
Wireless Network Technology, 36, 41
Bluetooth, 36, 39
Infrared, 40
Jini, 43
Wireless LAN, 38
networking application, 96
network-oriented positioning
future developments, 567
mobile phone location, 566
networks
2.5G/3G, 431
mobile positioning, 566
security, 617
NetworkSelection interface
javax.telephony.mobile, 354
newDatagram() method, 89
nextRecord(), 197
nodes
network security, 618
Nokia
mobile electronic transactions initiative, 628
non-invasive to recipient
SMS, 436
NOOP command
POP3, 101, 113, 118
Norman, Donald
design complexity reduction, 635
notifyDestroyed() method, 136, 142
MainMIDP class, 252
MIDlets, 161
n-tier client–server application model
Java 2 Micro Edition, 57
NTT DoCoMo iMode Technology
Java 2 Micro Edition, 41
numeration.hasMoreElements(), 197
NUMERIC
TextBox, 162
numRecords() method, 197
contact database design, 227
DBKJava class, 254
KJava, 255

O

Oak language, 12, 16
object classes
business model
time-sheet data entry case study, 69
ObjectInputStream
RMI, 398
ObjectOutputStream
RMI, 398
objects
administered, 453
Button
application conversion, 298
caches, 647
compared to components, 641
Connector
networking, 295
creation, 647
distribution, 394
Graphics
application conversion, 297
Menu
Connected Device Configuration, 271
multiple edit Field, 249
Order
RMI, 397
pools, 647
resusing, 647
Session, 454
observers, JavaPhone API events, 341
Omni cells
mobile positioning, 567
one directional transfer
synchronization, 309
onMessage(), 463, 468, 480
Open Services Gateway Initiative (OSGi)
Java 2 Micro Edition, 45
open() method
Connector class, 87, 88
contact database design, 227
DBKJava class, 253
DBMIDP, 242
Factory class, 105
OpenCard
smart card readers, 378
openDatabase() method
ContactDatabase class, 348, 356
openInputStream() method
HttpConnection, 204
openOutputStream() method
HttpConnection, 205
servlets, 419
openPrim() method
ConnectionBaseInterface, 93
server sockets, 95
openRecordStore(), 193
openXXXStream() helper methods, 88
operating systems
Java Card, 369
Windows for Smart Cards, 370
optimal code-levels, 646
optimal design, 644
optimal object creation, 647
options
synchronization, 329
Order interface
RMI, 400
Order object
RMI, 397
orderHotels() method
LocationHandler interface, 583, 588
LocationHandler main service method, 586
Orion
configuration update, 608
custom tags, 593
Orion application server
mobile positioning case study, 575, 576
OSGi (Open Services Gateway Initiative). See Open Services Gateway Initiative (OSGi)
outlines
low-level graphics, 186
OutputConnection interface, 87, 88
OutputStream
HttpConnection, 205

overlays
Palm, 495

P

P2P (Point-to-Point messaging). See Point- to-Point messaging (P2P or PTP)
package classes
MIDP implemetation, 513
paint() method
abstract classes, 509, 530
Canvas, 178
Graphics class, 179
KJava, 259
paintBorder()
abstract classes, 509, 530
paintInsertion() method
application conversion, 302
paintSlides() method
application conversion, 303
Palm
application development, 495
application testing, 150
Bidder client, 482
CLDC
ContentConnection interface, 231
connectivity, 151
creatorID, 253
emulators, 218
executable creation, 150
J2ME application development, 495
Java 2 Micro Edition, 27
JMS client, 484
KJava user interface, 252
list size, 260
messaging, 476
overlays, 495
protocol stack methods, 479
record deletion, 254, 256
running, 149
screen layout, 258
simulators, 482
Spotless API, 495
use of testhttp, 235
Palm database (PDB)
expert systems, 548
Palm devices
expert systems, 537, 561
Palm Memo, 548, 552, 559, 561
server connectivity, 406
time-sheet data entry
Java 2 Micro Edition architecture, 77
Palm emulator
Towers of Hanoi game, 533
Palm iBus//Mobile. See also iBus//Mobile. See also iBus
messaging, 476
Palm implementation
Hanoi game, 505
Palm Operating System Emulator (POSE)
expert systems, 539
Palm OS
emulater, 496, 497
Spotless implentation, 523
Palm Pilot
synchronization, 308, 330
synchronization code, 325
Palm Pilot device
Connected Limited Device Configuration, 295
Panasonic
mobile electronic transactions initiative, 628
panels
Insertion
application conversion, 292
slideList
application conversion, 291
parseAddress() method
DatagramService class, 355
Parser class
Simple API for XML, 551
participants
Factory Method, 504
Template Method, 505
PASS command
POP3, 100, 101, 107, 113, 117
PASSWORD
TextBox, 162
passwords
messaging, 476
messaging server administration, 461
smart card PIN, 375
smart cards, 380
WML login JSP, 599
pauseApp() method
launching and controlling MIDlets, 147
MainMIDP class, 252
MIDlets, 24, 137
rectangles, 182
Towers of Hanoi game, 514
PC
protocol stack methods, 479
PC/SC (Personal Computer/Smart Card). See Personal Computer/Smart Card (PC/SC)
PDB (Palm database). See Palm database (PDB)
PdbReader class
expert systems, 548
pen events
KJava user interface, 257
penDown() method
application conversion, 301
SelectScrollTextBox access, 259
Spotlets, 263
Towers of Hanoi game, 524
performance
compared to memory usage reduction, 646
component-based systems
performance, 641
periodic synchronization
disconnected access, 652
persistency
mobile positioning case study, 575, 584, 589
Personal Computer/Smart Card (PC/SC), 378
Personal Digital Assistants (PDA)
as connected limited device, 126
Java 2 Micro Edition, 27
Personal Identification Numbers (PINs), 617
Personal Profile, 122
Java 2 Micro Edition, 30
Personal Unblocking Key (PUK), 617

personalization
smart card file structure, 373
PersonalJava, 122
Connected Device Configuration
application conversion, 270, 271
emulation environment
application conversion, 274, 292, 293
EPOC operating system, 32
Java 2 Micro Edition, 30
JavaCheck, 272, 290
JavaPhone API, 336
PHONENUMBER
TextBox, 162
phones
mobile. See mobile phones
PIN
smart cards, 372, 374
pinging
mobile positioning, 567
PINs (Personal Identification Numbers). See Personal Identification Numbers (PINs)
PIX (Proprietary Application Identifier Extension). See Proprietary Application Identifier Extension (PIX)
pJava. See PersonalJava
PKI (Public Key Infrastructure). See Public Key Infrastructure (PKI)
platforms
CDC/CDLC-based
RMI, 403
playSound(), Alerts, 168
Pocket PC ActiveSync
synchronization, 329
Pocket PC devices
application conversion, 293
time-sheet data entry
Java 2 Micro Edition architecture, 77
pointerDragged(), 177
pointerPressed(), 177
pointerReleased(), 177
Point-to-Point (P2P or PTP) messaging, 448, 455
politeness
SMS, 436
pools
objects, 647
POP3 (Post Office Protocol), 100
application implementation, 102
commands, 101
proxy creation, 120
simple server, 113
PopProxy class, 120
PopReader class
e-mail client, 102, 105
PopServer class, 114
testing, 119
portability
design, 495
J2ME, 639
portable devices
expert systems, 537
porting
Connected Device Configuration, 293
Connected Limited Device Configuration, 294
K Virtual Machine, 129
ports
MIDP
contact databases, 242
Palm/KJava, 252
POSE (Palm Operating System Emulator). *See* Palm Operating System Emulator (POSE)
position areas visualization client
MPC emulator, 579
positioning
mobile. See mobile positioning
POST
networking, 203
POST data, 205
Post Office Protocol. See POP3
POST requests
database access, 425
servlet file writing, 419
power loss
Java Card applications, 383
Power Management package
JavaPhone API, architecture, 338
Power Monitor package
JavaPhone API, architecture, 338
javax.power.monitor, 360
Symbian implementation of JavaPhone API, 360
powerdown command
Java Card applets, 389
PowerMonitor class
getBatteryLevel() method, 361
getEstimatedSecondsRemaining() method, 361
javax.power.monitor, 360
usingExternalPowerSource() method, 361
PowerMonitorListener interface
javax.power.monitor, 360
powerup command
Java Card applets, 388
PowerWarningType class
javax.power.monitor, 360
PRECEDES
records, 198
pre-loaded data, 431
pre-personalization
smart card file structure, 373
presentation tier
mobile positioning case study, 579, 590
presentOnServer
synchronization, 322
pressed() method
application conversion, 301
prevention
security, 615
pre-verification
classes, 128, 149
previousRecord(), 197
primary keys
KJava record deletion, 256
readRecord(), 245
Record class, 223, 224
PrintWriter
servlets, 417
priority
Command classes, 160
privacy
SMS, 436
private fine grain methods
Towers of Hanoi game, 526
private methods
Towers of Hanoi game, 514
View, 527

process() method
Java Card applets, 385
Java Card application development, 383
processor cards, 369
processRequest() method
PopServer class, 115
production rules
expert systems, 538
profiles
Foundation
Java 2 Micro Edition, 32
Java 2 Micro Edition, 14, 15, 16
Java Remote Method Invocation
Java 2 Micro Edition, 32
Mobile Information Devices, 132, 157
Personal, 122
Java 2 Micro Edition, 30
RMI, 121
programming
device-based, 642
J2ME, 639
smart cards, 378
sockets, 90
UDP, 89
progress reports
user feedback, 430
project management
software design, 632
property values
JMS, 451
Proprietary Application Identifier Extension (PIX), 378
prosecution
security, 615
protection domains
security, 622
Protocol class
Connector instantiation, 87
server socket connection, 94
socket connection, 93
protocol creation
synchronization, 311
protocol stack methods, 479
protocols
creation methods, 312, 313
mobile positioning, 570
Simple Mail Transport protocol, 437
synchronization, 311
UDP, 441
Provider interface
getAddresses() method, 343
proxies
POP3, 120
proxy design patterns
RMI, 396
pseudo-business classes
ContactApp, 233
Psion devices
time-sheet data entry
Java 2 Micro Edition architecture, 77
PTP (Point-to-Point messaging). See Point-to-Point messaging (P2P or PTP)
Pub/Sub (Publish-and-Subscribe). See Publish-and-Subscribe (Pub/Sub)
public classes
MIDP implemetation, 513
Public Draft
Java Community Process, 47
public int countLines()
synchronization, 312
Public Key Infrastructure (PKI)
security, 625
public Line getLine ()
synchronization, 313
public long getTime()
synchronization, 313
public String getID()
synchronization, 313
public String getTitle()
synchronization, 312
public void addLine()
synchronization, 312
public void setID()
synchronization, 312
public void setTime()
synchronization, 313
public void setTitle()
synchronization, 312
Publish-and-Subscribe (Pub/Sub)
messaging, 449, 456
publishing
Pub/Sub messaging, 456
publishNewLot(), 464
PUK (Personal Unblocking Key). See Personal Unblocking Key (PUK)
purchasing agents
software design, 633

Q

qFactory
messaging, 479
Quartz devices
Symbian implementation of JavaPhone API, 339
queues
creation in JMSReplyTo Header, 455
creation in JNDI, 455
improvements, 492
Point to Point messaging, 448
receiver classes, 449
sender classes, 449
QUIT command
POP3, 101, 113, 115

R

RandomAccessConnection interface, 90
Read data command
smart cards, 376
read() method
database access, 425
MIDlet, 416
Web server communication, 408
readAllMail() method
MailClient, 110
readAttributes() method
XML expert systems, 555
readBoolean() method, 89
readContent() method
XML expert systems, 554
readDoc() method
XML expert systems, 553, 554

readEndElement() method
XML expert systems, 554
readers
smart card, 371, 377, 378
readLine() method, 421
connect(), 107
PopServer class, 115
readMessage(), 108
SimpleClient class, 98
readMail() method
MailClient, 110
readMessage() method
PopReader class, 108
readRecord() method
contact database design, 227
DBMIDP class, 245
readRecords() method
contact database design, 227
ContactApp, 236
DBMIDP class, 244
readStartElement() method
XML expert systems, 554
rebind() method
JNDI, 405
rebuild
records, 197
receive()
messaging, 444
receive() method, 89
DatagramService class, 355
receivers
JMS, 449
receiveSlide()
synchronization, 327
receiving messages, 453
P2P or PTP model, 455
Record class
databases, 223
Record Enumerator, 196
Record Management System (RMS) XE, 193
record store
creation, 193
deleting, 198
RecordComparator, 198
RecordFilter, 198
RecordListener events, 201
record-related methods
contact database design, 227
records
contact databases
fields of information, 221
creation, 193
deleting, 197
DBMIDP class, 246
filters, 198
modifying, 197
Palm/KJava databases, 254, 256
retrieval, 196
sorting, 198
synchronization, 201
RecordStore
DBMIDP class, 242
RecordStore events, 201
RecordStore.add(), 195
RecordStoreException, 194
recovery
device-based programming, 644
rectangles
draw, 181
fill, 181
round cornered, 185
refactoring
design, 645
reflection
CLDC, 624
eliminated KVM functionality, 127
register() method
application conversion, 299
Java Card applets, 384
registration index JSP
mobile positioning case study, 590
registration JSP
mobile positioning case study, 593
reliability
SMS, 437
Remote interfaces
mobile positioning case study, 582
Remote Method Invocation (RMI)
J2EE, 396, 400
Remote Method Invocation (RMI) Profile, 121
Java 2 Micro Edition, 32
Remote Method Invocation over the Internet Inter-ORB Protocol (RMI-IIOP)
J2EE, 396, 403, 433
rendering text, 190
repaint()
Canvas, 178
Repeat class
addExceptDate() method, 352
addRepeatDate() method, 352
javax.pim.calendar, 352
RepeatRule class
javax.pim.calendar, 352
request messages
mobile positioning, 570
request.getContentLength() method, 421
requesting resources
JNDI, 405
requirements
time-sheet data entry
Java 2 Micro Edition, 64, 66
resilience
Java 2 Micro Edition, 54
resolution
application conversion, 291
resource files
MIDlets, 142
resources
KVM, 213
responses
mobile positioning, 571
result processing
compared to network performance, 646
resuming interrupted data transfers, 431
RETR command
POP3, 101, 108, 113, 117
retrieval
records, 196
reverse usability
software design, 638
RID (National Registered Application Provider). See National Registered Application Provider (RID)

Ring
abstract class, 511
hierarchies, 504
Spotless implementation, 528
RMI (Remote Method Invocation). See Remote Method Invocation (RMI).
RMI-IIOP (Remote Method Invocation over the Internet Inter-ORB Protocol). See Remote Method Invocation over the Internet Inter-ORB Protocol (RMI-IIOP)
ROM images
Palm emulators, 218
round cornered rectangles, 185
Routfile.txt
MPC emulator, 573
RSET command
POP3, 101, 113, 118
Rule class
expert systems, 543
run
Palm JMS client, 484

S

SAAM (Software Architecture Analysis Method). See Software Architecture Analysis Method (SAAM)
safety
device-based programming, 644
Java 2 Micro Edition, 54
sales staff
software design, 633
sample knowledge
expert systems, 540
sandbox security, 128
sandboxes, 619
SAX (Simple API for XML). ***See*** **Simple API for XML (SAX)**
scalability, architecture
Java 2 Micro Edition, 54, 56
Schlumberger
CyberFlex, 390
SCIF (Secure Compartmented Information Facilities). See Secure Compartmented Information Facilities (SCIF)
screen design
time-sheet data entry case study, 70
Towers of Hanoi game, 498
screen navigation
Towers of Honoi game, 498
screen size
Palm OS emulator, 505
software design, 637
Wireless Toolkit, 505
screens
Alerts, 167
click depth, 636
dimension definition, 258
Displayable, 159
Forms, 169
Tickers, 175
UserInterfaceKJava class, 257
scrollable text boxes, 259
scrollable windows, 382
ScrollPane
GridLayout, 480
SDK
Java 2 Micro Edition, 17
Symbian EPOC, 32
MPS, 573
searching
contact database, 266
Secure Compartmented Information Facilities (SCIFs)
security, 617
secure computer systems, 615
secure systems, 614
security
access controllers, 616
availability, 616
Bluetooth technology, 628
CDC, 623
CLDC, 623, 624
communication channels, 626
communications networks, 625
component-based systems, 642
confidentiality, 616
Connected Device Configuration, 83
credit cards, 369
cryptography, 370
Denial of Service attacks, 615
detection, 615
digital attacks, 617
eliminated KVM functionality, 128
gateways, 627
integrity, 616
Java Virtual Machine (JVM), 619
MIDP, 624
mobile code, 618
mobile devices, 616, 620
mobile electronic transactions initiative, 628
Mobile Information Device Profile, 209
network communications, 625
Personal Unblocking Key, 617
PINs, 617
prevention, 615
prosecution, 615
protection domains, 622
Public Key Infrastructure, 625
sandboxes, 128, 619
Secure Compartmented Information Facilities, 617
secure computer systems, 615
secure systems, 614
smart cards, 369, 371, 391
SMS messages, 615
software, 619
time-sheet data entry
Java 2 Micro Edition architecture, 77
transactions, 627, 628
Transport Layer Security, 626
virtual machines, 619
WAP, 627
web servers, 627
Wireless Application Environment, 628
Wireless Application Protocol (WAP), 625
wireless networks, 617
Wireless Transport Security Layer, 626
worm programs, 618
security architecture
J2ME, 623
J2SE, 620
security managers
J2SE security architecture, 620, 622
SecurityException
socket programming, 91

seen variable
Slide class
application conversion, 284
select command
smart card programming, 379
select() method
Java Card application development, 383
Towers of Hanoi game, 529
selectedRecordPK class variable
ContactApp, 235
selectFile() command
smart cards, 377
SelectScrollTextBox class
KJava, 259
send() method
DatagramService class, 355
sendBytesLong(), 385
sendData(), 89
senders
JMS, 449
sending messages, 452
P2P or PTP model, 455
sendSlide()
synchronization, 327
separator characters
Record class, 223
user interface design, 228
sequence diagrams
time-sheet data entry case study, 73
server applications
Java 2 Enterprise Edition, 14
server architecture
messaging, 446
server code
synchronization, 320
server connectivity
MIDP, 406
server linking
Java 2 Micro Edition, 54
server sockets
Protocol, 94
servers
data integretity, 649
exchange synchronization, 318
JRun, 575, 576
Orion application, 575, 576
reading files from, 407
simple POP3, 113
synchronization, 315
synchronization code, 320
synchronization exchange, 318
synchronization flow charts, 315
ServerSocket
synchronization, 324
ServerSockets
acceptAndOpen() method, 88
implementing Connections, 91
service-oriented architecture
Java 2 Micro Edition, 61
serviceRepaints()
Canvas, 176
services
location based, 566
servlet instruction, 235
serviveRepaints()
Canvas, 178
servlets
ContactServlet class, 240
data formats, 431
database access, 422
deployment descriptor file (web.xml), 412
EJB access, 429
file writing, 418
filtered file reading, 414
JavaServer Pages, 431
server connectivity, 406
Session objects, 454
set() method
synchronization, 321
setClip(), 185
setColor (int.RGB), 180
setCurrentUser() method
UserProfile class, 351
setFont(), 188
setGrayScale, 180
setImplementation() method
UserProfile class, 351
setOutgoing()
Java Card applets, 384
setOutgoingLength()
Java Card applets, 384
setPK() method
contact database design, 225
setRecord() method, 197
KJava, 255
setStrokeStyle(), 186
setUIListener() method
UIListener interface, 230
UserInterface supporting method, 230
setVisible() method
application conversion, 281, 286
shared implementation
Hanoi game, 505
shared layers
high-level design, 502
short message service (SMS), 436
short term memory
conceptual models, 636
showNotify() events, 177
Siemans
mobile electronic transactions initiative, 628
sign-off meeting, use-case
time-sheet data entry, 68
SIM cards
smart cards, 368
Simple API for XML (SAX)
expert systems, 550
Simple Mail Transport Protocol (SMTP), 437
SIMPLEACTION
user interface design, 229
SimpleClient class
socket programming, 98
SimpleServer class
socket programming, 97
simulators
Palm, 482
SINGLEEDIT
user interface design, 229
single-key encryption systems, 625
size
font attributes, 188

size decreasement
applications, 209
skeleton objects
RMI, 396
Slide class
application conversion, 284
Connected Limited Device Configuration, 297
slide editors
application conversion, 274
Confirmation class, 285
Insertion class, 280, 292
Line class, 285
Status class, 287
Insert button
application conversion, 282
MouseListener interface
application conversion, 282
Synchronize button
application conversion, 279
Slide()
synchronization protocols, 311
SlideEditor class
application conversion, 276, 282
synchronization, 320
slideList panel
application conversion, 291
slides
synchronization, 325
SlideViewer class
application conversion, 282, 292
smart cards
Application Identifiers, 378
commands, 371
communication, 370, 372
construction, 368
definition, 368
deployment, 391
file structure, 372
file system, 373
listening to, 372
OpenCard, 378
operating systems, 369
PIN, 374
programming, 378
readers, 371, 377
security, 369, 371, 391
talking to, 371
using with Java, 377
working with real, 390
smart phones, 335
JavaPhone API, phone profile, 337
SmartPhone emulator
WML, 598
SMS (short message service). See short message service (SMS)
SMS messages
security, 615
SMTP (Simple Mail Transport Protocol). See Simple Mail Transport Protocol (SMTP)
sockets
basic application, 96
client, 87
implementing Connections, 91
programming, 90
Protocol, 93
RMI, 398
SimpleClient class, 98
SimpleServer class, 97
soft keys
Towers of Hanoi game, 502
soft masks
smart card file structure, 373
software
HotSync, 330
security, 619
Software Architecture Analysis Method (SAAM)
Java 2 Micro Edition, 75
software design
complexity management, 634
component-based, 640
considerations, 631
error handling, 638
J2ME, 631
metaphors, 640
patterns, 640
reverse usability, 638
small screens, 637
user-centered, 635
Softwired
iBus//Mobile libraries, 476
MIDP version of JMS, 486
SOLID lines, 186
Sony
mobile electronic transactions initiative, 628
sorts
records, 198
SPIN® questioning
Java 2 Micro Edition, 65
sponsors
software design, 632
Spotless API
Palm, 495
Spotless implementation
Factory Method, 528
Palm OS, 523
Spotless layers
Towers of Hanoi game, 503
Spotless Palm
Java 2 Micro Edition, 27
Spotlet, application conversion
Connected Limited Device Configuration, 297
register() method, 299
Spotlet class
CommandListener implementation, 263
constructor, 259
event handling methods, 263
KJava, 258
Towers of Hanoi game, 523
Spotlet model
Java 2 Micro Edition, 27
SpotletGameSettings class, 531
Spread Spectrum techniques
wireless communication, 37
SQL statements
servlets, 426
SSL package
JavaPhone API, architecture, 338
stand-alone applications
AuctionHouse, 465
stand-alone build J2MEWTK, 138
standards
lack in mobile positioning area, 570
mobile positioning protocols, 570
smart cards (ISO7816), 368, 370

Star–seven device (*7 device). See *7 device (Star–seven device)
start
MyHanoiGame class, 526
startAngle, arcs, 183
startApp() method
launching and controlling MIDlets, 147
MainMIDP class, 251
MIDlets, 24, 136
RecordStore, 194
rectangles, 182
Tickers, 175
Towers of Hanoi game, 514
Web server communication, 408
startElement() method
XML expert systems, 557
start-up time
compared to execution speed, 646
STAT command
POP3, 101, 107, 113, 117
static variables
button labels, 228
Status class
slide editors
application conversion, 287
Status Words
smart cards, 372
Stealth project
Java 2 Micro Edition, 11
Store networks
messaging, 437, 438
Stream
synchronization, 321
StreamConnection interface, 87, 88, 93
StreamConnectionNotifier, 88
acceptAndOpen() method, 96
SimpleServer class, 97
string searches
case sensitivity, 418
StringBuffer
HttpConnection, 204
StringItem
Forms, 170
strings
synchronization, 320
StringTokenizer class, 108
stringWidth()
fonts, 190
stub objects
RMI, 396
style
font attributes, 188
subclasses
Template Method, 505
subscribing
Pub/Sub messaging, 456
substringWidth()
fonts, 190
suites
MIDlet, 142
Sun Microsystems
Java 2 Micro Edition evolution, 11
Java Card, 367
Wireless Toolkit
Mobile Information Device Profile, 26
Sun Palm MIDP, 138
support staff
software design, 633
surrogate architecture
Jini network, 43
Swedish Yellow Pages service
mobile positioning, 568
Swing API
Connected Device Configuration
application conversion, 270
JavaCheck, 273, 290
List class
application conversion, 276
Symbian EPOC operating system
Personal Java, 32
Symbian implementation of JavaPhone API, 339
Address Book package, 344
sample application, 345
running examples, 350
AIF builder tool, 340
Calendar package, 352
accessing schedule information, 352
adding schedule information, 353
creating to-do lists, 353
Datagram package, 355
datagram addressing, 355
sending and receiving datagrams, 357
emulator file system, 339
EPOC operating system, 340
event handlers, 341
listeners, 341
Meta events, 341
observers, 341
JTAPI Core package, 341
JTAPI Mobile package, 354
Power Monitor package, 360
running Java aplications, 340
User Profile package, 351
Symbian Platform
Java and, 336
JavaPhone API, 336
Mobile Information Devices, 335
Symbian implementation of JavaPhone API, 339
Symbian SDK
JavaPhone API, 362
JTAPI implementation, 362
testing JTAPI with emulator, 363
symmetric encryption systems, 625
SyncBuilder, 331
synchronization, 307
client code, 325
code, 320
code for client, 325
code for server, 320
Palm Pilot, 308, 330
problems, 310
protocol creation, 311
protocol creation, 312, 313
protocols, 311
server code, 320
types, 308
using, 329
synchronization support
records, 201
Synchronize button
slide editors
application conversion, 279
synchronous communications, 446
SyncML, 331

system architecture
compared to system design, 633
messaging, 447
system design
compared to system architecture, 633
system properties
Connected Limited Device Configuration, 201
Mobile Information Device Profile, 202
system responsiveness
device-based programming, 643
system.getProperty(microedition.platform), 431
system-level APIs, 209
system-level design
infrastructure, 639

T

taglibs
Orion ejbtags, 593
telephone numbers
Address Book package, sample application, 345
inserting numbers, 345
updating numbers, 347
television set—top boxes
Java 2 Micro Edition evolution, 13
Template Method design pattern, 505
GameSettings abstract class, 512
templates
use-case
time-sheet data entry, 67
test tools
kvmkjava.exel, 533
testChar() method
XML expert systems, 553
testing
CLDC development environment, 148
e-mail client application, 112
Forms, 173
iBus installation, 481
Java Card Development Kit installation, 382
kAWT Palm application, 154
KJava applications, 149
KJava Palm application, 150
MPC emulator installation, 578
PopServer class, 119
text
anchor points, 188
low-level graphics, 188
render, 190
text boxes
scrollable, 259
text message construction, 452
TextArea
application conversion, 300
TextBox, 162
application conversion, 298
field display, 260
MIDlets, 136
TextField
Forms, 170
Java field editing, 261
UserInterfaceKJava class, 259
tf.loseFocus() method
stopping caret flashing, 264
tFactory
messaging, 479
ThinParser class
XML expert systems, 553
third generation (3G) networks. See 3G (third generation) networks
third party libraries, 212
thread groups
eliminated KVM functionality, 127
threading
CLDC/CDC, 430
threads
messaging, 480
three sector cells
mobile positioning, 567, 569
three-tier client–server application model
Java 2 Micro Edition, 56
tickerOn, 175
Tickers
Screen, 175
tiers
application
mobile positioning case study, 579
presentation
mobile positioning case study, 579, 590
time markers
mobile positioning, 572
Time of Arrival (TOA)
mobile positioning, 567
TIME variables
synchronization protocols, 311
timeliness
device-based programming, 643
timeouts
Alerts, 168
Connector.open() method, 88
timers, 206
TimerTask classes, 206
time-sheet data entry case study
Java 2 Micro Edition architecture, 63
Timestamp header
JMS, 464
timing advance
mobile positioning, 567, 569
titleEntry
application conversion, 298
TITLEHEIGHT, 258
titles
CHARHEIGHT, 260
KJava screen layout, 257
TLS1.0 (Transport Layer Security). See Transport Layer Security (TLS1.0)
TOA (Time of Arrival). See Time of Arrival (TOA)
toFields() function
contact database design, 226
tokenizers
POP3 commands, 116
Tomcat
setup, 411
tools
Mobile Information Device Profile, 211
TOP command
POP3, 101
topic
Publish-and-Subscribe messaging, 449
receiver classes, 449
sender classes, 449
toString() method
expert systems, 543, 544

Touchable interface
application conversion, 271, 274, 293
touchable screen design
time-sheet data entry case study, 71
Tower
abstract class, 509
hierarchies, 504
Spotless implementation, 528
Towers of Hanoi, 496
tracking queues
improvements, 492
transaction based synchronization, 309
transactions
security, 627, 628
translate(), 192
translations
low-level graphics, 192
transparent files
smart cards, 373
Transport Layer Security (TLS1.0)
security, 626
triangulation
mobile positioning, 567
trigger() method
expert systems, 544
trim() method
mobile positioning case study, 587
Truffle process
Connected Device Configuration
application conversion, 271
two-tier client–server application model
Java 2 Micro Edition, 55
two-way alternate communication
Java Card, 18

U

UDP (Undirected Datagram Protocol). See Undirected Datagram Protocol (UDP)
UDP data packets, 88
UDP programming
datagrams, 88
example, 89
UI (user interfaces). See user interfaces (UI)
ui class variable
ContactApp, 235
UIDL command
POP3, 101
UIListener interface, 230, 234
generic data, 250
interface definitions, 228
UIListener methods
UImode, 263
UImode
UIListener methods, 263
UML (Unified Modeling Language). See Unified Modeling Language (UML)
UMTS (Universal Mobile Telecommunications System). See Universal Mobile Telecommunications System (UMTS)
Undirected Datagram Protocol (UDP), 441
Unified Modeling Language (UML)
screen navigation, 498
uniqueID
synchronization, 313, 326
Universal Mobile Telecommunications System (UMTS)
3G (third generation) networks, 35
UnknownHostException
socket programming, 91
unpaint()
abstract classes, 509, 530
unsupported low-level interface methods
check methods, 192
update
synchronization, 321
update() method
expert systems, 545
updateItem() method
ContactDatabase class, 349
updating caches, 648
upload() method
Comms class, 231
UPLOAD_PATH
ContactApp, 235
uploadOK() method
ContactApp, 238
uploads
resuming, 431
URL
TextBox, 162
useBean
Orion custom tag, 593
use-case diagram
time-sheet data entry
Java 2 Micro Edition, 67
use-case sign-off meeting
time-sheet data entry
Java 2 Micro Edition, 68
use-case template
time-sheet data entry
Java 2 Micro Edition, 67
useHome
Orion custom tag, 593
user actions
design considerations, 638
USER command
POP3, 100, 101, 107, 113, 116
user feedback
device-based programming, 643
suggestions, 430
user interfaces (UI)
bidder, 469
contact databases, 228
design, 497
Displayable, 159
events, 160
expert systems, 538, 560
listener methods, 230, 234
logic, 503
messaging, 486
Mobile Information Device Profile, 157, 210
mobile positioning case study, 583
UIListener listener methods, 234
user names
messaging, 476
messaging server administration, 461
User Profile package
JavaPhone API, architecture, 338
Symbian implementation of JavaPhone API, 351
UserProfile class, 351
UserProfileImpl interface, 351
user-centered design, 635

user-defined class loaders
eliminated KVM functionality, 127
UserEJB class
mobile positioning case study, 589
UserHome interface
mobile positioning case study, 581
UserInterface
interface definition, 228
not allowed to be abstract class, 251
UserInterfaceKJava class
Palm/KJava port, 257
UserInterfaceMIDP class
class variables, 246
MIDP specific database class, 246
UserProfile class
getCurrentUser() method, 351
setCurrentUser() method, 351
setImplementation() method, 351
User Profile package, 351
UserProfileImpl interface
User Profile package, 351
usingExternalPowerSource() method
PowerMonitor class, 361
utilities
KVM, 129
utilities index JSP
mobile positioning case study, 595

V

validation
contact databases, 221
VALUEEND_ST, 228, 247, 260
variable length records
smart cards, 374
vCard standard
ContactCard class, 344
VCENTER
images, 190
Vector class
JMS, 459
Vector contents
synchronization, 320, 321, 325
Vectors
expert systems, 543, 546
Slide class
application conversion, 284
verification
class, 128
view()
records, 200
viewRecords(), 194
Views
Model–View–Controller system, 52
expert systems, 538
private methods, 515, 527
Spotless implementation, 525
viewSlide() method
application conversion, 283
virtual machine, 14. See also Java Virtual Machine. See also Kilo Virtual Machine (KVM)
Java 2 Micro Edition, 15, 21
K, 125, 126
security, 619
written in C, 84
VxWorks
Connected Device Configuration (CDC), 82

W

WAE (Wireless Application Environment). See Wireless Application Environment (WAE)
WAP (Wireless Application Protocol). See Wireless Application Protocol (WAP).
WAR (Web ARchive files)
hot deployment, 414
J2EE web application structure, 412
running servlets, 417
undeploying, 421
WARNING
Alert, 168
weak references
eliminated KVM functionality, 127
web application structure
(J2EE), 411
Web ARchive files (WAR). See WAR (Web ARchive files)
web servers
reading files from, 407
security, 627
web.xml file
deployment descriptor file, 412
mobile positioning case study, 610
WebRunner browser
Java 2 Micro Edition, 13
Welcome View
Towers of Hanoi game, 498, 499
WGS-84 (World Geodetic System 1984). See World Geodetic System 1984 (WGS-84)
width
fonts, 190
WindowListener interface
Confirmation class
application conversion, 286
Windows
IP address
tracking, 439
Windows 9x, 382
Windows for Smart Cards
operating system, 370
Wireless Application Environment (WAE), 628
Wireless Application Protocol (WAP)
Java 2 Micro Edition, 41
security, 625, 627
Wireless LAN
mobile devices, 38
Wireless Markup Language
Java 2 Micro Edition, 41
Wireless Network Technology
Java 2 Micro Edition, 36, 41
wireless networks
digital attacks, 617
disconnected access, 652
security, 617
wireless technology
design limitations, 496
Wireless Toolkit
class building, 523
Mobile Information Device Profile, 26
Towers of Hanoi, 496

Wireless Transport Security (WLTS) Layer, 626
Wizards
  Windows for Smart Cards, 370
WLTS (Wireless Transport Security). See Wireless Transport Security (WLTS)
WML
  mobile positioning case study, 598
WML do_login JSP
  mobile positioning case study, 600
WML error JSP
  mobile positioning case study, 606
WML index JSP
  mobile positioning case study, 601
WML login JSP
  mobile positioning case study, 598
WML search JSP
  mobile positioning case study, 604
workshop
  time-sheet data entry
    Java 2 Micro Edition architecture, 64, 75
World Geodetic System 1984 (WGS-84), 572
World Wide Web
  Java 2 Micro Edition evolution, 13
worm programs, 618
wrappers
  MailStore, 104
  Protocol, 93
write() method
  HttpConnection, 205
  servlets, 418
WRITE_DIRECTORY
  servlets, 421
writeChars() method, 89
writeln() method, 419
writeRecord() method
  contact database design, 227
  DBMIDP class, 245

X

Xerox Palo Alto Research Center, 618
XML
  contact databases, 221, 266
  data transfer, 431
  expert systems, 537, 542, 550, 561
    classes, 552
    knowledge base, 546
  Java 2 Enterprise Edition, 60
  mobile positioning case study descriptors, 608
  Mobile Positioning Center communication, 569
  mobile positioning request/response, 572
  SyncML, 331

PROFESSIONAL JAVA MOBILE

TOOL SELECTION

WEB TABLET
MOBILE PHONE
MP3 PLAYER
DIGITAL CAMERA
PALMTOP
LAPTOP

NOW SELECTED

PALMTOP

Wrox writes books for you. Any suggestions, or ideas about how you want information given in your ideal book will be studied by our team.
Your comments are always valued at Wrox.

Free phone in USA 800-USE-WROX
Fax (312) 893 8001

UK Tel.: (0121) 687 4100 Fax: (0121) 687 4101

Professional Java Mobile – Registration Card

Name ______________________

Address ______________________

____

____

____

City ____________ State/Region ____________

Country ____________ Postcode/Zip ____________

E-Mail ______________________

Occupation ______________________

How did you hear about this book?

- ❒ Book review (name) ______________
- ❒ Advertisement (name) ______________
- ❒ Recommendation ______________
- ❒ Catalog ______________
- ❒ Other ______________

Where did you buy this book?

- ❒ Bookstore (name) ____________ City ____________
- ❒ Computer store (name) ______________
- ❒ Mail order ______________
- ❒ Other ______________

What influenced you in the purchase of this book?

❒ Cover Design ❒ Contents ❒ Other (please specify):

____

How did you rate the overall content of this book?

❒ Excellent ❒ Good ❒ Average ❒ Poor

What did you find most useful about this book? ____________

____

____

What did you find least useful about this book? ____________

____

____

Please add any additional comments. ____________

____

____

What other subjects will you buy a computer book on soon?

____

What is the best computer book you have used this year?

____

Note: This information will only be used to keep you updated about new Wrox Press titles and will not be used for any other purpose or passed to any other third party.

3897 Check here if you DO NOT want to receive support for this book ■ **3897**

Note: If you post the bounce back card below in the UK, please send it to:

Wrox Press Limited, Arden House, 1102 Warwick Road,
Acocks Green, Birmingham B27 6HB. UK.

Computer Book Publishers

NO POSTAGE
NECESSARY
IF MAILED
IN THE
UNITED STATES

BUSINESS REPLY MAIL

FIRST CLASS MAIL PERMIT#64 CHICAGO, IL

POSTAGE WILL BE PAID BY ADDRESSEE

WROX PRESS INC.,
29 S. LA SALLE ST.,
SUITE 520
CHICAGO IL 60603-USA